Advances in Experimental Medicine and Biology

Volume 1131

More information about this series at http://www.springer.com/series/5584

Md. Shahidul Islam

Editor

Calcium Signaling

Second Edition

Volume 1

 Springer

Editor
Md. Shahidul Islam
Department of Clinical Science
and Education
Södersjukhuset, Karolinska Institutet
Stockholm, Sweden

Department of Emergency Care
and Internal Medicine
Uppsala University Hospital
Uppsala, Sweden

Calcium signaling contains a unique selection of chapters that cover a wide range of contemporary topics in this ubiquitous and diverse system of cell signaling.

ISSN 0065-2598   ISSN 2214-8019   (electronic)
Advances in Experimental Medicine and Biology
ISBN 978-3-030-12456-4   ISBN 978-3-030-12457-1   (eBook)
https://doi.org/10.1007/978-3-030-12457-1

Dedicated to the living memory of
Roger Yonchien Tsien
  (1952–2016)

Contents of Volume 1

Contents of Volume 2

Chapter 1
Calcium Signaling: From Basic to Bedside

Md. Shahidul Islam

Abstract Calcium signaling and its interacting networks are involved in mediating numerous processes including gene expression, excitation-contraction coupling, stimulus-secretion coupling, synaptic transmission, induction of synaptic plasticity, and embryonic development. Many structures, organelles, receptors, channels, calcium-binding proteins, pumps, transporters, enzymes, and transcription factors are involved in the generation and decoding of the different calcium signals in different cells. Powerful methods for measuring calcium concentrations, advanced statistical methods, and biophysical simulations are being used for modelling calcium signals. Calcium signaling is being studied in many cells, and in many model organisms to understand the mechanisms of many physiological processes, and the pathogenesis of many diseases, including cancers, diabetes, and neurodegenerative disorders. Studies in calcium signaling are being used for understanding the mechanisms of actions of drugs, and for discovery of new drugs for the prevention and treatment of many diseases.

Keywords Calcium signaling · Excitation-contraction coupling · Stimulus-secretion coupling · Calcium and gene expression · Calcium and diabetes · Calcium and cancer · Calcium channels · Calcium binding proteins · Calcium oscillations · Calcium pumps · Calcium-sensing receptor

Concentration of Ca^{2+} in the cytoplasm ($[Ca^{2+}]_c$) is >20, 000 times lower than that outside the cell. This is achieved by the Ca^{2+}-transporting ATPases, Na^+/Ca^{2+} exchangers, and Ca^{2+}-binding proteins. When activated by a variety of external stimuli, cells respond by an increase in the $[Ca^{2+}]_c$, in the form or Ca^{2+}-spikes and

M. S. Islam (✉)
Department of Clinical Science and Education, Södersjukhuset, Karolinska Institutet, Stockholm, Sweden

Department of Emergency Care and Internal Medicine, Uppsala University Hospital, Uppsala, Sweden
e-mail: Shahidul.Islam@ki.se

© Springer Nature Switzerland AG 2020
M. S. Islam (ed.), *Calcium Signaling*, Advances in Experimental Medicine and Biology 1131, https://doi.org/10.1007/978-3-030-12457-1_1

oscillations, which allosterically regulate many proteins leading to alterations of numerous processes like gene expression, meiotic resumption, gastrulation, somitogenesis, early embryonic development of different organs, left-right asymmetry, muscle contraction, exocytosis, synaptic transmission, and induction of synaptic plasticity, to name just a few.

The discovery of the method for measuring $[Ca^{2+}]_c$ by fluorescent Ca^{2+}-indicators [1] that are made temporarily membrane permeable [2] started a revolution in biology, and today many laboratories are using fluorescent Ca^{2+}-indicators for measuring $[Ca^{2+}]_c$. Since these indicators bind Ca^{2+}, it is necessary to assess to what extent the indicators attenuate the Ca^{2+} signal. At low concentration of the indicators, it is more likely that an un-attenuated Ca^{2+} signal is being measured, whereas at high concentration of the indicators, it is more likely that Ca^{2+}-flux is being measured [3]. Many investigators are using a variety of genetically encoded Ca^{2+} indicators suitable for measuring Ca^{2+} in different organelles and subcellular compartments [4]. The technology has advanced to the point that we can measure highly localized Ca^{2+} changes in femtoliter volume by high resolution laser microscopy. We can also measure the physiological changes in the membrane potential, and consequent changes in physiological Ca^{2+} currents, in living cells, in their native environment, by using fluorescence-based optical techniques. Pharmaceutical companies are using high- throughput fluorescence-based assays using a variety of Ca^{2+}-indicators, for screening of ion-channels and G-protein coupled receptor as drug targets, and for identifying novel lead compounds.

Plasma membrane Ca^{2+} transport ATPase (PMCA), Na^+/Ca^{2+} exchanger, and sarco/endoplasmic reticulum (SR/ER) Ca^{2+}-ATPase (SERCA) maintain $[Ca^{2+}]_c$ at a normal low level. More than 20 variants of PMCA, with different regulatory properties, cell-type-specific expressions, different localizations, and interactions with different signaling molecules, not only maintain Ca^{2+} homeostasis, but also shape the Ca^{2+} signals. Mutations or genetic variations in the PMCA genes have been associated with diseases like hypertension, preeclampsia, and neural disorders. Twelve isoforms of SERCA proteins encoded by three genes, are expressed in different patterns in different tissues. Impaired Ca^{2+} homeostasis and Ca^{2+} signaling caused by impaired functions of SERCA pumps have been implicated in the pathogenesis of Darier disease, and some neuropsychiatric and neurodegenerative disorders.

Cells contain numerous Ca^{2+}-binding proteins, some of which act as Ca^{2+}-sensors, others as Ca^{2+}-buffers, and some as both. Ca^{2+} is buffered by rapid binding to Ca^{2+}-binding proteins that vary in their Ca^{2+}-binding and -dissociation kinetics, their concentrations in different locations inside the cell, and in their diffusional mobility. Ca^{2+}-buffers make the $[Ca^{2+}]_c$ changes transient, and, thus, finely tune the timing and spatial extension of Ca^{2+} signaling.

Receptor activation increases incorporation of P^{32} into the phospholipids [5], due to activation of phospholipase C. Thirteen family members of phospholipase C, and their isoforms are regulated by numerous agonists in isozyme-selective manner; they perform distinct functions in signal transduction, and mediate a variety of cellular responses, in almost every cells of the body. Activation of phospholipase C leads to the formation of the second messengers inositol 1,4,5 trisphosphate, and diacylglycerol [6, 7], which activates protein kinase C [8]. Inositol 1,4,5 trisphosphate activates 1,4,5 trisphosphate receptor, and releases Ca^{2+} from the ER [9]. In addition to mediating Ca^{2+} release, the three 1,4,5 trisphosphate receptors that form homotetramers, and heterotetramers, interact with >100 proteins, and signaling molecules. The expression of different isoforms of 1,4,5 trisphosphate receptor in different tissues, and their interactions with other proteins add to the complexity and diversity of the regulation of Ca^{2+} signaling in mediating many processes including apoptosis, autophagy, and cancer development.

The critical components of excitation-contraction coupling are the ryanodine receptors, the L-type voltage-gated Ca^{2+} channels and the junctophilins. From evolutionary perspective, Ca^{2+}-induced-Ca^{2+} release (CICR) (as in the cardiomyocytes) is the earliest form of excitation-contraction coupling, whereas depolarization-induced Ca^{2+} release from SR (DICR) through the type 1 ryanodine receptor mediated by direct protein-protein interaction between the ryanodine receptor and the L-type voltage-gated Ca^{2+} channel, (as in skeletal muscle cells) is a later development in the vertebrates. Four types of junctophilins link the ER to the plasma membrane, and support both CICR and DICR. Ca^{2+} regulates the three ryanodine receptors both positively and negatively (depending on concentration), by binding to the receptors, directly or indirectly through the Ca^{2+}-binding proteins, both from the cytosolic side and the luminal side. Release of Ca^{2+} from the ER /SR where Ca^{2+} is present in the form or "Ca^{2+}-lattice" bound to proteins, is a highly regulated process that prevents depletion of Ca^{2+} stores and consequent activation of ER stress response.

The pyridine nucleotide metabolite cyclic ADP ribose (cADPR) releases Ca^{2+} from the intracellular stores by mechanisms that may involve activation of the ryanodine receptors, but it is not clear how this happens, and to which protein cADPR binds. In heart, cADPR increases the gain of CICR. Another pyridine nucleotide metabolite, NAADP also releases Ca^{2+} from some acidic lysosomal Ca^{2+} stores by mechanisms that involve activation of the two pore channels (TPC). It is not clear whether NAADP binds to proteins other than TPCs. Stimulation of β-adrenergic receptors increases formation of NAADP, which releases Ca^{2+} from the acidic stores by binding to TPCs, but numerous questions remain unanswered. Endosomes and lysosomes are acidic organelles that contain Ca^{2+} in high concentration in readily releasable form. Ca^{2+} is loaded into these stores by the actions of vacuolar H^{+}-ATPase, and Ca^{2+}/H^{+} exchange. Ca^{2+} can be released from these stores through the TRPML channel or the TPCs.

Receptor activation not only releases Ca^{2+} from the ER, but also leads to Ca^{2+} entry through the plasma membrane channels [10]. The stromal interaction molecule 1 (STIM1) senses the Ca^{2+} concentration in the ER Ca^{2+} stores, and it regulates the CRAC (Ca^{2+} release-activated Ca^{2+} channel) formed by Orai1, a highly calcium-selective channel located in the plasma membrane. STIM1 also regulates store-operated Ca^{2+} channel formed by TRPC1. Mutations in the STIM1/Orai1 have been associated with diseases like severe combined immune deficiency, Stormorken syndrome, and tubular aggregate myopathy. Disturbances in store-operated Ca^{2+} entry have been implicated in promoting angiogenesis, tumor growth, muscle differentiation, and progression from cardiac hypertrophy to heart failure.

Mitochondria plays important roles in Ca^{2+} signaling. Mitochondria-associated ER membrane (MAM) provides a mechanism for communication between the ER and the mitochondria. The Sigma-1 receptor, a chaperone protein located in the MAM, is involved in Ca^{2+} exchange between the ER and the mitochondria. Ca^{2+} enters into the mitochondria through the mitochondrial Ca^{2+}-uniporter. Efflux of Ca^{2+} from the mitochondria is mediated by the Na^+/Ca^{2+} exchanger. Ca^{2+} regulates mitochondrial respiration, and ATP synthesis, but mitochondrial Ca^{2+} overload triggers the apoptosis pathways. The Sigma-1 receptor is also located near the plasma membrane, and it interacts with many other proteins. Mutations of the Sigma-1 receptor may lead to diseases like amyotrophic lateral sclerosis, and distal hereditary neuropathy.

The discovery of the transient receptor potential channels was helped by the clues obtained from the photoreceptor cells of drosophila mutants [11, 12]. These channels act as molecular sensors, and they participate in numerous cellular processes. Many cells express many of the TRP channels, and these channels appear to be involved in the pathogenesis of many diseases. In lungs, TRPC3 appears to be involved in mediating airway hyper-responsiveness seen in asthma. Dysregulation of several TRP channels has been implicated in promoting cancer growth, metastasis, and in determining sensitivity or resistance to chemotherapy. Some TRP channels that regulate tumorogenesis or tumor progression, are themselves targets of specific microRNAs, which are expressed in many cancer cells, and which function in RNA silencing. Manipulation of the TRP/miRNA interactive network is a potential way to treat cancer.

Ca^{2+} signals are decoded by numerous proteins, including, many ion channels, enzymes, transcription factors, and exocytotic proteins, which can be activated or inactivated by Ca^{2+}. Activation of protein kinase C [8] and multifunctional calcium/calmodulin stimulated protein kinases (CaMK) by Ca^{2+}, mediates a variety of cellular processes. The CaMKs, which are expressed in numerous cells, are activated following a variety of stimuli. Specificity of the functions of these kinases is determined by "molecular targeting" mechanisms mediated by some specific binding proteins. CaMK-II remains active in proportion to the frequency and amplitude of the Ca^{2+} signals, and the activation persists for some time even after $[Ca^{2+}]_c$ returns to the normal basal level. CaMK-II plays important roles in decoding Ca^{2+} signals to activate specific events during the embryonic development. Ca^{2+} singling regulates expression of many genes by acting at the level of gene transcription, gene translation, regulation of alternative splicing, and by regulating the epi-genetic mechanisms. Alteration in the expression of many genes can alter the so called "Ca^{2+} homeostasome" [13].

The Ca^{2+}-microdomains comprised of Ca^{2+} channels, Ca^{2+}-activated Ca^{2+} channels, Ca^{2+}-buffers, and other molecules, are fundamental elements of Ca^{2+} signaling. Different simulation strategies including stochastic method, deterministic method, Gillespie's method, and hybrid methods in multi-scale simulations, have been used for modeling of the Ca^{2+} signaling systems. Ca^{2+} signals occur in the form of Ca^{2+} spikes, and Ca^{2+} oscillations, which are stochastic events. Advanced statistical approaches, and biophysical simulations are being used to obtain insight into the dynamics of Ca^{2+} oscillation, including the processes underlying the generation and decoding of the oscillations.

In biology, Ca^{2+} signaling is almost universal. Even bacteria use Ca^{2+} as a signal; they sense Ca^{2+} by using the so called two component regulatory systems consisting of a sensor kinase and a response element. To understand different biological phenomena, and many human diseases, Ca^{2+} signaling is being studied in many model organisms including *Drosophila melanogaster* [14], *Saccharomyces cerevisiae* [15], *Caenorhabditis elegans* [16], and zebrafish, and such researches have led to important discoveries. Study of Ca^{2+} imaging in the zebrafish has helped our understanding of the development processes, many other physiological processes, and the roles of disease-related genes in a vertebrate system.

Extracellular Ca^{2+} functions as charge carrier, and regulates neuromuscular excitability. Extracellular Ca^{2+} is sensed by a G-protein coupled receptor (calcium-sensing receptor), which regulates secretion of parathyroid hormone, and can be inhibited by the calcimimetic drug cinacalcet used in the treatment of hyper-parathyroidism. In the bone marrow high extracellular Ca^{2+} leads to predominant osteoblast formation by acting through calcium-sensing receptor. High Ca^{2+} in the bone marrow also inhibits the differentiation and the bone-resorbing function of the osteoclasts. Extracellular matrix is the largest Ca^{2+} store in animals. The macromolecules of extracellular matrix interact with the receptor on the plasma membrane, and by that way regulate the $[Ca^{2+}]_c$ through complex mechanisms. Ca^{2+} also controls cell-extracellular matrix interaction through focal adhesions.

Study of Ca^{2+} signaling is helpful in understanding the pathogenesis of many diseases including that of diabetes, and the neurodegenerative diseases, and in understanding the mechanisms of action of drugs used in the treatment of these diseases. Many calcium channel blockers are being extensively used in the treatment of hypertension and atrial fibrillation. Over 400 million people in the world have diabetes. Studies of Ca^{2+} signaling have increased our understanding of the mechanisms underlying stimulus-secretion coupling in the β-cells, which is impaired in type 2 diabetes. Some of the commonly used antidiabetic drugs act by altering Ca^{2+} signaling in the β-cells [17]. It is likely that studies of Ca^{2+}-signaling and its interacting networks, will lead to new breakthroughs that will increase our understanding of the molecular mechanisms of many cellular processes that we do not fully understand today.

References

1. Tsien RY, Pozzan T, Rink TJ (1982) Calcium homeostasis in intact lymphocytes: cytoplasmic free calcium monitored with a new, intracellularly trapped fluorescent indicator. J Cell Biol 94(2):325–334
2. Tsien RY (1981) A non-disruptive technique for loading calcium buffers and indicators into cells. Nature 290(5806):527–528

3. Neher E (1995) The use of fura-2 for estimating Ca buffers and Ca fluxes. Neuropharmacology 34(11):1423–1442
4. Miyawaki A, Llopis J, Heim R, McCaffery JM, Adams JA, Ikura M et al (1997) Fluorescent indicators for Ca^{2+} based on green fluorescent proteins and calmodulin. Nature 388(6645):882–887
5. Hokin MR, Hokin LE (1953) Enzyme secretion and the incorporation of P32 into phospholipides of pancreas slices. J Biol Chem 203(2):967–977
6. Berridge MJ (1984) Inositol trisphosphate and diacylglycerol as second messengers. Biochem J 220(2):345–360
7. Michell RH (1975) Inositol phospholipids and cell surface receptor function. Biochim Biophys Acta 415(1):81–47
8. Nishizuka Y (1992) Intracellular signaling by hydrolysis of phospholipids and activation of protein kinase C. Science 258(5082):607–614
9. Maeda N, Niinobe M, Mikoshiba K (1990) A cerebellar Purkinje cell marker P400 protein is an inositol 1,4,5-trisphosphate (InsP3) receptor protein. Purification and characterization of InsP3 receptor complex. EMBO J 9(1):61–67
10. Putney JW Jr (1986) A model for receptor-regulated calcium entry. Cell Calcium 7(1):1–12
11. Cosens DJ, Manning A (1969) Abnormal electroretinogram from a Drosophila mutant. Nature 224(5216):285–287
12. Montell C, Rubin GM (1989) Molecular characterization of the Drosophila trp locus: a putative integral membrane protein required for phototransduction. Neuron 2(4):1313–1323
13. Schwaller B (2012) The regulation of a cell's Ca^{2+} signaling toolkit: the Ca^{2+} homeostasome. Adv Exp Med Biol 740:1–25
14. Hardie RC, Raghu P (2001) Visual transduction in Drosophila. Nature 413(6852):186–193
15. Cui J, Kaandorp JA, Sloot PM, Lloyd CM, Filatov MV (2009) Calcium homeostasis and signaling in yeast cells and cardiac myocytes. FEMS Yeast Res 9(8):1137–1147
16. Orrenius S, Zhivotovsky B, Nicotera P (2003) Regulation of cell death: the calcium-apoptosis link. Nat Rev Mol Cell Biol 4(7):552–565
17. Islam MS (2014) Calcium signaling in the islets. In: Islam MS (ed) Islets of langerhans, 2nd edn. Springer, Dordrecht, pp 605–632

Chapter 2
Measuring Ca^{2+} in Living Cells

Joseph Bruton, Arthur J. Cheng, and Håkan Westerblad

Abstract Measuring free Ca^{2+} concentration ($[Ca^{2+}]$) in the cytosol or organelles is routine in many fields of research. The availability of membrane permeant forms of indicators coupled with the relative ease of transfecting cell lines with biological Ca^{2+} sensors have led to the situation where cellular and subcellular $[Ca^{2+}]$ is examined by many non-specialists. In this chapter, we evaluate the most used Ca^{2+} indicators and highlight what their major advantages and disadvantages are. We stress the potential pitfalls of non-ratiometric techniques for measuring Ca^{2+} and the clear advantages of ratiometric methods. Likely improvements and new directions for Ca^{2+} measurement are discussed.

Keywords Ca^{2+} · Laser confocal microscopy · Fluorescence · Ratiometric · Non-ratiometric

Changes in the free Ca^{2+} concentration ($[Ca^{2+}]$) inside a cell can fulfil many different roles. Local changes in near membrane $[Ca^{2+}]$ can modify channels in the plasma membrane while changes in mitochondrial $[Ca^{2+}]$ can help to promote ATP production. Changes in nuclear $[Ca^{2+}]$ are critical for modulating gene replication and temporal aspects of these changes may provide valuable clues. One of the challenges in the field of Ca^{2+} signaling is to monitor the sites, amplitude and duration of free Ca^{2+} changes in response to physiological stimuli. Earlier researchers relied on a variety of methods, including atomic absorption and radioactive $^{45}Ca^{2+}$ to monitor Ca^{2+} in samples and Ca^{2+} movements across membranes and the likely underlying uptake and release mechanisms. Typically cell fragments were isolated by centrifugation and then Ca^{2+} uptake and storage capacity of isolated cellular organelles were examined. These methods were useful in the detection of relatively slow Ca^{2+} changes (seconds to minutes) but were unable to follow the short-term, transient Ca^{2+} movements induced by neural or

J. Bruton (✉) · A. J. Cheng · H. Westerblad
Department of Physiology & Pharmacology, Karolinska Institutet, Stockholm, Sweden
e-mail: Joseph.Bruton@ki.se

© Springer Nature Switzerland AG 2020
M. S. Islam (ed.), *Calcium Signaling*, Advances in Experimental Medicine and Biology 1131, https://doi.org/10.1007/978-3-030-12457-1_2

hormonal stimuli. Nonetheless, they provided valuable information about Ca^{2+} in cells e.g. the majority of tissue Ca^{2+} exists as bound to the glycocalyx (extracellular cell coat, Borle [4]) and is essential for maintaining excitability of neurons and muscle cells. X-ray microanalysis or electron probe analysis was the most ambitious of these attempts looking at both cellular and subcellular changes in Ca^{2+} but even at its best, this technique reported only the result of a physiological stimulus and not what happened during the period of stimulation itself.

All of these earlier techniques looked at changes in total Ca^{2+} and could not distinguish between bound and free Ca^{2+}, but what is most relevant to physiologists is the free Ca^{2+} concentration. Free Ca^{2+} concentration in the cytosol is often written as $[Ca^{2+}]_i$, which can be confusing since the 'i' can be interpreted as meaning free or bound or both. In this review, $[Ca^{2+}]_i$ will be used to refer to the free cytosolic Ca^{2+} concentration. When muscle cells are electrically stimulated, free cytosolic $[Ca^{2+}]$ (i.e. $[Ca^{2+}]_i$) can increase more than tenfold in a few ms, whereas the intracellular $[Ca^{2+}]$ remains essentially constant. The transient increase in $[Ca^{2+}]_i$ is due to Ca^{2+} release from the sarcoplasmic reticulum (SR) into the cytosol and subsequent active removal from the cytosol. Thus, Ca^{2+} moves from one cellular compartment to another and back again and overall total intracellular $[Ca^{2+}]$ does not change.

Multiple bioluminescent and fluorescent Ca^{2+} indicators are now available to measure $[Ca^{2+}]$ in cells and subcellular regions. Published results focus often on amplitude and time course of the signal and gloss over the possible pitfalls of interpretation. Since many users are not experts and try to follow or modify methods described earlier, the likelihood of errors and misinterpretation of data has increased. Our focus in this chapter is to highlight what can and cannot be done with available Ca^{2+} indicators.

2.1   Earlier Attempts to Measure $[Ca^{2+}]$ Inside Cells

Measurements of $[Ca^{2+}]_i$ were rather complicated before the invention of the various fluorescent Ca^{2+} indicators that are commonplace today. An invaluable source of information about these methods is to be found here [3].

1. *Ca^{2+}-activated photoproteins*. In 1961, Osamu Shimomura spent a stressful summer mashing up the light organs distributed along the edge of the bell of many thousands of *Aequorea* jellyfishes trying to isolate and characterize what was responsible for the blue-green glow. These jellyfishes are pretty colorless in real life and do not spontaneously glow. However if they are poked or disturbed in the water, then a greenish bioluminescence is seen, localised only around the margins of the bell but not found anywhere else on the jellyfish's body. After many trials two proteins were isolated, the bioluminescent protein aequorin that glowed blue upon the addition of Ca^{2+} and the green fluorescent protein which in the living jellyfish produces green light because of resonant energy transfer

from aequorin. Shimomura was awarded the Nobel prize in Chemistry in 2008 for the green fluorescent protein discovery. Other bioluminescent proteins were subsequently isolated from other organisms (e.g. obelin, berovin) but none of them approached the versatility of aequorin either in their native form or with targeted mutations and thus they are hardly used today.

An advantage of aequorin is that as a bioluminescent molecule it does not require any external stimulating light and thus the background signal or noise is extremely low. On the other hand, the bioluminescence signal is quite small and measurement of the light emitted is not as easy as it is for other currently used fluorescent indicators. In practice, it is barely sensitive enough to following changes in resting [Ca^{2+}]$_i$. Even when aequorin is used to monitor changes in the high physiological range of [Ca^{2+}]$_i$ (0.5–10 μM) that are induced by electrical or chemical stimulation, there are difficulties in interpreting the light emission which increases as approximately the third power of [Ca^{2+}]. Translating aequorin light signal into actual values of [Ca^{2+}]$_i$ is complicated further by its consumption (i.e. the signal decreases over time) and since [Ca^{2+}]$_i$ differs within different regions of the cell (highest at release sites), the signal will be heavily dominated by the regions with the highest [Ca^{2+}]$_i$. The light emitted by aequorin in the presence of [Ca^{2+}]$_i$, will be influenced by Mg^{2+} and the ionic strength which can change markedly during intense stimulation. Moreover, it is sensitive to changes in pH especially below pH 7. It is useful to imagine the Ca^{2+}-activated photoproteins as being "precharged" and Ca^{2+} binding to an photoprotein molecule causes an energy-consuming reaction with emission of light that discharges the molecule. Each molecule emits light only once, which means that the light-emitting capacity declines over time but with experience and modelling, one can minimise this potentially confounding factor. In earlier days, the major problem with native photoproteins was getting them into a cell. In large cells this was achieved by microinjection which was not practical for smaller (< 20 μm) cells. Other loading techniques have been tried and of these, incubation combined with mild centrifugation seems to be the best. Once the sequence of aequorin was known, it became feasible to transfect cells and induce expression of recombinant aequorin. This works well with many cultured cells and in embryos but is problematic when one tries to induce expression in adult cells in culture or in a living animal. An advantage with this technique is that the aequorin gene can be modified and targeted to different cellular compartments (e.g. mitochondria or endoplasmic reticulum) and the Ca^{2+} binding properties of the proteins can be modified appropriately. Photoprotein-based methods to measure [Ca^{2+}] in organelles are useful because in some situations it is not possible to introduce other fluorescent probes into a subcellular compartment [1].

2. *Metallochromic Ca^{2+} dyes*. Murexide was the first of these and arsenazo III and antipylazo III followed soon afterwards. With these indicators, the light absorbance of the molecule is monitored by a photomultiplier and when [Ca^{2+}] increases, the light measured will decrease. The advantage of these dyes is that they are fast and therefore can detect rapid [Ca^{2+}]$_i$ transients. This is because

they display a relatively low Ca^{2+} affinity, which means that they readily can detect high $[Ca^{2+}]_i$ levels and show little Ca^{2+} buffering. However, there have some unwanted characteristics which include complex Ca^{2+}-binding properties, marked Mg^{2+} and pH sensitivity, and a tendency to bind to intracellular proteins. The metallochromic Ca^{2+} dyes do not easily enter intact cells, and therefore these dyes were usually microinjected. Today, with one exception, these dyes are seldom used by anyone except specialists looking at the kinetics and other properties of Ca^{2+} release in muscle cells. The exception is calcein a metallochromic indicator used since 1956 to look at calcium in minerals and salts. It is not sensitive to monitor resting $[Ca^{2+}]$ in unstimulated cells but has found a niche as a live live/dead cell indicator and looking at opening of the mitochondrial permeability transition pore.

3. *Ca^{2+}-selective microelectrodes*. Electrophysiological techniques were already used to probe channels in the plasma membrane and thus they could be readily adapted when suitable Ca^{2+} resin and ligands were produced by chemists. Double barrelled electrodes were adapted quite early on so that only one microelectrode impalement of the cell was necessary to measure both membrane potential and Ca^{2+} (the signal detected by the Ca^{2+} sensor includes both the membrane potential and the Ca^{2+} potential and thus, the membrane potential has to be subtracted). Ca^{2+}-selective electrodes are rather difficult to make since a special silane coat has to be applied to the glass first before the Ca^{2+}-selective ligand is loaded in the electrode [8]. Microelectrodes with tips less than 1 μm are used to minimise cell damage when the electrodes are inserted into cells. Ca^{2+}-selective microelectrodes have good selectivity for Ca^{2+} over other cations in the physiological range. They suffer from two drawbacks that have limited their use in Ca^{2+}-signalling. First they report the free $[Ca^{2+}]_i$ only in the vicinity of the microelectrode tip and second even under the best possible recording conditions, their response time is slow, on the order of seconds when changing between solutions containing different free $[Ca^{2+}]$. Thus, they are not able to follow the rapid $[Ca^{2+}]_i$ transients that occur in excitable cells such as muscle or neurons. Nonetheless, various groups have used them to report resting free $[Ca^{2+}]$ in both animal and plant cells as being 50 nM to 150 nM, slightly higher than was measured later with diffusible Ca^{2+} indicators and reflecting the fact that underneath and close to the plasma membrane, free $[Ca^{2+}]$ is higher than in the bulk of the cytosol.

2.2   Fluorescent Ca^{2+} Indicators

Many of the common Ca^{2+} indicators used today were derived from the Ca^{2+} chelator BAPTA developed by Roger Tsien and his co-workers [14]. The Ca^{2+} indicator molecule consists of two parts: the Ca^{2+}-binding cavity that changes its shape when Ca^{2+} binds to it and the scaffold part of the molecule giving the fluorescence changes in response to Ca^{2+} binding to or being released from the cavity. These indicators have high selectivity for Ca^{2+} over Mg^{2+} and other common monovalent cations and are relatively unaffected by modest changes in H$^+$. When Ca^{2+} binds into the Ca^{2+}-binding cavity, there are large absorbance and fluorescence changes. It should be remembered that even with a low affinity for Mg^{2+} and H$^+$, Ca^{2+} indicators can be affected by these ions in experiments that are designed to induce metabolic exhaustion and thus a rise in free Mg^{2+} or large changes in pH. Much work has gone into developing different Ca^{2+}-binding properties and fluorescent tails that are optimised to work in defined ranges of [Ca^{2+}] and with different types of detection systems.

Ca^{2+} indicators can be conveniently divided into two groups: single-wavelength non-ratiometric indicators and dual-wavelength ratiometric indicators. Indicators have absorption and emission spectra that have been well characterised in vitro and which apply in general to the behaviour of the molecules inside cells. Optimal excitation and emission wavelengths for individual indicators can generally be found in the papers where they were originally described and have been gathered here with additional details (https://www.thermofisher.com/se/en/home/life-science/cell-analysis/cell-viability-and-regulation/ion-indicators/calcium-indicators.html#crs).

Non-ratiometric indicators generally show very little fluorescence at low (<100 nM) [Ca^{2+}] but show up to a hundred-fold increase in fluorescence when [Ca^{2+}] increases maximally inside a cell so that the indicator becomes saturated with Ca^{2+}. The expectation that the light signal faithfully reflects [Ca^{2+}]$_i$, is probably true under ideal conditions. However, to be able to directly compare signals from different experiments the following requirements have to be fulfilled: (1) cells exposed to similar loading conditions will have similar concentrations of indicator; (2) indicators remain in the cytosol and do not leak or get pumped out of the cytosol; (3) cell volume remains constant and there is no change in cell thickness; (4) the cell does not move; (5) the indicator is not affected by repeated exposure to excitation light. Unfortunately all these requirements are almost never fulfilled and so data obtained with non-ratiometric indicators should be carefully assessed to avoid errors (see Fig. 2.1 and discussion below).

Ratiometric indicators have the advantage that the Ca^{2+}-free and Ca^{2+}-bound forms of the indicator have distinct peaks at different wavelengths. Thus, measurements can be made at the two separate peaks and combined into a ratio. The ratio is usually constructed so that the signal recorded at the wavelength where the fluorescence shows a maximum at high [Ca^{2+}] is divided by the signal recorded at the wavelength showing its maximum at low [Ca^{2+}]. Between the two wavelength peaks, there is an isosbestic point where the fluorescence does not depend on Ca^{2+}]. In some cases (e.g. measuring quenching of a dye by Mn^{2+}) measurements are best made at this isosbestic wavelength. The classical ratiometric indicator fura-2 requires excitation at two wavelengths while the emitted fluorescent light is measured at one wavelength ($\sim$510 nm). The isosbestic point for fura-2 excitation is $\sim$360 nm and with increasing [Ca^{2+}], the emitted light increases at shorter wavelengths and decreases at longer wavelengths. The ratio with maximal dynamic

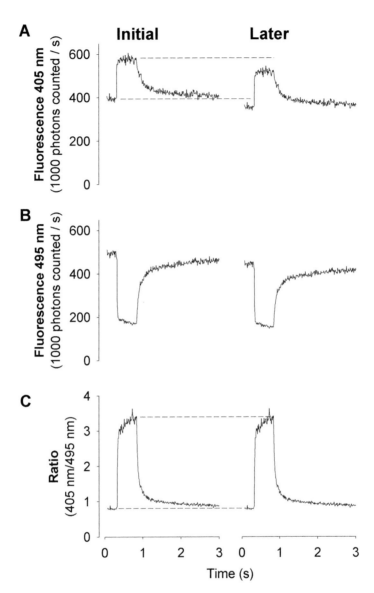

Fig. 2.1 Ratiometric indicators are best for experiments lasting hours. Indo-1 records obtained in a skeletal muscle cell stimulated to perform a tetanic contraction (70 Hz stimulation for 350 ms). Indo-1 was excited at 360 nm and the emitted light was measured simultaneously at 405 nm (**a**) and 495 nm (**b**) and the 405 nm/495 nm ratio was constructed (**c**). Over time, the ratio signal remained constant (dashed line in **c**) while the fluorescence intensity decreased for both the 405 nm (dashed line in **a**) and 495 nm signal. Note that the decline in the 405 nm trace seen in the right trace of the two shown in **a** is qualitatively similar to what would be seen if fluo-3 or another non-ratiometric indicator was used

range is then obtained by excitation below (~340 nm) divided by above (~380 nm) the isosbestic point. However, this requires continuous alteration between 340 nm and 380 nm excitation, which is technically troublesome, especially if rapid $[Ca^{2+}]_i$ transient are being measured. A simpler procedure is to measure the signal at the isosbestic point (360 nm) at regular intervals when constructing the ratio because the signal does not depend on $[Ca^{2+}]$. The preferred ratios will then be 340 nm/360 nm or 360 nm/380 nm, both of which will show an increase when $[Ca^{2+}]$ increases, albeit the ratio increase will not be as large as for the 340 nm/380 nm ratio. In our laboratory, we use the ratiometric dye indo-1 which requires excitation at only one wavelength and the emitted light be split into the $[Ca^{2+}]$-bound component (peaks about 400 nm) and the $[Ca^{2+}]$-free component (peaks about 475 nm).

The fundamental advantage of ratiometric over non-ratiometric indicators is exemplified in Fig. 2.1, which shows fluorescence records from a single skeletal muscle fiber at rest and during stimulation to produce a maximum contraction. Figure 2.1a shows the results as they would appear with a single wavelength indicator. As the experiments progressed, the fluorescent signal showed a general decline (probably representing pumping of the dye molecule out of the cell or transport by a non-specific anion transporter which can be blocked by probenecid or sulfinpyrazone), which might then be interpreted as a decrease in $[Ca^{2+}]_i$ both in the basal state and during contraction. However, the ratiometric indicator indo-1 was used in the experiment. In contrast to fura-2, this indicator is excited at one wavelength (~360 nm) and the emitted light is measured at two wavelengths (405 nm (increased signal with increasing $[Ca^{2+}]_i$) and 495 nm (decreased signal with increasing $[Ca^{2+}]_i$) in the depicted experiment). Figure 2.1b shows that there was a general decrease also in 495 nm signal as the experiment progressed. This means that there was no change in the 405 nm/495 nm ratio with time (Fig. 2.1c), which correctly reflects that there was no change in $[Ca^{2+}]_i$. The experimental traces in Fig. 2.1a show clearly that the signals from non-ratiometric indicators can result in completely erroneous conclusions if used without thinking. It should be noted that ratiometric indicators are not a cure for all problems. For instance, excessive UV light exposure can lead to qualitatively altered properties of the indicator (bleaching or inactivation), which cannot be corrected by ratioing [13].

2.3   Which Indicator Should One Use?

As outlined above, ratiometric indicators have clear advantages over non-ratiometric indicators and should be used whenever possible. Nowadays, visible-light laser scanning confocal microscopes are more common than any other Ca^{2+}-dedicated imaging systems meaning that a non-ratiometric indicator such as fluo-3/fluo-4 is often the first choice. Adding on a UV source to a microscope is reasonably straightforward and with suitable lens and filters, ratiometric indicators (i.e. fura-2 and indo-1 and their close relatives mag-fura-2 and mag-indo-1) could be used but this type of modification is rarely done.

In an ideal experiment, one would use an indicator which gives a fluorescence signal that shows large changes when $[Ca^{2+}]_i$ is changing and which is fast enough to follow the changes in $[Ca^{2+}]_i$ under study. However, the perfect indicator does not exist because some properties are difficult, or even impossible, to change. For instance, a Ca^{2+} indicator showing large changes in fluorescence with $[Ca^{2+}]_i$ changes in the low physiological range ($\sim$100 nM) is relatively slow and the opposite is also true. The relation between the intensity of the fluorescent signal (F) and $[Ca^{2+}]_i$ for a non-ratiometric indicator is given by the following equation (Eq. 2.1):

$$\left[Ca^{2+}\right]_i = K_d{}^* \ (F - F_{min}) / (F_{max} - F), \qquad (2.1)$$

where F_{min} and F_{max} mean the fluorescence intensity at virtually zero and saturating $[Ca^{2+}]_i$, respectively. K_d is the dissociation constant which in a plot of F against $[Ca^{2+}]_i$, will be the $[Ca^{2+}]$ where F is half-way between F_{min} and F_{max} and this is where the indicator displays its largest sensitivity. K_d is decided by an indicator's rates of Ca^{2+} binding (K_{on}) and dissociation (K_{off}), i.e. $K_d = K_{off}/K_{on}$. The on-rate constants of Ca^{2+}-binding are very fast and not that dissimilar whereas the rate that differs markedly between indicators is K_{off}. Accordingly, a slow indicator (low K_{off}) has a low K_d, which means that it is most sensitive at relatively low $[Ca^{2+}]_i$ and such indicators are therefore called high-affinity indicators. Conversely, a fast indicator has a high K_d and is referred to as a low-affinity indicator.

For ratiometric indicators, a slightly more complex equation describes the relation between fluorescence ratio (R) and $[Ca^{2+}]_i$ (Eq. 2.2):

$$\left[Ca^{2+}\right]_i = K_d{}^*\beta^* (R - R_{min}) / (R_{max} - R), \qquad (2.2)$$

where R_{min} and R_{max} is the fluorescence ratio at virtually zero and saturating $[Ca^{2+}]_i$, respectively. β is obtained by dividing the fluorescence intensity of the ratio's 2nd wavelength (denominator) acquired at virtually zero and saturating $[Ca^{2+}]_i$, respectively. Thus, the mid-point between R_{min} and R_{max} occurs at a $[Ca^{2+}]_i$ that equals $K_d * \beta$.

Figure 2.2 illustrates how the properties of two different Ca^{2+} indicators affect the change in fluorescence signal observed when $[Ca^{2+}]_i$ is changed in different concentration intervals. The comparison is between one high-affinity indicator, fura-2, and a low-affinity indicator, mag-fura-2. The name mag-fura-2 comes from the fact that it was designed to measure $[Mg^{2+}]$, but it has found its niche as a low-affinity Ca^{2+} indicator since $[Mg^{2+}]$ shows significant changes in the cytosol only when a cell is metabolically stressed by repetitive stimulation or exposed to poisons such as cyanide and its derivatives. $[Ca^{2+}]_i$ may vary dramatically between different physiological states. For instance, $[Ca^{2+}]_i$ peaks during contraction in a skeletal muscle cell may be up to 100-fold higher than resting $[Ca^{2+}]_i$. $[Ca^{2+}]_i$ is therefore often expressed as pCa or the negative $\log[Ca^{2+}]_i$ (analogous to the concept of pH). In Fig. 2.2, the 340 nm/380 nm ratio is shown for both indicators and β is set to

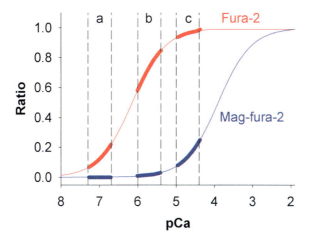

Fig. 2.2 High-affinity Ca²⁺indicators are more sensitive to stable changes in [Ca²⁺]ᵢin the normal physiological range. The relationships between $[Ca^{2+}]_i$ and fluorescence ratio (340 nm/380 nm excitation) are shown for the high-affinity indicator fura-2 and the low-affinity indicator mag-fura-2. $[Ca^{2+}]_i$ expressed as pCa $(-\log[Ca^{2+}]_i)$ in order to cover a larger range of concentrations. The thick lines are used to emphasise the differences between the two indicators at different $[Ca^{2+}]$ (**a**) 50–200 nM; (**b**) 1–4 μM; (**c**) 10–40 μM

4. This means that the mid-point between R_{min} and R_{max} occurs at a $[Ca^{2+}]_i$ of 0.56 μM for fura-2 (K_d assumed to be 0.14 μM) and 100 μM for mag-fura-2 (K_d assumed to be 25 μM). The interval (a) in Fig. 2.2 shows the change in ratio signal obtained when $[Ca^{2+}]_i$ is changed in the range of normal resting values, 50–200 nM. Here the fura-2 ratio signal shows a substantial increase, whereas mag-fura-2 ratio signal changes hardly at all. Thus, fura-2 can readily detect changes in basal $[Ca^{2+}]_i$, whereas mag-fura-2 is useless. The interval (b) in Fig. 2.2 (1–4 μM) would reflect $[Ca^{2+}]_i$ in cells that are activated. Again fura-2 is a rather sensitive indicator in this interval, whereas mag-fura-2 shows little change in the ratio signal. Finally the area marked (c) reflects $[Ca^{2+}]_i$ (10–40 μM) in a cell stimulated to maximal activation. In this case, fura-2 is saturated and changes little in the face of large concentration changes, whereas mag-fura-2 is clearly able to report changes in $[Ca^{2+}]_i$.

As a rule of thumb, all buffers are useful for detecting changes in an interval between about tenfold below and tenfold above the mid-point. Thus inside a cell where free $[Ca^{2+}]$ varies between 50 nM and 2–3 μM a suitable Ca²⁺ indicator which readily detects $[Ca^{2+}]_i$, at rest and during activation will have a K_d of 200–300 nM. This is also the $[Ca^{2+}]_i$ interval where Ca²⁺ most easily binds to the indicator, which has the potential to cause buffering problems. The noise in the detected fluorescent light signal decreases with increasing emitted light intensity. From this perspective it would be advantageous to have a large concentration of fluorescent indicator in a cell. However, a high concentration of indicator with a K_d in the physiological $[Ca^{2+}]_i$ range will buffer $[Ca^{2+}]_i$ markedly as illustrated in Fig. 2.3. When a relatively low concentration of indicator is present in the cell

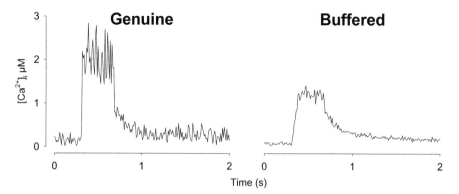

Fig. 2.3 Excessive cytosolic loading of Ca^{2+}indicator distorts [Ca^{2+}]$_i$transients. Typical records from the same skeletal muscle cell illustrate the real [Ca^{2+}]$_i$ response to 70 Hz tetanic stimulation (left trace) and the response as it looked after further injections of the Ca^{2+} indicator, indo-1, that caused buffering of the [Ca^{2+}]$_i$ transient (right trace). Note the reduced noise, the reduced amplitude and the slower rate of rise and decay in the buffered Ca^{2+} transient compared to the original record

("Genuine") a rapid and relatively large change in [Ca^{2+}]$_i$ is recorded but the signal contains some irregular fluctuations (noise). A markedly higher concentration of indicator ("Buffered") gives a far less noisy signal but the time course of the rise and fall of [Ca^{2+}]$_i$ is slowed and the amplitude of the change is less. Thus, with high-affinity Ca^{2+} indicators there is a delicate balance between introducing a sufficiently high indicator concentration to obtain records with an acceptable noise level and having so much indicator that cytosolic Ca^{2+} is markedly buffered, which leads to distorted [Ca^{2+}]$_i$ signals as well as altered cell signalling or function.

Figure 2.2 shows that a high-affinity Ca^{2+} indicator is better than a low-affinity indicator at monitoring changes in [Ca^{2+}]$_i$ in the normal physiological range. However, the diagram in Fig. 2.2 refers to stable or slowly changing [Ca^{2+}]$_i$. As discussed above, a trade-off of high Ca^{2+} sensitivity is that the indicator may be too slow to follow rapid changes in [Ca^{2+}]$_i$. In Fig. 2.4 this is illustrated for [Ca^{2+}]$_i$ transients in a skeletal muscle cell, the same would be true for any other excitable cell. The [Ca^{2+}]$_i$ transient resulting from a single stimulation pulse lasts for ~10 ms. Figure 2.4a shows such a [Ca^{2+}]$_i$ transient as recorded with the high-affinity indicator indo-1. However, the indicator is not fast enough to accurately follow the rapid changes in [Ca^{2+}]$_i$ and the recorded transient is too slow and the amplitude too low. In Fig. 2.4b the signal has been kinetically corrected [15] to take account of the properties of indo-1 and the [Ca^{2+}]$_i$ transient now better represents the true situation. While a low-affinity Ca^{2+} indicator could follow [Ca^{2+}]$_i$ transients more accurately and would therefore be preferable in experiments where rapid [Ca^{2+}]$_i$ changes are being studied, there is the drawback that the change in fluorescent signal is going to be small and hence difficult to measure. Figure 2.4c shows [Ca^{2+}]$_i$ as recorded by indo-1 during tetanic stimulation (70 Hz, 350 ms duration) of the

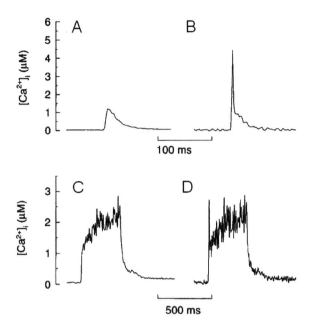

Fig. 2.4 Fast low-affinity Ca^{2+} indicators or kinetic correction of high-affinity Ca^{2+} indicator records are required to accurately portray rapid [Ca^{2+}]$_i$ transients. [Ca^{2+}]$_i$ records measured with indo-1 in a skeletal muscle cell in response to a single stimulation pulse (**a**) and a tetanus (**c**). This high-affinity indicator is too slow to accurately follow the most rapid changes in [Ca^{2+}]$_i$. Kinetic correction reveals a faster and larger [Ca^{2+}]$_i$ transient with the single stimulation pulse (**b**) and a [Ca^{2+}]$_i$ spike at start of the tetanus (**d**). (Figure adapted from Westerblad and Allen [15])

muscle cell; in Fig. 2.4d the record is corrected for the slow response of indo-1. It can be seen that the initial "spike" of [Ca^{2+}]$_i$ is missed without correction, but otherwise the records are rather similar. To sum up, Fig. 2.4 thus illustrates that problems with slow, high-affinity Ca^{2+} indicators are substantial when recording rapid [Ca^{2+}]$_i$ transients but much less so during more prolonged [Ca^{2+}]$_i$ changes. Thus, again there is a delicate balance between being able to measure large and rapid [Ca^{2+}]$_i$ changes (low-affinity indicators are preferable) and measure small prolonged changes (high-affinity indicators are better).

The signals recorded by the PMT or CCD are always transferred and stored on a computer and this means that the sampling rate can be high as one wishes. Sampling theorems are available which can be used empirically to decide what the optimal sampling rate is. As a general rule, we use a sampling rate at least tenfold faster (100 Hz to 1 kHz) than the expected fastest speed of [Ca^{2+}]$_i$ transients under study. It is worth remembering that using a high sampling rate means that less light signal is integrated for each time point and hence the noise level is higher with fast than with slow sampling. On the other hand, rapid or small [Ca^{2+}]$_i$ transients might be missed or distorted with a low sampling rate.

2.3.1   How Easy Is It to Get Indicators into Cells?

Indicators are charged molecules and do not easily pass lipid membranes. While many cells display endocytotic behaviour, we consider that the amount of indicator that can enter the cell by endocytosis during a couple of hours will be small and unlikely to be enough to make reproducible and meaningful measurements. Ca^{2+}-indicator can be introduced into cells by pressure injection or by electrophoreses. Electroporation of the cell membrane using very brief, high voltage pulses opens transient small pores in the cell membrane through which indicator molecules pass. All of these techniques require specialised equipment and some skill, but they maximise the likelihood that the indicator will be found only in the cytosol and not move into sub-cellular compartments, such as the mitochondria or sarco-endoplasmic reticulum.

Fortunately, there is a much easier method for introducing fluorescent indicators into the interior of single cells or tissue. The principle behind the method is that lipophilic groups (acetoxymethyl or acetate ester (AM) groups) are added to the charged indicator molecule. In this way the charges are hidden and the indicator complex becomes lipophilic and hence membrane-permeant. Once the complex has entered into the cytosol, cytoplasmic esterases gradually cut off the lipophilic groups and the free indicator molecule is then trapped in the cytosol and ready to detect $[Ca^{2+}]_i$. This simple method of loading fluorescent indicators into cells gives many a chance to investigate the regulation of $[Ca^{2+}]$ in their favourite cell. The lipophilic AM-indicator complex is typically dissolved in a mixture of dimethylsulfoxide (usually written as DMSO) and the detergent Pluronic to disperse the indicator molecules and aid cell loading. Typically, cells are exposed to the indicator (1–10 µM) for 10–30 min. After the loading period is finished, the cells are washed to remove residual extracellular AM-indicator and left for a further 30 minutes to ensure that all lipophilic groups have been cleaved off by cytoplasmic esterases allowing the indicator molecule to interact with Ca^{2+}. We have successfully used this basic loading protocol to detect electrically- or chemical-induced transient changes in $[Ca^{2+}]_i$ in myoblasts, myotubes and muscle fibres [12], pancreatic beta-cells [5], hippocampal neurons [10] and cardiomyocytes [11].

Loading of the lipophilic AM-indicator complex is not without problems. The quantity loaded into cells cannot be directly controlled. This leads to the risk of excessive loading and resultant buffering of $[Ca^{2+}]_i$, which affects Ca^{2+} homeostasis inside the cell and gives erroneous estimates of changes in $[Ca^{2+}]_i$ amplitude and time course as well as affecting Ca^{2+}-dependent cellular signalling. An additional problem with AM-indicators is that they may pass across intracellular membranes into organelles and report changes in $[Ca^{2+}]$ in this compartment in addition to changes in the cytosol. Our experience is that these problems seem to be minimised if cells are loaded at room temperature rather than at the higher physiological temperature of mammals. Unfortunately there is no single set of conditions that produces optimal loading of all cells and procedures needs to be optimised for each new cell type. For example, in our hands, indo-1 AM does not load into mouse cardiac myocytes but does load into rat cardiac myocytes. It should be noted that in tissues or densely coated cultured cells, indicator molecules can be trapped and cleaved by extracellular esterases to produce an indicator that reports extracellular [Ca^{2+}] and confounds the intracellular measurements [9].

For quantitative measurement within the cytosol of rapid or repeated transient [Ca^{2+}] in any intact neural or muscle cell, indo-1 is our first choice of fluorescent indicator. For slower changes lasting seconds or minutes, either indo-1 or fura-2 would be adequate. If one is interested only in the effect of a drug or other intervention and not the absolute numbers, then one could easily turn to fluo-3/fluo-4 as first choice indicators. For looking at intracellular organelles, the fluorescent indicator rhod-2 has been widely used to monitor changes in mitochondrial [Ca^{2+}] in neurones and muscles during and after stimulation by us and others. The low-affinity calcium indicator (K_{Ca} 90 μM) fluo-5 N has been used to monitor SR [Ca^{2+}] during repeated tetanic contractions. [Ca^{2+}] measurements can also be attempted using compartment-specific aequorin chimeras and other genetically engineered proteins [1, 7].

2.4   Equipment Overview

Typically, one uses the instruments that are available rather than those that are optimal for the task of measuring changes in [Ca^{2+}] inside a cell. The minimum needed to detect the fluorescence emitted from cells loaded with an indicator are a microscope with a light source to locate the cells and to excite the indicator, a detection device that is typically one or more photomultiplier tubes or a CCD camera and some recording or storage device. A simpler fluorometer-based system can be used if one is working with cell suspensions and is not interested in the response of individual cells. Filters are inserted into the light path to limit the wavelength and intensity of the light that excites the indicator and also to limit the wavelengths of the emitted light measured by the light detectors. The signals from the light detectors are generally digitised and stored on a computer. Newly purchased equipment dedicated to Ca^{2+} measurements is supplied with software controlling the various parameters related both to excitation wavelength and to detection of the emitted light that is more than capable of recording and performing a fast on-line analysis of signals.

The most important but often neglected part of the whole acquisition system is the light path and especially the objective lens. The lens is what allows one to magnify and focus on the cell or tissues. While magnification is important to see the sample, what is equally or more important is the ability of the lens to pass light of the appropriate wavelength and resolve fine specimen detail. The light collection effectiveness is described by the numerical aperture (N.A.) written on the lens casing. In general, one should have the lens with the highest N.A. possible (a more detailed description can be found here: http://micro.magnet.fsu.edu/primer/anatomy/numaperture.html). Lenses that are optimised to work with ultraviolet (UV) light are not optimal for visible light and vice versa. Lenses are exposed to the dust and moisture in the working environment unlike most of the other elements of the system, which are encased in protective housing. Even if an acquisition system is handled carefully, the lens is liable to become dirty from the particles floating in the air. If the lens requires oil or water for its proper operation, the combination of liquid and dust can lead rather quickly to the formation of a film coating the lens surface and the light path deteriorates. We use a superfusion system routinely in our experiments and over the years we have had a variety of problems ranging from leaks in aged tubing, overflow of liquid out of the recording chamber resulting in fluid on and inside the lens leading to a rapid deterioration of the signal. If not spotted quickly, this can lead to salt deposits on the lens or, in the worst case scenario, fluid entering the lens casing with a salt coating both outside and inside the lens. Problems of this kind are easily recognised as increased noise in the fluorescence signal and in the worse cases inability to focus on the cells or tissue. It should be routine to check the lens before and at the end of an experiment and to clean the lens with lens paper and an air spray before and after experiments or immediately one sees that solution has dripped on to the lens. If solution has dried and formed salt crystals on the lens, we use distilled water to rinse the salts away and ethanol is used finally to clear off residual water.

Nowadays, the most common types of detection set-ups are epifluorescence microscopy and scanning confocal microscopy. In epifluorescence microscopy, the whole sample consisting of a single cell or group of cells loaded with an indicator is excited by light of the appropriate wavelength and the photons emitted from the indicator are collected both from the sample section in focus (typically 0.3 μm with an objective lens with a high numerical aperture of 1.3) and also from above and below this plane of focus. Emitted light travels to one or more photomultiplier tubes or a CCD camera. Epifluorescent microscopy is used most commonly with ratiometric dyes such as indo-1 or fura-2 that are excited with light in the UV region. This type of set-up is ideal for measuring changes in $[Ca^{2+}]_i$ in virtually any cell type over extended periods of time while using mechanical, electrical or chemical stimulation. The area of interest can be limited to a single cell or data can be collected from a larger number of cells. While this method allows one to measure from the total volume of the cell, it is difficult (or with photomultiplier tubes basically impossible) to focus in on discrete areas of the cell and visualise events such as the entry of extracellular Ca^{2+} through surface membrane Ca^{2+} channels. However, when combined with special indicators, one can measure $[Ca^{2+}]$ changes in discrete organelles. For example, rhod-2 is a Ca^{2+}-indicator that loads preferentially into the mitochondria and indo-5 N has been used for measurements of $[Ca^{2+}]$ in the sarcoplasmic reticulum. Several groups including us have measured changes in $[Ca^{2+}]$ in the vicinity of the plasma membrane rather than in the bulk of the cytoplasm using an indicator moiety conjugated to fatty acid chains called FIP-18 which preferentially anchors into the surface membrane and measures $[Ca^{2+}]$ nearby (e.g. https://www.scbt.com/scbt/sv/product/ffp-18-am).

Confocal microscopy uses much the same hardware and software as that used in epifluorescence microscopy with two important additions: a laser acting as a point light source that excites the indicator and an adjustable diaphragm or pinhole in the emission pathway that when opened to its optimal size lets through light only from the focal plane, i.e. reducing light collection from cell regions outside the plane of focus. The fundamental advantage of the confocal microscope is that one can limit the focus to a very narrow section and thus measure discrete and rapid events such as localised release/entry of Ca^{2+} into the cytoplasm. While most confocal microscopes use lasers as light sources, this is not essential and the type of light source was not specified in the original patent (http://web.media.mit.edu/~minsky/papers/ConfocalMemoir.html).

Laser confocal microscopes come in three basic designs. These are (i) single photon laser scanning, (ii) the Nipkow or spinning disk, and (iii) two-photon versions.

(i) The single photon laser scanning confocal microscope is found in almost every biological/physiological institution. Most popular are those supplied by the major microscope manufacturers but nowadays for those who are technically proficient, it is possible to buy a confocal kit from the big optical suppliers (e.g. Thorlabs) and retrofit it to an existing microscope setup. In most systems, solid state lasers which have very precise and stable light emission and will work for many years have replaced Kr/Ar gas lasers. Physicists explain excitation of an indicator molecule as occurring when a single photon of the appropriate wavelength hits an indicator molecule and transiently lifts it from its ground state to a higher energy state. It remains in this higher energy state briefly (picoseconds) and then decays back to its original ground state by emitting a new photon with a longer wavelength than the original incident photon. An image of the sample is built up by moving a laser beam rapidly from one point to an adjacent point (pixel to pixel, typically dwelling a few to tens of μs on each pixel) along a horizontal line by means of a pair of mirrors (galvanometer-controlled or resonant-oscillating). A two dimensional image is built up by moving the laser beam vertically to a new line with a second pair of mirrors. The scanning and vertical movements are repeated until a full frame is obtained. This obviously takes a finite period of time and does not give an instantaneous view of what is happening in the cell. One can increase the scanning speed and obtain a full frame two to three times faster by reducing the "dwell time", i.e. the time for which the laser illuminates each pixel. The disadvantage of doing this is that the signal to noise ratio is reduced, which limits the ability to monitor small, spatially restricted changes in the fluorescence signal. If temporal resolution of a Ca^{2+} event in the cell is critical, the best approach is to abandon the two dimensional image acquisition approach and use the line scan mode instead. In this configuration, the laser beam scans the same line sequentially for a period of time. Line scans can be performed at over 1 kHz which is sufficiently fast to resolve even the most rapid change in local Ca^{2+} in a cell. The trade-off for the increased speed of data acquisition with line scanning is that only a single plane in a portion of the cell or tissue can be monitored. The line scan mode is extremely useful if one is trying to identify and characterise localised transient releases of Ca^{2+} from the sarcoplasmic reticulum in muscle or trying to localise the sites of Ca^{2+} entry in a neuron. Conversely, the full frame ("x-y mode") is best if one is trying to see what happens in the whole cell in response to a stimulus.

(ii) Spinning disk laser confocal microscopes use a spinning disk (rotating at several thousand revolutions per minute) with multiple pinholes (> 1000) through which parallel light beams pass. These beams excite the fluorescent indicator in the cell and the emitted light returns through a second collector disk with a matching pattern of microlenses to the detection device, which is normally a very sensitive CCD camera operated at low temperatures to minimise noise. The current generation of spinning disk confocals can easily acquire images at rates of up to 50 frames per second, which makes them suitable for visualizing temporal and spatial $[Ca^{2+}]$ changes in in a whole cell or cells rather than just a restricted line or set of lines using the line scan mode of a scanning confocal microscope. High frame rates generate large volumes of data but supplied software or ImageJ (download free from NIH) are sophisticated enough to select and analyze regions of interest only while masking data from uninteresting areas. The limited lack of popularity of these confocal microscopes may in part be due to the trade-off between spatial resolution and speed, i.e. greater spatial resolution generally requires a slower frame rate of acquisition and in part to the amount of incident light required that at best causes bleaching of the Ca^{2+} indicator only and at worst results cell damage and death.

(iii) The two- or multi-photon confocal microscopes overcome problems occurring when deeper parts of cells or tissues are being studied. Every microscope can be fitted with a motorised drive that accurately moves the plane of focus up or down in steps smaller than 1 μm. Thus, one can theoretically build up a three dimensional confocal image of a cell or tissue and check for possible hotspots or non-homogeneous change in $[Ca^{2+}]$ throughout a cell, tissue slice or cell culture in response to a stimulus. However, with a simple laser confocal scanning microscope, image quality deteriorates as one penetrates deeper into a cell or tissue. This impaired performance is due to the fact that a laser beam is a stream of photons that will excite any indicator molecule it meets as it travels to the plane of focus. Thus, a lens will receive photons not just from the plane of focus but also some photons that have been deflected into the light collection path following collision with proteins. As the distance from the region of interest to the lens increases, some photons from the focal plane of interest will be lost and photons from uninteresting regions will be collected.

Two-photon confocal microscopes minimises this problem by delivering the longer wavelength pulses required to excite indicator molecules only to a very confined region. The longer wavelength improves penetration depth into tissue which is especially important when looking at the behaviour of nerves in the brain or secretory cells in isolated parotid or pancreatic ducts. The beauty of the two photon technique is that excitation of an indicator molecule can only occur if two photons each with twice the wavelength and half of the energy of a single photon hit an indicator molecule. Indicator molecules hit by only one photon will not be excited. Longer wavelength light is less likely to cause damage to the cells. In a two-photon laser, the photons are sent out in femtosecond bursts. At the focal point, there is a high density of photons and the probability of two photons colliding with an indicator molecule is high. The major factor limiting more widespread usage of two-photon microscopy is the cost of the pulsed lasers themselves.

A final caution about experiments with lasers and intense light should be made. Children are routinely reminded to sunbathe in moderation and minimise prolonged exposure to ultraviolet light and avoid skin damage. The experience of seeing a cell start to bleb and die as one struggles to obtain the best record of [Ca^{2+}] transients highlights the fact that light energy is dangerous to cells. One should be aware that the energy that each photon of light contains may impact on the measurements being made and should try to limit the intensity of the light to the minimum possible. An additional problem is that intense light may produce photodegradation or photobleaching of Ca^{2+} indicators whereby the indicator is converted into a fluorescent but Ca^{2+} insensitive form that results in false measurements of resting and transient changes in [Ca^{2+}]$_i$ [13]. Again, the problem can be avoided by minimising the intensity and duration of light exposure.

2.5   Calibration of the Fluorescent Signal

Some kind of calibration is usually attempted in order to translate fluorescence signals into [Ca^{2+}]$_i$. Before any calibration is attempted, it is important to recognise that there is always some background signal in fluorescence systems, arising from the detectors themselves and because of imperfect filters and leakage of the excitation light to the detectors. Moreover, each tissue or cell will have an intrinsic or auto-fluorescence. The autofluorescence arises predominantly from proteins containing the amino acids tyrosine, tryptophan, and phenylalanine. The amount of background and intrinsic fluorescence depends on the excitation and emission wavelengths being used. It is necessary to measure the background and intrinsic fluorescence in a sample before loading the Ca^{2+} indicator and to subtract this value from all subsequent measurements. Failure to do this can have dramatic effects on the translation of the indicator signals into [Ca^{2+}]$_i$. Complete and accurate calibrations are generally tiresome or even impossible to perform on a single cell and some simplifications are usually made. This has led to an increased tendency to completely ignore calibrations and take the viewpoint that the fluorescence light intensity (F, non-ratiometric indicators) or ratio (R, ratiometric indicators) of Ca^{2+} indicators is linearly related to [Ca^{2+}]$_i$, which clearly is a severe oversimplification (e.g. see Fig. 2.2). Numerous papers erroneously state that [Ca^{2+}]$_i$ increased/decreased by x%, whereas what actually occurred was an increase/decrease in fluorescence intensity or ratio of x%, which can represent markedly different changes in $[Ca^{2+}]_i$. For instance, a minimal (<1%) change in fluorescence signal measured in a resting cell with a low-affinity indicator may represent a several-fold change in $[Ca^{2+}]_i$ (see Fig. 2.2). Similarly, a major increase in $[Ca^{2+}]_i$ may result in only a small increase in the fluorescence signal of a high-affinity indicator because the indicator was almost saturated with Ca^{2+} already before the increase.

Ca^{2+} indicators are affected by the surrounding protein and ionic environment and hence their properties inside a cell and in a test-tube will be markedly different. The relationship between fluorescence signals and $[Ca^{2+}]_i$ will also depend on the experimental setup. This means that all parameters in Eqs. 2.1 and 2.2 required to translate fluorescence signals into $[Ca^{2+}]_i$ should be established in the cell(s) using the same conditions and equipment as for the real experiments. This is of course easier said than done and some shortcuts are usually taken. In principle, the intracellular calibration is based on clamping $[Ca^{2+}]_i$ to a known value, without severe alterations of the cytosolic milieu, and then measure the fluorescence signal. The most important points to measure are at low/minimum $[Ca^{2+}]_i$, using EGTA or BAPTA to chelate Ca^{2+} to obtain F_{min} or R_{min}, and at saturating $[Ca^{2+}]_i$, to establish F_{max} or R_{max}. For ratiometric indicators, β is also obtained if R_{min} and R_{max} can be established without any major general decrease in fluorescence intensity. In addition, establishing K_d requires some intermediate $[Ca^{2+}]_i$. The reason why F_{min} or R_{min} and F_{max} or R_{max} are most important is because they set the limits between which the fluorescence signal can vary. Errors in measuring these parameters result in nonlinear errors when fluorescence signals are translated into $[Ca^{2+}]_i$. Erroneous estimates of F_{min} or R_{min} has the largest impact on the assessment of resting $[Ca^{2+}]_i$, whereas errors in F_{max} or R_{max} have the largest effects at high $[Ca^{2+}]_i$. On the other hand, K_d and β act as scaling factors and errors in these simply make the absolute changes in $[Ca^{2+}]_i$ smaller or larger, whereas relative changes during the course of an experiments are not affected.

Numerous methods have been used to perform a cytosolic calibration of $[Ca^{2+}]_i$. Most of these are based on introducing a strongly buffered solution with a set $[Ca^{2+}]$ to the cytosol. The solution can be introduced with methods similar to those described above for the introduction of the fluorescent indicator. An easy way of getting Ca^{2+} into cells is to add ionophores such as ionomycin or A23187 or even beta-escin to make the cell membrane leaky.

2.6   What Can We Hope for Now?

There have been marked improvements in the level of resolution. It was known and accepted for more than a century that separation of two objects closer than 250 nm in the horizontal plane was not possible with a standard single lens and light source. However, the use of two opposing and matched objective lenses and a complementary approach that relies on the photochemical properties of the indicators have led to at least a threefold improvement in both axial and horizontal resolution. While these technical improvements are still expensive to implement and are not yet generally available as ready to use equipment packages, it is likely that super-resolution fluorescence microscopy techniques will be used to image Ca^{2+} fluxes through groups of ion channels in the future (the clearest non-technical introduction is given in Hell [6]).

In recent years, different groups have further developed genetically encoded Ca^{2+} indicators (GECI's) and focussed on improving different aspects of their performance. The key to these developments was the recognition that the green fluorescent protein (GFP) found in jellyfish could be modified relatively easily to produce variants in various colours.

Green fluorescent protein GECI can be split into two broad groups. The first group are proteins that consist of a fusion of circularly permutated green fluorescent protein (GFP) or red fluorescent protein, a Ca^{2+}-binding protein (usually calmodulin or troponin C) and M13 (a short Ca^{2+}-CaM-binding peptide derived from from myosin light chain kinase that acts as a spacer). This shows weak fluorescence in the absence of Ca^{2+}. When Ca^{2+} binds there is change in its conformation and the protein construct now fluoresces brightly. The second group consists of the cameleons that rely on resonance energy transfer (FRET) to signal changes in [Ca^{2+}]. FRET works only if the two molecules making up the FRET pair are very close together (< 10 nm). Cameleons are a fusion of calmodulin binding Ca^{2+} to M13 and flanked on one side by a blue-shifted GFP and on the other side by a longer wavelength shifted GFP. When Ca^{2+} binds to calmodulin, the distance between the GFP molecules is altered and FRET efficiency increases. The cameleons are inherently ratiometric allowing one in theory at least to translate the FRET pair ratio into real [Ca^{2+}]. Since these complex proteins are genetically encoded, they have been targeted successfully to subcellular compartments. Interference from native forms of the Ca^{2+}-binding protein has been reduced through selective mutations. Their dynamic range has improved markedly but the maximum change of about 50% on average is markedly less than the classical fluorescent indicators such as indo-1 and fluo-3.

The 22 kDa bioluminescent protein aequorin and its prosthetic protein (coelenterazine) that is oxidised and released when Ca^{2+} binds have been massively re-engineered to optimise the properties of the photoprotein for monitoring of Ca^{2+} at different sites inside a cell [2]. Despite all the improvements, the inherent limitations of low light emission (one photon per aequorin versus hundreds of photons for other indicator molecules) and its consumption continue to make recording and interpretation of experiments difficult. It is difficult to see further improvements in this area.

Acknowledgment   Research reported from our laboratory was supported by the Swedish Research Council.

References

1. Agetsuma M, Matsuda T, Nagai T (2017) Methods for monitoring signaling molecules in cellular compartments. Cell Calcium 64:12–19
2. Alonso MT, Rodríguez-Prados M, Navas-Navarro P, Rojo-Ruiz J, García-Sancho J (2017) Using aequorin probes to measure Ca^{2+} in intracellular organelles. Cell Calcium 64:3–11
3. Blinks JR, Wier WG, Hess P, Prendergast FG (1982) Measurement of intracellular Ca^{2+} in living cells. Prog Biophys Mol Biol 40:1–114
4. Borle AB (1981) Control, modulation, and regulation of cell calcium. Rev Physiol Biochem Pharmacol 90:13–153
5. Bruton JD, Lemmens R, Shi CL, Persson-Sjögren S, Westerblad H, Ahmed M, Pyne NJ, Frame M, Furman BL, Islam MS (2003) Ryanodine receptors of pancreatic beta-cells mediate a distinct context-dependent signal for insulin secretion. FASEB J 17:301–303
6. Hell SW (2009) Microscopy and its focal switch. Nat Methods 6:24–32
7. Hossain MN, Suzuki K, Iwano M, Matsuda T, Nagai T (2018) Bioluminescent low-affinity Ca^{2+} indicator for ER with multicolor calcium imaging in single living cells. ACS Chem Biol 13:1862–1871
8. Hove-Madsen L, Baudet S, Bers DM (2010) Making and using calcium-selective mini- and microelectrodes. Methods Cell Biol 99:67–89
9. Jobsis PD, Rothstein EC, Balaban RS (2007) Limited utility of acetoxymethyl (AM) based intracellular delivery systems, in vivo: interference by extracellular esterases. J Microsc 226:74–81
10. Kloskowska E, Malkiewicz K, Winblad B, Benedikz E, Bruton JD (2008) APPswe mutation increases the frequency of spontaneous Ca^{2+}-oscillations in rat hippocampal neurons. Neurosci Lett 436:250–254
11. Llano-Diez M, Sinclair J, Yamada T, Zong M, Fauconnier J, Zhang SJ, Katz A, Jardemark K, Westerblad H, Andersson DC, Lanner JT (2016) The role of reactive oxygen species in β-adrenergic signaling in cardiomyocytes from mice with the metabolic syndrome. PLoS One 11(12):e0167090. https://doi.org/10.1371/journal.pone.0167090
12. Olsson K, Cheng AJ, Alam S, Al-Ameri M, Rullman E, Westerblad H, Lanner JT, Bruton JD, Gustafsson T (2015) Intracellular Ca^{2+}-handling differs markedly between intact human muscle fibers and myotubes. Skelet Muscle 162(3):285–293. https://doi.org/10.1186/s13395-015-0050-x
13. Scheenen WJ, Makings LR, Gross LR, Pozzan T, Tsien RY (1996) Photodegradation of indo-1 and its effect on apparent Ca^{2+} concentrations. Chem Biol 3:765–774
14. Tsien RY (1980) New calcium indicators and buffers with high selectivity against magnesium and protons: design, synthesis, and properties of prototype structures. Biochemistry 19:2396–2404
15. Westerblad H, Allen DG (1996) Intracellular calibration of the calcium indicator indo-1 in isolated fibers of *Xenopus* muscle. Biophys J 71:908–917

Chapter 3
High-Throughput Fluorescence Assays for Ion Channels and GPCRs

Irina Vetter, David Carter, John Bassett, Jennifer R. Deuis, Bryan Tay, Sina Jami, and Samuel D. Robinson

Abstract Ca^{2+}, Na^+ and K^+- permeable ion channels as well as GPCRs linked to Ca^{2+} release are important drug targets. Accordingly, high-throughput fluorescence plate reader assays have contributed substantially to drug discovery efforts and pharmacological characterization of these receptors and ion channels. This chapter describes some of the basic properties of the fluorescent dyes facilitating these assay approaches as well as general methods for establishment and optimisation of fluorescence assays for ion channels and G_q-coupled GPCRs.

Keywords High-throughput · High-content · Fluorescence imaging · G protein-coupled receptor · Voltage-gated ion channel · Ligand-gated ion channel · Assay development · Optimization · FLIPR

Abbreviations

ATP	adenosine triphosphate
Ca^{2+}	calcium ion
Ca_V and VGCC	Voltage-gated Ca^{2+} channels
DAG	diacylglycerol
FLIPR	Fluorescent Imaging Plate Reader
GPCR	G-protein coupled receptor
HTS	high throughput screening

I. Vetter (✉)
Institute for Molecular Bioscience, The University of Queensland, St. Lucia, QLD, Australia

School of Pharmacy, The University of Queensland, St. Lucia, QLD, Australia
e-mail: i.vetter@uq.edu.au

D. Carter · J. R. Deuis · B. Tay · S. Jami · S. D. Robinson
Institute for Molecular Bioscience, The University of Queensland, St. Lucia, QLD, Australia

J. Bassett
School of Pharmacy, The University of Queensland, St. Lucia, QLD, Australia

IP3	inositol-1,4,5,-triphosphate
LGCC	Ligand-gated Ca^{2+} channels
NCX	Na^{+}/Ca^{2+} exchanger
PIP_2	phosphatidylinositol 4, 5 bisphosphate
PMCA	Plasma Membrane Ca^{2+} ATPase
RyR	ryanodine receptors
SERCA	sarco/endoplasmic reticulum Ca^{2+} ATPase
EGTA	ethylene glycol-bis(2-aminoethylether)-N,N,N',N'-tetraacetic acid.
APTRA	2-aminophenol-N,N,O-triacetic acid
BAPTA	1,2-bis(o-aminophenoxy)ethane-N,N,N',N'-tetraacetic acid
K_d	dissociation constant
AM	acetoxymethyl
ER	endoplasmic reticulum
LED	light-emitting diode
CCD	charge-coupled device
EMCCD	Electron Multiplying Charge Coupled Device
ICDD	Intensified CCD
PDL	poly-D-lysine
PLL	poly-L-lysine
PLO	poly-L-ornithine
nAChR	nicotinic acetylcholine receptors
HEPES	4-(2-hydroxyethyl)-1-piperazineethanesulfonic acid
PAR2	protease-activated receptor 2
RFU	relative fluorescence unit
SBFI	Sodium-binding benzofuran isophthalate
PBFI	Potassium-binding benzofuran isophthalate
DRG	Dorsal Root Ganglion
RGS4	Regulator of G protein signalling 4
Na_V	voltage-gated sodium channel
MCU	mitochondrial Ca^{2+} uniporter
cAMP	cyclic adenosine monophosphate
CANDLES	Cyclic AMP iNdirect Detection by Light Emission from Sensor cells
GTP	guanosine triphosphate
TRP	Transient Receptor Potential
SK_{Ca}	small-conductance calcium-activated K^+ channel
IK_{Ca}	intermediate-conductance calcium-activated K^+ channel
BK_{Ca}	big-conductance calcium-activated K^+ channel
IRK	Inwardly-rectifying K^+ channel
TWIK	Tandem of pore domains in a Weakly Inward rectifying K^+ channel
TREK	TWIK-related K^+ channel

TASK	TWIK-related acid-sensitive K^+ channel
TALK	TWIK-related alkaline pH-activated K^+ channel
THIK	TWIK-related halothane-inhibited K^+ channel
TRESK	TWIK-related spinal cord K^+ channel
K_V or VGKC	voltage-gated K^+ channel
ANG-1 and ANG-2	Asante NaTRIUM Green-1 and -2
Di-4-ANEPPS	Pyridinium, 4-(2-(6(dibutylamino)-2-naphthalenyl)-1-(3-sulfopropyl)-hydroxide
$DiBAC_4(3)$	bis-(1,3-dibutylbarbituric acid) trimethine oxonol
FMP	FLIPR Membrane Potential
FRET	fluorescence resonance energy transfer
PeT	photoinduced electron transfer
CC2-DMPE	N-[6-chloro-7-hydroxycourmarin-3-carbonyl] dimyristroyl phosphatidyl ethanolamine
GEVI	genetically encoded voltage indicators
DPA	dipicrylamine
DiO	oxocyanine
LED	light-emitting diode
sCMOS	scientific complementary metal-oxide-semiconductor

3.1 Introduction

G-protein-coupled receptors and ion channels are, alongside kinases, the main protein families targeted by currently approved drugs [1]. Despite the considerable structural and functional diversity of these proteins, a common signalling mechanism involved in mediating some of their cellular effects is the directed flux of ions such as calcium (Ca^{2+}), sodium (Na^+) and potassium (K^+). Accordingly, over the past decades significant advances have been made towards the development of assays permitting high-throughput profiling of GPCRs or ion channels that are functionally coupled to ion flux. A particular focus of these efforts has been the development of high-throughput kinetic fluorescence plate reader assays for drug discovery and pharmacological characterisation of these targets, facilitated predominantly by fluorescent ion and membrane potential indicators (Fig. 3.1). Indeed, recent years have seen the development not only of fluorescent molecules and proteins capable of detecting physiologically relevant concentrations of Ca^{2+}, Na^+ and K^+, but also protons (H^+), chloride (Cl^-) and other halides which can detect the movement of these ions between subcellular compartments [2]. In combination with the development of sophisticated high-throughput and high-content plate readers incorporating fluid addition robots and kinetic read capabilities, these probes have significantly advanced our understanding of basic pharmacology of these important drug targets, not least by facilitating drug discovery programs that are accessible not only to pharmaceutical industry but also smaller academic groups.

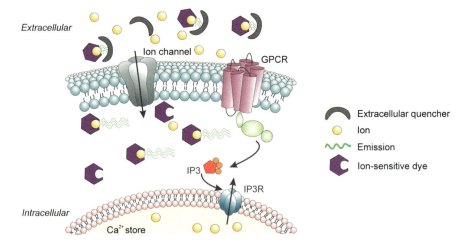

Fig. 3.1 Basic principle of fluorescent ion flux assays
Activation of ion channels and GPCRs leads to altered intracellular concentration of ions such as Ca^{2+}, Na^+ or K^+. The emission of fluorescent ion-sensitive dyes (e.g. Fluo-4, Sodium Green) increases in response to binding of their cognate ions (e.g. Ca^{2+}, Na^+). In no-wash format, extracellular signal is eliminated through incorporation of fluorescence quenchers in extracellular assay buffer, permitting quantification of intracellular ion accumulation in high-throughput format

While fluorescence approaches cannot recapitulate mechanistic insight into the activation and gating mechanics leading to ionic flux *via* voltage-gated ion channels, fluorescence-based screening approaches have nonetheless found application in bioactive discovery programs. For example, novel modulators of voltage-gated sodium (Na_V) channels have been isolated from venoms using high-throughput fluorescent plate reader assays [3–9], and fluorescence assays have also been used for identification and pharmacological characterisation of small molecule modulators of voltage-gated ion channels [10–13]. In addition, these techniques – which can be complementary to more traditional assays such as electrophysiological recordings – have provided novel insight into ligand-gated ion channel pharmacology as well as the physiological or pathological roles of these channels.

This review will discuss the properties of the most important Ca^{2+}, Na^+, K^+ and membrane potential dyes and fluorescent probes, as well as some of their applications in high-throughput and high-content kinetic fluorescence assays.

3.2 Ca^{2+} Signalling by Ion Channels and GPCRs

3.2.1 Calcium – A Universal Signalling Molecule

The calcium ion Ca^{2+} is often referred to as a 'universal" signalling molecule; indeed, most biological processes involve Ca^{2+} signalling in one form or another (for review, see [14, 15]). It is thus not surprising that Ca^{2+} is involved in diverse physiological functions ranging from differentiation, excitability and motility to apoptosis.

Because Ca^{2+} acts as a ubiquitous messenger molecule, a myriad of proteins are dedicated to its extrusion, chelation, sequestration and release, resulting in astonishingly precise temporal and spatial control of Ca^{2+} [14, 15]. At the cellular level, Ca^{2+} concentrations are extremely tightly controlled in the cytoplasm, where resting $[Ca^{2+}]$ is approximately 100 nM. It is maintained at this level by extrusion to the extracellular space through pumps such as Plasma Membrane Ca^{2+} ATPase (PMCA) and the Na^+/Ca^{2+} exchanger (NCX) [16]. Ca^{2+} is also sequestered into intracellular stores such as the endoplasmic or sarcoplasmic reticulum by the sarco/endoplasmic reticulum Ca^{2+} ATPase (SERCA) and the mitochondria by the mitochondrial Ca^{2+} uniporter (MCU) [17]. As a result, extracellular Ca^{2+} concentrations are significantly higher at approximately 1.8–2 mM, and a large Ca^{2+} reserve also is found in intracellular compartments, where Ca^{2+} is stored in protein-bound form and also occurs at relatively high concentrations as free Ca^{2+} [18].

To initiate signalling events, Ca^{2+} can be derived from the extracellular space, where voltage- or ligand-gated ion channels permit flow of this ion down its approximately 20,000-fold concentration gradient. Ca^{2+} can also be released from intracellular stores such as the endoplasmic reticulum through activation of inositol-1,4,5,-triphosphate (IP3) receptors and ryanodine receptors (RyR), resulting in a net increase in cytoplasmic Ca^{2+}. Accordingly, intracellular Ca^{2+} concentrations can rise several fold relatively to baseline during Ca^{2+} signalling events, a phenomenon that can be conveniently exploited for the development of fluorescence Ca^{2+} signalling assays.

3.2.2 GPCRs

Activation of some G-protein coupled receptors (GPCR), in particular those coupled to $G_{q/11}$, results in activation of phospholipase C which in turn facilitates cleavage of phosphatidylinositol 4, 5 bisphosphate (PIP_2) into 1,4,5-inositol trisphosphate (IP3) and diacylglycerol (DAG). IP3 then activates IP3 receptors located on the endoplasmic reticulum, causing release of Ca^{2+} into the cytoplasm [19]. While $G_{\alpha s}$ and $G_{\alpha i}$-coupled GPCR do not signal through Ca^{2+} physiologically, co-expression of chimeric or promiscuous G-proteins, such as $G_{\alpha 15/16}$ [20–22], can couple activation of these receptors to increases in intracellular Ca^{2+} and thus allows development of functional high throughput (HTS) assays based on Ca^{2+} imaging. In addition, fluorescence assays have also been developed for detection of second messengers that are modulated downstream of $G_{\alpha s}$ and $G_{\alpha i/o}$ activation, such as cAMP (for review, see [23]). While most of the techniques adapted for this purpose, such as time-resolved fluorescence resonance transfer, gene reporter assays, and fluorescence polarisation, are end-point assays, this limitation has been overcome by the development of genetically-encoded cAMP sensors that enable monitoring of real-time changes in intracellular cAMP concentrations (e.g. GloSensor™ cAMP (Promega); CANDLES (Cyclic AMP iNdirect Detection by Light Emission from Sensor cells); cADDis cAMP sensor (Montana Molecular) and CAMYEL, a cAMP sensor based on the BRET pair citrine-cp229 and *Renilla* luciferase which flank a human Epac1 (a cAMP-activated GTP exchange factor) protein) [24–26].

3.2.3 Ca^{2+}-Permeable Ion Channels

Voltage-gated Ca^{2+} channels (Ca_V), expressed in excitable cells, are large transmembrane proteins that undergo conformational changes in response to altered membrane potential [27]. As a result, activation of voltage-gated Ca^{2+} channels causes rapid influx of Ca^{2+} from the extracellular space, which controls processes such as muscle contractions or synaptic exocytosis. The properties of these channels can be exploited in the design of high throughput Ca^{2+} assays, where addition of extracellular KCl leads to membrane depolarization and thus channel opening [28, 29]. In addition, a plethora of Ca^{2+}-permeable transmembrane ion channels facilitate influx of extracellular Ca^{2+} along its concentration gradient in response to extra- or intracellular binding of ligands. These ligand-gated Ca^{2+} channels include, to name a few, ionotropic purinergic and glutamate receptors, nicotinic receptors and Transient Receptor Potential (TRP) channels and are indispensable to many physiological processes.

Accordingly, Ca^{2+} permeable ion channels and GPCRs linked to Ca^{2+} release are important drug targets, with modulation of Ca^{2+} signalling increasingly recognized as a valid therapeutic strategy in a range of diseases, including cardiac disease, neurological disorders such as Alzheimer's disease, and cancer [30–32].

3.3 Na^+-Permeable Ion Channels and Na^+ Signalling

3.3.1   Sodium – An Abundant Signalling Molecule

The sodium ion (Na^+) is the most abundant cation in the extracellular fluid, and as such, is essential for regulation of blood volume, osmotic equilibrium and cell membrane potential. At the cellular level, the concentration of Na^+ is predominantly maintained by the Na^+/K^+-ATPase, a membrane-bound protein that actively pumps Na^+ out of the cell in exchange for K^+, resulting in higher levels of extracellular Na^+ (145 mM) compared to intracellular Na^+ (10 mM) [33]. This maintains a negative membrane potential (-50 to -70 mV in neurons) that promotes an electrochemical gradient that drives Na^+ into the cell upon opening of Na^+-permeable channels [34]. Na^+-permeable channels are classified as either voltage-gated, which open in response to a change in membrane potential, or ligand-gated, which open upon binding of a ligand.

3.3.2 Voltage-Gated Na^+ Channels

Voltage-gated Na^+ channels (Na_Vs) are pore-forming membrane proteins that open in response to a positive change in membrane potential to allow Na^+ influx into the cell, resulting in the generation and propagation of action potentials in electrically excitable cells. In mammals, nine different Na_V subtypes have been described ($Na_V 1.1$–1.9), each having distinct tissue expression profiles and biophysical properties. $Na_V 1.1$, $Na_V 1.2$, $Na_V 1.3$ and $Na_V 1.6$ are expressed in the central nervous system, $Na_V 1.4$ is expressed in skeletal muscle, $Na_V 1.5$ is expressed in cardiac tissue, and $Na_V 1.1$, $Na_V 1.6$, $Na_V 1.7$, $Na_V 1.8$, $Na_V 1.9$ are expressed in the peripheral nervous system [35].

Despite the immense interest in Na_V modulators as therapeutic targets, high-throughput fluorescence assays of these channels have been comparatively limited, in part because of inherent difficulties in replicating the endogenous activation "signal" – a change in membrane voltage – in commonly available plate readers. Although allosteric modulators such as veratridine (activates $Na_V 1.1$–1.7) and deltamethrin (activates $Na_V 1.8$ & 1.9) can be used to elicit Na_V-mediated responses, they also alter channel kinetics such as reducing peak current and increasing sustained currents [4, 36]. This altered activity may interfere with screening assays and has the potential to generate false positive or negative results.

3.3.3 Ligand-Gated Na^+ Channels

Ligand-gated Na^+ channels are pore-forming membrane proteins that open in response to binding of a ligand to allow Na^+ influx into the cell. They can either be selective for Na^+, such as acid-sensing ion channels (ASICs) and epithelial sodium channels (eNaCs); or non-selective for Na^+, such as nicotinic acetylcholine receptors (nAChRs) or TRP channels, which can also conduct Ca^{2+} [37–39]. Indeed, the ionic selectivity of nAChR in particular is subtype-dependent, with the neuronal $\alpha 7$ nAChR being highly permeable to and selective for Ca^{2+}, while muscle nAChR subtypes are selective for Na^+ ions [37].

3.3.4 K^+-Permeable Ion Channels and K^+ Signalling

3.3.5 Potassium – A Comprehensive and Copious Counter to Calcium

K^+ flux generally opposes the function of Na^+ and Ca^{2+} signalling, with many (though not all) biological processes involving Ca^{2+} or Na^+ utilising K^+ as a physiological 'hand brake' to control and regulate signalling mechanisms. The intracellular K^+ concentration is typically high (140 mM) compared to the extracellular environment (4 mM), thus creating a pronounced outward gradient. K^+-permeable ion channels are the most numerous and functionally diverse ion channels in nature [40, 41]. In mammals, they can be divided into four major functional groups: calcium-activated K^+ channels, inwardly rectifying K^+ channels, two pore K^+ channels and voltage-gated K^+ channels, each of which can be further classified into sub-families with unique functional roles.

The calcium-activated K^+ channels consist of three families: small-conductance ($SK_{Ca}1, 2, 3$), intermediate conductance ($IK_{Ca}1$) and large/big-conductance (BK_{Ca}), all of which are widely expressed in many different tissues such as smooth muscle, epithelia, endothelium and neurons. In these tissues, K_{Ca} channels generally regulate excitability by restoring resting membrane potential, thus forming a negative feedback loop for Ca^{2+}-mediated signalling since these channels are activated by Ca^{2+}. For example, in cerebellar Purkinje neurons normal activity is reliant on both Ca^{2+} channels to initiate firing and the calcium-activated K^+ channels to regulate the firing [42].

Inwardly-rectifying K^+ (IRK) channels are the 'wild-card' in the K^+-permeable ion channel superfamily, as they allow the passage of external K^+ into the cell instead of moving K^+ ions out of the cell [43]. IRK channels can be classified into four functional groups: transporters, classical, G-protein gated (GIRK) and ATP-sensitive. As a result, the functional consequences of IRK channel activation, in particular for GIRK channels which functionally couple to GPCRs, are very diverse [43]. Overall, the function of these channels is highly dependent on the tissue in which they are expressed. For example, ATP-sensitive IRK channels aid in glucose homeostasis by regulating insulin release, whilst classical IRK channels facilitate passive movement of K^+ ions in order to orchestrate the electrical excitability of membranes needed for neuronal signalling [43].

Compared with other K^+-permeable ion channels, the two-pore K^+ channels (K_{2P}) remain open at a range of physiological voltages, although their activity is largely unaffected by voltage [44]. They are considered to contribute to leak and background K^+ currents and restore resting membrane potential [44]. K_{2P} channels include a wide array of functionally distinct families, such as the TWIK (Tandem of pore domains in a Weakly Inward rectifying K$^+$ channel), TREK (TWIK-related K^+ channel), TASK (TWIK-related acid-sensitive K^+ channel), TALK (TWIK-related alkaline pH-activated K^+ channel), THIK (TWIK-related halothan-inhibited K^+ channel), and TRESK (TWIK-related spinal cord K^+ channel) channels [44].

By far the largest of the four families, the voltage-gated K^+ channels (K_Vs or VGKCs) are comprised of 40 known individual subunits which can form functional homomeric or heteromeric tetramers, leading to incredible physiological variety [41]. As the name suggests, VGKCs activate and inactivate in response to changes in the membrane potential [41]. As such, they have a significant role in excitable cells where they contribute to the downstroke of an action potential. However, VGKCs are widely expressed in many tissues and have many physiological functions. For example, $K_V1.3$, a VGKC expressed in immune cells, is a key regulator of chronically activated effector T memory cells [45].

3.4 High-Throughput Plate Reader Assays: Fluorescent Sensors

3.4.1 Chemical Ca^{2+} Indicator Dyes

Assessment of calcium signalling has been greatly aided by the development of Ca^{2+} dyes which exhibit changes in their fluorescence spectra and/or intensity upon binding of free Ca^{2+} ions, enabling assessment of Ca^{2+} signals at the single cell level or in high-throughput format. Most of these dyes were developed from the Ca^{2+} chelators EGTA, APTRA or BAPTA, and incorporate a fluorophore with the characteristic ion binding groups of these molecules [46–50]. Continuous improvement of these compounds has resulted in a diverse array of dyes with unique properties (Table 3.1). Differences in their Ca^{2+} dissociation constant (K_d) and thus dynamic range, binding kinetics, photostability, sequestration into intracellular compartments, fluorescence quenching characteristics as well as excitation and emission wavelengths govern the usefulness of these compounds in a variety of applications [79]. In particular, the K_d of chemical Ca^{2+} indicator dyes should be carefully matched to the expected Ca^{2+} concentration in the cellular environment, with the useful range over which changes in Ca^{2+} are most reliably detected approximating 0.1–10 x K_d [80]. Thus, measurement of cytoplasmic Ca^{2+} events require high affinity Ca^{2+} dyes, while low affinity dyes will be useful in high Ca^{2+} cellular compartments such as the mitochondria or endoplasmic reticulum [81]. It is, however, important to take into consideration that the K_d of these compounds is affected by pH, temperature, viscosity, ionic strength, protein binding and the presence of other ions such as Mg^{2+} [48, 58, 82–84]. Accordingly, the actual intracellular K_d of these dyes is frequently several orders of magnitude higher than the K_d determined in vitro, and can be expected to vary depending on the cell type and even the cellular compartment assessed [79, 83].

In addition, the binding kinetics of fluorescent Ca^{2+} indicators can affect temporal resolution of Ca^{2+} signals [52, 53, 80]. Ca^{2+} signals are generally transient, so that the binding kinetics of the dye need to be significantly faster than the change in Ca^{2+} concentration if Ca^{2+} signals are to be resolved with sufficient temporal precision [52]. Dyes with slow binding kinetics thus lead to substantial inaccuracies, particularly with respect to the temporal resolution of Ca^{2+} signals. This problem is further compounded by the Ca^{2+} buffering properties displayed by these compounds, especially if present at sufficiently high concentrations [52, 53]. Thus, bright fluorescent dyes which enable reduction in concentration are often preferable to dyes that require higher concentrations in order to achieve sufficient signal strength and similarly, dyes with fast dissociation kinetics are preferable for transient Ca^{2+} signals and high throughput applications.

In addition to these key characteristics, the spectrometric properties of these compounds determine selection of fluorescent Ca^{2+} indicators for specific applications. Principally, chemical Ca^{2+} indicators can be divided into ratiometric and single wavelength dyes.

Table 3.1 Properties of fluorescent indicators

Dye	Excitation wavelength (nm)		Emission wavelength (nm)		K_d (nM)	F_{Max}/F_{min}	K_{on} 1/(M^{a}s)	K_{off} 1/s	References
	unbound	*ion-bound*	*unbound*	*ion-bound*					
Ca^{2+} dyes									
Quin-2	353	333	495	495	60	5–8			[49, 51]
Fura-2	363	335	512	505	135–258	13–25	4.0×10^8	103	[50–53]
Bis-Fura-2	366	338	511	504	370		5.5×10^8	257	[51, 53]
Fura-4F	366	336	511	505	770				[51, 54]
Fura-5F					400				[51]
Fura-6F	364	336	512	505	5300				[51, 54]
Fura-FF	364	335	510	506	5500				[51, 54]
Fura-PE3	364	335	508	500	250	18			[51, 55]
FFP18	364	335	475	408	331	7			[55, 56]
Fura-Red	472	436	657	637	140	5–12			[51, 57]
Mag-Fura-2	369	329	511	508	50,000	6–30	7.5×10^8	26,760	[53, 54]
Indo-1	346	330	485	410	250	20–80	9.4×10^8	180	[50, 58, 59]
Indo 1-PE3	346	330	475	408	260				[51]
Mag-Indo-1	349	328	480	390	35,000	12	2.3×10^5	>1000	[58, 60]
Fluo-3	503	506	526	526	400	40–100	9.2×10^8	587/186	[47, 58, 61]
Fluo-4	491	494	–	516	345	100		350	[48, 62]
Mag-Fluo-4	490	493	–	516	22,000				[48, 51, 63]
Fluo-5F	491	494	–	518	2300			300	[53, 62, 64, 65]
Fluo-5 N	491	494	–	516	90,000				[51, 53, 62, 64, 65]
Fluo-4FF	491	494	–	516	9700				[51, 53, 62, 64, 65]
Fluo-8	490			514	389	> 200			[66]
Fluo-8H	490			514	232	> 200			[66]

Fluo-8 L	490			514	1860	> 200			[66]
Rhod-2,	556	553	576	576	1000	14–100			[51]
X-Rhod-1	576	580	–	602	700	4–100			[51]
Rhod-FF	551		556		19,000				[51]
X-Rhod-5F	576	580	–	602	1600				[51]
X-Rhod-FF	568		605		17,000				[51]
Rhod-5 N	547	549	–	576	320,000				[51]
Calcium-Green-1	506	506	532	532	190	38	0.79×10^9	178	[53, 67]
Calcium-Green-2	506	503	536	536	550	60–100			[51]
Calcium-Green-5 N	506	506	532	532	4000–19,000				[68–70]
Oregon Green 488 BAPTA-1	494	494	523	523	170	14			[51, 71]
Oregon Green 488 BAPTA-2	494	494	523	523	580	100			[51, 71]
Oregon Green 488 BAPTA-6F	494	494	523	523	3000				[51, 71]
Oregon Green 488 BAPTA-5 N	494	494	521	521	20,000	44			[51]
Calcium crimson	590	589	615	615	185	2.5	0.86×10^9	232	[51, 67]
Calcium Ruby	579		598		30,000	32			[72]
Calcium Orange	549	549	575	576	185	3	0.51×10^9	233	[51, 67]
Na+ dyes									
Asante Natrium Green-1	488–517		540		92,000,000	29			[73]
Asante Natrium Green-2	488–517		540		20,000 000	29			[73]
SBFI	379	340	505	505	4000 000	0.08[a]			[73]

(continued)

Table 3.1 (continued)

Dye	Excitation wavelength (nm)		Emission wavelength (nm)		K_d (nM)	F_{Max}/F_{min}	K_{on} 1/(M^{a}s)	K_{off} 1/s	References
	unbound	*ion-bound*	*unbound*	*ion-bound*					
CoroNa Green	492	492	516	516	80,000,000	0.2^a			[73]
CoroNa Red	547	547	576	576	200,000,000				[73]
Sodium Green	506		532		6,000,000				[73]
Membrane potential/voltage-sensitive dyes									
DiBAC$_2$(3)	535		560		n/a				[74]
DiBAC$_4$(3)	492		516		n/a				[75]
FMP dye	530		565		n/a				[76]
CC2-DMPE	425		435		n/a				[77]
K$^+$dyes									
PBFI	390	340	500	500	8,000,000		2.7×10^8	1.4×10^9	[73, 78]
Asante Potassium Green-1	517	517	540	540	54,000,000				[73]
Asante Potassium Green-2	517	517	540	540	18,000,000				[73]

Fluorescence assays have been greatly aided by development of dyes that exhibit changes in fluorescence spectra and/or intensity upon binding of free Ca^{2+}, Na$^+$ or K$^+$ ions, or changes in membrane voltage. Ratiometric dyes exhibit a spectral shift, either in excitation or emission wavelength, upon ion binding, often in conjunction with altered fluorescence intensity, while binding of ions to single wavelength dyes elicits an increase in quantum efficiency, resulting in brighter fluorescence in the absence of spectral excitation or emission shifts. Differences in their dissociation constant (K_d) and thus dynamic range, binding kinetics (K_{on} and K_{off}), and fluorescence characteristics (emission and excitation maxima, F_{Max}/F_{Min}) govern the usefulness of these compounds in a variety of applications

^{a}quantum yield; *n/a* not applicable

3.4.2   Ratiometric Dyes

Ratiometric dyes exhibit a spectral shift, either in excitation or emission wavelength, upon Ca^{2+} binding, often in conjunction with altered fluorescence intensity. This effectively results in increased and decreased fluorescence intensity, respectively, at wavelengths on either side of the isosbestic point. Ratiometric dyes are advantageous for measurement of Ca^{2+} in application where uneven dye loading, dye leakage, photobleaching, compartmentalization, or cell thickness occur, as the fluorescence ratio is independent of the absolute signal strength, thus compensating for these variables [85]. However, these advantages come at the cost of increased photodamage to cells by excitation wavelengths in the ultraviolet range, increased cellular autofluorescence as well as decreased compatibility with caged compounds. Ratiometric dyes are generally poorly suited for high-throughput applications due to the need for equipment capable of dual excitation or emission monitoring. However, as more recent high-throughput plate readers incorporate optics that are, at least in principle, suitable for these ratiometric dyes, these will be discussed here for completeness.

3.4.2.1   Quin-2

Quin-2 is a first generation Ca^{2+} dye developed by the research group of Roger Tsien [49]. It exhibits low quantum yield and absorptivity, necessitating high dye concentrations to achieve adequate signal strength. This is turn leads to problems with Ca^{2+} buffering [86] and has resulted in this dye being largely superseded by newer derivatives.

3.4.2.2   Fura-2

Fura-2 is a dual excitation, single emission ratiometric dye and has become the Ca^{2+} indicator of choice for fluorescence microscopy, where it is more practical to use dual excitation wavelengths and maintain a single emission wavelength [87]. Upon binding of Ca^{2+}, the maximum fluorescence excitation wavelength of Fura-2 shifts from 362 nm to 335 nm, with an accompanying two-fold increase in fluorescence quantum efficiency [50]. In contrast, the fluorescence emission maxima of the free Fura-2 anion and Ca^{2+}-bound Fura-2 are, at 512 and 505 nm, virtually unaltered [50]. Thus, excitation of Fura-2 at 340 and 380 nm results in increased and decreased fluorescence, respectively, at an emission wavelength of $\sim$510 nm. The fluorescence ratio of 340/380 nm therefore increases with increasing concentrations of Ca^{2+}. With a K_d of approximately 135–258 nM, a K_{on} (1/(M.s)) of 4.0×10^8 and a K_{off} (1/s) of 103, Fura-2 and its derivatives are suitable for rapid, time-resolved measurement of cytoplasmic Ca^{2+} signals [50, 53]. In addition, Fura-2 has been reported to be more resistant to photobleaching than Indo-1 [88, 89], although it tends to be more susceptible to intracellular departmentalization [90].

3.4.2.3 Bis-Fura-2

Bis-Fura-2 consists of two fluorophores incorporated with one BAPTA molecule, resulting in brighter signal strength with a slightly reduced K_d (370 nM). With excitation and emission spectra identical to Fura-2, Bis-Fura-2 is particularly suitable for applications which require better signal or tolerate Ca^{2+} buffering poorly and thus require reduced dye concentrations. While on-rates are similar to Fura-2 with a K_{on} (1/(M.s)) of 5.5×10^8, off-rates are slightly higher for Bis-Fura-2 with a K_{off} of 257 (1/s) [53].

3.4.2.4 Fura-4F, Fura-5F, Fura-6F and Fura-FF

These analogues of Fura-2 exhibit similar excitation and emission spectra upon binding of Ca^{2+}, however, the K_d of these compounds has been significantly shifted by addition of one (Fura-4F, Fura-5F, Fura-6F) or two (Fura-FF) fluorine substitutes at varying positions. With K_d values of 400 nM (Fura-5F), 770 nM (Fura-4F), 5300 nM (Fura-6F) and 5500 nM (Fura-FF) [51, 54], these fluorescent Ca^{2+} indicators exhibit intermediate Ca^{2+} affinities and are useful for applications where Ca^{2+} concentrations >1 μM occur.

3.4.2.5 Fura-PE3 (Fura-2 LeakRes)

Fura-PE3 was developed from an analogue of BAPTA, Fura-FF6, by addition of a positive charge in order to improve cytosolic retention of the dye and minimize compartmental sequestration [55]. The spectral properties of Fura-PE3 are identical to Fura-2, but this dye avoids problems associated with uneven loading and dye leakage.

3.4.2.6 FFP18

FFP18 is similar to Fura-PE3 but incorporates a hydrophobic tail that targets this dye to lipids such as cell membranes [55]. The spectral properties of FFP18 are similar to Fura-2, with a slightly decreased K_d of 331 nM [55] and improved hydrophilicity compared to other membrane-associating Ca^{2+} indicators. Thus, FFP18 appears suitable for measurement of membrane-associated Ca^{2+} events [56].

3.4.2.7 Fura-Red

Fura-Red is a Fura-2 analogue excited by visible light, with excitation maxima at approximately 450–500 nm, depending on the presence of Ca^{2+}, and a very long-wave emission maximum at approximately 660 nm. Fura-Red fluorescence decreases upon binding of Ca^{2+}, and in addition, the relatively low quantum efficiency of Fura-Red necessitates use of higher concentrations to achieve an adequate fluorescence signal. The in vitro K_d of Fura-Red is similar to Fura-2 at approximately 140 nM, although the K_d of Fura-Red has been reported to be significantly higher ($\sim$1100–1600 nM) in myoplasm [57]. The large Stokes shift of Fura-Red permits simultaneous measurement of Ca^{2+} as well as other fluorophores excited at $\sim$488 nm. Accordingly, Fura-Red has been used for ratiometric Ca^{2+} measurement in conjunction with the single wavelength Ca^{2+} indicator Fluo-3 [91–93], although ratiometric imaging is also possible with Fura-Red alone using excitation wavelengths of 420/480 nm or 457/488 nm [57, 94].

3.4.2.8 Mag-Fura-2

Mag-Fura-2 (Furaptra) was, as the name suggests, originally developed to measure changes in Mg^{2+} concentration, and exhibits spectral properties similar to Fura-2.

Its propensity for intracellular departmentalization, in combination with its low affinity for Ca^{2+} with a K_d of approximately 50 μM [53, 95, 96], have seen application of this fluorescent indicator to measurement of Ca^{2+} in intracellular IP3-sensitive Ca^{2+} stores [97]. In addition, Mag-Fura-2 retains fast binding kinetics with a K_{on} (1/(M.s)) of 7.5 x 10^8 and particularly fast off-rates [53], enabling measurement of Ca^{2+} responses with little or no kinetic delay [98–100].

3.4.2.9 Indo-1

Like Fura-2, Indo-1 was developed as a BAPTA analogue by the research group of Roger Tsien [50]. However, in contrast to Fura-2, Indo-1 displays shifts in emission wavelength upon Ca^{2+} binding, with emission maxima of 485 nm and 410 nm in the absence and presence of Ca^{2+}, respectively [50]. Thus, this probe is generally more practical in flow cytometry applications, where it is easier to use a single excitation wavelength and monitor two emissions [101]. Indo-1 is also useful as it displays less compartmentalization than Fura-2, although it tends to photobleach more rapidly [90]. With a K_d of 250 nM, it displays slightly lower affinity for Ca^{2+} than Fura-2 and is useful for measurement of Ca^{2+} concentrations in the cytoplasmic range.

3.4.2.10 Indo-1-PE3 (Indo-1 LeakRes)

Like Fura-PE3, Indo-1-PE3 was developed as an Indo-1 analogue less prone to sequestration into intracellular compartments and dye extrusion [51]. Compared to the parent compound, Indo-1-PE3 displays the same spectral properties, but avoids problems with uneven loading, differences in cell thickness and uncontrolled loss of dye fluorescence due to extrusion or photobleaching [102, 103].

3.4.2.11 Mag-Indo-1

Mag-Indo-1 is a low affinity fluorescent Ca^{2+} indicator derived from Indo-1. Its spectral properties are virtually identical to its parent compound, except that they occur at significantly higher Ca^{2+} concentrations (K_d $\sim$35 μM) [58, 60, 80]. In combination with extremely fast kinetics [58], this compound is useful for measurement of Ca^{2+} kinetics in environments with high Ca^{2+} concentration.

3.4.3 Single Wavelength Ca^{2+} Dyes

In contrast to ratiometric fluorescent probes, binding of Ca^{2+} to single wavelength dyes elicits an increase in quantum efficiency, resulting in brighter fluorescence in the absence of spectral excitation or emission shifts. This eliminates the need for sophisticated equipment capable of dual excitation or dual emission monitoring and greatly simplifies experimental protocols. However, while single wavelength dyes generally exhibit large increases in fluorescence intensity upon binding of Ca^{2+}, brightness is also dependant on dye concentration. Thus, in addition to Ca^{2+} binding, fluorescence intensity of single wavelength Ca^{2+} probes is also affected by variables relating to the amount of dye present in cells. Most notably, differences in dye loading, extrusion, compartmentalization and photobleaching, as well as cell thickness and cellular environment can lead to apparent changes in dye concentration or fluorescence. Thus, because fluorescence intensity is the only measure for single wavelength dyes, quantitation of Ca^{2+}, particularly at the single cell level, tends to be less accurate than for ratiometric dyes. As an alternative to wavelength ratioing, time-based ratioing has been suggested as a viable strategy for single wavelength dyes, where the change in fluorescence intensity is expressed relative to a baseline fluorescence value [47, 104]. In circumstances where cell volume and shape changes as well as photobleaching have not been significant, this ratiometric Δ F/F value will approximate changes in Ca^{2+} [47]. Thus, single wavelength dyes have significantly advanced fluorescent Ca^{2+} imaging and are invaluable particularly for high-throughput assessment of Ca^{2+} responses.

3.4.3.1 Fluo-3

Fluo-3 was developed by Roger Tsien and his research group from the calcium chelator BAPTA conjugated with a xanthene chromophore [47]. Fluo-3 is excited by visible light, with absorption and emission maxima at 506 and 526 nm, respectively [47]. While the AM ester of Fluo-3 is virtually non-fluorescent, emission intensity of Fluo-3 increases approximately 40-fold in the presence of Ca^{2+} [47]. With a K_d of 400 nM, this makes Fluo-3 well suited for high resolution of cytosolic Ca^{2+} signals while at the same time being less prone to saturation and Ca^{2+} buffering at resting cytosolic Ca^{2+} than ratiometric dyes such as Fura-2 [61]. Fluo-3 – like all Ca^{2+}-dyes – is pH-sensitive, with an apparent pKa of 6.2, necessitating careful consideration of intracellular pH when measuring Ca^{2+} [47, 61]. In addition, Fluo-3 exhibits biphasic Ca^{2+} dissociation constants which, while also pH-dependent, are faster than Quin-2 and thus allow high time resolution of Ca^{2+} responses at neutral or physiological pH [61].

3.4.3.2 Fluo-4

Fluo-4 is a di-fluoro analogue of Fluo-3, and accordingly exhibits very similar spectral properties with absorption and emission maxima of 494 and 516 nm, respectively [48]. However, Fluo-4 is considerably brighter and more photostable than Fluo-3, and with a K_d of 345 nM [48], has become the dye of choice for measurement of cytosolic Ca^{2+} particularly in high throughput applications. Fluo-4 fluorescence, when excited at 488 nm, increases more than 100-fold upon binding of Ca^{2+}, permitting both use of lower dye concentrations and shorter loading times.

3.4.3.3 Mag-Fluo-4

Mag-Fluo-4 is a low affinity analogue of Fluo-4, and with a K_d of 22 μM, is particularly suitable for measurement of Ca^{2+} in the low μM – mM range. Accordingly, Mag-Fluo-4 has been used for measurement of Ca^{2+} responses in sarcoplasmic reticulum, and due to its affinity for Mg^{2+}, has also found applications in the measurement of intracellular Mg^{2+} [105, 106]. The spectral properties of Mag-Fluo-4, with an excitation maximum of 493 nm and an emission maximum of 516 nm, are very similar to Fluo-4 [48]. In addition, similar to Mag-Fura-2, Mag-Fluo-4 has fast dissociation kinetics which make this dye suitable for measurement of high resolution Ca^{2+} kinetics [63].

3.4.3.4 Fluo-5F, Fluo-5Cl, Fluo-5 N and Fluo-4FF

These mono-or di-substituted Fluo-4 analogues exhibit similar spectral properties to Fluo-4 upon binding of Ca^{2+}, however, the addition of one or two fluorine, chlorine or NO_2 substitutes results in significantly decreased Ca^{2+} affinity. The relatively high K_d values of Fluo-5F (2.3 µM), Fluo-5Cl (6.2 µM), Fluo-4FF (9.7 µM) and Fluo-5 N (90 µM) make these dyes suitable for measurement of Ca^{2+} under conditions which would result in saturation of higher affinity dyes [48]. The dissociation rate constant for Fluo-4 and Fluo-5F were determined as approximately 200–300 s^{-1} in vitro, making these dyes kinetically similar to Fura-2 [53, 62, 64, 65].

3.4.3.5 Fluo-8, Fluo-8H and Fluo-8 L

Fluo-8 and its analogues (AAT Bioquest) are a novel green-emitting Ca^{2+} probe (excitation/emission maxima ~490 nm/514 nm) that have been reported to be considerably brighter than Fluo-3 and even Fluo-4, resulting in improved signal-to-noise [107]. With K_d values of 389 nM (Fluo-8), 232 nM (Fluo-8H) and 1.86 µM (Fluo-8 L), these dyes may prove to be viable alternatives to Fluo-3, Fluo-4 and their analogues [66].

3.4.3.6 Rhod-2, X-Rhod-1and Low Affinity Derivatives

Rhod-2 was developed by the research group of Roger Tsien as a BAPTA analogue incorporating a rhodamine-like fluorophore [47]. This dye exhibits an excitation maximum of 553 nm and an emission maximum of 576 nm, with a K_d in the low µM range (~ 1.0 µM) [47]; while the analogue X-Rhod-1 has slightly shifted absorption and emission maxima (~580/602 nm, respectively) and a K_d of approximately 800 nM [108]. The cationic AM-esters of Rhod-2 and X-Rhod-1 can accumulate, using optimized loading protocols, in negatively charged mitochondria [109, 110], making Rhod-2, X-Rhod-1, and their derivatives useful for measurement of mitochondrial Ca^{2+} [108, 111, 112]. However, due to their relatively higher K_d values and thus lower potential for saturation in high Ca^{2+} environments, the low affinity Rhod-2 derivatives Rhod-FF (K_d 19 µM), X-Rhod-5F (K_d 1.9 µM), X-Rhod-FF (K_d 17 µM) and Rhod-5 N (K_d 320 µM) are preferred for this application [81, 113–115].

3.4.3.7 Calcium-Green-1 and Calcium-Green-2 Indicators

The Calcium-Green indicators display increased fluorescence emission intensity with little spectral shift upon binding of Ca^{2+} [67]. With peak excitation at ~507 nm, peak emission at ~530 nm and K_d values of approximately 200 nM and 550 nM for Calcium-Green-1 and Calcium-Green-2, respectively, these dyes are well suited to measurement of fast cytosolic Ca^{2+} responses [67].

In addition, Calcium-Green dyes have a higher quantum yield than Fluo-3, particularly at saturating Ca^{2+} concentrations, which improves brightness of these dyes at high Ca^{2+} concentrations and provides an excellent dynamic range [67]. In contrast to Fluo-3, Ca^{2+} dissociation from Calcium-Green dyes is mono- rather than bi-exponential with a K_{off} of approximately $180 \, s^{-1}$ [67]. Substitution of a NO_2 group in Calcium-Green-1 produced the low affinity analogue Calcium-Green-5 N, which has similar spectral properties but considerably lower Ca^{2+} affinity with a K_d of approximately 4–19 µM, making this dye less prone to saturation [68–70].

3.4.3.8   Oregon Green 488 BAPTA Indicators

Oregon Green 488 BAPTA-1, Oregon Green 488 BAPTA-2 and their derivatives are fluorinated analogues of Calcium-Green indicators that were developed to achieve increased excitation efficiencies by the 488 nm spectral line of an argon laser. Accordingly, peak excitation and emission wavelengths of these Oregon Ca^{2+} indicators are 494 nm/523 nm. Like the Calcium-green dyes, Oregon Green 488 BAPTA indicators are bright dyes with higher quantum efficiencies than Fluo-3. Oregon Green 488 BATPA-1 in particular, with a K_d of 170 nM, may thus be particularly well suited for measuring small changes in Ca^{2+} near resting cytosolic Ca^{2+}, while the K_d of Oregon Green 488 BAPTA-2, at $\sim$ 580 nM, more closely resembles the Ca^{2+} affinity of Fluo-4 [71]. Oregon Green 488 BAPTA-6F and Oregon Green 488 BAPTA-5 N are the 6'-fluorine and 5'-nitro analogues of Oregon Green 488 BAPTA-1 with reduced Ca^{2+} affinities (Oregon Green 488 BAPTA-6F K_d for Ca^{2+} $\sim$3 µM and Oregon Green 488 BAPTA-5 N K_d for Ca^{2+} $\sim$20 µM) and thus more suitable for measurement of larger Ca^{2+} responses.

3.4.3.9   Calcium Crimson, Calcium Ruby and Calcium Orange

Like other single wavelength dyes, Calcium Crimson, Ruby and Orange exhibit increased fluorescence emission intensity with little or no spectral shift in the presence of Ca^{2+}. As their names suggest, the excitation and emission spectra of these indicators are red-shifted with peak excitation/emission at 549/576 nm (Calcium Orange), 590/615 nm (Calcium Crimson) and 579/598 nm (Calcium Ruby) [72]. Calcium Ruby in particular displays large increases in fluorescence intensity upon Ca^{2+} binding, while all of these indicators are particularly useful where cellular autofluorescence is problematic [72]. The K_ds of these red-emitting Ca^{2+} indicators vary from $\sim$ 300 nM (Calcium Crimson) and $\sim$ 400 nM (Calcium Orange) to 30 µM (Calcium Ruby), allowing selection of Ca^{2+} dyes suitable for most imaging applications [67, 72].

3.4.4 Genetically Encoded Ca^{2+} Indicators

Genetically encoded Ca^{2+} indicators are protein-based sensors whose fluorescence changes in a concentration-dependent manner with alterations in intracellular Ca^{2+}. These indicators may be transiently or stably expressed and have been applied in a range of in vitro and in vivo models [116–120]. Modern genetically encoded Ca^{2+} indicators have improved properties compared to predecessors, which were limited by factors including pH sensitivity and low brightness [51]. One example is the GCaMP6 family of genetically encoded Ca^{2+} indicators, which have favourable characteristics such as a large signal to noise ratio, increased brightness and improved sensitivity comparable to leading small molecule fluorescent dyes [121]. This series has gained widespread use, particularly for the assessment of Ca^{2+} changes associated with neuronal activity using in vivo models [122–126]. Genetically encoded Ca^{2+} indicators are advantageous for certain applications compared to small molecule fluorescent dyes. This includes the repeated measurement of Ca^{2+} alterations occurring over a long time course (e.g. hours, days or weeks) [127], where small molecule Ca^{2+} indicators are not well suited due to sequestration and dye leakage that occurs over time. An example of the utility of genetically encoded over small molecule Ca^{2+} sensors is the use of GCaMP6f to monitor cortical neuron activity in marmosets over several months using a cranial window [128]. An additional advantage of genetically encoded Ca^{2+} indicators is their ability to be targeted to specific subcellular locations, which can enable investigations into Ca^{2+} dynamics in organelles such as the endoplasmic reticulum or mitochondria [129]. Additionally, the relationship between organellar and global cytosolic Ca^{2+} changes can be investigated using co-expression of an organelle-targeted and cytosolic genetically encoded Ca^{2+} indicator with distinct fluorescence spectra. For example, co-expression of GCaMP6s (cytosolic) with R-CEPIA*er* (targeted to endoplasmic reticulum) in MA104 cells following rotavirus infection identified cytosolic Ca^{2+} increases and corresponding endoplasmic reticulum Ca^{2+} depletion occurring over several hours [119]. Following the introduction of a genetically encoded Ca^{2+} indicator transgene into a model organism or stable cell line, further validation should be undertaken to ensure appropriate expression levels of the sensor and preservation of the parental phenotype [130]. Use of genetically encoded Ca^{2+} indicators will likely continue to increase with the continued development of spectrally distinct sensors enabling simultaneous assessment of Ca^{2+} changes with other cell features and further improvements in the properties of those Ca^{2+} sensors targeted to organelles.

3.4.5 Na$^+$ Dyes

Compared to the number of available fluorescent calcium indicators (∼40), there are currently only a limited number of useful fluorescent sodium indicators: SBFI (sodium-binding benzofuran isophthalate), Sodium Green, CoroNa Green and Red, as well as Asante NaTRIUM Green-1 and -2 (ANG-1 and ANG-2) [2, 73, 131]. These have been used widely as probes in fluorescence imaging with conventional and confocal microscopes and have only relatively recently been adapted for high-throughput format [132, 133]. With the exception of CoroNa Red – which is lipophilic and thus readily crosses the plasma membrane – Na^+ indicators are also available in their acetomethyl (AM) ester-conjugated forms which cross the plasma membrane and become trapped in the cytosol after cleavage by endogenous esterases (see below). In addition, the net positive charge of CoroNa Red leads to accumulation of the dye in mitochondria [2, 73, 131].

Like Fura-2, SBFI is a ratiometric dye that exhibits changes in fluorescence quantum yield as well as excitation characteristics upon binding of Na^+ ions. Generally, the selectivity of SBFI for Na^+ is reduced compared to the selectivity of Fura-2 for Ca^{2+}, although the spectral properties of the dyes are similar and the same optical filters and equipment can be used for both. A disadvantage of SBFI, as well as the related K^+ dye PBFI (Potassium-binding Benzofuran Isophthalate), is their sensitivity to ionic strength, pH and temperature, which necessitates calibration using gramicidin or palytoxin [2].

In contrast, Sodium Green, CoroNa Green and Red as well as Asante NaTRIUM Green-1 and 2 are single wavelength dyes with excitation peaks in the visible spectrum, making them particularly useful for high-throughput systems. However, although the quantum yield of these dyes (e.g. 0.2 for CoroNa Green; 0.64 for ANG-2) is considerably greater than that of SBFI (0.08 for SBFI), the dynamic range of fluorescent Na^+ dyes remains considerably lower than that of many Ca^{2+} dyes [2, 73]. These issues are compound by the relatively small Na^+ gradient across the membrane ($\sim$10-fold compared to the $\sim$10,000-fold difference in Ca^{2+} concentrations) which leads to a relatively small increase in intracellular Na^+ concentration in comparison with Ca^{2+} concentrations, which can increase several fold during signalling events. Accordingly, the signal-to-noise ratio for Na^+ detection is correspondingly – and inherently – low in comparison to other ions. Therefore, Na^+ channel activity is more commonly observed by electrophysiological means, or with surrogate measures such as membrane potential or voltage-sensitive dyes, as discussed below. In combination with difficulties relating to physiologically meaningful activation of Na_V channels – albeit these are also concerns with other voltage-gated channels such as K_V channels – these limitations have greatly hampered the use of Na^+ indicators for pharmacological characterisation of Na^+-selective channels, particularly those activated by changes in membrane voltage.

In practice, we have found only ANG-2 to have fluorescence characteristics compatible with high-throughput assay approaches of Na_V channels (unpublished data) and although TRPV1-expressing HEK293 cells loaded with CoroNa Green respond to capsaicin with an increase in fluorescence, these responses are considerably smaller than the corresponding changes in fluorescence in Fluo-4 loaded cells (Fig. 3.2).

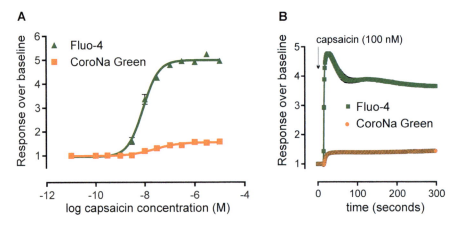

Fig. 3.2 Ca^{2+} and Na^+ responses elicited by capsaicin in cells stably expressing TRPV1
Methods: HEK293 cells stably expressing TRPV1 were generated as previously described [209, 210], plated at a density of 50,000 cells/well on PDL-coated, black-walled 96-well imaging plates and incubated in a 37 deg./5% CO_2 incubator for 24 h

Fluo-4-AM (Invitrogen) or CoroNa Green-AM (Invitrogen) were prepared as 5 mM stock solutions in DMSO and diluted to final concentration of 5 μM in physiological salt solution (PSS; composition see [195]). Cells were incubated with 50 μl of the loading solution in a 37 deg./5% CO2 incubator for 45 minutes. After removal of dye solution, cells were washed twice with PSS prior to addition of 100 μl PSS/well and transfer to the FLIPR^{TETRAPlus}

TRPV1-HEK293 cells were stimulated by addition of 50 μl capsaicin (final concentration: 10 μM – 0.1 nM) prepared as a 3 x concentrated stock solution in PSS and imaged at 1 Hz for 300 reads (excitation/emission: 470–495/515–575 nm) using a FLIPR^{TetraPlus} Fluorescent Plate reader

Raw fluorescence values were converted to response over baseline values using the analysis tool of ScreenWorks 3.1 by subtracting baseline reads (read 1–10) from subsequent fluorescent reads
Results: **(A)** Capsaicin-induced fluorescence responses were of significantly smaller magnitude in CoroNa Green-loaded cells (max. response over baseline 1.58 $\pm$ 0.02) compared with Fluo-4-loaded cells (max. response over baseline 4.96 $\pm$ 0.1); leading to an apparent decrease in the capsaicin EC_{50} in CoroNa Green-loaded cells (pEC_{50} 7.61 $\pm$ 0.07) compared with Fluo-4-loaded cells (pEC_{50} 8.07 $\pm$ 0.03). **(B)** Fluorescence responses to capsaicin (100 nM) differ in both magnitude and kinetics in Fluo-4- and CoroNa Green-loaded cells. Data presented as mean $\pm$ S.E.M. from n = 6 wells of the same plate. Error bars may be obscured by the data symbols

3.4.6 Membrane Potential and Voltage-Sensitive Dyes

Activity of Na^+-, K^+- and Ca^{2+}-permeable ion channels often leads to a change in membrane potential, or transmembrane charge distribution, which is most accurately measured using electrophysiological techniques. However, despite advances in high-throughput electrophysiology approaches (reviewed in [134]), these techniques remain technically challenging and relatively expensive. As the intracellular ion concentration may not be an accurate reflection of membrane potential or cell excitability, alternative imaging approaches are needed to measure membrane depolarisation. With the exception of Ca^{2+}, the flux of ions required to depolarise a cell of 50 μm diameter by 60 mV ($\sim$ 30 million ions) may not appreciably change the intracellular or extracellular ion concentration. Thus, the development of membrane potential dyes has greatly aided high-throughput assessment of cell excitability. These dyes generally incorporate into the lipid membrane of cells and exhibit a change in fluorescence spectra and/or intensity upon depolarisation. Examples include the fast responding styryl dyes, which have high temporal resolution (3 ms) but limited sensitivity; oxonol and carbocyanine dyes and their derivatives, including bis-(1,3-dibutylbarbituric acid) trimethine oxonol ($DiBAC_4(3)$); and FLIPR Membrane Potential (FMP) Dye which displays slower kinetics and moderate sensitivity. The use of two probes allows more sensitive detection of membrane potential changes via the measurement of fluorescence resonance energy transfer (FRET). More recently, the use of photoinduced electron transfer (PeT) has driven the development of dyes such as FluoVolt, which is both fast, sensitive and displays a higher fluorescent yield. Finally, there have been considerable advances in the effective use of genetically encoded membrane potential indicators. Membrane potential is an attractive output for high-throughput screening as it is sensitive, generic, and applicable to living cells.

Although membrane potential dyes act via different mechanisms, sensitivity to depolarisation is dependent on having a dynamic interaction with phospholipid bilayers. Generally, styryl, oxonol, and carbocyanine dyes move across lipid bilayers dependent on the Nernstian distribution during depolarisation [135, 136]. The direction depends on their charge, but results in a change in the concentration of intracellular and extracellular dye. The interactions of dye, membrane or cytosolic molecules results in changes to the emission spectra and/or quantum yield underpinning their usefulness as voltage-sensitive dyes. As the membrane potential of mitochondria can be as high as -180 mV, cationic dyes are less suitable for measuring depolarisation in the plasma membrane. However, they have been employed in limited circumstances to discriminate mitochondrial membrane potential [137, 138]. FMP Dye and $DiBAC_4(3)$ possess lipophilic tails, and negatively charged head groups, allowing these dyes to embed in cell membranes and act as voltage sensors. The negatively charged dyes preferentially sequester to the extracellular leaflet of a hyperpolarised cell membrane (inside the cell is negatively charged) and the presence of an extracellular quencher limits emissions in this state. Upon depolarisation, dye molecules are recruited to the cytosolic side of the plasma membrane, escaping the quencher, and increasing emissions. Despite sharing a mechanism of action, FLIPR Membrane Potential Dye display superior temporal and membrane potential sensitivity to $DiBAC_4(3)$ [76, 139].

In contrast, FRET relies on the dynamic interactions of the emission donor and/or acceptor within the membrane. The most commonly used system utilises a membrane bound coumarin derivative, N-[6-chloro-7-hydroxycourmarin-3-carbonyl] dimyristroyl phosphatidyl ethanolamine, CC2-DMPE) that acts as a FRET donor, co-loaded with a mobile voltage-sensitive oxonol acceptor, such as $DiBAC_4(3)$ [77, 140] . Other FRET systems co-load two oxonol derivatives (e.g. $DiBAC_1(3)$ and $DiBAC_1(5)$) into the plasma membrane as a donor/acceptor pair, respectively. Recording concurrent emission wavelength maxima from both molecules after excitation of the donor results in transfer of emissions to the acceptor, proportional to the proximity of the two molecules. As the position of at least one of the donor/acceptor pair in the membrane is dependent on cell polarity, this allows the ratiometric measurement of membrane potential [141]. The catalogue of voltage-dependent dyes is ever expanding, with new "VoltageFluors" being developed to overcome the shortcomings of previously designed dyes. More recently, experimenters have utilised photoinduced electron transfer (PeT), by tethering a dichlorosulfoflorescein reporter to a dimethylaniline electron donor via a phenylenevinylene molecular "wire" [142]. The fluorescent reporter, embedded in the extracellular membrane leaflet is quenched by the electron donor when the cell is hyperpolarised. Depolarisation of the membrane decreases the dipole moment of the dye, allowing increased emissions from the fluorescein, proportional to the voltage difference across the membrane. Due to these slight differences in mechanism of action, voltage-sensitive dyes may require elaborate, highly sensitive detection equipment and/or labour intensive tissue preparation, while others are more useful for high-throughput screening applications.

In addition to dyes, there are many genetically encoded voltage indicators (GEVI) that have been developed over the past 20 years. There are two main structurally distinct GEVI. The first to be developed exploited the voltage-dependent structural rearrangement of voltage-sensing domains from potassium channels [143], sodium channels [144], phosphatases [145, 146], and proton channels [147]. These are generally tethered to one or more fluorescent molecules and the structural shift during depolarisation changes some spectral property of the fluorophore/s. In addition to those based on voltage-sensing domains are a contingent of GEVI engineered from microbial opsins [148]. These rely on membrane depolarisation changing the protonation of the opsin's retinal cofactor, resulting in changes to absorption or emission intensity from a conjugated fluorophore [149]. All GEVI offer similar advantages to calcium indicators, and can be selectively and stably expressed in cells and animal models, making this approach advantageous for in vivo imaging as well as imaging over multiple days. Some benefit has also been derived from combining GEVI and traditional voltage-sensitive dyes to assess membrane potential [150]. As GEVI have extensively reviewed in recent years, this chapter will not discuss them in detail, however comprehensive comparison can be found elsewhere [151–153].

When designing experiments for membrane potential assays, it is important to consider limitations in temporal resolution. Depending on the cell type, the duration of an action potential ranges from 1–2 ms in neurons, 2–5 ms in skeletal muscle cells, and 200–400 ms in cardiac muscle cells. With the exception of styryl dyes, PeT dyes, and some GEVIs, all membrane potentials dyes $t_{1/2}$ fall in the range between 1 s and >60 s, and are thus not capable of resolving the membrane potential changes characteristic of an action potential [139]. However, paramount to these considerations is the system in which compounds are being screened – many in vitro cell systems are not capable of firing action potentials and so the inability to resolve single action potentials is not necessarily a limitation.

Membrane potential dyes impart several advantages over ion-sensing dyes. One benefit of membrane potential dyes is the high stability, with stable signals being reported 4–15 h after loading [139, 154]. Unfortunately, due to their relatively insolubility in water, dye loading may require the addition of a pluronic surfactant, which may interfere with interactions between test compounds and molecular targets. Each cell system requires dye concentration titration with the aim of optimal loading to increase signal-to-noise ratio. This can be particularly laborious when using FRET-based assays, as multiple compounds require loading [140]. In addition, membrane potential dyes can be particularly prone to artefacts, with alterations in pH, quantity of cytosolic proteins and/or RNA, and membrane integrity leading to changes in fluorescence that can be misinterpreted as changes in membrane potential. Likewise, molecules – particularly charged compounds – can directly interact with dye molecules, and positive hits should be screened at least once in a cell free system or better yet, in the absence of the suspected molecular target.

3.4.6.1 Styryl Dyes

Styryl dyes offer the temporal resolution to accurately measure the onset of a single action potential and low background fluorescence [155]. One example is pyridinium, 4-(2-(6(dibutylamino)-2-naphthalenyl)-1-(3-sulfopropyl)-hydroxide (di-4 ANEPPS), which has been used in a number of different in vitro membrane potential measurements including red blood cells, squid giant axon, cardiomyocytes, and isolated membrane vesicles [156]. The excitation and emission wavelengths are dependent on the cell system, but ranges between 520 ± 20 nm (Ex) and >640 nm (Em). However, the relative fluorescence change upon depolarisation is limited (<10% with depolarisations of more than 100 mV) and the dye has proven incompatible with some cultured cell lines [157, 158]. Other styryl dyes, such as Di-2-ANEPEQ must be microinjected into cells, making them capable of distinguishing sub-cellular depolarisation [159]. Thus, stryryl dyes are well suited for use with highly optimized, low noise equipment, but ill-suited for high-throughput screening.

3.4.6.2 Carbocyanine Derivatives

Indo-, thia-, and oxo-cyanine dyes were some of the first potentiometric dyes used. Oxocyanine (DiO) $C_6(3)$, the most commonly used for this application, has excitation/emission maxima of 484/501 nm [160]. However, due to its high affinity for mitochondria and other intracellular organelles, its application for measurement of plasma membrane potential is limited. Another cyanine derivative, mercocyanine 540, was also initially used as a membrane potential probe, but its high phototoxicity limits its application in living cells [161].

3.4.6.3   Oxonol Derivatives

A diverse family of oxonol derivatives with a variety of fluorescent properties have been produced. Two examples of oxonol derivatives with potentiometric sensitivity are $DiBAC_2(3)$ and $DiBAC_4(3)$ with excitation/emission maxima of 537/554 nm and 492/516 nm, respectively. Due to their overall negative charge, these dye molecules are excluded from mitochondrial membranes, thus reducing noise. $DiBAC_4(3)$ is often used with an extracellular quencher (*e.g.* bromophenol blue) allowing a substantial improvement in signal to noise ratio when compared to other oxonol and carbocyanine derivatives. $DiBAC_4(3)$ also shows other advantages over $DiBAC_2(3)$ with more than $2 \times$ fluorescence yield upon depolarisation and faster kinetics ($t_{1/2} = 30$ s) [139]. However, the slow kinetics of activation leaves the dye susceptible to interference from temperature and pH changes as longer read times may be required to capture differences in membrane depolarisation [162]. Additionally, loading cells with $DiBAC_4(3)$ requires surfactant, and additional dye must be washed off before reading, making the application of these dyes difficult for poorly adherent cells.

3.4.6.4   FLIPR Membrane Potential Dye

The FMP dye is excited at 530 nm with emissions measured at 565 nm. There is the option of FMP dye being bundled with two proprietary quenchers, "red" and "blue", which must be optimised to individual assay conditions. FMP displays a relative fluorescence change of more than four times that of $DiBAC_4(3)$ during depolarisation [139]. This is accompanied by enhanced kinetics over oxonol and carbocyanine derivatives ($t_{1/2} = 8$ s). The inclusion of a proprietary quencher offers more stability to changing temperatures, making them useful during short or long time-course experiments. The loading of cells is streamlined, with fast loading times (< 30 min) and the absences of wash steps making these dyes useful for poorly adherent cells while enhancing the reproducibility of assays [163, 164].

3.4.6.5   Fret-Based Membrane Potential Dyes

The most commonly used commercially available FRET system for measurement of membrane potential is based on the $CC2$-DMPE/$DiBAC_2(3)$ donor/acceptor pair. The donor is excited at 405 ± 15 nm with emission monitored at both 460 nm and 580 nm, leading to a ratiometric readout that is relatively resistant to changes in pH, temperature and bleaching [165]. While ratiometric systems offer the most sensitive measurement of membrane potential available, more complex loading procedures are required [139]. These include optimisation of the FRET donor concentration for each cell system assayed, and multiple wash steps which may limit application when cells are poorly adherent. Another fluorescent molecule, dipicrylamine (DPA, Abs 406 nM) has been paired with DiO for the measurement of membrane voltage potential by FRET [166]. This system is applicable to two-photon imaging, allowing voltage measurement from deep within tissue slices [167].

3.4.6.6 FluoVolt™

Described as "molecular wires", these fluorophores display the favourable properties of both fast and slow membrane potential dyes [168]. The fast temporal dynamics allow the dye to change fluorescence and resolve single action potentials while displaying a greater relative fluorescence yield upon depolarisation (approx. 25% per 100 mV). As the conjugated fluorescein shares an excitation/emission maxima with GFP, standard filter sets can be used for imaging. Furthermore, the molecule can be applied for high throughput screening, and is easily loaded into cells with the use of a proprietary surfactant [169]. Due to emission overlap with other biologically useful molecules, a far-red fluorescent equivalent has also been developed [170]. Thus, one can record the membrane potential change of a cell, while simultaneously using GCAMP6 to record the resulting calcium spike, for example.

3.4.7 K^+ Dyes

Similar to Na^+-sensitive dyes, the repertoire of K^+ dyes is relatively limited. PBFI comprises a benzofuranyl fluorophore that is linked to a crown ether chelator [2] which confers some modest selectivity for K^+ ions, although the K_d is also affected by pH, temperature, ionic strength and the presence of Na^+ ions. Like SBFI and Fura-2, PBFI is a ratiometric dye with excitation/emission peaks of 336/557 and 338/507 nm in the K^+-free and K^+-bound state, respectively. Although the selectivity of PBFI for K^+ is sufficient for the quantification of intracellular K^+ concentrations in vitro – owing to the relatively higher concentration compared to Na^+ – the dye properties of PBFI make it poorly suited to high-throughput applications. In addition, dyes such as PBFI are less useful for high-throughput screening because of their broad excitation spectra, which can overlap with many optically active compounds in drug screening libraries [171].

More recently, single wavelength K^+ dyes – including Asante Potassium Green-1 and -2 – have been reported [73]. Although these have not been widely used as yet, at least in principle they could be useful for high-throughput and/or high-content applications in light of their more favourable properties, such as excitation/emission peaks of ~515/540 nm and more favourable loading [172].

Historically, rubidium ion (Rb^+) efflux was used to determine K^+ channel activity by incubating cells expressing the K^+ channel of interest in a Rb^+ rich buffer and examining the movement of Rb^+ using radioactive Rb^+ isotopes or Rb^+ ions in conjunction with atomic absorption spectroscopy (AAS) or as its commonly known, "flame photometry", to quantify the amount of Rb^+ [171, 173]. In addition to Rb^+, many K^+-permeable channels are also thallium (Tl^+) permeable. Although permeability of K^+-permeable ion channels families and subtypes to Tl^+ has not been assessed systematically, the pore structure is relatively conserved and it is reasonable to assume that these channels are likely Tl^+ permeable [171]. Accordingly, a change in fluorescence occurs when Tl^+ enters a cell loaded with a Tl^+-sensitive dye via K^+ channels, thus providing a surrogate measure of K^+ channel activity [171].

3.4.8 Dye Loading

Several methods for introduction of Ca^{2+}, Na^+ and K^+ dyes into the cell cytoplasm have been developed; these include ATP-induced permeabilization, electroporation, hypo-osmotic shock, cationic liposomes, chelators mediating dye uptake through pinocytosis, microinjection as well as loading of dyes coupled to acetoxymethyl (AM) ester [58, 82, 83, 97, 174–177]. Of these, the acetoxymethyl (AM) ester loading technique has become popular due to its simplicity, ease of use and low toxicity and is particularly well suited to high-throughput applications.

3.4.9 AM Ester Loading Technique

As free poly-anionic, large fluorescent probes are unable to passively cross the cell membrane, fluorescent dyes can be conjugated to lipophilic acetoxymethyl (AM) groups to render them membrane-permeable. Once in the cytoplasm, ubiquitous esterases hydrolyse these derivatized indicators, which again become unable to passively cross the plasma membrane, thus effectively trapping the free fluorescent probe. As an additional advantage, these AM derivatives are often non-fluorescent, thus reducing or eliminating fluorescence from non-hydrolysed extracellular dye.

3.4.10 No-Wash Extracellular Quenchers

Because physiologically, extracellular Ca^{2+} and Na^+ concentrations are high, fluorescence signals from the extracellular compartment generally need to be excluded to enable measurement of the often relatively small changes in cytosolic ion concentration. This can be achieved either by physically removing extracellular dyes by media or buffer exchange ("washing"), or alternatively by incorporation of fluorescence quenchers in the extracellular media. Quenchers that have been used successfully include Trypan Blue, haemoglobin and Brilliant Black [178, 179]. In addition, several of these quenchers or "no-wash" kits are now commercially available, and they can provide significant improvements in assay performance particularly for cells that are only weakly adherent, or for cells that are prone to dye extrusion. However, as the composition and nature of these quenchers is largely proprietary information, it can be difficult to assess potential interference of quenchers with assays or to design protocols for assay optimization [179].

3.4.11   Problems with AM Ester Loading

While the AM ester loading technique is undoubtedly one of the most widely used and easiest approaches to introducing fluorescent dyes into cells, a number of issues – including limited dye solubility, sequestration, incomplete hydrolysis and dye extrusion – can limit the applications of this approach. These issues are typically dye- and also cell-specific, and perhaps best understood (or appreciated) for fluorescent Ca^{2+} dyes. Specific considerations relating to each of these issues are thus described below in the context of Ca^{2+} dyes, although similar problems may also arise for other indicators.

Solubility By virtue of increased lipophilicity, many AM esters are poorly soluble in aqueous solutions, thus necessitating inclusion of dispersants such as pluronic acid in the loading media. However, satisfactory dye loading can be achieved in the absence of pluronic acid for several dyes (including e.g. Fluo-4, Fura-2 and ANG-2) provided they are prepared as 100x to 1000x stock solutions in dimethylsulfoxide and carefully stored frozen as aliquots to avoid water absorption and thus precipitation. Individual optimisation of loading conditions is thus advisable.

Sequestration Once introduced into the cell cytoplasm, fluorescent Ca^{2+} dyes start to accumulate into intracellular membrane-bound vacuoles and organelles such as the endoplasmic reticulum (ER) and mitochondria in a process commonly referred to as sequestration or compartmentalization [180]. This process is, however, not restricted to the AM ester loading technique and may result in increasing baseline fluorescence readings as the free dye accumulates in high Ca^{2+} intracellular compartments, as well as accompanying decreases in cytosolic Ca^{2+} responses due to dye loss. To minimize dye sequestration, loading with the lowest AM ester concentration that produces reliable Ca^{2+} signals, as well as loading for the shortest possible time is beneficial. Loading and imaging cells at room temperature rather than 37 °C can also help reduce dye sequestration, although restricting recordings to approximately 30 minutes largely avoids this problem. In addition, while for most applications, retention of the dye in the cell cytoplasm is desirable, dye sequestration can be exploited to assess calcium levels in organelles [181–184].

Incomplete AM Ester Hydrolysis Residual cytosolic non-hydrolysed dyes, due to insufficient intracellular esterase activity or failure to completely remove AM ester dyes, can lead to signal artefacts, most notably an apparent decrease in fluorescence response, as AM esters tend to be non-fluorescent [185, 186]. In addition, efficiency of ester hydrolysis can be highly variable and often depends on the cell type; incubation at 37 °C generally improves ester hydrolysis but optimal conditions usually have to be determined empirically. In contrast, excessive extracellular ester hydrolysis leads to poor dye loading and as a result poor fluorescence signals, while extracellular hydrolysed probes tend to provide high fluorescence background [187].

Dye Extrusion Extrusion of hydrolysed intracellular Ca^{2+} probes by cellular anion transporters results in decreased available dye concentrations and thus, decreased signal strength. This problem is not restricted to the AM-loading technique, and can be a particular problem in certain cell types. Dye extrusion or leakage can be minimized by incorporation of anion transport inhibitors such as probenecid or sulphinpyrazone [180]. However, these compounds can alter cellular function and should thus be used with caution. Loading cells and measuring fluorescence a quickly as possible, as well as performing experiments at room temperature rather than 37 °C generally also aid in minimising dye extrusion [80].

3.5 Assay Platforms

3.5.1 High-Throughput Fluorescence Plate Readers

A number of plate reader platforms suitable for fluorescence assays in high-throughput format are available commercially. The industry-leading instruments are set apart from lower throughput instruments by the capacity to dispense liquids and measure fluorescence emission from 96,384 or 1536 wells simultaneously with high temporal resolution. This in turn negates some of the problems associated with conventional fluorescence imaging that arise from uneven dye loading, extrusion, intracellular compartmentalization or photobleaching, as loading and imaging conditions are constant across all wells. Thus, high-throughput imaging plate readers such as the FLIPR^{TetraPlus} (Molecular Devices), Hamamatsu FDSS7000EX (Hamamatsu), WaveFront Biosciences PanOptic (WaveFront Biosciences) and CellLux (Perkin Elmer) are well-suited for the primary identification of drug leads as well as detailed pharmacological characterization of compounds. Because functional responses are measured, pharmacological characterization of full or partial agonists as well as competitive and non-competitive antagonists at a range of targets can be accomplished.

While the precise specifications differ between these instruments, all combine sophisticated liquid handling systems with optics that permit real-time fluorescence readings before, during, and after compound addition to enable characterisation of kinetic responses. Desirable attributes of these high-throughput platforms include high-precision robotics in 96-, 384- or 1534-well format that permit custom configuration of aspiration and dispense height and speed to minimize disruption of cell monolayers and addition artefacts. The ability to program multiple additions is crucial for testing antagonists as well as agonists in the same experiment, with additional reagents plates permitting assessment of more complex responses such as Ca^{2+} influx through Orai-1 channels following store depletion [188].

The incorporation of state-of-the-art excitation and detection systems – such as two sets of customisable LED banks and an EMCCD cooled charge-coupled device (CCD) camera for fluorescence detection, or optionally an ICCD intensified CCD camera for the FLIPR^{TetraPlus}, or a variable-wattage white light xenon lamp with Hamamatsu fluorescence/luminescence camera for the FDSS7000EX – permit the versatile design of experiments. Measurements from an entire plate can be taken in as little as sub-second intervals, with up to 800 (FLIPR^{TetraPlus}) or 4000 (Hamamatsu FDSS7000EX) reads enabling prolonged real-time pre-incubation with antagonists as well as kinetics recordings for even the slowest fluorescent responses.

In addition, an increasing number of available emission filters, with wavelengths ranging from 340 nm to > 650 nm, permit selection of fluorescent dyes that are most suitable for individual applications. An important consideration is the number of excitation/emission filter sets that each instrument can be configured to, which in case of the FLIPR^{TetraPlus} is currently limited to 2 and 3, respectively.

While most high-throughput instruments include optional temperature-control, typically permitting heating from ambient to $\sim$ 40 °C, most assays perform well at ambient temperature and both the speed and precision of plate heating are insufficient for modulation of most temperature-sensitive pharmacological targets, although this feature could be useful if assaying poorly soluble compounds.

3.5.2 Single-Cell Imaging

In high-throughput plate reader assays, it is the sum of individual cellular responses in each well that is recorded and analysed. It is this relatively simple experimental output that has made this technology particularly amenable to high-throughput applications. However, for some studies, multiple readouts may be desirable. Such assays, where multiple cellular or subcellular responses or components are recorded individually and in parallel, have been described as "high-content" [189]. For example, changes in intracellular Ca^{2+} can be quantified simultaneously – in individual cells – with other cellular and subcellular features or events such as cell morphology, cell death or nuclear translocation [127].

High-content imaging assays are often considered as target-agnostic and provide information on cellular phenotypes – associated with one, or many, targets and signalling pathways – through the analysis of fluorescence microscopy images in high-throughput format. However, in the context of this review, the term "high-content imaging" will be used for single-cell, live imaging approaches that permit analysis of individual cellular responses rather than population responses that are assessed in high-throughput plate reader assays.

For high-content imaging assays, cells can be prepared and loaded in the same manner as for high-throughput imaging assays, but the imaging instrumentation and software used is necessarily different. At its most basic, cells are imaged by videomicroscopy using an inverted fluorescent microscope equipped with an appropriate light source, excitation and emission filters and a camera. However, while this approach may be "high-content" in the sense that many single cells can be captured simultaneously, it is limited in its throughput as typically a single experiment (often from individual coverslips or wells on which cells were plated) is conducted. Nevertheless, because this type of single cell imaging provides information that is complementary to high-throughput approaches, it has found use in multiple applications. One of these is the systematic characterisation of distinct functional populations in heterogeneous cell populations such as primary cell cultures [190, 191]. In this application, researchers exploit a suite of well-defined pharmacological tools to functionally profile cell populations — an application that is complementary to immunocytochemistry or RNAseq-based techniques. In cases where a cell population is already well-defined, a second application becomes possible: defining the cellular mechanism of action of a compound of interest [192–194]. An example of this application is shown in Fig. 3.3, where high-content imaging of cultured DRG neurons was used to define the cellular mechanism of action of two algogenic toxins. Using a similar high-content imaging approach, the Na_V activator Pacific Ciguatoxin-1 was shown to selectively activate TRPA1-expressing sensory neurons in cultured DRG neurons [195]. While these example compounds act as agonists, it is worth mentioning that this type of experiment is equally relevant for the investigation of antagonists: a cellular response can be elicited by the application of a well-defined agonist in the presence/absence of the compound of interest and cellular responses recorded in the same manner. The information acquired through this type of high-content imaging assays, particularly when coupled with complementary immunocytochemistry, RNAseq or gene-knockout data, can be valuable in determining the molecular mechanism of action of a compound of interest.

In more recent years, single cell imaging has been adapted to automated platforms that permit imaging from cells plated in multi-well plates – so-called high-content plate readers, examples of which will be discussed in the following paragraphs – and have opened the door to other applications such as the parallel recording and quantitation of spatial redistribution of cellular or subcellular targets and/or individual cell and organelle morphology [189].

Temperature control and CO_2 regulation is available in some systems (e.g. ImageXpress Micro (Molecular Devices), Operetta CLS LIVE (PerkinElmer)) which permits live cell imaging over extended durations, such as time-lapse measurement of Ca^{2+} changes in cells undergoing cellular events. For example, using an ImageXpress Micro, global $[Ca^{2+}]_{CYT}$ increases were identified in the subset of MDA-MB-231 breast cancer cells undergoing cell death within 6 h of ceramide treatment [196]. The Kinetic Image Cytometer (Vala) can also be used for timelapse high content imaging and is capable of rapid image capture particularly suited to transient cellular events occurring over milliseconds. This instrument also has the option for electrical stimulation of cells, useful for applications such as neuronal screening. Wide-field imaging is commonly used for high-content systems, although confocal imaging is also available. The Opera Phenix (PerkinElmer) is an example of a confocal high content imaging system, which can also be configured with four sCMOS cameras to enable increased frame rates, low signal to noise ratio

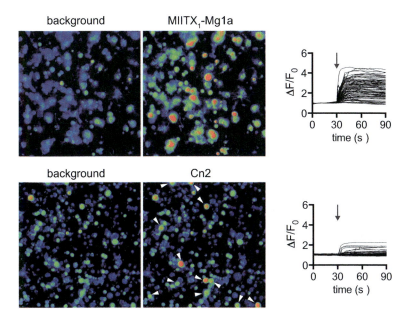

Fig. 3.3 High-content calcium imaging of the effects of algogenic toxins on DRG cells
Methods: Experiments involving animals were approved by the University of Queensland Animal ethics committee. DRGs from 6–8 week male C57BL/6 mice were dissociated as described [211], plated on a poly-D-lysine-coated 96-well culture plate and maintained in a 37 deg./5% CO_2 incubator overnight
Fluo-4-AM (Invitrogen) was prepared according to the manufacturer's instructions. Cells were loaded for 30 min in a 37 deg./5% CO_2 incubator, followed by 30 min equilibration at room temperature. After loading, the dye-containing solution was replaced with assay solution (1x Hanks' balanced salt solution, 20 mM HEPES), and the plate transferred to the inverted microscope setup
Fluorescence corresponding to $[Ca^{2+}]_i$ of 100–150 DRG cells per experiment was monitored in parallel using a Nikon Ti-E Deconvolution inverted microscope, equipped with a Lumencor Spectra LED Lightsource. Baseline fluorescence was monitored for 30 s, after which assay solution was replaced with $MIITX_1$-Mg1a (10 μM in assay solution; panel A) or Cn2 (500 nM in assay solution; panel B).
Raw fluorescence values were converted to $\Delta F/F_0$ (fluorescence minus baseline fluorescence, divided by baseline fluorescence)
Results: $MIITX_1$-Mg1a is the major peptide component of the venom of the giant red bull ant, *Myrmecia gulosa*. The toxin exerts algogenic activity in vivo, and causes a sustained increase in intracellular Ca^{2+} concentration in neuronal as well as non-neuronal cells obtained from a culture of dissociated dorsal root ganglia. This activity is suggestive of a widespread cellular target, and indeed, $MIITX_1$-Mg1a acts on cellular lipid bilayers where it causes a leak in membrane ion conductance, which, in neurons, triggers depolarization [211]. In contrast, the algogenic Na_V1.6-selective Cn2 – a major peptide component of the venom from the scorpion *Centruroides noxious* – caused a sustained increase in intracellular calcium concentration only in a subset of large diameter neurons (labelled with an arrowhead), indicative of a molecular target expressed only in this neuronal population. Snapshots shown are before and after addition of each compound. Each trace corresponds a single cell in each field of view. Arrows indicate toxin addition

and parallel measurement in four colour channels simultaneously. The ImageXpress Micro has the option for automated liquid addition, a feature that can be used for the assessment of transient Ca^{2+} changes following compound addition. However, as for automated high-throughput imaging systems – and in contrast to traditional setups that permit perfusion imaging – high-content imaging in microwell plates typically does not permit removal of compounds and is limited to cumulative additions. High-content imaging systems can also vary in number of objectives, filter configurations and ability to measure bright-field in addition to fluorescence (e.g. IN Cell Analyzer 6000 (GE Healthcare), CellInsight CX7 (Thermo Fisher Scientific)). A key challenge of high-content imaging approaches is the inherent generation of large datasets, which typically requires advanced computing solutions for analysis including large data storage capacity. Most instruments are packaged with analysis software, with open-source applications such as ImageJ and CellProfiler also used widely. For comprehensive reviews on high-content imaging in drug discovery, including live-cell and fluorescent Ca^{2+} imaging, see e.g. [197–200].

3.5.3 High-Throughput Electrophysiology Platforms

For assessment of ion channel function, in particular voltage-gated channels, electrophysiological assays remain the technique of choice as they assess the flux of ions across cell membranes in real time. However, these approaches are not only very low-throughput, but also require highly technically skilled operators and are thus generally limited to detailed characterisation of few compounds. A number of automated high-throughput electrophysiology platforms have been developed in recent years to overcome some of these limitations. These include the IonFlux 16, HT and Mercury (FluxIon Biosciences), IonWorks.

Barracuda Plus (Molecular Devices), IonWorks Quattro (Molecular Devices), PatchXpress 7000A (Molecular Devices), SyncroPatch (96 and 384/768 PE) (Nanion), Patchliner (Nanion), Port-a-Patch (Nanion), QPatch 8, 16 and HT (Sophion), QPatch II (Sophion), Qube (Sophion), Patch-Server (Multichannel Systems GmbH) as well as Robocyte2 for automated recording from oocytes (Multichannel Systems GmbH).

Most of these systems operate on the principle of planar patch-clamping, where cell membranes are patched though a small opening on a plate or chip which can be configured to achieve recordings from 4 to 384 cells in parallel. While the specific advantages and disadvantages of these systems are discussed in detail elsewhere [134, 201, 202], the arguably more physiologically relevant read-out of high-throughput electrophysiological assays is offset by the relatively greater purchasing and operating cost of these systems.

3.6 Technical Considerations

3.6.1 Design and Optimization of Fluorescence Assays

High-content and high-throughput fluorescence assays can be performed on cells either heterologously or endogenously expressing ion channels and GPCRs of interest [203]. While endogenously expressed targets may provide more physiologically relevant data owing to co-expression appropriate auxiliary subunits, there is usually little control over subtypes present or expression levels [203]. In contrast, heterologously expressed ion channels and GPCRs allow control of both subtype expression as well as selection of cells with appropriate target expression levels, and are thus often the favoured approach particularly for primary identification of novel lead compounds.

High-throughput fluorescence assays are generally possible with both adherent and suspension cell lines, though adherent cell lines tend to produce less addition artefacts. Suspension cell lines often require use of no-wash kits using extracellular quenchers, while this is optional for adherent cell lines. While commercial no-wash kits are considerably more expensive, they also require less time due to omission of washing steps and can improve assay performance particularly for poorly adherent cell lines. In addition, it is possible to use readily available dyes for in-house optimisation of no-wash protocols, which may be considerably cheaper than commercial kits which typically include quenchers of undisclosed composition.

The most important aspect of a successful high-content or high-throughput assays is the quality of cells. For adherent cell lines used in high-throughput applications, this optimally requires a 90–95% confluent monolayer of cells. As a rule of thumb, over-confluent cells tend to produce better assays than sub-confluent or patchy cells, although changes in cell morphology that occur as a result of confluency need to be considered. For example, some cancer cell lines differentiate when over-confluent, and receptor expression can also vary with cell confluency. In contrast, high-content assays ideally require well-defined "single" cells that grow evenly, do not clump and are not overly confluent. For both approaches, firmly adherent cells are desirable so no disruption of the cell layer occurs even with multiple washing or liquid addition steps. If cells can be even partially dislodged from their tissue culture flask by mechanical means, cell adhesion probably needs to be optimized for successful imaging assays.

Cell adherence can be improved by coating plates with poly-D-lysine (PDL), poly-L-lysine (PLL), collagen, Matrigel, laminin, poly-L-ornithine (PLO) or similar substances. Adhesion-promoting plates such as Cell^{BIND} (Corning) plates can also help, though improvement appears generally less dramatic than with coating. Typically, ideal plating conditions as well as coating substances and procedures need to be optimised according to specific cell, assay and research requirements. Generally, cells should be plated at least overnight, although cell viability and morphology is often improved with plating several days prior to the assay.

3.6.2   Agonists

In order to design successful fluorescence assays, cellular responses need to be elicited by addition of suitable agonists; the choice of agonist is often crucial to the success of the assay and will need to be determined carefully.

In the case of ligand-gated ion channels or GPCRs, these will generally consist of endogenous or exogenous ligands. For example, addition of nicotine or acetylcholine could be utilized to elicit Ca^{2+} responses mediated through nicotinic acetylcholine receptors (nAChR) [203]. Similarly, membrane depolarization can be induced through addition of KCl in order to activate voltage-gated Ca^{2+} channels, though in the case of N-type Ca_V ($Ca_v2.2$), addition of extracellular Ca^{2+} is required to elicit sufficiently robust responses [28]. It may also be necessary to co-express inward rectifier K^+ channels to adequately control membrane potential [204], or to include allosteric modulators to delay inactivation or enhance signalling, as is the case for $\alpha7$ nAChR [203]. A more difficult application is high-throughput assays for voltage-gated Na^+ or K^+ channels; the former typically require use of allosteric modulators such as veratridine or deltamethrin which may skew the pharmacology of these channels in undesirable, or unexpected, ways.

3.6.3   Limitations

Fluorescence assays can be used for drug screening or the primary identification of novel drug leads, as well as for detailed pharmacological characterization of known or novel agonists and antagonists. However, while the most commonly used protocol involves addition of antagonists first, followed by addition of agonists, this setup can lead to ambiguous pharmacological profiles due to the kinetics of receptor/channel binding and the elicited fluorescence response. For example, the kinetics of Ca^{2+} responses measured using fluorescent approaches result from a combination of the binding rate constants of the Ca^{2+} dye, the binding kinetics of agonists and antagonists used, the rate of inactivation or desensitization of receptors and ion channels as well as extrusion and sequestration of Ca^{2+} by pumps such as PMCA or SERCA [205, 206]. This complex interplay may lead to functional profiles resembling irreversible rather than reversible agonism [207]. In addition, physical factors such as mixing of test compounds in the well – which in turn can be affected by fluid volume, addition speed, and the physicochemical properties of the compounds – may not only affect the kinetics but also the magnitude of the observed responses. While some of these difficulties are decreased by addition to, and measurement from, an entire multi-well plate simultaneously, it nonetheless is important to remember that fluorescent responses remain a surrogate measure of ion accumulation rather than ionic currents or receptor activation. In the case of membrane potential, Na^+ or K^+ responses, some of these issues are compounded by the unfavourable dye kinetics and, as discussed earlier, the need for non-physiological or allosteric agonists.

Nonetheless, response kinetics obtained using fluorescence imaging approaches can provide valuable information. For example, activation of IP3 receptors result in Ca^{2+} kinetics that are often quite distinct from those of voltage- or ligand ion channels. GPCR activation leads to relatively slow, concentration-dependent increases in intracellular Ca^{2+}, which peak approximately 5–20 s after ligand addition and return to baseline within $\sim$ 100–180 s. In contrast, ion channels often display extremely rapid increases in Ca^{2+}, which, depending on the desensitization kinetics of the channel as well as the Ca^{2+} load, may or may not return to baseline (see e.g. Fig. 3.2). Thus, when using fluorescence imaging for pharmacological characterization of compounds, careful design and validation of individual assays is essential.

3.7   Future Directions

Development of the first fluorescence plate readers incorporating liquid handling robotics for the real-time measurement of intracellular Ca^{2+} concentrations began more than 20 years ago and has revolutionised screening of ion channel and GPCR drug targets. The long-standing industry-leader of these instruments – the FLIPR – was undoubtedly developed due to the availability of ever-improving fluorescent Ca^{2+} sensors, an area pioneered by the Nobel prize-winning Roger Tsien and colleagues [47]. While the miniaturisation and development of true high-throughput capacities of these platforms were driven predominantly by industry needs relating to high-throughput screening campaigns, these advances have also led to several advantages for mechanistic and pharmacological studies. These include for example the ability to obtain fluorescence reads from all wells in parallel, thus eliminating or at least considerably reducing assay errors arising from inconsistent dye loading or sequestration effects; as well as the ease with which large datasets can be collected. Thus, in conjunction with an ever-increasing repertoire of fluorescent indicators, optical capabilities and access to high-throughput platforms outside of large pharmaceutical companies, fluorescence assays are evolving to become invaluable tools addressing basic biological research questions. For example, using FLIPR-based fluorescence Ca^{2+} imaging, Regulator of G protein signalling 4 (RGS4) was found to negatively regulate muscarinic receptor 3 (M3R)-mediated Ca^{2+} signalling in pancreatic β-cells, leading to reduced glucose-stimulated insulin secretion, thus identifying novel mechanisms and drug targets for treatment of diabetes [208].

However, while high-throughput fluorescence assays are increasingly used as tools across many areas of research, these applications also highlight some of the limitations of these assays. Currently, most kinetic fluorescence plate reader assays are poorly suited to drug discovery or mechanistic studies of targets that are activated, or regulated, by temperature (in particular cool temperatures), mechanical stimuli, or membrane voltage. Nonetheless, high-throughput fluorescence assays remain an important and easily accessible methodology that is particularly powerful when combined with complementary approaches such as single cell imaging, patch-clamp electrophysiology or multi-electrode arrays, as these techniques can inform on signalling events at the subcellular, channel and network excitability level, which cannot (yet) be captured by existing high-throughput fluorescence platforms. The utility and versatility of high-throughput fluorescence assays will undoubtedly continue to improve, particularly if instrumentation with routine capabilities for control of membrane voltage and temperature become available.

References

1. Santos R et al (2017) A comprehensive map of molecular drug targets. Nat Rev Drug Discov 16(1):19–34
2. Johnson I, Spence MTZ (eds) (2010) *Molecular probes handbook, a guide to fluorescent probes and labeling technologies*, 11th edn. Invitrogen, Carlsbad
3. Cardoso FC et al (2017) Modulatory features of the novel spider toxin mu-TRTX-Df1a isolated from the venom of the spider Davus fasciatus. Br J Pharmacol 174(15):2528–2544
4. Deuis JR et al (2016) Development of a muO-Conotoxin analogue with improved lipid membrane interactions and potency for the analgesic Sodium Channel NaV1.8. J Biol Chem 291(22):11829–11842
5. Deuis JR et al (2017) Pharmacological characterisation of the highly NaV1.7 selective spider venom peptide Pn3a. Sci Rep 7:40883
6. Jin AH et al (2015) delta-Conotoxin SuVIA suggests an evolutionary link between ancestral predator defence and the origin of fish-hunting behaviour in carnivorous cone snails. Proc Biol Sci 282(1811):pii: 20150817
7. Klint JK et al (2015) Seven novel modulators of the analgesic target NaV 1.7 uncovered using a high-throughput venom-based discovery approach. Br J Pharmacol 172(10):2445–2458
8. Vetter I et al (2012) Isolation, characterization and total regioselective synthesis of the novel muO-conotoxin MfVIA from Conus magnificus that targets voltage-gated sodium channels. Biochem Pharmacol 84(4):540–548
9. Vetter I et al (2012) Characterisation of Na(v) types endogenously expressed in human SH-SY5Y neuroblastoma cells. Biochem Pharmacol 83(11):1562–1571
10. Benjamin ER et al (2006) State-dependent compound inhibition of Nav1.2 sodium channels using the FLIPR Vm dye: on-target and off-target effects of diverse pharmacological agents. J Biomol Screen 11(1):29–39
11. Liu K et al (2010) High-throughput screening for Kv1.3 channel blockers using an improved FLIPR-based membrane-potential assay. J Biomol Screen 15(2):185–195
12. Trivedi S et al (2008) Cellular HTS assays for pharmacological characterization of Na(V)1.7 modulators. Assay Drug Dev Technol 6(2):167–179
13. Zhao F et al (2016) Development of a rapid throughput assay for identification of hNav1.7 antagonist using unique efficacious sodium channel agonist, antillatoxin. Mar Drugs 14(2):pii: E36
14. Clapham DE (2007) Calcium signaling. Cell 131(6):1047–1058
15. Berridge MJ, Lipp P, Bootman MD (2000) The versatility and universality of calcium signalling. Nat Rev Mol Cell Biol 1(1):11–21
16. Brini M, Carafoli E (2011) The plasma membrane Ca^{2+} ATPase and the plasma membrane sodium calcium exchanger cooperate in the regulation of cell calcium. Cold Spring Harb Perspect Biol 3(2):pii: a004168
17. Pathak T, Trebak M (2018) Mitochondrial Ca^{2+} signaling. Pharmacol Ther 192:112–123

18. Bygrave FL, Benedetti A (1996) What is the concentration of calcium ions in the endoplasmic reticulum? Cell Calcium 19(6):547–551
19. Berridge MJ (1993) Inositol trisphosphate and calcium signalling. Nature 361(6410):315–325
20. Liu AM et al (2003) Galpha(16/z) chimeras efficiently link a wide range of G protein-coupled receptors to calcium mobilization. J Biomol Screen 8(1):39–49
21. Zhu T, Fang LY, Xie X (2008) Development of a universal high-throughput calcium assay for G-protein- coupled receptors with promiscuous G-protein Galpha15/16. Acta Pharmacol Sin 29(4):507–516
22. Kostenis E, Waelbroeck M, Milligan G (2005) Techniques: promiscuous Galpha proteins in basic research and drug discovery. Trends Pharmacol Sci 26(11):595–602
23. Vasudevan NT (2017) cAMP assays in GPCR drug discovery. Methods Cell Biol 142:51–57
24. Jiang LI et al (2007) Use of a cAMP BRET sensor to characterize a novel regulation of cAMP by the sphingosine 1-phosphate/G13 pathway. J Biol Chem 282(14):10576–10584
25. Trehan A et al (2014) CANDLES, an assay for monitoring GPCR induced cAMP generation in cell cultures. Cell Commun Signal 12:70
26. Matthiesen K, Nielsen J (2011) Cyclic AMP control measured in two compartments in HEK293 cells: phosphodiesterase K(M) is more important than phosphodiesterase localization. PLoS One 6(9):e24392
27. Catterall WA (2000) Structure and regulation of voltage-gated Ca^{2+} channels. Annu Rev Cell Dev Biol 16:521–555
28. Benjamin ER et al (2006) Pharmacological characterization of recombinant N-type calcium channel (Cav2.2) mediated calcium mobilization using FLIPR. Biochem Pharmacol 72(6):770–782
29. Belardetti F et al (2009) A fluorescence-based high-throughput screening assay for the identification of T-type calcium channel blockers. Assay Drug Dev Technol 7(3):266–280
30. Monteith GR et al (2007) Calcium and cancer: targeting Ca^{2+} transport. Nat Rev Cancer 7(7):519–530
31. Duncan RS et al (2010) Control of intracellular calcium signaling as a neuroprotective strategy. Molecules 15(3):1168–1195
32. Talukder MA, Zweier JL, Periasamy M (2009) Targeting calcium transport in ischaemic heart disease. Cardiovasc Res 84(3):345–352
33. Suhail M (2010) Na, K-ATPase: Ubiquitous multifunctional transmembrane protein and its relevance to various pathophysiological conditions. J Clin Med Res 2(1):1–17
34. Davidson S et al (2014) Human sensory neurons: membrane properties and sensitization by inflammatory mediators. Pain 155(9):1861–1870
35. Catterall WA, Goldin AL, Waxman SG (2005) International Union of Pharmacology. XLVII. Nomenclature and structure-function relationships of voltage-gated sodium channels. Pharmacol Rev 57(4):397–409
36. Zhang XY et al (2018) Veratridine modifies the gating of human voltage-gated sodium channel Nav1.7. Acta Pharmacol Sin 39(11):1716–1724
37. Albuquerque EX et al (2009) Mammalian nicotinic acetylcholine receptors: from structure to function. Physiol Rev 89(1):73–120
38. Hanukoglu I, Hanukoglu A (2016) Epithelial sodium channel (ENaC) family: phylogeny, structure-function, tissue distribution, and associated inherited diseases. Gene 579(2):95–132
39. Kweon HJ, Suh BC (2013) Acid-sensing ion channels (ASICs): therapeutic targets for neurological diseases and their regulation. BMB Rep 46(6):295–304
40. Kuang Q, Purhonen P, Hebert H (2015) Structure of potassium channels. Cell Mol Life Sci 72(19):3677–3693
41. Grizel AV, Glukhov GS, Sokolova OS (2014) Mechanisms of activation of voltage-gated potassium channels. Acta Nat 6(4):10–26
42. Womack MD, Chevez C, Khodakhah K (2004) Calcium-activated potassium channels are selectively coupled to P/Q-type calcium channels in cerebellar Purkinje neurons. J Neurosci 24(40):8818–8822

43. Hibino H et al (2010) Inwardly rectifying potassium channels: their structure, function, and physiological roles. Physiol Rev 90(1):291–366
44. Feliciangeli S et al (2015) The family of K2P channels: salient structural and functional properties. J Physiol 593(12):2587–2603
45. Chiang EY et al (2017) Potassium channels Kv1.3 and KCa3.1 cooperatively and compensatorily regulate antigen-specific memory T cell functions. Nat Commun 8:14644
46. Otten PA, London RE, Levy LA (2001) A new approach to the synthesis of APTRA indicators. Bioconjug Chem 12(1):76–83
47. Minta A, Kao JP, Tsien RY (1989) Fluorescent indicators for cytosolic calcium based on rhodamine and fluorescein chromophores. J Biol Chem 264(14):8171–8178
48. Gee KR et al (2000) Chemical and physiological characterization of fluo-4 Ca^{2+}-indicator dyes. Cell Calcium 27(2):97–106
49. Tsien RY (1980) New calcium indicators and buffers with high selectivity against magnesium and protons: design, synthesis, and properties of prototype structures. Biochemistry 19(11):2396–2404
50. Grynkiewicz G, Poenie M, Tsien RY (1985) A new generation of Ca^{2+} indicators with greatly improved fluorescence properties. J Biol Chem 260(6):3440–3450
51. Whitaker M (2010) Genetically encoded probes for measurement of intracellular calcium. Methods Cell Biol 99:153–182
52. Kao JP, Tsien RY (1988) Ca^{2+} binding kinetics of fura-2 and azo-1 from temperature-jump relaxation measurements. Biophys J 53(4):635–639
53. Naraghi M (1997) T-jump study of calcium binding kinetics of calcium chelators. Cell Calcium 22(4):255–268
54. Wokosin DL, Loughrey CM, Smith GL (2004) Characterization of a range of fura dyes with two-photon excitation. Biophys J 86(3):1726–1738
55. Vorndran C, Minta A, Poenie M (1995) New fluorescent calcium indicators designed for cytosolic retention or measuring calcium near membranes. Biophys J 69(5):2112–2124
56. Etter EF et al (1996) Near-membrane $[Ca^{2+}]$ transients resolved using the Ca^{2+} indicator FFP18. Proc Natl Acad Sci U S A 93(11):5368–5373
57. Kurebayashi N, Harkins AB, Baylor SM (1993) Use of fura red as an intracellular calcium indicator in frog skeletal muscle fibers. Biophys J 64(6):1934–1960
58. Lattanzio FA Jr (1990) The effects of pH and temperature on fluorescent calcium indicators as determined with Chelex-100 and EDTA buffer systems. Biochem Biophys Res Commun 171(1):102–108
59. Westerblad H, Allen DG (1996) Intracellular calibration of the calcium indicator indo-1 in isolated fibers of Xenopus muscle. Biophys J 71(2):908–917
60. Launikonis BS et al (2005) Confocal imaging of $[Ca^{2+}]$ in cellular organelles by SEER, shifted excitation and emission ratioing of fluorescence. J Physiol 567(Pt 2):523–543
61. Eberhard M, Erne P (1989) Kinetics of calcium binding to fluo-3 determined by stopped-flow fluorescence. Biochem Biophys Res Commun 163(1):309–314
62. Goldberg JH et al (2003) Calcium microdomains in aspiny dendrites. Neuron 40(4):807–821
63. Hollingworth S, Gee KR, Baylor SM (2009) Low-affinity Ca^{2+} indicators compared in measurements of skeletal muscle Ca^{2+} transients. Biophys J 97(7):1864–1872
64. Scott R, Rusakov DA (2006) Main determinants of presynaptic Ca^{2+} dynamics at individual mossy fiber-CA3 pyramidal cell synapses. J Neurosci 26(26):7071–7081
65. Woodruff ML et al (2002) Measurement of cytoplasmic calcium concentration in the rods of wild-type and transducin knock-out mice. J Physiol 542(Pt 3):843–854
66. Falk S, Rekling JC (2009) Neurons in the preBotzinger complex and VRG are located in proximity to arterioles in newborn mice. Neurosci Lett 450(3):229–234
67. Eberhard M, Erne P (1991) Calcium binding to fluorescent calcium indicators: calcium green, calcium orange and calcium crimson. Biochem Biophys Res Commun 180(1):209–215
68. Stout AK, Reynolds IJ (1999) High-affinity calcium indicators underestimate increases in intracellular calcium concentrations associated with excitotoxic glutamate stimulations. Neuroscience 89(1):91–100

69. Rajdev S, Reynolds IJ (1993) Calcium green-5N, a novel fluorescent probe for monitoring high intracellular free Ca^{2+} concentrations associated with glutamate excitotoxicity in cultured rat brain neurons. Neurosci Lett 162(1–2):149–152
70. Eilers J et al (1995) Calcium signaling in a narrow somatic submembrane shell during synaptic activity in cerebellar Purkinje neurons. Proc Natl Acad Sci U S A 92(22):10272–10276
71. Agronskaia AV, Tertoolen L, Gerritsen HC (2004) Fast fluorescence lifetime imaging of calcium in living cells. J Biomed Opt 9(6):1230–1237
72. Gaillard S et al (2007) Synthesis and characterization of a new red-emitting Ca^{2+} indicator, calcium ruby. Org Lett 9(14):2629–2632
73. Teflabs (2011) Fluorescent ion indicator handbook, vol 1–44. Texas Teflabs, Austin
74. Sguilla FS, Tedesco AC, Bendhack LM (2003) A membrane potential-sensitive dye for vascular smooth muscle cells assays. Biochem Biophys Res Commun 301(1):113–118
75. Brauner T, Hulser DF, Strasser RJ (1984) Comparative measurements of membrane potentials with microelectrodes and voltage-sensitive dyes. Biochim Biophys Acta 771(2):208–216
76. Baxter DF et al (2002) A novel membrane potential-sensitive fluorescent dye improves cell-based assays for ion channels. J Biomol Screen 7(1):79–85
77. Adams DS, Levin M (2012) Measuring resting membrane potential using the fluorescent voltage reporters DiBAC4(3) and CC2-DMPE. Cold Spring Harb Protoc 2012(4):459–464
78. Meuwis K et al (1995) Photophysics of the fluorescent K^+ indicator PBFI. Biophys J 68(6):2469–2473
79. Thomas D et al (2000) A comparison of fluorescent Ca^{2+} indicator properties and their use in measuring elementary and global Ca^{2+} signals. Cell Calcium 28(4):213–223
80. Paredes RM et al (2008) Chemical calcium indicators. Methods 46(3):143–151
81. Yasuda R et al (2004) *Imaging calcium concentration dynamics in small neuronal compartments.* Sci STKE 2004(219):pl5
82. Oliver AE et al (2000) Effects of temperature on calcium-sensitive fluorescent probes. Biophys J 78(4):2116–2126
83. O'Malley DM, Burbach BJ, Adams PR (1999) Fluorescent calcium indicators: subcellular behavior and use in confocal imaging. Methods Mol Biol 122:261–303
84. Poenie M (1990) Alteration of intracellular Fura-2 fluorescence by viscosity: a simple correction. Cell Calcium 11(2–3):85–91
85. Dustin LB (2000) Ratiometric analysis of calcium mobilization. Clin Appl Immunol Rev 1(1):5–15
86. Hesketh TR et al (1983) Duration of the calcium signal in the mitogenic stimulation of thymocytes. Biochem J 214(2):575–579
87. O'Connor N, Silver RB (2007) Ratio imaging: practical considerations for measuring intracellular Ca^{2+} and pH in living cells. Methods Cell Biol 81:415–433
88. Becker PL, Fay FS (1987) Photobleaching of fura-2 and its effect on determination of calcium concentrations. Am J Phys 253(4 Pt 1):C613–C618
89. Scheenen WJ et al (1996) Photodegradation of indo-1 and its effect on apparent Ca^{2+} concentrations. Chem Biol 3(9):765–774
90. Wahl M, Lucherini MJ, Gruenstein E (1990) Intracellular Ca^{2+} measurement with Indo-1 in substrate-attached cells: advantages and special considerations. Cell Calcium 11(7):487–500
91. Floto RA et al (1995) IgG-induced Ca^{2+} oscillations in differentiated U937 cells; a study using laser scanning confocal microscopy and co-loaded fluo-3 and fura-red fluorescent probes. Cell Calcium 18(5):377–389
92. Lipp P, Niggli E (1993) Ratiometric confocal Ca^{2+}-measurements with visible wavelength indicators in isolated cardiac myocytes. Cell Calcium 14(5):359–372
93. Schild D, Jung A, Schultens HA (1994) Localization of calcium entry through calcium channels in olfactory receptor neurones using a laser scanning microscope and the calcium indicator dyes Fluo-3 and Fura-red. Cell Calcium 15(5):341–348
94. Lohr C (2003) Monitoring neuronal calcium signalling using a new method for ratiometric confocal calcium imaging. Cell Calcium 34(3):295–303

95. Martinez-Zaguilan R, Parnami J, Martinez GM (1998) Mag-Fura-2 (Furaptra) exhibits both low (microM) and high (nM) affinity for Ca^{2+}. Cell Physiol Biochem 8(3):158–174
96. Zhao M, Hollingworth S, Baylor SM (1996) Properties of tri- and tetracarboxylate Ca^{2+} indicators in frog skeletal muscle fibers. Biophys J 70(2):896–916
97. Hofer AM (2005) Measurement of free $[Ca^{2+}]$ changes in agonist-sensitive internal stores using compartmentalized fluorescent indicators. Methods Mol Biol 312:229–247
98. Claflin DR et al (1994) The intracellular Ca^{2+} transient and tension in frog skeletal muscle fibres measured with high temporal resolution. J Physiol 475(2):319–325
99. Konishi M et al (1991) Myoplasmic calcium transients in intact frog skeletal muscle fibers monitored with the fluorescent indicator furaptra. J Gen Physiol 97(2):271–301
100. Berlin JR, Konishi M (1993) Ca^{2+} transients in cardiac myocytes measured with high and low affinity Ca^{2+} indicators. Biophys J 65(4):1632–1647
101. MacFarlane AWt, Oesterling JF, Campbell KS (2010) Measuring intracellular calcium signaling in murine NK cells by flow cytometry. Methods Mol Biol 612:149–157
102. Takahashi A et al (1999) Measurement of intracellular calcium. Physiol Rev 79(4):1089–1125
103. Overholt JL et al (2000) HERG-like potassium current regulates the resting membrane potential in glomus cells of the rabbit carotid body. J Neurophysiol 83(3):1150–1157
104. Smith SJ, Augustine GJ (1988) Calcium ions, active zones and synaptic transmitter release. Trends Neurosci 11(10):458–464
105. Lee S, Lee HG, Kang SH (2009) Real-time observations of intracellular Mg2+ signaling and waves in a single living ventricular myocyte cell. Anal Chem 81(2):538–542
106. Shmigol AV, Eisner DA, Wray S (2001) Simultaneous measurements of changes in sarcoplasmic reticulum and cytosolic. J Physiol 531(Pt 3):707–713
107. Bioquest A (2011) Quest fluo-8™ calcium reagents and screen quest™ Fluo-8 NW calcium assay kits. [cited 2011]
108. Gerencser AA, Adam-Vizi V (2005) Mitochondrial Ca^{2+} dynamics reveals limited intramitochondrial Ca^{2+} diffusion. Biophys J 88(1):698–714
109. Tao J, Haynes DH (1992) Actions of thapsigargin on the Ca^{2+}-handling systems of the human platelet. Incomplete inhibition of the dense tubular Ca^{2+} uptake, partial inhibition of the Ca^{2+} extrusion pump, increase in plasma membrane Ca^{2+} permeability, and consequent elevation of resting cytoplasmic Ca^{2+}. J Biol Chem 267(35):24972–24982
110. Trollinger DR, Cascio WE, Lemasters JJ (1997) Selective loading of Rhod 2 into mitochondria shows mitochondrial Ca^{2+} transients during the contractile cycle in adult rabbit cardiac myocytes. Biochem Biophys Res Commun 236(3):738–742
111. Davidson SM, Yellon D, Duchen MR (2007) Assessing mitochondrial potential, calcium, and redox state in isolated mammalian cells using confocal microscopy. Methods Mol Biol 372:421–430
112. Gerencser AA, Adam-Vizi V (2001) Selective, high-resolution fluorescence imaging of mitochondrial Ca^{2+} concentration. Cell Calcium 30(5):311–321
113. Pologruto TA, Yasuda R, Svoboda K (2004) Monitoring neural activity and $[Ca^{2+}]$ with genetically encoded Ca^{2+} indicators. J Neurosci 24(43):9572–9579
114. David G, Talbot J, Barrett EF (2003) Quantitative estimate of mitochondrial $[Ca^{2+}]$ in stimulated motor nerve terminals. Cell Calcium 33(3):197–206
115. Simpson AW (2006) Fluorescent measurement of $[Ca^{2+}]c$: basic practical considerations. Methods Mol Biol 312:3–36
116. Palmer AE et al (2004) Bcl-2-mediated alterations in endoplasmic reticulum Ca^{2+} analyzed with an improved genetically encoded fluorescent sensor. Proc Natl Acad Sci U S A 101(50):17404–17409
117. McCombs JE, Gibson EA, Palmer AE (2010) Using a genetically targeted sensor to investigate the role of presenilin-1 in ER Ca^{2+} levels and dynamics. Mol BioSyst 6(9):1640–1649

118. Kuchibhotla KV et al (2008) Abeta plaques lead to aberrant regulation of calcium homeostasis in vivo resulting in structural and functional disruption of neuronal networks. Neuron 59(2):214–225
119. Perry JL et al (2015) Use of genetically-encoded calcium indicators for live cell calcium imaging and localization in virus-infected cells. Methods 90:28–38
120. Ouzounov DG et al (2017) In vivo three-photon imaging of activity of GCaMP6-labeled neurons deep in intact mouse brain. Nat Methods 14(4):388–390
121. Chen TW et al (2013) Ultrasensitive fluorescent proteins for imaging neuronal activity. Nature 499(7458):295–300
122. Cichon J, Gan WB (2015) Branch-specific dendritic Ca^{2+} spikes cause persistent synaptic plasticity. Nature 520(7546):180–185
123. Sheffield ME, Dombeck DA (2015) Calcium transient prevalence across the dendritic arbour predicts place field properties. Nature 517(7533):200–204
124. Sun W et al (2016) Thalamus provides layer 4 of primary visual cortex with orientation- and direction-tuned inputs. Nat Neurosci 19(2):308–315
125. Falkner S et al (2016) Transplanted embryonic neurons integrate into adult neocortical circuits. Nature 539(7628):248–253
126. Lee KS, Huang X, Fitzpatrick D (2016) Topology of ON and OFF inputs in visual cortex enables an invariant columnar architecture. Nature 533(7601):90–94
127. Bassett JJ, Monteith GR (2017) Genetically encoded calcium indicators as probes to assess the role of calcium channels in disease and for high-throughput drug discovery. Adv Pharmacol 79:141–171
128. Sadakane O et al (2015) Long-term two-photon calcium imaging of neuronal populations with subcellular resolution in adult non-human primates. Cell Rep 13(9):1989–1999
129. Suzuki J, Kanemaru K, Iino M (2016) Genetically encoded fluorescent indicators for Organellar calcium imaging. Biophys J 111(6):1119–1131
130. Tian L, Hires SA, Looger LL (2012) Imaging neuronal activity with genetically encoded calcium indicators. Cold Spring Harb Protoc 2012(6):647–656
131. Schreiner AE, Rose CR (2012) Quantitative imaging of intracellular sodium. In: Mendez-Vilas A (ed) Current microscopy contributions to advances in science and technology. Formatex Research Center, Badajoz, pp 119–129
132. O'Donnell GT et al (2011) *Evaluation of the sodium sensing dye asante natrium green 2 in a voltage-gated sodium channel assay in 1536-well format.* Merck & Co., Inc., Whitehouse Station
133. Antonia B et al (2016) *Overcoming historical challenges of Nav1.9 voltage gated sodium channel as a drug discovery target for treatment of pain.* Icagen, Durham
134. Priest BT et al (2004) Automated electrophysiology assays. In: Sittampalam GS et al (eds) Assay guidance manual. Eli Lilly & Company and the National Center for Advancing Translational Sciences, Bethesda
135. Sims PJ et al (1974) Studies on the mechanism by which cyanine dyes measure membrane potential in red blood cells and phosphatidylcholine vesicles. Biochemistry 13(16):3315–3330
136. Bashford CL, Chance B, Prince RC (1979) Oxonol dyes as monitors of membrane potential. Their behavior in photosynthetic bacteria. Biochim Biophys Acta 545(1):46–57
137. Rottenberg H, Wu S (1998) Quantitative assay by flow cytometry of the mitochondrial membrane potential in intact cells. Biochim Biophys Acta 1404(3):393–404
138. Huang SG (2002) Development of a high throughput screening assay for mitochondrial membrane potential in living cells. J Biomol Screen 7(4):383–389
139. Wolff C, Fuks B, Chatelain P (2003) Comparative study of membrane potential-sensitive fluorescent probes and their use in ion channel screening assays. J Biomol Screen 8(5):533–543
140. Gonzalez JE, Maher MP (2002) Cellular fluorescent indicators and voltage/ion probe reader (VIPR) tools for ion channel and receptor drug discovery. Receptors Channels 8(5–6):283–295

141. Dunlop J et al (2008) Ion channel screening. Comb Chem High Throughput Screen 11(7):514–522
142. Woodford CR et al (2015) Improved PeT molecules for optically sensing voltage in neurons. J Am Chem Soc 137(5):1817–1824
143. Siegel MS, Isacoff EY (1997) A genetically encoded optical probe of membrane voltage. Neuron 19(4):735–741
144. Ataka K, Pieribone VA (2002) A genetically targetable fluorescent probe of channel gating with rapid kinetics. Biophys J 82(1 Pt 1):509–516
145. Dimitrov D et al (2007) Engineering and characterization of an enhanced fluorescent protein voltage sensor. PLoS One 2(5):e440
146. Murata Y et al (2005) Phosphoinositide phosphatase activity coupled to an intrinsic voltage sensor. Nature 435(7046):1239–1243
147. Kang BE, Baker BJ (2016) Pado, a fluorescent protein with proton channel activity can optically monitor membrane potential, intracellular pH, and map gap junctions. Sci Rep 6:23865
148. Kralj JM et al (2011) Optical recording of action potentials in mammalian neurons using a microbial rhodopsin. Nat Methods 9(1):90–95
149. Maclaurin D et al (2013) Mechanism of voltage-sensitive fluorescence in a microbial rhodopsin. Proc Natl Acad Sci U S A 110(15):5939–5944
150. Mutoh H, Akemann W, Knopfel T (2012) Genetically engineered fluorescent voltage reporters. ACS Chem Neurosci 3(8):585–592
151. St-Pierre F, Chavarha M, Lin MZ (2015) Designs and sensing mechanisms of genetically encoded fluorescent voltage indicators. Curr Opin Chem Biol 27:31–38
152. Xu Y, Zou P, Cohen AE (2017) Voltage imaging with genetically encoded indicators. Curr Opin Chem Biol 39:1–10
153. Yang HH, St-Pierre F (2016) Genetically encoded voltage indicators: opportunities and challenges. J Neurosci 36(39):9977–9989
154. Whiteaker KL et al (2001) Validation of FLIPR membrane potential dye for high throughput screening of potassium channel modulators. J Biomol Screen 6(5):305–312
155. Muller W, Windisch H, Tritthart HA (1986) Fluorescent styryl dyes applied as fast optical probes of cardiac action potential. Eur Biophys J 14(2):103–111
156. Loew LM et al (1992) A naphthyl analog of the aminostyryl pyridinium class of potentiometric membrane dyes shows consistent sensitivity in a variety of tissue, cell, and model membrane preparations. J Membr Biol 130(1):1–10
157. Fluhler E, Burnham VG, Loew LM (1985) Spectra, membrane binding, and potentiometric responses of new charge shift probes. Biochemistry 24(21):5749–5755
158. Gross D, Loew LM (1989) Fluorescent indicators of membrane potential: microspectrofluorometry and imaging. Methods Cell Biol 30:193–218
159. Canepari M et al (2010) Imaging inhibitory synaptic potentials using voltage sensitive dyes. Biophys J 98(9):2032–2040
160. Waggoner AS (1979) Dye indicators of membrane potential. Annu Rev Biophys Bioeng 8:47–68
161. Picaud S, Wunderer HJ, Franceschini N (1988) 'Photo-degeneration' of neurones after extracellular dye application. Neurosci Lett 95(1–3):24–30
162. Yamada A et al (2001) Usefulness and limitation of DiBAC4(3), a voltage-sensitive fluorescent dye, for the measurement of membrane potentials regulated by recombinant large conductance Ca^{2+}−activated K^+ channels in HEK293 cells. Jpn J Pharmacol 86(3):342–350
163. Joesch C et al (2008) Use of FLIPR membrane potential dyes for validation of high-throughput screening with the FLIPR and microARCS technologies: identification of ion channel modulators acting on the GABA(A) receptor. J Biomol Screen 13(3):218–228
164. Molinski SV et al (2015) Facilitating structure-function studies of CFTR modulator sites with efficiencies in mutagenesis and functional screening. J Biomol Screen 20(10):1204–1217
165. Maher MP, Wu NT, Ao H (2007) pH-Insensitive FRET voltage dyes. J Biomol Screen 12(5):656–667

166. Bradley J et al (2009) Submillisecond optical reporting of membrane potential in situ using a neuronal tracer dye. J Neurosci 29(29):9197–9209
167. Fink AE et al (2012) Two-photon compatibility and single-voxel, single-trial detection of subthreshold neuronal activity by a two-component optical voltage sensor. PLoS One 7(8):e41434
168. Miller EW et al (2012) Optically monitoring voltage in neurons by photo-induced electron transfer through molecular wires. Proc Natl Acad Sci U S A 109(6):2114–2119
169. Bedut S et al (2016) High-throughput drug profiling with voltage- and calcium-sensitive fluorescent probes in human iPSC-derived cardiomyocytes. Am J Physiol Heart Circ Physiol 311(1):H44–H53
170. Huang YL, Walker AS, Miller EW (2015) A Photostable silicon rhodamine platform for optical voltage sensing. J Am Chem Soc 137(33):10767–10776
171. Weaver CD et al (2004) A thallium-sensitive, fluorescence-based assay for detecting and characterizing potassium channel modulators in mammalian cells. J Biomol Screen 9(8):671–677
172. Rimmele TS, Chatton JY (2014) A novel optical intracellular imaging approach for potassium dynamics in astrocytes. PLoS One 9(10):e109243
173. Terstappen GC (2004) Nonradioactive rubidium ion efflux assay and its applications in drug discovery and development. Assay Drug Dev Technol 2(5):553–559
174. Roe MW, Lemasters JJ, Herman B (1990) Assessment of Fura-2 for measurements of cytosolic free calcium. Cell Calcium 11(2–3):63–73
175. Tsien RY (1981) A non-disruptive technique for loading calcium buffers and indicators into cells. Nature 290(5806):527–528
176. Williams DA, Bowser DN, Petrou S (1999) Confocal Ca^{2+} imaging of organelles, cells, tissues, and organs. Methods Enzymol 307:441–469
177. Johnson I (1998) Fluorescent probes for living cells. Histochem J 30(3):123–140
178. Cronshaw DG et al (2006) Evidence that phospholipase-C-dependent, calcium-independent mechanisms are required for directional migration of T-lymphocytes in response to the CCR4 ligands CCL17 and CCL22. J Leukoc Biol 79(6):1369–1380
179. Mehlin C, Crittenden C, Andreyka J (2003) No-wash dyes for calcium flux measurement. BioTechniques 34(1):164–166
180. Di Virgilio F, Steinberg TH, Silverstein SC (1990) Inhibition of Fura-2 sequestration and secretion with organic anion transport blockers. Cell Calcium 11(2–3):57–62
181. Vetter I et al (2008) Rapid, opioid-sensitive mechanisms involved in transient receptor potential vanilloid 1 sensitization. J Biol Chem 283(28):19540–19550
182. Kabbara AA, Allen DG (2001) The use of the indicator fluo-5N to measure sarcoplasmic reticulum calcium in single muscle fibres of the cane toad. J Physiol 534(Pt 1):87–97
183. Rehberg M et al (2008) A new non-disruptive strategy to target calcium indicator dyes to the endoplasmic reticulum. Cell Calcium 44(4):386–399
184. Solovyova N, Verkhratsky A (2002) Monitoring of free calcium in the neuronal endoplasmic reticulum: an overview of modern approaches. J Neurosci Methods 122(1):1–12
185. Oakes SG et al (1988) Incomplete hydrolysis of the calcium indicator precursor fura-2 pentaacetoxymethyl ester (fura-2 AM) by cells. Anal Biochem 169(1):159–166
186. Gillis JM, Gailly P (1994) Measurements of $[Ca^{2+}]i$ with the diffusible Fura-2 AM: can some potential pitfalls be evaluated? Biophys J 67(1):476–477
187. Jobsis PD, Rothstein EC, Balaban RS (2007) Limited utility of acetoxymethyl (AM)-based intracellular delivery systems, in vivo: interference by extracellular esterases. J Microsc 226(Pt 1):74–81
188. Azimi I et al (2017) Evaluation of known and novel inhibitors of Orai1-mediated store operated Ca^{2+} entry in MDA-MB-231 breast cancer cells using a fluorescence imaging plate reader assay. Bioorg Med Chem 25(1):440–449
189. Buchser W et al (2004) *Assay development guidelines for image-based high content screening, high content analysis and high content imaging*. In: Sittampalam GS et al (eds) Assay guidance manual. Eli Lilly & Company and the National Center for Advancing Translational Sciences, Bethesda

190. Teichert RW et al (2012) Characterization of two neuronal subclasses through constellation pharmacology. Proc Natl Acad Sci U S A 109(31):12758–12763
191. Teichert RW et al (2012) Functional profiling of neurons through cellular neuropharmacology. Proc Natl Acad Sci U S A 109(5):1388–1395
192. Siemens J et al (2006) Spider toxins activate the capsaicin receptor to produce inflammatory pain. Nature 444(7116):208–212
193. Imperial JS et al (2014) A family of excitatory peptide toxins from venomous crassispirine snails: using Constellation Pharmacology to assess bioactivity. Toxicon 89:45–54
194. Robinson SD et al (2015) Discovery by proteogenomics and characterization of an RF-amide neuropeptide from cone snail venom. J Proteome 114:38–47
195. Vetter I et al (2012) Ciguatoxins activate specific cold pain pathways to elicit burning pain from cooling. EMBO J 31(19):3795–3808
196. Bassett JJ et al (2018) Assessment of cytosolic free calcium changes during ceramide-induced cell death in MDA-MB-231 breast cancer cells expressing the calcium sensor GCaMP6m. Cell Calcium 72:39–50
197. Harrill JA (2018) Human-derived neurons and neural progenitor cells in high content imaging applications. Methods Mol Biol 1683:305–338
198. Esner M, Meyenhofer F, Bickle M (2018) Live-cell high content screening in drug development. Methods Mol Biol 1683:149–164
199. Adams CL, Sjaastad MD (2009) Design and implementation of high-content imaging platforms: lessons learned from end user-developer collaboration. Comb Chem High Throughput Screen 12(9):877–887
200. Shumate C, Hoffman AF (2009) Instrumental considerations in high content screening. Comb Chem High Throughput Screen 12(9):888–898
201. McManus OB (2014) HTS assays for developing the molecular pharmacology of ion channels. Curr Opin Pharmacol 15:91–96
202. Picones A et al (2016) Contribution of automated technologies to ion channel drug discovery. Adv Protein Chem Struct Biol 104:357–378
203. Gleeson EC et al (2015) Inhibition of N-type calcium channels by fluorophenoxyanilide derivatives. Mar Drugs 13(4):2030–2045
204. Dai G et al (2008) A high-throughput assay for evaluating state dependence and subtype selectivity of Cav2 calcium channel inhibitors. Assay Drug Dev Technol 6(2):195–212
205. Redondo PC et al (2005) Collaborative effect of SERCA and PMCA in cytosolic calcium homeostasis in human platelets. J Physiol Biochem 61(4):507–516
206. Brini M et al (2000) Effects of PMCA and SERCA pump overexpression on the kinetics of cell Ca^{2+} signalling. EMBO J 19(18):4926–4935
207. Vetter I (2012) Development and optimization of FLIPR high throughput calcium assays for ion channels and GPCRs. Adv Exp Med Biol 740:45–82
208. Ruiz de Azua I et al (2010) RGS4 is a negative regulator of insulin release from pancreatic beta-cells in vitro and in vivo. Proc Natl Acad Sci U S A 107(17):7999–8004
209. Vetter I et al (2006) The mu opioid agonist morphine modulates potentiation of capsaicin-evoked TRPV1 responses through a cyclic AMP-dependent protein kinase A pathway. Mol Pain 2:22
210. Vetter I et al (2008) Mechanisms involved in potentiation of transient receptor potential vanilloid 1 responses by ethanol. Eur J Pain 12(4):441–454
211. Samuel D, Robinson SD et al (2018) A comprehensive portrait of the venom of the giant red bull ant Myrmecia gulosa reveals a hyperdiverse hymenopteran toxin gene family. Sci Adv 4:eaau4640

Chapter 4
Imaging Native Calcium Currents in Brain Slices

Karima Ait Ouares, Nadia Jaafari, Nicola Kuczewski, and Marco Canepari

Abstract Imaging techniques may overcome the limitations of electrode techniques to measure locally not only membrane potential changes, but also ionic currents. Here, we review a recently developed approach to image native neuronal Ca^{2+} currents from brain slices. The technique is based on combined fluorescence recordings using low-affinity Ca^{2+} indicators possibly in combination with voltage sensitive dyes. We illustrate how the kinetics of a Ca^{2+} current can be estimated from the Ca^{2+} fluorescence change and locally correlated with the change of membrane potential, calibrated on an absolute scale, from the voltage fluorescence change. We show some representative measurements from the dendrites of CA1 hippocampal pyramidal neurons, from olfactory bulb mitral cells and from cerebellar Purkinje neurons. We discuss the striking difference in data analysis and interpretation between Ca^{2+} current measurements obtained using classical electrode techniques and the physiological currents obtained using this novel approach. Finally, we show how important is the kinetic information on the native Ca^{2+} current to explore the potential molecular targets of the Ca^{2+} flux from each individual Ca^{2+} channel.

K. Ait Ouares · N. Jaafari
Univ. Grenoble Alpes, CNRS, LIPhy, Grenoble, France

Laboratories of Excellence, Ion Channel Science and Therapeutics, France

N. Kuczewski
Centre de Recherche en Neurosciences de Lyon, INSERM U1028/CNRS UMR5292, Université Lyon1, Lyon, France

M. Canepari (✉)
Univ. Grenoble Alpes, CNRS, LIPhy, Grenoble, France

Laboratories of Excellence, Ion Channel Science and Therapeutics, France

Institut National de la Santé et Recherche Médicale (INSERM), Paris, France

Laboratoire Interdisciplinaire de Physique (UMR 5588), St Martin d'Hères cedex, France
e-mail: marco.canepari@univ-grenoble-alpes.fr

© Springer Nature Switzerland AG 2020
M. S. Islam (ed.), *Calcium Signaling*, Advances in Experimental Medicine and Biology 1131, https://doi.org/10.1007/978-3-030-12457-1_4

Keywords Calcium currents · Calcium imaging · Voltage sensitive dyes imaging · CA1 hippocampal pyramidal neuron · Olfactory bulb mitral cell · Purkinje neuron · Brain slices · Action potential · Synaptic potential · Biophysical modeling

4.1 Introduction

Optical measurements have been historically designed to monitor the electrical activity of the nervous system, a task where the use of electrode techniques has clear limitations [1]. In the last two decades, the development of new organic voltage sensitive dyes (VSD), in parallel with the progress of devices to excite and detect fluorescence [2], allowed optical recordings of sub-cellular membrane potential (V_m) changes <1 mV with a signal-to-noise ratio (S/N) comparable to that of patch clamp recordings [3]. This achievement suggested that voltage imaging can be used to investigate voltage-dependent proteins, in particular voltage-gated ion channels, in their physiological environment. The principal function of an ion channel is to allow an ion flux through a membrane, i.e. to produce an ionic current. Thus, the study of the biophysics of ion channels is routinely performed by measurements of ionic currents in single-electrode or two-electrode voltage clamp [4]. A way to investigate the biophysics of isolated native ion channels is to perform excised patches from ex-vivo membranes [5]. Alternatively, ion channels can be expressed in foreign cells such as oocytes or mammalian cell lines [6] and studied by using patch clamp techniques [7]. Yet, the physiological role and function of voltage-gated ion channels must be investigated in their natural environment, i.e. in their native cellular compartment and during physiological changes of V_m. To this purpose, the voltage clamp electrode approach has serious limitations for several reasons. First, the ionic current is measured by maintaining the cell at a given artificial V_m and even if the cell is dynamically clamped the V_m change is never a physiological signal [8]. Second, the current measured with the electrode is the summation of the filtered currents from all different cellular regions, including remote regions where V_m is unclamped, and no information is available on the site of origin of the current [9]. Third, different ionic currents contribute to the physiological change of V_m producing a functional coupling among the different ion channels [10]. Thus, a single native ionic current must be pharmacologically isolated from the total current mediated by the other channels, but the block of these channels will make the V_m change non-physiological.

In the last few years, we designed a novel approach to measure physiological Ca^{2+} currents from neurons in brain slices [11]. The method is based on fast Ca^{2+} optical measurements using low-affinity indicators that can be combined with sequential [12] or simultaneous [13] V_m optical recordings. The latter measurements can be calibrated in mV [14] using cell-specific protocols. Individual cells are loaded with Ca^{2+} and V_m indicators using a patch clamp recording. In contrast to voltage-clamp current measurements, the current approach permits independent recordings of the V_m change and of the Ca^{2+} influx, i.e. the study of voltage gating during physiological V_m changes. Since the Ca^{2+} current is reconstructed by the measurement of Ca^{2+} locally binding to an indicator, this approach provides information on channels in different areas of the cell with a spatial resolution as good as the optical recording allows. Finally, the Ca^{2+} current is recorded without blocking all Na^+ and K^+ channels that are necessary to produce the physiological V_m change. The principle of obtaining an optical measurement of a fast Ca^{2+} current is based on the analysis of the dye-Ca^{2+} binding reaction in a cell, a scenario initially studied by Kao and Tsien [15]. According to their theoretical estimates and to our recent empirical measurements [16], the relaxation time of the dye-Ca^{2+} binding reaction is less than 200 µs for low-affinity indicators with equilibrium constant (K_D) $\geq$ 10 µM such as Oregon Green BAPTA-5N (OG5N, $K_D = 35$ µM, [17]) or Fura-FF ($K_D = 10$ µM, [18]). Therefore, a fast Ca^{2+} current with duration of a few milliseconds can be reliably tracked by low-affinity indicators if fluorescence is acquired at sufficiently high speed. The goal of this methodological article is to provide an exhaustive tool for those scientists aiming at performing this type of measurement. The next section addresses in detail the problem of extracting the Ca^{2+} current kinetics from Ca^{2+} fluorescence measurements under different cellular buffering conditions. The following section is devoted to the technical aspects of how to set up combined V_m and Ca^{2+} optical measurements and to calibrate V_m signals on an absolute scale. We then illustrate some examples of combined V_m and Ca^{2+} current measurements and we finally discuss how to correctly interpret the results and how to use this information to significantly advance our knowledge on Ca^{2+} channels function. All data shown here were from experiments performed at the Laboratoire Interdisciplinaire de Physique and approved by the Isere prefecture (Authorisation n. 38 12 01). These experiments were performed at 32–34 °C using brain slices from 21 to 40 postnatal days old C57Bl6 mice of both genders.

4.2 Extracting Ca^{2+} Current Kinetics from Ca^{2+} Fluorescence Measurements

4.2.1 Biophysical Foundations of Ca^{2+} Currents Imaging

An optical measurement of a Ca^{2+} signal is ultimately a measurement of the Ca^{2+} indicator bound to Ca^{2+} ions, which is proportional to the Ca^{2+} fractional change of fluorescence ($\Delta F/F_0$) if the indicator is not saturated. If the kinetics of the Ca^{2+}-binding reaction of the indicator is slower than the kinetics of the Ca^{2+} source, and imaging is performed at higher rate, the time-course of Ca^{2+} $\Delta F/F_0$ essentially tracks the kinetics of the chemical reaction. Alternatively, if the kinetics of the Ca^{2+}-

binding reaction is faster than the kinetics of the Ca^{2+} source, the Ca^{2+} $\Delta F/F_0$ signal tracks the kinetics of the Ca^{2+} source. It follows that the equilibration (or relaxation) time of the Ca^{2+}-indicator binding reaction is a crucial variable to use the technique to investigate the biophysics and the physiology of the Ca^{2+} source. The relaxation of the Ca^{2+}- binding reactions for early indicators was studied by Kao and Tsien [15] who established that the rate of association for all these molecules is limited by diffusion leading to an association constant of $\sim 6 \cdot 10^8$ M^{-1} s^{-1}. Thus, both the equilibrium constant (K_D) and the equilibrium time are determined by the dissociation constant, i.e. the lower is the affinity of the indicator the shorter is its equilibrium time. We have empirically demonstrated that indicators with $K_D \geq 10$ μM such as OG5N or FuraFF have relaxation time < 200 μs [16]. Since the kinetics of activation and deactivation of voltage-gated Ca^{2+} channels (VGCCs) during physiological changes of V_m (for instance action potentials), is governed by the kinetics of the V_m transient, it follows that the relaxation time for those indicators is shorter than the duration of the Ca^{2+} influx. Hence, since Ca^{2+} binds to the indicator linearly in time, the Ca^{2+} $\Delta F/F_0$ is proportional to the integral of the Ca^{2+} influx, i.e. to the integral of the Ca^{2+} current. In the cell, however, Ca^{2+} simultaneously binds to proteins that form the endogenous buffer and this binding is competing with the binding to the indicator. An endogenous buffer can be, in principle, at least as fast as the indicator in equilibrating. In this case, only a fraction of Ca^{2+} is bound to the indicator, but this fraction is proportional to the total Ca^{2+} entering the cell and therefore to the integral of the Ca^{2+} current. Alternatively, an endogenous buffer can equilibrate over a time scale that is longer than the duration of the Ca^{2+} current. In this case, Ca^{2+} first binds to the dye and later to the endogenous buffer, implying that part of Ca^{2+} moves from the indicator to the endogenous buffer during its relaxation time. Under this condition, the Ca^{2+} $\Delta F/F_0$ is not linear with the integral of the Ca^{2+} current over this time scale. To clarify this important concept we make use of two simple computer simulations shown in Fig. 4.1, produced by a model that takes into account the chemical reactions as well as an extrusion mechanism re-establishing the initial Ca^{2+} conditions over a time scale >100 ms. We analyse what hypothetically can happen if a Ca^{2+} current with Gaussian shape occurs in a cell filled with 2 mM OG5N. In the first simulation (Fig. 4.1a), the cell has only 1 mM of a fast endogenous buffer behaving with the same association constant of the indicator and $K_D = 10$ μM. In the second simulation (Fig. 4.1b), the cell has additional 400 μM of a slower endogenous buffer with association rate ~ 3 times slower than that of the indicator and $K_D = 0.2$ μM. In the first case, the time derivative of the Ca^{2+} $\Delta F/F_0$ signal matches the kinetics of the Ca^{2+} current (Fig. 4.1a). In contrast, in the presence of the slower buffer, the time derivative of the Ca^{2+} $\Delta F/F_0$ signal has a negative component and does not match the kinetics of the Ca^{2+} current (Fig. 4.1b). In the next two paragraphs, we present the analysis strategies that can be applied to extract the kinetics of Ca^{2+} currents from Ca^{2+} imaging recordings.

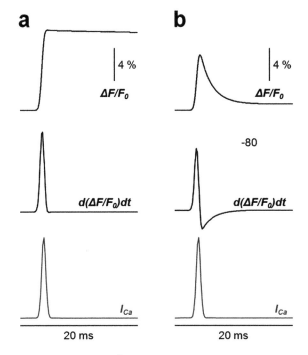

Fig. 4.1 Simulation of hypothetical Ca^{2+} $\Delta F/F_0$ signals from 2 mM OG5N in the presence of endogenous Ca^{2+} buffers. (**a**) Simulation of $\Delta F/F_0$ signal (top trace) following a Ca^{2+} current with simple Gaussian kinetics (I_{Ca}, gray trace on the bottom) in the presence of 1 mM of a fast endogenous Ca^{2+} buffer with same association constant of the Ca^{2+} indicator ($5.7 \cdot 10^8$ M^{-1} s^{-1}) and $K_D = 10$ μM. The kinetics of the $\Delta F/F_0$ time derivative (middle trace) matches the kinetics of the Ca^{2+} current. (**b**) Same as in the previous panel but in this case in the presence of additional 400 μM of a slower buffer with association constant equal to $2 \cdot 10^8$ M^{-1} s^{-1} and $K_D = 0.2$ μM. The kinetics of the $\Delta F/F_0$ time derivative (middle trace) does not matches the kinetics of the Ca^{2+} current

4.2.2   The Case of Linearity Between Ca^{2+} Influx and Ca^{2+} Fluorescence Changes

The proteins expressed in a cell determine whether or not the time course of the Ca^{2+} $\Delta F/F_0$ signal is linear with the kinetics of the Ca^{2+} current. As previously demonstrated [11], in the case of linear behaviour, the Ca^{2+} $\Delta F/F_0$ signal must reach its peak and remain constant for a few milliseconds afterwards, i.e. for the entire duration of the current. As shown in the simulation of Fig. 4.1a, the kinetics of Ca^{2+} extrusion producing a slow decrease of the Ca^{2+} $\Delta F/F_0$ signal has negligible effect on the time derivative. Thus, the estimate of the Ca^{2+} current kinetics is reliably obtained by the calculation of the time derivative of the Ca^{2+} $\Delta F/F_0$ signal. This calculation, however, requires the signal noise to be smaller than the signal change between two consecutive samples. The classical way to achieve this necessary condition is to apply to the Ca^{2+} $\Delta F/F_0$ signal a "smoothing algorithm", i.e. a temporal filter that reduces the noise of the signal with minimal distortion of its kinetics. At 20 kHz acquisition rate, we have found that the Savitky-Golay algorithm [19] is an optimal filtering tool permitting noise reduction of the signal without significant temporal distortion using time-windows of up to 20–30 samples [11]. The applicability of this strategy has however limitations, i.e. if the signal or the region of measurement are too small, or if the light is too dim, the smoothing of the signal might not be sufficient to reduce the noise down to the level permitting calculation of the time derivative. In this case, the alternative strategy to apply consists in fitting the raw or the filtered Ca^{2+} $\Delta F/F_0$ signal with a model function obtaining a noiseless curve that mimics the time course of the Ca^{2+} $\Delta F/F_0$ signal. A simple choice of function that resembles the time course of the Ca^{2+} $\Delta F/F_0$ transient is the sigmoid. In particular, we found that the product of three sigmoid functions always provides an excellent fit of the Ca^{2+} $\Delta F/F_0$ signal associated with a backpropagating action potential in CA1 hippocampal pyramidal neurons [16]. As shown in the example of Fig. 4.2a both strategies are faithful in correctly calculating the time derivative of the $\Delta F/F_0$ signal. In this example, a CA1 hippocampal pyramidal neuron was filled with 2 mM OG5N and the dendritic Ca^{2+} $\Delta F/F_0$ signal associated with a backpropagating action potential was recorded at 20 kHz and averaged over 16 trials. This high sampling frequency was necessary to avoid signal aliasing and therefore distortion of the kinetics of the current. The filtering strategy is the straightforward approach that enables the calculation of the time derivative, but it produces a curve with noise. The noise can be reduced (if possible) by increasing the number of trials to average or by enlarging the dendritic area from where fluorescence is averaged. The fitting strategy is less direct but it produces a noiseless curve and it is therefore the only possible approach when the noise of the Ca^{2+} $\Delta F/F_0$ signal is above a certain level, as quantitatively estimated in an original report [16]. In particular, this is the case when the current must be extracted from single trials or when the recording is obtained from small or relatively dim regions.

4.2.3   The Case of Nonlinearity Between Ca^{2+} Influx and Ca^{2+} Fluorescence Changes

The method of estimating the kinetics of a Ca^{2+} current by calculating the Ca^{2+} $\Delta F/F_0$ time derivative fails when Ca^{2+} unbinds from the indicator over a time scale that is longer than the current duration, but sufficiently short to distort the estimate of Ca^{2+} influx dynamics by fluorescence measurement. In other words, this method fails when the Ca^{2+} $\Delta F/F_0$ signal decays rapidly, after correction for bleaching, generating a negative component in its time derivative. Such a situation occurs, for example, where slow buffering is produced by Calbindin-D28k [20, 21] and Parvalbumin [22, 23]. As shown in the example of Fig. 4.2b, the Ca^{2+}

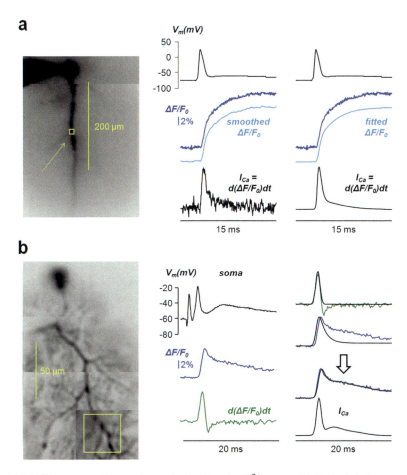

Fig. 4.2 Different strategies to estimate the kinetics of a Ca^{2+} current (**a**) On the left, fluorescence image of CA1 hippocampal pyramidal neuron filled with 2 mM OG5N with a small region of the apical dendrite outlined and indicated by the arrow. On the right, somatic V_m change associated with an action potential (top black traces) and Ca^{2+} $\Delta F/F_0$ signal in the indicated region (blue trace). The $\Delta F/F_0$ signal is either smoothed with a Savitky-Golay algorithm (left) or fitted with a 3-sigmoid function (right). The Ca^{2+} current kinetics (I_{Ca}) is then estimated by calculating the time derivative of the processed $\Delta F/F_0$ signal. The kinetics of the current is the same using the two strategies, but the curve obtained with the strategy of data fitting is noiseless. Data, recorded at 20 kHz, were from averages of 16 trials. (**b**) On the left, fluorescence image of PN filled with 2 mM OG5N with square region of interest outlined. On the right, somatic V_m change associated with a climbing fibre EPSP (top-left black trace) and Ca^{2+} $\Delta F/F_0$ signal in the indicated region (blue traces). The time derivative of the Ca^{2+} $\Delta F/F_0$ signal (green traces) does not match the kinetics of the current. To estimate the kinetics of the current we use a strategy that consists in matching the result of a computer simulation to the Ca^{2+} $\Delta F/F_0$ signal using an optimised two-buffer model [24]. We start from the Gaussian function fitting the rising phase of the $\Delta F/F_0$ time derivative (top-right black trace). We then correct the current with three additional Gaussian components until a match of the computer simulation with the Ca^{2+} $\Delta F/F_0$ signal is obtained (process indicated by the arrow. The curve producing this match (I_{Ca}, bottom-right black trace) is the estimate of the Ca^{2+} current kinetics. Data, recorded at 5 kHz, were from averages of 4 trials

$\Delta F/F_0$ signal associated with a climbing fibre excitatory postsynaptic potential (EPSP), recorded at 5 kHz from a dendritic region and averaged over four trials, decays rapidly after its maximum resulting in a negative component of its time derivative. The distortion from the linear behaviour produced by the slow buffers can be compensated by taking into account the kinetics of Ca^{2+} unbinding from the indicator. We have recently developed a successful method to achieve this goal [24]. The strategy is based on fitting the decay time of the Ca^{2+} $\Delta F/F_0$ signal with the result of a computer simulation of a model with a slow buffer. Initially the input current is the Gaussian function fitting the rising phase of the time derivative (that is still a good approximation of the initial part of the current). The kinetic parameters and the concentration of the slow buffer are set to obtain the best fit of the decay phase of the Ca^{2+} $\Delta F/F_0$ signal. Then, the kinetics of the Ca^{2+} current is obtained as summation of four Gaussian functions that maximise the match between the result of the computer simulation and the experimental Ca^{2+} $\Delta F/F_0$ signal. Although this new method provides only an indirect approximation of the kinetics of the Ca^{2+} current, this information is crucial at understanding the activation and deactivation of different types of VGCCs. In the dendrites of PNs, for instance, different Ca^{2+} current kinetics components are associated with the activation of P/Q-type VGCCs [25] and T-type Ca^{2+} channels [26] that can be in principle separated by pharmacological block of one component. Thus, the extrapolation of a curve that approaches the kinetics of the Ca^{2+} current can be used to quantitatively investigate the variability of channels activation at different dendritic sites, the modulation of channel activation due to physiological activity or to pharmacological action. Finally, it is important to say that such a strategy can be extended to estimate slower Ca^{2+} currents where the fitting procedure can be applied to the slower decay time due to Ca^{2+} extrusion [27].

4.3 Combining Membrane Potential and Ca^{2+} Imaging

4.3.1 Setting up Combined Voltage and Ca^{2+} Fluorescence Measurements

To combine V_m and Ca^{2+} optical measurements, the VSD and the Ca^{2+} indicator must have minimal overlap in the emission spectra. Water soluble voltage indicators with different excitation and emission spectra have been recently developed [28]. In particular, the red-excitable and IR emitting VSD ANBDQPTEA (or PY3283) is suitable for coupling with other optical techniques [29]. Nevertheless, the most used VSDs for single cell applications are still JPW3028 [30] and the commercially available JPW1114 [18]. These indicators have wide excitation spectrum in the blue/green region and they emit mainly in the red region. We have previously demonstrated that both indicators can be optimally combined with Fura indicators that are excited in the UV region and emit in the short green region [12]. In

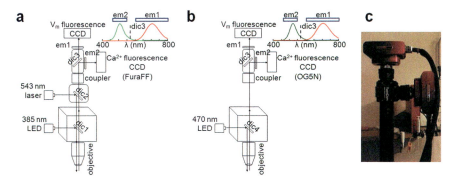

Fig. 4.3 Configurations and camera for combined voltage and Ca^{2+} fluorescence measurements. (**a**) Schematic drawing of the apparatus for simultaneous voltage and Ca^{2+} imaging using the VSD JPW1114 and the Ca^{2+} indicator FuraFF [13]; 385 nm LED light via the epifluorescence port of a commercial microscope is reflected by a 506 nm long-pass dichroic mirror (dic1); 543 nm laser light via the top of the microscope is reflected by a dual-band dichroic mirror transmitting wavelengths between 493 nm and 530 nm and wavelengths longer than 574 nm (dic2); the fluorescence images of the two dyes are demagnified and separated by a 565 nm long-pass dichroic mirror (dic3); The V_m image and the Ca^{2+} images are filtered by a 610 nm long-pass (em1) and by a 510 ± 42 nm band-pass filter (em2) respectively, then acquired by two CCD cameras; the emission spectra of FuraFF (green) and JPW1114 (red) are shown on the top-right. (**b**) Schematic drawing of the apparatus for simultaneous voltage and Ca^{2+} imaging using the VSD JPW1114 and the Ca^{2+} indicator OG5N [11]; 470 nm LED light via the epifluorescence port of a commercial microscope is reflected by a 495 nm long-pass dichroic mirror (dic4); the fluorescence images of the two dyes are demagnified and processed as in the previous configuration; the emission spectra of OG5N (green) and JPW1114 (red) are shown on the top-right. (**c**) "The dual NeuroCCD camera designed by RedshirtImaging for this type of measurement

this case, VSDs were excited at 543 nm using a laser and Fura indicators were excited at 385 nm using a light emitting diode (LED) as shown in the scheme of Fig. 4.3a. Alternatively, simultaneous voltage and Ca^{2+} imaging can be achieved using Oregon Green, Calcium Green or Fluo Ca^{2+} indicators using blue light (470–490 nm) to excite both VSDs and Ca^{2+} indicators [31]. Simultaneous imaging of JPW1114 and OG5N was adopted to obtain the first combined measurement of V_m and Ca^{2+} currents using the configuration of Fig. 4.3b. This type of measurement, however, has several disadvantages. First, OG5N fluorescence has a small tail component in the red region [31] which can be negligible or not depending on the ratio of the two dyes at each site as well as on the ratio between the two signals. Thus, for example, it works in proximal dendrites of CA1 pyramidal neurons for signals associated with action potentials [11], where V_m fluorescence is stronger than Ca^{2+} fluorescence, but it does not in distal dendrites of cerebellar Purkinje neurons (PNs, data not shown), where V_m fluorescence is weaker than Ca^{2+} fluorescence. A second disadvantage is that the JPW1114 signal at 470 nm excitation is ~4 times smaller than that at 532 nm excitation. If simultaneous recordings are not critical, one can replace them with sequential recordings obtained by alternating 470 nm and 532 nm excitation as used in a recent study [32]. Finally, a third disadvantage is that JPW1114 absorbs more in the blue range than in the green range, i.e. it exhibits toxic effects after fewer exposures. A crucial technical aspect to take into consideration while setting up combined voltage and Ca^{2+} fluorescence measurements is the ability to record the two signals simultaneously at high speed. To this purpose, the company RedShirtImaging (Decantur, GA) has developed a dual-head version of the SMQ NeuroCCD (Fig. 4.3c). This camera permits simultaneous image acquisitions from both heads at 5–20 kHz, i.e. at the required speed. A demagnifier developed by Cairn Research Ltd. (Faversham, UK) allows adjusting the size of the image before it is split in two images at the emission wavelengths of the two dyes. Thus, the alignment of the two heads of the camera allows obtaining, at each precise region of interest, the V_m and the Ca^{2+} signal.

4.3.2 Calibrating Membrane Potential Fluorescence Transients

The calibration of V_m optical signals on an absolute scale (in mV) is crucial to analyse the gating of Ca^{2+} channels at the same locations where Ca^{2+} recordings are performed. This is not, however, straightforward. Indeed, the fractional change of VSD fluorescence is proportional to V_m [33], but the linear coefficient between these two quantities depends on the ratio between the inactive dye and the active dye that varies from site to site. The inactive dye is bound to membranes that do not change potential and contributes only to the resting fluorescence, while the active dye is bound to the plasma membrane and contributes to the resting fluorescence, but also carries the signal. In particular, in experiments utilising intracellular application of the dye, inactive dye is the dye that binds to intracellular membranes and organelles. Since the sensitivity of recording varies from site to site, a calibration can be achieved only if a calibrating electrical signal that has known amplitude at all locations is available. Such a signal is different in different systems. In mitral cells of the olfactory bulb, the amplitude of an action potential is the same in the whole apical dendrite and it can be used to create a sensitivity profile of the measuring system [34]. Another type of calibrating electrical signal can be a slow electrical change spreading with minimal attenuation over relatively long distances. Such a signal can be used to reliably calibrate VSD signals in PNs [18]. An example of this type of calibration is reported in Fig. 4.4a. Starting from the resting V_m, that we assume nearly uniform over the entire cell, long current hyperpolarising or depolarising current pulses are injected to the soma *via* the patch pipette and the change in V_m is recorded. As shown by direct dendritic patch recording, the dendrite is hyperpolarised by the same amount of the soma [35]. Thus, the measurement of somatic hyperpolarisation can be used as voltage reference to calibrate the dendritic VSD fractional change of fluorescence (VSD $\Delta F/F_0$) optical signal, as shown in Fig. 4.4a. In contrast, a depolarisation step attenuates along the dendrite. A third type of calibrating signal is a uniform depolarisation over the entire dendritic tree

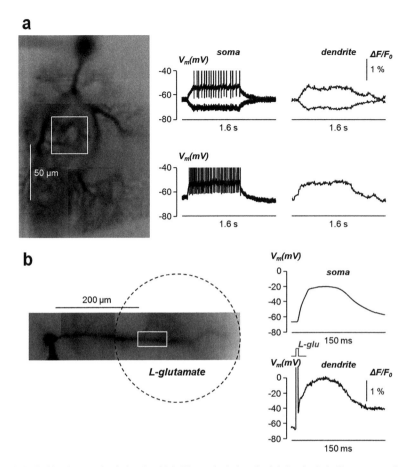

Fig. 4.4 Calibration methods for dendritic V_m optical signals. (**a**) On the left, fluorescence image of PN with square region of interest outlined. On the right, somatic V_m change associated with long hyperpolarising or depolarising steps and associated dendritic VSD $\Delta F/F_0$ signals in the region of interest; the hyperpolarising step spreads to the dendrites with negligible attenuation [35] and is used to calibrate the VSD $\Delta F/F_0$ signals; the weak depolarising step on the top also spreads with minimal attenuation, but the associated somatic action potentials do not propagate into the dendrites; in contrast, the strong depolarising step on the bottom attenuates. (**b**) On the left, fluorescence image of CA1 hippocampal pyramidal neuron with rectangular region, 200–250 μm from the soma, outlined. On the right, somatic V_m change associated with L-glutamate photorelease from MNI-glutamate [14] in the area limited by the dotted line and in the presence of 1 μM tetrodotoxin blocking action potentials; the dendritic VSD $\Delta F/F_0$ signal is reported on the bottom; the saturating L-glutamate concentration depolarises the illuminated area from the resting V_m (~ -70 mV) to the reversal potential of AMPA receptors (0 mV). All calibrations were from single trials

using L-glutamate photolysis from 4-Methoxy-7-nitroindolinyl-caged-L-glutamate (MNI-glutamate) [14]. This calibration procedure is applicable to all membrane expressing a relatively large number of glutamate receptors, i.e. to dendrites with high densities of excitatory synapses. The calibration is based on the principle that if the ionotropic glutamate receptor becomes the dominant conductance in a particular neuronal compartment, its reversal potential will determine the membrane potential of the compartment. Thus, in the area where dominance of glutamate receptor conductance is obtained, the resulting V_m change will be the same and can be used to calibrate VSD signals. An example of this protocol to calibrate backpropagating action potentials in CA1 hippocampal pyramidal neurons is shown in Fig. 4.4b. The VSD $\Delta F/F_0$ signal associated with the backpropagating AP at different sites of the apical dendrites is variable and cannot be directly correlated with the absolute change of V_m. In the presence of 1 μM TTX, to block action potentials, L-glutamate is photoreleased to saturate glutamate receptors over the whole field of view. Since the recording is performed starting from the resting V_m, the size of the VSD $\Delta F/F_0$ corresponds to this potential in the whole illuminated area where V_m reaches the reversal potential of 0 mV. Thus, this information is used to extrapolate the V_m at each dendritic site.

4.4 Examples of Combined Voltage and Ca^{2+} Current Imaging

4.4.1 Ca^{2+} Currents Associated with Backpropagating Action Potentials in CA1 Hippocampal Pyramidal Neurons and in Olfactory Bulb Mitral Cells

In many neurons, action potentials generated in the axon hillock adjacent to the soma do not only propagate along the axon to reach neurotransmitter release terminals, but also backpropagate throughout dendrites to signal cell activation at the sites where the neuron receives the synaptic inputs. At least part of this information is given by the fast Ca^{2+} transients produced by activation of VGCCs caused by the dendritic depolarisation associated with the action potential. The analysis that can be performed using the present imaging method is therefore crucial at understanding signal processing in individual neurons, as well as the specific role and function of the diverse VGCCs activated in dendrites. The propagation of the action potential and the consequent activation of VGCCs may be very different in different neuronal systems. In CA1 hippocampal pyramidal neurons, action potentials attenuate along the dendrite and activate both high-voltage activated (HVA) and low-voltage activated (LVA) VGCCs [36, 37]. We have very recently demonstrated that HVA-VGCCs and LVA-VGCCs operate synergistically to stabilise Ca^{2+} signals during burst firing [32]. Somatic and dendritic action potentials, at nearly physiological temperature, have 1–4 ms duration as in the example shown in Fig. 4.5a. In agreement with this evidence, the kinetics of the Ca^{2+} current is similar to that of the action potential, with a peak delayed by a few hundred milliseconds from the peak of the action potential. In total contrast to the CA1 hippocampal pyramidal neuron, in

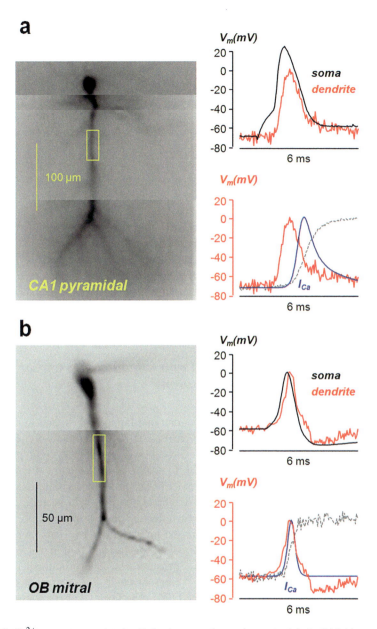

Fig. 4.5 Ca^{2+} currents associated with backpropagating action potentials in CA1 hippocampal pyramidal neurons and in olfactory bulb (OB) mitral cells (**a**) On the left, fluorescence image of CA1 hippocampal pyramidal neuron filled with JPW1114 and 2 mM OG5N with a region of the apical dendrite outlined. On the right, somatic V$_m$ change associated with an action potential (black trace) in the soma and in the dendritic region. The associated Ca^{2+} current kinetics, obtained with the fitting strategy from the raw Ca^{2+} signal (dashed gray trace), is shown in the bottom. (**b**) On the left, fluorescence image of OB cell filled with JPW1114 and 2 mM OG5N with a region of the principal dendrite outlined. On the right, somatic V$_m$ change associated with an action potential in the soma (black trace) and in the dendritic region (red trace). The associated Ca^{2+} current kinetics, obtained with the fitting strategy from the raw Ca^{2+} signal (dashed gray trace), is shown in the bottom (blue trace) superimposed to the dendritic action potential (red trace). Data, recorded at 20 kHz, were from averages of 4 trials. All experiments were performed at 32–34 °C

olfactory bulb mitral cells the action potential does not attenuate along the dendrites [38]. In addition, as shown in the representative example of Fig. 4.5b, the somatic and dendritic action potential at near physiological temperature (32–34 °C) has duration <1 ms. Thus, in this system, the activation and deactivation of VGCCs is also faster leading to a Ca^{2+} current with shorter duration and shorter delay from the V_m waveform peak. This preliminary comparison between the two cases indicates that the role of VGCCs, activated by the action potential, is different in different systems. For example, Ca^{2+} currents are delayed by ~100 μs in presynaptic terminals where the function of this signal is to trigger neurotransmitter release [39]. Here, the kinetics of the Ca^{2+} current was obtained by calculating the time derivative of the Ca^{2+} $\Delta F/F_0$ signal fit [16]. VGCCs contribute to the shape of the action potential directly and indirectly by activating K^+ channels, but also provide a precise time-locked Ca^{2+} transient capable to select fast-activated Ca^{2+} binding proteins. The possibility to locally investigate, using combined V_m and Ca^{2+} current optical measurements, the physiological occurrence of Ca^{2+} signals mediated by VGCCs will contribute enormously, in the near future, to the understanding of complex signal processing in neurons.

4.4.2 Ca^{2+} Currents Associated with Climbing Fibre EPSPs in Cerebellar Purkinje Neurons

In contrast to pyramidal neurons of the cortex and hippocampus, and to olfactory bulb mitral cells, somatic/axonal action potentials in PNs do not actively propagate in the dendrites [40]. The dendrites of PNs, however, express P/Q-type HVA-VGCCs [25] and T-type LVA-VGCCs [26] that are activated by the dendritic depolarisation produced by climbing fibre EPSPs. As shown in the example of Fig. 4.6, the shape of the dendritic V_m calibrated in Fig. 4.4a is quite different in the soma and in the dendrite, mainly reflecting the absence of Na^+ action potentials in the dendrite. In this system, the low-affinity Ca^{2+} indicator used to estimate the Ca^{2+} current was Fura-FF, since the larger Ca^{2+} signal produced by OG5N contaminated the optical V_m measurement. The prominent dendritic depolarisation produces a biphasic Ca^{2+} current, which is in this case obtained by applying our recent generalised method [24]. The fast and sharp component is nearly concomitant to the short period in which $V_m > -40$ mV and it is therefore likely mediated by HVA-VGCCs. The slower and more persistent component is instead mostly concomitant to the whole depolarisation transient and is therefore likely mediated by LVA-VGCCs, as demonstrated by selectively blocking T-type VGCCs (unpublished data not shown). The analysis of Ca^{2+} signalling associated with the climbing fibre EPSP is crucial for the understanding of synaptic plasticity in PNs [41]. Yet, while the role of the Ca^{2+} transient associated with the climbing fibre EPSP has been postulated to be auxiliary to the principal Ca^{2+} signal mediated by parallel fibre EPSPs, these first measurements of the Ca^{2+} current

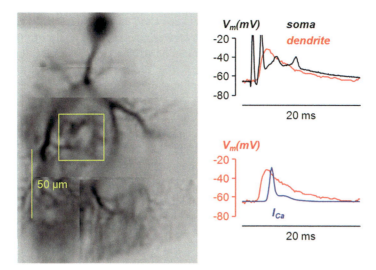

Fig. 4.6 Ca^{2+} currents associated with climbing fibre EPSPs in cerebellar Purkinje neurons. On the left, fluorescence image of the same PN of Fig. 4.4a filled with JPW1114 and 2 mM Fura-FF with a region of the dendrite outlined. On the right, somatic V_m change associated with climbing fibre EPSP (black trace) in the soma and in the dendritic region. The associated Ca^{2+} current kinetics, obtained with the generalised method [24], is shown in the bottom. Data, recorded at 5 kHz, were from averages of 4 trials. Experiments were performed at 32–34 °C

kinetics elucidate a quite precise timing of occurrence of the Ca^{2+} source that may eventually provide a less ambiguous explanation of the precise function of this spread dendritic depolarisation. In summary, the examples illustrated in this section show the potentials of this novel optical method in providing physiological information not available by using electrode techniques.

4.5   Data Interpretation and Future Prospective

The method described here has been developed to overcome the limitations of single-electrode or two-electrode voltage clamp techniques permitting the analysis of physiological Ca^{2+} currents of native Ca^{2+} channels. Indeed, in contrast to patch-clamp recordings, these Ca^{2+} optical currents can be measured in conditions of a physiological change of V_m and the measured currents are confined to the sites where they are recorded, as shown in apical dendrites of hippocampal pyramidal neurons [32]. The additional information on local V_m change, necessary to correlate the behaviour of the conductance with its biophysical properties, is obtained by combining VSD imaging. In cases of linear behaviour between Ca^{2+} influx and Ca^{2+} fluorescence changes the kinetics of the Ca^{2+} current can be extracted by calculating the time derivative of the Ca^{2+} $\Delta F/F_0$ signal using low-affinity Ca^{2+} indicators [11,

16]. In the case of nonlinear behaviour between Ca^{2+} influx and Ca^{2+} fluorescence, produced by Ca^{2+}-binding proteins with slower kinetics with respect to the Ca^{2+} current, the kinetics of the Ca^{2+} current can be still correctly estimated by taking into account the faster unbinding of Ca^{2+} from the low-affinity indicator [24]. In this last section we address the question of how data, obtained using this technique, should be interpreted. In Ca^{2+} current recordings from channels expressed in heterologous systems using voltage clamp, V_m is controlled artificially and its change is therefore independent of the channel deactivation. Under physiological conditions, Ca^{2+} channels contribute to the V_m change directly, through the ion flux, and indirectly by regulating other conductances. It follows that the channel deactivation changes the V_m waveform. We have shown that in CA1 hippocampal pyramidal neurons this phenomenon produces a modulation of LVA-VGCCs by HVA-VGCCs [32]. More in general, a Ca^{2+} current mediated by diverse VGCCs is always the result of a synergy among all different ion channels contributing to the V_m waveform. It follows that in a Ca^{2+} current optical measurement, a single component of the current cannot be extracted simply by blocking the underlying channel, since this block may affect the residual current as well. This evidence has important implications in the study of transgenic animals carrying Ca^{2+} channel mutations. In this case, a certain phenotype is likely to result from the combined modification of function of many different channels, rather than from the specific Ca^{2+} influx component, making the study of these animals as models for disease challenging. In summary, the investigation of the role and function of individual Ca^{2+} channels must be performed in the global context of activation of all channels participating to the local V_m waveform.

Another important aspect of data interpretation is the relation of the kinetics of Ca^{2+} current with the putative molecular targets of Ca^{2+} ions entering the cell. While importance is normally given to possible molecular coupling between the Ca^{2+} channel and the Ca^{2+} binding protein, the kinetics of the Ca^{2+} current can be a potent selector of the molecular pathway which is activated. To illustrate this important concept we make use of computer simulations using the same theoretical framework for simple Ca^{2+}-binding dynamics that we already used in the past [42]. We imagine the possible activation of two proteins: a "fast" protein with $K_{ON} = 5.7 \cdot 10^8$ M^{-1} s^{-1} and $K_D = 10$ μM, expressed at the concentration of 500 μM; and a "slow" protein with $K_{ON} = 4 \cdot 10^8$ M^{-1} s^{-1} and $K_D = 0.4$ μM, expressed at the concentration of 100 μM. In the first case, shown in Fig. 4.7a, the cell is receiving a fast Ca^{2+} current with ~ 2 ms total duration which binds first to the fast protein and later to the slow protein. In the second case, shown in Fig. 4.7b, the cell is receiving a slower Ca^{2+} current that is smaller in amplitude but that carries approximately the same amount of Ca^{2+}. In this case the slow protein binds to Ca^{2+} with a slower kinetics but the amount of the fast protein binding to Ca^{2+} is less than half with respect to the first case. These simulations indicate that the ability to activate for a molecular pathway triggered by the fast protein strongly depends on the kinetics of the Ca^{2+} current. Thus, the approach described here should drastically improve our understanding of the physiological function of Ca^{2+} channels by providing the possibility to explore the biophysics of

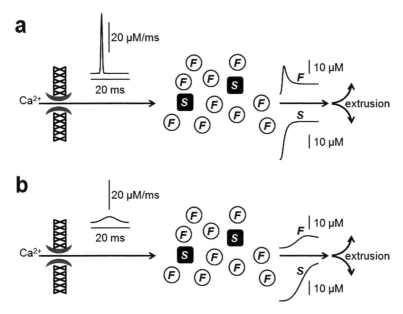

Fig. 4.7 Simulated activation of two different Ca^{2+}-binding proteins by Ca^{2+} currents. (**a**) In a cell containing 500 μM of a fast (*F*) buffer with $K_{ON} = 5.7 \cdot 10^8$ M^{-1} s^{-1} and $K_D = 10$ μM, and 100 μM of a slow (*S*) buffer with $K_{ON} = 4 \cdot 10^8$ M^{-1} s^{-1} and $K_D = 0.4$ μM, the curves on the right report the binding to Ca^{2+} of the *F* and *S* proteins following the fast Ca^{2+} current reported on the left. (**b**) Same as in the previous panel but following the slow Ca^{2+} current reported on the left

native channels during physiological activity locally within the complex neuronal architecture. The examples of combined V_m and Ca^{2+} current optical measurements from CA1 hippocampal pyramidal neurons, olfactory bulb mitral cells and PNs reported here are representative of the types of exploration that can be performed using this novel approach.

Acknowledgment This work was supported by the *Agence Nationale de la Recherche* through three grants: (1) Grant *WaveFrontImag*, program number ANR-14-CE17–0006-01; (2) Labex *Ion Channels Science and Therapeutics*, program number ANR-11-LABX-0015; (3) National Infrastructure France Life Imaging "Noeud Grenoblois"; and by the *Federation pour la recherché sur le Cerveau* (FRC) through the grant *Espoir en tête* (in partnership with Rotary France).

References

1. Braubach O, Cohen LB, Choi Y (2015) Historical overview and general methods of membrane potential imaging. Adv Exp Med Biol 859:3–26
2. Davies R, Graham J, Canepari M (2013) Light sources and cameras for standard in vitro membrane potential and high-speed ion imaging. J Microsc 251:5–13

3. Canepari M, Willadt S, Zecevic D, Vogt KE (2010) Imaging inhibitory synaptic potentials using voltage sensitive dyes. Biophys J 98:2032–2040
4. Sakmann B, Neher E (1986) Patch clamp techniques for studying ionic channels in excitable membranes. Annu Rev Physiol 46:455–472
5. Gray R, Johnston D (1985) Rectification of single GABA-gated chloride channels in adult hippocampal neurons. J Neurophysiol 54:134–142
6. Lester HA (1988) Heterologous expression of excitability proteins: route to more specific drugs? Science 241:1057–1063
7. Guy HR, Conti F (1990) Pursuing the structure and function of voltage-gated channels. Trends Neurosci 13:201–206
8. Antic SD (2016) Simultaneous recordings of voltage and current waveforms from dendrites. J Physiol 594:2557–2558
9. Williams SR, Mitchell SJ (2008) Direct measurement of somatic voltage clamp errors in central neurons. Nat Neurosci 11:790–798
10. Hodgkin AL, Huxley AF (1952) Currents carried by sodium and potassium ions through the membrane of the giant axon of Loligo. J Physiol 116:449–472
11. Jaafari N, De Waard M, Canepari M (2014) Imaging fast calcium currents beyond the limitations of electrode techniques. Biophys J 107:1280–1288
12. Canepari M, Vogt K, Zecevic D (2008) Combining voltage and calcium imaging from neuronal dendrites. Cell Mol Neurobiol 58:1079–1093
13. Vogt KE, Gerharz S, Graham J, Canepari M (2011a) High-resolution simultaneous voltage and Ca^{2+} imaging. J Physiol 589:489–494
14. Vogt KE, Gerharz S, Graham J, Canepari M (2011) Combining membrane potential imaging with L-glutamate or GABA photorelease. PLoS One 6:e24911
15. Kao JP, Tsien RY (1988) Ca^{2+} binding kinetics of fura-2 and azo-1 from temperature-jump relaxation measurements. Biophys J 53:635–639
16. Jaafari N, Marret E, Canepari M (2015) Using simultaneous voltage and calcium imaging to study fast Ca^{2+} channels. Neurophotonics 2:021010
17. Canepari M, Odgen D (2006) Kinetic, pharmacological and activity-dependent separation of two Ca^{2+} signalling pathways mediated by type 1 metabotropic glutamate receptors in rat Purkinje neurons. J Physiol 573:65–82
18. Canepari M, Vogt KE (2008) Dendritic spike saturation of endogenous calcium buffer and induction of postsynaptic cerebellar LTP. PLoS One 3:e4011
19. Savitzky A, Golay MJE (1964) Smoothing and differentiation of data by simplified least squares procedures. Anal Chem 36:1627–1639
20. Nägerl UV, Novo D, Mody I, Vergara JL (2000) Binding kinetics of calbindin-D(28k) determined by flash photolysis of caged Ca^{2+}. Biophys J 79:3009–3018
21. Airaksinen MS, Eilers J, Garaschuk O, Thoenen H, Konnerth A, Meyer M (1997) Ataxia and altered dendritic calcium signalling in mice carrying a targeted nullmutation of the calbindin D28k gene. Proc Natl Acad Sci U S A 94:1488–1493
22. Lee SH, Schwaller B, Neher E (2000) Kinetics of Ca^{2+} binding to parvalbumin in bovine chromaffin cells: implications for $[Ca^{2+}]$ transients of neuronal dendrites. J Physiol 525:419–432
23. Schmidt H, Stiefel KM, Racay P, Schwaller B, Eilers J (2003) Mutational analysis of dendritic Ca^{2+} kinetics in rodent Purkinje cells: role of parvalbumin and calbindin D28k. J Physiol 551:13–32
24. Ait Ouares K, Jaafari N, Canepari M (2016) A generalised method to estimate the kinetics of fast Ca^{2+} currents from Ca^{2+} imaging experiments. J Neurosci Methods 268:66–77
25. Usowicz MM, Sugimori M, Cherksey B, Llinás R (1992) P-type calcium channels in the somata and dendrites of adult cerebellar Purkinje cells. Neuron 9:1185–1199
26. Isope P, Hildebrand ME, Snutch TP (2012) Contributions of T-type voltage-gated calcium channels to postsynaptic calcium signaling within Purkinje neurons. Cerebellum 11:651–665

27. Miyakawa H, Lev-Ram V, Lasser-Ross N, Ross WN (1992) Calcium transients evoked by climbing fiber and parallel fiber synaptic inputs in guinea pig cerebellar Purkinje neurons. J Neurophysiol 68:1178–1189

28. Yan P, Acker CD, Zhou WL, Lee P, Bollensdorff C, Negrean A, Lotti J, Sacconi L, Antic SD, Kohl P, Mansvelder HD, Pavone FS, Loew LM (2012) Palette of fluorinated voltage-sensitive hemicyanine dyes. Proc Natl Acad Sci U S A 109:20443–20448

29. Willadt S, Canepari M, Yan P, Loew LM, Vogt KE (2014) Combined optogenetics and voltage sensitive dye imaging at single cell resolution. Front Cell Neurosci 8:311

30. Antic SD (2003) Action potentials in basal and oblique dendrites of rat neocortical pyramidal neurons. J Physiol 550:35–50

31. Bullen A, Saggau P (1998) Indicators and optical configuration for simultaneous high-resolution recording of membrane potential and intracellular calcium using laser scanning microscopy. Pflügers Arch 436:788–796

32. Jaafari N, Canepari M (2016) Functional coupling of diverse voltage-gated Ca^{2+} channels underlies high fidelity of fast dendritic Ca^{2+} signals during burst firing. J Physiol 594:967–983

33. Loew LM, Simpson LL (1981) Charge-shift probes of membrane potential: a probable electrochromic mechanism for p-aminostyrylpyridinium probes on a hemispherical lipid bilayer. Biophys J 34:353–365

34. Djurisic M, Antic S, Chen WR, Zecevic D (2004) Voltage imaging from dendrites of mitral cells: EPSP attenuation and spike trigger zones. J Neurosci 24:6703–6714

35. Roth A, Häusser M (2001) Compartmental models of rat cerebellar Purkinje cells based on simultaneous somatic and dendritic patch-clamp recordings. J Physiol 535:445–472

36. Spruston N, Schiller Y, Stuart G, Sakmann B (1995) Activity-dependent action potential invasion and calcium influx into hippocampal CA1 dendrites. Science 268:297–300

37. Canepari M, Djurisic M, Zecevic D (2007) Dendritic signals from rat hippocampal CA1 pyramidal neurons during coincident pre- and post-synaptic activity: a combined voltage- and calcium-imaging study. J Physiol 580:463–484

38. Bischofberger J, Jonas P (1997) Action potential propagation into the presynaptic dendrites of rat mitral cells. J Physiol 504:359–365

39. Sabatini BL, Regerh WG (1996) Timing of neurotransmission at fast synapses in the mammalian brain. Nature 384:170–172

40. Stuart G, Häusser M (1994) Initiation and spread of sodium action potentials in cerebellar Purkinje cells. Neuron 13:703–712

41. Vogt KE, Canepari M (2010) On the induction of postsynaptic granule cell-Purkinje neuron LTP and LTD. Cerebellum 9:284–290

42. Canepari M, Mammano F (1999) Imaging neuronal calcium fluorescence at high spatio-temporal resolution. J Neurosci Methods 87:1–11

Chapter 5
Molecular Diversity of Plasma Membrane Ca^{2+} Transporting ATPases: Their Function Under Normal and Pathological Conditions

Luca Hegedűs, Boglárka Zámbó, Katalin Pászty, Rita Padányi, Karolina Varga, John T. Penniston, and Ágnes Enyedi

Abstract Plasma membrane Ca^{2+} transport ATPases (PMCA1-4, *ATP2B1-4*) are responsible for removing excess Ca^{2+} from the cell in order to keep the cytosolic Ca^{2+} ion concentration at the low level essential for normal cell function. While these pumps take care of cellular Ca^{2+} homeostasis they also change the duration and amplitude of the Ca^{2+} signal and can create Ca^{2+} gradients across the cell. This is accomplished by generating more than twenty PMCA variants each having the character – fast or slow response, long or short memory, distinct interaction partners and localization signals – that meets the specific needs of the particular cell-type in which they are expressed. It has become apparent that these pumps are essential to normal tissue development and their malfunctioning can be linked to different pathological conditions such as certain types of neurodegenerative and heart diseases, hearing loss and cancer. In this chapter we summarize the complexity of PMCA regulation and function under normal and pathological conditions with particular attention to recent developments of the field.

L. Hegedűs
Department of Thoracic Surgery, Ruhrlandklinik, University Clinic Essen, Essen, Germany

B. Zámbó
Research Centre for Natural Sciences, Institute of Enzymology, Hungarian Academy of Sciences, Budapest, Hungary

K. Pászty
Department of Biophysics, Semmelweis University, Budapest, Hungary

R. Padányi · K. Varga · Á. Enyedi (✉)
2nd Department of Pathology, Semmelweis University, Budapest, Hungary
e-mail: enyedi.agnes@med.semmelweis-univ.hu

J. T. Penniston
Department of Neurosurgery, Massachusetts General Hospital, Boston, MA, USA

© Springer Nature Switzerland AG 2020
M. S. Islam (ed.), *Calcium Signaling*, Advances in Experimental Medicine and Biology 1131, https://doi.org/10.1007/978-3-030-12457-1_5

Keywords Plasma membrane Ca^{2+} ATPase (PMCA) · *ATP2B1-4* · Alternative splice · Calmodulin · Phosphatidylinositol-4, 5-bisphosphate · Actin cytoskeleton · Ca^{2+} signal · Genetic variation · Altered expression · Pathological condition

Abbreviations

AD	Alzheimer's disease
ATP	adenosine triphosphate
CaM	calmodulin
CaMKII	calcium/calmodulin-dependent protein kinase II
CASK	calcium/calmodulin-dependent serine protein kinase
CBS	calmodulin binding sequence
ER	endoplasmic reticulum
ERK	extracellular-signal regulated kinase
HDAC	histone deacetylase
HER2	human epidermal growth factor receptor 2
HUVEC	human umbilical vein endothelial cell
IP3	inositol 1,4,5-trisphosphate
IP_3R	inositol 1,4,5-trisphosphate receptor
IS	immunological synapse
MAGUK	membrane-associated guanylate kinase
MLEC	mouse lung endothelial cells
NFAT	nuclear factor of activated T-cell
NHERF2	Na^+/H^+ exchanger regulatory factor 2
nNOS	neural nitric oxide synthase
PIP2	phosphatidylinositol-4,5- bisphosphate
PKC	protein kinase C
PKA	protein kinase A
PMCA	plasma membrane Ca^{2+} ATPases
POST	partner of STIM
PSD-95	post synaptic density protein 95
RANKL	nuclear factor κB ligand
RASSF1	Ras association domain-containing protein 1
RBC	red blood cell
SCD	sickle cell disease
SERCA	sarco/endoplasmic reticulum Ca^{2+} ATPases
SNP	small nucleotide polymorphisms
SOCE	store operated Ca^{2+} entry
SPCA	secretory-pathway Ca^{2+} ATPase
STIM	stromal interacting molecule
TGF	transforming growth factor
TM domain	transmembrane domain

TSA	trichostatin A
VEGF	vascular endothelial growth factor
VSMC	vascular smooth muscle cell

5.1   Introduction

The plasma membrane Ca^{2+} transport ATPase (PMCA protein, *ATP2B gene*) was first described as a Ca^{2+} extrusion pump in red blood cells by Hans J. Schatzmann in 1966 [1]. It became evident that this pump is an essential element of the Ca^{2+} signaling toolkit, and that it plays a vital role in maintaining Ca^{2+} homeostasis in all mammalian cells [2]. After the first discovery of the PMCA many years were spent on identifying its regulators (for example calmodulin and acidic phospholipids) before it was cloned and sequenced at around the time when sequences for many of the other P-type ATPase family members also became available [3, 4]. Further structure-function studies concentrated on the PMCAs unique C-terminal regulatory region (often called the C-tail) and identified there calmodulin and PDZ-domain binding sequence motifs, a built-in inhibitor sequence, phosphorylation sites for protein kinases and a localization signal [5]. It became apparent that PMCAs comprise a P-type ATPase sub-family, encoded by four separate genes *ATP2B1-4* [6, 7] from which alternative splicing generates more than 20 variants with distinct biochemical characteristics that make them suitable to perform specific cellular functions [8, 9]. By now it is well documented that PMCAs are not simply Ca^{2+} extrusion pumps but by changing their abundance and variant composition, having different activation kinetics, locale and partners, they can actively modulate the Ca^{2+} signal in space and time, and hence affect Ca^{2+} mediated signaling events downstream. The PMCA variants are expressed in a tissue and cell type specific manner and many of them have specific function. Although, in the past decades these pumps have been extensively characterized their importance is rather underestimated. This is because only recently we gathered more information on their involvement in diseases such as cancer, neurological disorders, hearing loss and others. In this book chapter, therefore, we will summarize briefly the long known basic characteristics of these pumps paying more attention to the most recent findings on their roles under normal and pathophysiological conditions.

5.2   Structural Features of the PMCA

PMCAs (*ATP2B1-4 gene*) belong to the P-type ATPase family and share basic structural and catalytic features with them. The closest relatives of the PMCAs are the sarco/endoplasmic reticulum type Ca^{2+} pumps (SERCAs, *ATP2A1-3*) with an overall 30 % sequence homology between PMCA4 and SERCA1 [10]. Homology modeling using the SERCA1 structure as a template [11–13] has revealed four major domains shared with SERCA1, and a relatively large unstructured C-terminal region (30–130 residues depending on the isoforms and their variants), which is unique

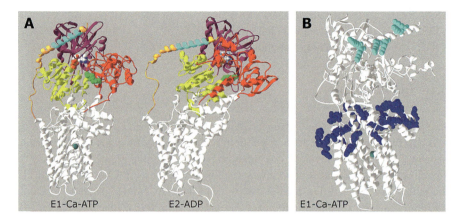

Fig. 5.1 Structural model of the PMCA in the E1-Ca-ATP and E2-ADP conformations.
Structures of several intermediates in the enzyme cycle have been determined for SERCA [11–
13]. Based on those, models of the intermediates have been constructed for PMCA4b (lacking
90 residues from the C-tail). (**a**): The models show 2 of the intermediates, E1 with Ca and ATP
(SERCA PDB 1VFP and 1T5S) and E2 with ADP (SERCA PDB 2C88). In the latter, the Ca
has been ejected into the extracellular space. They are colored as follows: A domain *red*; P
domain *yellow*; N domain *purple*; stalk; insert and transmembrane domains *white*; C-tail *straw*;
Calmodulin-binding domain *cyan*; Ca^{2+} *metallic blue-green*. (**b**): The positively charged residues
of the PIP$_2$ binding regions are colored. The blue collar and the insert *blue*, the calmodulin-binding
domain (CBD) *cyan*. The CBD would have the potential of releasing from the conformation shown
and lying on the surface of the membrane in a PIP$_2$-rich region

to the PMCAs (Fig. 5.1a) (for a review see also [14]). The M-domain consists
of 10 trans-membrane spanning helices that provide the coordinating ligands for
the binding of one cytosolic Ca^{2+} ion to be transported. The N-domain binds an
ATP molecule of which the terminal phosphate is transferred to a highly conserved
aspartate in the P-domain forming a high-energy acyl-phosphate intermediate. As a
result of these events hydrolysis of one ATP molecule provides sufficient energy
to translocate one Ca^{2+} ion through the membrane [15] that is coupled to H^+
transport in return with a $Ca^{2+}:H^+$ ratio of 1:2 [16]. The A-domain coordinates
the movements of the other three domains during the E1-E2 transition to complete
a full reaction cycle [17]. While the catalytic domains N, P and the M-domain
are largely conserved between the PMCAs the C-tail and the A-domain – where
alternative splicing generates substantial sequence divergence – vary substantially.
These variations in the C-tail and A-domain can generate PMCA proteins with
distinct characteristics [18, 19].

The Blue Collar In contrast to the endoplasmic reticulum-resident SERCA pump
a cluster of positively charged residues were found at the intracellular near-
membrane region of the PMCA forming four binding pockets for the phosphorylated
inositol ring of PIP2 (phosphatidylinositol-4,5-bisphosphate) [20], in addition to the
previously determined linear PIP2 binding sequences near the A splice-site region
at the A-domain [21, 22] and the calmodulin binding sequence at the C-tail [23].
Figure 5.1b shows a blue collar formed from the four PIP2 binding pockets and the

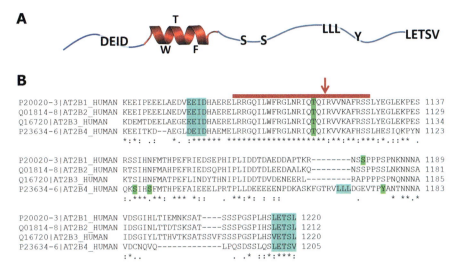

A

B

```
                        KEEIPEEELAEDVEEIDHAERELRRGQILWFRGLNRIQTQIRVVNAFRSSLYEGLEKPES
P20020-3|AT2B1_HUMAN    KEEIPEEELAEDVEEIDHAERELRRGQILWFRGLNRIQTQIRVVNAFRSSLYEGLEKPES 1137
Q01814-8|AT2B2_HUMAN    KEEIPEEELNEDVEEIDHAERELRRGQILWFRGLNRIQTQIRVVKAFRSSLYEGLEKPES 1129
Q16720|AT2B3_HUMAN      KDEMTDEELAEGEEEIDHAERELRRGQILWFRGLNRIQTQIRVVKAFRSSLYEGLEKPES 1134
P23634-6|AT2B4_HUMAN    KEEITKD--AEGLDEIDHAEMELRRGQILWFRGLNRIQTQIKVVKAFHSSLHESIQKPYN 1123
                        *:*: .:   *. :****** ******************:**:**:***:*.::** .

P20020-3|AT2B1_HUMAN    RSSIHNFMTHPEFRIEDSEPHIPLIDDTDAEDDAPTKR--------NSSPPPSPNKNNNA 1189
Q01814-8|AT2B2_HUMAN    RTSIHNFMAHPEFRIEDSQPHIPLIDDTDLEEDAALKQ--------NSSPPSSLNKNNSA 1181
Q16720|AT2B3_HUMAN      KTSIHNFMATPEFLINDYTHNIPLIDDTDVDENEERL---------RAPPPPSPNQNNNA 1185
P23634-6|AT2B4_HUMAN    QKSIHSFMTHPEFAIEEELPRTPLLDEEEEENPDKASKFGTRVLLLDGEVTPYANTNNNA 1183
                        :.***.**: *** *::  . **:*: : ::         .       * **.*

P20020-3|AT2B1_HUMAN    VDSGIHLTIEMNKSAT----SSSPGSPLHSLETSL 1220
Q01814-8|AT2B2_HUMAN    IDSGINLTTDTSKSAT----SSSPGSPIHSLETSL 1212
Q16720|AT2B3_HUMAN      IDSGIYLTTHVTKSATSSVFSSSPGSPLHSVETSL 1220
P23634-6|AT2B4_HUMAN    VDCNQVQ------------LPQSDSSLQSLETSV 1205
                        :*..                . .* ::*:***:
```

Fig. 5.2 C-tail of the "b" splice variants of PMCA1-4. (a) Schematic representation of the C-tail of PMCA4b emphasizing important sequence motifs highlighted below. Calmodulin-binding domain is colored *burgundy*. **(b)** An alignment of C-terminal sequences of "b" splice forms of PMCA1-4 demonstrates that the variants may have distinct regulatory features (i.e. the di-leucine motif in PMCA4b) however; some sequence motifs (caspase 3 sites and the PDZ-binding tails) are relatively conserved. These motifs are colored *cyan*. The PKA, PKC and tyrosine kinase phosphorylation sites are highlighted in *green* and the calmodulin-binding sequence is marked *burgundy*. The arrow indicates where alternative splice changes the sequence in the other splice variants of PMCA1-4

linear lipid binding region of the A domain around the stalk region of the PMCA. This arrangement of positively charged residues follows the positive inside role, which is quite common in plasma membrane proteins and often involved in PIP2 binding [24, 25].

The C-Tail The C-tail, which is also known as the main regulatory unit of these pumps, is the most characterized although the least conserved region of the PMCAs (Fig. 5.2). A major portion of this region is structurally disordered [5], containing multiple recognition sites: a DxxD caspase cleavage site [26, 27], a calmodulin-binding domain (CBD) with an overlapping auto-inhibitory region and acidic lipid binding side chains [3], several protein kinase phosphorylation sites [28, 29], a di-leucine-like localization signal [30] and a PDZ-domain-binding sequence motif at the C-terminus [31]. Some of these motifs are present in nearly all PMCAs (caspase 3 cleavage sites, CBD) while others are specific to certain variants; for instance the di-leucine-like motif is specific to PMCA4b whereas the PDZ-binding motif is present in all "b" splice variants. However, specificity of the PDZ binding may vary because the terminal amino acid is Val in PMCA4b but Leu in PMCA1-3. As an example the sodium-hydrogen exchange regulatory cofactor NHERF2 interacts with PMCA2b but not with PMCA4b [32].

Ca²⁺-Calmodulin Binding is critical for PMCA function. Early studies identified a 28 residue long sequence at the C-tail of PMCA4b that could bind Ca^{2+}-calmodulin with high affinity. Extensive kinetic [33, 34] and NMR [35] studies with a peptide (c28) representing the complete 28-residue sequence region have revealed two anchor sites Phe-1110 and Trp-1093 in a relative position of 18 and 1, and two steps of Ca^{2+}-calmodulin binding in an anti-parallel manner (Fig. 5.3). In the first step the C-terminal lobe of calmodulin binds the N-terminal Trp-1093, followed by the second step, which is binding of the C-terminal Phe-1110 to the N-terminal lobe of calmodulin. As a result, calmodulin wraps around the c28 peptide that adopts an α-helix with its anchors buried in the hydrophobic pockets of the two distinct CaM lobes. This model correlates well with an earlier NMR structure of Ca^{2+}-calmodulin with a shorter c20 peptide lacking the second anchor Phe-1110 [36]. In that case the peptide could bind only to the C-terminal lobe of calmodulin, which retained its extended structure, as is expected (Fig. 5.3).

The w Insert Another structurally less defined region of the molecule is the sequence that couples the A domain to the third membrane spanning helix. An

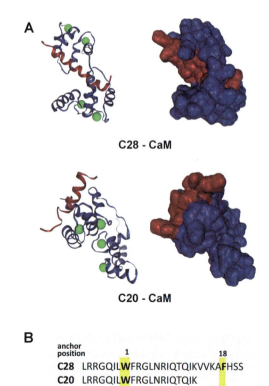

Fig. 5.3 NMR structure of calmodulin in complex with calmodulin binding peptides. (a): Structures of C28-calmodulin (https://www.rcsb.org/structure/2KNE) and C20-calmodulin complexes (https://www.rcsb.org/structure/1CFF). C20 and C28 correspond to the appropriate calmodulin binding sequence of PMCA4b. Colors: calmodulin *blue*; peptide *burgundy*; green spheres correspond to the 4 Ca^{2+} bound to CaM. **(b):** Sequences of the peptides C20 and C28 with the anchors 18-1

alternative splice at splice site A changes the structure of this region by including or excluding a single exon, producing the x and z variants of the isoforms [37], however, no functional significance has been linked to these changes. In PMCA2, however, additional variations exist in which two more exons can be inserted generating the PMCA2 y and w forms. Importantly, the w insert – which is a 44-residue long sequence – is essential for targeting PMCA2 to the apical compartment of polarized cells.

5.3 Regulation of PMCA Expression and Function

PMCAs are encoded by four separate genes (*ATP2B1-4*) located at distinct chromosomes: 12q21–23, 3p25.3, Xq28 and 1q25–q32, respectively [8]. Two major alternative splice options at splice sites A and C of the primary transcripts of each *ATP2B* gene have the potential of generating >30 PMCA protein variants, however, only 20 of them have been identified in different tissues [38, 39]. In addition, mutations, single nucleotide polymorphisms and posttranslational modifications further increase PMCA variations. It is not surprising that to keep the level of calcium within a suitable range in the cytoplasm of different cell types with very different function tight regulation of PMCAs is required at the transcriptional, splicing, translational and protein levels.

5.3.1 Regulation at the Transcription Level

Transcriptional regulation of *ATP2B* genes is complex and still not well understood. The intricate regulatory structure of the promoter and enhancer regions of the genes allows the fine-tuning of each PMCA's transcription during embryonic development, in various tissues, as well as upon various stimuli. It been shown that in mouse smooth muscle cells Atp2b1 expression during G1/S phase is reduced via c-myb binding to the promoter region of the gene [40]. This transcription factor is also involved in the down-regulation of Atp2b1 in differentiating B-lymphocytes [41]. The active form of vitamin D induces the transcription of *ATP2B1* in various tissues and cell types [42–45]. *ATP2B2* gene has four alternative promoters and alternatively spliced 5' exons, which showed higher expression and different promoter usage in mammary gland compared to neuronal cells [46]. EGR1 can bind to a specific region in the CpG island of the *ATP2B2* gene and controls the α-type promoter activity, which is specific to brain and auditory cells [47]. The *ATP2B4* gene contains an enhancer in the intron 1, which has an essential role in the erythroid differentiation, but has no effect in other cell types [48]. From these studies it appears that PMCAs possess general and specific transcription factor binding sites and regions, which only play role under certain conditions, under proper stimulus or differentiation state of the given cell type.

5.3.2 Regulation at the Protein Level

Auto-Inhibition PMCA activity is determined by the presence of an auto-inhibitory unit at the C-tail, which largely overlaps with the calmodulin-binding sequence [49]. This inhibitory unit binds to the N- and A-domains interfering with Ca^{2+} binding to the catalytic sites, and slowing down the reaction cycle by inhibiting the movements of the cytosolic domains [23]. The extent of the auto-inhibition differs from one isoform to the other and is affected by the alternative splice at splice site C [50–52]. As a result, PMCA4b is the only truly inactive pump at resting cytosolic Ca^{2+} ion concentration while all the other pumps are partially active, as determined in cell free systems.

Activation by Caspase 3 The auto-inhibitory C-tail is removed by the executor protease caspase 3 during apoptosis. Caspase 3 cleaves PMCA4b at a canonic caspase 3 cleavage site (DEID) just upstream of the CBD-auto-inhibitory sequence removing the complete auto-inhibitory region [26, 27]. While there has been a long debate on whether caspase 3 activates or inhibits PMCA4b during apoptosis [53] it is conceivable that deleting the auto-inhibitor should result in a gain-of- function pump [54], however, the overall outcome could depend on the given cell type, stimulus and conditions that need further studies.

Activation with Ca^{2+}-Calmodulin A functionally important feature of the PMCA variants is the difference in their activation with Ca^{2+}-calmodulin that determines the rate by which they can respond to the incoming Ca^{2+} signal, and equally important is the length of time during which they remain active after the stimulus [55]. Since pump and calmodulin compete for CBD-autoinhibitor it is expected that a strong pump-CBD-auto-inhibitor interaction will result not only in a low basal activity but also in a slow activation rate. Indeed, PMCA4b has both the lowest basal activity and the slowest activation with calmodulin among the isoforms (slow pump, $T_{1/2}$ is about 1 min) [56]. Although, PMCA4b is activated slowly its inactivation rate is even slower (long memory, remains active for about 20 min) because calmodulin remains bound to the pump for a long period of time [57]. An alternative splice that creates a shorter version of PMCA4 changes the response of the pump to Ca^{2+} completely so that PMCA4a binds Ca^{2+}-calmodulin quickly (fast pump, T1/2 is about 20 s) but then calmodulin dissociates also quickly, resulting in a fast responding pump that remains active for a relatively short period of time (short memory, active for less than a minute) [34]. It is important to note, that PMCA4a also has a relatively high basal activity suggesting weak interaction between pump and auto-inhibitor. All other forms – variants of PMCA2 and PMCA3 – that have been characterized are fast responding pumps having slow inactivation rates (long memory), and as mentioned above they also have relatively high activity even without activators [50, 57].

Activation with Acidic Phospholipids Acidic lipids like PS and the PIPs – PI, PIP and PIP2 – can activate the pump and the amount of activation is augmented as the negative charge of the phospholipid head group increases [58]. It has been demonstrated that both the CBD and the linear basic sequence in the A-domain are involved in this type of activation [21–23]. It has been suggested that changes in the lipid composition may affect PMCA activity and that PMCAs might be more active in PIP2-rich lipid rafts [59]. Recently, it was demonstrated that the activity of the PMCA is also modulated by neutral phospholipids. The activity of PMCA4b was optimal when it was reconstituted in a 1,2-dimyristoyl-*sn*-glycero-3-phosphocholine (DMPC) bilayer of approximately 24 Å thickness [60]. Molecular simulation studies have revealed that in DMPC several lysine and arginine residues at the extracellular surface are exposed to the medium while in a thicker layer of 1,2-dioleoyl-sn-glycero-3-phosphocholine (DOPC) these residues are embedded in the hydrophobic core that could explain the reduced activity observed in DOPC.

Regulation by the Actin Cytoskeleton First it was shown that PMCAs interact with F-actin in activated platelets and they are associated with the F-actin rich cytoskeleton at or near the filopodia [61, 62]. Later it was documented that the purified PMCA4b can bind both monomeric and filamentous actin and while actin monomers activate the pump, F-actin may inhibit its function [63, 64]. These results were confirmed by using live HEK cells expressing isoforms PMCA2 and PMCA4 [65]. Based on these findings it has been suggested that PMCA can regulate actin dynamics through a series of feed-back regulations by lowering Ca^{2+} concentration in its vicinity and promoting actin polymerization, which in turn switches off the PMCA function allowing increase in Ca^{2+} levels and hence actin de-polymerization [66].

5.4 Function of the PMCAs in the Living Cell

It is quite remarkable how the above described diverse structural and biochemical characteristics of the PMCA proteins are translated into specific physiological functions in the different cell types. Distinct kinetics of the PMCAs are transcribed into distinct Ca^{2+} signaling properties while additional structural diversity between the PMCAs determines their localization and interaction patterns with different scaffolding and signaling molecules resulting in unique PMCA variant-specific cellular function (Table 5.2).

5.4.1 PMCAs Shape the Ca^{2+} Signal

It has been widely accepted that PMCAs play a role in the decay phase of the store-operated Ca^{2+} entry (SOCE). However, expression of PMCAs with distinct kinetic properties (see also Table 5.1 and Fig. 5.4) – fast or slow, with or without memories – resulted not only in a faster decay of the signal but also in very different Ca^{2+} signaling patterns in HEK and HeLa cells [67]. While the "slow with

Table 5.1 Distinct kinetic properties and distribution of the PMCAs

Basal activity	Activation with CaM	Memory	Pump variant	Cells, tissues
High	Fast	Short	PMCA4a	Smooth muscle, heart, sperm
High	Fast	Long	PMCA2b	Neuron, mammary gland
			PMCA2a	Neuron, cochlear hair cells
Low	Slow	Long	PMCA4b	Erythrocyte, breast, colon heart, kidney, HUVEC, melanoma

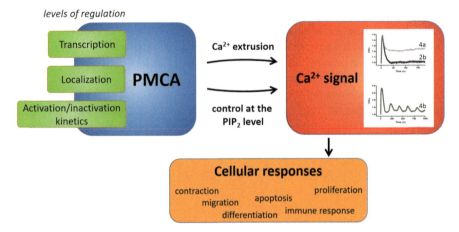

Fig. 5.4 Schematic representation of the role of PMCAs in Ca^{2+} signaling. The abundance of PMCA is regulated at the transcriptional and protein levels. Localization is affected by specific sequence motifs and interaction with other proteins. The activity is regulated by proteins (such as calmodulin) and acidic lipids. This can result in the modulation of the Ca^{2+} signal at two levels: (i) Ca^{2+} extrusion; (ii) IP_3-induced Ca^{2+} release. The resultant Ca^{2+} signal then might be translated to distinct cell responses

memory" PMCA4b induced Ca^{2+} oscillation after the first spike, the C-terminal splice variant of the PMCA4 isoform – the "fast without memory" PMCA4a – responds quickly to the incoming Ca^{2+} but then since it becomes inactivated also quickly the signal returns to an elevated level without oscillation. PMCA2b – a fast pump with memory – allows only short Ca^{2+} spikes and Ca^{2+} concentration always returns to the basal level quickly. It was also demonstrated that in addition to shaping the SOCE mediated Ca^{2+} signal PMCAs also control the formation of IP3 by controlling the availability of the signaling PIP2 molecules, and hence regulate the release of Ca^{2+} from the stores [20] (Fig. 5.4). It is important to note that the Ca^{2+} signal can also be altered through additional cell type-specific regulatory mechanisms of the PMCA. During T-cell activation, for example, it was shown that the activity of PMCA4b is inhibited by the interaction with the ER Ca^{2+} sensor protein STIM1 [68] and its partner scaffold protein POST [69] resulting in a more sustained elevation in intracellular Ca^{2+} concentration.

5.4.2 Cell Type Specific Expression of the PMCAs

Homozygous deletion of the *ATP2B1* gene in mice is lethal suggesting that PMCA1 is the housekeeping isoform [70]. The other isoforms PMCA2-4 are expressed at different stages of development [8]. The slow PMCA4b variant is present in erythrocytes, T lymphocytes and in epithelial cells but also abundantly expressed in the heart and smooth muscle cells [39]. PMCA4a is expressed in the brain and it is the only PMCA isoform present in the sperm tail [71]. Altered expression of *ATP2B4* in mice was associated with arrhythmias, cardiac hypertrophy and heart failure. Deletion of both copies of *ATP2B4* in mice caused male infertility [70, 72]. Interestingly, in activated sperm cells the pattern of the Ca^{2+} signal is similar to that seen in the PMCA4a expressing Hela cells [73]. Ca^{2+} pumps (PMCA1 and PMCA4) were shown to contribute to sustained Ca^{2+} oscillations in human mesenchymal stem cells [74] and airway smooth muscle cells [75].

The fast pumps PMCA2 and PMCA3 are abundantly expressed in excitable tissues such as the brain and skeletal muscle [76, 77]. The PMCA2w/a and PMCA2w/b forms are found in vestibular hair cells and in Purkinje neurons of the cerebellum where they can react quickly to the fast signals induced by the voltage-gated Ca^{2+} channels. A specific form PMCA2w/b is also expressed in the lactating mammary gland. Knock down of the *ATP2B2* gene induced ataxia, deafness [78] and reduced Ca^{2+} concentration in the milk [79]. These are just a few examples demonstrating how variations in PMCA expression contribute to cell-type specific functions (see more details in refs (39, 55, 76, 77) and in Table 5.1.

5.4.3 Polarized Expression of the PMCA

To perform their cellular function it is also important to target PMCA proteins to the appropriate membrane compartment. This is accomplished by intrinsic localization signals and/or by interaction with other proteins in a cell-type specific manner. In many cases these characteristics of the PMCAs are sensitive to alternative splicing. For example, the di-leucine-like localization motif is unique to the "b" splice variant of PMCA4 that was shown to direct this pump to endocytic vesicles in non-confluent epithelial cells [30]. Hence, PMCA4b localizes to the plasma membrane only in fully confluent differentiated cells where it can be stabilized and/or modulated by other interacting molecules [80]. Most recently, basigin/CD147 was identified as a novel interacting protein that may serve as a subunit of the PMCA [81]. It was demonstrated in a variety of cell types that PMCA1-4 interacts with basigin in the ER, which is essentially involved in functional targeting PMCAs to the plasma membrane.

PMCA proteins are localized to specific membrane compartments in polarized cells where they contribute to trans-cellular Ca^{2+} fluxes. While the lateral compartment seems to be the default place, in some cell types PMCAs localize apically.

PMCA2 for example can be directed to the apical compartment by an alternative splice option at site A that introduces a 44-residue long "w" sequence at the region that connects the A and TM domains [37]. The resultant PMCA2w/b and PMCA2w/a variants have very specific functions in the lactating mammary gland [79] and the stereocilia of hair cells [78, 82, 83] where PMCA2w/b is responsible for milk Ca^{2+} while PMCA2w/a contributes to hearing, respectively. The "b" splice variant of PMCA2w might be connected through PDZ-interactions with the scaffold protein NHERF2 to the actin cytoskeleton by which it is immobilized in the apical membrane [84]. In contrast, PMCA2w/a, which is lacking the PDZ-interacting tail, is very mobile, trafficking in and out of the stereocilia of hair cells [85]. In parotid gland acinar cells PMCA4b was found in the apical membrane compartment and its localization was modulated by PDZ-interaction with Homer2 [86]. In the same cells PMCA1 was also found in the apical membrane but only when it was phosphorylated by PKA [87]. PMCA4b plays an important role in the immune synapse where it is targeted to specific signaling micro domains beneath the mitochondria where it is actively involved in Ca^{2+} handling during T-cell activation controlling Ca^{2+} influx through the CRAC channels [88].

Polarized distribution of PMCA was also found in migrating cells. In collectively migrating human umbilical vein endothelial cells (HUVEC) PMCA located to the front of the cells by which it contributed to the front-to-rear Ca^{2+} gradient essential for directed cell migration [89]. In addition, downregulation of PMCA4 increased while its overexpression decreased cell migration in a wound-healing assay of HUVECs [90]. These data are in line with the latest finding demonstrating that PMCA4b interferes with cell migration of a highly motile BRAF mutant melanoma cell line [91]. These examples highlight the importance of PMCA targeting and demonstrate that different interacting partners may change the location of PMCAs resulting in distinct cellular functions (see Table 5.2).

5.4.4   Interaction of PMCAs with Signaling Molecules

Through interactions with other proteins PMCAs can influence downstream signaling events (Table 5.2). In many cases they influence the activity of the interacting signaling molecule by reducing the Ca^{2+} concentration in its vicinity. One example is the interaction of PMCA2 and PMCA4 with the Ca^{2+}-CaM dependent phosphatase calcineurin through their catalytic domain that was found to reduce the activity of the nuclear factor activated T-cell (NFAT) pathway [92, 93]. Inhibition of this interaction increased Fas-ligand expression and apoptosis in breast cancer cells [94], while PMCA4b overexpression in endothelial cells reduced VEGF initiated cell migration and angiogenesis [95]. Another example for this type of interaction was described between PMCA4b and calcium/calmodulin-dependent serine protein kinase (CASK) in rat brain and kidney where PMCA4b binds CASK through its C-terminal PDZ binding motif [96]. CASK together with Tbr-1 induces T-element dependent transcription; however, this is strongly decreased upon interaction with

Table 5.2 PMCA interactions

Interacting partner	Isoform preference	Type/place of interaction	Cell type	Effect	References
Effect of interacting on PMCA					
14-3-3ε	PMCA1, PMCA3, PMCA4	N-terminal	HeLa, CHO	Inhibition of PMCA activity	[233, 234]
G-actin (short oligomers)	PMCA4	Putative actin-binding site	Erythrocyte	Increasing PMCA activity	[63, 64, 235]
F-actin	PMCA4	Putative actin-binding site	Erythrocyte	Inhibition of PMCA activity	[64]
G/F-actin	PMCA4b	Through CLP36 and α-actinin	Platelet		[62, 236]
Calmodulin	All PMCA forms	Calmodulin-binding domain CaMBS1, CaMBS2	All kinds of cells	RELIEVING autoinhibition of the pump, increasing both Ca2+ affinity and pump rate	[237, 238]
CASK	PMCA4b	Via PDZ interaction	Mouse sperm	Reduced PMCA activity	[97]
Homer1	PMCA1-4 b-splice forms	Via PDZ interaction	Hippocampal neurons, MDCK, COS-7	Increasing PMCA activity	[239, 240]
Homer2	PMCA1b	Homer-binding	Mouse parotid gland acinar cells, HEK293	Inhibition of PMCA activity	[86]
	PMCA4b				

(continued)

Table 5.2 (continued)

Interacting partner	Isoform preference	Type/place of interaction	Cell type	Effect	References
Effect of PMCA on the interacting partner					
Calcineurin	PMCA4b	Main intracellular loop (amino acids 501–575)	HEK293, MCF-7, HUVEC, endothelial cells, cardiomyocytes	Inhibition of the calcineurin/NFAT signal transduction pathway	[92, 93, 95, 204]
	PMCA2(b)	Amino acids 462–684	Breast cancer cells, MCF-7, ZR-75-1, MDA-MB-231, T47D,HEK293	Inhibition of the calcineurin/NFAT signal transduction pathway	[93, 94]
CASK	PMCA4b	Via PDZ interaction	Rat brain and kidney, HEK293	Inhibition of CASK function	[96]
CD147 (basigin)	PMCA4		Jurkat E6.1	Negatively regulates IL-2 expression	[231]
nNOS	PMCA4b	Via PDZ interaction	HEK293, neuro-2a, cardiomyocytes, sperm	Decrease nNOS activity, decrease NO production	[197, 241, 242]
eNOS	PMCA4	Catalytic intracellular loop	Endothelial cells, sperm	Decrease eNOS activity, decrease NO production	[243]
Ras-associated factor-1	PMCA4b	Main intracellular loop (region 652–748)	HEK293, cardiomyocytes	Modulating the EGF-mediated Ras signaling pathways	[244]
α1-syntrophin	PMCA1b, PMCA4b	Main intracellular loop (region 652–748)	HEK293, heart	Inhibit NOS-1-mediated NO production	[200]

Changing localization of PMCA

CLP36	PMCA4b	Via PDZ interaction	Platelet	Connect PMCA to the cytoskeleton	[62]
mGluR1–Homer3–IP3R1	PMCA2	via PDZ interaction	Purkinje neurons	Component of the mGluR1–Homer3–IP3R1 signaling complex	[245]
MAGUK family members: PSD95, PSD93, SAP97 and SAP102	PMCA2b, PMCA4b	Via PDZ interaction	MDCK, hippocampal neurons	Recruit PMCA to specific membrane domains	[31, 80, 246]
	Only PMCA4b	Via PDZ interaction			
neuroplastin or basigin	PMCA1-4	Heteromeric complexes of PMCAs1–4 and either neuroplastin or basigin	Brain, kidney	Stability and effective surface trafficking of PMCA	[81]
NHERF2	PMCA2(w)b	Via PDZ interaction	MDCK	NHERF2 anchors PMCA2w/b to the apical actin filaments via ezrin	[32, 84]
	PMCA1b	Via PDZ interaction	HT-29	Scaffolding and maintaining PMCA at the cell membrane	[247]
PISP	All PMCA b-splice forms	Via PDZ interaction	MDCK	Sorting of PMCAs to and from the plasma membrane	[248]

PMCA4b in HEK cells. Interestingly, CASK and PMCA4b interaction was also found in mouse sperm where CASK inhibited the activity of the pump resulting in an increased Ca^{2+} level and ultimately decreased motility of the sperm [97]. Several other interactions between PMCA proteins and their partners were described that influence downstream signaling events such as interactions with nNOS in the heart, CD147 in T-cells, STIM and POST in the immune synapse or with F- and G-actin. These results demonstrate that besides maintaining the low intracellular calcium level PMCAs are also important signaling molecules modulating the outcome of a variety of cell-type specific functions.

5.5 PMCAs in Disease Pathogenesis

PMCA proteins have been associated with several diseases in humans. Since many isoforms have highly specialized, cell type specific function alterations in their expression, localization, regulation or activity may contribute to the development of distinct pathological conditions (Table 5.3) [98]. Alterations of the PMCAs have been described in cardiovascular diseases, neurodegenerative disorders and cancer [99, 100]. More recently genetic variations in the *ATP2B* genes were also linked to certain pathological conditions.

5.5.1 Diseases Related to Genetic Variations in ATP2B1-4

ATP2B1 Small nucleotide polymorphisms (SNPs) found in the *ATP2B1* gene were associated with hypertension [101, 102], coronary artery disease [103–105] and early onset preeclampsia [106]. Preeclampsia is a disorder during pregnancy and it is characterized by high blood pressure and proteinuria. Reduced Ca^{2+}-ATPase activity of myometrium and the placental trophoblast was described in preeclamptic women [107], and a decreased expression of PMCA1 and PMCA4 in preeclamptic placental tissue was also found [108] suggesting a pivotal role of PMCAs in calcium homeostasis and transport through the placenta. The susceptibility to hypertension resulting in elevated blood pressure was linked to SNP rs11105378 in *ATP2B1* that was suggested to decrease PMCA1 expression in human umbilical artery smooth muscle cells [109]. In patients with chronic kidney disease, SNPs in ATB2B1 were associated with coronary atherosclerosis and myocardial infarction [105].

ATP2B2 SNPs in the *ATP2B2* gene were associated with autism in both European and Chinese population [110, 111]. Also, a missense mutation of PMCA2 (V586M) was shown to exacerbate the effect of the mutation in cadherin-23 leading to hearing loss [112, 113] in good accordance with the finding that ablation or missense mutations in PMCA2 cause deafness in mice [83, 114].

Table 5.3 PMCA related diseases

PMCA/*ATP2B*	Diseases associated with genetic variation in the *ATP2B1-4* genes	Diseases associated with altered expression, localization or activity of the PMCA proteins
PMCA1 (*ATP2B1*)	Hypertension	Multiple sclerosis
	Coronary artery disease	Reduced bone mineral density
	Myocardial infarction	Oral cancer
	Early onset preeclampsia	Ovarian cancer
PMCA2 (*ATP2B2*)	Hereditary deafness	Parkinson's disease
	Autism	Type 1 and type 2 diabetes
		Breast cancer
PMCA3 (*ATP2B3*)	X-linked cerebellar ataxia	Multiple sclerosis
	Aldosterone producing adenomas	
PMCA4 (*ATP2B4*)	Familial spastic paraplegia	Cardiac hypertrophy
	Developmental dysplasia of the hip	Hypertension
	Malaria resistance	Sickle cell disease
		Alzheimer's disease
		Chronic kidney disease
		Diabetes
		Adult idiopathic scoliosis
		Colon cancer
		Breast cancer
		Melanoma

ATP2B3 Missense mutation in the *ATP2B3* gene was found in patients with X-linked congenital cerebellar ataxia in two separate cases, in which the ability of the pump to decrease intracellular Ca^{2+} concentration after stimulation was compromised [115, 116]. Later it was demonstrated that the G1107D replacement altered both activation and auto-inhibition of this pump at low Ca^{2+} levels [117]. Mutations in the *ATP2B3* gene were also identified in some aldosterone producing adenomas (APA), and were linked to elevated aldosterone production compared with wild type APAs [118, 119]. In cellular models it was demonstrated that impaired PMCA3 function resulted in elevated intracellular Ca^{2+} levels and consequently increased aldosterone synthase production in the cells [120].

ATP2B4 Missense mutation in the *ATP2B4* gene was found in one family with familial spastic paraplegia that causes lower limb spasticity and weakness in patients [121]. Later it was shown that overexpression of the mutant PMCA4 protein in human neuroblastoma cells increased the resting cytosolic Ca^{2+} concentration and elevated the maximal Ca^{2+} surge after stimulation relative to the wild type pump [122]. Rear heterozygous variants in the *ATP2B4* and the *HSPG2* genes were described in a family with developmental dysplasia of the hip and based on *in silico* analysis an epistatic interaction was suggested between the genes [123]. SNPs in the *ATP2B4* gene were related to resistance against severe malaria that will be discussed in detail in the next chapter.

5.5.2   PMCAs in Red Blood Cell Related Diseases

PMCAs were among the first proteins described – and later characterized – in the membrane of red blood cells [124–126]. Since mature red cells (RBCs) are easily accessible, and have no internal membrane organelles involved in Ca^{2+} homeostasis, they have become important model cells for the examination of the enzymatic activity and kinetic parameters of the plasma membrane-bound PMCA protein [22, 127, 128]. Two isoforms have been identified in the RBC surface, PMCA1b and PMCA4b, of which PMCA4b appeared to be the most abundant [129–132]. These high affinity calcium pumps are responsible for maintaining the exceptionally low total Ca^{2+} content of red cells [133–135]. They have a crucial role in balancing cell calcium during shear stress in the microcirculation [136], volume control [137, 138] and in senescence and programmed cell death [131, 139, 140] of RBCs. Under certain pathological conditions – such as hereditary hemolytic anemia, malaria and diabetes mellitus – the intracellular Ca^{2+} levels in RBCs are altered [135, 141], therefore, the role of PMCAs in these cases emerges.

In Hereditary Hemolytic Anemia Ca^{2+} transport has a particular importance. In case of sickle cell anemia (SCD) and thalassemia, atypical hemoglobin (such as HbS) polymerization and deoxygenating processes lead to membrane deformation and activation of the mechanosensitive stretch-activated cation channel PIEZO1 [142]. As a result, Ca^{2+} permeability of these atypical RBCs increases. Subsequent stochastic activation of the Gardos or Ca^{2+}-sensitive potassium channel can lead to sickling and dehydration of red cells in SCD patients [131, 138, 143, 144]. It was found that PMCA inhibition is also involved in the maintenance of the high Ca^{2+} concentration needed for sickle cell dehydration [145, 146].

Severe Malaria is one of the most studied infectious diseases worldwide [147, 148]; however, the molecular mechanisms underlying the survival and growth of the parasite in the human body are still not fully understood. As a result of co-evolution of human and *Plasmodium* species, many alleles preserved in our genome, which provide some degree of protection against malaria infection [149, 150]. Majority of these alleles are important in the erythroid stage of the parasite [150] when it binds to the uninfected RBC, invades it and grows inside the red cells. The firstly described genetic factors linked to malaria protection were the hemoglobin genes [151, 152], but there are several other red cell related genetic variants involved in the susceptibility to malaria [148] including ABO blood group [153, 154], G6PD [151, 155], glycophorin genes [156, 157], CR1 [158], band 3 protein (*SLC4A1*) [157], pyruvate kinase (Pklv) [159], basigin [160] and ABCB6 [161]. It was recently discovered that PMCAs present in RBCs are involved in the survival and growth of the parasite and some variations in the *ATP2B4* (encoding PMCA4) gene may lead to malaria resistance [162–165].

The latest genome wide association (GWA) [163, 164] and multicenter [165] studies have shown that the *ATP2B4* gene also carries a haplotype that is involved in malaria protection and this haplotype showed association with red blood cell traits such as mean corpuscular hemoglobin concentration (MCHC) [166]. According to Lessard et al. [167], this haplotype is located in the enhancer region of the protein, and the complete deletion of this region lead to complete loss of PMCA expression in some erythroid related cell lines, while in case of some other cell lines the deletion does not cause any change in its expression. It is also described [168] that this haplotype leads to reduced expression of PMCA4b in RBCs, but this change is not associated with any additional physiological conditions, probably because this genome region is only essential in erythrocyte development. It is also notable, that this haplotype is much more frequent in malaria-endemic than in malaria-free countries (NCBI and CDC databases). While the relationship between these variations in the *ATP2B4* gene and malaria susceptibility is apparent, the exact function of the PMCA in the parasite's lifecycle within RBCs is still not known [169]. There are controversial data [170] whether the parasitophorous vacuolar membrane (PVM), surrounding the parasite inside the RBCs, contains host membrane proteins [171] or they are excluded from it [172]. Although, the locale of the PMCA during RBC phase of the parasite lifecycle has not been determined, it has been suggested that PMCA remains in the vacuolar membrane, and the parasite may use this protein to maintain a sufficiently high concentration of Ca^{2+} within the vacuolar membrane to proliferate [162]. Thus, selective inhibition of the PMCA may offer a potential new treatment option for malaria in the future.

Diabetes In poorly controlled diabetic patients increased glycosylation and decreased Ca^{2+}-ATPase activity were detected [173]. In another study, oral glucose administration to healthy subjects also decreased the activity of the RBC Ca^{2+}-ATPase [174] Similar results were obtained when protein glycosylation and Ca^{2+}-ATPase activity were measured in membranes from normal erythrocytes pre-incubated with glucose [175]. It has also been shown that the activity of the pump decreases with cell age, however, this effect was independent of the patients' glucose level indicating that glycation could not be responsible for the age dependent decline in pump's activity [176].

5.5.3   PMCAs Linked to Neuronal Disorders and Other Diseases

Although, in several diseases no genetic alterations in the *ATP2B* genes have been identified, modified expression, altered activity or de-regulation of one or more PMCA isoforms could be associated with the disorder. For example, PMCAs have an important role in the brain where they have been linked to certain neurodegenerative disorders [100]. In Alzheimer's disease (AD) deposits of amyloid β-peptide are extensively formed and it was suggested that activation of the amyloidogenic pathway was associated with the remodeling of neuronal Ca^{2+} signaling [177]. First it was found that Ca^{2+} dependence of PMCAs was different in membrane vesicles prepared from human AD brains as compared to non-AD brains [178]. Later amyloid β-peptide aggregates were shown to bind to PMCA and inhibit its activity in the absence of calmodulin [179]. Furthermore, microtubule-associated regulatory protein tau, that is hyperphosphorylated and forms neurofibrillary tangles in AD, has been shown to interact with PMCA, as well, and inhibited its activity [180].

Altered activity of PMCA proteins in human brain tissue was also proposed in Parkinson's disease (PD) [181]. In an in vitro model of PD in neuroblastoma cells it was found that the resting cytosolic Ca^{2+} concentration was elevated while PMCA2 expression was decreased leading to decreased cell survival [182]. Alterations in the expression of PMCAs were also found in multiple sclerosis (MS), an inflammatory, demyelinating and neurodegenerative disorder of the central nervous system. In gene microarray analysis of brain lesions from MS patients both PMCA1 and PMCA3 expression was found to be downregulated compared to control [183]. Down-regulation of PMCA2 expression was also described in rats with experimental autoimmune encephalomyelitis (EAE), an animal model of MS. Interestingly, after disease recovery PMCA2 expression was restored in the animals, while in mouse models with chronic EAE PMCA2 level remained low throughout the disease course [184, 185].

Expression of PMCA4b has been shown to be increased in platelets from patient with both type I and type II diabetes compared to control; this might contribute to increased thrombus formation in diabetic patients [186]. In cellular models it was found that PMCA2 plays an important role in the regulation of pancreatic β-cell proliferation, survival and insulin secretion [187–189]. An analysis of PMCA expression in rat pancreatic islets showed that PMCA1 and PMCA4 are expressed in all islet cells while PMCA3 is present only in the β-cells [190]. In fructose rich diet induced insulin resistant rats PMCA expression was altered in the islet cells resulting in reduced total activity. This caused an elevation in the intracellular calcium level that contributes to the compensatory elevated insulin secretion in response to glucose [191]. Alterations in PMCA activity were related to kidney diseases, as well. Decreased PMCA activity and concomitantly increased cytosolic Ca^{2+} concentration was described in red blood cells of children with chronic kidney disease [192]. Furthermore, in patients with idiopathic hypercalciuria PMCA activity of the erythrocytes was increased compared to controls [193].

5.5.4 PMCA4 in Heart Diseases

During cardiac relaxation SERCA and NCX proteins are mainly responsible for Ca^{2+} removal and PMCA4 acts primarily as a signaling molecule in the heart. It plays a role in the regulation of cardiac β-adrenergic response, hypertrophy and heart failure [194]. β-adrenergic stimulation can initiate neural nitric oxide synthase (nNOS) activity and NO production in cardiac myocytes [195] while nNOS regulates contractility and oxygen radical production [196]. It was demonstrated that PMCA4b can directly interact with the Ca^{2+} sensitive nNOS molecule through its C-terminal PDZ binding motif and it decreases nNOS activity by reducing the Ca^{2+} concentration in its vicinity [197]. In cardiac specific PMCA4b transgenic mice nNOS activity was reduced compared to WT animals and that caused a decreased responsiveness to β-adrenergic stimulation [198]. This interaction might play an important role in remodeling after myocardial infarction (MI). In mice, after induction of MI, nNOS and its adaptor protein CAPON (carboxy-terminal PDZ ligand of NOS1) relocate to caveolae where they make a complex also with PMCA and this way possibly protect the cardiomyocytes from calcium overload. In mice lacking nNOS the redistribution does not happen [199].

PMCA4 also forms a ternary complex in cardiac cells with α-1 syntrophin and nNOS [200]. A mutation in α-1 syntrophin (A390V-SNTA1) was found in patients with long QT syndrome and it was demonstrated that the mutation resulted in the disruption of the interaction with PMCA4. This led to increased nNOS activation and late sodium current causing arrhythmias [201]. Interestingly, in a GWAS study a mutation in CAPON was found to be associated with QT interval variations [202] and variants of the *ATP2B4* gene were associated with congenital ventricular arrhythmia [203].

PMCA4 can also influence cardiac hypertrophy. It is well established that the calcineurin-NFAT pathway is activated during cardiac hypertrophy and it was found that PMCA4 is able to inhibit this pathway through direct binding of calcineurin [92]. In mice overexpressing PMCA4 in the heart both the NFAT-calcineurin signaling and hypertrophy were reduced, while the mice lacking PMCA4 were more susceptible to hypertrophy [204]. Furthermore, after induction of experimental myocardial infarction in mice overexpression of PMCA4 in cardiomyocytes reduced infarct expansion, cardiac hypertrophy and heart failure [205]. However, deletion of PMCA4 in cardiac fibroblasts also prevented cardiac hypertrophy in mice. In the absence of PMCA4, intracellular Ca^{2+} level was elevated in the fibroblasts enhancing secreted frizzled related protein 2 (sFRP2) production and secretion which reduced Wnt signaling in the neighboring cardiomyocytes [206]. Interestingly, overexpression of PMCA4 in arterial smooth muscle cells in mice caused an increase in blood pressure through the inhibition of nNOS [207].

5.5.5 The Role of PMCAs in the Intestine and Bone Mineralization

PMCA1 plays a crucial role in the transcellular Ca^{2+} absorption both in the duodenum and in the large bowel. Its expression is induced by vitamin D metabolite 1,25-(OH)2D3 and by estrogens, as well [208]. In mice it was demonstrated that high bone density correlated with PMCA expression and mucosal to serosal Ca^{2+} transport in the duodenum [209]. Treatment of mice with 1,25-(OH)2D3 strongly increased PMCA1 mRNA level in the duodenum [210] while selective deletion of PMCA1 in the intestinal absorptive cells caused reduced whole body bone mineral density and lower serum Ca^{2+} level [211]. Furthermore, in ovariectomized rats a negative Ca^{2+} balance was induced and this was associated with decreased PMCA1 mRNA expression in an estrogen dependent manner [212], a model for postmenopausal osteoporosis. Interestingly, in biopsies of ulcerative colitis patients reduced PMCA1 expression was also found [213].

PMCAs play an important role in the regulation of bone mineral density already during development. The expression level of PMCA3 in the placenta correlates with neonatal bone mineral content [214] while during lactation PMCA2 expression is strongly induced in the mammary epithelium and it provides Ca^{2+} into the breast milk that is required for the normal bone development of the offspring. In PMCA2-null mice the Ca^{2+} content of the milk was 60% less than in the wild type mice [79]. PMCA isoforms 1, 2 and 4 were described in human osteoblasts, and PMCA1 and PMCA4 in osteoclasts. In osteoblasts of patient with adolescent idiopathic scoliosis expression of PMCA4 was found to be downregulated [215]. During osteoclast differentiation PMCA4 was shown to have an anti-osteoclastogenic effect on one hand by reducing NF-κB ligand–induced Ca^{2+} oscillations, on the other hand by decreasing NO synthesis in the cells [216]. However in mature osteoclast PMCA had an anti-apoptotic effect on the cells. Furthermore, in premenopausal women PMCA4b level showed correlation with high peak bone mass.

5.5.6 Altered PMCA Expression Linked to Tumorigenesis

Ca^{2+} plays an important role in the regulation of many cellular processes such as proliferation, migration or cell death. In tumorous cells these processes are strongly altered and changes in the expression or activity of Ca^{2+} handling molecules in several cancer types have been described. These modifications can result in altered resting Ca^{2+} level in the cellular compartments and can change the spatial and temporal characteristics of the intracellular calcium transients [217].

Alterations in the expression of PMCA proteins have been described in several cancer types. In colorectal cancer a decrease in PMCA4 expression was found during the multistep carcinogenesis of the human colon [218]. In normal human colon mucosa samples PMCA4 was present both at the mRNA and protein levels, however, in high grade adenomas, adenocarcinomas and lymph node metastases the protein expression strongly decreased. Interestingly, the PMCA4 mRNA level was not altered in the samples. Furthermore, after spontaneous differentiation of the colorectal cancer cell line Caco-2 the expression of PMCA4 strongly increased, and treatment with the histone deacetylase (HDAC) inhibitor Trichostatin A induced differentiation and PMCA4 expression in several gastric and colon cancer cell lines [219, 220]. PMCA1 was also found in colon cancer cells and its expression increased after 1,25-(OH)2D3 treatments, however, this was not accompanied by a change in cellular differentiation [221].

Expression of PMCA proteins was also analyzed in breast cancer. In normal breast epithelium PMCA4 is abundantly present [222], while PMCA2 expression is induced only in the lactating mammary glands. In breast cancer cell lines it was found that the mRNA level of PMCA1 and PMCA2 is increased compared to non-tumorigenic human breast epithelial cell lines [223, 224], while PMCA4 expression is downregulated [222]. In human breast cancer samples PMCA2 mRNA level showed association with higher tumor grade and docetaxel resistance in patients. In a tissue microarray analysis of 652 primary breast tumors PMCA2 expression showed positive correlation with lymph node metastasis and human epidermal growth factor receptor 2 (HER2) positivity. Furthermore, overexpression of PMCA2 in breast cancer cells reduced their sensitivity to apoptosis [225]. It was suggested that PMCA2 regulates HER2 signaling in breast cancer cells and knocking down PMCA2 inhibits HER2 mediated cell growth [226]. In another study PMCA2 expression was found in 9% of 96 breast tumors with various histological subtypes and there was no association with grade or hormone receptor status. However, higher PMCA2 expression was described in samples with basal histological subtype. It was also demonstrated that downregulation of PMCA2 level decreased breast cancer cell proliferation and increased the sensitivity to doxorubicin treatment [227]. While PMCA2 expression is upregulated in certain breast cancer cells, PMCA4 level seems to be downregulated. In MCF-7 breast cancer cells treatment with HDAC inhibitors or with phorbol 12-myristate 13-acetate (PMA) strongly induced PMCA4b expression and this effect was coupled with increased Ca^{2+} clearance from the cells [222].

Altered PMCA protein levels were described in melanomas. In melanoma cell lines with different BRAF and NRAS mutational status PMCA4 and PMCA1 isoforms were detected. Mutant BRAF specific inhibitor treatment selectively increased PMCA4b expression in BRAF mutant melanoma cells and this was coupled with faster Ca^{2+} clearance and strong inhibition of migration [91]. When PMCA4b was overexpressed in a BRAF mutant melanoma cell line A375, it strongly reduced the migratory and metastatic capacity of the cells both in vitro and in vivo, while it did not influence their proliferation rate. Furthermore, HDAC inhibitor treatment increased the expression of both PMCA4b and PMCA1 in melanoma cell lines independently from their BRAF mutational status [228]. Similarly to BRAF inhibitor treatment, HDAC inhibition also increased Ca^{2+} clearance and reduced the migratory activity of the highly motile A375 melanoma cells. These results suggested that PMCA4b plays an important role in the regulation of melanoma cell motility, and its expression is under epigenetic control.

PMCA1 was also found to be epigenetically downregulated in human oral cancer. PMCA1 expression was reduced both in primary oral squamous cell carcinomas (OSCCs) and in oral premalignant lesions (OPLs) compared to normal tissue. In OSCC derived cell lines it was demonstrated that decreased PMCA1 level was caused by the increased DNA methylation in the promoter region of PMCA1 [229].

The emerging role of PMCAs in the regulation of the immune response might also be considered in the treatment of malignant diseases. Immune checkpoint inhibitors are relatively new but promising treatment options in cancer therapy that are able to enhance cytotoxic T-cell activation by blocking the negative regulatory signals coming from tumor cells [230]. Recently, it was found that PMCA4 interacts with Ig-like glycoprotein CD147 upon T-cell activation and this interaction is necessary for the immunosuppressive effect of CD147 through the decrease of IL-2 production [231]. CD147 was shown to participate in the development and progression of several cancer types including malignant melanomas, and antibodies targeting CD147 are under development [232]. All these results show that remodeling of the activity and expression of PMCA proteins play an important role in altered cancer cell growth, motility, and in T-cell activation during the immune response to cancer cells that might influence therapy response, as well.

5.6   Conclusion

PMCAs comprise a big family of Ca^{2+} transport ATPases including four separate genes *(ATP2B1-4)* from which more than twenty different protein variants are transcribed. The variants have different regulatory properties, and hence they respond differently to the incoming Ca^{2+} signal, differ in their sub-plasma membrane localization and interact with different signaling molecules. The expression, and thus the abundance of the variants are also tightly regulated in a development and cell-type specific manner, by processes not yet very well understood. In the past we studied many aspects of the biochemical characteristics of these pumps, but we still know very little on how their transcription and translation are regulated and how stable the proteins are in the plasma membrane. Our main goal, therefore, should be to study further these mechanisms particularly because alterations in the PMCA expression and genetic variations in the *ATP2B* genes have been linked to several diseases such as cardiovascular and neurodegenerative disorders, and cancer. Understanding PMCA pathophysiology and learning more about the consequences of PMCA dysfunction may help finding ways to predict, prevent and/or cure such diseases.

Acknowledgement The authors are supported by grants from the Hungarian Scientific Research Funds NKFIH K119223 and FIKP-EMMI (AE).

References

1. Schatzmann HJ (1966) ATP-dependent Ca++-extrusion from human red cells. Experientia 22(6):364–365
2. Berridge MJ, Bootman MD, Roderick HL (2003) Calcium signalling: dynamics, homeostasis and remodelling. Nat Rev Mol Cell Biol 4(7):517–529

3. Penniston JT, Enyedi A (1998) Modulation of the plasma membrane Ca^{2+} pump. J Membr Biol 165(2):101–109
4. Strehler EE (1990) Plasma membrane Ca^{2+} pumps and Na$^+$/Ca^{2+} exchangers. Semin Cell Biol 1(4):283–295
5. Padanyi R, Paszty K, Hegedus L, Varga K, Papp B, Penniston JT et al (2016) Multifaceted plasma membrane Ca^{2+} pumps: from structure to intracellular Ca^{2+} handling and cancer. Biochim Biophys Acta 1863(6 Pt B):1351–1363
6. Axelsen KB, Palmgren MG (1998) Evolution of substrate specificities in the P-type ATPase superfamily. J Mol Evol 46(1):84–101
7. Thever MD, Saier MH Jr (2009) Bioinformatic characterization of p-type ATPases encoded within the fully sequenced genomes of 26 eukaryotes. J Membr Biol 229(3):115–130
8. Strehler EE, Zacharias DA (2001) Role of alternative splicing in generating isoform diversity among plasma membrane calcium pumps. Physiol Rev 81(1):21–50
9. Krebs J (2015) The plethora of PMCA isoforms: alternative splicing and differential expression. Biochim Biophys Acta 1853(9):2018–2024
10. Green NM (1989) ATP-driven cation pumps: alignment of sequences. Biochem Soc Trans 17(6):972
11. Toyoshima C, Mizutani T (2004) Crystal structure of the calcium pump with a bound ATP analogue. Nature 430(6999):529–535
12. Sorensen TL, Moller JV, Nissen P (2004) Phosphoryl transfer and calcium ion occlusion in the calcium pump. Science 304(5677):1672–1675
13. Jensen AM, Sorensen TL, Olesen C, Moller JV, Nissen P (2006) Modulatory and catalytic modes of ATP binding by the calcium pump. EMBO J 25(11):2305–2314
14. Sweadner KJ, Donnet C (2001) Structural similarities of Na,K-ATPase and SERCA, the Ca^{2+}-ATPase of the sarcoplasmic reticulum. Biochem J 356(Pt 3):685–704
15. Toyoshima C (2009) How Ca^{2+}-ATPase pumps ions across the sarcoplasmic reticulum membrane. Biochim Biophys Acta 1793(6):941–946
16. Thomas RC (2009) The plasma membrane calcium ATPase (PMCA) of neurones is electroneutral and exchanges 2 H$^+$ for each Ca^{2+} or Ba^{2+} ion extruded. J Physiol 587(2):315–327
17. Carafoli E, Brini M (2000) Calcium pumps: structural basis for and mechanism of calcium transmembrane transport. Curr Opin Chem Biol 4(2):152–161
18. Strehler EE, Treiman M (2004) Calcium pumps of plasma membrane and cell interior. Curr Mol Med 4(3):323–335
19. Di Leva F, Domi T, Fedrizzi L, Lim D, Carafoli E (2008) The plasma membrane Ca^{2+} ATPase of animal cells: structure, function and regulation. Arch Biochem Biophys 476(1):65–74
20. Penniston JT, Padanyi R, Paszty K, Varga K, Hegedus L, Enyedi A (2014) Apart from its known function, the plasma membrane Ca^{2+}ATPase can regulate Ca^{2+} signaling by controlling phosphatidylinositol 4,5-bisphosphate levels. J Cell Sci 127(Pt 1):72–84
21. Filoteo AG, Enyedi A, Penniston JT (1992) The lipid-binding peptide from the plasma membrane Ca^{2+} pump binds calmodulin, and the primary calmodulin-binding domain interacts with lipid. J Biol Chem 267(17):11800–11805
22. Enyedi A, Flura M, Sarkadi B, Gardos G, Carafoli E (1987) The maximal velocity and the calcium affinity of the red cell calcium pump may be regulated independently. J Biol Chem 262(13):6425–6430
23. Brodin P, Falchetto R, Vorherr T, Carafoli E (1992) Identification of two domains which mediate the binding of activating phospholipids to the plasma-membrane Ca^{2+} pump. Eur J Biochem 204(2):939–946
24. Wang K, Sitsel O, Meloni G, Autzen HE, Andersson M, Klymchuk T et al (2014) Structure and mechanism of Zn2+-transporting P-type ATPases. Nature 514(7523):518–522
25. Hansen SB (2015) Lipid agonism: the PIP2 paradigm of ligand-gated ion channels. Biochim Biophys Acta 1851(5):620–628
26. Paszty K, Verma AK, Padanyi R, Filoteo AG, Penniston JT, Enyedi A (2002) Plasma membrane Ca^{2+}ATPase isoform 4b is cleaved and activated by caspase-3 during the early

phase of apoptosis. J Biol Chem 277(9):6822–6829

27. Schwab BL, Guerini D, Didszun C, Bano D, Ferrando-May E, Fava E et al (2002) Cleavage of plasma membrane calcium pumps by caspases: a link between apoptosis and necrosis. Cell Death Differ 9(8):818–831

28. Enyedi A, Elwess NL, Filoteo AG, Verma AK, Paszty K, Penniston JT (1997) Protein kinase C phosphorylates the "a" forms of plasma membrane Ca^{2+} pump isoforms 2 and 3 and prevents binding of calmodulin. J Biol Chem 272(44):27525–27528

29. Enyedi A, Verma AK, Filoteo AG, Penniston JT (1996) Protein kinase C activates the plasma membrane Ca^{2+} pump isoform 4b by phosphorylation of an inhibitory region downstream of the calmodulin-binding domain. J Biol Chem 271(50):32461–32467

30. Antalffy G, Paszty K, Varga K, Hegedus L, Enyedi A, Padanyi R (2013) A C-terminal di-leucine motif controls plasma membrane expression of PMCA4b. Biochim Biophys Acta 1833(12):2561–2572

31. DeMarco SJ, Strehler EE (2001) Plasma membrane Ca^{2+}-atpase isoforms 2b and 4b interact promiscuously and selectively with members of the membrane-associated guanylate kinase family of PDZ (PSD95/Dlg/ZO-1) domain-containing proteins. J Biol Chem 276(24):21594–21600

32. DeMarco SJ, Chicka MC, Strehler EE (2002) Plasma membrane Ca^{2+} ATPase isoform 2b interacts preferentially with Na^+/H^+ exchanger regulatory factor 2 in apical plasma membranes. J Biol Chem 277(12):10506–10511

33. Penniston JT, Caride AJ, Strehler EE (2012) Alternative pathways for association and dissociation of the calmodulin-binding domain of plasma membrane Ca^{2+}-ATPase isoform 4b (PMCA4b). J Biol Chem 287(35):29664–29671

34. Caride AJ, Filoteo AG, Penniston JT, Strehler EE (2007) The plasma membrane Ca^{2+} pump isoform 4a differs from isoform 4b in the mechanism of calmodulin binding and activation kinetics: implications for Ca^{2+} signaling. J Biol Chem 282(35):25640–25648

35. Juranic N, Atanasova E, Filoteo AG, Macura S, Prendergast FG, Penniston JT et al (2010) Calmodulin wraps around its binding domain in the plasma membrane Ca^{2+} pump anchored by a novel 18-1 motif. J Biol Chem 285(6):4015–4024

36. Elshorst B, Hennig M, Forsterling H, Diener A, Maurer M, Schulte P et al (1999) NMR solution structure of a complex of calmodulin with a binding peptide of the Ca^{2+} pump. Biochemistry 38(38):12320–12332

37. Chicka MC, Strehler EE (2003) Alternative splicing of the first intracellular loop of plasma membrane Ca^{2+}-ATPase isoform 2 alters its membrane targeting. J Biol Chem 278(20):18464–18470

38. Strehler EE (2013) Plasma membrane calcium ATPases as novel candidates for therapeutic agent development. J Pharm Pharm Sci 16(2):190–206

39. Strehler EE (2015) Plasma membrane calcium ATPases: from generic Ca^{2+} sump pumps to versatile systems for fine-tuning cellular Ca(2.). Biochem Biophys Res Commun 460(1):26–33

40. Afroze T, Husain M (2000) c-Myb-binding sites mediate G(1)/S-associated repression of the plasma membrane Ca^{2+}-ATPase-1 promoter. J Biol Chem 275(12):9062–9069

41. Habib T, Park H, Tsang M, de Alboran IM, Nicks A, Wilson L et al (2007) Myc stimulates B lymphocyte differentiation and amplifies calcium signaling. J Cell Biol 179(4):717–731

42. Zelinski JM, Sykes DE, Weiser MM (1991) The effect of vitamin D on rat intestinal plasma membrane CA-pump mRNA. Biochem Biophys Res Commun 179(2):749–755

43. Cai Q, Chandler JS, Wasserman RH, Kumar R, Penniston JT (1993) Vitamin D and adaptation to dietary calcium and phosphate deficiencies increase intestinal plasma membrane calcium pump gene expression. Proc Natl Acad Sci U S A 90(4):1345–1349

44. Glendenning P, Ratajczak T, Dick IM, Prince RL (2000) Calcitriol upregulates expression and activity of the 1b isoform of the plasma membrane calcium pump in immortalized distal kidney tubular cells. Arch Biochem Biophys 380(1):126–132

45. Glendenning P, Ratajczak T, Dick IM, Prince RL (2001) Regulation of the 1b isoform of the plasma membrane calcium pump by 1,25-dihydroxyvitamin D3 in rat osteoblast-like cells. J

Bone Miner Res 16(3):525–534
46. Silverstein RS, Tempel BL (2006) Atp2b2, encoding plasma membrane Ca^{2+}-ATPase type 2, (PMCA2) exhibits tissue-specific first exon usage in hair cells, neurons, and mammary glands of mice. Neuroscience 141(1):245–257
47. Minich RR, Li J, Tempel BL (2017) Early growth response protein 1 regulates promoter activity of alpha-plasma membrane calcium ATPase 2, a major calcium pump in the brain and auditory system. BMC Mol Biol 18(1):14
48. Lessard S, Gatof ES, Beaudoin M, Schupp PG, Sher F, Ali A et al (2017) An erythroid-specific ATP2B4 enhancer mediates red blood cell hydration and malaria susceptibility. J Clin Invest 127(8):3065–3074
49. Verma AK, Enyedi A, Filoteo AG, Penniston JT (1994) Regulatory region of plasma membrane Ca^{2+} pump. 28 residues suffice to bind calmodulin but more are needed for full auto-inhibition of the activity. J Biol Chem 269(3):1687–1691
50. Caride AJ, Filoteo AG, Penheiter AR, Paszty K, Enyedi A, Penniston JT (2001) Delayed activation of the plasma membrane calcium pump by a sudden increase in Ca^{2+}: fast pumps reside in fast cells. Cell Calcium 30(1):49–57
51. Enyedi A, Vorherr T, James P, McCormick DJ, Filoteo AG, Carafoli E et al (1989) The calmodulin binding domain of the plasma membrane Ca^{2+} pump interacts both with calmodulin and with another part of the pump. J Biol Chem 264(21):12313–12321
52. Ba-Thein W, Caride AJ, Enyedi A, Paszty K, Croy CL, Filoteo AG et al (2001) Chimaeras reveal the role of the catalytic core in the activation of the plasma membrane Ca^{2+} pump. Biochem J 356(Pt 1):241–245
53. Bruce JIE (2018) Metabolic regulation of the PMCA: role in cell death and survival. Cell Calcium 69:28–36
54. Paszty K, Antalffy G, Penheiter AR, Homolya L, Padanyi R, Ilias A et al (2005) The caspase-3 cleavage product of the plasma membrane Ca^{2+}-ATPase 4b is activated and appropriately targeted. Biochem J 391(Pt 3):687–692
55. Strehler EE, Caride AJ, Filoteo AG, Xiong Y, Penniston JT, Enyedi A (2007) Plasma membrane Ca^{2+} ATPases as dynamic regulators of cellular calcium handling. Ann N Y Acad Sci 1099:226–236
56. Caride AJ, Elwess NL, Verma AK, Filoteo AG, Enyedi A, Bajzer Z et al (1999) The rate of activation by calmodulin of isoform 4 of the plasma membrane Ca^{2+} pump is slow and is changed by alternative splicing. J Biol Chem 274(49):35227–35232
57. Caride AJ, Penheiter AR, Filoteo AG, Bajzer Z, Enyedi A, Penniston JT (2001) The plasma membrane calcium pump displays memory of past calcium spikes. Differences between isoforms 2b and 4b. J Biol Chem 276(43):39797–39804
58. Missiaen L, Raeymaekers L, Wuytack F, Vrolix M, de Smedt H, Casteels R (1989) Phospholipid-protein interactions of the plasma-membrane Ca^{2+}-transporting ATPase. Evidence for a tissue-dependent functional difference. Biochem J 263(3):687–694
59. Zaidi A, Adewale M, McLean L, Ramlow P (2018) The plasma membrane calcium pumps-The old and the new. Neurosci Lett 663:12–17
60. Pignataro MF, Dodes-Traian MM, Gonzalez-Flecha FL, Sica M, Mangialavori IC, Rossi JP (2015) Modulation of plasma membrane Ca^{2+}-ATPase by neutral phospholipids: effect of the micelle-vesicle transition and the bilayer thickness. J Biol Chem 290(10):6179–6190
61. Dean WL, Whiteheart SW (2004) Plasma membrane Ca^{2+}-ATPase (PMCA) translocates to filopodia during platelet activation. Thromb Haemost 91(2):325–333
62. Bozulic LD, Malik MT, Powell DW, Nanez A, Link AJ, Ramos KS et al (2007) Plasma membrane Ca^{2+}-ATPase associates with CLP36, alpha-actinin and actin in human platelets. Thromb Haemost 97(4):587–597
63. Dalghi MG, Fernandez MM, Ferreira-Gomes M, Mangialavori IC, Malchiodi EL, Strehler EE et al (2013) Plasma membrane calcium ATPase activity is regulated by actin oligomers through direct interaction. J Biol Chem 288(32):23380–23393
64. Vanagas L, de La Fuente MC, Dalghi M, Ferreira-Gomes M, Rossi RC, Strehler EE et al (2013) Differential effects of G- and F-actin on the plasma membrane calcium pump activity.

Cell Biochem Biophys 66(1):187–198
65. Dalghi MG, Ferreira-Gomes M, Montalbetti N, Simonin A, Strehler EE, Hediger MA et al (2017) Cortical cytoskeleton dynamics regulates plasma membrane calcium ATPase isoform-2 (PMCA2) activity. Biochim Biophys Acta 1864(8):1413–1424
66. Dalghi MG, Ferreira-Gomes M, Rossi JP (2017) Regulation of the plasma membrane calcium ATPases by the actin cytoskeleton. Biochem Biophys Res Commun
67. Paszty K, Caride AJ, Bajzer Z, Offord CP, Padanyi R, Hegedus L et al (2015) Plasma membrane Ca^{2+}-ATPases can shape the pattern of Ca^{2+} transients induced by store-operated Ca^{2+} entry. Sci Signal 8(364):ra19
68. Ritchie MF, Samakai E, Soboloff J (2012) STIM1 is required for attenuation of PMCA-mediated Ca^{2+} clearance during T-cell activation. EMBO J 31(5):1123–1133
69. Krapivinsky G, Krapivinsky L, Stotz SC, Manasian Y, Clapham DE (2011) POST, partner of stromal interaction molecule 1 (STIM1), targets STIM1 to multiple transporters. Proc Natl Acad Sci U S A 108(48):19234–19239
70. Okunade GW, Miller ML, Pyne GJ, Sutliff RL, O'Connor KT, Neumann JC et al (2004) Targeted ablation of plasma membrane Ca^{2+}-ATPase (PMCA) 1 and 4 indicates a major housekeeping function for PMCA1 and a critical role in hyperactivated sperm motility and male fertility for PMCA4. J Biol Chem 279(32):33742–33750
71. Schuh K, Cartwright EJ, Jankevics E, Bundschu K, Liebermann J, Williams JC et al (2004) Plasma membrane Ca^{2+} ATPase 4 is required for sperm motility and male fertility. J Biol Chem 279(27):28220–28226
72. Prasad V, Okunade GW, Miller ML, Shull GE (2004) Phenotypes of SERCA and PMCA knockout mice. Biochem Biophys Res Commun 322(4):1192–1203
73. Lefievre L, Nash K, Mansell S, Costello S, Punt E, Correia J et al (2012) 2-APB-potentiated channels amplify CatSper-induced Ca^{2+} signals in human sperm. Biochem J 448(2):189–200
74. Kawano S, Otsu K, Shoji S, Yamagata K, Hiraoka M (2003) Ca^{2+} oscillations regulated by $Na(+)$-Ca^{2+} exchanger and plasma membrane Ca^{2+} pump induce fluctuations of membrane currents and potentials in human mesenchymal stem cells. Cell Calcium 34(2):145–156
75. Chen YF, Cao J, Zhong JN, Chen X, Cheng M, Yang J et al (2014) Plasma membrane Ca^{2+}-ATPase regulates Ca^{2+} signaling and the proliferation of airway smooth muscle cells. Eur J Pharmacol 740:733–741
76. Prasad V, Okunade G, Liu L, Paul RJ, Shull GE (2007) Distinct phenotypes among plasma membrane Ca^{2+}-ATPase knockout mice. Ann N Y Acad Sci 1099:276–286
77. Cali T, Brini M, Carafoli E (2018) The PMCA pumps in genetically determined neuronal pathologies. Neurosci Lett 663:2–11
78. Ficarella R, Di Leva F, Bortolozzi M, Ortolano S, Donaudy F, Petrillo M et al (2007) A functional study of plasma-membrane calcium-pump isoform 2 mutants causing digenic deafness. Proc Natl Acad Sci U S A 104(5):1516–1521
79. Reinhardt TA, Lippolis JD, Shull GE, Horst RL (2004) Null mutation in the gene encoding plasma membrane Ca^{2+}-ATPase isoform 2 impairs calcium transport into milk. J Biol Chem 279(41):42369–42373
80. Padanyi R, Paszty K, Strehler EE, Enyedi A (2009) PSD-95 mediates membrane clustering of the human plasma membrane Ca^{2+} pump isoform 4b. Biochimica et Biophysica Acta 1793(6):1023–1032
81. Schmidt N, Kollewe A, Constantin CE, Henrich S, Ritzau-Jost A, Bildl W et al (2017) Neuroplastin and basigin are essential auxiliary subunits of plasma membrane Ca^{2+}-ATPases and key regulators of Ca^{2+} clearance. Neuron 96(4):827–38 e9
82. Grati M, Aggarwal N, Strehler EE, Wenthold RJ (2006) Molecular determinants for differential membrane trafficking of PMCA1 and PMCA2 in mammalian hair cells. J Cell Sci 119(Pt 14):2995–3007
83. Spiden SL, Bortolozzi M, Di Leva F, de Angelis MH, Fuchs H, Lim D et al (2008) The novel mouse mutation Oblivion inactivates the PMCA2 pump and causes progressive hearing loss. PLoS Genet 4(10):e1000238

84. Padanyi R, Xiong Y, Antalffy G, Lor K, Paszty K, Strehler EE et al (2010) Apical scaffolding protein NHERF2 modulates the localization of alternatively spliced plasma membrane Ca^{2+} pump 2B variants in polarized epithelial cells. J Biol Chem 285(41):31704–31712

85. Grati M, Schneider ME, Lipkow K, Strehler EE, Wenthold RJ, Kachar B (2006) Rapid turnover of stereocilia membrane proteins: evidence from the trafficking and mobility of plasma membrane Ca^{2+}-ATPase 2. J Neurosci 26(23):6386–6395

86. Yang YM, Lee J, Jo H, Park S, Chang I, Muallem S et al (2014) Homer2 protein regulates plasma membrane Ca^{2+}-ATPase-mediated Ca^{2+} signaling in mouse parotid gland acinar cells. J Biol Chem 289(36):24971–24979

87. Baggaley E, McLarnon S, Demeter I, Varga G, Bruce JI (2007) Differential regulation of the apical plasma membrane Ca^{2+}-ATPase by protein kinase A in parotid acinar cells. J Biol Chem 282(52):37678–37693

88. Quintana A, Pasche M, Junker C, Al-Ansary D, Rieger H, Kummerow C et al (2011) Calcium microdomains at the immunological synapse: how ORAI channels, mitochondria and calcium pumps generate local calcium signals for efficient T-cell activation. EMBO J 30(19):3895–3912

89. Tsai FC, Seki A, Yang HW, Hayer A, Carrasco S, Malmersjo S et al (2014) A polarized Ca^{2+}, diacylglycerol and STIM1 signalling system regulates directed cell migration. Nat Cell Biol 16(2):133–144

90. Kurusamy S, Lopez-Maderuelo D, Little R, Cadagan D, Savage AM, Ihugba JC et al (2017) Selective inhibition of plasma membrane calcium ATPase 4 improves angiogenesis and vascular reperfusion. J Mol Cell Cardiol 109:38–47

91. Hegedus L, Garay T, Molnar E, Varga K, Bilecz A, Torok S et al (2017) The plasma membrane Ca^{2+} pump PMCA4b inhibits the migratory and metastatic activity of BRAF mutant melanoma cells. Int J Cancer 140(12):2758–2770

92. Buch MH, Pickard A, Rodriguez A, Gillies S, Maass AH, Emerson M et al (2005) The sarcolemmal calcium pump inhibits the calcineurin/nuclear factor of activated T-cell pathway via interaction with the calcineurin A catalytic subunit. J Biol Chem 280(33):29479–29487

93. Holton M, Yang D, Wang W, Mohamed TM, Neyses L, Armesilla AL (2007) The interaction between endogenous calcineurin and the plasma membrane calcium-dependent ATPase is isoform specific in breast cancer cells. FEBS Lett 581(21):4115–4119

94. Baggott RR, Mohamed TM, Oceandy D, Holton M, Blanc MC, Roux-Soro SC et al (2012) Disruption of the interaction between PMCA2 and calcineurin triggers apoptosis and enhances paclitaxel-induced cytotoxicity in breast cancer cells. Carcinogenesis 33(12):2362–2368

95. Baggott RR, Alfranca A, Lopez-Maderuelo D, Mohamed TM, Escolano A, Oller J et al (2014) Plasma membrane calcium ATPase isoform 4 inhibits vascular endothelial growth factor-mediated angiogenesis through interaction with calcineurin. Arterioscler Thromb Vasc Biol 34(10):2310–2320

96. Schuh K, Uldrijan S, Gambaryan S, Roethlein N, Neyses L (2003) Interaction of the plasma membrane Ca^{2+} pump 4b/CI with the Ca^{2+}/calmodulin-dependent membrane-associated kinase CASK. J Biol Chem 278(11):9778–9783

97. Aravindan RG, Fomin VP, Naik UP, Modelski MJ, Naik MU, Galileo DS et al (2012) CASK interacts with PMCA4b and JAM-A on the mouse sperm flagellum to regulate Ca^{2+} homeostasis and motility. J Cell Physiol 227(8):3138–3150

98. Stafford N, Wilson C, Oceandy D, Neyses L, Cartwright EJ (2017) The plasma membrane calcium ATPases and their role as major new players in human disease. Physiol Rev 97(3):1089–1125

99. Giacomello M, De Mario A, Scarlatti C, Primerano S, Carafoli E (2013) Plasma membrane calcium ATPases and related disorders. Int J Biochem Cell Biol 45(3):753–762

100. Hajieva P, Baeken MW, Moosmann B (2018) The role of Plasma Membrane Calcium ATPases (PMCAs) in neurodegenerative disorders. Neurosci Lett 663:29–38

101. Johnson T, Gaunt TR, Newhouse SJ, Padmanabhan S, Tomaszewski M, Kumari M et al (2011) Blood pressure loci identified with a gene-centric array. Am J Hum Genet 89(6):688–700

102. Kato N, Takeuchi F, Tabara Y, Kelly TN, Go MJ, Sim X et al (2011) Meta-analysis of genome-wide association studies identifies common variants associated with blood pressure variation in east Asians. Nat Genet 43(6):531–538

103. Weng L, Taylor KD, Chen YD, Sopko G, Kelsey SF, Bairey Merz CN et al (2016) Genetic loci associated with nonobstructive coronary artery disease in Caucasian women. Physiol Genomics 48(1):12–20

104. Lu X, Wang L, Chen S, He L, Yang X, Shi Y et al (2012) Genome-wide association study in Han Chinese identifies four new susceptibility loci for coronary artery disease. Nat Genet 44(8):890–894

105. Ferguson JF, Matthews GJ, Townsend RR, Raj DS, Kanetsky PA, Budoff M et al (2013) Candidate gene association study of coronary artery calcification in chronic kidney disease: findings from the CRIC study (Chronic Renal Insufficiency Cohort). J Am Coll Cardiol 62(9):789–798

106. Wan JP, Wang H, Li CZ, Zhao H, You L, Shi DH et al (2014) The common single-nucleotide polymorphism rs2681472 is associated with early-onset preeclampsia in Northern Han Chinese women. Reprod Sci 21(11):1423–1427

107. Carrera F, Casart YC, Proverbio T, Proverbio F, Marin R (2003) Preeclampsia and calcium-ATPase activity of plasma membranes from human myometrium and placental trophoblast. Hypertens Pregnancy 22(3):295–304

108. Hache S, Takser L, LeBellego F, Weiler H, Leduc L, Forest JC et al (2011) Alteration of calcium homeostasis in primary preeclamptic syncytiotrophoblasts: effect on calcium exchange in placenta. J Cell Mol Med 15(3):654–667

109. Tabara Y, Kohara K, Kita Y, Hirawa N, Katsuya T, Ohkubo T et al (2010) Common variants in the ATP2B1 gene are associated with susceptibility to hypertension: the Japanese Millennium Genome Project. Hypertension 56(5):973–980

110. Yang W, Liu J, Zheng F, Jia M, Zhao L, Lu T et al (2013) The evidence for association of ATP2B2 polymorphisms with autism in Chinese Han population. PLoS One 8(4):e61021

111. Prandini P, Pasquali A, Malerba G, Marostica A, Zusi C, Xumerle L et al (2012) The association of rs4307059 and rs35678 markers with autism spectrum disorders is replicated in Italian families. Psychiatr Genet 22(4):177–181

112. Schultz JM, Yang Y, Caride AJ, Filoteo AG, Penheiter AR, Lagziel A et al (2005) Modification of human hearing loss by plasma-membrane calcium pump PMCA2. N Engl J Med 352(15):1557–1564

113. Bortolozzi M, Mammano F (2018) PMCA2 pump mutations and hereditary deafness. Neurosci Lett 663:18–24

114. Street VA, McKee-Johnson JW, Fonseca RC, Tempel BL, Noben-Trauth K (1998) Mutations in a plasma membrane Ca^{2+}-ATPase gene cause deafness in deafwaddler mice. Nat Genet 19(4):390–394

115. Zanni G, Cali T, Kalscheuer VM, Ottolini D, Barresi S, Lebrun N et al (2012) Mutation of plasma membrane Ca^{2+} ATPase isoform 3 in a family with X-linked congenital cerebellar ataxia impairs Ca^{2+} homeostasis. Proc Natl Acad Sci U S A 109(36):14514–14519

116. Cali T, Lopreiato R, Shimony J, Vineyard M, Frizzarin M, Zanni G et al (2015) A novel mutation in isoform 3 of the plasma membrane Ca^{2+} pump impairs cellular Ca^{2+} homeostasis in a patient with cerebellar ataxia and laminin subunit 1alpha mutations. J Biol Chem 290(26):16132–16141

117. Cali T, Frizzarin M, Luoni L, Zonta F, Pantano S, Cruz C et al (2017) The ataxia related G1107D mutation of the plasma membrane Ca^{2+} ATPase isoform 3 affects its interplay with calmodulin and the autoinhibition process. Biochim Biophys Acta 1863(1):165–173

118. Williams TA, Monticone S, Schack VR, Stindl J, Burrello J, Buffolo F et al (2014) Somatic ATP1A1, ATP2B3, and KCNJ5 mutations in aldosterone-producing adenomas. Hypertension 63(1):188–195

119. Kitamoto T, Suematsu S, Yamazaki Y, Nakamura Y, Sasano H, Matsuzawa Y et al (2016) Clinical and steroidogenic characteristics of aldosterone-producing adenomas with ATPase or CACNA1D gene mutations. J Clin Endocrinol Metab 101(2):494–503

120. Tauber P, Aichinger B, Christ C, Stindl J, Rhayem Y, Beuschlein F et al (2016) Cellular pathophysiology of an adrenal adenoma-associated mutant of the plasma membrane Ca^{2+}-ATPase ATP2B3. Endocrinology. 157(6):2489–2499

121. Li M, Ho PW, Pang SY, Tse ZH, Kung MH, Sham PC et al (2014) PMCA4 (ATP2B4) mutation in familial spastic paraplegia. PLoS One 9(8):e104790

122. Ho PW, Pang SY, Li M, Tse ZH, Kung MH, Sham PC et al (2015) PMCA4 (ATP2B4) mutation in familial spastic paraplegia causes delay in intracellular calcium extrusion. Brain Behav 5(4):e00321

123. Basit S, Albalawi AM, Alharby E, Khoshhal KI (2017) Exome sequencing identified rare variants in genes HSPG2 and ATP2B4 in a family segregating developmental dysplasia of the hip. BMC Med Genet 18(1):34

124. Schatzmann HJ, Rossi JL (1971) (Ca^{2+} + Mg2+)-activated membrane ATPases in human red cells and their possible relations to cation transport. Biochimica et 75659:379–392

125. Wolf HU (1972) Studies on a Ca^{2+}-dependent ATPase of human erythrocyte membranes – effects of Ca^{2+} and H$^+$. Biochimica et Biophysica Acta 66:361–375

126. Sarkadi B (1980) Active calcium transport in human red cells. Biochimica et Biophysica Acta 4:159–190

127. Schatzmann HJ (1975) Active calcium transport and Ca^{2+}-Activated ATPase in human red cells. Curr Topics Membr Transport 6:125–168

128. Strehler EE (1991) Recent advances in the molecular characterization of plasma membrane Ca^{2+} pumps. J Membr Biol 120:1–15

129. Borke JL, Minami J, Verma A, Penniston JT, Kumar R (1987) Monoclonal antibodies to human erythrocyte membrane Ca++-Mg++ adenosine triphosphatase pump recognize an epitope in the basolateral membrane of human kidney distal tubule cells. J Clin Invest 80:1225–1231

130. Caride AJ, Filoteo AG, Enyedi A, Verma AK, Penniston JT (1996) Detection of isoform 4 of the plasma membrane calcium pump in human tissues by using isoform-specific monoclonal antibodies. Biochem J 316:353–359

131. Bogdanova A, Makhro A, Wang J, Lipp P, Kaestner L (2013) Calcium in red blood cells – a perilous balance. Int J Mol Sci 14:9848–9872

132. Pasini EME, Kirkegaard M, Mortensen P, Lutz HU, Thomas AW, Mann M (2006) In-depth analysis of the membrane and cytosolic proteome of red blood cells. Blood 108:791–801

133. Harrison D, Long C (1968) The calcium content of human erythrocytes. J Physiol 199:367–381

134. Schatzmann HJ (1973) Dependence on calcium concentration and stoichiometry of the calcium pump in human red cells. J Physiol 235:551–569

135. Tiffert T, Bookchin RM, Lew VL (2003) Calcium homeostasis in normal and abnormal human red cells. In: Red cell membrane transport in health and disease. Springer, Berlin/Heidelberg, pp 373–405

136. Larsen FL, Katz S, Roufogalis BD, Brooks DE (1981) Physiological shear stresses enhance the Ca^{2+} permeability of human erythrocytes. Nature 294:667–668

137. Lew VL, Daw N, Perdomo D, Etzion Z, Bookchin RM, Tiffert T (2003) Distribution of plasma membrane Ca^{2+} pump activity in normal human red blood cells. Distribution 102:4206–4213

138. Lew VL, Tiffert T, Etzion Z, Perdomo D, Daw N, Macdonald L et al (2005) Distribution of dehydration rates generated by maximal Gardos-channel activation in normal and sickle red blood cells. Blood 105:361–367

139. Lew VL, Daw N, Etzion Z, Tiffert T, Muoma A, Vanagas L et al (2007) Effects of age-dependent membrane transport changes on the homeostasis of senescent human red blood

cells. Blood 110:1334–1342

140. Lew VL, Tiffert T (2017) On the mechanism of human red blood cell longevity: roles of calcium, the sodium pump, PIEZO1, and gardos channels. Front Physiol 8:977

141. Hertz L, Huisjes R, Llaudet-Planas E, Petkova-Kirova P, Makhro A, Danielczok JG, et al (2017) Is increased intracellular calcium in red blood cells a common component in the molecular mechanism causing anemia? Front Physiol 8

142. Vandorpe DH, Xu C, Shmukler BE, Otterbein LE, Trudel M, Sachs F et al (2010) Hypoxia activates a Ca^{2+}-permeable cation conductance sensitive to carbon monoxide and to GsMTx-4 in human and mouse sickle erythrocytes. PLoS One. 5(1):e8732

143. Gibson JS, Ellory JC (2002) Membrane transport in sickle cell disease. Blood Cells Mol Dis 28:303–314

144. Lew VL, Ortiz OE, Bookchin RM (1997) Stochastic nature and red cell population distribution of the sickling-induced Ca^{2+} permeability. J Clin Invest 99(11):2727–2735

145. Etzion Z, Tiffert T, Bookchin RM, Lew VL (1993) Effects of deoxygenation on active and passive Ca^{2+} transport and on the cytoplasmic Ca^{2+} levels of sickle cell anemia red cells. J Clin Investig 92:2489–2498

146. Lew VL, Bookchin RM (2005) Ion transport pathology in the mechanism of sickle cell dehydration. Physiol Rev 85(1):179–200

147. Wassmer SC, Taylor TE, Rathod PK, Mishra SK, Mohanty S, Arevalo-Herrera M et al (2015) Investigating the pathogenesis of severe malaria: a multidisciplinary and cross-geographical approach. Am J Trop Med Hyg 93:42–56

148. Marquet S (2018) Overview of human genetic susceptibility to malaria: from parasitemia control to severe disease. Infect Genet Evol 66:399–409

149. Min-Oo G, Gros P (2005) Erythrocyte variants and the nature of their malaria protective effect. Cell Microbiol 7:753–763

150. Williams TN (2006) Human red blood cell polymorphisms and malaria. Curr Opin Microbiol 9:388–394

151. Gilles HM, Fletcher KA, Hendrickse RG, Linder R, Reddy S, Allan N (1967) Glucose-6-phosphate-dehydrogenase deficiency, sickling, and malaria in African children in South Western Nigeria. Lancet 289(7482):138–140

152. Hill AVS, Allsopp CEM, Kwiatkowski D, Anstey NM, Twumasi P, Rowe PA et al (1991) Common West African HLA antigens are associated with protection from severe malaria. Nature 352:595–600

153. Lell B, May J, Schmidt-Ott RJ, Lehman LG, Luckner D, Greve B et al (1999) The role of red blood cell polymorphisms in resistance and susceptibility to malaria. Clin Infect Dis 28:794–799

154. Fischer PR, Boone P (1998) Short report: severe malaria associated with blood group. Am J Trop Med Hyg 58:122–123

155. Shah SS, Rockett KA, Jallow M, Sisay-Joof F, Bojang KA, Pinder M et al (2016) Heterogeneous alleles comprising G6PD deficiency trait in West Africa exert contrasting effects on two major clinical presentations of severe malaria. Malar J 15:1–8

156. Pasvol G, Wainscoat JS, Weatherall DJ (1982) Erythrocytes deficient in glycophorin resist invasion by the malarial parasite Plasmodium falciparum. Nature 297:64–66

157. Patel SS, King CL, Mgone CS, Kazura JW, Zimmerman PA (2004) Glycophorin C (Gerbich Antigen Blood Group) and Band 3 Polymorphisms in Two Malaria Holoendemic Regions of Papua New Guinea. Am J Hematol 75:1–5

158. Teeranaipong P, Ohashi J, Patarapotikul J, Kimura R, Nuchnoi P, Hananantachai H et al (2008) A functional single-nucleotide polymorphism in the CR1 promoter region contributes to protection against cerebral malaria. J Infect Dis 198:1880–1891

159. Durand PM, Coetzer TL (2008) Pyruvate kinase deficiency protects against malaria in humans. Haematologica 93:939–940

160. Crosnier C, Bustamante LY, Bartholdson SJ, Bei AK, Theron M, Uchikawa M et al (2011) Basigin is a receptor essential for erythrocyte invasion by Plasmodium falciparum. Nature 480(7378):534–537

161. Egan ES, Weekes MP, Kanjee U, Manzo J, Srinivasan A, Lomas-Francis C et al (2018) Erythrocytes lacking the Langereis blood group protein ABCB6 are resistant to the malaria parasite Plasmodium falciparum. Commun Biol 1(1):45

162. Gazarini ML, Thomas AP, Pozzan T, Garcia CRS (2003) Calcium signaling in a low calcium environment: how the intracellular malaria parasite solves the problem. J Cell Biol 161:103–110

163. Timmann C, Thye T, Vens M, Evans J, May J, Ehmen C et al (2012) Genome-wide association study indicates two novel resistance loci for severe malaria. Nature 489:443–446

164. Bedu-Addo G, Meese S, Mockenhaupt FP (2013) An ATP2B4 polymorphism protects against malaria in pregnancy. J Infect Dis 207:1600–1603

165. Rockett KA, Clarke GM, Fitzpatrick K, Hubbart C, Jeffreys AE, Rowlands K et al (2014) Reappraisal of known malaria resistance loci in a large multicenter study. Nat Genet 46:1197–1204

166. Li J, Glessner JT, Zhang H, Hou C, Wei Z, Bradfield JP et al (2013) GWAS of blood cell traits identifies novel associated loci and epistatic interactions in Caucasian and African-American children. Hum Mol Genet 22:1457–1464

167. Lessard S, Stern EN, Beaudoin M, Schupp PG, Sher F, Ali A et al (2017) An erythroid – specific enhancer of ATP2B4 mediates red blood cell hydration and malaria susceptibility. J Clin Investig 1:1–10

168. Zambo B, Varady G, Padanyi R, Szabo E, Nemeth A, Lango T et al (2017) Decreased calcium pump expression in human erythrocytes is connected to a minor haplotype in the ATP2B4 gene. Cell Calcium 65:73–79

169. Tiffert T, Staines HM, Ellory JC, Lew VL (2000) Functional state of the plasma membrane Ca^{2+} pump in Plasmodium falciparum-infected human red blood cells. J Physiol 525(Pt 1):125–134

170. Spielmann T, Montagna GN, Hecht L, Matuschewski K (2012) Molecular make-up of the Plasmodium parasitophorous vacuolar membrane. Int J Med Microbiol 302:179–186

171. Lauer S, VanWye J, Harrison T, McManus H, Samuel BU, Hiller NL et al (2000) Vacuolar uptake of host components, and a role for cholesterol and sphingomyelin in malarial infection. EMBO J 19:3556–3564

172. Dluzewski AR, Fryer PR, Griffiths S, Wilson RJ, Gratzer WB (1989) Red cell membrane protein distribution during malarial invasion. J cell Sci 92:691–699

173. Gonzalez Flecha FL, Castello PR, Caride AJ, Gagliardino JJ, Rossi JP (1993) The erythrocyte calcium pump is inhibited by non-enzymic glycation: studies in situ and with the purified enzyme. Biochem J 293(Pt 2):369–375

174. Davis FB, Davis PJ, Nat G, Blas SD, MacGillivray M, Gutman S et al (1985) The effect of in vivo glucose administration on human erythrocyte Ca^{2+}-ATPase activity and on enzyme responsiveness in vitro to thyroid hormone and calmodulin. Diabetes 34(7):639–646

175. Gonzalez Flecha FL, Bermudez MC, Cedola NV, Gagliardino JJ, Rossi JP (1990) Decreased Ca2(+)-ATPase activity after glycosylation of erythrocyte membranes in vivo and in vitro. Diabetes 39(6):707–711

176. Bookchin RM, Etzion Z, Lew VL, Tiffert T (2009) Preserved function of the plasma membrane calcium pump of red blood cells from diabetic subjects with high levels of glycated haemoglobin. Cell Calcium 45(3):260–263

177. Berridge MJ (2010) Calcium hypothesis of Alzheimer's disease. Pflugers Arch 459(3):441–449

178. Berrocal M, Marcos D, Sepulveda MR, Perez M, Avila J, Mata AM (2009) Altered Ca^{2+} dependence of synaptosomal plasma membrane Ca^{2+}-ATPase in human brain affected by Alzheimer's disease. FASEB J 23(6):1826–1834

179. Berrocal M, Sepulveda MR, Vazquez-Hernandez M, Mata AM (2012) Calmodulin antagonizes amyloid-beta peptides-mediated inhibition of brain plasma membrane Ca^{2+}-ATPase. Biochim Biophys Acta 1822(6):961–969

180. Berrocal M, Corbacho I, Vazquez-Hernandez M, Avila J, Sepulveda MR, Mata AM (2015) Inhibition of PMCA activity by tau as a function of aging and Alzheimer's neuropathology. Biochim Biophys Acta 1852(7):1465–1476

181. Zaidi A (2010) Plasma membrane Ca-ATPases: targets of oxidative stress in brain aging and neurodegeneration. World J Biol Chem 1(9):271–280
182. Brendel A, Renziehausen J, Behl C, Hajieva P (2014) Downregulation of PMCA2 increases the vulnerability of midbrain neurons to mitochondrial complex I inhibition. Neurotoxicology 40:43–51
183. Lock C, Hermans G, Pedotti R, Brendolan A, Schadt E, Garren H et al (2002) Gene-microarray analysis of multiple sclerosis lesions yields new targets validated in autoimmune encephalomyelitis. Nat Med 8(5):500–508
184. Nicot A, Kurnellas M, Elkabes S (2005) Temporal pattern of plasma membrane calcium ATPase 2 expression in the spinal cord correlates with the course of clinical symptoms in two rodent models of autoimmune encephalomyelitis. Eur J Neurosci 21(10):2660–2670
185. Kurnellas MP, Donahue KC, Elkabes S (2007) Mechanisms of neuronal damage in multiple sclerosis and its animal models: role of calcium pumps and exchangers. Biochem Soc Trans 35(Pt 5):923–926
186. Chaabane C, Dally S, Corvazier E, Bredoux R, Bobe R, Ftouhi B et al (2007) Platelet PMCA- and SERCA-type Ca^{2+} -ATPase expression in diabetes: a novel signature of abnormal megakaryocytopoiesis. J Thromb Haemost 5(10):2127–2135
187. Souza KLA, Elsner M, Mathias PCF, Lenzen S, Tiedge M (2004) Cytokines activate genes of the endocytotic pathway in insulin-producing RINm5F cells. Diabetologia 47(7):1292–1302
188. Jiang L, Allagnat F, Nguidjoe E, Kamagate A, Pachera N, Vanderwinden JM et al (2010) Plasma membrane Ca^{2+}-ATPase overexpression depletes both mitochondrial and endoplasmic reticulum Ca^{2+} stores and triggers apoptosis in insulin-secreting BRIN-BD11 cells. J Biol Chem 285(40):30634–30643
189. Pachera N, Papin J, Zummo FP, Rahier J, Mast J, Meyerovich K et al (2015) Heterozygous inactivation of plasma membrane Ca^{2+}-ATPase in mice increases glucose-induced insulin release and beta cell proliferation, mass and viability. Diabetologia 58(12):2843–2850
190. Garcia ME, Del Zotto H, Caride AJ, Filoteo AG, Penniston JT, Rossi JP et al (2002) Expression and cellular distribution pattern of plasma membrane calcium pump isoforms in rat pancreatic islets. J Membr Biol 185(1):17–23
191. Alzugaray ME, Garcia ME, Del Zotto HH, Raschia MA, Palomeque J, Rossi JP et al (2009) Changes in islet plasma membrane calcium-ATPase activity and isoform expression induced by insulin resistance. Arch Biochem Biophys 490(1):17–23
192. Polak-Jonkisz D, Purzyc L, Laszki-Szczachor K, Musial K, Zwolinska D (2010) The endogenous modulators of Ca^{2+}-Mg2+-dependent ATPase in children with chronic kidney disease (CKD). Nephrol Dial Transplant 25(2):438–444
193. Bianchi G, Vezzoli G, Cusi D, Cova T, Elli A, Soldati L et al (1988) Abnormal red-cell calcium pump in patients with idiopathic hypercalciuria. N Engl J Med 319(14):897–901
194. Cartwright EJ, Oceandy D, Austin C, Neyses L (2011) Ca^{2+} signalling in cardiovascular disease: the role of the plasma membrane calcium pumps. Sci China Life Sci 54(8):691–698
195. Queen LR, Ferro A (2006) Beta-adrenergic receptors and nitric oxide generation in the cardiovascular system. Cell Mol Life Sci 63(9):1070–1083
196. Cartwright EJ, Oceandy D, Neyses L (2009) Physiological implications of the interaction between the plasma membrane calcium pump and nNOS. Pflugers Arch 457(3):665–671
197. Schuh K, Uldrijan S, Telkamp M, Rothlein N, Neyses L (2001) The plasmamembrane calmodulin-dependent calcium pump: a major regulator of nitric oxide synthase I. J Cell Biol 155(2):201–205

198. Mohamed TM, Oceandy D, Prehar S, Alatwi N, Hegab Z, Baudoin FM et al (2009) Specific role of neuronal nitric-oxide synthase when tethered to the plasma membrane calcium pump in regulating the beta-adrenergic signal in the myocardium. J Biol Chem 284(18):12091–12098

199. Beigi F, Oskouei BN, Zheng M, Cooke CA, Lamirault G, Hare JM (2009) Cardiac nitric oxide synthase-1 localization within the cardiomyocyte is accompanied by the adaptor protein, CAPON. Nitric Oxide 21(3-4):226–233

200. Williams JC, Armesilla AL, Mohamed TM, Hagarty CL, McIntyre FH, Schomburg S et al (2006) The sarcolemmal calcium pump, alpha-1 syntrophin, and neuronal nitric-oxide synthase are parts of a macromolecular protein complex. J Biol Chem 281(33):23341–23348

201. Ueda K, Valdivia C, Medeiros-Domingo A, Tester DJ, Vatta M, Farrugia G et al (2008) Syntrophin mutation associated with long QT syndrome through activation of the nNOS-SCN5A macromolecular complex. Proc Natl Acad Sci U S A. 105(27):9355–9360

202. Arking DE, Pfeufer A, Post W, Kao WH, Newton-Cheh C, Ikeda M et al (2006) A common genetic variant in the NOS1 regulator NOS1AP modulates cardiac repolarization. Nat Genet. 38(6):644–651

203. Dewey FE, Grove ME, Priest JR, Waggott D, Batra P, Miller CL et al (2015) Sequence to medical phenotypes: a framework for interpretation of human whole genome DNA sequence data. PLoS Genet. 11(10):e1005496

204. Wu X, Chang B, Blair NS, Sargent M, York AJ, Robbins J et al (2009) Plasma membrane Ca^{2+}-ATPase isoform 4 antagonizes cardiac hypertrophy in association with calcineurin inhibition in rodents. J Clin Invest. 119(4):976–985

205. Sadi AM, Afroze T, Siraj MA, Momen A, White-Dzuro C, Zarrin-Khat D et al (2018) Cardiac-specific inducible overexpression of human plasma membrane Ca^{2+} ATPase 4b is cardioprotective and improves survival in mice following ischemic injury. Clin Sci (Lond). 132(6):641–654

206. Mohamed TM, Abou-Leisa R, Stafford N, Maqsood A, Zi M, Prehar S et al (2016) The plasma membrane calcium ATPase 4 signalling in cardiac fibroblasts mediates cardiomyocyte hypertrophy. Nat Commun. 7:11074

207. Gros R, Afroze T, You XM, Kabir G, Van Wert R, Kalair W et al (2003) Plasma membrane calcium ATPase overexpression in arterial smooth muscle increases vasomotor responsiveness and blood pressure. Circ Res. 93(7):614–621

208. Perez AV, Picotto G, Carpentieri AR, Rivoira MA, Peralta Lopez ME, Tolosa de Talamoni NG (2008) Minireview on regulation of intestinal calcium absorption. Emphasis on molecular mechanisms of transcellular pathway. Digestion. 77(1):22–34

209. Armbrecht HJ, Boltz MA, Hodam TL (2002) Differences in intestinal calcium and phosphate transport between low and high bone density mice. Am J Physiol Gastrointest Liver Physiol. 282(1):G130–G136

210. Lee SM, Riley EM, Meyer MB, Benkusky NA, Plum LA, DeLuca HF et al (2015) 1,25-Dihydroxyvitamin D3 Controls a Cohort of Vitamin D Receptor Target Genes in the Proximal Intestine That Is Enriched for Calcium-regulating Components. J Biol Chem. 290(29):18199–18215

211. Ryan ZC, Craig TA, Filoteo AG, Westendorf JJ, Cartwright EJ, Neyses L et al (2015) Deletion of the intestinal plasma membrane calcium pump, isoform 1, Atp2b1, in mice is associated with decreased bone mineral density and impaired responsiveness to 1, 25-dihydroxyvitamin D3. Biochem Biophys Res Commun 467(1):152–156

212. Dong XL, Zhang Y, Wong MS (2014) Estrogen deficiency-induced Ca balance impairment is associated with decrease in expression of epithelial Ca transport proteins in aged female rats. Life Sci. 96(1-2):26–32

213. Wu F, Dassopoulos T, Cope L, Maitra A, Brant SR, Harris ML et al (2007) Genome-wide gene expression differences in Crohn's disease and ulcerative colitis from endoscopic pinch biopsies: insights into distinctive pathogenesis. Inflamm Bowel Dis. 13(7):807–821

214. Martin R, Harvey NC, Crozier SR, Poole JR, Javaid MK, Dennison EM et al (2007) Placental calcium transporter (PMCA3) gene expression predicts intrauterine bone mineral accrual. Bone. 40(5):1203–1208

215. Bredoux R, Corvazier E, Dally S, Chaabane C, Bobe R, Raies A et al (2006) Human platelet Ca^{2+}-ATPases: new markers of cell differentiation as illustrated in idiopathic scoliosis. Platelets. 17(6):421–433

216. Kim HJ, Prasad V, Hyung SW, Lee ZH, Lee SW, Bhargava A et al (2012) Plasma membrane calcium ATPase regulates bone mass by fine-tuning osteoclast differentiation and survival. J Cell Biol. 199(7):1145–1158

217. Prevarskaya N, Ouadid-Ahidouch H, Skryma R, Shuba Y (2014) Remodelling of Ca^{2+} transport in cancer: how it contributes to cancer hallmarks? Philos Trans R Soc Lond B Biol Sci. 369(1638):20130097

218. Ruschoff JH, Brandenburger T, Strehler EE, Filoteo AG, Heinmoller E, Aumuller G et al (2012) Plasma membrane calcium ATPase expression in human colon multistep carcinogenesis. Cancer Invest. 30(4):251–257

219. Ribiczey P, Tordai A, Andrikovics H, Filoteo AG, Penniston JT, Enouf J et al (2007) Isoform-specific up-regulation of plasma membrane Ca^{2+} ATPase expression during colon and gastric cancer cell differentiation. Cell Calcium. 42(6):590–605

220. Aung CS, Kruger WA, Poronnik P, Roberts-Thomson SJ, Monteith GR (2007) Plasma membrane Ca^{2+}-ATPase expression during colon cancer cell line differentiation. Biochem Biophys Res Commun. 355(4):932–936

221. Ribiczey P, Papp B, Homolya L, Enyedi A, Kovacs T (2015) Selective upregulation of the expression of plasma membrane calcium ATPase isoforms upon differentiation and 1,25(OH)2D3-vitamin treatment of colon cancer cells. Biochem Biophys Res Commun. 464(1):189–194

222. Varga K, Paszty K, Padanyi R, Hegedus L, Brouland JP, Papp B et al (2014) Histone deacetylase inhibitor- and PMA-induced upregulation of PMCA4b enhances Ca^{2+} clearance from MCF-7 breast cancer cells. Cell calcium. 55(2):78–92

223. Lee WJ, Roberts-Thomson SJ, Holman NA, May FJ, Lehrbach GM, Monteith GR (2002) Expression of plasma membrane calcium pump isoform mRNAs in breast cancer cell lines. Cell Signal 14(12):1015–1022

224. Lee WJ, Roberts-Thomson SJ, Monteith GR (2005) Plasma membrane calcium-ATPase 2 and 4 in human breast cancer cell lines. Biochem Biophys Res Commun 337(3):779–783

225. VanHouten J, Sullivan C, Bazinet C, Ryoo T, Camp R, Rimm DL et al (2010) PMCA2 regulates apoptosis during mammary gland involution and predicts outcome in breast cancer. Proc Natl Acad Sci U S A. 107(25):11405–11410

226. Jeong J, VanHouten JN, Dann P, Kim W, Sullivan C, Yu H et al (2016) PMCA2 regulates HER2 protein kinase localization and signaling and promotes HER2-mediated breast cancer. Proc Natl Acad Sci U S A. 113(3):E282–E290

227. Peters AA, Milevskiy MJ, Lee WC, Curry MC, Smart CE, Saunus JM et al (2016) The calcium pump plasma membrane Ca^{2+}-ATPase 2 (PMCA2) regulates breast cancer cell proliferation and sensitivity to doxorubicin. Sci Rep. 6:25505

228. Hegedus L, Padanyi R, Molnar J, Paszty K, Varga K, Kenessey I et al (2017) Histone deacetylase inhibitor treatment increases the expression of the plasma membrane Ca^{2+} pump PMCA4b and inhibits the migration of melanoma cells independent of ERK. Front Oncol. 7:95

229. Saito K, Uzawa K, Endo Y, Kato Y, Nakashima D, Ogawara K et al (2006) Plasma membrane Ca^{2+} ATPase isoform 1 down-regulated in human oral cancer. Oncol Rep. 15(1):49–55

230. Farkona S, Diamandis EP, Blasutig IM (2016) Cancer immunotherapy: the beginning of the end of cancer? BMC Med. 14:73

231. Supper V, Schiller HB, Paster W, Forster F, Boulegue C, Mitulovic G et al (2016) Association of CD147 and calcium exporter PMCA4 uncouples IL-2 expression from early TCR signaling. J Immunol. 196(3):1387–1399

232. Hu X, Su J, Zhou Y, Xie X, Peng C, Yuan Z et al (2017) Repressing CD147 is a novel therapeutic strategy for malignant melanoma. Oncotarget. 8(15):25806–25813

233. Rimessi A, Coletto L, Pinton P, Rizzuto R, Brini M, Carafoli E (2005) Inhibitory interaction of the 14-3-3{epsilon} protein with isoform 4 of the plasma membrane Ca^{2+}-ATPase pump. J Biol Chem. 280(44):37195–37203

234. Linde CI, Di Leva F, Domi T, Tosatto SC, Brini M, Carafoli E (2008) Inhibitory interaction of the 14-3-3 proteins with ubiquitous (PMCA1) and tissue-specific (PMCA3) isoforms of the plasma membrane Ca^{2+} pump. Cell Calcium. 43(6):550–561

235. Vanagas L, Rossi RC, Caride AJ, Filoteo AG, Strehler EE, Rossi JP (2007) Plasma membrane calcium pump activity is affected by the membrane protein concentration: evidence for the involvement of the actin cytoskeleton. Biochim Biophys Acta. 1768(6):1641–1649

236. Zabe M, Dean WL (2001) Plasma membrane Ca^{2+}-ATPase associates with the cytoskeleton in activated platelets through a PDZ-binding domain. J Biol Chem. 276(18):14704–14709

237. James P, Maeda M, Fischer R, Verma AK, Krebs J, Penniston JT et al (1988) Identification and primary structure of a calmodulin binding domain of the Ca^{2+} pump of human erythrocytes. J Biol Chem. 263(6):2905–2910

238. Cali T, Brini M, Carafoli E (2017) Regulation of cell calcium and role of plasma membrane calcium ATPases. Int Rev Cell Mol Biol. 332:259–296

239. Sgambato-Faure V, Xiong Y, Berke JD, Hyman SE, Strehler EE (2006) The Homer-1 protein Ania-3 interacts with the plasma membrane calcium pump. Biochem Biophys Res Commun. 343(2):630–637

240. Salm EJ, Thayer SA (2012) Homer proteins accelerate Ca^{2+} clearance mediated by the plasma membrane Ca^{2+} pump in hippocampal neurons. Biochem Biophys Res Commun. 424(1):76–81

241. Oceandy D, Cartwright EJ, Emerson M, Prehar S, Baudoin FM, Zi M et al (2007) Neuronal nitric oxide synthase signaling in the heart is regulated by the sarcolemmal calcium pump 4b. Circulation. 115(4):483–492

242. Olli KE, Li K, Galileo DS, Martin-DeLeon PA (2018) Plasma membrane calcium ATPase 4 (PMCA4) co-ordinates calcium and nitric oxide signaling in regulating murine sperm functional activity. J Cell Physiol. 233(1):11–22

243. Holton M, Mohamed TM, Oceandy D, Wang W, Lamas S, Emerson M et al (2010) Endothelial nitric oxide synthase activity is inhibited by the plasma membrane calcium ATPase in human endothelial cells. Cardiovasc Res. 87(3):440–448

244. Armesilla AL, Williams JC, Buch MH, Pickard A, Emerson M, Cartwright EJ et al (2004) Novel functional interaction between the plasma membrane Ca^{2+} pump 4b and the proapoptotic tumor suppressor Ras-associated factor 1 (RASSF1). J Biol Chem. 279(30):31318–31328

245. Kurnellas MP, Lee AK, Li H, Deng L, Ehrlich DJ, Elkabes S (2007) Molecular alterations in the cerebellum of the plasma membrane calcium ATPase 2 (PMCA2)-null mouse indicate abnormalities in Purkinje neurons. Mol Cell Neurosci. 34(2):178–188

246. Kim E, DeMarco SJ, Marfatia SM, Chishti AH, Sheng M, Strehler EE (1998) Plasma membrane Ca^{2+} ATPase isoform 4b binds to membrane-associated guanylate kinase (MAGUK) proteins via their PDZ (PSD-95/Dlg/ZO-1) domains. J Biol Chem. 273(3):1591–1595

247. Kruger WA, Yun CC, Monteith GR, Poronnik P (2009) Muscarinic-induced recruitment of plasma membrane Ca^{2+}-ATPase involves PSD-95/Dlg/Zo-1-mediated interactions. J Biol Chem 284(3):1820–1830

248. Goellner GM, DeMarco SJ, Strehler EE (2003) Characterization of PISP, a novel single-PDZ protein that binds to all plasma membrane Ca^{2+}-ATPase b-splice variants. Ann N Y Acad Sci 986:461–471

Chapter 6
A Role for SERCA Pumps in the Neurobiology of Neuropsychiatric and Neurodegenerative Disorders

Aikaterini Britzolaki, Joseph Saurine, Benjamin Klocke, and Pothitos M. Pitychoutis

Abstract Calcium (Ca^{2+}) is a fundamental regulator of cell fate and intracellular Ca^{2+} homeostasis is crucial for proper function of the nerve cells. Given the complexity of neurons, a constellation of mechanisms finely tunes the intracellular Ca^{2+} signaling. We are focusing on the sarco/endoplasmic reticulum (SR/ER) calcium (Ca^{2+})-ATPase (SERCA) pump, an integral ER protein. SERCA's well established role is to preserve low cytosolic Ca^{2+} levels ($[Ca^{2+}]_{cyt}$), by pumping free Ca^{2+} ions into the ER lumen, utilizing ATP hydrolysis. The SERCA pumps are encoded by three distinct genes, *SERCA1-3*, resulting in 12 known protein isoforms, with tissue-dependent expression patterns. Despite the well-established structure and function of the SERCA pumps, their role in the central nervous system is not clear yet. Interestingly, SERCA-mediated Ca^{2+} dyshomeostasis has been associated with neuropathological conditions, such as bipolar disorder, schizophrenia, Parkinson's disease and Alzheimer's disease. We summarize here current evidence suggesting a role for SERCA in the neurobiology of neuropsychiatric and neurodegenerative disorders, thus highlighting the importance of this pump in brain physiology and pathophysiology.

Keywords SERCA · Calcium · Central nervous system · Bipolar disorder · Schizophrenia · Alzheimer's disease · Parkinson's disease

6.1 Introduction

Calcium (Ca^{2+}) is a critical and universal regulator of cell fate [1–3]. While Ca^{2+} is crucial for the electrophysiological properties of all cells, it also serves as a prominent second messenger triggering a cascade of intracellular molecular

A. Britzolaki · J. Saurine · B. Klocke · P. M. Pitychoutis (✉)
Department of Biology & Center for Tissue Regeneration and Engineering at Dayton (TREND), University of Dayton, Dayton, OH, USA
e-mail: ppitychoutis1@udayton.edu

© Springer Nature Switzerland AG 2020   131
M. S. Islam (ed.), *Calcium Signaling*, Advances in Experimental Medicine and Biology 1131, https://doi.org/10.1007/978-3-030-12457-1_6

processes [1, 4–8]. Neurons are no exception to this; Ca^{2+} is pivotal for their survival and function, and disruptions of intracellular Ca^{2+} homeostasis may elicit neuropathology [9–14]. Given the innate complexity of neurons and the importance of Ca^{2+} in maintaining proper neuronal function, nerve cells have developed an intricate Ca^{2+} signaling network. A variety of channels, pumps, exchangers and proteins ensure a finely-tuned handling of the intraneuronal Ca^{2+} distribution [1, 4, 15–18]. In this review, we are focusing on the sarco/endoplasmic reticulum (SR/ER) calcium (Ca^{2+})-ATPase (SERCA) pump, an integral ER protein. The SERCA pumps are major regulators of intracellular Ca^{2+} homeostasis, facilitating the influx of Ca^{2+} in the ER lumen, thus regulating the levels of free Ca^{2+} in the cytosol [2, 19]. SERCAs belong to the family of P-type ATPases which includes a variety of membrane pumps that utilize ATP hydrolysis and a phosphorylated enzyme intermediate to transport ions across cellular membranes [20–22]. Found in all eukaryotic cells, three distinct genes, *SERCA1-3* (or *ATPA1-3* in humans), encode SERCA, producing 12 known protein isoforms, mainly *via* alternate splicing [23]. Interestingly, although all SERCA isoforms have a highly conserved structure, their expression patterns, affinity for Ca^{2+} and turnover rates may differ [23–30].

The SERCA structure and function have been recently reviewed [26]. It is well-established that SERCA is composed of a 1000-amino-acid-long single 100 kDa polypeptide chain [29, 31–33]. Much of the current knowledge on the structure of the SERCA pumps was based on SERCA1a isoform crystallography studies [34]. In fact, the folded SERCA protein resides on the ER membrane, and it is comprised of three cytosolic domains (A, N and P), one short luminal loop, and ten transmembrane α-helices (M1-M10). The extension of four transmembrane α-helices (M2-M5) results in the formation of the three cytosolic domains; Ca^{2+} binding and release are mediated by the actuator (A) domain, the ATP-binding cavity is formed by the nucleotide-binding (N) domain, whereas the high-energy phosphorylation intermediate product is formed in the phosphorylation (P) domain [34–39]. Moreover, the transmembrane α-helices are crucial for the formation of the Ca^{2+} channel (M2, M5, M5 and M8), and facilitate the Ca^{2+} transportation across the ER membrane (M4-M6) [34, 36, 40, 41]. This organization is highly conserved amongst all SERCA isoforms with differences mainly detected in the C-terminus [26, 31, 42–47] (Fig. 6.1). Despite the plethora of the SERCA splice variants, the universal role of SERCA entails the pumping of Ca^{2+} into the ER, resulting in decreased free cytosolic Ca^{2+} levels $([Ca^{2+}]_{cyt})$ and maintained internal Ca^{2+} storages. Specifically, SERCA couples the active transport of two Ca^{2+} ions at the expense of one ATP molecule throughout a cycle of conformational alterations between two biochemical states (E1/E2) [19, 24, 48–55] (Fig. 6.2). At the E1 state, once Ca^{2+} and ATP bind to the high-affinity cytosolic sites, ATP hydrolysis is triggered, and the high energy intermediate state is formed. Subsequently, ADP is released and SERCA is phosphorylated, leading to the conformational change of the transmembrane domain (E2 state). During this transition, Ca^{2+} is shortly occluded from both the cytosol and the ER lumen. At the E2 state, Ca^{2+} is exposed to the ER lumen and its binding affinity to SERCA is very low, resulting in its release into

SERCA isoform	Protein size	Carboxyl terminus sequence
SERCA1a	994aa; 109.3kDa	LDEILKFVARNYLEG
SERCA1b	1001aa; 110.5kDa	LDEILKFVARNYLEDPEDERRK
SERCA2a	997aa; 109.7kDa	LYVEPLP/LIFQITPLLNVTQWLMVLKISLPVILMDETLKFVARNYLEP/AILE
SERCA2b	1042aa; 114.8kDa	LYVEPLP/LIFQITPLLNVTQWLMVLKISLPVILMDETLKFVARNYLEP/GKECVQPATKS CSFSACTDGISWPFVLLIMPLVIYVYSTDTNFSDMFWS
SERCA2c	999aa; 109.9kDa	LYVEPLP/LIFQITPLLNVTQWLMVLKISLPVILMDETLKFVARNYLEP/VLSSEL
SERCA2d	1007aa; 110.6kDa	LYVEPLP/VSGWVGLGTSHLLPGEAGGVTRLPCVS/AHLPDHTAERDPVADGAENLLA RDSHG
SERCA3a	999aa; 109.2kDa	SRNHMH/EEMSQK
SERCA3b	1043aa; 113.9kDa	SRNHMH/ACLYPGLLRTVSQAWSRQPLTTSWTPDHTGRNEPEVSAGNRVESPVCTS D
SERCA3c	1029aa; 112.4kDa	SRNHMH/ACLYPGLLRTVSQAWSRQPLTTSWTPDHTGLASLKK
SERCA3d	1044aa; 114.1kDa	SRNHMH/ACLYPGLLRTVSQAWSRQPLTTSWTPDHTGARDTASSRCQSCSEREEAGK K
SERCA3e	1052aa; 114.9kDa	SRNHMH/ACLYPGLLRTVSQAWSRQPLTTSWTPDHTGLASLGQGHSIVSLSELLREG GSREEMSQK
SERCA3f	1033aa; 112.6kDa	SRNHMH/GPGTQHRLAVRAAQRGRKQGRNEPEVSAGNRVESPVCTSD

Fig. 6.1 The primary sequence of the carboxyl termini of all known human SERCA protein isoforms. SERCA pumps have a highly conserved structure, but differences are detected in the carboxyl termini amongst protein isoforms. Variants that are encoded by the same gene tend to be less different. The splice sites are marked with slashes

the ER lumen. Simultaneously, 2-3 H^+ ions bind to the pump. Afterwards, SERCA gets dephosphorylated and returns to its E1 state, while releasing H^+ into the cytosol [19, 36, 37, 39, 48, 56, 57]. Of note, ATP-binding is independent of whether Ca^{2+} is bound to SERCA, but Ca^{2+} is essential for the enzymatic cycle to proceed [36, 37, 58]. Taken together, these conformational changes mediate the SERCA-dependent Ca^{2+} shuttling from the cytosol into the ER.

Interestingly, certain SERCA functions are isoform-dependent, with different splice-variants presenting slight differences in the affinity for Ca^{2+} and turnover rate [26, 28, 31, 59–61]. Several studies have revealed that the different SERCA isoforms present distinct expression profiles, suggesting tissue-specific functions. Two SERCA1 isoforms have been identified so far; SERCA1a and SERCA1b. The gene encoding these protein isoforms, *SERCA1*, is expressed in both the neonatal and the adult period [46, 62], but alternative splicing determines which isoform will be expressed in each developmental period [46, 62, 63]. Hence, SERCA1a is considered the adult isoform, and SERCA1b the neonatal isoform [46, 62–64]. In addition to temporal expression, SERCA1 expression is also tissue-dependent. To date, functional SERCA1 protein expression has been found primarily in the fast-twitch skeletal muscle fibers [46, 62], while studies have also shown low expression of SERCA1 in the Purkinje neurons of the cerebellum [65].

Similarly, SERCA2 expression is spatially regulated in the different tissues of the body [27, 29, 43, 44, 66, 67]. To date, four SERCA2 splice variants (SERCA2a-

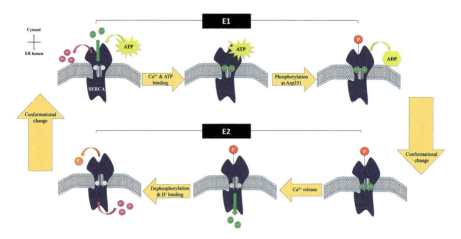

Fig. 6.2 SERCA-mediated Ca^{2+} transport across the ER membrane. SERCA facilitates the transport of two Ca^{2+} ions across the ER membrane in the expense of one ATP molecule, through a cycle of conformational changes (E1/E2 states). Upon Ca^{2+} and ATP binding (E1 state), SERCA gets phosphorylated due to ATP hydrolysis and changes conformation (E2 state). This conformational change allows Ca^{2+} release in the ER lumen. Afterwards, phosphate cleavage leads to dephosphorylation of SERCA, while two to three H^+ are coupled. SERCA dephosphorylation will reinforce a second conformational change to the initial E1 state. H^+ are released and SERCA is ready to bind new Ca^{2+} ions

d) have been identified with tissue-specific expression [26, 27, 42–44, 68–70]. Specifically, SERCA2a mRNA and protein have been detected in slow-twitch skeletal muscle, as well as in cardiac muscle fibers [44, 66, 67]. Furthermore, SERCA2a expression has been identified in the brain, albeit at low levels, and it is confined in the granular cells of cerebellar Purkinje neurons, as well as in the giant cells of the brainstem reticular formation [71–73]. In contrast, SERCA2b is constitutively expressed in all tissues including slow skeletal muscle fibers, smooth muscle cells, cardiac muscle fibers, neurons and astrocytes [27, 43, 44, 74]. SERCA2b has also been found to be the only isoform expressed in neuronal microsomes, synaptic plasma membrane vesicles, and synaptosomes [75]. The extensive expression pattern of SERCA2b in the CNS has been recently reviewed [26]. Given its constitutive expression, SERCA2b is considered an ER housekeeping enzyme, crucial for maintaining intracellular Ca^{2+} homeostasis [27, 44, 64]. About 15 years after the discovery of the SERCA2a and SERCA2b isoforms, a third splice variant, SERCA2c, was identified in hematopoietic, mesenchymal and epithelial cells, as well as in the brain at low levels [30, 31, 42]. Moreover, the mRNA expression of a fourth splice variant, SERCA2d, has been detected in skeletal muscle fibers [68].

The expression pattern of the third SERCA-encoding gene, *SERCA3*, was first discovered using northern blot analysis in rat tissues [64]. To date, at least six different SERCA3 isoforms (SERCA3a-f) have been identified with a complex expression pattern in many tissue types and in a wide variety of species [60, 61,

64, 76–79]. The SERCA3a, SERCA3b, and SERCA3c protein isoforms have been detected in human platelets and human immortalized T-lymphocytes, known as T-lymphoid Jurkat cells [78]. The most distributed SERCA3 isoforms are SERCA3a, SERCA3b and SERCA3d; their expression has been found in a wide variety of tissue types, including the brain [60]. Specifically, SERCA3a and SERCA3b mRNA expression has been identified in the brain, heart, lung, pancreas, liver and placenta [60]. SERCA3d mRNA has also been found in the same tissue types as well as in skeletal muscle fibers [60]. The expression of SERCA3e is rather confined to human lung and pancreatic tissues [60]. Further studies have revealed the expression of a sixth SERCA3 variant, SERCA3f, in hematopoietic and non-muscle cell lines, as well as in all human cell and tissue types [61]. Importantly, immunoblotting and immunohistochemical experiments have indicated the expression of SERCA3 protein in the rat cerebral cortex and cerebellar Purkinje neurons, as well as in the human choroid plexus [79–81].

6.2   The SERCA Pumps: Potential Indicators of Brain Pathophysiology?

Taking into consideration that SERCA plays a pivotal role in preserving intracellular Ca^{2+} homeostasis, and that specific SERCA isoforms are expressed in the brain, several studies have explored the role of this important Ca^{2+} regulator in brain pathophysiology. In this context we conduct a comprehensive review of contemporary experimental evidence suggesting a central role for SERCA in the pathophysiology of neuropsychiatric and neurodegenerative disorders. Moreover, we discuss the potential of SERCA pumps as molecular targets for the development of novel pharmacotherapies to combat such debilitating disorders.

6.3   A Role for SERCA in the Pathophysiology of Neuropsychiatric Disorders

Neuropsychiatric disorders such as schizophrenia (SZ) and bipolar disorder (BD) are characterized by high lifetime prevalence and early onset, with deleterious long-time effects on public health [82–85]. Of note, neuropsychiatric disorders account for approximately 70% of disabilities worldwide, with SZ and BD contributing approximately 7% each [86]. SZ is a clinically and genetically heterogenous neuropsychiatric disorder, that affects approximately 1% of the general population, and is associated with hallucinations, delusions and profound cognitive impairment [87, 88]. BD is a debilitating chronic mood disorder with a complex clinical and genetic background. About 1% of the general population is affected by BD, with episodes of depression, mania and hypomania being the characteristic symptoms of the disease [89, 90]. Despite the devastating effects of these disorders and the rapid increase in incidence worldwide, their underlying pathophysiology is yet elusive. Interestingly, although SZ and BD are classified as distinct diagnostic categories, the two disorders greatly overlap in clinical presentation and genetic liability, as extensively discussed in recent association studies [91], highlighting the importance of understanding the common mechanisms underlying both disorders.

Intracellular Ca^{2+} signaling is believed to play a vital role in SZ etiology. In fact Ca^{2+} has been proposed as the common mechanism underlying SZ pathology [92, 93]. According to this theory, SZ symptomatology rises due to the disruption of the intracellular Ca^{2+} homeostasis, and the subsequent dysfunction of Ca^{2+}-mediated signal transduction processes [93]. The role of Ca^{2+} in SZ was first proposed by Jimerson et al., as a positive correlation was evidenced between psychotic episodes and increased cerebrospinal fluid (CSF) Ca^{2+} levels in SZ patients [92]. Since then, growing evidence has further supported the central role of Ca^{2+}-signaling in the pathogenesis of SZ, as discussed in several reviews [26, 93–96].

In addition to SZ, altered intracellular Ca^{2+} signaling has also been proposed as a potential mechanism underlying BD pathophysiology. In fact, ex vivo platelet studies have indicated that intracellular Ca^{2+} mobilization is dysregulated in BD. More specifically, an increase in serotonin- or thrombin-induced Ca^{2+} mobilization, as well as elevation of intracellular Ca^{2+} levels, have been observed in platelets harvested from BD patients [97–102]. Importantly, common pharmacological therapies for BD (i.e., lithium and valproate) have been reported to enhance the expression of proteins essential for Ca^{2+} sequestration, further supporting a possible role for Ca^{2+} in BD pathogenesis [103–109].

Disruption of intracellular Ca^{2+} homeostasis is crucial in the emergence of neuropsychiatric pathology. Indeed, recent Genome Wide Association Studies (GWAS) and meta-analyses have indicated the association of altered Ca^{2+} channel activity with SZ and BD [110–112]. More specifically, the expression of genes encoding several types of voltage-gated Ca^{2+} (Ca_v) channels have been linked to both SZ and BD [113–117]. Of note, Ca_v channels are central in regulating Ca^{2+} influx into neurons, and Ca_v channel blockers have been examined in clinical trials as potential therapeutic approach for SZ, highlighting the importance of altered Ca^{2+}-signaling mechanisms in the pathogenesis of this disorder [118]. Thus, efforts have focused on deciphering the possible role of Ca^{2+}-regulating mechanisms in SZ and BD.

6.3.1 Early Association Between SERCA2 Mutations and Psychosis in Darier's Disease Patients

The implication of SERCA in BD and SZ was first introduced almost two decades ago and since then, growing evidence has continued to support the importance of SERCA in the pathogenesis of these disorders. An early association between

SERCA and neuropsychiatric pathophysiology was made in patients with Darier's Disease (DD) [14, 119]. DD or Darier-White disease, also known as *keratosis follicularis*, is a rare autosomal dominant skin disorder with the presence of characteristic warty papules and keratotic plaques in seborrheic areas [120]. It is a highly penetrating disorder, with variable expressivity and early onset, affecting 1 in 100,000 individuals in the general population [120–123]. Mutations in the *ATP2A2* (i.e. SERCA2) gene on chromosome 12q23-24.1 have been identified as the underlying mechanism of DD pathogenesis, with at least 253 unique *ATP2A2* variants being reported in 353 familiar or sporadic DD cases throughout the years [124]. At least 75% of the reported mutations are unique to the affected family, many are *de novo* mutations and only a few are common between families [125]. Evidence indicates that the majority of *ATP2A2* mutations are missense (51%), while all mutations are dispersed throughout the gene without hotspots [121, 124–126]. It is also believed that the *ATP2A2* mutations lead to DD phenotype through haploinsufficiency, and that DD symptomatology occurs independently of the SERCA2 mutation type [124, 126]. Several studies have indicated that *ATP2A2* mutations might affect SERCA2 protein expression and function, as well as proteasomal degradation [24, 127–129]. Specifically, some *ATP2A2* mutants have shown reduced SERCA2 expression, subsequent decreased Ca^{2+} ATPase activity, Ca^{2+} turnover and phosphorylation rates [24, 127–129]. Other *ATP2A2* mutants only reduce Ca^{2+} affinity and sensitivity to feedback inhibition by ER Ca^{2+} [129]. Moreover, frameshift deletions and nonsense mutations have been reported to promote apoptosis by increasing proteasomal degradation [127]. Further in vitro experiments using missense, nonsense and deletion *ATPA2* variants have indicated that ER luminal Ca^{2+} is depleted in keratinocytes, leading to abnormal $[Ca^{2+}]_{cyt}$ [130, 131]. Additional experiments with cultured cells from DD patients have revealed abrogated post-translational protein trafficking to plasma membrane and induced ER stress, caused by diminished ER Ca^{2+} [132, 133]. More importantly, ER stress induced by SERCA-blockade in primary human keratinocytes has been associated with increased keratinocyte differentiation, further confirming the fundamental role of SERCA2 in DD [134].

Interestingly, as extensively discussed in recent reviews [26, 124, 135, 136], DD patients present with increased life-time prevalence of several neuropsychiatric disorders, including: major depression disorder (MDD-30%), BD (4%), epilepsy (3%), SZ (1%) and cognitive disabilities (4%) [137–143]. These early observations along with the fact that the ubiquitous expression of SERCA2b had been identified in the brain [144], raised the question whether skin and brain disorders occur independently, or due to a SERCA2 pleiotropic effect. Early studies in DD patients have associated mutations in the SERCA2 ATP-binding domain with dysthymia, hinge domain mutations with MDD and BD, and transmembrane domain mutations with epilepsy, MDD and cognitive disability [14]. Other studies have supported the implication of SERCA2 mutations in the neuropsychopathology of DD patients. Identified SERCA2 mutations affecting the phosphorylation, stalk, hinge or transduction domains, and the transmembrane M6/M7 loop, have been reported in clinical DD cases with concurrent mental disease and vegetative growth

[121]. More recent clinical reports have further supported the association between SERCA2 dysfunction and schizophrenic or bipolar symptomatology in DD patients by identifying novel or recurrent *ATP2A2* mutations [145–147]. Specifically, the same missense mutation in the stalk domain of the SERCA2 pump has been identified in two different clinical DD cases with concurrent SZ [145, 147], while an altered-splicing mutation in the acceptor site has been characterized in another DD patient with concurrent BD [146].

6.3.2 Altered SERCA Activity Associated with the Pathogenesis of Schizophrenia and Bipolar Disorder

The role of SERCA in SZ and BD pathophysiology has been further discussed in a variety of preclinical and clinical studies. Early studies have investigated the role of SERCA in the pathophysiology of BD; platelets derived from BD patients presented a higher increase in $[Ca^{2+}]_{cyt}$ upon SERCA blockade by thapsigargin when compared to healthy controls, indicating altered SERCA activity in BD [148]. Other studies have attempted to explore the role of SERCA in SZ using a common animal model for SZ, the *Df(16)1/+* mouse, carrying SZ-related 22q11 microdeletions. These *de novo* deletions are variable in size and occur on the 22q11.21 to 22q11.23 chromosomal region in humans [149–153], leading to haploinsufficiency of several genes [149, 154–160]. The syndrome associated with these microdeletions is known as the 22q11 deletion syndrome (22q11DS) or DiGeorge syndrome, and it affects 1in 4000 to 1 in 6000 individuals worldwide [161–165]. DiGeorge symptomatology includes mild to moderate cognitive deficits, intellectual and learning disabilities, with a progressive worsening of cognitive function [166–170]. Interestingly, DiGeorge patients present high life-time prevalence to SZ or psychotic spectrum disorders (25%) [171–176]. Moreover, DiGeorge-related and idiopathic psychosis are not distinguished from each other, presenting similar demographics, age-onset and symptomatology [177, 178]. Therefore, *Df(16)1/+* mice are widely used to study the mechanisms underlying SZ pathophysiology in vivo. In vivo and in vitro experiments have suggested profound neuromolecular and behavioral alterations in *Df(16)1/+* mice, such as hippocampus-dependent spatial memory deficits and enhanced long-term potentiation (LTP) at the Schaffer collateral synapses [13]. More specifically, presynaptic SERCA2 expression has been found to be enhanced in the hippocampus, subsequently increasing presynaptic glutamate release and altering Ca^{2+} dynamics in the axon terminal [13]. Additional studies in *Df(16)1/+* mice have indicated that SERCA2 expression is altered in a brain-specific manner, and that these alterations extend to other brain regions beyond the hippocampus, such as the cortex and cerebellum [179]. Interestingly, later studies have proposed that the microRNA-processing gene *Dgcr8* regulates SZ-related SERCA2 expression [179, 180]. Earls et al. identified the loss of two miRNAs, miR-25 and miR-185, in the brain of *Df(16)1/+* mice, leading to abnormal

LTP increase [179]. According to the study, miR-25 and miR-185 affect Ca^{2+} influx into the ER by targeting SERCA2. Hence, loss of these two miRNAs has been linked to significantly elevated presynaptic SERCA2 levels of the hippocampus. Strikingly, presynaptic injection of either miR-25 and miR-185, restored SERCA2 activity and rescued the LTP abnormalities, introducing novel targets for the development of pharmacotherapies for the treatment of SZ [179]. More importantly, the same study demonstrated that SERCA2 was upregulated in the prefrontal cortex and the hippocampus of SZ patients post-mortem, further supporting the mechanistic link between SZ, DiGeorge syndrome and SERCA2. These data collectively led the investigators to hypothesize that SERCA2 upregulation results in perturbation of presynaptic Ca^{2+} signaling that may ultimately lead to the cognitive deficits observed in SZ [179].

In agreement with previous studies, a recent GWAS assay identified *ATP2A2* as a risk gene for SZ [181]. In addition, a meta-analysis classified *ATP2A2* as a SZ risk gene that could synergistically contribute to the enhanced generation of thalamic delta oscillations, associated with both negative and positive symptoms of SZ [182]. Furthermore, recent studies using another SZ-relevant animal model, the Neonatal Lesion in Ventral Hippocampus (NLVH) rat, have identified decreased SERCA2 expression in the prefrontal cortex of NLVH rats [183]. Genis-Mendoza et al. used microarrays to assess the global transcriptomic profiles of the prefrontal cortex, hippocampus and nucleus accumbens in juvenile and adult NVLH rats. The investigators reported a marked decline in *ATP2A2* expression only in the prefrontal cortex of adult NVLH rats, but not in their juvenile counterparts. These data led to the hypothesis that decreased *ATP2A2* expression could result in elevated $[Ca^{2+}]_{cyt}$, thus activating Ca^{2+}-dependent transcription factors and kinases that could potentially exert genomic effects contributing to the cognitive defects observed in SZ [183, 184].

Adding to the proposed role of SERCA in the neurobiology of SZ, a very interesting study used a ketamine-induced pharmacological model of experimental psychosis to demonstrate brain region-specific alterations in SERCA3 expression [185]. Specifically, Lisek et al. treated rats repeatedly (5 consecutive days) with the non-competitive N-methyl-D-aspartate receptor (NMDAR) antagonist, ketamine (30 mg/kg); this protocol has been found to induce profound psychosis-like neurochemical, behavioral and neuroanatomical alterations in the rat brain [94, 186, 187]. The authors first observed that ketamine-treated rats presented severe stereotypic behavior (i.e., increased cumulative turning, weaving and bobbing), as assessed by open field (OF) tests. Interestingly, a marked increase of $[Ca^{2+}]_{cyt}$ in the striatum, cerebellum, cortex and hippocampus of ketamine-treated rats was reported. However, a positive correlation between hyperlocomotion and elevated $[Ca^{2+}]_{cyt}$ was only established in cortical and striatal neurons [185]. Moreover, SERCA3 expression was found decreased in the cortex, but increased in the cerebellum, hippocampus and striatum of ketamine-treated rats. Correlation studies linked the abnormal elevation of $[Ca^{2+}]_{cyt}$ to increased SERCA3 expression, suggesting that overexpression of SERCA acted as a compensatory mechanism for the increased $[Ca^{2+}]_{cyt}$ [185].

Taken together, clinical and preclinical evidence point to altered SERCA activity in SZ and BD. SERCA dysregulation most likely affects downstream events, such as synaptic glutamate release and LTP in excitatory neurons, likely leading to characteristic neuropsychiatric symptomatology [13, 179, 180]. Evidently, SERCA pumps synergistically drive neuropsychiatric pathology, reinforcing the notion that this protein family might be an important target for the development of future pharmacological approaches in treating schizophrenia and bipolar disorder.

6.4 A Proposed Implication of SERCA Pump Dysregulation in Neurodegenerative Disorders

6.4.1 Ca^{2+} Aberrations in Alzheimer's Disease and Parkinson's Disease

Alzheimer's disease (AD) and Parkinson's disease (PD) are the two most common progressive neurodegenerative disorders amongst aging populations with high morbidity and mortality [188–190]. The majority of AD and PD cases are sporadic, but a few early-onset cases have been attributed to familial mutations [188, 190–193]. Interestingly, neuronal loss and malformed protein aggregations are characteristic features in both AD and PD. The pathophysiological hallmarks of PD include the loss of dopamine (DA) neurons in the substantia nigra *pars compacta* (SNc) as well as the presence of intraneuronal Lewy bodies, resulting in motor deficits manifested as bradykinesia, resting tremor and rigidity [194–196]. Likewise, AD is characterized by progressive cognitive loss and dementia due to the onset of extracellular senile amyloid (Aβ) plaques, intraneuronal neurofibrillary tangles (NFT), as well as the detrimental loss of neurons and synapses predominately in the hippocampus and the cerebral cortex [197–199].

Despite the distinct clinical presentation, recent studies have revealed a crosstalk in the molecular mechanisms underlying AD and PD pathophysiology, suggesting that common molecular mechanisms drive aging brain pathology [200–203]. As extensively reviewed recently [200], oxidative stress and the generation of free radicals could be caused by abnormal accumulation of proteins with altered conformation (i.e., proteinopathy). Subsequently, proteinopathy results in impaired proteasomal degradation of toxic proteins and subsequent proteotoxicity observed in AD and PD [200, 204–208]. Additionally, proteotoxicity has been associated with mitochondrial dysfunction in both disorders. Evidence has indicated that abnormal protein aggregation could result in disruption of bioenergetics and important mitochondrial functions, such as mitophagy and fusion/fission, and to a subsequent increase of free radicals that may trigger cell apoptosis [200, 209–215].

Taking into consideration that the molecular pathogenetic mechanisms underlying AD and PD appear to be highly similar, great effort has focused on revealing whether these two disorders share common mechanisms that reinforce a cross-talk. Interestingly, decreased ER Ca^{2+} levels and subsequent increased $[Ca^{2+}]_{cyt}$ have been associated with ER-stress induced apoptosis and neurodegeneration [198, 216–220]. As discussed in several reviews [26, 217, 221–225], evidence indicates an intricate interplay between intracellular Ca^{2+} homeostasis, amyloid metabolism, neurotransmission and plasticity in the pathophysiology of AD. Increase in intracellular Ca^{2+} concentrations is considered to be secondary to the characteristic AD lesions, including the accumulation Aβ and the hyperphosphorylation of TAU protein [11, 226–230]. The aftermath of such complex interactions is believed to be rendering the neurons more vulnerable to Ca^{2+} overload, causing a major remodeling of the neural Ca^{2+} circuits, that leads to apoptosis and progressive cognitive deterioration [11, 225, 231–233]. Many in vitro and in vivo studies have supported an association between dysfunctional amyloid precursor protein (APP), Aβ plaque deposition and Ca^{2+} overload with dendritic degeneration, leading to impaired spino-dendritic Ca^{2+} signaling and disrupted plasticity [234–237].

Similarly, intracellular Ca^{2+} dyshomeostasis is believed to be critical for the elicitation of PD-related histopathological features [238–241]. The possible implication of intraneuronal Ca^{2+} signaling in PD originated by the association between increased PD-related neuronal vulnerability and decreased levels of the Ca^{2+}-buffering protein, CB-D$_{28k}$, as well as between high $[Ca^{2+}]_{cyt}$ and a two- to threefold increase in DA levels in the cytosol of SNc DAergic neurons [242, 243]. Subsequent studies supported the involvement of Ca^{2+} in PD, as overexpression of familial PD-related proteins, such as DJ-1, Parkin and α-synuclein (αSYN) enhanced the ER-mitochondrial Ca^{2+} sequestration [244, 245]. More importantly, ER stress caused by the disruption of Ca^{2+} signaling and buffering in the SNc DAergic neurons has been linked to abnormal levels of αSYN and Parkin [239, 242, 246–249]. Thus, it seems that an intricate interplay exists between elevated $[Ca^{2+}]_{cyt}$ and DA levels, as well as αSYN overexpression in triggering the selective apoptosis of SNc DAergic neurons. Aggregation of αSYN has been shown to increase $[Ca^{2+}]_{cyt}$ and caspase-3-mediated cell death in cultured neurons and human neuroblastoma cell lines [250, 251]. Interestingly, it seems that αSYN aggregates regulate $[Ca^{2+}]_{cyt}$ in a biphasic fashion [252–257]; in vitro experiments in primary hippocampal neurons from αSYN-transgenic mice overexpressing αSYN, showed that the initial decline in $[Ca^{2+}]_{cyt}$ was followed by an increase 10 days later, suggesting a sophisticated a-SYN-dependent regulation of intracellular Ca^{2+} signaling in PD [257]. In addition, Ca^{2+} binding to the αSYN carboxyl terminus has been shown to accelerate the accumulation of the annular toxic form of αSYN in vitro [247, 258]. It has been also demonstrated that elevated intracellular Ca^{2+} and oxidative stress cooperatively stimulate αSYN accumulation. Goodwin et al. reported that when oxidative stress is induced in 1321N1 cells in vitro by treatment with the SERCA inhibitor thapsigargin, accumulation of αSYN aggregates increases dramatically [259]. In addition, αSYN aggregates might associate with ER in vivo, as demonstrated by experiments using transgenic mice overexpressing WT or mutant αSYN and human brain tissue from PD patients [260]. According to Colla et al. a small subset of αSYN resides in the ER lumen leading to protein aggregation. Additionally, the initial appearance of aggregated αSYN precedes onset and increases with the progression of the disease, while ER-stress attenuation reduces αSYN aggregation and rescues disease phenotype [260]. Such observations, directly associated αSYN aggregation with ER stress and α-synucleinopathy, highlighting the importance of further investigation of ER-stress-associated mechanisms underlying PD pathology. Most importantly, clinical studies with hypertensive patients have suggested that Ca^{2+} channel blockers could decrease the risk for PD, thus suggesting intracellular Ca^{2+} regulation as a novel pharmacological approach for PD [261–265].

In the context of unraveling the molecular mechanisms underlying Ca^{2+} dysregulation in AD and PD pathogenesis, the role of SERCA in neurodegeneration has also been investigated. It is well established that SERCA pumps play an important role in oxidative stress, the generation of free radicals, and mitochondrial dysfunction, as aberrations in SERCA function lead to ER stress. Compelling studies have suggested that pharmacological activation and/or overexpression of SERCA may alleviate ER stress [266–268]. For instance, Park et al. infected mouse embryonic fibroblasts with an adenovirus encoding SERCA2b and challenged the cells with thapsigargin to induce ER stress, by blocking SERCA. Interestingly, the overexpression of SERCA2b resulted in the alleviation of ER stress and the increase of ER folding capacity [267]. ER stress-induced apoptosis may also be reduced by pharmacological activation of SERCA, as indicated by recent in vitro experiments in human embryonic cells [268]. In this study, ER stress was induced by H_2O_2 exposure in SERCA2b-expressing HEK293 cells, and viability assays were conducted in the presence and absence of a small allosteric specific SERCA activator (i.e., CDN1163). CDN1163-induced SERCA activation dramatically decreased H_2O_2-induced apoptosis, suggesting CDN1163 as a potential pharmacological treatment for ER stress-mediated cell death. SERCA activity maintains ER function and intracellular Ca^{2+} balance, ultimately preventing the activation of apoptotic pathways. Thus, it is very likely that SERCA is associated with proteotoxicity present in AD and PD.

6.4.2 Evidence for Altered SERCA Function in Alzheimer's Disease

In the quest of deciphering the role of SERCA in AD, several studies have sought for a possible correlation between fundamental disease features and SERCA activity. Indeed, studies have proposed an association between SERCA and proteins that are central in AD pathogenesis, namely presenilins (PS). Over 200 reported mutations on the genes encoding for PS1, PS2 and APP have been suggested as causatives of 30–50% of all familial AD (FAD) cases [269–273]. Of note, PS resides on the ER membrane until it is cleaved to its N- and C-terminal fragments. These fragments then travel to the plasma membrane where they act as γ-secretase units, subsequently cleaving APP to form Aβ peptides [274–276]. Given the important role of γ-secretase in cleaving APP, PS mutations disturbing APP processing, result in the formation of cytotoxic Aβ peptides. In vitro studies utilizing mouse embryonic fibroblasts and *Xenopus laevis* oocytes introduced the concept of SERCA physically associating with PS [9]. Green et al. proposed that SERCA2 function was directly relying on the presence of PS, as shown by the elevated levels of SERCA2b in double-PS null-mutant (PSDKO) neuronal cells. Surprisingly, the levels of resting $[Ca^{2+}]_{cyt}$ was elevated while the ER Ca^{2+} load was diminished in the absence of PS, indicating impaired SERCA activity that led to a compensatory increase in SERCA expression. Their hypothesis was confirmed as SERCA2b levels were rescued in PS mutant cells in the presence of either PS1 or PS2. In addition, siRNA knockdown of SERCA2b in CHO cells mimicked the Ca^{2+} phenotype of PSDKO cells, further supporting their initial observations. The study also highlighted the importance of PS in Ca^{2+} sequestration by overexpressing PS1, PS2 and SERCA2b in *X. laevis* oocytes; cytosolic Ca^{2+} sequestration was accelerated in all cells, with PS2 overxpression showing the most robust effect. Interestingly, pharmacological blockade of SERCA2 by thapsigargin prevented the PS-mediated accelerated ER filling. Further immunohistochemical and immunoprecipitation experiments on WT mouse fibroblasts showed that SERCA2b and PS1/PS2 colocalize on the ER membrane and that PS1/PS2 specifically bind to SERCA2b. Moreover, the implication of SERCA2b in Aβ production was suggested in the same study; overexpression of SERCA2b in APP-expressing CHO cells resulted in elevated Aβ production, while siRNA knockdown of SERCA2b caused a drastic decline in Aβ40 and Aβ42 production [9].

Adding to the notion that SERCA and PS physically associate, in vitro experiments in human epithelial, glioma and neuroblastoma cell lines (HEK293, KNS-42 and GOTO) indicated the formation of a complex between SERCA2 and the PS1 holoprotein [277]. Jin et al. proposed that the PS1 holoprotein may act as an anti-apoptotic agent under stress by upregulating SERCA2, and thus preventing the disruption of intracellular Ca^{2+} homeostasis [277]. Further co-immunoprecipitation data, derived from untransfected and stably PS1-expressing cells, demonstrated the interaction between SERCA2b and the PS1 holoprotein. In addition, immunohisto-chemical analysis revealed that SERCA2b and PS1-holoprotein co-localize [277]. The importance of the interaction between PS and SERCA in AD pathogenesis was further supported by in vitro studies using HeLa cells, embryonic cells (HEK293), human neuroblastoma cells (SH-SY5Y) and fibroblasts expressing PS2 [278]. Brunello et al. showed that both WT PS2 and the FAD-related PS2-T122R mutation could reduce ER Ca^{2+} uptake by affecting SERCA2 activity. Indeed, overexpression of SERCA2b in PS2-T122R-expressing SH-SY5Y cells resulted in full restoration of ER Ca^{2+} uptake rate and steady-state levels, highlighting the inhibitory activity of PS2 on SERCA activity [278].

Apart from the PS-SERCA interaction, recent studies have also indicated another possible interaction between SERCA and a novel sporadic AD-associated protein, known as the Ca^{2+} homeostasis modulator 1 (CALHM1) [218]. CALHM1 is preferentially expressed in the nervous system and is located on the neuronal plasma and ER membranes, while it is proposed to increase Aβ production by disrupting intracellular Ca^{2+} homeostasis [279]. Intriguingly, earlier in vitro experiments in CALHM1-transfected HEK-293 cells, showed that CALHM1 provokes ER stress by affecting the influx and efflux of Ca^{2+} into ER [218]; The CALHM1-transfected cells presented a drastic increase in the plasma membrane permeability to Ca^{2+}, whereas ER Ca^{2+} accumulation at the steady-state was significantly decreased. Such a reduction in ER Ca^{2+} uptake was initially hypothesized to be attributed to either enhanced ER Ca^{2+} leak by IP_3R and RyR, or diminished ER Ca^{2+} pumping by SERCA2. Further measurements of Ca^{2+} levels revealed that CALHM1 promoted ER Ca^{2+} leak, as indicated by the increased efflux of Ca^{2+} from the ER. Despite the increase in both plasma membrane-mediated Ca^{2+} influx and ER Ca^{2+} leak, the diminished intraluminal ER Ca^{2+} suggested that Ca^{2+} pumping into the ER might also be impaired. Further experiments using plasma membrane-permeabilized cells indicated that ER uptake was altered, as both Ca^{2+} transport capacity and Ca^{2+} affinity had been significantly reduced. Interestingly, the combined decline in ER pumping and increase in ER leak resulted to a 6-7-fold decrease in steady-state ER Ca^{2+} levels at physiological $[Ca^{2+}]_{cyt}$. Subsequently, Gallego-Sandin et al. proposed that the aftermath of such alterations in the ER Ca^{2+} levels could be the activation of store-operated Ca^{2+} entry, justifying the enhanced plasma membrane- mediated Ca^{2+} influx and the $[Ca^{2+}]_{cyt}$. In agreement with the notion that diminished ER Ca^{2+} promotes ER stress, the same study demonstrated the activation of ER stress-related genes in CALHM1-transfected cells. The derived data further supported the importance of maintaining SERCA activity and the detrimental consequences of dysfunctional SERCA in intracellular Ca^{2+} homeostasis, leading to AD-associated neuronal cell death [218].

Another intriguing pull-down assay employing human post-mortem brains identified that SERCA2 interacts with FE65, a cytosolic adapter protein as well as an APP-binding partner [280]. Of note, the FE65-APP interaction is believed to be critical for APP processing, apoptosis and neurite outgrowth [281–285]. Among other identified candidate proteins, SERCA2 was confirmed to interact with FE65, as supported by further co-immunoprecipitation assays from HEK293 cells overexpressing both SERCA2 and FE65. Notably, SERCA2 was found to be upregulated in primary hippocampal neurons derived from FE65/FE65L1 DKO mice, whereas HEK293 cells were more prone to thapsigargin-induced ER stress when FE65 was knocked down. According to Nensa et al. FE65 is possibly involved in the regulation of intracellular Ca^{2+} homeostasis, through its interaction with key components on the ER membrane. Taken together, it was suggested that the regulation of SERCA activity by PS may be attributed to the increased availability of APP intracellular domains (AICD) upon γ-secretase-mediated APP cleavage. Subsequently, the free AICDs bind to FE65 (or FE65 changes conformation), leading to further interaction with binding partners, such as SERCA2 [280].

As demonstrated by in vitro data, SERCA possibly holds an important role in AD pathology. Recently, promising in vivo experiments proposed that SERCA activation could alleviate AD-like pathology in mice [286]. Krajnak et al. employed the allosteric SERCA activator, CDN1163, which had been previously shown to increase ER Ca^{2+} uptake and SERCA activity in vivo [268]. To test whether CDN1163 had a therapeutic effect, double-transgenic APPSwe/PSEN1dE9 mice (i.e., transgenic mice carrying both mutant APP and PS1) were administered CDN1163 (10 mg/kg) for 4 weeks and were then subjected to a series of behavioral tests to assess learning and memory (i.e., Morris Water Maze; MWM), depressive-like behavior (i.e., tail suspension test; TST) and motor coordination (i.e., Rotarod). Strikingly, CDN1163-treated mice presented enhanced memory retention, better motor coordination and an antidepressant-like phenotype, as compared to vehicle-treated controls. These data collectively suggested SERCA as a therapeutic target in AD pathogenesis and CDN1163 as a promising drug candidate [286].

6.4.3 SERCA Dysregulation as a Proposed Mechanism Underlying Parkinson's Disease Pathophysiology

Similar to AD, SERCA activity has recently been linked to PD pathophysiology, further supporting the pivotal role of the pump in neurodegeneration. Growing evidence suggests that the intricate interplay between SERCA activity, αSYN aggregation and ER stress in neurons could be the main driving force of PD pathogenesis. Of note, several reviews have discussed the multifunctional properties of αSYN, including the regulation of synaptic transmission, synaptic function and plasticity, as well as the organization of membrane activity through interaction with intracellular membranes and lipid surfaces [287, 288]. Furthermore, synucleinopathies have been established as the genetic component of PD, as extensively discussed in a recent excellent review by Nussbaum; autosomal-dominant familial PD cases have been attributed to mutations on the αSYN-encoding *SNCA* gene, while *SNCA* gene variations have been linked with sporadic PD cases [287]. Previous studies have associated abnormal αSYN aggregation with ER stress, intracellular Ca^{2+} signaling alterations and subsequent neuronal cell death (reviewed in [289]). However, the association of SERCA with αSYN, and its subsequent implication in PD pathophysiology, has been only recently investigated. An interesting study showed that 6-hydroxy-DA (6-OHDA)-mediated oxidative stress affects SERCA2 activity in rats, and that physical exercise may rescue the PD phenotype, supporting the idea that SERCA dysregulation might be implicated in PD pathogenesis [290]. Prior to striatal stereotaxic injections of 6-OHDA, rats were challenged with an incremental running program on a treadmill for a total period of 8 weeks. One-week post-surgery, all animals were subjected to the Rotational test to assess ipsiversive and contraversive movement, and to detect any rotational asymmetry. Even though both untrained and trained 6-OHDA-lesioned rats presented high asymmetry, the trained

6-OHDA-lesioned group presented a marked decline in number of rotations when compared to their untrained counterparts, indicating that physical exercise could have a protective effect against PD-mediated motor deficits. Interestingly, a significant increase in αSYN levels accompanied by a decline in SERCA2 was observed in untrained 6-OHDA-lesioned rats as opposed to the non-lesioned group, suggesting a correlation between αSYN accumulation and SERCA2 downregulation. However, trained 6-OHDA-lesioned rats presented full restoration of αSYN and SERCA2 protein levels, further highlighting the potential neuroprotective effect of physical exercise. Taken together, the investigators proposed that abnormal accumulation of αSYN could lead to intracellular Ca^{2+} dyshomeostasis possibly *via* downregulation of SERCA2 activity, and that physical exercise could indirectly prevent these events most likely by reducing ER stress through restoring SERCA activity [290].

It has been previously shown that αSYN aggregation leads to biphasic alterations in $[Ca^{2+}]_{cyt}$ [252–256]. The hypothesized SERCA/αSYN interaction has been recently confirmed in vitro; aggregated, but not monomeric, αSYN specifically binds to SERCA, stimulating the activity of the pump and affecting downstream processes that drive intracellular Ca^{2+} dysregulation [257]. Betzer et al. investigated whether αSYN physically associates with SERCA using SH-SY5Y cells and primary hippocampal neurons from αSYN-overexpressing transgenic mice, human brain tissue and *C. elegans*. Interestingly, co-immunoprecipitation experiments showed that oligomeric αSYN preferentially binds to the E1 conformation of SERCA, indicating a high degree of structural specificity for the SERCA/αSYN interaction. Further, Proximity Ligation Assay (PLA) experiments in SH-SY5Y cells showed that the two proteins are in close proximity, confirming the proposed interaction, while inhibiting the abnormal aggregation of αSYN abolished the SERCA/αSYN interaction. Taken together, these data confirmed that SERCA specifically binds to oligomeric αSYN. In vitro biochemical experiments also revealed that aggregated αSYN increase SERCA-dependent ATP hydrolysis, thus accelerating Ca^{2+} transfer across the ER membrane [257]. Of note, β-synuclein (βSYN) and Tau proteins did not influence SERCA activity, highlighting the high specificity of this interaction. Further in vitro experiments using SERCA1a-containing microsomes showed that only αSYN aggregates (and not monomeric αSYN species) increased SERCA-dependent Ca^{2+} uptake and SERCA's dephosphorylation rate, suggesting that the aggregated protein form regulates the function of SERCA. Interestingly, cyclopiazonic acid (CPA)-mediated SERCA inhibition in SH-SY5Y cells resulted to the normalization of the elevated $[Ca^{2+}]_{cyt}$, reducing αSYN-mediated apoptosis via the reduction of total αSYN levels by 10% and αSYN/SERCA interaction by 50%. In the same study, it was revealed that αSYN aggregation may enhance dendritic loss in dopaminergic neurons by 40% in *C. elegans*, and that this could be rescued by CPA-induced SERCA inhibition [257]. Betzer et al. also reported that SERCA and αSYN oligomers interact in the brain of patients affected by PD, Lewy body dementia and multiple system atrophy (MSA). Taken together, the investigators proposed that at the early stages of disease αSYN accumulation most likely stimulates SERCA activity *via* direct interaction, affecting three key SERCA functions; Ca^{2+} pumping, ATP hydrolysis and dephosphorylation. Subsequently Ca^{2+} uptake into the ER is increased and $[Ca^{2+}]_{cyt}$ is diminished, ultimately altering Ca^{2+} homeostasis and promoting cell death [257].

Strikingly, in vivo experiments have further supported the crucial role of SERCA activity in PD pathology [10]. Dahl demonstrated that the SERCA activator CDN1163 could alleviate dyskinesia in the 6-OHDA-lesioned rat model of PD. Adult male Wistar rats were unilaterally injected with 6-OHDA in the left SNc, and on day 11 post-surgery they were administered either L-DOPA (i.e., a current PD drug treatment)- or the SERCA activator CDN1163. Once the animals were pharmacologically challenged, they were subjected to three standardized akinesia tests; the Initiation Time (IT) test, the stepping test, and the cylinder test. Of note, the same tests were also conducted prior to surgery to establish a baseline response. As expected 6-OHDA-treated rats showed a delay in stepping initiation in the IT test that was reversed by L-DOPA treatment. Importantly, CDN1163 treatment also reversed 6-OH-DA-induced delay in stepping initiation. The stepping test was then employed to evaluate the degree of contralateral limb akinesia by measuring the number of adjusting steps the animal performed while moving sideways [291]. As expected, the lesioned animals presented a significantly reduced number of adjusting steps, while treatment with either L-DOPA or CDN1163 completely rescued this phenotype. Finally, the cylinder test aimed to assess spontaneous forelimb lateralization by scoring for akinesia of the contralateral forelimb *via* measuring the number of contacts made with the cylinder wall [291, 292]. The 6-OHDA lesioned group showed a drastically decreased number of contacts. Treatment with L-DOPA restored forelimb lateralization to almost baseline levels, while CDN1163-treated rats presented with a significantly increased number of contacts, restoring forelimb lateralization to approximately 50% of baseline. Observing that administration of CDN1163 exerted similar therapeutic effects to L-DOPA treatment, the study proposed SERCA activation as a potential therapeutic target for PD.

6.5   Conclusions

It is well established that Ca^{2+} is a fundamental signaling molecule for cell survival and function, and that SERCA is a gatekeeper of intracellular Ca^{2+} homeostasis. Aberrations in the finely-tuned intraneuronal Ca^{2+} homeostasis may impose detrimental effects, leading to the emergence of brain pathology. Despite the extensive knowledge on SERCA distribution and function, the exact regulatory networks and mechanisms that operate in the brain are still elusive. In this context, we have presented current evidence regarding SERCA's involvement in the pathophysiology of neuropsychiatric and neurodegenerative disorders (for summary, see Fig. 6.3). Although only DD has been directly linked to SERCA2 mutations, compelling data supports that brain pathology is caused by alterations in SERCA activity, and that SERCA synergistically contributes to neuronal pathogenesis. It is yet unclear to what extent SERCA is involved and what upstream or downstream events participate in determining the fate of nerve cells. However, targeting SERCA activity could

SERCA isoform	SERCA modification	Study model	Associated disease
SERCA2	Missense mutation in stalk domain	Human tissue	DD with concurrent SZ (Takeichi, Sugiura et al. 2016, Noda, Takeichi et al. 2016)
SERCA2	Altered-splicing mutation in acceptor site	Human tissue	DD with concurrent BD (Nakamura et al., 2016)
SERCA2	Downregulation	Human platelets	BD (Hough, Lu et al. 1999)
SERCA2	Upregulation in hippocampus, cortex and cerebellum	Df(16)1/+ mice Human brain tissue	DiGeorge-related SZ (Earls, Bayazitov et al. 2010, Earls, Fricke et al. 2012)
SERCA2	Downregulation in prefrontal cortex	NLVH rats	SZ (Genis-Mendoza, Gallegos-Silva et al. 2018)
SERCA3	Downregulation in cortex/ Upregulation in hippocampus, striatum and cerebellum	Ketamine-induced psychotic rats	SZ (Lisek, Boczek et al. 2016)
SERCA2	Upregulation	PSDKO mouse neuronal cells	AD (Green, Demuro et al. 2008)
		APP-expressing CHO cells	
SERCA2	Downregulation	PS2-expressing SH-SY5Y cells CALHM1-transfected cells APPSwe/PSEN1dE9 mice	AD (Brunello, Zampese et al. 2009, Gallego-Sandin, Alonso et al. 2011, Krajnak and Dahl, 2018)
SERCA2	Upregulation	Primary hippocampal neurons of FE65/FE65L1 DKO mice	AD (Nensa, Neumann et al. 2014)
SERCA2	Downregulation	6-OHDA lesioned rats	PD (Tuon et al., 2012, Dahl, 2017)
SERCA2	Upregulation	αSYN-overexpressing cell lines	PD (Betzer et al., 2018)

Fig. 6.3 Clinical and preclinical experimental evidence of altered SERCA expression and/or activity in neuropsychiatric and neurodegenerative disorders. SERCA2 mutations have been reported in Darier's disease (DD) patients with concurrent schizophrenia (SZ) or bipolar disease (BD). SERCA2 and SERCA3 alterations have been associated with SZ, BD, Alzheimer's disease (AD) and Parkinson's disease (PD)

reveal more protein interactions, shedding light on the neuromolecular circuitry involved in brain pathophysiology. Most importantly, advancing knowledge on the neuronal function of SERCA may contribute to the future development of safer and more effective therapeutic strategies to combat such incapacitating disorders.

Acknowledgements A.B. was supported by the *University of Dayton (UD) Graduate School* and by the *UD Office for Graduate Affairs* through the *Graduate Student Summer Fellowship* (GSSF) Program. J.S. was supported by a *Barry Goldwater Scholarship in Excellence and Education Award*, a *Biology Department Lancaster-McDougall Award, a CAS Dean Fellowship*, the *Berry Summer Thesis Institute*, and the *UD Honors Program*. P.M.P. was supported by an *inaugural STEM Catalyst grant and Start-up funding from UD*, as well as by *Research Council Seed Grants (RCSG) from the University of Dayton Research Institute (UDRI)*; this work was supported by the National Institute Of Neurological Disorders and Stroke of the National Institutes of Health under award number R03NS109836 (to P.M.P.). Funding sponsors had no further role in study design; in the collection, analysis and interpretation of data; in the writing of the report; and in the decision to submit the article for publication.

Conflict of Interest None.

References

1. Brini M et al (2014) Neuronal calcium signaling: function and dysfunction. Cell Mol Life Sci 71(15):2787–2814
2. Berridge MJ, Lipp P, Bootman MD (2000) The versatility and universality of calcium signalling. Nat Rev Mol Cell Biol 1(1):11–21
3. Orrenius S, Zhivotovsky B, Nicotera P (2003) Regulation of cell death: the calcium–apoptosis link. Nat Rev Mol Cell Biol 4(7):552–565
4. Berridge MJ, Bootman MD, Roderick HL (2003) Calcium signalling: dynamics, homeostasis and remodelling. Nat Rev Mol Cell Biol 4(7):517–529
5. Zucker RS (1999) Calcium-and activity-dependent synaptic plasticity. Curr Opin Neurobiol 9(3):305–313
6. Lyons MR, West AE (2011) Mechanisms of specificity in neuronal activity-regulated gene transcription. Prog Neurobiol 94(3):259–295
7. Neher E, Sakaba T (2008) Multiple roles of calcium ions in the regulation of neurotransmitter release. Neuron 59(6):861–872
8. Carafoli E (2003) The calcium-signalling saga: tap water and protein crystals. Nat Rev Mol Cell Biol 4(4):326–332
9. Green KN et al (2008) SERCA pump activity is physiologically regulated by presenilin and regulates amyloid beta production. J Gen Physiol 132(2):i1
10. Dahl R (2017) A new target for Parkinson's disease: small molecule SERCA activator CDN1163 ameliorates dyskinesia in 6-OHDA-lesioned rats. Bioorg Med Chem 25(1):53–57
11. LaFerla FM (2002) Calcium dyshomeostasis and intracellular signalling in Alzheimer's disease. Nat Rev Neurosci 3(11):862–872
12. Bezprozvanny I, Mattson MP (2008) Neuronal calcium mishandling and the pathogenesis of Alzheimer's disease. Trends Neurosci 31(9):454–463
13. Earls LR et al (2010) Dysregulation of presynaptic calcium and synaptic plasticity in a mouse model of 22q11 deletion syndrome. J Neurosci 30(47):15843–15855
14. Jacobsen NJ et al (1999) ATP2A2 mutations in Darier's disease and their relationship to neuropsychiatric phenotypes. Hum Mol Genet 8(9):1631–1636
15. Grienberger C, Konnerth A (2012) Imaging calcium in neurons. Neuron 73(5):862–885
16. Fucile S (2004) Ca^{2+} permeability of nicotinic acetylcholine receptors. Cell Calcium 35(1):1–8
17. Berridge MJ, Bootman MD, Lipp P (1998) Calcium – a life and death signal. Nature 395(6703):645–648
18. Schwaller B (2010) Cytosolic Ca^{2+} buffers. Cold Spring Harb Perspect Biol 2(11):a004051
19. Vandecaetsbeek I et al (2009) Structural basis for the high Ca^{2+} affinity of the ubiquitous SERCA2b Ca^{2+} pump. Proc Natl Acad Sci 106(44):18533–18538
20. Morth JP et al (2010) A structural overview of the plasma membrane Na^+,K^+-ATPase and H^+-ATPase ion pumps. Nat Rev Mol Cell Biol 12:60
21. Toyoshima C (2008) Structural aspects of ion pumping by Ca^{2+}-ATPase of sarcoplasmic reticulum. Arch Biochem Biophys 476(1):3–11
22. Bublitz M et al (2010) In and out of the cation pumps: P-type ATPase structure revisited. Curr Opin Struct Biol 20(4):431–439
23. Bobe R et al (2005) How many Ca^{2+} ATPase isoforms are expressed in a cell type? A growing family of membrane proteins illustrated by studies in platelets. Platelets 16(3–4):133–150
24. Dode L et al (2003) Dissection of the functional differences between sarco (endo) plasmic reticulum Ca^{2+}-ATPase (SERCA) 1 and 2 isoforms and characterization of Darier disease (SERCA2) mutants by steady-state and transient kinetic analyses. J Biol Chem 278: 47877–47889
25. Dode L et al (2002) Dissection of the functional differences between sarco (endo) plasmic reticulum Ca^{2+}-ATPase (SERCA) 1 and 3 isoforms by steady-state and transient kinetic analyses. J Biol Chem 277(47):45579–45591

26. Britzolaki A et al (2018) The SERCA2: a gatekeeper of neuronal calcium homeostasis in the brain. Cell Mol Neurobiol 38:981–994
27. Lytton J, MacLennan DH (1988) Molecular cloning of cDNAs from human kidney coding for two alternatively spliced products of the cardiac Ca^{2+}-ATPase gene. J Biol Chem 263(29):15024–15031
28. Lytton J et al (1992) Functional comparisons between isoforms of the sarcoplasmic or endoplasmic reticulum family of calcium pumps. J Biol Chem 267(20):14483–14489
29. MacLennan DH et al (1985) Amino-acid sequence of a Ca^{2+} + Mg2+–dependent ATPase from rabbit muscle sarcoplasmic reticulum, deduced from its complementary DNA sequence. Nature 316(6030):696–700
30. Dally S et al (2010) Multiple and diverse coexpression, location, and regulation of additional SERCA2 and SERCA3 isoforms in nonfailing and failing human heart. J Mol Cell Cardiol 48(4):633–644
31. Dally S et al (2006) Ca^{2+}-ATPases in non-failing and failing heart: evidence for a novel cardiac sarco/endoplasmic reticulum Ca^{2+}-ATPase 2 isoform (SERCA2c). Biochem J 395(2):249–258
32. MacLennan DH (1970) Purification and properties of an adenosine triphosphatase from sarcoplasmic reticulum. J Biol Chem 245(17):4508–4518
33. Toyoshima C, Inesi G (2004) Structural basis of ion pumping by Ca^{2+}-ATPase of the sarcoplasmic reticulum. Annu Rev Biochem 73:269–292
34. Toyoshima C et al (2000) Crystal structure of the calcium pump of sarcoplasmic reticulum at 2.6 Å resolution. Nature 405(6787):647–655
35. Abu-Abed M et al (2002) Characterization of the ATP-binding domain of the sarco (endo) plasmic reticulum Ca^{2+}-ATPase: probing nucleotide binding by multidimensional NMR. Biochemistry 41(4):1156–1164
36. Toyoshima C, Nomura H (2002) Structural changes in the calcium pump accompanying the dissociation of calcium. Nature 418(6898):605–611
37. Smolin N, Robia SL (2015) A structural mechanism for calcium transporter headpiece closure. J Phys Chem B 119(4):1407–1415
38. Brini M, Carafoli E, Cali T (2017) The plasma membrane calcium pumps: focus on the role in (neuro)pathology. Biochem Biophys Res Commun 483(4):1116–1124
39. Møller JV et al (2005) The structural basis for coupling of Ca^{2+} transport to ATP hydrolysis by the sarcoplasmic reticulum Ca^{2+}-ATPase. J Bioenerg Biomembr 37(6):359–364
40. Guerini D (1998) The Ca^{2+} pumps and the Na^+/Ca^{2+} exchangers. Biometals 11(4):319–330
41. Zhang P et al (1998) Structure of the calcium pump from sarcoplasmic reticulum at 8-A resolution. Nature 392(6678):835–839
42. Gelebart P et al (2003) Identification of a new SERCA2 splice variant regulated during monocytic differentiation. Biochem Biophys Res Commun 303(2):676–684
43. Gunteski-Hamblin AM, Greeb J, Shull GE (1988) A novel Ca^{2+} pump expressed in brain, kidney, and stomach is encoded by an alternative transcript of the slow-twitch muscle sarcoplasmic reticulum Ca-ATPase gene. Identification of cDNAs encoding Ca^{2+} and other cation-transporting ATPases using an oligonucleotide probe derived from the ATP-binding site. J Biol Chem 263(29):15032–15040
44. Lytton J et al (1989) Molecular cloning of the mammalian smooth muscle sarco(endo)plasmic reticulum Ca^{2+}-ATPase. J Biol Chem 264(12):7059–7065
45. Zarain-Herzberg A, MacLennan D, Periasamy M (1990) Characterization of rabbit cardiac sarco (endo) plasmic reticulum Ca2 (+)-ATPase gene. J Biol Chem 265(8):4670–4677
46. Brandl CJ et al (1986) Two Ca^{2+} ATPase genes: homologies and mechanistic implications of deduced amino acid sequences. Cell 44(4):597–607
47. Dally S et al (2009) Multiple and diverse coexpression, location, and regulation of additional SERCA2 and SERCA3 isoforms in nonfailing and failing human heart. J Mol Cell Cardiol 48:633–644
48. Periasamy M, Kalyanasundaram A (2007) SERCA pump isoforms: their role in calcium transport and disease. Muscle Nerve 35(4):430–442

49. Hao L, Rigaud J-L, Inesi G (1994) Ca^{2+}/H^+ countertransport and electrogenicity in proteoliposomes containing erythrocyte plasma membrane Ca-ATPase and exogenous lipids. J Biol Chem 269(19):14268–14275

50. Yu X et al (1993) H^+ countertransport and electrogenicity of the sarcoplasmic reticulum Ca^{2+} pump in reconstituted proteoliposomes. Biophys J 64(4):1232–1242

51. Salvador JM et al (1998) Ca^{2+} transport by reconstituted synaptosomal ATPase is associated with H^+ countertransport and net charge displacement. J Biol Chem 273(29):18230–18234

52. Brini M, Carafoli E (2009) Calcium pumps in health and disease. Physiol Rev 89(4):1341–1378

53. Hasselbach W, Makinose M (1961) The calcium pump of the "relaxing granules" of muscle and its dependence on ATP-splitting. Biochem Z 333:518–528

54. Lee C-H et al (2002) Ca^{2+} oscillations, gradients, and homeostasis in vascular smooth muscle. Am J Phys Heart Circ Phys 282(5):H1571–H1583

55. Dyla M et al (2018) Dynamics of P-type ATPase transport cycle revealed by single-molecule FRET. Biophys J 114(3):559a

56. Carafoli E, Brini M (2000) Calcium pumps: structural basis for and mechanism of calcium transmembrane transport. Curr Opin Chem Biol 4(2):152–161

57. Olesen C et al (2004) Dephosphorylation of the calcium pump coupled to counterion occlusion. Science 306(5705):2251–2255

58. Mueller B et al (2004) SERCA structural dynamics induced by ATP and calcium. Biochemistry 43(40):12846–12854

59. Verboomen H et al (1994) The functional importance of the extreme C-terminal tail in the gene 2 organellar Ca^{2+}-transport ATPase (SERCA2a/b). Biochem J 303(Pt 3):979–984

60. Martin V et al (2002) Three novel Sarco/endoplasmic reticulum Ca^{2+}-ATPase (SERCA) 3 isoforms expression, regulation, and function of the members of the SERCA3 family. J Biol Chem 277(27):24442–24452

61. Bobe R et al (2004) Identification, expression, function, and localization of a novel (sixth) isoform of the human sarco/endoplasmic reticulum Ca^{2+} ATPase 3 gene. J Biol Chem 279(23):24297–24306

62. Brandl CJ et al (1987) Adult forms of the Ca^{2+} ATPase of sarcoplasmic reticulum. Expression in developing skeletal muscle. J Biol Chem 262(8):3768–3774

63. Korczak B et al (1988) Structure of the rabbit fast-twitch skeletal muscle Ca^{2+}-ATPase gene. J Biol Chem 263(10):4813–4819

64. Burk SE et al (1989) cDNA cloning, functional expression, and mRNA tissue distribution of a third organellar Ca^{2+} pump. J Biol Chem 264(31):18561–18568

65. Wu KD et al (1995) Localization and quantification of endoplasmic reticulum ca(2+)-ATPase isoform transcripts. Am J Phys 269(3 Pt 1):C775–C784

66. Wuytack F et al (1989) Smooth muscle expresses a cardiac/slow muscle isoform of the $Ca^{2+}-$transport ATPase in its endoplasmic reticulum. Biochem J 257(1):117–123

67. Lompre A-M et al (1989) Characterization and expression of the rat heart sarcoplasmic reticulum Ca^{2+}-ATPase mRNA. FEBS Lett 249(1):35–41

68. Kimura T et al (2005) Altered mRNA splicing of the skeletal muscle ryanodine receptor and sarcoplasmic/endoplasmic reticulum Ca^{2+}-ATPase in myotonic dystrophy type 1. Hum Mol Genet 14(15):2189–2200

69. Miller KK et al (1991) Localization of an endoplasmic reticulum calcium ATPase mRNA in rat brain by in situ hybridization. Neuroscience 43(1):1–9

70. Sepulveda MR, Hidalgo-Sanchez M, Mata AM (2004) Localization of endoplasmic reticulum and plasma membrane Ca^{2+}-ATPases in subcellular fractions and sections of pig cerebellum. Eur J Neurosci 19(3):542–551

71. Baba-Aissa F et al (1996) Distribution of the organellar Ca^{2+} transport ATPase SERCA2 isoforms in the cat brain. Brain Res 743(1–2):141–153

72. Campbell AM, Wuytack F, Fambrough DM (1993) Differential distribution of the alternative forms of the sarcoplasmic/endoplasmic reticulum Ca^{2+}-ATPase, SERCA2b and SERCA2a, in the avian brain. Brain Res 605(1):67–76

73. Plessers L et al (1991) A study of the organellar Ca2(+)-transport ATPase isozymes in pig cerebellar Purkinje neurons. J Neurosci 11(3):650–656
74. Morita M, Kudo Y (2010) Growth factors upregulate astrocyte $[Ca^{2+}]i$ oscillation by increasing SERCA2b expression. Glia 58(16):1988–1995
75. Salvador JM et al (2001) Distribution of the intracellular Ca^{2+}-ATPase isoform 2b in pig brain subcellular fractions and cross-reaction with a monoclonal antibody raised against the enzyme isoform. J Biochem 129(4):621–626
76. Dode L et al (1996) cDNA cloning, expression and chromosomal localization of the human sarco/endoplasmic reticulum Ca (2)-ATPase 3 gene. Biochem J 318(2):689–699
77. Wuytack F et al (1994) A sarco/endoplasmic reticulum Ca^{2+}-ATPase 3-type Ca^{2+} pump is expressed in platelets, in lymphoid cells, and in mast cells. J Biol Chem 269(2):1410–1416
78. KOVÁCS T et al (2001) All three splice variants of the human sarco/endoplasmic reticulum Ca^{2+}-ATPase 3 gene are translated to proteins: a study of their co-expression in platelets and lymphoid cells. Biochem J 358(3):559–568
79. Ait-Ghezali L et al (2014) Loss of endoplasmic reticulum calcium pump expression in choroid plexus tumours. Neuropathol Appl Neurobiol 40(6):726–735
80. Pottorf W et al (2001) Function of SERCA mediated calcium uptake and expression of SERCA3 in cerebral cortex from young and old rats. Brain Res 914(1–2):57–65
81. Baba-Aïssa F et al (1996) Purkinje neurons express the SERCA3 isoform of the organellar type Ca^{2+}−transport ATPase. Mol Brain Res 41(1–2):169–174
82. Kessler RC et al (2007) Age of onset of mental disorders: a review of recent literature. Curr Opin Psychiatry 20(4):359–364
83. Kessler RC et al (2005) Lifetime prevalence and age-of-onset distributions of DSM-IV disorders in the National Comorbidity Survey Replication. Arch Gen Psychiatry 62(6):593–602
84. Murray CJ, Lopez AD (1997) Global mortality, disability, and the contribution of risk factors: global burden of disease study. Lancet 349(9063):1436–1442
85. Murray CJ, Lopez AD (1997) Alternative projections of mortality and disability by cause 1990–2020: global burden of disease study. Lancet 349(9064):1498–1504
86. Whiteford HA et al (2013) Global burden of disease attributable to mental and substance use disorders: findings from the global burden of disease study 2010. Lancet 382(9904):1575–1586
87. Ohi K et al (2017) Cognitive clustering in schizophrenia patients, their first-degree relatives and healthy subjects is associated with anterior cingulate cortex volume. NeuroImage Clin 16:248–256
88. Purcell S, International Schizophrenia Consortium, Wray NR, Stone JL, Visscher PM, O'Donovan MC, Sullivan PF, Sklar P (2009) Common polygenic variation contributes to risk of schizophrenia and bipolar disorder. Nature 460:748–752
89. Grunze H (2015) Bipolar disorder. In: Neurobiology of brain disorders. Elsevier, pp 655–673
90. Miller S et al (2016) Mixed depression in bipolar disorder: prevalence rate and clinical correlates during naturalistic follow-up in the Stanley bipolar network. Am J Psychiatr 173(10):1015–1023
91. Allardyce J et al (2018) Association between schizophrenia-related polygenic liability and the occurrence and level of mood-incongruent psychotic symptoms in bipolar disorder. JAMA Psychiat 75(1):28–35
92. Jimerson DC et al (1979) CSF calcium: clinical correlates in affective illness and schizophrenia. Biol Psychiatry 14:37–51
93. Lidow MS (2003) Calcium signaling dysfunction in schizophrenia: a unifying approach. Brain Res Rev 43(1):70–84
94. Lajtha A, Tettamanti G, Goracci G (2009) Handbook of neurochemistry and molecular neurobiology. Springer, New York
95. Hertzberg L, Domany E (2018) *Commentary:* Integration of gene expression and GWAS results supports involvement of calcium signaling in Schizophrenia. J Ment Health Clin Psychol 2(3):5–7

96. Woon PS et al (2017) CACNA1C genomewide supported psychosis genetic variation affects cortical brain white matter integrity in Chinese patients with schizophrenia. J Clin Psychiatry 75(11):e1284–e1290

97. Dubovsky SL et al (1991) Elevated platelet intracellular calcium concentration in bipolar depression. Biol Psychiatry 29(5):441–450

98. Kusumi I, Koyama T, Yamashita I (1992) Thrombin-induced platelet calcium mobilization is enhanced in bipolar disorders. Biol Psychiatry 32(8):731–734

99. Kusumi I, Koyama T, Yamashita I (1994) Serotonin-induced platelet intracellular calcium mobilization in depressed patients. Psychopharmacology 113(3–4):322–327

100. Suzuki K et al (2003) Altered 5-HT-induced calcium response in the presence of staurosporine in blood platelets from bipolar disorder patients. Neuropsychopharmacology 28(6):1210–1214

101. Suzuki K et al (2004) Effects of lithium and valproate on agonist-induced platelet intracellular calcium mobilization: relevance to myosin light chain kinase. Prog Neuro-Psychopharmacol Biol Psychiatry 28(1):67–72

102. Suzuki K et al (2001) Serotonin-induced platelet intracellular calcium mobilization in various psychiatric disorders: is it specific to bipolar disorder? J Affect Disord 64(2–3):291–296

103. Manji HK, Moore GJ, Chen G (2000) Clinical and preclinical evidence for the neurotrophic effects of mood stabilizers: implications for the pathophysiology and treatment of manic–depressive illness. Biol Psychiatry 48(8):740–754

104. Chen G et al (1999) The mood-stabilizing agents Lithium and valproate RobustlIncrease the levels of the neuroprotective protein bcl-2 in the CNS. J Neurochem 72(2):879–882

105. Chen R-W, Chuang D-M (1999) Long term lithium treatment suppresses p53 and Bax expression but increases Bcl-2 expression A prominent role in neuroprotection against excitotoxicity. J Biol Chem 274(10):6039–6042

106. Manji HK et al (1996) Regulation of signal transduction pathways by mood-stabilizing agents: implications for the delayed onset of therapeutic efficacy. J Clin Psychiatry 57:34–46. discussion 47-8

107. Manji HK, Moore GJ, Chen G (1999) Lithium at 50: have the neuroprotective effects of this unique cation been overlooked? Biol Psychiatry 46(7):929–940

108. Manji HK, Moore GJ, Chen G (2000) Lithium up-regulates the cytoprotective protein Bcl-2 in the CNS in vivo: a role for neurotrophic and neuroprotective effects in manic depressive illness. J Clin Psychiatry 61:82–96

109. Wang J-F, Bown C, Young LT (1999) Differential display PCR reveals novel targets for the mood-stabilizing drug valproate including the molecular chaperone GRP78. Mol Pharmacol 55(3):521–527

110. Bhat S et al (2012) CACNA1C (Cav1.2) in the pathophysiology of psychiatric disease. Prog Neurobiol 99(1):1–14

111. Gonzalez S et al (2013) Suggestive evidence for association between L-type voltage-gated calcium channel (CACNA1C) gene haplotypes and bipolar disorder in Latinos: a family-based association study. Bipolar Disord 15(2):206–214

112. Isaac C, Januel D (2016) Neural correlates of cognitive improvements following cognitive remediation in schizophrenia: a systematic review of randomized trials. Socioaffect Neurosci Psychol 6(1):30054

113. Li W et al (2018) A molecule-based genetic association approach implicates a range of voltage-gated calcium channels associated with schizophrenia. Am J Med Genet B Neuropsychiatr Genet 177(4):454–467

114. Schizophrenia Working Group of the Psychiatric Genomics, C et al (2014) Biological insights from 108 schizophrenia-associated genetic loci. Nature 511:421–427

115. Li Z et al (2017) Genome-wide association analysis identifies 30 new susceptibility loci for schizophrenia. Nat Genet 49(11):1576–1583

116. Ferreira MAR et al (2008) Collaborative genome-wide association analysis supports a role for ANK3 and CACNA1C in bipolar disorder. Nat Genet 40(9):1056–1058

117. Sklar P et al (2011) Large-scale genome-wide association analysis of bipolar disorder identifies a new susceptibility locus near ODZ4. Nat Genet 43(10):977–983
118. Lencz T, Malhotra A (2015) Targeting the schizophrenia genome: a fast track strategy from GWAS to clinic. Mol Psychiatry 20(7):820–826
119. Craddock N et al (1994) Familial cosegregation of major affective disorder and Darier's disease (keratosis follicularis). Br J Psychiatry 164(3):355–358
120. Burge SM, Wilkinson JD (1992) Darier-white disease: a review of the clinical features in 163 patients. J Am Acad Dermatol 27(1):40–50
121. Ringpfeil F et al (2001) Darier disease–novel mutations in ATP2A2 and genotype-phenotype correlation. Exp Dermatol 10(1):19–27
122. Judge M, McLean W, Munro C (2010) Disorders of keratinization. In: Rook's textbook of dermatology, vol 1, pp 1–122
123. Munro C (1992) The phenotype of Darier's disease: penetrance and expressivity in adults and children. Br J Dermatol 127(2):126–130
124. Nellen RGL et al (2017) Mendelian disorders of cornification caused by defects in intracellular calcium pumps: mutation update and database for variants in ATP2A2 and ATP2C1 associated with Darier disease and Hailey–Hailey disease. Hum Mutat 38(4):343–356
125. Green EK et al (2013) Novel ATP2A2 mutations in a large sample of individuals with D arier disease. J Dermatol 40(4):259–266
126. Sakuntabhai A et al (1999) Mutations in ATP2A2, encoding a Ca^{2+} pump, cause Darier disease. Nat Genet 21(3):271–277
127. Ahn W et al (2003) Multiple effects of SERCA2b mutations associated with Darier's disease. J Biol Chem 278(23):20795–20801
128. Sato K et al (2004) Distinct types of abnormality in kinetic properties of three Darier disease-causing sarco (endo) plasmic reticulum Ca^{2+}-ATPase mutants that exhibit normal expression and high Ca^{2+} transport activity. J Biol Chem 279(34):35595–35603
129. Miyauchi Y et al (2006) Comprehensive analysis of expression and function of 51 sarco (endo) plasmic reticulum Ca^{2+}-ATPase mutants associated with Darier disease. J Biol Chem 281(32):22882–22895
130. Foggia L et al (2006) Activity of the hSPCA1 Golgi Ca^{2+} pump is essential for Ca^{2+}−mediated Ca^{2+} response and cell viability in Darier disease. J Cell Sci 119(4):671–679
131. Leinonen P et al (2005) Keratinocytes cultured from patients with Hailey–Hailey disease and Darier disease display distinct patterns of calcium regulation. Br J Dermatol 153(1):113–117
132. Hetz C (2012) The unfolded protein response: controlling cell fate decisions under ER stress and beyond. Nat Rev Mol Cell Biol 13(2):89–102
133. Savignac M et al (2014) SERCA2 dysfunction in Darier disease causes endoplasmic reticulum stress and impaired cell-to-cell adhesion strength: rescue by Miglustat. J Investig Dermatol 134(7):1961–1970
134. Celli A et al (2011) Endoplasmic reticulum Ca^{2+} depletion activates XBP1 and controls terminal differentiation in keratinocytes and epidermis. Br J Dermatol 164(1):16–25
135. Engin B et al (2015) Darier disease: a fold (intertriginous) dermatosis. Clin Dermatol 33(4):448–451
136. Suryawanshi H et al (2017) Darier disease: a rare genodermatosis. J Oral Maxillofac Pathol 21(2):321–321
137. Cheour M et al (2009) Darier's disease: an evaluation of its neuropsychiatric component. Encéphale 35(1):32–35
138. Wang SL et al (2002) Darier's disease associated with bipolar affective disorder: a case report. Kaohsiung J Med Sci 18(12):622–626
139. Gordon-Smith K et al (2010) The neuropsychiatric phenotype in Darier disease. Br J Dermatol 163(3):515–522
140. Jones I et al (2002) Evidence for familial cosegregation of major affective disorder and genetic markers flanking the gene for Darier's disease. Mol Psychiatry 7(4):424–427

141. Cederlöf M et al (2015) The association between Darier disease, bipolar disorder, and schizophrenia revisited: a population-based family study. Bipolar Disord 17(3):340–344
142. Cederlöf M et al (2015) Intellectual disability and cognitive ability in Darier disease: Swedish nation-wide study. Br J Dermatol 173(1):155–158
143. Dodiuk-Gad R et al (2016) Darier disease in Israel: combined evaluation of genetic and neuropsychiatric aspects. Br J Dermatol 174(3):562–568
144. Baba-Aissa F et al (1998) Distribution and isoform diversity of the organellar Ca^{2+} pumps in the brain. Mol Chem Neuropathol 33(3):199–208
145. Takeichi T et al (2016) Darier's disease complicated by schizophrenia caused by a novel ATP2A2 mutation. Acta Derm Venereol 96(7):993–994
146. Nakamura T et al (2016) Loss of function mutations in ATP2A2 and psychoses: a case report and literature survey. Psychiatry Clin Neurosci 70(8):342–350
147. Noda K et al (2016) Novel and recurrent ATP2A2 mutations in Japanese patients with Darier's disease. Nagoya J Med Sci 78(4):485–492
148. Hough C et al (1999) Elevated basal and thapsigargin-stimulated intracellular calcium of platelets and lymphocytes from bipolar affective disorder patients measured by a fluorometric microassay. Biol Psychiatry 46(2):247–255
149. Scambler PJ (2000) The 22q11 deletion syndromes. Hum Mol Genet 9(16):2421–2426
150. McDonald-McGinn DM et al (2001) Phenotype of the 22q11.2 deletion in individuals identified through an affected relative: cast a wide FISHing net! Genet Med 3(1):23–29
151. Schreiner MJ et al (2013) Converging levels of analysis on a genomic hotspot for psychosis: insights from 22q11. 2 deletion syndrome. Neuropharmacology 68:157–173
152. Chun S et al (2014) Specific disruption of thalamic inputs to the auditory cortex in schizophrenia models. Science 344(6188):1178–1182
153. Kobrynski LJ, Sullivan KE (2007) Velocardiofacial syndrome, DiGeorge syndrome: the chromosome 22q11.2 deletion syndromes. Lancet 370(9596):1443–1452
154. Yagi H et al (2003) Role of TBX1 in human del22q11.2 syndrome. Lancet 362(9393):1366–1373
155. Papangeli I, Scambler P (2013) The 22q11 deletion: DiGeorge and velocardiofacial syndromes and the role of TBX1. Wiley Interdisc Rev Dev Biol 2(3):393–403
156. Devaraju P et al (2017) Haploinsufficiency of the 22q11. 2 microdeletion gene Mrpl40 disrupts short-term synaptic plasticity and working memory through dysregulation of mitochondrial calcium. Mol Psychiatry 22(9):1313–1326
157. Shi H, Wang Z (2018) Atypical microdeletion in 22q11 deletion syndrome reveals new candidate causative genes: a case report and literature review. Medicine 97(8):e9936
158. Ellegood J et al (2014) Neuroanatomical phenotypes in a mouse model of the 22q11.2 microdeletion. Mol Psychiatry 19(1):99–107
159. Mukai J et al (2015) Molecular substrates of altered axonal growth and brain connectivity in a mouse model of schizophrenia. Neuron 86(3):680–695
160. Karpinski BA et al (2014) Dysphagia and disrupted cranial nerve development in a mouse model of DiGeorge (22q11) deletion syndrome. Dis Model Mech 7(2):245–257
161. Oskarsdottir S, Vujic M, Fasth A (2004) Incidence and prevalence of the 22q11 deletion syndrome: a population-based study in Western Sweden. Arch Dis Child 89(2):148–151
162. Chow EW et al (2006) Neurocognitive profile in 22q11 deletion syndrome and schizophrenia. Schizophr Res 87(1):270–278
163. Pulver AE et al (1994) Psychotic illness in patients diagnosed with velo-cardio-facial syndrome and their relatives. J Nerv Ment Dis 182(8):476–477
164. Botto LD et al (2003) A population-based study of the 22q11.2 deletion: phenotype, incidence, and contribution to major birth defects in the population. Pediatrics 112(1 Pt 1):101–107
165. Bassett AS et al (2011) Practical guidelines for managing patients with 22q11.2 deletion syndrome. J Pediatr 159(2):332–339.e1
166. Bearden CE et al (2001) The neurocognitive phenotype of the 22q11. 2 deletion syndrome: selective deficit in visual-spatial memory. J Clin Exp Neuropsychol 23(4):447–464

167. Eliez S et al (2000) Young children with Velo-cardio-facial syndrome (CATCH-22). Psychological and language phenotypes. Eur Child Adolesc Psychiatry 9(2):109–114
168. Swillen A et al (2000) Chromosome 22q11 deletion syndrome: update and review of the clinical features, cognitive-behavioral spectrum, and psychiatric complications. Am J Med Genet A 97(2):128–135
169. Gothelf D et al (2007) Developmental trajectories of brain structure in adolescents with 22q11. 2 deletion syndrome: a longitudinal study. Schizophr Res 96(1):72–81
170. Rauch A et al (2006) Diagnostic yield of various genetic approaches in patients with unexplained developmental delay or mental retardation. Am J Med Genet A 140(19):2063–2074
171. Bassett AS, Chow EW (2008) Schizophrenia and 22q11.2 deletion syndrome. Curr Psychiatry Rep 10(2):148–157
172. Fung WL et al (2010) Elevated prevalence of generalized anxiety disorder in adults with 22q11.2 deletion syndrome. Am J Psychiatry 167(8):998
173. Karayiorgou M, Simon TJ, Gogos JA (2010) 22q11.2 microdeletions: linking DNA structural variation to brain dysfunction and schizophrenia. Nat Rev Neurosci 11:402–416
174. Schneider M et al (2014) Psychiatric disorders from childhood to adulthood in 22q11. 2 deletion syndrome: results from the international consortium on brain and behavior in 22q11. 2 deletion syndrome. Am J Psychiatr 171(6):627–639
175. Jonas RK, Montojo CA, Bearden CE (2014) The 22q11.2 deletion syndrome as a window into complex neuropsychiatric disorders over the lifespan. Biol Psychiatry 75(5):351–360
176. Green T et al (2009) Psychiatric disorders and intellectual functioning throughout development in velocardiofacial (22q11. 2 deletion) syndrome. J Am Acad Child Adolesc Psychiatry 48(11):1060–1068
177. Bassett AS et al (2003) The schizophrenia phenotype in 22q11 deletion syndrome. Am J Psychiatr 160(9):1580–1586
178. Tang SX et al (2017) The psychosis Spectrum in 22q11.2 deletion syndrome is comparable to that of nondeleted youth. Biol Psychiatry 82(1):17–25
179. Earls LR et al (2012) Age-dependent microRNA control of synaptic plasticity in 22q11 deletion syndrome and schizophrenia. J Neurosci 32(41):14132–14144
180. Earls LR, Zakharenko SS (2014) A synaptic function approach to investigating complex psychiatric diseases. Neuroscientist 20(3):257–271
181. Schizophrenia_Working_Group_of_the_Psychiatric_Genomics-Consortium (2014) Biological insights from 108 schizophrenia-associated genetic loci. Nature 511(7510):421–427
182. Richard EA et al (2017) Potential synergistic action of 19 schizophrenia risk genes in. Schizophr Res 180:64–69
183. Genis-Mendoza A et al (2018) Comparative analysis of gene expression profiles involved in calcium signaling pathways using the NLVH animal model of schizophrenia. J Mol Neurosci 64(1):111–116
184. Hagenston AM, Bading H (2011) Calcium signaling in synapse-to-nucleus communication. Cold Spring Harb Perspect Biol:a004564
185. Lisek M et al (2016) Regional brain dysregulation of Ca^{2+}—handling systems in ketamine-induced rat model of experimental psychosis. Cell Tissue Res 363(3):609–620
186. Stefani MR, Moghaddam B (2005) Transient N-methyl-D-aspartate receptor blockade in early development causes lasting cognitive deficits relevant to schizophrenia. Biol Psychiatry 57(4):433–436
187. Neill JC et al (2010) Animal models of cognitive dysfunction and negative symptoms of schizophrenia: focus on NMDA receptor antagonism. Pharmacol Ther 128(3):419–432
188. Chakrabarti S et al (2015) Metabolic risk factors of sporadic Alzheimer's disease: implications in the pathology, pathogenesis and treatment. Aging Dis 6(4):282–299
189. Davie CA (2008) A review of Parkinson's disease. Br Med Bull 86:109–127
190. Nussbaum RL, Ellis CE (2003) Alzheimer's disease and Parkinson's disease. N Engl J Med 348(14):1356–1364

191. Warner TT, Schapira AH (2003) Genetic and environmental factors in the cause of Parkinson's disease. Ann Neurol 53(S3):S16–S25
192. Cookson MR, Xiromerisiou G, Singleton A (2005) How genetics research in Parkinson's disease is enhancing understanding of the common idiopathic forms of the disease. Curr Opin Neurol 18(6):706–711
193. Gilks WP et al (2005) A common LRRK2 mutation in idiopathic Parkinson's disease. Lancet 365(9457):415–416
194. Forno LS (1996) Neuropathology of Parkinson's disease. J Neuropathol Exp Neurol 55(3):259–272
195. Braak H et al (2003) Staging of brain pathology related to sporadic Parkinson's disease. Neurobiol Aging 24(2):197–211
196. Gandhi S, Wood NW (2005) Molecular pathogenesis of Parkinson's disease. Hum Mol Genet 14(18):2749–2755
197. Sudo H et al (2001) Secreted Abeta does not mediate neurotoxicity by antibody-stimulated amyloid precursor protein. Biochem Biophys Res Commun 282(2):548–556
198. Mattson MP (2007) Calcium and neurodegeneration. Aging Cell 6(3):337–350
199. Mattson MP (2004) Pathways towards and away from Alzheimer's disease. Nature 430(7000):631–639
200. Ganguly G et al (2017) Proteinopathy, oxidative stress and mitochondrial dysfunction: cross talk in Alzheimer's disease and Parkinson's disease. Drug Des Devel Ther 11:797–810
201. Xie A et al (2014) Shared mechanisms of neurodegeneration in Alzheimer's disease and Parkinson's disease. Biomed Res Int 2014:648740
202. Bonda DJ et al (2011) The mitochondrial dynamics of Alzheimer's disease and Parkinson's disease offer important opportunities for therapeutic intervention. Curr Pharm Des 17(31):3374–3380
203. Perier C, Vila M (2012) Mitochondrial biology and Parkinson's disease. Cold Spring Harb Perspect Med 2(2):a009332
204. Sen CK, Packer L (1996) Antioxidant and redox regulation of gene transcription. FASEB J 10(7):709–720
205. Aiken CT et al (2011) Oxidative stress-mediated regulation of proteasome complexes. Mol Cell Proteomics 10(5):R110.006924
206. Pajares M et al (2015) Redox control of protein degradation. Redox Biol 6:409–420
207. Baillet A et al (2010) The role of oxidative stress in amyotrophic lateral sclerosis and Parkinson's disease. Neurochem Res 35(10):1530–1537
208. Zhou C, Huang Y, Przedborski S (2008) Oxidative stress in Parkinson's disease: a mechanism of pathogenic and therapeutic significance. Ann NY Acad Sci 1147(1):93–104
209. Protter D, Lang C, Cooper AA (2012) alphaSynuclein and mitochondrial dysfunction: a pathogenic partnership in parkinson's disease? Parkinsons Dis 2012:829207
210. Banerjee K et al (2010) Alpha-synuclein induced membrane depolarization and loss of phosphorylation capacity of isolated rat brain mitochondria: implications in Parkinson's disease. FEBS Lett 584(8):1571–1576
211. Wang X et al (2009) Impaired balance of mitochondrial fission and fusion in Alzheimer's disease. J Neurosci 29(28):9090–9103
212. Devi L et al (2006) Accumulation of amyloid precursor protein in the mitochondrial import channels of human Alzheimer's disease brain is associated with mitochondrial dysfunction. J Neurosci 26(35):9057–9068
213. Mounsey RB, Teismann P (2011) Mitochondrial dysfunction in Parkinson's disease: pathogenesis and neuroprotection. Parkinson's Dis 2011
214. Pacelli C et al (2011) Mitochondrial defect and PGC-1α dysfunction in parkin-associated familial Parkinson's disease. Biochim Biophys Acta (BBA) - Mol Basis Dis 1812(8):1041–1053
215. Wen Y et al (2011) Alternative mitochondrial electron transfer as a novel strategy for neuroprotection. J Biol Chem, 2011:jbc. M110. 208447

216. Stutzmann GE, Mattson MP (2011) Endoplasmic reticulum Ca^{2+} handling in excitable cells in health and disease. Pharmacol Rev:pr. 110.003814
217. Woods NK, Padmanabhan J (2012) Neuronal calcium signaling and Alzheimer's disease. In: Islam MS (ed) Calcium signaling. Springer Netherlands, Dordrecht, pp 1193–1217
218. Gallego-Sandin S, Alonso MT, Garcia-Sancho J (2011) Calcium homoeostasis modulator 1 (CALHM1) reduces the calcium content of the endoplasmic reticulum (ER) and triggers ER stress. Biochem J 437(3):469–475
219. Verkhratsky A (2005) Physiology and pathophysiology of the calcium store in the endoplasmic reticulum of neurons. Physiol Rev 85(1):201–279
220. Verkhratsky A (2002) The endoplasmic reticulum and neuronal calcium signalling. Cell Calcium 32(5–6):393–404
221. Corona C et al (2011) New therapeutic targets in Alzheimer's disease: brain deregulation of calcium and zinc. Cell Death Dis 2(6):e176
222. Mattson MP (2010) ER calcium and Alzheimer's disease: in a state of flux. Sci Signal 3(114):pe10–pe10
223. Magi S et al (2016) Intracellular calcium dysregulation: implications for Alzheimer's disease. Biomed Res Int 2016:1–14
224. Egorova P, Popugaeva E, Bezprozvanny I (2015) Disturbed calcium signaling in spinocerebellar ataxias and Alzheimer's disease. Semin Cell Dev Biol 40:127–133
225. Berridge MJ (2013) Dysregulation of neural calcium signaling in Alzheimer disease, bipolar disorder and schizophrenia. Prion 7(1):2–13
226. Mattson MP (1990) Antigenic changes similar to those seen in neurofibrillary tangles are elicited by glutamate and Ca^{2+} influx in cultured hippocampal neurons. Neuron 4(1):105–117
227. Mattson MP et al (1993) Comparison of the effects of elevated intracellular aluminum and calcium levels on neuronal survival and tau immunoreactivity. Brain Res 602(1):21–31
228. Kurbatskaya K et al (2016) Upregulation of calpain activity precedes tau phosphorylation and loss of synaptic proteins in Alzheimer's disease brain. Acta Neuropathol Commun 4(1):34
229. Mattson MP et al (1992) Beta-amyloid peptides destabilize calcium homeostasis and render human cortical neurons vulnerable to excitotoxicity. J Neurosci 12(2):376–389
230. Mattson MP, Tomaselli KJ, Rydel RE (1993) Calcium-destabilizing and neurodegenerative effects of aggregated β-amyloid peptide are attenuated by basic FGF. Brain Res 621(1):35–49
231. Khachaturian ZS (1989) Calcium, membranes, aging and Alzheimer's disease: introduction and overview. Ann NY Acad Sci 568(1):1–4
232. Shankar GM et al (2007) Natural oligomers of the Alzheimer amyloid-β protein induce reversible synapse loss by modulating an NMDA-type glutamate receptor-dependent signaling pathway. J Neurosci 27(11):2866–2875
233. Thibault O, Gant JC, Landfield PW (2007) Expansion of the calcium hypothesis of brain aging and Alzheimer's disease: minding the store. Aging Cell 6(3):307–317
234. Kamenetz F et al (2003) APP processing and synaptic function. Neuron 37(6):925–937
235. Cirrito JR et al (2003) In vivo assessment of brain interstitial fluid with microdialysis reveals plaque-associated changes in amyloid-β metabolism and half-life. J Neurosci 23(26):8844–8853
236. Kuchibhotla KV et al (2008) Aβ plaques lead to aberrant regulation of calcium homeostasis in vivo resulting in structural and functional disruption of neuronal networks. Neuron 59(2):214–225
237. Abramov E et al (2009) Amyloid-β as a positive endogenous regulator of release probability at hippocampal synapses. Nat Neurosci 12(12):1567–1576
238. Surmeier DJ et al (2017) Calcium and Parkinson's disease. Biochem Biophys Res Commun 483(4):1013–1019
239. Calì T, Ottolini D, Brini M (2011) Mitochondria, calcium, and endoplasmic reticulum stress in Parkinson's disease. Biofactors 37(3):228–240

240. Chan CS, Gertler TS, Surmeier DJ (2009) Calcium homeostasis, selective vulnerability and Parkinson's disease. Trends Neurosci 32(5):249–256
241. Surmeier DJ, Guzman JN, Sanchez-Padilla J (2010) Calcium, cellular aging, and selective neuronal vulnerability in Parkinson's disease. Cell Calcium 47(2):175–182
242. Mosharov EV et al (2009) Interplay between cytosolic dopamine, calcium, and α-synuclein causes selective death of substantia nigra neurons. Neuron 62(2):218–229
243. Nedergaard S, Flatman J, Engberg I (1993) Nifedipine-and omega-conotoxin-sensitive Ca^{2+} conductances in guinea-pig substantia nigra pars compacta neurones. J Physiol 466(1):727–747
244. Calì T et al (2013) Enhanced parkin levels favor ER-mitochondria crosstalk and guarantee Ca^{2+} transfer to sustain cell bioenergetics. Biochim Biophys Acta (BBA) - Mol Basis Dis 1832(4):495–508
245. Calì T et al (2012) α-Synuclein controls mitochondrial calcium homeostasis by enhancing endoplasmic reticulum-mitochondria interactions. J Biol Chem 287(22):17914–17929
246. Sandebring A et al (2009) Parkin deficiency disrupts calcium homeostasis by modulating phospholipase C signalling. FEBS J 276(18):5041–5052
247. Nath S et al (2011) Raised calcium promotes α-synuclein aggregate formation. Mol Cell Neurosci 46(2):516–526
248. Rcom H et al (2014) Interactions between calcium and alpha-Synuclein in neurodegeneration. Biomol Ther 4(3)
249. Surmeier DJ (2007) Calcium, ageing, and neuronal vulnerability in Parkinson's disease. Lancet Neurol 6(10):933–938
250. Danzer KM et al (2007) Different species of α-synuclein oligomers induce calcium influx and seeding. J Neurosci 27(34):9220–9232
251. Hettiarachchi NT et al (2009) α-Synuclein modulation of Ca^{2+} signaling in human neuroblastoma (SH-SY5Y) cells. J Neurochem 111(5):1192–1201
252. El-Agnaf OM et al (2004) A strategy for designing inhibitors of α-synuclein aggregation and toxicity as a novel treatment for Parkinson's disease and related disorders. FASEB J 18(11):1315–1317
253. Rockenstein E et al (2002) Differential neuropathological alterations in transgenic mice expressing α-synuclein from the platelet-derived growth factor and Thy-1 promoters. J Neurosci Res 68(5):568–578
254. Vekrellis K et al (2011) Pathological roles of α-synuclein in neurological disorders. Lancet Neurol 10(11):1015–1025
255. Vekrellis K et al (2009) Inducible over-expression of wild type α-synuclein in human neuronal cells leads to caspase-dependent non-apoptotic death. J Neurochem 109(5):1348–1362
256. Kragh CL et al (2009) α-Synuclein aggregation and Ser-129 phosphorylation-dependent cell death in oligodendroglial cells. J Biol Chem 284(15):10211–10222
257. Betzer C et al (2018) Alpha-synuclein aggregates activate calcium pump SERCA leading to calcium dysregulation. EMBO Rep 19:e44617
258. Lowe R et al (2004) Calcium (II) selectively induces α-synuclein annular oligomers via interaction with the C-terminal domain. Protein Sci 13(12):3245–3252
259. Goodwin J et al (2013) Raised calcium and oxidative stress cooperatively promote alpha-synuclein aggregate formation. Neurochem Int 62(5):703–711
260. Colla E et al (2012) Accumulation of toxic α-synuclein oligomer within endoplasmic reticulum occurs in α-synucleinopathy in vivo. J Neurosci 32(10):3301–3305
261. Hurley MJ et al (2013) Parkinson's disease is associated with altered expression of CaV1 channels and calcium-binding proteins. Brain 136(7):2077–2097
262. Pasternak B et al (2012) Use of calcium channel blockers and Parkinson's disease. Am J Epidemiol 175(7):627–635
263. Ritz B et al (2010) L-type calcium channel blockers and Parkinson disease in Denmark. Ann Neurol 67(5):600–606
264. Becker C, Jick SS, Meier CR (2008) Use of antihypertensives and the risk of Parkinson disease. Neurology 70(16 Part 2):1438–1444

265. Marras C et al (2012) Dihydropyridine calcium channel blockers and the progression of parkinsonism. Ann Neurol 71(3):362–369
266. Lytton J, Westlin M, Hanley MR (1991) Thapsigargin inhibits the sarcoplasmic or endoplasmic reticulum Ca-ATPase family of calcium pumps. J Biol Chem 266(26):17067–17071
267. Park SW et al (2010) Sarco (endo) plasmic reticulum Ca^{2+}-ATPase 2b is a major regulator of endoplasmic reticulum stress and glucose homeostasis in obesity. Proc Natl Acad Sci 2010:12044
268. Kang S et al (2015) Small molecular allosteric activator of the sarco/endoplasmic reticulum Ca^{2+}-ATPase (SERCA) attenuates diabetes and metabolic disorders. J Biol Chem, 2015:p. jbc. M115. 705012
269. Tanzi RE et al (1987) Amyloid beta protein gene: cDNA, mRNA distribution, and genetic linkage near the Alzheimer locus. Science 235(4791):880–884
270. Sherrington R et al (1995) Cloning of a gene bearing missense mutations in early-onset familial Alzheimer's disease. Nature 375(6534):754–760
271. Levy-Lahad E et al (1995) Candidate gene for the chromosome 1 familial Alzheimer's disease locus. Science 269(5226):973–977
272. Rogaev EI et al (1995) Familial Alzheimer's disease in kindreds with missense mutations in a gene on chromosome 1 related to the Alzheimer's disease type 3 gene. Nature 376:775
273. Cruts M, Theuns J, Van Broeckhoven C (2012) Locus-specific mutation databases for neurodegenerative brain diseases. Hum Mutat 33(9):1340–1344
274. Wolfe MS et al (1999) Two transmembrane aspartates in presenilin-1 required for presenilin endoproteolysis and γ-secretase activity. Nature 398(6727):513–517
275. Annaert WG et al (1999) Presenilin 1 controls γ-secretase processing of amyloid precursor protein in pre-Golgi compartments of hippocampal neurons. J Cell Biol 147(2):277–294
276. De Strooper B et al (1998) Deficiency of presenilin-1 inhibits the normal cleavage of amyloid precursor protein. Nature 391(6665):387–390
277. Jin H et al (2010) Presenilin-1 holoprotein is an interacting partner of sarco endoplasmic reticulum calcium-ATPase and confers resistance to endoplasmic reticulum stress. J Alzheimers Dis 20(1):261–273
278. Brunello L et al (2009) Presenilin-2 dampens intracellular ca(2+) stores by increasing ca(2+) leakage and reducing ca(2+) uptake. J Cell Mol Med 13(9b):3358–3369
279. Dreses-Werringloer U et al (2008) A polymorphism in CALHM1 influences Ca^{2+} homeostasis, Abeta levels, and Alzheimer's disease risk. Cell 133(7):1149–1161
280. Nensa FM et al (2014) Amyloid beta a4 precursor protein-binding family B member 1 (FE65) interactomics revealed synaptic vesicle glycoprotein 2A (SV2A) and sarcoplasmic/endoplasmic reticulum calcium ATPase 2 (SERCA2) as new binding proteins in the human brain. Mol Cell Proteomics 13(2):475–488
281. Ikin AF et al (2007) A macromolecular complex involving the amyloid precursor protein (APP) and the cytosolic adapter FE65 is a negative regulator of axon branching. Mol Cell Neurosci 35(1):57–63
282. Cao X, Südhof TC (2001) A transcriptively active complex of APP with Fe65 and histone acetyltransferase Tip60. Science 293(5527):115–120
283. Pietrzik CU et al (2004) FE65 constitutes the functional link between the low-density lipoprotein receptor-related protein and the amyloid precursor protein. J Neurosci 24(17):4259–4265
284. Kinoshita A et al (2002) The γ secretase-generated carboxyl-terminal domain of the amyloid precursor protein induces apoptosis via Tip60 in H4 cells. J Biol Chem 277(32):28530–28536
285. Santiard-Baron D et al (2005) Expression of human FE65 in amyloid precursor protein transgenic mice is associated with a reduction in β-amyloid load. J Neurochem 93(2):330–338
286. Krajnak K, Dahl R (2018) A new target for Alzheimer's disease: a small molecule SERCA activator is neuroprotective in vitro and improves memory and cognition in APP/PS1 mice. Bioorg Med Chem Lett 28(9):1591–1594
287. Nussbaum RL (2018) Genetics of synucleinopathies. Cold Spring Harb Perspect Med 8(6):a024109

288. Lashuel HA et al (2013) The many faces of α-synuclein: from structure and toxicity to therapeutic target. Nat Rev Neurosci 14(1):38–48
289. Ghiglieri V, Calabrese V, Calabresi P (2018) Alpha-Synuclein: from early synaptic dysfunction to neurodegeneration. Front Neurol 9:295
290. Tuon T et al (2012) Physical training exerts neuroprotective effects in the regulation of neurochemical factors in an animal model of. Neuroscience 227:305–312
291. Lundblad M et al (2002) Pharmacological validation of behavioural measures of akinesia and dyskinesia in a rat model of Parkinson's disease. Eur J Neurosci 15(1):120–132
292. Jouve L et al (2010) Deep brain stimulation of the center median–parafascicular complex of the thalamus has efficient anti-parkinsonian action associated with widespread cellular responses in the basal ganglia network in a rat model of Parkinson's disease. J Neurosci 30(29):9919–9928

Chapter 7
Cytoplasmic Calcium Buffering: An Integrative Crosstalk

Juan A. Gilabert

Abstract Calcium (Ca^{2+}) buffering is part of an integrative crosstalk between different mechanisms and elements involved in the control of free Ca^{2+} ions persistence in the cytoplasm and hence, in the Ca^{2+}-dependence of many intracellular processes. Alterations of Ca^{2+} homeostasis and signaling from systemic to subcellular levels also play a pivotal role in the pathogenesis of many diseases.

Compared with Ca^{2+} sequestration towards intracellular Ca^{2+} stores, Ca^{2+} buffering is a rapid process occurring in a subsecond scale. Any molecule (or binding site) with the ability to bind Ca^{2+} ions could be considered, at least in principle, as a buffer. However, the term Ca^{2+} buffer is applied only to a small subset of Ca^{2+} binding proteins containing acidic side-chain residues.

Ca^{2+} buffering in the cytoplasm mainly relies on mobile and immobile or fixed buffers controlling the diffusion of free Ca^{2+} ions inside the cytosol both temporally and spatially. Mobility of buffers depends on their molecular weight, but other parameters as their concentration, affinity for Ca^{2+} or Ca^{2+} binding and dissociation kinetics next to their diffusional mobility also contribute to make Ca^{2+} signaling one of the most complex signaling activities of the cell.

The crosstalk between all the elements involved in the intracellular Ca^{2+} dynamics is a process of extreme complexity due to the diversity of structural and molecular elements involved but permit a highly regulated spatiotemporal control of the signal mediated by Ca^{2+} ions. The basis of modeling tools to study Ca^{2+} dynamics are also presented.

Keywords Ca^{2+} buffering · Mobile buffers · Immobile buffers · Modeling Ca^{2+} signaling

J. A. Gilabert (✉)
Department of Pharmacology and Toxicology, Faculty of Veterinary Medicine, Complutense University of Madrid, Madrid, Spain
e-mail: jagilabe@ucm.es

© Springer Nature Switzerland AG 2020
M. S. Islam (ed.), *Calcium Signaling*, Advances in Experimental Medicine and Biology 1131, https://doi.org/10.1007/978-3-030-12457-1_7

7.1 Introduction

In the previous chapters the reader has had the opportunity to explore the properties and characteristics making of calcium ions (Ca^{2+}) the essential and more versatile messenger in the cells. But, how its concentration is so tightly regulated within the cells and for what reason?

Ca^{2+} ions are involved in many processes along the vital cycle of the cells. Notwithstanding, they also can be cytotoxic for living organisms across the entire phylogenetic tree (from bacteria to eukaryotic cells) making necessary an universal Ca^{2+} homeostasis system [1].

Ancestral cells probably faced to low Ca^{2+} levels in the prehistoric alkaline ocean, which millennia later became acidified and Ca^{2+} in the seawater started gradually to increase [2]. Thus, primitive cells had to face a massive, constant and toxic Ca^{2+} gradient. The appearance of a plasma membrane in the primitive cell was the border between extra- and intracellular spaces providing permeability features to enable a tight control of this cation concentration in the cytoplasm, with low concentrations inside against high concentrations in the extracellular milieu. This generates a huge gradient both in terms of concentration and of net charge considering the negative net charge of the intracellular milieu.

Moreover, since primitive cells successful mechanisms were developed very early in the evolution process to precisely regulate the cellular concentrations of free and bound/sequestered Ca^{2+} both in time and spatial dimensions. These mechanisms are essentially the same in prokaryote and eukaryote organisms: a low permeability of the cell membrane depending on influx mechanisms, a high intracellular buffering capacity, and an effective removal system [3, 4]. A significant difference in the mechanisms of buffering in eukaryotic cells is the presence of a nucleus and several sets of intracellular organelles, which can divide the cytoplasm into specialized compartments, with distinct mechanisms and capacities of Ca^{2+} handling [3].

At the cell level, Ca^{2+} homeostasis determines that basal cytosolic Ca^{2+} concentration ($[Ca^{2+}]_c$) is set at around 100 nM (10^{-7} M). This is crucial for a proper signaling process mediated by Ca^{2+} which as other effective signals, must be fast, with an adequate magnitude or dynamic range (to exceed a threshold) and finite in spatial and/or temporal terms.

This resting $[Ca^{2+}]_c$ works as a threshold to switch on any signaling processes mediated by Ca^{2+} against an electrochemical gradient due to a 10,000–20,000 times higher concentration in the extracellular milieu (around 1–2 mM). This threshold is kept by different mechanisms in a continuous and well-orchestrated crosstalk involving active transport expending energy (i.e. ATP-dependent) of Ca^{2+} out of the cell or into the organelles, antiport systems, Ca^{2+} buffering (mobile and immobile buffers) and ion condensation [5].

Ca^{2+} buffering in the cytoplasm relies on mobile or immobile buffers controlling the diffusion of free Ca^{2+} ions inside the cytosol both temporally and spatially. In other words, Ca^{2+} changes can occur in the whole cell space or be restricted to smaller areas around those elements involved in Ca^{2+} fluxes.

The importance of Ca^{2+} control mechanisms can be also seen at intercellular and multicellular levels in higher organisms. Thus, Ca^{2+} homeostasis also operates in the extracellular fluid where is influenced by dietary intake, Ca^{2+} absorption in the small intestine, exchange to and from the bones, and by excretion of Ca^{2+} in the urine. So, control of Ca^{2+} homeostasis in the cell and between cells underlies different physiological processes at the whole organism level [6].

Next, we will review these mechanisms with special attention to the role of cytoplasmic Ca^{2+} buffering in the generation of different spatiotemporal Ca^{2+} signals and its physiological relevance at the single cell (eukaryote) level.

7.2 Ca^{2+} Buffering: An Overview

The control of $[Ca^{2+}]_c$ at the resting level ($\sim 10^{-7}$ M range) involves several mechanisms, including Ca^{2+} influx (from the extracellular space), Ca^{2+} release (from internal Ca^{2+} stores), Ca^{2+} sequestration (towards internal Ca^{2+} stores), Ca^{2+} efflux (to the extracellular space) or Ca^{2+} buffering. Hence, Ca^{2+} persistence in the cytosol and derived Ca^{2+}-mediated actions are determined by two main processes: Ca^{2+} removal and Ca^{2+} diffusion (Fig. 7.1). Note that occasionally some of the efflux mechanisms can increase cytosolic Ca^{2+} as "slippage" of Ca^{2+} through Ca^{2+}-ATPases and the reverse-mode action of Na^+/Ca^{2+} exchanger) [7].

Ca^{2+} removal from the cytosol results from the combined action of Ca^{2+} sequestration and Ca^{2+} efflux, while Ca^{2+} diffusion is mainly determined by Ca^{2+} buffering. Ca^{2+} buffering is the rapid binding of Ca^{2+} entering cytosolic space to different cellular binding sites. Ca^{2+} buffering is an important process in Ca^{2+} signaling because it has been estimated that only about 1–5% of Ca^{2+} entering the cell remains as free Ca^{2+}, its physiologically active form [8–12].

In 1992, Neher and Augustine showed that Ca^{2+} buffering was a rapid process (time scale in subsecond range) kinetically distinct of Ca^{2+} sequestration, which is slower occurring in a tens of seconds scale. They also were able to determine using a combination of fura-2 microfluorimetry and Ca^{2+} current measurements in single adrenal chromaffin cells, the Ca^{2+} binding capacity of cytoplasm (κ_s) (bound Ca^{2+} over free Ca^{2+}) [13]. The κ_s value was approximately 75, which did not change during prolonged whole-cell recording (in a dialyzed cell with a disrupted membrane). Thus, they concluded that the majority of cellular Ca^{2+} binding sites were immobile [14].

Immobile buffers are represented by molecules of high molecular weight or Ca^{2+} binding sites anchored to intracellular structures. On the contrary, mobile buffers are molecules of low molecular size, typically less than 20–25 kDa, comprising soluble proteins or small organic anions and metabolites, like ATP. Their contribution can be difficult to estimate in some experiments involving whole-cell recordings where washout phenomena lead to the loss of some of these small molecules. In the presence of mobile buffers, immobile buffers increase the complexity of the spatiotemporal signaling repertoires, depending on the relative affinities, kinetics, and concentrations of the different buffers [15].

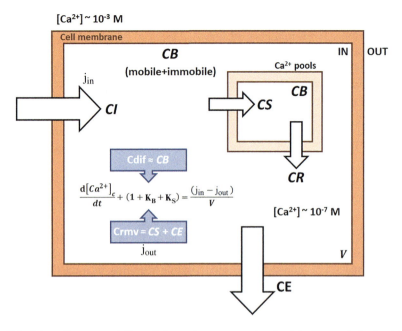

Fig. 7.1 Schematic representation of the main processes involved in the control of calcium concentration in the cytosol ([Ca^{2+}]$_c$) of a eukaryotic cell. *CI* calcium influx, *CR* calcium release, *CS* calcium sequestration, *CE* calcium efflux, and *CB* calcium buffering (mobile and immobile buffers). Ca^{2+} diffusion (Cdif) is mainly determined by cytoplasmic *CB*. Ca^{2+}*pools*, mainly represent the endoplasmic or sarcoplasmic reticulum and mitochondria. The combined action of these fluxes determines how [Ca^{2+}]$_c$ changes with time to generate different spatiotemporal signaling patterns. A basic transient in a cell of volume V and a Ca^{2+} influx (j_{in}) as consequence of a stimulus will produce an increase of [Ca^{2+}]$_c$ (see text for details). These Ca^{2+} ions will be partitioned into the cytoplasm between the endogenous Ca^{2+} buffer component and, in common experimental conditions, the exogenous Ca^{2+} indicator/buffer with constant Ca^{2+}-binding ratios κ_S and κ_B, respectively. Ca^{2+} removal from the cytoplasm (Crmv) (j_{out}) mainly reflects a combined action of *CS* and *CE*

Mobile buffers are estimated to have a Ca^{2+} binding capacity about one tenth of the cytoplasm Ca^{2+} binding capacity. One important mobile Ca^{2+} buffer is ATP, a highly mobile and effective Ca^{2+} chelator. ATP concentration in the cytosol is estimated to be around 2–3 mM, of which 0.4 mM is in a free form [16, 17].

But, which is the mobility of buffers in the cell? When a signal opens a Ca^{2+} route (i.e., a channel or a receptor) a fast Ca^{2+} flux reach the cytoplasm from extracellular space (or from organelles) driven by a large electrochemical gradient. More than 95% of Ca^{2+} are immediately bound to buffers within a distance of 10–50 nm from the focal point of Ca^{2+} entry [18, 19]. A mobile Ca^{2+} buffer will act to disperse such local domains of elevated [Ca^{2+}]$_c$ whereas fixed Ca^{2+} buffers will tend to prolong them.

Therefore, commonly used exogenous buffers/chelators for experimental purposes (EGTA, BAPTA or fluorescent Ca^{2+} indicators like fura-2) must be also considered since they compete with endogenous buffers increasing the transport of Ca^{2+} across the cell [20].

The presence of mobile and immobile buffers greatly reduces the diffusion spread of Ca^{2+}. The effective diffusion constant of free Ca^{2+} is between 200–300 $\mu m^2/s$, this being reduced more than 20 times (<16 $\mu m^2/s$) by cytoplasmic buffering [21, 22]. This fact was already described by Hodgkin and Keynes in 1957 in squid giant axon, where the retardation factor of radioactivity-labeled Ca^{2+} was about 40 [8]. However, the timescale of those experiments was of minutes so that Ca^{2+} buffering and Ca^{2+} sequestration could be equilibrated and hence the effects of buffers overestimated.

One parameter that offers an intuitive idea of Ca^{2+} diffusion is the mean distance covered by Ca^{2+} ions in one dimension. This distance can be calculated as $(2D_{eff}t)^{1/2}$, where D_{eff} is the effective diffusion coefficient for Ca^{2+} and t is the uptake time constant.

A stationary state situation (determining the basal level, usually around 100 nM) will be reached when Ca^{2+} influx or Ca^{2+} release to the cytoplasm space equals Ca^{2+} efflux plus Ca^{2+} buffering and sequestration into organelles [23, 24].

The results of simulations of the spatial and temporal pattern of Ca^{2+} changes following stimulation depend very much on assumptions regarding mobility of buffers [25, 26]. Thus, it is important to have experimental data on the mobility of cellular Ca^{2+} buffers. The result, showed by Neher and Augustine [13], that there is very little mobile buffer, would mean that the addition of even minute amounts of an exogenous mobile Ca^{2+} buffer, such as fura-2 (as in Ca^{2+} imaging experiments), should alter the temporal pattern of Ca^{2+} redistribution.

7.2.1 Ca^{2+} Buffers

Any molecule (or process) with the ability to bind Ca^{2+} ions could be considered, at least in principle, as a buffer. Thus, many molecules with several negatively charged groups can act as Ca^{2+} chelators. However, the term Ca^{2+} buffer is applied only to a small subset of Ca^{2+} binding proteins containing acidic side-chain residues [27, 28].

Ca^{2+} binding proteins can be found in the cytosol (as soluble proteins); but also, inside organelles (intraluminal proteins) [29] like the endoplasmic reticulum (ER) or as intrinsic proteins in membranes (plasma or organelle membranes) (see *Calcium Binding Proteins* chapter on this volume). The first of these proteins to be described was troponin C [30].

Different protein families bind Ca^{2+} through different structural motifs as EF-hands [31] or other well conserved Ca^{2+} binding domains founded in several proteins families (C2 domain proteins, annexins) [27, 32]. However, the term Ca^{2+}

buffer is applied only to a small subset of proteins of the EF-hand family, including parvalbumins (α and β isoforms), calbindin-D9k (CB-D9k), calbindin D-28k and calretinin in the cytosol [28].

In the ER, main intraluminal Ca^{2+} buffers are calsequestrin and calreticulin (which can also operate in the cytosol). These organellar Ca^{2+} buffers play a significant role as modulators in a dynamic network of organellar Ca^{2+} signaling [29].

Other Ca^{2+} binding proteins play a role as Ca^{2+} sensors more than as Ca^{2+} buffers due to its low concentration in the cell. Ca^{2+} ions bind to Ca^{2+} sensors inducing a conformational change, which permits them to interact with specific targets in a Ca^{2+}-regulated manner. A prototype of Ca^{2+} sensor is calmodulin [33].

On the other hand, depending on their diffusion characteristics, buffers can be considered as mobile or immobile. In addition, their Ca^{2+} binding and dissociation rate constants cover a wide range from slow buffers (with constant values about $1\ s^{-1}$) to fast buffers (constant values about $100\ s^{-1}$) [34].

The obvious consequence of the presence of Ca^{2+} buffers is the Ca^{2+} buffering capacity of the cytoplasm, which will be directly related to the concentration and spatial location of the Ca^{2+} buffers. However, other parameters as affinity for Ca^{2+}, Ca^{2+} binding and dissociation kinetics, and diffusional mobility also contribute to make Ca^{2+} signaling one of the most complex signaling activities of the cell.

7.2.2 Intracellular Concentration of Ca^{2+} Buffers

Intracellular or endogenous Ca^{2+} buffering capacity is directly related to the concentration of Ca^{2+} buffers located in the cytosol [19]. However, this unitless parameter is quite variable amongst different cells or even in the same type of cell depending of host tissue, species or clonality (e.g., PC12 *vs* adrenal chromaffin cells in Table 7.1) [19, 27, 35]. However, a minimal value could be 15 [15].

A useful approximation to endogenous Ca^{2+} buffering properties is to estimate the ratio of changes of buffer-bound Ca^{2+} over changes of free Ca^{2+}

Table 7.1 Values of Ca2+ binding ratio (κ) in different cell types

Cell type	κ	References
Motor neurons	40–50	[37, 41]
Adrenal chromaffin cells (PC12 cells)	40–75 (268)	[13, 14, 35]
Hippocampal neurons (excitatory/inhibitory)	60/150	[42]
Dopaminergic neurons	110–179	[40]
Smooth muscle (coronary artery)	150	[43]
Cerebellar Purkinje cells	900–2000	[38]
Pancreatic acinar cells	1500–2000	[39]

More details (as experimental conditions of measurements) can be founded in the respective original references. More values of Ca2+ binding ratios and their quantification for different types of neurons can be founded in Matthews and Dietrich [15]

($\kappa = d[BCa^{2+}]/d[Ca^{2+}] \approx [B]/K_d$) [15, 19, 36]. We can use this Ca^{2+} binding ratio to compare the values of κ of different cells ranging from low (below 50) for motor neurons [37] to very high (around 1000–2000) as that found in cerebellar Purkinje neurons [38] or pancreatic acinar cells [39] (Table 7.1). It indicates that lower values are probably needed in cells with a rapid Ca^{2+} signaling, as occurs in many neurons types [40]. In other words, a high buffering capacity means that very few Ca^{2+} will remain free following an action potential or other Ca^{2+}-generating event.

7.2.3 Ca^{2+} Binding and Kinetics by Ca^{2+} Buffers

EF-hands motifs are Ca^{2+}-binding sites with different selectivity and affinity for Ca^{2+} and Mg^{2+} [31, 44]. The Ca^{2+}-specific sites display affinities for Ca^{2+} from 10^{-3} to 10^{-7} M and significant lower ones for Mg^{2+} (10^{-1} to 10^{-2} M). The Ca^{2+} and Mg^{2+} sites bind Ca^{2+} with high (10^{-7} to 10^{-9} M) and Mg^{2+} with moderate (10^{-3} to 10^{-5} M) affinities (see [28]).

The majority of Ca^{2+} buffers have values of dissociation constants in the low micromolar range, such that, in a resting cell, Ca^{2+} buffers are mostly in a Ca^{2+}-free form. Another parameter to be considered is the kinetics of this binding. Ca^{2+} buffers considered as fast have rates $>10^8$ M^{-1} s^{-1} (as CB-D9k and troponin C and the synthetic buffers BAPTA or Fura-2), while those considered as slow have rates around 10^6 M^{-1} s^{-1} (as parvalbumins and the synthetic buffer EGTA). It is the presence of Mg^{2+} (0.5–1.0 mM in physiological conditions) that determines the Ca^{2+}-binding due to the slow Mg^{2+} off rate. Thus, in an experimental setting in the absence of Mg^{2+}, the on-rate of Ca^{2+} binding to parvalbumins is very rapid (around 10^8 M^{-1} s^{-1}) [45]. Usually, endogenous Ca^{2+} buffers possess several Ca^{2+} binding sites with different affinities and kinetics. The ratio between high:intermediate affinity sites can be 3:1 or 2:2 [46]. Likewise, Ca^{2+} binding sites (EF-hands) show allosterism in function of the occupied sites resulting in a non-linear Ca^{2+} buffering [47, 48], therefore increasing the complexity and versatility of the signaling mediated by Ca^{2+}.

7.2.4 Intracellular Mobility of Ca^{2+} Buffers

Ca^{2+} buffering is, at least in mechanistic terms a different process that Ca^{2+} sequestration. The majority of Ca^{2+} entering into the cell will be rapidly bound by Ca^{2+} buffers and later sequestered into Ca^{2+} storing organelles by slower processes [13]. Mobility of Ca^{2+} buffers is an important determinant of Ca^{2+} signaling since a mobile buffer will disperse a local increment of Ca^{2+} [15]. However, not only mobile buffers can alter the time course and spatial distribution of the Ca^{2+} signal. Immobile or fixed buffers (which also include some mobile Ca^{2+} binding proteins that may change their mobility upon binding Ca^{2+}) have been proposed to participate in the generation of the repertoire of intracellular Ca^{2+} signaling [15, 49].

The mobility of a Ca^{2+} buffer is expressed by its diffusion coefficient, which is proportional to the molecular weight but also influenced by other factors. Molecular diffusion is a complex process determined by a dynamic and out of equilibrium environment as the cytoplasm. The rate of the diffusion of a molecule or particle is also a function of temperature (thermal Brownian movement), viscosity and its mass (i.e. radius) (see [50]). Therefore, the same molecule can have different diffusion coefficient values in different cellular compartments as occurs with parvalbumin that exhibits around 12 μm^2/s in axons, somata and nuclei [51] versus 43 μm^2/s in dendrites [52].

7.3   Physiological Relevance of Cytosolic Ca^{2+} Buffering

Ca^{2+} signaling can be observed and described at different levels of life organization ranging from the whole organism to the subcellular level, but they are closely related [53, 54]. Most of Ca^{2+} in higher organisms is bound to bones and teeth forming hydroxyapatite. In humans, from a total amount of Ca^{2+} of approximately 1250 g, only a few grams are in the extracellular and intracellular fluids, this level is controlled by the slow Ca^{2+} movements in and out of the bone deposits [55] under the influence of vitamin D and parathyroid hormone regulatory actions. Serum concentration of Ca^{2+} is quite variable in animals ranging from 1 to 15 mM. In man, normal range is 2.1–2.6 mM, but levels out of this range can produce uncontrollable muscle spasms and cardiac alterations [56].

As discussed above, extracellular (plasma) levels will determine the electrochemical gradient with respect to intracellular space and Ca^{2+} fluxes from the extracellular fluid to inside the cells are both signals and sources of Ca^{2+} for subcellular organelles as the ER, mitochondrion or nucleus. Thus, the extracellular pool acts as a large reservoir of free Ca^{2+}.

Intracellular Ca^{2+} homeostasis is finely tuned by different regulatory systems making of Ca^{2+} a crucial cation in the cellular physiopathology and cell fate. An increase of $[Ca^{2+}]_c$ is the key signal to initiate many physiological actions as synaptic transmission, muscle contraction, hormone secretion or gene expression [57, 58]. Moreover, Ca^{2+} signaling is also present throughout the life history of the cell from its birth (mitosis) to death (apoptosis) [23, 59]. Alterations of Ca^{2+} homeostasis and signaling from systemic to subcellular levels also play a pivotal role in the pathogenesis of many diseases [60–67].

Ca^{2+} signaling does not act solely by "on-off" changes in their concentration at the entire cytoplasm. Cells have developed a complex code of signals based on the modulation of Ca^{2+} concentration on different spatial and temporal basis [66]. Thus, Ca^{2+} signals can be graded from unitary and spatially located signals, with a very limited spread through the cell, to whole cell signals as Ca^{2+} waves or Ca^{2+}

oscillations with repetitive Ca^{2+} changes travelling across the cell over longer time periods. These global signals are possible due to a coordinated activity of Ca^{2+} entry, Ca^{2+} release and Ca^{2+} sequestration and finely tuned by Ca^{2+} buffering.

However, a signal to be effective as messenger must be finite. Ca^{2+} buffering and Ca^{2+} sequestration are mechanisms to make intracellular Ca^{2+} changes transient, permitting the cell to recover the basal values so that get ready again for a new signaling round.

In general, fast responses are mediated by rapid and highly localized Ca^{2+} spikes as in neurons during synaptic transmission. By contrast, repetitive Ca^{2+} transients or waves are involved in slower intracellular processes, like in astrocytes. Moreover, neuronal excitability can be modulated by Ca^{2+} buffering changing from a spike-based pattern to a bursting signaling [67].

7.3.1 Spatiotemporal Signaling

Intracellular Ca^{2+} signaling, as occurs with other second messengers-mediated signaling is coded by changes in amplitude and frequency. Likewise, the quality and quantity of an extracellular incoming signal are reflected in different domains in the cell [68]. Amplitude may be proportional to the strength of the stimulus while the frequency to its strength and quality. Localized increases of Ca^{2+} in restricted domains represents an additional way of coding the signal. The combination of all these three characteristics makes possible a wide variety of Ca^{2+} signals and slight modifications of the amplitude or the temporal and spatial features of Ca^{2+} signals can trigger deleterious processes involved in multiple pathogenic states, such as cancer, inflammation, heart failure, and neurodegeneration [58].

The elementary phenomena involved in Ca^{2+} dynamics at a cytosolic level can be the opening of a single channel in the plasma membrane or in an intracellular Ca^{2+}-store organelle leading to Ca^{2+} entry or Ca^{2+} release. Such events can act as starters of whole cell events (i.e. Ca^{2+} sparks and Ca^{2+} induced Ca^{2+} release from the sarcoplasmic reticulum). Many other different elementary Ca^{2+} signaling events have been described from many cell and tissue types [69].

In an elementary event, the opening is usually brief and leads to a small and local increase of Ca^{2+} concentration. This local increase results in the formation of submicron sharp Ca^{2+} concentration profile in the vicinity of the channel [70]. The temporal collapse of these Ca^{2+} domains after channel closing is believed to be achieved in the microsecond time scale.

Ca^{2+} microdomains are restrained by strong buffering and slow diffusion. The microdomain's size is a function of several parameters as the conductance and the opening duration of the channel (how many ions can pass through the channel pore by time unit) and the electrochemical driving force for Ca^{2+} (potential difference and free Ca^{2+} concentration on both sides of the channel) but it is strongly influenced by the properties of Ca^{2+} buffers.

7.3.2 Ca^{2+} Buffering and Organelles

In addition to Ca^{2+} buffering, Ca^{2+} pumping significantly influences cellular Ca^{2+} signaling, albeit in a slower time scale. Ca^{2+} diffusion throughout the cell is not only restrained by Ca^{2+} buffering, Ca^{2+} pumping (Ca^{2+} fluxes against the electrochemical gradients) removes Ca^{2+} from the cytosol towards the extracellular space (by plasma membrane Ca^{2+}-ATPases and Na^+/Ca^{2+} exchanger) or into the ER (by sarco/endoplasmic reticulum Ca^{2+}-ATPases or SERCA pumps). Therefore, Ca^{2+} can be sequestered into the mitochondria (by action of the mitochondrial Ca^{2+} uniporter). In the resting state or for small $[Ca^{2+}]_c$ changes the dominant pumping fluxes are into the ER and to the extracellular space, whereas larger Ca^{2+} signals (micromolar range) involves mitochondrial participation [71].

SERCA pumps are in charge of Ca^{2+} sequestration from the cytosol to inside of the ER, where Ca^{2+} binds to intraluminal Ca^{2+} binding proteins with high capacity (10 mol per mol of protein) but low affinity ($K_D \approx 1$ mM) which permit ER to store vast amounts of Ca^{2+} which can rapidly be exchanged with the cytosol. Many ER proteins bind Ca^{2+}, including calreticulin, protein disulfide isomerase (PDI), glucose regulated protein 94 (Grp94), immunoglobulin binding protein (BiP), and ERp57. In the sarcoplasmic reticulum the most abundant Ca^{2+} binding protein is calsequestrin. Sarcolumenin, a histidine-rich protein, junctin, junctate, and triadin are unique to their membrane providing buffering and structural support [72].

ER contribution to Ca^{2+} signaling is also mediated by inositol 1,4,5-trisphosphate (InsP3) and ryanodine receptors as well as by passive leak channels. An example of this coordinated exchange between ER Ca^{2+} pools and $[Ca^{2+}]_c$ is the propagation of Ca^{2+} waves observed in mature oocytes during activation [73].

Organellar buffers also play a multifunctional role in a variety of processes, including protein folding, regulation of apoptosis, and regulating Ca^{2+} release pathways [29]. Thus, organellar Ca^{2+} dynamics is also dependent both structural and functional relationships between different organelles and their respective buffers.

The total Ca^{2+} concentration in the ER has been estimated between 5 and 50 mM [74]. The majority of this Ca^{2+} is bound to proteins with a low affinity ($K_d = 1$–4 mM) like calreticulin/calsequestrin and other intraluminal proteins with additional function as chaperones. The high intraluminal content of Ca^{2+} and the low affinity of ER Ca^{2+} buffers suggest that free Ca^{2+} concentration inside the ER could be in the micromolar range (300–800 mM, depending on cell type) [75].

Mitochondria is another crucial organelle in Ca^{2+} homeostasis in the cell, being capable of store substantial amounts of Ca^{2+}. Moreover, Ca^{2+} play an important role linking Ca^{2+} signaling with mitochondrial energetic status (through the production of ATP necessary, among other functions, for active Ca^{2+} transport) and cell death by apoptosis [76].

The Ca^{2+} entry way to the mitochondria is the uniporter located in the organelle's inner membrane. This mitochondrial Ca^{2+} uniporter is a selective Ca^{2+} channel with low affinity (K_d estimated at $\sim$10–50 μM) and high conductance [77].

Mitochondria becomes the predominant system of Ca^{2+} sequestration when the level of $[Ca^{2+}]_c$ is well above that reached during normal cell activation (around 1 μM) when work at saturation [78].

An additional mechanism of Ca^{2+} entry into the mitochondria known as rapid uptake mode or RaM was described in hepatocytes [79]. RaM functions could be to create a brief but high free Ca^{2+} elevation inside mitochondria, which may be enough to activate metabolic reactions with small amounts of Ca^{2+} uptake that are not capable to open mitochondrial membrane permeability transition pore [80]. Thus, mitochondria are the decoder between intensity of signaling and metabolic activity till apoptosis is induced when Ca^{2+} homeostasis is lost.

A final interesting point to understand Ca^{2+} dynamics in the cytoplasm is the crosstalk between organelles and plasma membrane. A good example of it is the mechanism, described by Putney in 1986 [81], known as capacitative calcium entry in which plasma membrane Ca^{2+} channels (like Ca^{2+} release activated Ca^{2+} or CRAC channels) open after depletion of ER Ca^{2+} stores to slowly replenish their resting Ca^{2+} levels [82].

This communication involves different signaling molecules and/or close interactions between subcellular structures. Ca^{2+} entry through store-operated Ca^{2+} channels involve two main proteins. The stromal interaction molecule 1 (STIM1), an ER-located Ca^{2+}-sensing protein and, Orai1, a pore-forming subunit located at the plasma membrane. After depletion of Ca^{2+} stores, STIM1 multimerizes and redistributes into discrete sites close to the plasma membrane and, STIM1 couples to and stimulates Orai, initiating the Ca^{2+} entry to replenish the empty stores [83].

The high-resolution imaging techniques have permitted to confirm that the mitochondrial network inside the cells can be very close or even in contact with the ER membrane and/or plasma membrane channels. Also mitochondria could be able to sense large, but spatially limited Ca^{2+} increments derived from InsP3 receptor activation or other Ca^{2+} fluxes from plasma membrane channels as voltage-dependent calcium channels or store-operated calcium entry [84–86]. But, mitochondria can also increase Ca^{2+} buffering capacity by a local release of ATP which acts as an endogenous highly mobile and effective Ca^{2+} chelator. Thus, mitochondria located close to CRAC channels in T lymphocytes can regulate slow Ca^{2+}-dependent inactivation of the Ca^{2+} current through these channels by increasing the Ca^{2+} buffering capacity beneath the plasma membrane, mainly through the release of ATP [17].

7.4 Elements for Modeling Ca^{2+} Buffering and Signaling

Ca^{2+} dynamics can be seen as a process of extreme complexity due to the diversity of molecular and structural elements (i.e., pumps, channels, Ca^{2+} buffers, organelles, ...) involved [23] which can be analyzed at several levels from unitary or elementary events (restricted to subcellular spaces) to global events (as waves or oscillations) occurring at the whole cell level.

First attempts to model intracellular Ca^{2+} dynamics focused in extracellular Ca^{2+} entry pathways (i.e. channels) considering the cytoplasm as a uniform medium. Later, more descriptive models were proposed to explain the activation kinetics of a single channel or the spatial arrangements of a group of channels, the role of mobile and immobile buffers or the alterations produced by experimentally-added exogenous buffers.

In general, the inositol 1,4,5-trisphosphate receptor (InsP3R), the ryanodine receptor, and the SERCA pumps govern most Ca^{2+} exchanges between the cytosol and the ER. When combined in a cellular model, they can explain many observations regarding signal-induced Ca^{2+} oscillations and waves. On the other hand, modeling also permits a predictive analysis of experimental results and to study the role of different elements in a complex system (see an excellent review by Dupont [87] about the relative contribution of the different elements in Ca^{2+} dynamics).

Mathematics behinds Ca^{2+} modeling can be hard for some of us (a comprehensive review about this issue have been written by Martin Falcke [88]). For it, only the main elements (variables) and their role in the control of cytoplasmic Ca^{2+} will be shortly described below.

A valuable approach to study the properties and contribution of buffers to Ca^{2+} signaling has been the microfluorimetry combined with single-cell electrophysiological techniques using Ca^{2+} indicators (as fura-2) [89]. They have also permitted to study the contribution of exogenous Ca^{2+} buffers (i.e. fluorescent Ca^{2+} probes) competing with endogenous Ca^{2+} buffers [19]. The method known as "added-buffer method" (see next section) allows to study endogenous buffer capacity by adding a competing exogenous buffer. Recently, new experimental approaches using minimal Ca^{2+} buffer (the low affinity Ca^{2+} indicator Fura-6F) concentrations, in order not to overwhelm endogenous Ca^{2+} buffers, have permitted more realistic intracellular conditions and to obtain more accurate values of parameters for modelling Ca^{2+} dynamics in neurons [90].

The basis to understand complex signals as Ca^{2+} waves or oscillations is the measurement of Ca^{2+} fluxes through a single channel and the subsequent estimation of its contribution to the global Ca^{2+} signal. The free Ca^{2+} increase due to a single channel opening (elementary events as blips or quarks) will depend mainly on the magnitude of the Ca^{2+} current and the buffering characteristics of the cytoplasm.

In early works, the aim was to estimate the cytosolic Ca^{2+} rise near the inner side of a single Ca^{2+} channel [91, 92] or a cluster of single channels [93]. These studies also allowed to make inferences about processes controlled by local Ca^{2+} signals like Ca^{2+}-dependent inactivation of Ca^{2+} channels [17], activation of Ca^{2+}-dependent potassium channels [94] or neurosecretion mediated by Ca^{2+} regulated exocytosis [95, 96].

The Ca^{2+} signal generated by a single channel is known as a nanodomain, which produces a Ca^{2+} elevation until 50 nm away from the channel pore in neurons [97]. Thus, any potential Ca^{2+}-dependent process controlled by a nanodomain should have a Ca^{2+} sensor in such distance range as occurs in rapid neurotransmitter release or the Ca^{2+} dependent modulation of ionic channels commented above.

Interestingly, small groups or clusters of channels can lead to the summation of nanodomains producing stronger signals called microdomains. These microdomains imply Ca^{2+} sensors that are placed within a fraction of a micrometer from the Ca^{2+} channel cluster center, to allow detection of summed signals.

The spread of Ca^{2+} during elemental events is the process most directly affected by buffer properties [26]. Thus, many efforts have been done to understand the cytosolic buffer dynamics and their influence in Ca^{2+} signaling from elementary events. However, Ca^{2+} nano or microdomains occurs in spatial and temporal scales that requires highly sensitive optical imaging techniques [98]. An alternative is the theoretical approach to study the influence of buffering on Ca^{2+} microdomains formation and diffusion [13, 16, 20, 99, 100]. At this level some other phenomena can be considered, as the stochastic behavior or the spatial grouping of Ca^{2+} permeating channels forming clusters [34], much more complex and demanding in order to design a reliable mathematical approach.

Despite several technological and methodological improvements in microfluorometry, such as the development of two-photon and super-resolution microscopy, we remain unable to trace Ca^{2+} dynamics over longer periods of time limiting our knowledge to their relevance in different physiological (or pathological) conditions [58].

An increased level of complexity of Ca^{2+} signaling is represented by global phenomena as waves or oscillations. They reflect the coordinated activity of release and diffusion processes involving several fluxes among different cellular compartments and organelles. These fluxes are usually visualized as a periodic behavior of $[Ca^{2+}]_c$ which spread across the cell and between cells. Ca^{2+} waves were first observed in fertilized fish oocytes [101], now we know different patterns from one-way linear displacement to spiral waves with a synchronous or asynchronous behavior depending of cell types [102, 103]. Ca^{2+} wave propagation can be modified by changing InsP3, buffer concentration, mitochondrial Ca^{2+} uptake or overexpressing SERCA pumps (see [88]). Repetitive Ca^{2+} spikes were instead first time observed in agonist-stimulated hepatocytes [104].

Ca^{2+} oscillations reflect an exchange of Ca^{2+} between the buffers and the ER mediated by InsP3R and the SERCA [25] with a dependence on external Ca^{2+} in most cells. Ca^{2+} oscillations and waves can be found in many cells from intra- to an intercellular (coordinated and cooperative responses in multicellular systems) level of signaling [105]. These transient events permit signaling based on frequency instead of amplitude; hence avoiding prolonged exposures to high Ca^{2+} concentrations potentially toxic for cells.

An additional challenge in the development of models is represented by the stochastic or deterministic nature of the signals involved in Ca^{2+} dynamics [87]. Some groups have concluded that Ca^{2+} dynamics remains as stochastic process even at the cellular level, mainly because of the poor communication between Ca^{2+}-releasing channels due to the low diffusivity of Ca^{2+} inside the cytoplasm [106–108].

The mathematical description of global phenomena can be very complex (in function of the number of variables considered, see [109]). However, Ca^{2+} signaling involves a complex system with many elements and multiple interactions where modeling is a useful tool to understand the spatiotemporal behavior of Ca^{2+} signaling in a particular system.

There are three basic types of models: qualitative, phenomenological and quantitative or mechanistic. Qualitative models are presented in a diagrammatic rather than in mathematical form; they are easy to do and can serve to support an initial hypothesis. Phenomenological models are based and expressed in a mathematical form to explain the experimental observations. A main disadvantage of this sort of models is that widely differing mathematical expressions can behave in a comparable way. Finally, the quantitative or mechanistic models are based as far as possible on known mechanisms and experimentally validated parameters [110].

7.4.1 Ca^{2+}-Binding Ratio

Several experimental approaches have been employed to estimate Ca^{2+} fluxes and free Ca^{2+} concentration in cells being the most popular a combination of fluorescent imaging using Ca^{2+} indicator dyes with the electrophysiological measurements [42]. But, these Ca^{2+} indicators are exogenous Ca^{2+} buffers that contribute to cytoplasmic Ca^{2+} binding capacity.

In 1995, E. Neher developed the "added buffer method" to estimate the endogenous Ca^{2+} buffer [19]. It consists to measure the Ca^{2+} transients elicited by voltage or drug stimulation in the presence of different concentrations of an exogenous Ca^{2+} indicator. By extrapolating to zero concentration of added exogenous indicator it is possible to estimate the endogenous Ca^{2+} buffer.

In practice, to calculate κ_S we need to measure changes in both the fluorescent signal and in the total calcium entering the cytosol (valid for brief time intervals and small incremental elevations in $[Ca^{2+}]_c$). As explained before, κ_S can be estimated by different approaches like the analysis of Ca^{2+} signal or the amount of Ca^{2+} bound to buffer (see [18, 19] for alternative methods and problems measuring Ca^{2+}-binding ratio). Some limitations of the "added buffer method" regarding to the relative contributions of mobile and immobile buffers to the total buffering capacity are widely described in Matthews and Dietrich [15].

7.4.2 Calculating Changes in Free Ca^{2+} Concentration

The most basic scenario to model Ca^{2+} transients is to consider that they occur in a single compartment [13]. Consider a cell (or subcellular location, e.g. a dendritic segment) with a volume V and a Ca^{2+} influx (j_{in}) induced by a stimulus that produces an increase in the total Ca^{2+} concentration. Ca^{2+} will be partitioned between the endogenous Ca^{2+} buffer component (S) in the cytoplasm and the exogenous Ca^{2+} indicator/buffer (B) with constant Ca^{2+}-binding ratios κ_S and κ_B, respectively. Ca^{2+} removal (j_{out}) is modeled as a linear extrusion mechanism with rate constant γ. The next equation describes the kinetics of Ca^{2+} changes (with conservation of total Ca^{2+}) [36]:

$$\frac{d[Ca^{2+}]_i}{dt} + \frac{d[BCa]}{dt} + \frac{d[SCa]}{dt} = \frac{(j_{in} - j_{out})}{V} \tag{7.1}$$

where [BCa] is the concentration of a mobile buffer (such as fura-2 or some other exogenous buffer) in its Ca^{2+}-bound form, [SCa] is the concentration of fixed (endogenous) Ca^{2+} buffer in the Ca^{2+}-bound form, and V is the accessible volume of the cell (or compartment).

The Eq. 7.1 can be expressed as a function of calcium binding capacities of κ_B and κ_S as

$$\frac{d[Ca^{2+}]_i}{dt} (1 + \kappa_B + \kappa_S) = \frac{(j_{in} - j_{out})}{V}$$

where

$$\kappa_B = \frac{d[BCa]}{d[Ca^{2+}]_i}$$

and

$$\kappa_S = \frac{d[SCa]}{d[Ca^{2+}]_i}$$

Thus, in a typical patch-clamp experiment we need to know the proportion of total current carried by Ca^{2+} ions, the accessible cell volume, and the cellular Ca^{2+} binding ratio to calculate the free Ca^{2+} concentration.

CalC ("Calcium Calculator") is a free modeling tool for simulating intracellular Ca^{2+} diffusion and buffering developed by Prof. Victor Matveev and available for download at https://web.njit.edu/matveev/calc.html [111]. CalC solves continuous reaction-diffusion partial differential equations describing the entry of Ca^{2+} into a volume through point-like channels, and its diffusion, buffering and binding to calcium "receptors".

The diffusion of calcium within a three-dimensional space in the presence of multiple buffers of mixed mobility is a more complex phenomenon, which it is described by a set of partial differential equations, which can be linearized to isolate individual interactions of Ca^{2+} with buffers at the nano and microdomain levels [16, 112].

An additional simplification assumes that interactions between Ca^{2+} and buffers are instantaneous ('Rapid Buffer Approximation') and that the spatiotemporal localization of Ca^{2+} depends only on the diffusion coefficients and affinities of the various buffers [20, 99, 100]. This yields a very useful analytical expression describing Ca^{2+} diffusion in the presence of multiple buffers using a new, smaller diffusion coefficient of calcium (*Dapp*), depending on the number, amount and mobility of the calcium buffers present (see [15] for details):

$$Dapp = D_{Ca} \frac{\left(1 + \frac{D_{mobile}}{D_{Ca}} \kappa_{mobile}\right)}{(1 + \kappa_{mobile} + \kappa_{immobile})}$$

where, D_{Ca} is the diffusion coefficient of free Ca^{2+} in the cytosol, D_{mobile} is the diffusion coefficient of mobile buffers, and κ_{mobile} and $\kappa_{immobile}$ are the Ca^{2+} buffering capacities of mobile and immobile buffers, respectively. This equation dictates that adding a mobile buffer can accelerate Ca^{2+} diffusion (increasing *Dapp*), but only if the mobile buffer's diffusion coefficient is larger than the *Dapp* of the system in the absence of the mobile buffer.

References

1. Dominguez DC (2004) Calcium signalling in bacteria. Mol Microbiol 54:291–297
2. Kazmierczak J, Kempe S, Kremer B (2013) Calcium in the early evolution of living systems: a biohistorical approach. Curr Org Chem 17:1738–1750
3. Case RM, Eisner D, Gurney A, Jones O, Muallem S, Verkhratsky A (2007) Evolution of calcium homeostasis: from birth of the first cell to an omnipresent signalling system. Cell Calcium 42:345–350
4. Domínguez DC, Guragain M, Patrauchan M (2015) Calcium binding proteins and calcium signaling in prokaryotes. Cell Calcium 57:151–165
5. Ripoll C, Norris V, Thellier M (2004) Ion condensation and signal transduction. BioEssays 26:549–557
6. Bronner F (2001) Extracellular and intracellular regulation of calcium homeostasis. Sci World J. https://doi.org/10.1100/tsw.2001.489
7. Harzheim D, Roderick HL, Bootman MD (2010) Chapter 117 – Intracellular calcium signaling. In: Bradshaw RA, Dennis EA (eds) Handbook of cell signal, 2nd edn. Academic, San Diego, pp 937–942
8. Hodgkin AL, Keynes RD (1957) Movements of labelled calcium in squid giant axons. J Physiol 138:253–281
9. Smith SJ, Zucker RS (1980) Aequorin response facilitation and intracellular calcium accumulation in molluscan neurones. J Physiol 300:167–196
10. Gorman AL, Thomas MV (1980) Intracellular calcium accumulation during depolarization in a molluscan neurone. J Physiol 308:259–285
11. McBurney RN, Neering IR (1985) The measurement of changes in intracellular free calcium during action potentials in mammalian neurons. J Neurosci Methods 13:65–76
12. Ahmed Z, Connor JA (1988) Calcium regulation by and buffer capacity of molluscan neurons during calcium transients. Cell Calcium 9:57–69
13. Neher E, Augustine GJ (1992) Calcium gradients and buffers in bovine chromaffin cells. J Physiol 450:273–301
14. Zhou Z, Neher E (1993) Mobile and immobile calcium buffers in bovine adrenal chromaffin cells. J Physiol 469:245–273
15. Matthews EA, Dietrich D (2015) Buffer mobility and the regulation of neuronal calcium domains. Front Cell Neurosci 9:1–11

16. Naraghi M, Neher E (1997) Linearized buffered Ca^{2+} diffusion in microdomains and its implications for calculation of $[Ca^{2+}]$ at the mouth of a calcium channel. J Neurosci 17:6961–6973
17. Montalvo GB, Artalejo AR, Gilabert JA (2006) ATP from subplasmalemmal mitochondria controls Ca^{2+}-dependent inactivation of CRAC channels. J Biol Chem 281:35616–35623
18. Augustine GJ, Neher E (1992) Calcium requirements for secretion in bovine chromaffin cells. J Physiol 450:247–271
19. Neher E (1995) The use of fura-2 for estimating Ca buffers and Ca fluxes. Neuropharmacology 34:1423–1442
20. Wagner J, Keizer J (1994) Effects of rapid buffers on Ca^{2+} diffusion and Ca^{2+} oscillations. Biophys J 67:447–456
21. Gabso M, Neher E, Spira ME (1997) Low mobility of the Ca^{2+} buffers in axons of cultured Aplysia neurons. Neuron 18:473–481
22. Allbritton NL, Meyer T, Stryer L (1992) Range of messenger action of calcium ion and inositol 1,4,5-trisphosphate. Science 258:1812–1815
23. Berridge MJ, Lipp P, Bootman MD (2000) The versatility and universality of calcium signalling. Nat Rev Mol Cell Biol 1:11–21
24. Berridge MJ, Bootman MD, Roderick HL (2003) Calcium: calcium signalling: dynamics, homeostasis and remodelling. Nat Rev Mol Cell Biol 4:517–529
25. Sala F, Hernández-Cruz A (1990) Calcium diffusion modeling in a spherical neuron. Relevance of buffering properties. Biophys J 57:313–324
26. Nowycky MC, Pinter MJ (1993) Time courses of calcium and calcium-bound buffers following calcium influx in a model cell. Biophys J 64:77–91
27. Schwaller B (2010) Chapter 120 – Ca^{2+} buffers. In: Bradshaw RA, Dennis EA (eds) Handbook of cell signal, 2nd edn. Academic, San Diego, pp 955–962
28. Schwaller B (2010) Cytosolic Ca^{2+} buffers. Cold Spring Harb Perspect Biol 2:a004051
29. Prins D, Michalak M (2011) Organellar calcium buffers. Cold Spring Harb Perspect Biol 3:a004069
30. Ebashi S (1963) Third component participating in the superprecipitation of "natural actomyosin". Nature 200:1010
31. Gifford JL, Walsh MP, Vogel HJ (2007) Structures and metal-ion-binding properties of the Ca^{2+}-binding helix–loop–helix EF-hand motifs. Biochem J 405:199–221
32. Bindreither D, Lackner P (2009) Structural diversity of calcium binding sites. Gen Physiol Biophys 28 Spec No Focus:F82–F88
33. Chin D, Means AR (2000) Calmodulin: a prototypical calcium sensor. Trends Cell Biol 10:322–328
34. Falcke M (2003) Buffers and oscillations in intracellular Ca^{2+} dynamics. Biophys J 84: 28–41
35. Duman JG, Chen L, Hille B (2008) Calcium transport mechanisms of PC12 cells. J Gen Physiol 131:307–323
36. Mathias RT, Cohen IS, Oliva C (1990) Limitations of the whole cell patch clamp technique in the control of intracellular concentrations. Biophys J 58:759–770
37. Lips MB, Keller BU (1998) Endogenous calcium buffering in motoneurones of the nucleus hypoglossus from mouse. J Physiol 511:105–117
38. Fierro L, Llano I (1996) High endogenous calcium buffering in Purkinje cells from rat cerebellar slices. J Physiol 496:617–625
39. Mogami H, Gardner J, Gerasimenko OV, Camello P, Petersen OH, Tepikin AV (1999) Calcium binding capacity of the cytosol and endoplasmic reticulum of mouse pancreatic acinar cells. J Physiol 518:463–467
40. Foehring RC, Zhang XF, Lee JCF, Callaway JC (2009) Endogenous calcium buffering capacity of substantia nigral dopamine neurons. J Neurophysiol 102:2326–2333
41. Palecek J, Lips MB, Keller BU (1999) Calcium dynamics and buffering in motoneurones of the mouse spinal cord. J Physiol 520:485–502

42. Lee S-H, Rosenmund C, Schwaller B, Neher E (2000) Differences in Ca^{2+} buffering properties between excitatory and inhibitory hippocampal neurons from the rat. J Physiol 525:405–418
43. Ganitkevich VYa, Isenberg G (1995) Efficacy of peak Ca^{2+} currents (ICa) as trigger of sarcoplasmic reticulum Ca^{2+} release in myocytes from the guinea-pig coronary artery. J Physiol 484: 287–306
44. Kawasaki H, Kretsinger RH (2017) Structural and functional diversity of EF-hand proteins: evolutionary perspectives. Protein Sci 26:1898–1920
45. Lee S-H, Schwaller B, Neher E (2000) Kinetics of Ca^{2+} binding to parvalbumin in bovine chromaffin cells: implications for $[Ca^{2+}]$ transients of neuronal dendrites. J Physiol 525: 419–432
46. Nägerl UV, Novo D, Mody I, Vergara JL (2000) Binding kinetics of calbindin-D(28k) determined by flash photolysis of caged Ca^{2+}. Biophys J 79:3009–3018
47. Schwaller B (2009) The continuing disappearance of "pure" Ca^{2+} buffers. Cell Mol Life Sci 66:275–300
48. Faas GC, Schwaller B, Vergara JL, Mody I (2007) Resolving the fast kinetics of cooperative binding: Ca^{2+} buffering by calretinin. PLoS Biol. https://doi.org/10.1371/journal.pbio.0050311
49. Matthews EA, Schoch S, Dietrich D (2013) Tuning local calcium availability: cell-type-specific immobile calcium buffer capacity in hippocampal neurons. J Neurosci 33:14431–14445
50. Brangwynne CP, Koenderink GH, MacKintosh FC, Weitz DA (2008) Cytoplasmic diffusion: molecular motors mix it up. J Cell Biol 183:583–587
51. Schmidt H, Arendt O, Brown EB, Schwaller B, Eilers J (2007) Parvalbumin is freely mobile in axons, somata and nuclei of cerebellar Purkinje neurones. J Neurochem 100:727–735
52. Schmidt H, Brown EB, Schwaller B, Eilers J (2003) Diffusional mobility of parvalbumin in spiny dendrites of cerebellar Purkinje neurons quantified by fluorescence recovery after photobleaching. Biophys J 84:2599–2608
53. Williams RJP (1998) Calcium: outside/inside homeostasis and signalling. Biochim Biophys Acta Mol Cell Res 1448:153–165
54. Williams RJP (2006) The evolution of calcium biochemistry. Biochim Biophys Acta Mol Cell Res 1763:1139–1146
55. Carafoli E (1987) Intracellular calcium homeostasis. Annu Rev Biochem 56:395–433
56. Soar J, Perkins GD, Abbas G et al (2010) European Resuscitation Council guidelines for resuscitation 2010 Section 8. Cardiac arrest in special circumstances: electrolyte abnormalities, poisoning, drowning, accidental hypothermia, hyperthermia, asthma, anaphylaxis, cardiac surgery, trauma, pregnancy, electrocution. Resuscitation 81:1400–1433
57. Carafoli E (2002) Calcium signaling: a tale for all seasons. Proc Natl Acad Sci 99:1115–1122
58. Giorgi C, Danese A, Missiroli S, Patergnani S, Pinton P (2018) Calcium dynamics as a machine for decoding signals. Trends Cell Biol 28:258–273
59. Krebs J, Michalak M (eds) (2007) Calcium: a matter of life or death, vol 41, 1st edn. Elsevier Science, Amsterdam
60. Peacock M (2010) Calcium metabolism in health and disease. Clin J Am Soc Nephrol 5: S23–S30
61. Chan CS, Gertler TS, Surmeier DJ (2009) Calcium homeostasis, selective vulnerability and Parkinson's disease. Trends Neurosci 32:249–256
62. Blair HC, Schlesinger PH, Huang CL-H, Zaidi M (2007) Calcium signalling and calcium transport in bone disease. Subcell Biochem 45:539–562
63. Feske S (2007) Calcium signalling in lymphocyte activation and disease. Nat Rev Immunol 7:690–702
64. Duchen MR, Verkhratsky A, Muallem S (2008) Mitochondria and calcium in health and disease. Cell Calcium 44:1–5
65. Lloyd-Evans E, Waller-Evans H, Peterneva K, Platt FM (2010) Endolysosomal calcium regulation and disease. Biochem Soc Trans 38:1458–1464

66. Bootman MD, Collins TJ, Peppiatt CM et al (2001) Calcium signalling—an overview. Semin Cell Dev Biol 12:3–10
67. Roussel C, Erneux T, Schiffmann SN, Gall D (2006) Modulation of neuronal excitability by intracellular calcium buffering: from spiking to bursting. Cell Calcium 39:455–466
68. Lipp P, Niggli E (1996) A hierarchical concept of cellular and subcellular Ca^{2+}-signalling. Prog Biophys Mol Biol 65:265–296
69. Niggli E, Shirokova N (2007) A guide to sparkology: the taxonomy of elementary cellular Ca^{2+} signaling events. Cell Calcium 42:379–387
70. Stern MD (1992) Buffering of calcium in the vicinity of a channel pore. Cell Calcium 13: 183–192
71. Saris NE, Carafoli E (2005) A historical review of cellular calcium handling, with emphasis on mitochondria. Biochemistry (Mosc) 70:187–194
72. Lee D, Michalak M (2010) Membrane associated Ca^{2+} buffers in the heart. BMB Rep 43:151–157
73. Kaneuchi T, Sartain CV, Takeo S, Horner VL, Buehner NA, Aigaki T, Wolfner MF (2015) Calcium waves occur as Drosophila oocytes activate. Proc Natl Acad Sci 112:791–796
74. Meldolesi J, Pozzan T (1998) The endoplasmic reticulum Ca^{2+} store: a view from the lumen. Trends Biochem Sci 23:10–14
75. Alvarez J, Montero M, García-Sancho J (1999) Subcellular Ca^{2+} dynamics. Physiology 14:161–168
76. Gunter TE, Gunter KK, Sheu SS, Gavin CE (1994) Mitochondrial calcium transport: physiological and pathological relevance. Am J Physiol Cell Physiol 267:C313–C339
77. Mishra J, Jhun BS, Hurst S, O-Uchi J, Csordás G, Sheu S-S (2017) The mitochondrial Ca^{2+} Uniporter: structure, function and pharmacology. Handb Exp Pharmacol 240:129–156
78. Alonso MT, Villalobos C, Chamero P, Alvarez J, García-Sancho J (2006) Calcium microdomains in mitochondria and nucleus. Cell Calcium 40:513–525
79. Sparagna GC, Gunter KK, Sheu S-S, Gunter TE (1995) Mitochondrial calcium uptake from physiological-type pulses of calcium. A description of the rapid uptake mode. J Biol Chem 270:27510–27515
80. Buntinas L, Gunter KK, Sparagna GC, Gunter TE (2001) The rapid mode of calcium uptake into heart mitochondria (RaM): comparison to RaM in liver mitochondria. Biochim Biophys Acta Bioenerg 1504:248–261
81. Putney JW (1986) A model for receptor-regulated calcium entry. Cell Calcium 7:1–12
82. Stathopulos PB, Ikura M (2017) Store operated calcium entry: from concept to structural mechanisms. Cell Calcium 63:3–7
83. Zhou Y, Meraner P, Kwon HT, Machnes D, Oh-hora M, Zimmer J, Huang Y, Stura A, Rao A, Hogan PG (2010) STIM1 gates the store-operated calcium channel ORAI1 in vitro. Nat Struct Mol Biol 17:112–116
84. Rizzuto R, Brini M, Murgia M, Pozzan T (1993) Microdomains with high Ca^{2+} close to IP3-sensitive channels that are sensed by neighboring mitochondria. Science 262:744–747
85. Rizzuto R, Pinton P, Carrington W, Fay FS, Fogarty KE, Lifshitz LM, Tuft RA, Pozzan T (1998) Close contacts with the endoplasmic reticulum as determinants of mitochondrial Ca^{2+} responses. Science 280:1763–1766
86. Gilabert JA, Bakowski D, Parekh AB (2001) Energized mitochondria increase the dynamic range over which inositol 1,4,5-trisphosphate activates store-operated calcium influx. EMBO J 20:2672–2679
87. Dupont G (2014) Modeling the intracellular organization of calcium signaling. Wiley Interdiscip Rev Syst Biol Med 6:227–237
88. Falcke M (2004) Reading the patterns in living cells—the physics of Ca^{2+} signaling. Adv Phys 53:255–440
89. Daub B, Ganitkevich VY (2000) An estimate of rapid cytoplasmic calcium buffering in a single smooth muscle cell. Cell Calcium 27:3–13

90. Lin K-H, Taschenberger H, Neher E (2017) Dynamics of volume-averaged intracellular Ca^{2+} in a rat CNS nerve terminal during single and repetitive voltage-clamp depolarizations. J Physiol 595:3219–3236

91. Chad JE, Eckert R (1984) Calcium domains associated with individual channels can account for anomalous voltage relations of CA-dependent responses. Biophys J 45:993–999

92. Neher E (1986) Concentration profiles of intracellular calcium in the presence of a diffusible chelator. In: Klee M, Neher E, Singer W, Heinemann U (eds) Calcium electrogenesis neuronal functioning. Springer, Berlin, pp 80–96

93. Fogelson AL, Zucker RS (1985) Presynaptic calcium diffusion from various arrays of single channels. Implications for transmitter release and synaptic facilitation. Biophys J 48:1003–1017

94. Fakler B, Adelman JP (2008) Control of KCa channels by calcium nano/microdomains. Neuron 59:873–881

95. Klingauf J, Neher E (1997) Modeling buffered Ca^{2+} diffusion near the membrane: implications for secretion in neuroendocrine cells. Biophys J 72:674–690

96. Pedersen MG, Tagliavini A, Cortese G, Riz M, Montefusco F (2017) Recent advances in mathematical modeling and statistical analysis of exocytosis in endocrine cells. Math Biosci 283:60–70

97. Augustine GJ, Santamaria F, Tanaka K (2003) Local calcium signaling in neurons. Neuron 40:331–346

98. Demuro A, Parker I (2006) Imaging single-channel calcium microdomains. Cell Calcium 40:413–422

99. Smith GD, Wagner J, Keizer J (1996) Validity of the rapid buffering approximation near a point source of calcium ions. Biophys J 70:2527–2539

100. Neher E (1998) Usefulness and limitations of linear approximations to the understanding of Ca++ signals. Cell Calcium 24:345–357

101. Gilkey JC, Jaffe LF, Ridgway EB, Reynolds GT (1978) A free calcium wave traverses the activating egg of the medaka, Oryzias latipes. J Cell Biol 76:448–466

102. Jaffe LF (2008) Calcium waves. Philos Trans R Soc Lond Ser B Biol Sci 363:1311–1317

103. Jaffe LF (1993) Classes and mechanisms of calcium waves. Cell Calcium 14:736–745

104. Woods NM, Cuthbertson KSR, Cobbold PH (1987) Agonist-induced oscillations in cytoplasmic free calcium concentration in single rat hepatocytes. Cell Calcium 8:79–100

105. MacQuaide N, Dempster J, Smith GL (2007) Measurement and modeling of Ca^{2+} waves in isolated rabbit ventricular cardiomyocytes. Biophys J 93:2581–2595

106. Thul R, Bellamy TC, Roderick HL, Bootman MD, Coombes S (2008) Calcium oscillations. Adv Exp Med Biol 641:1–27

107. Williams GSB, Molinelli EJ, Smith GD (2008) Modeling local and global intracellular calcium responses mediated by diffusely distributed inositol 1,4,5-trisphosphate receptors. J Theor Biol 253:170–188

108. Skupin A, Kettenmann H, Falcke M (2010) Calcium signals driven by single channel noise. PLoS Comput Biol. https://doi.org/10.1371/journal.pcbi.1000870

109. Schuster S, Marhl M, Höfer T (2002) Modelling of simple and complex calcium oscillations. Eur J Biochem 269:1333–1355

110. Sneyd J, Keizer J, Sanderson MJ (1995) Mechanisms of calcium oscillations and waves: a quantitative analysis. FASEB J 9:1463–1472

111. Matveev V, Sherman A, Zucker RS (2002) New and corrected simulations of synaptic facilitation. Biophys J 83:1368–1373

112. Zador A, Koch C (1994) Linearized models of calcium dynamics: formal equivalence to the cable equation. J Neurosci 14:4705–4715

Chapter 8
An Update to Calcium Binding Proteins

Jacobo Elíes, Matilde Yáñez, Thiago M. C. Pereira, José Gil-Longo, David A. MacDougall, and Manuel Campos-Toimil ⓘ

Abstract Ca^{2+} binding proteins (CBP) are of key importance for calcium to play its role as a pivotal second messenger. CBP bind Ca^{2+} in specific domains, contributing to the regulation of its concentration at the cytosol and intracellular stores. They also participate in numerous cellular functions by acting as Ca^{2+} transporters across cell membranes or as Ca^{2+}-modulated sensors, i.e. decoding Ca^{2+} signals. Since CBP are integral to normal physiological processes, possible roles for them in a variety of diseases has attracted growing interest in recent years. In addition, research on CBP has been reinforced with advances in the structural characterization of new CBP family members. In this chapter we have updated a previous review on CBP, covering in more depth potential participation in physiopathological processes and candidacy for pharmacological targets in many diseases. We review intracellular CBP that contain the structural EF-hand domain: parvalbumin, calmodulin, S100 proteins, calcineurin and neuronal Ca^{2+} sensor proteins (NCS). We also address intracellular CBP lacking the EF-hand domain: annexins, CBP within intracellular Ca^{2+} stores (paying special attention to calreticulin and calsequestrin), proteins that

Authors Jacobo Elíes and Matilde Yáñez have contributed equally for this chapter

J. Elíes
Pharmacology and Experimental Therapeutics, Faculty of Life Sciences, University of Bradford, Bradford, UK

M. Yáñez · J. Gil-Longo · M. Campos-Toimil (✉)
Pharmacology of Chronic Diseases (CD Pharma), Centro de Investigación en Medicina Molecular y Enfermedades Crónicas (CIMUS), Universidad de Santiago de Compostela, Santiago de Compostela, Spain
e-mail: manuel.campos@usc.es

T. M. C. Pereira
Pharmaceutical Sciences Graduate Program, Vila Velha University (UVV), Vila Velha, ES, Brazil

Federal Institute of Education, Science and Technology (IFES), Vila Velha, ES, Brazil

D. A. MacDougall
Research and Enterprise, University of Huddersfield, Huddersfield, UK

© Springer Nature Switzerland AG 2020
M. S. Islam (ed.), *Calcium Signaling*, Advances in Experimental Medicine and Biology 1131, https://doi.org/10.1007/978-3-030-12457-1_8

contain a C2 domain (such as protein kinase C (PKC) or synaptotagmin) and other proteins of interest, such as regucalcin or proprotein convertase subtisilin kexins (PCSK). Finally, we summarise the latest findings on extracellular CBP, classified according to their Ca^{2+} binding structures: (i) EF-hand domains; (ii) EGF-like domains; (iii) ɣ-carboxyl glutamic acid (GLA)-rich domains; (iv) cadherin domains; (v) Ca^{2+}-dependent (C)-type lectin-like domains; (vi) Ca^{2+}-binding pockets of family C G-protein-coupled receptors.

Keywords Annexins · Ca^{2+} sensors · Calcineurin · Calmodulin · Calreticulin · EF-hand domain · Parvalbumin · Protein kinase C · S100 proteins · Synaptotagmin

8.1   Introduction

Calcium (Ca^{2+}) is a ubiquitous and highly versatile intracellular signal that operates over a wide spatial and temporal range to regulate many different cellular processes [1]. However, cellular Ca^{2+} overload can be cytotoxic and therefore a homeostatic system is necessary to regulate ionic balance. Calcium binding proteins (CBP), grouped together into a very large and heterogeneous family, not only regulate Ca^{2+} homeostasis but also control numerous Ca^{2+} signalling pathways [2]. Those that regulate Ca^{2+} levels are mainly membrane proteins (Ca^{2+} pumps) which maintain low cytosolic free Ca^{2+} concentrations under resting conditions (~100 nM) to avoid calcium precipitation or excess of Ca^{2+} signal activity. Other CBP regulate a plethora of cellular functions by interacting with and modulating a wide range of proteins.

This review presents an update on discoveries pertaining to intracellular CBP with or without EF-hand domains, as well as extracellular CBP, in health and disease, based on our previous review [3]. Here we focus on CBP cellular functions and diseases associated with mutations and dysregulation of CPB, but not on the structural and Ca^{2+}-binding affinities of the CPB previously discussed [3].

We have also gathered evidence of how recently-developed experimental tools such as genetically encoded Ca^{2+} indicators (GECI), mutagenesis studies combined with in vivo calcium imaging, optogenetics and chemogenetics, have contributed to a better understanding of cellular Ca^{2+} signalling.

8.2   Intracellular Ca^{2+} Binding Proteins with EF-Hand Domains

The superfamily of EF-hand proteins includes a large number of members which share a common structural motif consisting of two alpha helices oriented perpendicular to each other (Fig. 8.1). The loop integrated in this sequence can

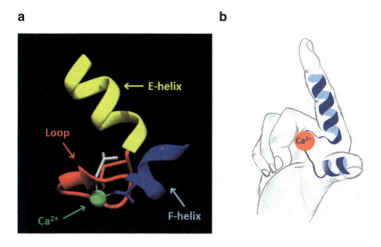

Fig. 8.1 (**a**) Representation of a EF-hand motif constituted by two alpha helices (E and F) perpendicularly placed and linked by a short loop region that facilitates Ca^{2+} binding. (**b**) The spatial arrangement of the EF-hand motif mimics the spread thumb and the index finger of a human right hand

accommodate Ca^{2+} or Mg^{2+} with distinct geometries and the affinity for these ions is a determining factor for the function of the protein [4].

Commonly, EF-hand motifs occur in adjacent pairs giving rise to different structural and functional proteins. Conformational changes induced in EF-hand regulatory proteins usually lead to an increase in enzymatic activity or in signal transduction between cellular compartments [5]. Meanwhile, structural EF-hand domains play an important role in calcium buffering in the cytosol [3].

8.2.1 Parvalbumin Family Proteins

Traditionally, parvalbumin (PV) was considered a cytosolic Ca^{2+}-binding protein acting as a slow-onset Ca^{2+} buffer that modulates the shape of Ca^{2+} transients in fast-twitch muscles and a subpopulation of neurons. However, PV is also widely expressed in non-excitable cells like distal convoluted tubule (DCT) cells of the kidney, where it might act as an intracellular Ca^{2+} shuttle facilitating transcellular Ca^{2+} resorption by influencing mitochondrial Ca^{2+} buffering [6]. Two isoforms, alpha and beta parvalbumin, exist in vertebrates and are associated with several calcium-mediated cellular activities and physiological processes.

Parvalbumin-expressing GABAergic interneurons (PVIs), present in different brain regions, play a role in short-term synaptic plasticity [7], high frequency neuronal synchronization, maintaining a proper excitatory/inhibitory balance, and voluntary movement tasks [8]. Their activity supports critical developmental trajectories, sensory and cognitive processing, and social behaviour [9–13]. Alterations in PVIs are commonly observed in post mortem brains of schizophrenia patients and are reported in bipolar and autism spectrum disorders [14]. Furthermore, dysregulation of PV (and calretinin) is associated with the development of affective disorders [15], anxiety, and fear extinction [16].

The use of designer receptors exclusively activated by designer drugs (DREADD) technology, a selective noninvasive chemogenetic approach, allowed the characterization of specific and reversible activation of PVIs in subregions of the hippocampus associated with social behaviour such as the dentate gyrus (DG) [17].

PVIs can generate feedforward inhibition that opposes seizure spread in both experimental models [18] and patients [19]. Moreover, in vivo optogenetic studies revealed that PVI activation interrupts spontaneous ongoing seizures [20–22].

Importantly, distribution of CBP (regulators of intracellular Ca^{2+} levels) is associated with functionally distinct neuronal subpopulations (with different neurotransmitter profiles), suggesting that CBP can serve as anatomical (and potentially functional) markers of locomotor network as recently demonstrated by Ca^{2+} imaging studies in zebra fish [23].

8.2.2   Calmodulin Family Proteins

The calmodulin family, represented by calmodulin, troponin C and essential and regulatory myosin light chains (ELC and RLC of myosin), is one of the most extensively characterized sets of the EF-hand Ca^{2+} sensor proteins. Calmodulin is a ubiquitous Ca^{2+} sensor molecule encoded by 3 distinct genes, CALM1-3. CALM modulates the activity of various proteins including ion channels [24–26] which play important roles in the generation and profile of cardiomyocyte action potentials. Mutagenesis studies in cardiac cells have identified that mutations in CALM genes are associated with severe early-onset of congenital long-QT syndrome (LQTS) [27–29], and idiopathic ventricular fibrillation [30]. A recent study of the calmodulin interactome characterised a pivotal role for this CBP in invadopodia formation associated with the invasive nature of glioblastoma multiforme (GBM) cells [31].

Troponin C, as a part of the troponin complex, is present in all striated muscle, being the protein trigger that initiates myocyte contraction [32]. Two isoforms of this protein have been described: fast skeletal muscle troponin C, which is activated by Ca^{2+} binding to two low-affinity sites on the N-terminal domain, and slow skeletal (and cardiac) muscle troponin C, which is activated by Ca^{2+} binding to a single affinity site [33].

ELC and RLC of myosin bind to the neck region (approximately 70 amino acids) of myosin heavy chain; the neighbouring head region contains the globular catalytic domain, responsible for binding to actin and hydrolysing of ATP [34].

Phosphorylation of RLC generates a structural signal transmitted between myosin molecules in the thick filament and finally to the thin filaments (actin), which forms the basis of contractile regulation in cardiac muscle [35].

8.2.3 S100 Family Proteins

The S100 proteins, a family of Ca^{2+}-binding cytosolic proteins expressed exclusively in vertebrates, constitute the major family of EF-hand calcium sensor proteins. S100 proteins are characterised by the presence of a unique S100-specific Ca^{2+}-binding loop named "pseudo EF-hand" [36], often involved in the formation of homo- or heterodimers, and the ability to bind other divalent metals such as Zn^{2+} and Cu^{2+} [37, 38].

S100 proteins have a wide range of intracellular and extracellular functions through regulating calcium balance, cell apoptosis, migration, differentiation, energy metabolism and inflammation [39–41].

Initial research showed that S100 proteins are involved in cell growth, division and differentiation [38, 42–46]. Nonetheless, more recent contributions have demonstrated the role of S100 proteins in cell migration and invasion [39, 40, 47, 48], neuronal plasticity [49, 50], cartilage repair [51], inflammation [41], and several types of cancer including lung [52], ovarian [53], pancreatic [54], and melanoma [55, 56].

The S100 family of calcium-binding proteins are gaining importance as both potential molecular key players and biomarkers in the aetiology, progression, manifestation, and therapy of neoplastic disorders, including lymphoma, pancreatic cancer and malignant melanoma [57]. For example, S100A2 is downregulated in melanoma, whilst S100A1, S100A4, S100A6, S100A13, S100B, and S100P are upregulated [56].

8.2.4 Calcineurin

Calcineurin is classified as a calmodulin-dependent serine/threonine phosphatase and is ubiquitously expressed in lower and higher eukaryotes [58]. Calcineurin plays a pivotal role in the information flow from local or global Ca^{2+} signals to effectors that control immediate cellular responses and alter gene transcription. It is a heterodimeric protein consisting of a catalytic A subunit (CNA), which is highly homologous to protein phosphatases 1 and 2, and a regulatory B subunit (CNB), that contains four EF-hand motifs and binds to CNA to regulate its phosphatase activity even in the absence of Ca^{2+} [59]. A malfunction in calcineurin-NFAT signalling can engender several pathologies, such as cardiac hypertrophy, autoimmune diseases, osteoporosis, and neurodegenerative diseases [60–62].

8.2.5 Neuronal Ca²⁺ Sensor (NCS) Proteins

Frequenin was the first NCS protein to be discovered and was originally designated NCS-1, due to its distribution in neuronal cell types [63]. Fourteen NCS proteins have since been identified, and classified into A-E subgroups on the basis of their amino acid sequences, including NCS-1, hippocalcin, neurocalcin-δ, VILIP1-3, recoverin, GCAP1-3 and KChIP1-4 [64]. Most are expressed only in neurons, where they have different roles in the regulation of neuronal function [65]. For instance, direct interaction of NCS1 and calneuron-1 with the GPCR cannabinoid CB1 Receptor (CB1R) determines cAMP/Ca^{2+} crosstalk which regulates the action of endocannabinoids [66]. Dysregulation of NCS proteins have been observed in several CNS disorders, such as Alzheimer's disease, schizophrenia, and cancer [67–69].

8.3 Intracellular Ca²⁺ Binding Proteins Without EF-Hand Domains

An important and heterogeneous group of proteins capable of binding Ca $^{2+}$ but lacking the EF-hand domain are also found within eukaryotic cells. Important functional roles have been described for most of them, which has led to many being considered new potential therapeutic targets in various pathologies (see below). This is the case for annexins, whose role in different types of cancer, among other pathologies, is currently under investigation. Other CBP, such as calreticulin and calsequestrin, perform a fundamental role in Ca^{2+} homeostasis, since they fix the ion within the endoplasmic/sarcoplasmic reticulum (ER/SR), facilitating the role of these organelles as intracellular Ca^{2+} reservoirs. In addition, there are other proteins, such as regucalcin, calcium-and integrin-binding protein 1 (CIB1), proprotein convertases, or those that share a Ca^{2+}-binding domain named C2, whose Ca^{2+} binding ability serves as a regulatory mechanism for several cellular functions.

8.3.1 Annexins

Annexins, also known as lipocortins, are a multigene superfamily of Ca^{2+}-dependent phospholipid- and membrane-binding proteins. We have previously reviewed their classification and structure [3]. Briefly, annexins are classified into 5 families (A–E) [70–72]. The annexin A family, which is common to humans and other vertebrates, contains 12 members (annexin A1–11 and A13). Their structure consists of a highly conserved COOH-terminal core domain and a NH_2-terminal region that shows marked diversity and regulates membrane association and interaction with protein ligands [72, 73] (Fig. 8.2).

Calreticulin

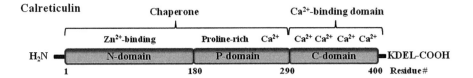

Annexins

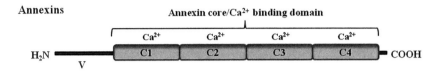

Classic PKCs

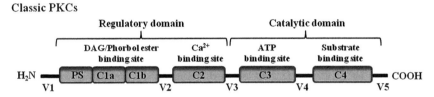

Fig. 8.2 Domain structures of calreticulin, annexins and classic PKCs. Calreticulin contains a KDEL signal that is responsible for the retrieval of ER proteins from the Golgi complex; *C* conserved regions, *V* variable regions, *PS* pseudosubstrate binding site. For more details, see main text

Upon binding and activation by Ca^{2+}, annexins undergo a conformational change which allows them to bind negatively charged membrane phospholipids and form a ternary complex bridging adjacent membranes [74].

Annexins can participate in a large number of cellular processes including anti-coagulation, anti-inflammation, endocytosis and exocytosis, membrane aggregation and fusion, regulation of membrane dynamics and organization, signal transduction, cell division, cell proliferation, differentiation, growth regulation and apoptosis (for detailed review see e.g. [75, 76]). Accordingly, their involvement in different disease states, including cancer, diabetes and inflammatory pathologies, has been extensively studied, and they have consequently become promising pharmacological targets [75–77].

Annexin A1 in particular has received much interest from multiple perspectives in the search for novel therapies. It participates in many important biological processes that encompass cell migration, recruitment, permeability, apoptosis, phagocytosis and proliferation [78, 79]. It makes up 2–4% of the total cytosolic protein in some cell types and can also be found in the nucleus [80]. The expression level of annexin A1 is modified in various tumours, suggesting a potential influence on tumorigenic and metastatic processes [81, 82] and has been found in breast cancer and squamous cancer cells [83, 84]. Annexin A1 may act as a modulator of inflammation and inflammatory pain [85, 86]. In fact, annexin 1 replicates anti-inflammatory effects due to its potential for glucocorticoid sparing in rheumatoid arthritis [87]. As a result of these properties, annexin A1-based therapies could be used in myocardial ischaemia-reperfusion injury, limiting neutrophil infiltration and preserving both cardiomyocyte viability and left ventricular contractile function, thereby offering a novel route for treating myocardial inflammatory disorders [88].

Annexin A2 also fulfils a wide range of biological functions both at the plasma membrane and within multiple intracellular compartments [89]. It regulates tumour cell adhesion, proliferation, invasion, metastasis and tumour neovasculogenesis, thus playing a crucial role in tumour development, which in turn means it is a potential therapeutic target for efficient molecular-based strategies in tumour treatment [90] as well as a putative cancer biomarker [91, 92]. Other pathologies in which annexin A2 is identified as a potential target are ulcerative colitis [93], thrombosis and lupus [94].

Concerning annexin A3, its function either as a tumour suppressor or as a tumour promoter candidate for different cancers depending on the types of tumour cells and tissues has been investigated [95].

Annexin A4 is considered a promising therapeutic target for the treatment of platinum-resistant cancers [96]. Furthermore, studies on the function of this protein on tumour tissues have great potential importance not only for understanding cancer progression but also for developing diagnostic and therapeutic approaches [97].

Similarly, annexin A5 contributes to several aspects of tumour progression and drug resistance in certain types of cancer, so it can be used as a therapeutic target for broad applications in the diagnosis, treatment, and prognosis of tumours [98]. Other important functions that have been described for annexin A5 are its role in cell membrane repair [99, 100] and in the maintenance of pregnancy since it acts as an immunomodulator and anticoagulant at the level of the placenta [101].

Finally, other annexins with the potential to be pharmacological targets are: annexin A6 (which regulates converging steps of autophagy and endocytic trafficking in hepatocytes [102]; annexin A7 (abnormal expression of the Anxa7 gene is associated with several pathologies, including glioblastoma, melanoma, urinary bladder transitional cell carcinoma and Hodgkin lymphoma ovarian carcinoma [103]; and annexin A11 (dysregulation and mutation of this protein are involved in systemic autoimmune diseases and sarcoidosis, and are associated with the development, chemoresistance and recurrence of cancers [104].

8.3.2 Ca^{2+} Binding Proteins at Intracellular Ca^{2+} Stores

The SR and ER of eukaryotic cells act as intracellular reservoirs of readily-releasable Ca^{2+} ions. Several CBP within these organelles are essential for this function. Calreticulin, BiP/Grp78, glucose-regulated protein 94 (Grp 94) and protein disulfide isomerase (PDI) are CBP that participate in smooth muscle SR/ER-dependent Ca^{2+} homeostasis and act as protein chaperones [105, 106].

Amongst these CBP, calreticulin is arguably the most important. This protein consists of three distinct structural and functional domains (Fig. 8.2) [3, 107]. In addition to SR/ER, it is also present on the membrane of other subcellular organelles, cell surface and in the extracellular environment where it contributes to different physiological and pathological processes [108]. Calreticulin regulates Ca^{2+} uptake and release within the ER and mitochondria [109, 110], and it is a central component of the folding quality control system of glycoproteins [111, 112].

Outside of the ER, calreticulin also regulates critical biological functions including cell adhesion, gene expression, and RNA stability [107]. New roles of calreticulin in the extracellular space such as an involvement in cutaneous wound healing and possible diagnostic applications of calreticulin in blood or urine have also emerged [108].

A possible role for calreticulin in the development of several human pathologies, including congenital arrhythmias and some cancers (including oral, esophagus, breast, pancreas, gastric, colon, bladder, prostate, vagina, ovarian and neuroblastoma) has been described [107, 109, 112, 113]. Mutations in the calreticulin gene, for example, are present in myeloproliferative neoplasms [114]. Thus, the multifaceted properties of calreticulin may in the future provide a means to treat a number of diseases [115].

BiP/Grp78 is an immunoglobulin binding protein that can bind Ca^{2+} at relatively low stoichiometry [116]. As a consequence, its over-expression increases the Ca^{2+} pool available for transfer to mitochondria. Grp94 protein is a high affinity CBP whose inhibition induces Ca^{2+}-mediated apoptosis [117]. PDI is a major CBP of the ER that is also involved in protein folding and isomerization [118]. Increasing evidence suggests that PDI supports the survival and progression of several cancers. However, no PDI inhibitor has been approved for clinical use [119].

The storage and rapid release of Ca^{2+} from the skeletal and cardiac SR reservoirs has been associated with calsequestrin [120, 121], as previously reviewed [3]. Both calsequestrin 1 and calsequestrin 2 subtypes act as SR luminal sensors for skeletal or cardiac ryanodine receptors together with the proteins triadin and junctin [122, 123] (Fig. 8.3). Calsequestrin communicates changes in the luminal Ca^{2+} concentration to the cardiac ryanodine receptor channel [124] and it has been demonstrated that a lack of this protein causes important structural changes in the SR and alters the storage and release of appropriate levels of SR Ca^{2+} [125]. Alterations in calsequestrin expression are an underlying cause of cardiac complications. For example, loss of calsequestrin 2 causes abnormal SR Ca^{2+} release and selective interstitial fibrosis in the atrial pacemaker complex (which disrupts sinoatrial node pacemaking but enhances latent pacemaker activity) and creates conduction abnormalities and increased susceptibility to atrial fibrillation [126]. Also, a change in calsequestrin expression is involved in the pathogenesis of Duchenne progressive muscular dystrophy [127]. In addition, calsequestrin-1 knockout mice suffer episodes of exertional/environmental heatstroke when exposed to strenuous exercise and environmental heat, and aerobic training significantly reduces mortality rate by lowering oxidative stress [128].

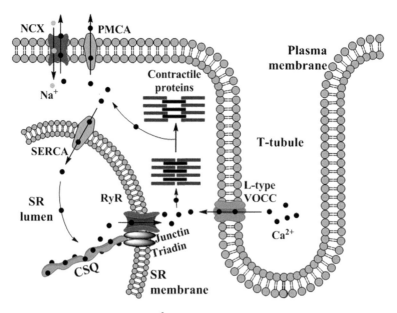

Fig. 8.3 Calsequestrin (CSQ) binds Ca^{2+} within the SR of cardiac myocytes. It is attached to the ryanonide receptor through the junctin-triadin protein complex. *NCX* Na^{2+}/Ca^{2+} exchanger, *PMCA* plasma membrane Ca^{2+} ATPase, *VOCC* voltage-operated Ca^{2+} channel, *SERCA* sarcoplasmic/endoplasmic reticulum Ca^{2+} ATPase, *RyR* ryanodine receptor

8.3.3 Ca^{2+} Binding Proteins with C2 Domains

The C2 domain is a Ca^{2+}-binding motif which also has the ability to bind phospholipids, inositol polyphosphates and some other intracellular proteins. It was first identified as Conserved Domain 2 of the PKC kinase family, hence its name [129, 130]. It has been found in well over two hundred different proteins, making it the second most common lipid binding domain [3, 131] and it mediates a wide range of intracellular processes, such as membrane trafficking, generation of lipid-second messengers, activation of GTPases, and control of protein phosphorylation [132]. Some of the best-known proteins that contain a C2 domain are PKC, synaptotagmins, phospholipase C (PLC) and phospholipase A (PLA).

PKC is composed of a large family of lipid-activated enzymes that regulate the function of other proteins via phosphorylation of serine and threonine residues (Fig. 8.2). Multiple isoforms of PKC, ten of them found in mammals, can exist in the cytosol in a soluble form or bind to the plasma membrane, participating in many functions, such as intracellular signalling, secretion, cell growth and differentiation [133, 134].

PKC activity depends on the presence of several lipid cofactors and Ca^{2+}, although these requirements vary for different isozymes. Three categories have been established for mammalian PKCs: classical, novel and atypical isoforms, the last lacking the C2 domain [134, 135]. Classical isozymes contain a C2 type I Ca^{2+}- and phospholipid-binding domain and require Ca^{2+} in order to stabilize interaction with the plasma membrane [133]. On the other hand, novel isozymes contain a variant of the C2 domain (type 2) that lacks key Ca^{2+}-coordinating residues; as a result, novel isozymes are not sensitive to Ca^{2+} [135]. Therefore, classical PKC isoforms are activated by Ca^{2+} and diacylglycerol (DAG), whereas novel PKC isoforms are activated by DAG, but not Ca^{2+}. Atypical PKC isoforms require neither Ca^{2+} nor DAG for activation [136].

A role for different PKC isozymes has been described in a large number of diseases, including cardiac pathologies, cancer, dermatological diseases, lung and kidney diseases, autoimmune diseases, neurological diseases and some others (for detailed review see e.g. [137]).

Synaptotagmins are a family of transmembrane Ca^{2+} sensors found in synaptic vesicles and in secretory granules of endocrine cells that have a main role in exocytosis [2, 138, 139]. They bind Ca^{2+} with low affinity by means of two C2 domains: C2A and C2B [140]. The most studied protein in this family is synaptotagmin I, which functions as a sensor for evoked, synchronous neurotransmitter release in neurons [141].

Otoferlin is another transmembrane protein with a C2 domain that binds Ca^2 and membranes and triggers the fusion of neurotransmitter-filled vesicles with the plasma membrane, in conjunction with a specific ensemble of molecular machinery proteins [142]. Its significance in priming and fusion of synaptic vesicles during sound encoding is clear, and mutations in otoferlin cause human deafness [143].

Other proteins involved in the modification of lipids that also contain functional Ca^{2+}-binding C2 domains are phosphoinositide-specific PLC, which liberates IP_3 and DAG in response to mitogenic signals that raise intracellular Ca^{2+} levels [144] and PLA2, which liberates arachidonic acid from glycerophospholipids to initiate production of leukotrienes and prostaglandins, potent mediators of inflammation [145].

8.3.4 Other Intracellular CBP Without EF-Hand Domains

Regucalcin is a CBP with multiple physiological functions that has been localised to the cell nucleus and cytoplasm, as well as in the mitochondrial fraction [146]. It is involved in aging, participating in brain calcium signalling [147]. Also, it may be a key molecule in lipid metabolic disorder and diabetes [148]. In addition, regucalcin is implicated in carcinogenesis and the enhancement of regucalcin gene expression may reveal preventive and therapeutic effects in the progression of cancer cells. Thus, it has been suggested that targeting the regucalcin gene could be a useful tool in cancer therapy [149, 150].

Proprotein convertase subtilisin kexins (PCSK) are a family of CBP that activate other proteins. Nine subtypes of PCSKs with varying functions and tissue distributions have been described [151]. Some of these proteins have been studied as possible therapeutic targets. For example, inhibition of PCSK9 holds considerable promise as a therapeutic option for decreasing cardiovascular disease risk, since it plays an important role in the regulation of cholesterol homeostasis [152]. Also, furin has been the most intensively researched member of the family with regard to tumor regulation, promotion, and progression [153] and there are numerous patents related to the uses of this protein and its inhibitors as therapeutics [154].

8.4   Extracellular Ca^{2+} Binding Proteins

The extracellular Ca^{2+} (Ca^{2+}_o) concentration in mammals is maintained at about 1.2 mM [155]. Deviations in this concentration lead to severe pathological malfunctions. Different organs and hormones must cooperate to regulate the uptake, excretion and recycling of calcium in the body and, as a consequence, serum Ca^{2+} concentration. To regulate Ca^{2+}_o homeostasis in the body, cells must be equipped with the ability to monitor its level. In this regard, several extracellular Ca^{2+} binding proteins (ECBP) have been discovered and their roles investigated. The study of extracellular Ca^{2+} as a messenger has probably been hampered by the generally accepted (although erroneous) view that the Ca^{2+} concentration in the extracellular space does not fluctuate, as well as by the technical difficulties inherent in measuring spatial and temporal changes in extracellular Ca^{2+} concentration [156].

If the Ca^{2+} concentration in the extracellular space did not differ from that measured in serum, Ca^{2+} would not play an extracellular messenger role. ECBP would only use Ca^{2+} for static roles (e.g. formation of active sites in enzymes or active conformations in receptors, structure protein stabilization and formation of supramolecular structures with other proteins or carbohydrates) or to maintain extracellular Ca^{2+} homeostasis.

Despite the constant concentration of serum Ca^{2+}, there are clear demonstrations that Ca^{2+} levels of interstitial fluids in many tissues differ from those usually measured in serum and, do indeed fluctuate (e.g. fluctuations are very likely to occur during intracellular Ca^{2+} signalling events) [156, 157]. Therefore, extracellular Ca^{2+} could fulfil a more dynamic function as a 'first messenger' in extracellular signal transduction pathways and contributor to autocrine/paracrine cell-to-cell communication. This role for Ca^{2+}_o is possible only if ECBP undergo a conformational change in response to physiological fluctuations in extracellular Ca^{2+} concentrations; this would allow interactions with, and modulation of, specific target proteins.

ECBP are modulators of numerous cellular functions (e.g. neuronal signalling, blood-clotting, complement activation, cell-cell interactions, cell-matrix interactions, receptor-ligand interactions, Ca^{2+} transport and Ca^{2+} homeostasis, cardiovascular remodelling, tumour cell migration and cancer metastasis regulation, gene expression, apoptosis, and more) and may serve as important therapeutic targets (see the following sections). In this section we review the main ECBP, focussing on some proteins that could be Ca^{2+}_o sensors and mediators of Ca^{2+}_o signalling. We have grouped ECBP by shared Ca^{2+}-binding domain structures (also see Table 8.1).

8.4.1 Extracellular Ca^{2+} Binding Proteins with EF-Hand Domains

Ca^{2+}-binding proteins containing a pair of EF-hand motifs are present within cells as described in Sect. 8.2, and in the extracellular environment (or matrix). EF-hand proteins are widely found in animal genomes and distributed throughout the cell [158, 159]. These proteins are fundamental for many cellular functions and at the same time are associated with neuronal diseases, cardiac arrhythmias, cancer and autism [160]. Physiologically, all EF-hand proteins can be divided into two groups: (1) calcium buffers, controlling the level of free cytosolic Ca^{2+} or (2) calcium sensors, acting to translate the signal to various responses [160, 161]. In this context, some examples of these proteins will be explained below.

Osteonectin, also known as BM-40 or SPARC (secreted protein acidic and rich in cysteine), is a 32-kDa calcium-binding glycoprotein matrix protein that serves as the prototype of the osteonectin family [162]. Other members of this family are hevin, QR1, testicans 1-3, tsc 36, SMOC-1 and SMOC-2. The osteonectin family is characterized by a follistatin-like, cysteine rich domain and a C-terminal module with two EF-hand Ca^{2+}-binding domains; each EF-hand domain is predicted to bind one Ca^{2+} ion [163]. The EF-hand pair is very similar to those of intracellular EF-hand proteins such as calmodulin. The affinity of the Ca^{2+}-binding domain for Ca^{2+} is high; for example, osteonectin binds 2 Ca^{2+} with a K_{d1} of 490 nM and K_{d2} of 26 nM [164]. The osteonectin-like proteins modulate cell function by interacting with cell-surface receptors, metalloproteinases, growth factors and other bioeffector molecules and proteins of the matrix such as collagens, secreted by many types of cells, e.g. endothelial cells, fibroblasts, and fragments of megakaryocyte-platelets [162, 165]. Different lines of evidence link osteonectin-like proteins with human cancer progression [166, 167].

Another example is the FKBP65 (or FKBP10) protein. Its molecular structure includes four PPIase–FKBP type domains that are involved in activation of the release of ER Ca^{2+} stores. Moreover, its expression coincides with increased expression of tropoelastin and type I collagen expression [168, 169]. Although not all of its biological roles are known, emerging genetic studies have shown that, in humans, mutation of this protein might cause a form of osteogenesis imperfecta, a brittle bone disease resulting from deficient secretion of mature type I collagen [168, 169].

Table 8.1 Main extracellular Ca^{2+}- binding domain structures

Group and calcium-binding domain structure	Main proteins	Possible role of Ca^{2+}- binding to proteins
EF-hand domains	Osteonectin family: osteonectin, hevin, QR1, testicans 1–3, tsc 36, SMOC-1 and SMOC-2, FKBP65 (or FKBP10)	Formation of binding sites for extracellular ligands
		Ca^{2+} signal transmission?
EGF-like domains	EGF protein, Neuregulin, Transforming growth factor α, Coagulation factors VII, IX and X, protein C and protein S	Induction of protein conformation required for biological activity
	Fibrillin	Stabilization of proteins
	Notch and delta receptors	Ca^{2+} signal transmission?
	LDL receptors	
γ-Carboxyl glutamic acid-rich domains	Coagulation factors II, VII, IX and X, protein C and protein Z	Anchoring of proteins to membrane
	Osteocalcin, matrix GLA protein and periostin, telopeptide of type I collagen, bone alkaline phosphatase	Activation of proteins
	Growth arrest-specific protein 6, matrix Gla protein-MGP	
Cadherin domains	Cadherin family: classical cadherins, protocadherins, and atypical cadherins (Fat, Dachsous, and Flamingo)	Modulation of mechanical integrity and mechanotransduction capability of proteins (using an adaptor complex, cadherins connect to the cytoskeleton)
		Ca^{2+} signal transmission
(C)-type lectin-like domains	Selectins	Modulation of ligand binding
	Mannose receptor family	Stabilization of proteins
	Dendritic cell-specific ICAM-3 grabbing non-integrin molecule	Cell-cell adhesion
	Several collectins (e.g. mannose-binding protein)	Apoptotic process
Ca^{2+}-binding pockets of family C GPCRs	Ca^{2+}-sensing receptor	Change of receptor conformation
	Metabotropic glutamate receptors	Ca^{2+} signal transmission through G-proteins
	GABA$_B$ receptors	

All Ca^{2+}-binding domain structures have in common a highly negative surface potential usually associated with Asp or Glu residues

8.4.2   Extracellular Ca^{2+} Binding Proteins with EGF-Like Domains

The epidermal growth factor (EGF)-like domain is one of the most widely distributed disulfide-containing domains in nature and is involved in multiple cellular regulations [170]. Its name derives from the epidermal growth factor where it was first described [171]. Normally, EGF domains contain six cysteine residues that form disulphide bridges. Although many EGF-like domains participate in Ca^{2+}-dependent processes by responding to local Ca^{2+} concentrations, very little detail has been resolved concerning how this regulation is programmed at the molecular level [170]. What is known is that a subset of EGF-like domains also contains a Ca^{2+}-binding domain with a wide range of high Ca^{2+} affinities (K$_d$ from 0.1 mM to nM values) and these represent prevalent extracellular Ca^{2+}-binding sites [172–179].

Among the proteins containing Ca^{2+}-binding EGF-like domains are those involved in cell growth (e.g. EGF, neuregulin and transforming growth factor α), blood coagulation, fibrinolysis and the complement system (e.g. factors VII, IX and X, protein C and protein S), matrix proteins (e.g. laminin, fibrillin and nidogen) and cell surface receptors (e.g. selectins, low density lipoprotein receptor and Notch receptor and their homologues) [3, 176, 177]. The coagulation enzymes, factors VII, IX and X and protein C, all have two EGF-like domains, whereas the cofactor of activated protein C, protein S, has four EGF-like domains in tandem [178]. On the other hand, fibrillin, low density lipoprotein receptor and the developmentally important receptor Notch have numerous EGF-like domains in tandem that might mediate heterophilic interactions with other family members based on binding between the EGF-like repeats of adjacent receptors [176, 179–181].

Additionally, recent studies have demonstrated that Ca^{2+} binding to an EGF-like domain is important to orient neighbouring domains and to induce the protein conformation required for biological activity [170, 178, 182]. For example, Ca^{2+} binding to an EGF-like domain contributes to protein stability [181, 182]. Moreover, the EGF-like domains seem to be involved in protein-protein interactions, receptor-ligand interactions and blood coagulation [175, 181]. Interestingly, some studies have demonstrated that mutations in EGF-like domains might be involved in hemophilia B, Marfan syndrome or hypercholesterolemia (due to mutation of factor IX, fibrillin and low-density lipoprotein receptor, respectively) [172, 183, 184].

8.4.3   Extracellular Ca^{2+} Binding Proteins with ɣ-Carboxyl Glutamic Acid-Rich Domains

Several human proteins have a ɣ-carboxyl glutamic acid (GLA)-rich domain that binds Ca^{2+}. These proteins play key roles in the regulation of blood coagulation (factors II-prothrombin- VII, IX, X, protein C, protein S, and protein Z), bone metabolism (osteocalcin, matrix GLA protein, periostin, telopeptide of type I collagen and bone alkaline phosphatase) and vascular biology (growth arrest-specific protein 6, matrix Gla protein-MGP) [185–188]. It is important to emphasize that in addition to the GLA-rich domain, several blood-clotting proteins also have EGF-like domains as described above. The GLA-rich domain consists of approximately 45 amino acids, of which the 10–12 glutamic acids are carboxylated to GLA by a vitamin K dependent carboxylase [189–191]. In coagulation factors VII, IX and X and protein C, the GLA-rich domain occupies the N-terminal half of the molecule and is followed by two EGF-like domains. Regarding protein S, the GLA-rich domain also occupies the N-terminal half of the molecule and is followed by a thrombin-sensitive domain and four EGF-like domains [172, 192]. According to different structural studies, it is known that the number of Ca^{2+} ions associated with the GLA domains seems to be variable. For example, in the GLA domain of human FVIIa, nine of the ten GLA residues bind seven Ca^{2+} [193, 194].

More specifically, Ca^{2+} binding to blood-clotting proteins is required for initiation of the coagulation cascade at sites of injury. The GLA domains of most coagulation factors have similar Ca^{2+} affinities, the average K_d being $\sim$0.5 mM [195, 196]. Thus, coagulation factors should generally be in the Ca^{2+}-saturated form in the extracellular space and Ca^{2+} should play a structural role rather a regulatory role. However, local fluctuations of Ca^{2+} levels have been described and a regulatory role of Ca^{2+} may occur in some circumstances [156]. For example, 4-hydroxycoumarin anticoagulants are currently used in clinical practice because they indirectly inhibit the vitamin K dependent carboxylation of several blood-clotting proteins; as a consequence, clotting proteins cannot bind Ca^{2+} and they cannot participate in the coagulation cascade [155, 197]. However, this intervention may not be free of risks. Since γ-carboxylated coagulation proteins are potent inhibitors of vascular calcification, some recent studies have suggested that anticoagulants such as warfarin may accelerate vascular calcification in addition to frequent or irregular heavy bleeding [198–200]. Although there are already studies demonstrating this effect, clinical data need to be supplemented with controlled studies to confirm this plausible hypothesis [200].

Another protein involved in arterial calcification is MGP, a 10-kDa vitamin K-dependent extracellular matrix protein. Although the normal physiological process by which MGP inhibits vascular calcification is still unknown, recent evidence demonstrates that MGP regulates vascular calcification by binding to and inhibiting bone morphogenic protein 2 and preventing the deposition of calcium phosphate in vascular matrix [201–203]. The association with vascular calcification and atherosclerotic disease may be justified by several single nucleotide polymorphisms (SNPs) of the MGP gene, compromising its function [203, 204].

8.4.4 Extracellular Ca^{2+} Binding Proteins with Cadherin Domains

As their name implies, cadherins are Ca^{2+}-dependent adherent receptors that mediate adhesion between cells (extracellular domains of cadherins from adjacent cells form trans bonds). Moreover, they sense and transmit several extracellular signals to the inside of the cell [205]. The cadherin family includes classical cadherins, desmosomal cadherins, protocadherins and atypical cadherins. All are single membrane-spanning proteins with the exception of flamingo, an atypical cadherin that is a seven-pass membrane protein. As we have previously reviewed [3], the different types of cadherins have diverse structures, but all possess Ca^{2+}-binding extracellular repeats of the same protein chain and may act as Ca^{2+} sensors that respond to external Ca^{2+} fluctuations [206, 207].

Once a cadherin has been activated by an appropriate signal, the cytoplasmic tails connect to the cytoskeleton using an adaptor complex formed by three proteins (p120 catenin, β-catenin and α-catenin) and, as a consequence, several important intracellular signalling pathways are modulated [205]. Cadherins are involved in development, morphogenesis, synaptogenesis, differentiation and carcinogenesis (for detailed review see e.g. [206]).

Because of their essential biological functions, cadherins are being investigated as drug targets, especially in oncology. For example, neural (N)-cadherin is involved in angiogenesis and the maintenance of blood vessel stability [208]. Since tumour growth depends on an adequate blood supply, it could be affected by N-cadherin antagonists [209]. Also, E-cadherin plays an important role in epithelial cell adhesion and the loss of its function is a major contributor to cancer progression because most solid tumours arise from epithelial tissue [210, 211].

8.4.5 Extracellular Ca^{2+} Binding Proteins with Ca^{2+}-Dependent (C)-Type Lectin-Like Domains

The superfamily of proteins containing C-type lectin-like domains (CTLDs) is a large group of extracellular proteins with diverse functions of biomedical interest. Curiously, the term 'C-type lectin' was introduced to distinguish a group of Ca^{2+}-dependent (C-type) carbohydrate-binding (lectin) animal lectins from the other (Ca^{2+}-independent) type [212].

CTLD are present not only in C-type lectins, but also in other extracellular proteins. They are found in more than 1000 proteins and represent a ligand-binding motif that is not necessarily restricted to binding sugars [213]. C-type lectins are found either as transmembrane proteins (e.g. selectins, the mannose receptor family and the dendritic cell-specific ICAM-3 grabbing non-integrin molecule) or as secreted soluble proteins (e.g. mannose-binding protein and other collectins) [3, 212]. The C-type lectin fold is rare and complex. It presents a compact domain of approximately 120 amino acid residues with a double-looped ('loop-in-a-loop'), two-stranded antiparallel β-sheet formed by the amino-and carboxy-terminal residues connected by two α-helices and a three-stranded antiparallel β-sheet [213]. This secondary loop is involved in Ca^{2+}-dependent carbohydrate binding, the main CTLD function, and in interactions with other ligands [214]. Four Ca^{2+}-binding sites are consistently detected in the CTLD domain, but depending on the particular CTLD sequence and on experimental conditions, zero, one, two or three sites are occupied [212, 214]. Ca^{2+} is involved in ligand binding to C-type lectins and also serves to stabilize their molecular structure [215, 216]. Since changes in Ca^{2+} concentration dramatically enhance carbohydrate binding by C-type lectins, it has been suggested that physiological fluctuations of extracellular Ca^{2+} have a regulatory effect on ligand binding by C-type lectins [216, 217]. Additionally, C-type lectins are involved in many cell surface carbohydrate recognition events, e.g. cell-to-cell contact and recognition of pathogens, cell-to-cell adhesion and apoptosis [215, 218, 219].

8.4.6 Ca^{2+}-Binding Pockets of Family C G-Protein-Coupled Receptors

Family C of the superfamily of G protein-coupled receptors (GPCRs) includes the Ca^{2+}-sensing receptor (CaR), eight metabotropic glutamate receptors (mGluRs), two $GABA_B$ receptors, three taste (T1R) receptors, one L-amino acid receptor (GPRC6A), and five orphan receptors [220]. The function of some of these receptors is altered in response to small fluctuations in Ca^{2+}_o concentrations, suggesting Ca^{2+} sensing and Ca^{2+} signalling regulator capabilities.

8.4.6.1 Ca^{2+}-Sensing Receptor (CaR; Also Named CaSR or CaS)

CaR was initially identified in bovine parathyroid glands as a plasma membrane-located GPCR and was the first GPCR recognised to have Ca^{2+} as its major physiological agonist. The main function originally ascribed to CaR was the control of Ca^{2+}_o concentrations by modulating parathyroid hormone (PTH) secretion. Later, CaR was also identified on the surface of cells from multiple tissues, e.g. bone, thyroid, kidney, nervous system, gastrointestinal system, ovary, mamma, prostate, blood vessels and heart cells, pancreas as well as monocytes, macrophages, dendritic and intestine and haematopoietic stem cells [173, 221]. Furthermore, it was found that other agonists and allosteric modulators interact with CaR and influence its response to Ca^{2+}, and that CaR has several non-calciotropic roles.

CaR has a long extracellular amino-terminal domain of 612 amino acids called a Venus flytrap module, containing a ligand binding pocket that gives the receptor the necessary sensitivity to detect small fluctuations in external Ca^{2+} concentration.

According to Silve et al. [222], the Ca^{2+}-binding site in CaR hosts a set of polar residues directly involved in Ca^{2+} coordination (Ser-170, Asp-190, Gln-193, Ser-296, and Glu-297), and an additional set of residues that completes the coordination sphere of the cation (Phe-270, Tyr-218, and Ser-147). The Venus flytrap module is followed by a seven transmembrane helix cassette (250 amino acids) and a C-terminal domain (216 amino acids) [223]. CaR activity is modulated by changes in Ca^{2+}_o concentrations occurring in the mM range. CaR is coupled through G-proteins, preferentially $G\alpha q/11$, $G\alpha i/o$, $G\alpha s$ and $G\alpha 12/13$ subunits, to intracellular signal transduction pathways involving phospholipase C, cytosolic phospholipase A_2 and various MAP kinase proteins [155, 156, 224].

An increase in serum Ca^{2+} concentration negatively regulates PTH secretion via CaR. The vital role of CaR in maintaining normal systemic Ca^{2+} homeostasis is illustrated by more than 230 different germline mutations in CaR that cause hypocalcemic and hypercalcemic disorders [225] and by mouse knockout models. Complete ablation of functional CaR is lethal, resulting in severe skeletal demineralization, extremely high Ca^{2+} serum concentration, growth defects and ultrastructural changes in the epidermis [157].

In vitro experiments have revealed that CaR on the cell surface responds to Ca^{2+} exported from the same or a neighbouring cell during Ca^{2+} signalling events; Ca^{2+}_o might thus act as an autocrine/paracrine messenger via CaR [157]. CaR not only responds to Ca^{2+}, but also to other agonists (inorganic divalent and trivalent cations, polyamines, aminoglycoside antibiotics, basic polypeptides) and to allosteric modulators (L-amino acids, glutathione analogs, small molecule calcimimetics, small molecule calcilytics) that influence the effect of Ca^{2+} [156, 224]. CaR can discriminate between different ligands and selectively activate one signalling transduction pathway in response to a particular ligand (a phenomenon known as biased agonism) [224].

It is therefore not surprising that CaR is important for processes other than maintenance of systemic Ca^{2+} homeostasis, e.g. regulation of hormone and fluid secretion, activities of various ion channels, synaptic transmission and neuronal activity via multiple signaling pathways, gene expression, programmed cell death (apoptosis), cellular proliferation, development and more [155, 226, 227].

Activators of the Ca^{2+} binding site in CaR (orthosteric agonists and also termed type I calcimimetics), synthetic allosteric activators of CaR (termed type II calcimimetics) and inhibitors of CaR (termed calcilytics) have been developed for clinical use. Strontium, which can be considered a type I calcimimetic, is currently used in osteoporosis in the form of strontium ranelate. Cinnacalcet, a type II calcimimetic, is currently used in primary and secondary hyperparathyroidism and parathyroid carcinoma. Calcilytics failed in previous clinical trials as anabolic therapies for the treatment of osteoporosis but are now being evaluated for the treatment of some cases of hypocalcemic and hypercalciuric disorders [228]. Sustained elevations of PTH, as occurs in hyperparathyroid states, have a net catabolic effect on bone, favouring resorption, whereas short bursts are anabolic, favouring formation [229]. Therefore, continuous administration of a calcimimetic or intermittent administration of a calcilytic could promote anabolic over catabolic actions.

Because Ca^{2+} and CaR are implicated in many physiopathological conditions, the use of calcilytics or CaR allosteric activators have been proposed for the treatment of a range of disorders [223]. For instance, the former could be beneficial for the treatment of Alzheimer's disease, osteolytic breast cancer, pulmonary hypertension, asthma and other inflammatory lung disorders, whilst the latter could be useful in inflammatory bowel disease, secretory diarrhoea, hypertension, and colon cancer.

8.4.6.2 Metabotropic Glutamate Receptors (mGluRs) and GABA$_B$ Receptors

Metabotropic glutamate receptors are family C GPCRs and structurally similar to CaR. On the basis of sequence homology to CaR and several experimental results, it was postulated that some mGluRs can respond to external Ca^{2+} fluctuations [157, 220]. mGluRs are expressed principally in the brain, where the levels of external Ca^{2+} are highly dynamic. A Ca^{2+}-binding site in the mGluR1α extracellular domain was recently identified: it comprises Asp-318, Glu-325 and Asp-322 and the carboxylate side chain of Glu-701 [230]. External Ca^{2+} has been proposed: (i) to directly activate mGluRs; (ii) to increase sensitivity of mGluRs to L-glutamate; (iii) to modulate mGluRs synergistically with L-glutamate; and (iv) to modulate actions of allosteric drugs targeting mGluRs [220, 230]. External Ca^{2+} also modulates other type C GPCRs, namely GABA$_B$ receptors, although it is not thought that this occurs directly [155]. The numerous clinical effects in which these receptors are involved have been the subject of a large number of studies and reviews (see e.g. [231–234]).

8.5 Concluding Remarks

Over the few last decades, research on CBP has resulted in a profusion of important advances. Crystallography, molecular biology, microscopy and other techniques have widened our knowledge on the sequence, structure, and functionality of CBP. Many new proteins equipped with the ability to bind Ca^{2+} have been discovered and there has been a great effort to rationally classify them. Currently, much of the research in this field is oriented towards understanding the physiological roles of these proteins. This is why ever more CBP are emerging as potential targets in therapeutic approaches for a staggering array of diseases. Nonetheless, differentiating between structural and regulatory roles for Ca^{2+} binding (especially in the case of extracellular CBP) remains a significant challenge. Also, although progress has been made, determining which proteins act exclusively as Ca^{2+} sensors and which act uniquely as Ca^{2+} buffers is still not straightforward and, indeed, for some CBP, both functions are observed. A deeper understanding of the structure and function of CBP, together with a better definition of their physiological role, will allow us to further investigate targeting of these proteins in novel therapeutic strategies.

Author Contribution JE was responsible for the writing of Sect. 8.2; he also participated in the drafting of the Introduction and Concluding Remarks sections. MY was responsible for the writing of Sect. 8.2. TMCP and JGL were responsible for the writing of Sect. 8.4. DAM contributed throughout, provided focus and flow to the various sections of the review, and oversaw/edited written English. MCT was responsible for the writing of Sect. 8.3; he also participated in the writing of the Introduction and Concluding Remarks sections and in the coordination of all authors.

References

1. Berridge MJ, Bootman MD, Roderick HL (2003) Calcium signalling: dynamics, homeostasis and remodelling. Nat Rev Mol Cell Biol 4:517–529
2. Carafoli E, Santella L, Branca D, Brini M (2001) Generation, control, and processing of cellular calcium signals. Crit Rev Biochem Mol Biol 36:107–260
3. Yáñez M, Gil-Longo J, Campos-Toimil M (2012) Calcium binding proteins. Adv Exp Med Biol 740:461–482
4. Lewit-Bentley A, Réty S (2000) EF-hand calcium-binding proteins. Curr Opin Struct Biol 10:637–643
5. Skelton NJ, Kordel J, Akke M, Forsen S, Chazin WJ (1994) Signal transduction versus buffering activity in Ca^{2+}-binding proteins. Nat Struct Biol 1:239–245
6. Henzi T, Schwaller B (2015) Antagonistic regulation of parvalbumin expression and mito-chondrial calcium handling capacity in renal epithelial cells. PLoS One 10:e0142005
7. Caillard O, Moreno H, Schwaller B, Celio MR, Marty A (2000) Role of the calcium-binding protein parvalbumin in short-term synaptic plasticity. Proc Natl Acad Sci 97:13372–13377
8. Estebanez L, Hoffmann D, Voigt BC, Poulet JFA (2017) Parvalbumin-expressing GABAergic neurons in primary motor cortex signal reaching. Cell Rep 20:308–318
9. Yizhar O, Fenno LE, Prigge M, Schneider F, Davidson TJ, O'Shea DJ et al (2011) Neocortical excitation/inhibition balance in information processing and social dysfunction. Nature 477:171–178
10. Inan M, Petros TJ, Anderson SA (2013) Losing your inhibition: linking cortical GABAergic interneurons to schizophrenia. Neurobiol Dis 53:36–48
11. Zikopoulos B, Barbas H (2013) Altered neural connectivity in excitatory and inhibitory cortical circuits in autism. Front Hum Neurosci 7:609
12. Hu H, Gan J, Interneurons JP (2014) Fast-spiking, parvalbumin(+) GABAergic interneurons: from cellular design to microcircuit function. Science 345:1255263
13. Hashemi E, Ariza J, Rogers H, Noctor SC, Martínez-Cerdeño V (2017) The number of parvalbumin-expressing interneurons is decreased in the medial prefrontal cortex in autism. Cereb Cortex 27:1931–1943
14. Steullet P, Cabungcal JH, Coyle J, Didriksen M, Gill K, Grace AA et al (2017) Oxidative stress-driven parvalbumin interneuron impairment as a common mechanism in models of schizophrenia. Mol Psychiatry 22:936–943
15. Brisch R, Bielau H, Saniotis A, Wolf R, Bogerts B, Krell D et al (2015) Calretinin and parvalbumin in schizophrenia and affective disorders: a mini-review, a perspective on the evolutionary role of calretinin in schizophrenia, and a preliminary post-mortem study of calretinin in the septal nuclei. Front Cell Neurosci 9:393

16. Soghomonian JJ, Zhang K, Reprakash S, Blatt GJ (2017) Decreased parvalbumin mRNA levels in cerebellar purkinje cells in autism. Autism Res 10:1787–1796

17. Zou D, Chen L, Deng D, Jiang D, Dong F, McSweeney C et al (2016) DREADD in parvalbumin interneurons of the dentate gyrus modulates anxiety, social interaction and memory extinction. Curr Mol Med 16:91–102

18. Cammarota M, Losi G, Chiavegato A, Zonta M, Carmignoto G (2013) Fast spiking interneuron control of seizure propagation in a cortical slice model of focal epilepsy. J Physiol 591:807–822

19. Schevon CA, Weiss SA, McKhann G Jr, Goodman RR, Yuste R, Emerson RG et al (2012) Evidence of an inhibitory restraint of seizure activity in humans. Nat Commun 3:1060

20. Krook-Magnuson E, Armstrong C, Oijala M, Soltesz I (2013) On-demand optogenetic control of spontaneous seizures in temporal lobe epilepsy. Nat Commun 4:1376

21. Paz JT, Davidson TJ, Frechette ES, Delord B, Parada I, Peng K et al (2013) Closed-loop optogenetic control of thalamus as a tool for interrupting seizures after cortical injury. Nat Neurosci 16:64–70

22. Sessolo M, Marcon I, Bovetti S, Losi G, Cammarota M, Ratto GM et al (2015) Parvalbumin-positive inhibitory interneurons oppose propagation but favor generation of focal epileptiform activity. J Neurosci 35:9544–9557

23. Berg EM, Bertuzzi M, Ampatzis K (2018) Complementary expression of calcium binding proteins delineates the functional organization of the locomotor network. Brain Struct Funct 223:2181–2196

24. Shamgar L, Ma L, Schmitt N, Haitin Y, Peretz A, Wiener R et al (2006) Calmodulin is essential for cardiac IKS channel gating and assembly: impaired function in long-QT mutations. Circ Res 98:1055–1063

25. Ciampa EJ, Welch RC, Vanoye CG, George AL Jr (2011) KCNE4 juxtamembrane region is required for interaction with calmodulin and for functional suppression of KCNQ1. J Biol Chem 286:4141–4149

26. Chang A, Abderemane-Ali F, Hura GL, Rossen ND, Gate RE, Minor DL Jr (2018) A calmodulin c-lobe Ca^{2+}-dependent switch governs Kv7 channel function. Neuron 97:836–852

27. Crotti L, Johnson CN, Graf E, De Ferrari GM, Cuneo BF, Ovadia M et al (2013) Calmodulin mutations associated with recurrent cardiac arrest in infants. Circulation 127:1009–1017

28. Makita N, Yagihara N, Crotti L, Johnson CN, Beckmann BM, Roh MS et al (2014) Novel calmodulin mutations associated with congenital arrhythmia susceptibility. Circ Cardiovasc Genet 7:466–474

29. Yamamoto Y, Makiyama T, Harita T, Sasaki K, Wuriyanghai Y, Hayano M et al (2017) Allele-specific ablation rescues electrophysiological abnormalities in a human iPS cell model of long-QT syndrome with a CALM2 mutation. Hum Mol Genet 26:1670–1677

30. Marsman RF, Barc J, Beekman L, Alders M, Dooijes D, van den Wijngaard A et al (2014) A mutation in CALM1 encoding calmodulin in familial idiopathic ventricular fibrillation in childhood and adolescence. J Am Coll Cardiol 63:259–266

31. Li T, Yi L, Hai L, Ma H, Tao Z, Zhang C et al (2018) The interactome and spatial redistribution feature of Ca^{2+} receptor protein calmodulin reveals a novel role in invadopodia-mediated invasion. Cell Death Dis 9:292

32. McDonald KS (2018) Jack-of-many-trades: discovering new roles for troponin C. J Physiol 596(19):4553–4554. https://doi.org/10.1113/JP276790

33. Gillis TE, Marshall CR, Tibbits GF (2007) Functional and evolutionary relationships of troponin C. Physiol Genomics 32:16–27

34. Trybus KM (1994) Role of myosin light chains. J Muscle Res Cell Motil 15:587–594

35. Kampourakis T, Sun YB, Irving M (2016) Myosin light chain phosphorylation enhances contraction of heart muscle via structural changes in both thick and thin filaments. Proc Natl Acad Sci U S A 113:E3039–E3047

36. Pechere JF (1968) Muscular parvalbumins as homologous proteins. Comp Biochem Physiol 24:289–295

37. Nishikawa T, Lee IS, Shiraishi N, Ishikawa T, Ohta Y, Nishikimi M (1997) Identification of S100b protein as copper-binding protein and its suppression of copper-induced cell damage. J Biol Chem 272:23037–23041
38. Fritz G, Botelho HM, Morozova-Roche LA, Gomes CM (2010) Natural and amyloid self-assembly of S100 proteins: structural basis of functional diversity. FEBS J 277:4578–4590
39. Gross SR, Sin CG, Barraclough R, Rudland PS (2014) Joining S100 proteins and migration: for better or for worse, in sickness and in health. Cell Mol Life Sci 71:1551–1579
40. Donato R, Sorci G, Giambanco I (2017) S100A6 protein: functional roles. Cell Mol Life Sci 74:2749–2760
41. Xia C, Braunstein Z, Toomey AC, Zhong J, Rao X (2018) S100 proteins as an important regulator of macrophage inflammation. Front Immunol 8:1908
42. Zimmer DB, Wright Sadosky P, Weber DJ (2003) Molecular mechanisms of S100-target protein interactions. Microsc Res Tech 60:552–559
43. Eckert RL, Broome AM, Ruse M, Robinson N, Ryan D, Lee K (2004) S100 proteins in the epidermis. J Invest Dermatol 123:23–33
44. Donato R, Sorci G, Riuzzi F, Arcuri C, Bianchi R, Brozzi F et al (2009) S100B's double life: intracellular regulator and extracellular signal. Biochim Biophys Acta 1793:1008–1022
45. He H, Li J, Weng S, Li M, Yu Y (2009) S100A11: diverse function and pathology corresponding to different target proteins. Cell Biochem Biophys 55:117–126
46. Sherbet GV (2009) Metastasis promoter S100A4 is a potentially valuable molecular target for cancer therapy. Cancer Lett 280:15–30
47. Naz S, Ranganathan P, Bodapati P, Shastry AH, Mishra LN, Kondaiah P (2012) Regulation of S100A2 expression by TGF-beta-induced MEK/ERK signalling and its role in cell migration/invasion. Biochem J 447:81–91
48. Donato R, Cannon BR, Sorci G, Riuzzi F, Hsu K, Weber DJ et al (2013) Functions of S100 proteins. Curr Mol Med 13:24–57
49. Brockett AT, Kane GA, Monari PK, Briones BA, Vigneron PA, Barber GA et al (2018) Evidence supporting a role for astrocytes in the regulation of cognitive flexibility and neuronal oscillations through the Ca^{2+} binding protein S100β. PLoS One 13:e0195726
50. Sakatani S, Seto-Ohshima A, Shinohara Y, Yamamoto Y, Yamamoto H, Itohara S et al (2008) Neuralactivity-dependent release of S100β from astrocytes enhances kainate-induced gamma oscillations in vivo. J Neurosci 28:10928–10936
51. Diaz-Romero J, Nesic D (2017) S100A1 and S100B: calcium sensors at the cross-roads of multiple chondrogenic pathways. J Cell Physiol 232:1979–1987
52. Wang T, Huo X, Chong Z, Khan H, Liu R, Wang T (2018) A review of S100 protein family in lung cancer. Clin Chim Acta 476:54–59
53. Tian T, Li X, Hua Z, Ma J, Liu Z, Chen H et al (2017) S100A1 promotes cell proliferation and migration and is associated with lymph node metastasis in ovarian cancer. Discov Med 23:235–245
54. Chen X, Liu X, Lang H, Zhang S, Luo Y, Zhang J (2015) S100 calcium-binding protein A6 promotes epithelial-mesenchymal transition through b-catenin in pancreatic cancer cell line. PLoS One 10:e0121319
55. Belter B, Haase-Kohn C, Pietzsch J (2017) Biomarkers in malignant melanoma: recent trends and critical perspective. In: Ward WH, Farma JM (eds) Cutaneous melanoma: etiology and therapy. Codon Publications, Brisbane
56. Bresnick AR, Weber DJ, Zimmer DB (2015) S100 proteins in cancer. Nat Rev Cancer 15:96–109
57. Tesarova P, Kalousova M, Zima T, Tesar V (2016) HMGB1, S100 proteins and other RAGE ligands in cancer-markers, mediators and putative therapeutic targets. Biomed Pap Med Fac Univ Palacky Olomouc Czech Repub 160:1–10
58. Rusnak F, Mertz P (2000) Calcineurin: form and function. Physiol Rev 80:1483–1521
59. Li J, Jia Z, Zhou W, Wei Q (2009) Calcineurin regulatory subunit B is a unique calcium sensor that regulates calcineurin in both calcium-dependent and calcium-independent manner. Proteins 77:612–623

60. Li H, Rao A, Hogan PG (2011) Interaction of calcineurin with substrates and targeting proteins. Trends Cell Biol 21:91–103
61. Parra V, Rothermel BA (2017) Calcineurin signaling in the heart: the importance of time and place. J Mol Cell Cardiol 103:121–136
62. Shah SZ, Hussain T, Zhao D, Yang L (2012) A central role for calcineurin in protein misfolding neurodegenerative diseases. Cell Mol Life Sci 74:1061–1074
63. Pongs O, Lindemeier J, Zhu XR, Theil T, Engelkamp D, Krah-Jentgens I et al (1993) Frequenin – a novel calcium-binding protein that modulates synaptic efficacy in the Drosophila nervous system. Neuron 11:15–28
64. Burgoyne RD, Haynes LP (2015) Sense and specificity in neuronal calcium signalling. Biochim Biophys Acta 1853:1921–1932
65. Burgoyne RD (2007) Neuronal calcium sensor proteins: generating diversity in neuronal Ca^{2+} signalling. Nat Rev Neurosci 8:182–193
66. Angelats E, Requesens M, Aguinaga D, Kreutz MR, Franco R, Navarro G (2018) Neuronal calcium and cAMP cross-talk mediated by cannabinoid CB1 receptor and EF-hand calcium sensor interactions. Front Cell Dev Biol 6:67
67. Braunewell KH (2005) The darker side of Ca^{2+} signaling by neuronal Ca^{2+}-sensor proteins: from Alzheimer's disease to cancer. Trends Pharmacol Sci 26:345–351
68. Kaeser PS, Regehr WG (2014) Molecular mechanisms for synchronous, asynchronous, and spontaneous neurotransmitter release. Annu Rev Physiol 76:333–363
69. Romanov RA, Alpár A, Hökfelt T, Harkany T (2017) Molecular diversity of corticotropin-releasing hormone mRNA-containing neurons in the hypothalamus. J Endocrinol 232:R161–R172
70. Gerke V, Moss SE (2002) Annexins: from structure to function. Physiol Rev 82:331–371
71. Moss SE, Morgan RO (2004) The annexins. Genome Biol 5:219
72. Rescher U, Gerke V (2004) Annexins -unique membrane binding proteins with diverse functions. J Cell Sci 117:2631–2639
73. Mishra S, Chander V, Banerjee P, Oh JG, Lifirsu E, Park WJ et al (2011) Interaction of annexin A6 with alpha actinin in cardiomyocytes. BMC Cell Biol 12:7
74. Santamaria-Kisiel L, Rintala-Dempsey AC, Shaw GS (2006) Calcium-dependent and-independent interactions of the S100 protein family. Biochem J 396:201–214
75. Raynal P, Pollard HB (1994) Annexins: the problem of assessing the biological role for a gene family of multifunctional calcium- and phospholipid-binding proteins. Biochim Biophys Acta 1197:63–93
76. Gerke V, Creutz CE, Moss SE (2005) Annexins: linking Ca2þ signalling to membrane dynamics. Nat Rev Mol Cell Biol 6:449–461
77. Fatimathas L, Moss SE (2010) Annexins as disease modifiers. Histol Histopathol 25:527–532
78. D'Acunto CW, Gbelcova H, Festa M, Ruml T (2014) The complex understanding of annexin A1 phosphorylation. Cell Signal 26:173–178
79. Leoni G, Nusrat A (2016) Annexin A1: shifting the balance towards resolution and repair. Biol Chem 397:971–979
80. Christmas P, Callaway J, Fallon J, Jones J, Haigler HT (1991) Selective secretion of annexin-1, a protein without a signal sequence, by the human prostate-gland. J Biol Chem 266:2499–2507
81. Guo C, Liu S, Sun M (2013b) Potential role of Anxa1 in cáncer. Future Oncol 9:1773–1793
82. Tu Y, Johnstone CN, Stewart AG (2017) Annexin A1 influences in breast cancer: controversies on contributions to tumour, host and immunoediting processes. Pharmacol Res 119:278–288
83. Boudhraa Z, Bouchon B, Viallard C, D'Incan M, Degoul F (2016) Annexin A1 localization and its relevance to cancer. Clin Sci 130:205–220
84. Yan Tu Y, Johnstone CN, Stewart AG (2017) Annexin A1 influences in breast cancer: Controversies on contributions to tumour, host and immunoediting processes. Pharmacol Res 119:278–288

85. Chen L, Lv F, Pei L (2014) Annexin 1: a glucocorticoid-inducible protein that modulates inflammatory pain. Eur J Pain 18:338–347
86. Sugimoto MA, Vago JP, Teixeira MM, Sousa LP (2016) Annexin A1 and the resolution of inflammation: modulation of neutrophil recruitment, apoptosis, and clearance. J Immunol Res 2016:8239258
87. Yang YH, Morand E, Leech M (2013) Annexin A1: potential for glucocorticoid sparing in RA. Nat Rev Rheumatol 9:595–603
88. Qin C, Yang YH, May L, Gao X, Stewart AG, Tu Y et al (2015) Cardioprotective potential of annexin-A1 mimetics in myocardial infarction. Pharmacol Ther 148:47–65
89. Luo M, Hajjar KA (2013) Annexin A2 system in human biology: cell surface and beyond. Semin Thromb Hemost 39:338–346
90. Xu XH, Pan W, Kang LH, Feng H, Song YQ (2015) Association of annexin A2 with cancer development. Oncol Rep 33:2121–2128
91. Wang C, Lin C (2014a) Annexin A2: its molecular regulation and cellular expression in cancer development. Dis Markers 2014:308976
92. Liu X, Ma D, Jing X, Wang B, Yang W, Qiu W (2015) Overexpression of ANXA2 predicts adverse outcomes of patients with malignant tumors: a systematic review and meta-analysis. Med Oncol 32:392
93. Tanida S, Mizoshita T, Ozeki K, Katano T, Kataoka H, Kamiya T et al (2015) Advances in refractory ulcerative colitis treatment: a new therapeutic target, Annexin A2. World J Gastroenterol 21:8776–8786
94. Cañas F, Simonin L, Couturaud F, Renaudineau Y (2015) Annexin A2 autoantibodies in thrombosis and autoimmune diseases. Thromb Res 135:226–230
95. Wu N, Liu S, Guo C, Hou Z, Sun MZ (2013) The role of annexin A3 playing in cancers. Clin Transl Oncol 15:106–110
96. Matsuzaki S, Serada S, Morimoto A, Ueda Y, Yoshino K, Kimura T et al (2014) Annexin A4 is a promising therapeutic target for the treatment of platinum-resistant cancers. Expert Opin Ther Targets 18:403–414
97. Wei B, Guo C, Liu S, Sun MZ (2015) Annexin A4 and cancer. Clin Chim Acta 447:72–78
98. Peng B, Guo C, Guan H, Liu S, Sun MZ (2014) Annexin A5 as a potential marker in tumors. Clin Chim Acta 427:42–48
99. Bouter A, Carmeille R, Gounou C, Bouvet F, Degrelle SA, Evain-Brion D et al (2015) Review: Annexin-A5 and cell membrane repair. Placenta 36:S43–S49
100. Carmeille R, Degrelle SA, Plawinski L, Bouvet F, Gounou C, Evain-Brion D, Brisson AR, Bouter A (2015) Annexin-A5 promotes membrane resealing in human trophoblasts. Biochim Biophys Acta 1853:2033–2044
101. Udry S, Aranda F, Latino O, de Larrañaga G (2013) Annexins and recurrent pregnancy loss. Medicina (B Aires) 73:495–500
102. Enrich C, Rentero C, Grewal T (2017) Annexin A6 in the liver: from the endocytic compartment to cellular physiology. Biochim Biophys Acta 1864:933–946
103. Guo C, Liu S, Greenaway F, Sun MZ (2013a) Potential role of annexin A7 in cancers. Clin Chim Acta 423:83–89
104. Wang J, Guo C, Liu S, Qi H, Yin Y, Liang R et al (2014b) Annexin A11 in disease. Clin Chim Acta 431:164–168
105. Coe H, Michalak M (2009) Calcium binding chaperones of the endoplasmic reticulum. Gen Physiol Biophys 28:F96–F103
106. Gutierrez T, Simmen T (2018) Endoplasmic reticulum chaperones tweak the mitochondrial calcium rheostat to control metabolism and cell death. Cell Calcium 70:64–75
107. Lu YC, Weng WC, Lee H (2015) Functional roles of calreticulin in cancer biology. Biomed Res Int 2015:526524
108. Zamanian M, Veerakumarasivam A, Abdullah S, Rosli R (2013) Calreticulin and cancer. Pathol Oncol Res 19:149–154
109. Nakamura K, Zuppini A, Arnaudeau S, Lynch J, Ahsan I, Krause R et al (2001) Functional specialization of calreticulin domains. J Cell Biol 154:961–972

110. Arnaudeau S, Frieden M, Nakamura K, Castelbou C, Michalak M, Demaurex N (2002) Calreticulin differentially modulates calcium uptake and release in the endoplasmic reticulum and mitochondria. J Biol Chem 277:46696–46705
111. Trombetta ES (2003) The contribution of N-glycans and their processing in the endoplasmic reticulum to glycoprotein biosynthesis. Glycobiology 13:77R–91R
112. Gelebart P, Opas M, Michalak M (2005) Calreticulin, a Ca^{2+}-binding chaperone of the endoplasmic reticulum. Int J Biochem Cell Biol 37:260–266
113. Zhu N, Wang Z (1999) Calreticulin expression is associated with androgen regulation of the sensitivity to calcium ionophore-induced apoptosis in LNCaP prostate cancer cells. Cancer Res 59:1896–1902
114. Clinton A, McMullin MF (2016) The Calreticulin gene and myeloproliferative neoplasms. J Clin Pathol 69:841–845
115. Eggleton P, Bremer E, Dudek E, Michalak M (2016) Calreticulin, a therapeutic target? Expert Opin Ther Targets 20:1137–1147
116. Lievremont JP, Rizzuto R, Hendershot L, Meldolesi J (1997) BiP: a major chaperone protein of the endoplasmic reticulum lumen, plays a direct and important role in the storage of the rapidly exchanging pool of Ca^{2+}. J Biol Chem 272:30873–30879
117. Taiyab A, Sreedhar AS, Rao CM (2009) Hsp90 inhibitors: GA and 17AAG, lead to ER stress-induced apoptosis in rat histiocytoma. Biochem Pharmacol 78:142–152
118. Narindrasorasak S, Yao P, Sarkar B (2003) Protein disulfide isomerase, a multifunctional protein chaperone, shows copper-binding activity. Biochem Biophys Res Commun 311:405–414
119. Xu S, Sankar S, Neamati N (2014) Protein disulfide isomerase: a promising target for cancer therapy. Drug Discov Today 19:222–240
120. MacLennan DH, Wong PT (1971) Isolation of a calcium-sequestering protein from sarcoplasmic reticulum. Proc Natl Acad Sci U S A 68:1231–1235
121. Novák P, Soukup T (2011) Calsequestrin distribution, structure and function, its role in normal and pathological situations and the effect of thyroid hormones. A review. Physiol Res 60:439–452
122. Gyorke I, Hester N, Jones LR, Gyorke S (2004) The role of calsequestrin, triadin, and junctin in conferring cardiac ryanodine receptor responsiveness to luminal calcium. Biophys J 86:2121–2128
123. Qin J, Valle G, Nani A, Nori A, Rizzi N, Priori SG et al (2008) Luminal Ca^{2+} regulation of single cardiac ryanodine receptors: insights provided by calsequestrin and its mutants. J Gen Physiol 131:325–334
124. Gaburjakova M, Bal NC, Gaburjakova J, Periasamy M (2013) Functional interaction between calsequestrin and ryanodine receptor in the heart. Cell Mol Life Sci 70:2935–2945
125. Paolini C, Quarta M, Nori A, Boncompagni S, Canato M, Volpe P et al (2007) Reorganized stores and impaired calcium handling in skeletal muscle of mice lacking calsequestrin-1. J Physiol 583:767–784
126. Glukhov AV, Kalyanasundaram A, Lou Q, Hage LT, Hansen BJ, Belevych AE et al (2015) Calsequestrin 2 deletion causes sinoatrial node dysfunction and atrial arrhythmias associated with altered sarcoplasmic reticulum calcium cycling and degenerative fibrosis within the mouse atrial pacemaker complex1. Eur Heart J 36:686–697
127. Pertille A, de Carvalho CL, Matsumura CY, Neto HS, Marques MJ (2010) Calcium-binding proteins in skeletal muscles of the mdx mice: potential role in the pathogenesis of Duchenne muscular dystrophy. Int J Exp Pathol 91:63–71
128. Guarnier FA, Michelucci A, Serano M, Pietrangelo L, Pecorai C, Boncompagni S et al (2018) Aerobic training prevents heatstrokes in calsequestrin-1 knockout mice by reducing oxidative stress. Oxidative Med Cell Longev 2018:4652480
129. Nishizuka Y (1998) The molecular heterogeneity of protein kinase C and its implications for cellular regulation. Nature 334:661–665
130. Kikkawa U, Kishimoto A, Nishizuka Y (1989) The protein kinase C family: heterogeneity and its implications. Annu Rev Biochem 58:31–44

131. Stahelin RV (2009) Lipid binding domains: more than simple lipid effectors. J Lipid Res 50:S299–S304
132. Nalefski EA, Falke JJ (1996) The C2 domain calcium-binding motif: structural and functional diversity. Protein Sci 5:2375–2390
133. Steinberg SF (2008) Structural basis of protein kinase C isoform function. Physiol Rev 88:1341–1378
134. Newton AC (2010) Protein kinase C: poised to signal. Am J Physiol Endocrinol Metab 298:E395–E402
135. Newton AC, Johnson JE (1998) Protein kinase C: a paradigm for regulation of protein function by two membrane-targeting modules. Biochim Biophys Acta 1376:155–172
136. Breitkreutz D, Braiman-Wiksman L, Daum N, Denning MF, Tennenbaum T (2007) Protein kinase C family: on the crossroads of cell signaling in skin and tumor epithelium. J Cancer Res Clin Oncol 133:793–808
137. Mochly-Rosen D, Das K, Grimes KV (2012) Protein kinase C, an elusive therapeutic target? Nat Rev Drug Discov 11:937–957
138. Perin MS, Fried VA, Mignery GA, Jahn R, Sudhof TC (1990) Phospholipid binding by a synaptic vesicle protein homologous to the regulatory region of protein kinase C. Nature 345:260–263
139. Jackman SL, Turecek J, Belinsky JE, Regehr WG (2016) The calcium sensor synaptotagmin 7 is required for synaptic facilitation. Nature 529:88–91
140. Fernandez I, Araç D, Ubach J, Gerber SH, Shin O, Gao Y et al (2001) Three-dimensional structure of the synaptotagmin 1 C2B-domain: synaptotagmin 1 as a phospholipid binding machine. Neuron 32:1057–1069
141. Fernández-Chacón R, Königstorfer A, Gerber SH, García J, Matos MF, Stevens CF et al (2001) Synaptotagmin I functions as a calcium regulator of release probability. Nature 410:41–49
142. Johnson CP, Chapman ER (2010) Otoferlin is a calcium sensor that directly regulates SNARE-mediated membrane fusion. J Cell Biol 191:187–197
143. Pangrsic T, Reisinger E, Moser T (2012) Otoferlin: a multi-C2 domain protein essential for hearing. Trends Neurosci 35:671–680
144. Bunney TD, Katan M (2011) PLC regulation: emerging pictures for molecular mechanisms. Trends Biochem Sci 36:88–96
145. Lee JC, Simonyi A, Sun AY, Sun GY (2011) Phospholipases A2 and neural membrane dynamics: implications for Alzheimer's disease. J Neurochem 116:813–819
146. Marques R, Maia CJ, Vaz C, Correia S, Socorro S (2014) The diverse roles of calcium-binding protein regucalcin in cell biology: from tissue expression and signalling to disease. Cell Mol Life Sci 71:93–111
147. Yamaguchi M (2012) Role of regucalcin in brain calcium signaling: involvement in aging. Integr Biol (Camb) 4:825–837
148. Yamaguchi M, Murata T (2013) Involvement of regucalcin in lipid metabolism and diabetes. Metab Clin Exp 62:1045–1051
149. Yamaguchi M (2013) Suppressive role of regucalcin in liver cell proliferation: involvement in carcinogenesis. Cell Prolif 46:243–253
150. Yamaguchi M (2015) Involvement of regucalcin as a suppressor protein in human carcinogenesis: insight into the gene therapy. J Cancer Res Clin Oncol 141:1333–1341
151. Seidah NG, Chrétien M (1999) Proprotein and prohormone convertases: a family of subtilases generating diverse bioactive polypeptides. Brain Res 848:45–62
152. Bergeron N, Phan BA, Ding Y, Fong A, Kraussn RM (2015) Proprotein convertase subtilisin/kexin type 9 inhibition: a new therapeutic mechanism for reducing cardiovascular disease risk. Circulation 132:1648–1666
153. Kadio B, Yaya S, Basak A, Djè K, Gomes J, Mesenge C (2016) Calcium role in human carcinogenesis: a comprehensive analysis and critical review of literature. Cancer Metastasis Rev 35:391–411

154. Couture F, Kwiatkowska A, Dory YL, Day R (2015) Therapeutic uses of furin and its inhibitors: a patent review. Expert Opin Ther Pat 25:379–396
155. Brown EM, MacLeod RJ (2001) Extracellular calcium sensing and extracellular calcium signaling. Physiol Rev 81:239–297
156. Colella M, Gerbino A, Hofer AM, Curci S (2016) Recent advances in understanding the extracellular calcium-sensing receptor. F1000Res 5:2535
157. Hofer AM (2005) Another dimension to calcium signaling: a look at extracellular calcium. J Cell Sci 118:855–862
158. Chazin WJ (2011) Relating form and function of EF-hand calcium binding proteins. Acc Chem Res 44:171–179
159. Martínez J, Cristóvão JS, Sánchez R, Gasset M, Gomes CM (2018) Preparation of amyloidogenic aggregates from EF-hand β-parvalbumin and S100 proteins. Methods Mol Biol 1779:167–179
160. Denessiouk K, Permyakov S, Denesyuk A, Permyakov E, Johnson MS (2014) Two structural motifs within canonical EF-hand calcium-binding domains identify five different classes of calcium buffers and sensors. PLoS One 9:e109287
161. Murphy-Ullrich JE, Sage EH (2014) Revisiting the matricellular concept. Matrix Biol 37:1–14
162. Wang H, Workman G, Chen S, Barker TH, Ratner BD, Sage EH et al (2006) Secreted protein acidic and rich in cysteine (SPARC/osteonectin/BM-40) binds to fibrinogen fragments D and E, but not to native fibrinogen. Matrix Biol 25:20–26
163. Bradshaw AD (2012) Diverse biological functions of the SPARC family of proteins. Int J Biochem Cell Biol 44:480–448
164. Busch E, Hohenester E, Timpl R, Paulsson M, Maurer P (2000) Calcium affinity, cooperativity, and domain interactions of extracellular EF-hands present in BM-40. J Biol Chem 275:25508–25515
165. Papapanagiotou A, Sgourakis G, Karkoulias K, Raptis D, Parkin E, Brotzakis P et al (2018) Osteonectin as a screening marker for pancreatic cancer: a prospective study. J Int Med Res 46:2769–2779
166. Podhajcer OL, Benedetti L, Girotti MR, Prada F, Salvatierra E, Llera AS (2008) The role of the matricellular protein SPARC in the dynamic interaction between the tumor and the host. Cancer Metastasis Rev 27:523–537
167. Vaz J, Ansari D, Sasor A, Andersson R (2015) SPARC: a potential prognostic and therapeutic target in pancreatic cancer. Pancreas 44:1024–1035
168. Murphy LA, Ramirez EA, Trinh VT, Herman AM, Anderson VC, Brewster JL (2011) Endoplasmic reticulum stress or mutation of an EF-hand Ca^{2+}-binding domain directs the FKBP65 rotamase to an ERAD-based proteolysis. Cell Stress Chaperones 16:607–619
169. Ishikawa Y, Holden P, Bächinger HP (2017) Heat shock protein 47 and 65-kDa FK506-binding protein weakly but synergistically interact during collagen folding in the endoplasmic reticulum. J Biol Chem 292:17216–17224
170. Wang CK, Ghani HA, Bundock A, Weidmann J, Harvey PJ, Edwards IA et al (2018) Calcium-mediated allostery of the EGF fold. ACS Chem Biol 13:1659–1667
171. Engel J (1989) EGF-like domains in extracellular matrix proteins: localized signals for growth and differentiation? FEBS Lett 251:1–7
172. Stenflo J, Stenberg Y, Muranyi A (2000) Calcium-binding EGF-like modules in coagulation proteinases: function of the calcium ion in module interactions. Biochim Biophys Acta 1477:51–63
173. Krebs J, Heizmann CW (2007) Calcium-binding proteins and the EF-hand principle. In: Krebs J, Michalak M (eds) Calcium: a matter of life or death. Elsevier, Amsterdam, pp 51–93
174. Rose-Martel M, Smiley S, Hincke MT (2015) Novel identification of matrix proteins involved in calcitic biomineralization. J Proteome 116:81–96
175. Hu P, Luo BH (2018) The interface between the EGF1 and EGF2 domains is critical in integrin affinity regulation. J Cell Biochem 119(9):7264–7273. https://doi.org/10.1002/jcb.26921

176. Balzar M, Briaire-de Bruijn IH, Rees-Bakker HA, Prins FA, Helfrich W, de Leij L et al (2001) Epidermal growth factor-like repeats mediate lateral and reciprocal interactions of Ep-CAM molecules in homophilic adhesions. Mol Cell Biol 21:2570–2580

177. Saha S, Boyd J, Werner JM, Knott V, Handford PA, Campbell ID et al (2001) Solution structure of the LDL receptor EGF-AB pair: a paradigm for the assembly of tandem calcium binding EGF domains. Structure 9:451–456

178. Wildhagen KC, Lutgens E, Loubele ST, ten Cate H, Nicolaes GA (2011) The structure-function relationship of activated protein C. Lessons from natural and engineered mutations. Thromb Haemost 106:1034–1045

179. Robertson I, Jensen S, Handford P (2011) TB domain proteins: evolutionary insights into the multifaceted roles of fibrillins and LTBPs. Biochem J 433:263–276

180. Andersen OM, Dagil R, Kragelund BB (2013) New horizons for lipoprotein receptors: communication by β-propellers. J Lipid Res 54:2763–2774

181. Jensen SA, Handford PA (2016) New insights into the structure, assembly and biological roles of 10-12 nm connective tissue microfibrils from fibrillin-1 studies. Biochem J 473:827–838

182. Pena F, Jansens A, van Zadelhoff G, Braakman I (2010) Calcium as a crucial cofactor for low density lipoprotein receptor folding in the endoplasmic reticulum. J Biol Chem 285:8656–8664

183. Wang F, Li B, Lan L, Li L (2015) C596G mutation in FBN1 causes Marfan syndrome with exotropia in a Chinese family. Mol Vis 21:194–200

184. Garvie CW, Fraley CV, Elowe NH, Culyba EK, Lemke CT, Hubbard BK et al (2016) Point mutations at the catalytic site of PCSK9 inhibit folding, autoprocessing, and interaction with the LDL receptor. Protein Sci 25:2018–2027

185. Cranenburg EC, Schurgers LJ, Vermeer C (2007) Vitamin K: the coagulation vitamin that became omnipotent. Thromb Haemost 98:120–125

186. Cristiani A, Maset F, De Toni L, Guidolin D, Sabbadin D, Strapazzon G, Moro S, De Filippis V, Foresta C (2014) Carboxylation-dependent conformational changes of human osteocalcin. Front Biosci 19:1105–1116

187. Palta S, Saroa R, Palta A (2014) Overview of the coagulation system. Indian J Anaesth 58:515–523

188. Zhao D, Wang J, Liu Y, Liu X (2015) Expressions and clinical significance of serum bone Gla-protein, bone alkaline phosphatase and C-terminal telopeptide of type I collagen in bone metabolism of patients with osteoporosis. Pak J Med Sci 31:91–94

189. Maurer P, Hohenester E, Engel J (1996) Extracellular calcium-binding proteins. Curr Opin Cell Biol 8:609–617

190. Viegas CS, Simes DC, Laizé V, Williamson MK, Price PA, Cancela ML (2008) Gla-rich protein (GRP), a new vitamin K-dependent protein identified from sturgeon cartilage and highly conserved in vertebrates. J Biol Chem 283:36655–36664

191. Tie JK, Carneiro JD, Jin DY, Martinhago CD, Vermeer C, Stafford DW (2016) Characterization of vitamin K-dependent carboxylase mutations that cause bleeding and nonbleeding disorders. Blood 127:1847–1855

192. Persson E, Madsen JJ, Olsen OH (2014) The length of the linker between the epidermal growth factor-like domains in factor VIIa is critical for a productive interaction with tissue factor. Protein Sci 23:1717–1727

193. Ohkubo YZ, Tajkhorshid E (2008) Distinct structural and adhesive roles of Ca^{2+} in membrane binding of blood coagulation factors. Structure 16:72–81

194. Sumarheni S, Hong SS, Josserand V, Coll JL, Boulanger P, Schoehn G et al (2014) Human full-length coagulation factor X and a GLA domain-derived 40-merpolypeptide bind to different regions of the adenovirus serotype 5 hexoncapsomer. Hum Gene Ther 25:339–349

195. Hansson K, Stenflo J (2005) Post-translational modifications in proteins involved in blood coagulation. J Thromb Haemost 3:2633–2648

196. Egorina EM, Sovershaev MA, Osterud B (2008) Regulation of tissue factor procoagulant activity by post-translational modifications. Thromb Res 122:831–837

197. Czogalla KJ, Watzka M, Oldenburg J (2015) Structural modeling insights into human VKORC1 phenotypes. Nutrients 7:6837–6851
198. Zhang YT, Tang ZY (2014) Research progress of warfarin-associated vascular calcification and its possible therapy. J Cardiovasc Pharmacol 63:76–82
199. Kapustin AN, Schoppet M, Schurgers LJ, Reynolds JL, McNair R, Heiss A et al (2017) Prothrombin loading of vascular smooth muscle cell-derived exosomes regulates coagulation and calcification. Arterioscler Thromb Vasc Biol 37:e22–e32
200. Siltari A, Vapaatalo H (2018) Vascular calcification, vitamin K and warfarin therapy – possible or plausible connection? Basic Clin Pharmacol Toxicol 122:19–24
201. Wallin R, Cain D, Hutson SM, Sane DC, Loeser R (2000) Modulation of the binding of matrix Gla protein (MGP) to bone morphogenetic protein-2 (BMP-2). Thromb Haemost 84:1039–1044
202. Lomashvili KA, Wang X, Wallin R, O'Neill WC (2011) Matrix Gla protein metabolism in vascular smooth muscle and role in uremic vascular calcification. J Biol Chem 286:28715–28722
203. Sheng K, Zhang P, Lin W, Cheng J, Li J, Chen J (2017) Association of Matrix Gla protein gene (rs1800801, rs1800802, rs4236) polymorphism with vascular calcification and atherosclerotic disease: a meta-analysis. Sci Rep 7:8713
204. Luo G, Ducy P, McKee MD, Pinero GJ, Loyer E, Behringer RR et al (1997) Spontaneous calcification of arteries and cartilage in mice lacking matrix GLA protein. Nature 386:78–81
205. Klezovitch O, Vasioukhin V (2015) Cadherin signaling: keeping cells in touch. F1000Res 4(F1000 Faculty Rev):550
206. Halbleib JM, Nelson WJ (2006) Cadherins in development: cell adhesion, sorting, and tissue morphogenesis. Genes Dev 20:3199–3214
207. Oroz J, Valbuena A, Vera AM, Mendieta J, Gomez-Puertas P, Carrion-Vazquez M (2011) Nanomechanics of the cadherin ectodomain: "canalization" by Ca^{2+} binding results in a new mechanical element. J Biol Chem 286:9405–9418
208. Gaengel K, Genové G, Armulik A, Betsholtz C (2009) Endothelial-mural cell signalling in vascular development and angiogenesis. Arterioscler Thromb Vasc Biol 29:630–638
209. Blaschuk OW (2015) N-cadherin antagonists as oncology therapeutics. Philos Trans R Soc B 370:20140039
210. Jeanes A, Gottardi CJ, Yap AS (2008) Cadherins and cancer: how does cadherin dysfunction promote tumor progression? Oncogene 27:6920–6929
211. Wong SHM, Fang CM, Chuah LH, Leong CO, Ngai SC (2018) E-cadherin: Its dysregulation in carcinogenesis and clinical implications. Crit Rev Oncol Hematol 121:11–22
212. Zelensky AN, Gready JE (2005) The C-type lectin-like domain superfamily. FEBS J 272:6179–6217
213. Varki A, Cummings RD, Esko JD, Stanley P, Hart GW, Aebi M et al (2017) Essentials of glycobiology, 3rd edn. Cold Spring Harbor Laboratory Press, New York
214. Aretz J, Wamhoff EC, Hanske J, Heymann D, Rademacher C (2014) Computational and experimental prediction of human C-type lectin receptor druggability. Front Immunol 5:323
215. Cambi A, Koopman M, Figdor CG (2005) How C-type lectins detect pathogens. Cell Microbiol 7:481–488
216. Abdian PL, Caramelo JJ, Ausmees N, Zorreguieta A (2013) RapA2 is a calcium-binding lectin composed of two highly conserved cadherin-like domains that specifically recognize *Rhizobium leguminosarum* acidic exopolysaccharides. J Biol Chem 288:2893–2904
217. Bravo R, Parra V, Gatica D, Rodriguez AE, Torrealba N, Paredes F et al (2013) Endoplasmic reticulum and the unfolded protein response: dynamics and metabolic integration. Int Rev Cell Mol Biol 301:215–290
218. Cambi A, Figdor C (2009) Necrosis: C-type lectins sense cell death. Curr Biol 19:R375–R378
219. Bellande K, Bono JJ, Savelli B, Jamet E, Canut H (2017) Plant lectins and lectin receptor-like kinases: how do they sense the outside? Int J Mol Sci 18:E1164
220. Zou J, Jiang JY, Yang JJ (2017) Molecular basis for modulation of metabotropic glutamate receptors and their drug actions by extracellular Ca^{2+}. Int J Mol Sci 18:672

221. Peterlik M, Kállay E, Cross HS (2013) Calcium nutrition and extracellular calcium sensing: relevance for the pathogenesis of osteoporosis, cancer and cardiovascular diseases. Nutrients 5:302–327
222. Silve C, Petrel C, Leroy C, Bruel H, Mallet E, Rognan D et al (2005) Delineating a Ca^{2+} binding pocket within the venus flytrap module of the human calcium-sensing receptor. J Biol Chem 280:37917–37923
223. Lopez-Fernandez I, Schepelmann M, Brennan SC, Yarova PL, Riccardi D (2015) The calcium-sensing receptor: one of a kind. Exp Physiol 100:1392–1399
224. Conigrave AD, Ward DT (2013) Calcium-sensing receptor (CaSR): pharmacological properties and signaling pathways. Best Pract Res Clin Endocrinol Metab 27:315–331
225. Hannan FM, Babinsky VN, Thakker RV (2016) Disorders of the calcium-sensing receptor and partner proteins: insights into the molecular basis of calcium homeostasis. J Mol Endocrinol 57:R127–R142
226. Riccardi D, Kemp PJ (2012) The calcium-sensing receptor beyond extracellular calcium homeostasis: conception, development, adult physiology, and disease. Rev Physiol 74:271–297
227. Jones BL, Smith SM (2016) Calcium-sensing receptor: a key target for extracellular calcium signaling in neurons. Front Physiol 7:116
228. Nemeth EF, Shoback D (2013) Calcimimetic and calcilytic drugs for treating bone and mineral-related disorders. Best Pract Res Clin Endocrinol Metab 27:373–384
229. Steddon SJ, Cunningham J (2005) Calcimimetics and calcilytics -fooling the calcium receptor. Lancet 365:2237–2239
230. Jiang Y, Huang Y, Wong HC, Zhou Y, Wang X, Yang J et al (2010) Elucidation of a novel extracellular calcium-binding site on metabotropic glutamate receptor 1{alpha} (mGluR1{alpha}) that controls receptor activation. J Biol Chem 285:33463–33474
231. Samardzic J (ed) (2018) GABA and glutamate. New developments in neurotransmission research. Intech Open, London
232. Willard SS, Koochekpour S (2013) Glutamate, glutamate receptors, and downstream signaling pathways. Int J Biol Sci 9:948–959
233. Jacobson LH, Vlachou S, Slattery DA, Li X, Cryan JF (2018) The gamma-aminobutyric acid B receptor in depression and reward. Biol Psychiatry 83:963–976
234. Benarroch EE (2012) GABAB receptors: structure, functions, and clinical implications. Neurology 78:578–584

Chapter 9
Phospholipase C

Colin A. Bill and Charlotte M. Vines

Abstract Phospholipase C (PLC) family members constitute a family of diverse enzymes. Thirteen different family members have been cloned. These family members have unique structures that mediate various functions. Although PLC family members all appear to signal through the bi-products of cleaving phospholipids, it is clear that each family member, and at times each isoform, contributes to unique cellular functions. This chapter provides a review of the current literature on PLC. In addition, references have been provided for more in-depth information regarding areas that are not discussed including tyrosine kinase activation of PLC. Understanding the roles of the individual PLC enzymes, and their distinct cellular functions, will lead to a better understanding of the physiological roles of these enzymes in the development of diseases and the maintenance of homeostasis.

Keywords Phospholipase C family · G protein-coupled receptors · Phosphatidylinositol 4 · 5 – bisphosphate · Diacylglycerol · Inositol 1 · 4 · 5 – triphosphate · Calcium · Isoform · Structure · Ubiquitous expression · Multiple functions

9.1 Discovery

In 1953, it was reported that the addition of acetylcholine or carbamylcholine to pancreatic cells led to the production of phospholipids [1]. In these studies, ^{32}P was used to detect a sevenfold increase in the levels of phospholipids in the samples treated with the drugs, when compared with control slices, which had remained un-stimulated. Although unrecognized at that time, this was the first evidence of

C. A. Bill · C. M. Vines (✉)
Department of Biological Sciences, Border Biomedical Research Center, The University of Texas at El Paso, El Paso, TX, USA
e-mail: cvines@utep.edu

© Springer Nature Switzerland AG 2020
M. S. Islam (ed.), *Calcium Signaling*, Advances in Experimental Medicine and Biology 1131, https://doi.org/10.1007/978-3-030-12457-1_9

the presence of phospholipase C (PLC) function in cells. More than 20 years later, in 1975, it was shown that impure preparations of PLC could be used to cleave phosphatidylinositol [2]. In 1981, the first purified preparation of PLC was isolated [3]. A couple of years later it was found that the inositol 1,4,5 trisphosphate (IP$_3$) generated from the cleavage of phosphatidyl inositol 4,5 bisphosphate (also known as PI (4,5)P$_2$ or PIP$_2$) could induce the release of Ca^{2+} from intracellular stores [4] (Figs. 9.1 and 9.2). This important observation provided new insight into the function of PLC in living organisms. Eventually, the PLCβ, PLCγ, PLCδ, PLCε, PLCη and PLCζ cDNAs were cloned [5–10]. Although PIP$_2$ is a minor phospholipid in the plasma membrane, it plays a central role in regulating a host of cellular processes. PLC is activated following stimulation of cells by either tyrosine kinase receptors, T-cell receptors, B-cell receptors, Fc receptors, integrin adhesion proteins or G protein-coupled receptors via cognate ligands including neurotransmitters, histamine, hormones and growth factors [11–15]. Signaling through PLC family members regulates diverse functions, which will be outlined within this chapter. In addition, we will discuss PLC mediated signaling, common structural domains found in this family of enzymes, current knowledge about the isoforms and areas that have yet to be explored.

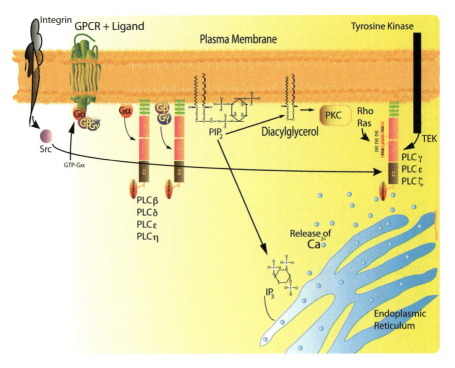

Fig. 9.1 Different effectors activate signaling through PLC to induce cleavage of PIP$_2$ to yield diacylglycerol and inositol triphosphate

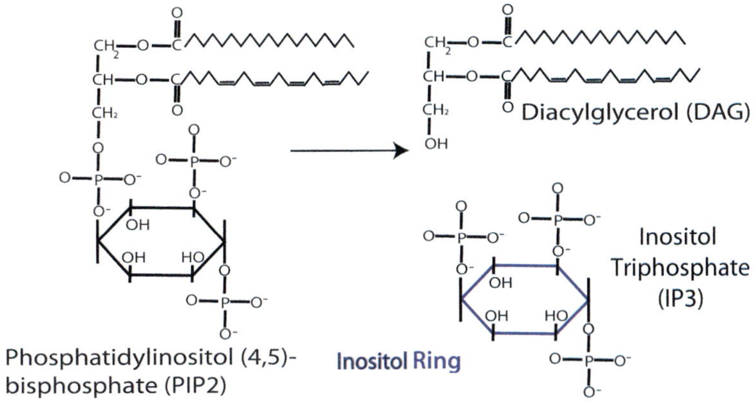

Fig. 9.2 PLC family members cleave PIP$_2$ to produce diacylglycerol and inositol triphosphate

9.2 Cleavage of PIP$_2$ and Signaling

PLC is a cytoplasmic protein that controls the levels of PIP$_2$ in cells by localizing within or outside of lipid rafts in the plasma membrane and catalyzing the hydrolysis of phosphorylated forms of phosphatidyl inositol in response to cellular stimuli (Figs. 9.1 and 9.2). These enzymes have been reported to increase the rate of lysis of phosphatidyl inositol >1000 s^{-1} at 30 °C at low concentrations of substrate, but is likely to reach rates of >5000 s^{-1} (as reviewed by [16]). Therefore, targeting of PLC to the plasma membrane plays a critical role in the functioning of this enzyme. The preferred substrate of PLC is PIP$_2$, a relatively uncommon phospholipid in the plasma membrane, followed by phosphatidyl inositol phosphate (PIP), and then phosphatidyl inositol (PI). Cleavage of PIP$_2$ leads to the generation of two products. One product, diacylglycerol (DAG), activates the calcium dependent protein kinase C (PKC), which then phosphorylates downstream effectors such as AKT to activate an array of cellular functions including regulating cell proliferation, cell polarity, learning, memory and spatial distribution of signals [17, 18]. DAG, which remains membrane bound, can then be cleaved to produce another signaling molecule, arachidonic acid. The second product of PLC action on PIP$_2$, IP$_3$ is a small water-soluble molecule, which diffuses away from the membrane, and through the cytosol to bind to IP$_3$ receptors on the endoplasmic reticulum inducing the release of Ca^{2+} from intracellular stores found within the organelle [4]. In turn, the cytoplasmic calcium levels are quickly elevated and cause the characteristic calcium spike that signals cell activation. Once the endoplasmic reticulum stores have been used up, they are replenished through the store-operated calcium channels. Ca^{2+} activates downstream transcription factors resulting in a plethora of gene activation pathways. In this way, signaling through PLC regulates proliferation, differentiation, fertilization, cell division, growth, sensory transduction, modification of gene expression, degranulation, secretion and motility [15, 19–26].

9.3   Structure of PLC

There are thirteen different PLC family members that can be subdivided into six classes, β, γ, δ, ε, η and ζ (Fig. 9.3). Different isoforms have been discovered in a wide range of species including mouse, rat and cattle. PLC-like isozymes have been found in *Drosophila melanogaster*, *Glycine max* (soybean), *Arabidopsis thaliana*, *Saccharomyces cerevisiae* and *Schizosaccharomyces pombe* [27, 28]. Overall, there is a low level of amino acid conservation between the family members; however, the similarity of the pleckstrin homology domains, the EF hand motifs, the X and Y domains and the C2 domains is greater than 40–50% [15]. Since these domains are common to all organisms they might represent a minimum requirement for a functioning PLC [29]. With the exception of the PH domain, which is not expressed on PLCζ, each family member shares all of the core domains. A description of each domain follows:

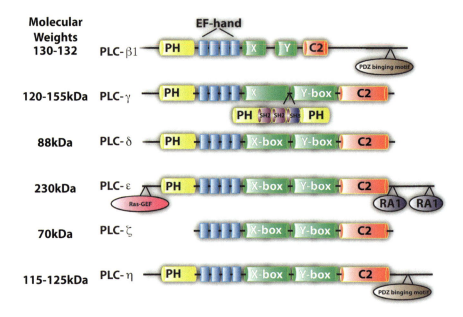

Fig. 9.3 Unique domains found in individual family members include the following: Post Synaptic density (PSD-95), Drosophila disc large tumor suppressor (DLgA) and Zonula Occludens-1 (ZO-1) (PDZ), src homology 2 (SH2)

9.3.1 Pleckstrin Homology (PH) Domains

As mentioned, with the exception of PLCζ, all PLC family members have N-terminal pleckstrin homology (PH) domains which consists of approximately 120 amino acids, and is the eleventh most common domain in the human genome. PH domains are found in a large number of distinct protein families involved in signal transduction [30]. PH domains can mediate recruitment of the PLC family members to the plasma membrane via phosphoinositides. Computer simulations and crystal structures of the PH domain found in kindlins, proteins which co-activate integrin adhesion proteins, have revealed that PH domains consist of 7 beta sheets and an alpha helix, and that the beta sheets form the PIP_2 binding site [31]. Surface plasmon resonance studies have revealed a 1mM affinity for PIP_2 within lipid bilayers.

Notably, membrane binding of PLCδ to PIP_2 is blocked by high levels of intracellular Ca^{2+} in hepatocytes due to generation of phosphoinositides [32]. This may also be due to the ability of Ca^{2+} to regulate the conformation of the headgroup of PIP_2 [33]. Unlike the PH domain of PLCδ1, which uses the PH domain to bind to the PIP_2 in the membrane, the PH domain of PLCβ2, cannot bind to phosphoinositides [34].

PLCγ contains 2 PH domains, one in the N-terminus and a C-terminal split PH domain. This PH domain of PLCγ is unique, since it is split between two tandem Src homology domains [35]. Early on, it was found that the carboxy-terminal region of the PH domains of PLCγ, PLCβ2 and PLCβ3 control the binding of heterotrimeric G protein βγ subunits to PLC following activation of G protein-coupled receptors [36, 37]. Interestingly, the binding of $G_{\beta\gamma}$ to the PH domain, and the binding of $G_{\beta\gamma}$ to Gα are mutually exclusive [36]. Therefore, this competition for binding to $G_{\beta\gamma}$ implicates PLC activation in preventing the regeneration of the Gα/$G_{\beta\gamma}$ heterotrimeric G proteins. In this way [34], PLC activation may regulate the signaling of proteins that are turned on in response to stimulation of G protein-coupled receptors. Additionally, downstream of SDF1α (CXCL12) binding to the G protein-coupled receptor CXC chemokine receptor 4 (CXCR4), the PH domains of PLCε1 promote lipase independent activation of Rap1, which leads to β2-integrin-mediated recruitment and adhesion of T-cells to sites of inflammation [38]. Overall, from these observations it can be inferred that PH domains have multiple roles in regulating the signaling via PLC.

In contrast to PLC signaling through heterotrimeric G-protein, it should be noted that Rap1 which belongs to the Rap-family of small GTPases and Ras-family small GTPases are also involved in PLC signaling. Rap and Ras are small, closely related GTP binding proteins. While Rap is an important factor in cell junctions and cell adhesion, Ras is linked to cell proliferation and survival [39]. Both of these small, monomeric G proteins also play critical roles in signaling through PLC as will be discussed below:

9.3.1.1 EF-Hand Motifs

The EF-hand motifs are helix-loop-helix motifs present in a number of calcium-binding proteins, such as myosin, calmodulin, calreticulin and troponin [40]. EF-hand motifs were first described for PLC when the crystal structure analysis of PLCδ1 revealed the characteristic helix-loop-helix motifs [41]. Within PLC, the EF-hand is part of the catalytic core that consists of an EF-hand, the X and Y and the C2 domains ([41] and see below). Upon binding to Ca^{2+}, the structure of PLC is stabilized as the EF-hand motifs undergo a conformational change to activate calcium-regulated functions, by exposing sites that become ligands for other proteins [42]. For example, in PLCβ, the EF-hands contain sites that mediate association with subunits of heterotrimeric G proteins, while in PLCγ, the EF-hands contain regions that lead to binding of tyrosine kinases [43]. Independent of the Ca^{2+} concentration, deletion of the EF-hands in the enzyme reduces PLC function, [44]; however, binding of Ca^{2+} to the EF-hand motifs can promote binding of PLC to PIP_2 via the PH domain. Lacking a PH domain, PLCζ may bind to membrane PIP_2 via cationic residues in the EF-hand [45] as well as the X-Y linker (as reviewed [46]).

9.3.1.2 X and Y Domains

So far, only PLCδ1 and PLCβ2 have been crystallized and their structures analyzed [34, 41]. The X and Y domains consist of approximately 300 amino acids and lie at the C-terminus of the EF-hand motifs. These domains consist of alternating α-helices and β-sheets that form αβαβαβαβ motif with a triosephosphate isomerase (TIM) barrel-like structure [41]. The X-region, containing all of the catalytic residues, is somewhat conserved across the PLC family members [27, 41]. The X-region forms one half of the TIM-barrel like structure. Within the X-region lies histidine residues that support the generation of the 1,2 cyclic inositol 4,5-bisphosphate [47]. The catalytic activity of this domain increases as the concentration of Ca^{2+} rises from 0.01μM to 10μM. Mutational analysis of rat PLCδ1 revealed that histidine[311] and histidine[356], which are crucial for catalyzing the hydrolysis of PIP_2, have an important role within the X domain [47]. These residues are well conserved in PLC family members [47].

Structurally, the Y-domain (residues 489–606) forms the other half of the TIM-barrel-like architecture. This eightfold barrel structure is almost always found within an enzyme that regulates metabolism [48], although the functions of the enzymes are quite diverse. With the exception of an extended loop connection between the β5 and β6 strand, instead of a helix, this domain forms the second half of the TIM-barrel-like structure. This Y-domain is important for substrate recognition and regulates the preference of PLC for PIP_2, PIP and PI [49, 50].

PLCγ contains a unique region that splits the X and Y domains. This region contains the split PH domains at the ends and the middle consists of two N-terminal src homology (SH2) domains followed by an SH3 domain. The SH2 domains provide docking sites for tyrosine kinase growth factor receptors such as the platelet derived growth factor receptors (PDGFRs) and the epidermal growth factor receptors (EGFRs) to promote activation of this PLC family member [51–53]. The binding of tyrosine kinase receptors to PLCγ results in phosphorylation and activation of the enzymes [54, 55]. The SH3 domain directs the cellular localization of signaling proteins such as dynamin and the actin cytoskeleton. In addition, the SH3 domains have been found to mediate nerve growth factor-induced cell proliferation through activation of a guanine nucleotide exchange factor for phosphoinositide 3 kinase (PI3K) [56, 57].

9.3.1.3   C2 Domains

C2 domains are formed from about 120 amino acids [58] and can be found in more than 40 different proteins [41]. These motifs have several binding targets and have been implicated in signal transduction and membrane interactions. The C2 domains found within PLC family members are formed by an eight-stranded anti-parallel β-sandwich [41]. There are between three and four C2 domains found within PLCδ family members. In combination with Ca^{2+}, the C2 domain mediates the binding of PLCδ1 to anionic phospholipids to mediate signal transduction and membrane trafficking [43]. C2 domains have common structural motifs, which are found in PKCβ, rabphilin 3A [59, 60], and synaptotagmin I [61]. High cooperativity of calcium-dependent phospholipid binding sites implies that there are multiple sites that bind Ca^{2+}, which function synergistically [43].

C2 domains belong to the non-continuous Ca^{2+}-binding sites in which the Ca^{2+}-binding pockets are found far from each other in the amino acid sequence. In contrast EF-hands have binding pockets for Ca^{2+} produced by a stretch of continuous amino acids in the primary sequence [62, 63]. Functionally, the EF-hand motif, the most common Ca^{2+} binding motif in proteins, may compete for binding to Ca^{2+} with the C2 domains. The affinity of the EF-hand for Ca^{2+} is within the nanomolar to millimolar range, which overlaps the micromolar to millimolar binding constants of C2 domains [64, 65]. This broad affinity of C2 domains for $[Ca^{2+}]$ reflects the diversity of the functions of proteins containing the C2 domains over a wide range of calcium concentrations [66–68].

9.3.1.4   PDZ Domains

PDZ (Post synaptic density (PSD)-95, Drosophila disc large tumor suppressor (DlgA), and Zonula occludens-1 protein (zo-1)) regions are separate from C2 domains, and are found in the C-terminal tails of PLCβ and PLCη lipases (Fig. 9.1) [58]. The PDZ domains are formed by 5 of 6 β-strands and 2 or 3 α-helices [69]. This common structural motif is found in many signaling proteins, where it functions as a scaffold for large molecular complexes [70]. In this way, the motif links many proteins to signaling from the cytoskeletal membranes. It has been postulated that each PLCβ form may be used by different G protein-coupled receptors in regulating signaling events [71]. The sequences within the last five amino acids of the C-terminus are thought to regulate the specificity of the interaction of PLC with the Gα or Gβγ subunits [72].

9.4 Roles of Each PLC

As mentioned, there are six PLC family members (β, γ, δ, ε, η and ζ) consisting of thirteen different PLCs identified based on structure (Fig. 9.3) and activation mechanism. There is no alpha form of PLC, since the protein that was originally described as the α form turned out to be a protein disulfide isomerase without phospholipase activity [73]. Under most conditions, PLC is a cytoplasmic protein that moves to the plasma membrane. Its role within the membrane lipid rafts is somewhat controversial. For instance, PLC has been shown to accumulate within lipid rafts that consist of cholesterol, sphingomyelin and ceramide, *Xenopus* egg activation, catalyzing the hydrolysis of PIP$_2$ within these frog eggs [16, 74]. In contrast, PLC associates with the tyrosine kinase HER2 within non-raft domains in ovarian cancer cells [72]. In eggs and in ovarian cancer cells, PLC catalyzes hydrolysis of PIP$_2$ to promote classic functions (Fig. 9.2 and 9.3). With the exception of PLCγ2, there have been splice variants reported for each PLC isoform (as reviewed by [28, 44]). For PLC a different gene encodes each isoform. The diversity of the PLC isoforms is created with splice variants. PLC isoforms are quite distinct in regard to tissue distribution, cell localization, expression and regulation. PLCβ and PLCγ are typically activated by extracellular stimuli and are termed first line PLC's, whereas PLCδ, ε, η and ζ are activated by intracellular stimuli and known as secondary PLC's [75]. For the purposes of this chapter, we will focus on the general properties described for each isoform.

9.4.1 PLCβ$_{1,2,3,4}$

There are four isoforms of PLCβ that range in size from 130 kDa for PLCβ4, 140 kDa for PLCβ2, 150 kDa for PLCβ1 and 152 kDa for PLCβ3. In addition, splice variants have been reported for each of these isoforms [76–78]. The PLCβ subfamily consists of a well-conserved core structure with an N-terminal PH domain, four EF-hands, a split X + Y catalytic domain, C2 domain and an extended C-terminal domain (Fig. 9.3). The catalytic domain being the most conserved domain of all PLC's isozymes with a substrate preference for PIP2 over PIP and PI [79]. PLCβ family members show distinct tissue expression and G protein regulation. PLCβ1 and PLCβ3 are ubiquitously expressed, whereas PLCβ2 and PLCβ4 are found only in hematopoietic and neuronal tissues, respectively [80]. These well-characterized isoforms of PLC are classically activated by G protein-coupled receptors and their catalytic activity is entirely dependent upon Ca^{2+}. All four PLCβ isoforms are activated by $G\alpha_q$ subunit. PLCβ2 and PLCβ3 can also be activated by βγ subunits of the $G\alpha_{i/o}$ family of G proteins and by small GTPases such as Rac and Cdc42 (Figs. 9.1 and 9.3). In addition, PLCβ's are GTPase-activating proteins (GAPs) for the $G\alpha_q$ proteins that activate them [80, 81]. While $G\alpha_q$, $G\alpha_{11}$, and $G\alpha_{16}$ can activate PLCβ1, PCLβ2 and PLCβ3 family members [82]. In this case, the G protein-coupled receptor is stimulated by binding to its ligand, undergoing a conformational change to release $G\alpha_q$ or $G\alpha_{i/o}$ and Gβ/γ [81, 83, 84]. PLCβ is recruited to the membranes through interactions with Gβγ, but not Gαq [85]. In addition, PLCβ is recruited only through specific Gα subunits and the Gβγ subunits. These studies demonstrate that the PLC family members respond not only to Gα, but to Gβγ as well [37, 86]. Phosphoinositide-specific-phospholipase C β (PLCβ) is the main effector of $G\alpha_q$ stimulation that is coupled to receptors binding acetylcholine, dopamine, bradykinin, angiotensin II, other hormones and neurotransmitters [87].

The PLCβ family members have an additional 450 amino acid residues in the C-terminus (Fig. 9.3). While all PLCβ family members have been found in the nucleus, PLCβ1 is the major nuclear PLC [88–90]. Within this C-terminal 450 amino acid region, lies the greatest dissimilarity between PLC family members. In this region of the PLCβ1a and 1b splice variants is a nuclear localization signal, which directs localization of PLCβ1 isoforms, mostly to the nucleus while a nuclear export signal allows PLCβ1a to remain in the cytosol [77]. The likely consequence of DAG generation inside the nucleus is activation of nuclear PKC [91, 92]. Nuclear PLCβ1 regulates the cell cycle by modulating cyclin levels with cells overexpressing PLCβ1 producing increased levels of Cyclin D3 and a higher percentage of cells in S phase, in an erythroleukemia cell line [92, 93]. The binding site for Gαq is found within a region that mediates activation of Gαq by regulator of G protein signaling 4 (RGS4) and G alpha interacting protein (GAIP), which are GTPase-activating proteins (GAPs) [94]. This binding site blocks activation of PLCβ [95]. PLCβ1 is expressed at high levels in the cerebral cortex, retina, hippocampus and cardiomyocytes [96–98].

As mentioned, the expression of PLCβ2, which shares 48% identity with PLCβ1, appears to be restricted to cells of the hematopoietic lineages [99]. PLCβ2 can be activated by Rac, a member of the Rho-family of kinases [100]. The PH domain of PLCβ2 mediates binding of active forms of Rac (Rac1, Rac2 and Rac3), which leads to activation [101]. In contrast to PLCβ1 and PLCβ2, PLCβ3 lacks 10–20 amino acids within its C-terminus [102], although the significance of this difference is unknown. This PLC isoform is expressed by liver, brain and parotid gland [102].

PLCβ1 and PLCβ4 are expressed within the brain including the cerebral cortex, amygdala, hippocampus, and olfactory bulb and are thought to be involved in brain development and synaptic plasticity [91, 103–105]. Mis-regulation of PLCβ1 and/or PLCβ4 have been linked to several brain conditions such as schizophrenia, epilepsy, depression, Alzheimer's disease, bipolar disease and Huntington's disease [105–107]. In addition, studies of PLCβ1$^{-/-}$ mice revealed roles for PLCβ1 in regulating vision and central nervous system homeostasis and loss of PLCβ1 can lead to seizures and sudden death [108].

PLCβ1 plays important roles in cell differentiation, particularly in osteogenesis, hematopoiesis and myogenesis [79, 80, 109]. At least for myogenic differentiation, PLCβ1 signaling involves inositol polyphosphate multikinase and β -catenin as downstream effectors. By means of c-jun binding to cyclin D3 promoter, the activation of PLC β 1 pathway determines cyclin D3 accumulation and muscle cell differentiation [110]. Also, PLCβ participates in the differentiation and activation of immune cells involved in both the innate and adaptive immune systems including, macrophages, neutrophils, mast cells, T cells and B cells [79]. Consistent with a role of PLCβ3 in neutrophil development, it was reported that PLCβ3$^{-/-}$ mice develop myeloproliferative neoplasm with increased mature neutrophils [80].

A role for PLCβ in several cancers has been proposed. Recently, it has been reported that PLCβ2 acts as a negative regulator of triple negative breast cancer since up-regulation in invasive triple negative breast cancer cells was sufficient to lower the expression of surface antigens required for malignancy and to reduce the number of cells with a stem-like phenotype suggesting that enhancing PLCβ2 expression is a potential therapy for triple negative breast cancer [111]. Similarly, a high expression of PLCβ1 was associated with an enhanced long-term survival of patients with a proneural subtype high grade gliomas [112] and patients affected by myelodysplastic syndromes showed a reduced propensity to develop acute myeloid leukemia when the expression of nuclear PLCβ1 was reduced [91].

9.4.2   PLCγ$_{1,2}$

There are two isoforms of PLCγ, PLCγ1 and PLCγ2. PLCγ1 is ubiquitously expressed, and operates downstream of tyrosine kinase growth factor receptors such as vascular endothelial growth factor (VEGF), fibroblast growth factor (FGF), PDGF and EGF, whereas PLCγ2 is primarily expressed in hematopoietic cell lineages, often functioning downstream of immune cell receptors (Fig. 9.1 and [113, 114]). PLCγ subtypes are primarily activated by receptor tyrosine kinases (RTKs). Both PLCγ1 [115] and PLCγ2 can be activated by adhesion receptors, such as integrins [116]. PLCγ1 signaling acts via direct interactions with other signaling molecules via SH domains, as well as its lipase activity [117]. Some PLCγ signaling via nonreceptor tyrosine kinases has been reported [118, 119], including the B-cell receptor and via the Spleen tyrosine kinase (Syk)-activated PLCγ2 signaling in T cells [120] or osteogenic differentiation of bone marrow stromal cells [121]. PLCγ has important roles in differentiation, proliferation, transformation, calcium flux and tumorigenesis [22, 25, 122, 123]. In addition, it has been shown that PLC γ1 is activated by Src tyrosine kinase in Xenopus [124].

PLCγ can regulate proliferation by functions that are independent of its lipase activity. One example is that DNA synthesis does not require phospholipase function, but instead is regulated through the SH3 recruitment of a Ras exchange factor, SOS1 [125]. In addition to the PH domain found in the N-terminus, these PLCγ family members have a second PH domain, which is split into an N-terminal domain of the PH domain that flanks two SH2 domains, followed by an SH3 domain and a C-terminal PH domain (Fig. 9.3). This C-terminal is thought to bind directly to the TRPC3 calcium channel, which then leads to agonist-induced calcium entry into the cell [35]. In addition, Vav1, c-Cbl and Slp76, via interactions with either the SH3 domain or the C-terminal SH2 domain are also required to help stabilize the recruitment of PLCγ1to the plasma membrane [126]. PLCγ2 and PKC are important upstream signals of macrophage-colony stimulating factor (M-CSF) and granulocyte-colony stimulating factor (G-CSF) that regulate myelopoiesis through cytokine production. These pathways activate ERK1/2, NFAT and JAK1/STAT-3 pathways [127]. PLCγ isoforms have been reported to be expressed in several innate immune cell types, including natural killer cells, macrophages, neutrophils and mast cells [128–131]. PLCγ activates the innate immune system by regulating respiratory bursts, phagocytosis, cell adhesion, and cell migration. PLCγ also modulates the inflammatory response by controlling Toll-like receptor-mediated signaling [132]. T cells express more PLCγ1 than PLCγ2 and PLCγ1 is activated by ligation of the T cell antigen receptor [126] and recruitment of PLCγ1 by Linker of Activated T cells (LAT) to the plasma membrane [133]. Phosphorylated LAT, in turn, serves as the primary docking site for the amino terminal SH2 domain of PLCγ1 to the membrane [134, 135]. All three SH domains of PLCγ1, however, are required to stabilize association of PLCγ1 with LAT, which is required to activate T cells [126]. Following engagement of the TCR, PLCγ1 production of DAG leads to activation of not only PKC, but also Ras guanyl releasing protein (GRP)-dependent signaling events [136, 137].

PLCγ1 is also activated by certain G protein-coupled receptors. We have shown that PLCγ1 can be activated following stimulation by the G protein-coupled receptor, C–C chemokine Receptor 7, a $G\alpha_{i/o}$ receptor, to mediate activation of β1 integrin, heterodimeric adhesion receptors [138]. In addition, PLCγ1 and PLCγ2 are both activated by the angiotensin and bradykinin G protein coupled receptors.

Homozygous disruption of PLCγ1 in a mouse model revealed that this PLC plays an essential role in growth and development [139]. In the absence of PLCγ1, the mice die at day E9.0, although until that stage of development the embryos appear normal. This mouse model revealed that although other PLCγ family members might be available, the role of PLCγ1 is essential and is not compensated by another PLC. In contrast, homozygous deletion of PLCγ2 leads to defects in platelet functions that are stimulated through β1 and β3 integrin adhesion proteins [140, 141]. PLCγ2 plays an essential role in B cell development, and function [20, 26]. Similar to PLCβ2, Rac, a member of the Rho-family of GTPases, can bind to and activate PLCγ2 [100]. This PLC family member can be activated through interactions with growth factor receptors, via phosphorylated tyrosines within their cytoplasmic tails via their intracellular tyrosine activation motifs (ITAMs). PLCγ2 also regulates calcium oscillations induced by the transcription factor, Nuclear Factor of Activated T cells (NFAT). Additionally, the SH2 domains can mediate activation of this receptor.

A role for PLCγ in neural development and certain neurological condition has become increasingly evident. PLCγ1 is highly expressed in the brain and is required for normal neuronal development and activation [114]. Since deregulation of PLCγ1 activation in response to brain derived neuronal factor can alter calcium influx and actin rearrangements that control neuronal migration, this PLC has been linked to diverse neurological disorders, including epilepsy, Huntington's disease and depression [114]. In this case mis-regulation of PLCγ1 function has been observed in animal models of Huntington's disease [142]. Moreover, genomic analysis has revealed a PLCγ2 variant that appears to be protective against Alzheimer's disease, possibly acting via microglia-modulated immune responses [143]. Other physiological roles for PLCγ are provided by recent evidence suggesting that PLCγ1 activates Akt-mediated Notch1 signaling, which is required for intima formation of blood vessels, and also plays a role in influenza viral entry into human epithelial cells [144, 145].

PLCγ1 is often mutated and highly expressed in several cancers being involved in tumorigenic processes including migration, invasion and in some cases, proliferation (as reviewed by [146]). Moderately to poorly differentiated breast tumors showed significantly higher levels of PLCγ1, compared with well differentiated tumors [147, 148]. Also, three distinct mutations in PLCγ2 were described in patients with chronic lymphocytic leukemia that were resistant to Ibrutinib treatment [148]. Indeed, studies have shown that mutated DNA sequences associated with human cancers and autoimmune diseases are well conserved between PLCγ1 und PLCγ2 and these mutations are gain-of-function effectors that destabilize normal regulatory signaling [149].

9.4.3   PLCδ$_{1, 3, 4}$

There are three identified isoforms of PLCδ with similar amino acid sequences that are highly evolutionary conserved from lower to higher eukaryotes [150]. PLCδ family members are activated by levels of calcium that are normally found in the cytoplasm (10^{-7}M to 10^{-5}M), making them one of the most calcium sensitive PLC isoforms [151, 152]. While PLCδ1 is localized to the cytoplasm in quiescent cells, this PLC isoform shuttles between the nucleus and the cytoplasm in active cells [153]. Human PLCδ4 was found to be primarily nuclear in human adipose derived mesenchymal stem cells [154]. Depletion of PLCδ1 leads to a block in the cell cycle [155]. PLCδ family members are thought to have a role in potentiating calcium signaling [151]. This form of PLC is similar to non-mammalian forms of PLC [15, 156] PLCδ1 can be activated by $G_{i/o}$ and Ga_q following stimulation of G protein-coupled receptors [157]. PLCδ is involved in regulating the activation of the actin cytoskeleton. Studies using PLCδ knockout mice have shown that PLCδ1 is required for maintenance of skin homeostasis; a recent study suggested that PLCδ1 is required for epidermal barrier integrity [158], whereas PLCδ3 regulates microvilli genesis within the intestine and the directed migration of neurons in the cerebral cortex of developing brains [159, 160]. Knockout of both PLCδ1 and PLCδ3 resulted in embryonic lethality [161].

Similar to PLCγ1, mis-regulation of PLCδ1 has been linked to Alzheimer's disease [162]. Interestingly, this enzyme function is inhibited by sphingomyelin, a membrane lipid that is found in high concentrations in neurons. PLCδ1 is also mis-regulated in rat models of hypertension [163]. In addition, a decrease in PLCδ1 downregulation in cystic fibrosis cells resulted in dysregulation of Transient Receptor Potential Vanilloid 6 channel activity leading to an increase in the constitutive calcium influx, exacerbating cystic fibrosis effects [164].

PLCδ1 is expressed at high levels in hair follicles. Homozygous deletion of PLCδ1 leads to hair loss [165, 166]. The hair loss was due to an increase in leukocytes, specifically macrophages, neutrophils and T cells within the hair follicle [166]. Homozygous deletion of *Plcδ3* or *Plcδ4* had no apparent affect and the mice appeared normal.

During fertilization, a transient increase in Ca^{2+} precedes egg activation. Like other forms of PLC, this isoform appears to play a role in fertilization. Notably, PLCδ4$^{-/-}$ male mice are sterile [167, 168]. Even when PLCδ4$^{-/-}$ sperm were injected into eggs, few viable embryos developed. These studies implicate this family member in the regulation of fertilization [167]. In the same study, sperm isolated from PLCδ4 knockout mice were found to be inferior to sperm isolated from wild type mice in that the Ca^{2+} oscillations in these mice were delayed or did not occur at all [167].

Similar to several other PLC's, PLCδ's role in carcinogenesis is controversial. In one study, high expression levels of PLCδ significantly correlated with a shorter disease-free survival of patients with poorly-differentiated breast tumors suggesting a possible role as a tumor promoter [147]. In contrast, an unrelated study found that downregulation of PLCδ1 in breast cancers induced cell migration and invasion in an in vitro assay by inhibiting the phosphorylation of ERK1/2, suggesting a role as a tumor suppressor [169]. In support of the tumor suppressor effects, another study in colorectal cancer revealed that expression of PLCδ1, as shown by immunohistochemistry, was down-regulated in colorectal cancer samples, which was also linked to suppression of ERK1/2 phosphorylation [170] and increased autophagy of the colorectal cancer cells [171]. These results are in line with the concept that PLCδ1 may function as a tumor promoter or as tumor suppressor [147], and it is clear that further studies are needed to clarify the role of PLCδ in carcinogenesis.

9.4.4 PLCε

PLCε is the largest of the PLC family members with an apparent molecular weight of ~230 kDa and was originally described in 1998 as a Let-60 Ras binding protein [172]. Two splice variants of PLCε have been reported, termed PLCε1a and PLCε1b that are widely expressed, but distinct roles for these variants have not been described [173]. PLCε is expressed at the highest levels in the heart, liver and lung, but can also be found in the skeletal muscle, spleen brain, lungs, kidneys, pancreas, testis, uterus, thymus and intestine [7, 174, 175]. This class of PLC, which was originally identified in *Caenorhabditis elegans*, and was later cloned in humans [7, 172, 174, 175]. The Ras-associated (RA) domains consist of approximately 100 amino acids that interact directly with the Ras-family GTPases, Ras [7, 175] and Rho [176]. A point mutation at a lysine residue in the RA2 domain of PLCε is sufficient to prevent Ras binding of the enzyme in a GTP-dependent manner [7]. Subsequently, it was found that PLCε could also be activated by the $G\alpha_{12}$ and $G\beta/\gamma$ released by activated G protein-coupled receptors [175, 177]. Later, it was shown that hydrolysis of Golgi-associated phosphatidylinositol 4- phosphate (PI4P) in cardiac myocytes is mediated by $G\beta\gamma$ via the RA2 and N-terminal CDC25 and cysteine-rich domains [178, 179]. G protein-coupled receptors that activate PLCε include the adrenergic and PGE receptors. At the same time $G\alpha s$ has been shown to stimulate activation of PLCε [180] while $G\alpha_{12}$ and $G\alpha_{13}$ can activate RhoA which can stimulate PLCε [180, 181]. Not only is this PLC family member activated by Ras and RhoA, it can also function as a guanine nucleotide exchange factor (GEF) for the Ras superfamily of GTPases [175]. In a contrasting study, the CDC25 domain of PLCε was found to serve as a GEF for Rap1 but not for other Ras family members [182]. These characteristics of PLCε reveal that this enzyme can be activated not only by subunits of heterotrimeric G proteins, but also by small GTPases.

This ability of PLCε to be regulated by both Ras and Rho suggest that it can contribute to both proliferation and to migration. More interestingly, since PLCβ can be activated by Rho, both PLC family members may work together to regulate signal transduction pathways that are activated following stimulation of cells by Rho to control cell migration. Similarly, since PLCε can be regulated by Ras, a downstream effector of PLCγ signaling following activation of growth factor receptors such as the EGF receptor, the signaling pathways may work together to promote proliferation. The ability of PLCε to coordinate signaling through these pathways points to regulatory mechanisms that may be more complex than originally thought.

Since PLCε can regulate inflammatory ligands for G protein-coupled receptors, it was suggested that PLCε may protect against ischemia/reperfusion injuries [183]. In contrast, in a separate study it was shown that PLCε is often upregulated in patients with heart failure [184] and recently it was shown that chronic activation of this isoform leads to cardiac hypertrophy [178]. Additionally, PLCε-null mice have abnormal development of aortic and pulmonary valves [185]. The role of PLCε in carcinogenesis is controversial, although the enzyme is thought to play important roles in the regulation of cancer development and progression, possibly acting as either an oncogene or tumor suppressor depending upon the type of tumor [186, 187]. Inflammatory processes induced by PLCε are thought to be involved in the progression towards cancer [188]. Mutation analysis of the PLCE1 gene landscape via The Cancer Genome Atlas (TCGA) database showed that PLCE1 is an often-mutated gene in several types of cancer, in particular digestive tract cancer such as gastric cancer and esophageal squamous carcinoma, but also including skin cancer, lung cancer and head and neck cancers [187].

9.4.5 PLCη₁,₂

PLCη consists of two members that are the most recently discovered PLC's and are most closely related to PLCδ subtype [189]. The sequence homology between PLCη1 and PLCη2 are ~50% similar. PLCη1 has an apparent molecular weight of 115 kDa in mouse and humans, while PLCη2 is larger at 125 kDa. PLCη can be activated by G protein-coupled receptors and RTK's [190] with PLC activity amplified by both intracellular Ca^{2+} mobilization and extracellular Ca^{2+} entry [191]. PLCη sequence analysis showed a novel EF-hand domain including a non-canonical EF-loop 2 sequence that is responsible for the enhanced binding of Ca^{2+} and enhanced hydrolysis of PIP_2 [189]. The PLCη1 and PLCη2, isoforms are localized to the brain and neurons and are extremely sensitive to changes in calcium levels within the physiological range [8, 9, 192, 193]. Like PLCδ, this form of PLC responds to the 100 nM calcium concentrations found inside the cell [194]. However, PLCη is more sensitive than PLCδ [8] and PLCη can modulate a sustained Ca^{2+} release via production of IP_3 [189].

PLCη2 is expressed in the infant brain, specifically in the hippocampus, cerebral cortex and olfactory bulb [9], where it may play an important role in calcium mobilization required for axon growth and retraction, growth cone guidance, the generation of synapses and neurological responses [9]. In humans, loss of the human chromosomal region, which encodes PLCη2 has been linked to mental retardation [195] and role for PLCη2 in neurite growth has been postulated [196]. Alzheimer's disease has been linked to altered calcium homeostasis within neurons of the central nervous system with calcium accumulation occurring in disease affected neuronal cells [197]. Since PLCη is expressed in these same regions of the brain, a potential role for PLCη in Alzheimer's disease pathogenesis has been postulated [197].

9.4.6 PLCζ

PLCζ is the smallest of the mammalian PLC family members with a molecular mass of ~70 kDa in humans and ~74 kDa in mice [10, 198]. Interestingly, studies have shown PLC-like activities in plants with non-specific PLC hydrolyzing membrane phospholipids like phosphatidylcholine (PC) and phosphatidylethanolamine and another PLC with structural similarities to PLCζ [29]. In mammals, PLCζ expression has been confined to sperm heads [10, 198, 199] where it serves to activate eggs during fertilization [10, 200]. Subsequent studies have also identified further mammalian orthologues of PLCζ in human, hamster, monkey, and horse sperm [201, 202]. Although some studies suggested the possibility that a post-acrosomal sheath WW domain-binding protein, termed PAWP, could be responsible for eliciting Ca^{2+} oscillations at egg activation [203–205], more recent studies now convincingly suggest that PAWP is not required to stimulate Ca^{2+} oscillations during egg activation, while strong evidence supports PLCζ as a soluble sperm factor responsible for the Ca^{2+} oscillations [206–210].

In line with its key role as a sperm factor, PLCζ generally localizes to distinct regions of the sperm head in mammals [211–213]. In humans, three distinct populations of PLCζ within the sperm head have been determined in the acrosomal, equatorial and post-acrosomal regions [211, 214–216]. Although this is the only isoform of PLC identified, which lacks the N-terminal PH domain, it shares the closest homology with PLCδ1 [217]. The absence of the PH domain demonstrates that presence is not required for membrane localization of PLCζ. It is unclear, however, how PLCζ targets the plasma membrane in the absence of the PH domain. There is some indication that the C2 domain may contribute to targeting PLCζ to membrane-bound PIP_2. Following fusion of sperm with the egg, PLCζ is released into an egg, which until that point, is arrested at the second meiotic division. Ca^{2+} oscillations that mediate activation of an egg are due to IP_3 mediated Ca^{2+} release. The presence of PLCζ within the cytoplasm leads to Ca^{2+} oscillations, which are classically observed during activation of the egg and release from the meiotic arrest [218]. In addition, immuno-depletion of PLCζ suppresses Ca^{2+} release. After the egg is fertilized the Ca^{2+} oscillations end when the pronuclei merge [219, 220]. Sperm from infertile men who are unable to activate eggs have been reported to exhibit reduced or abolished types of PLCζ [214, 216, 221]. Also, the proportion of sperm expressing PLCζ correlates with fertilization rates following intracytoplasmic sperm injection making PLCζ a diagnostic marker of fertilization [75].

9.5   Methods to Inhibit PLC

There are several chemical inhibitors that can be used to block PLC function. A commonly used pan inhibitor, 1-[6-((17β-3-methoxyestra-1,3,5(10)trien-17-yl)amino)hexyl)-1H-pyrrole-2,5-dione, (U73122), of phospholipase C, is thought to function by blocking translocation of the enzyme to the membrane [222]. For example, using 2 μM U73122 in contrast to the control U73343, we found that stimulation of CCR7 through one of its ligands, CCL21 [138], but not CCL19 promoted PLC dependent migration of T cells via β1 integrin adhesion proteins. In the same study, we were able to determine that the PLCγ1 isoform regulated migration by preventing CCL21 directed migration with targeted siRNA. This data suggests that one G protein-coupled receptor can activate PLCγ1 through two different ligands to control migration in T cells. In this case we speculate that PLCγ1 mediates integrin activation through inside-out signaling leading to activation of β1-integrins.

Recently, it has been shown that U73122 forms covalent associations with human PLCβ3, when the phospholipase is associated with mixed micelles [223]. While U73122 has been used as a pan inhibitor of PLC in numerous studies [21, 138, 224–228], in the study by Klein et al., instead of inhibiting PLC, U73122 activated human PLCγ1, human PLCβ2 and human PLCβ3, which had been incorporated into micelles to differing magnitudes. Since the PLC used in these studies was in a purified form, it is unclear, how U73122 functions to regulate the extent of PLC activation. In a second study, 1 μM U73122 was found to directly inhibit G protein activated inwardly rectifying potassium channels. This was in contrast to a second PLC inhibitor, 2-Nitro-4-carboxyphenylN,N -diphenylcarbamate (NCDC), which did not have that effect [229]. NCDC, however, is also thought to have non-specific effects that are not related to PLC functions [230].

It should also be noted that in rabbit parietal cells, use of the U73122 led to a number of unexpected effects including mis-regulation of Ca^{2+} mobilization, and acid secretion induced by an agonist. Of equal concern, the negative control U73343 blocked acid secretion [231]. Therefore, this PLC inhibitor when used, should be used with caution.

Similarly, there are at least three other known inhibitors and two activators of PLC, yet they are not specific. These inhibitors include O-(Octahydro-4,7-mthano-1H-iden-5-yl)carbonopotassium dithioate, [232], Edelfosine [233] and RHC 80267 (O,O'[1,6-Hexanediyl*bis*(iminocarbonyl)]dioxime cyclohexanone) [234]. The two activators are *m*-3M3FBS (2,4,6-Trimethyl-*N*-[3-(trifluoromethyl)phenyl]benzenesulfonamide), and the ortho version *o*-3M3FBS [235].

Heterozygous deletion of a specific PLC family member via siRNA, however, can yield targeted results [138]. As mentioned, in these studies, PLCγ1 specific siRNA was used to confirm the role of this PLC isoform in the regulation of β1 integrins during the adhesion of primary T cells. In the future it may be advisable to determine the specific PLC family member involved in a cellular response, by using siRNAs. More recently the discovery of Clustered Regularly Spaced Short Palindromic Repeats-Cas9 (CRISPR Cas9) technology, which was originally described in bacterial systems, allows for long-term targeted disruption or in some cases activation of specific genes [236, 237]. This technology, will likely be used to target specific PLC isoforms in the future.

The highly specific 3-phosphoinositide-dependent protein kinase 1(PDK1) inhibitor 2-O-benzyl-myo-inositol 1,3,4,5,6-pentakisphosphate (2-O-Bn-InsP5) can also block PLCγ1 dependent cell functions such as EGFR-induced phosphorylation of PLCγ1. This interaction takes place through the PH domain of PDK1. The loss of phosphorylation blocks PLCγ1 activity and downstream the cell migration and invasion [238], and has been considered as a lead compound for an anti-metastatic drug.

9.6 Future Directions

9.6.1 Hierarchy of Isozymes

It is unclear how the different isoforms of PLC are activated in cells receiving multiple stimuli from different receptors. With thirteen identified isoforms, expressed in multiple cell types, it will be important to define how the different signaling events that are linked to each isoform are controlled. Since PLC activation leads to release of IP$_3$ and DAG in response to activation, it will be important to determine how cells discriminates between multiple PLC signals to determine the hierarchy, intensity and duration of signaling events. As mentioned, PLCβ2 and PLCγ2 are activated by Rac while PLCε is activated by RhoA. These observations suggest that key regulators of cell motility function through different PLC family members, and may have pivotal roles in defining where and when a cell migrates.

PLC enzymes are found in every cell in the body, where they play critical roles in regulating diverse cellular responses (as reviewed in [28]). As mentioned, some family members serve as scaffolds for other signaling proteins, while others can serve as GAPs or GEFs, for secondary signaling proteins. Other PLCs function to amplify the Ca^{2+} oscillations in the cell. Certain PLC family members can travel to the nucleus to control signaling there. With PLC family members playing key roles in numerous cell functions, it will be important to define how each PLC is regulated and how the cellular environment affects the duration and intensity of the response.

References

1. Hokin MR, Hokin LE (1953) Enzyme secretion and the incorporation of P32 into phospholipides of pancreas slices. J Biol Chem 203(2):967–977
2. Michell RH, Allan D (1975) Inositol cyclis phosphate as a product of phosphatidylinositol breakdown by phospholipase C (Bacillus cereus). FEBS Lett 53(3):302–304
3. Takenawa T, Nagai Y (1982) Effect of unsaturated fatty acids and Ca^{2+} on phosphatidylinositol synthesis and breakdown. J Biochem 91(3):793–799
4. Streb H et al (1983) Release of Ca^{2+} from a nonmitochondrial intracellular store in pancreatic acinar cells by inositol-1,4,5-trisphosphate. Nature 306(5938):67–69
5. Suh PG et al (1988) Inositol phospholipid-specific phospholipase C: complete cDNA and protein sequences and sequence homology to tyrosine kinase-related oncogene products. Proc Natl Acad Sci U S A 85(15):5419–5423
6. Suh PG et al (1988) Cloning and sequence of multiple forms of phospholipase C. Cell 54(2):161–169
7. Kelley GG et al (2001) Phospholipase C(epsilon): a novel Ras effector. EMBO J 20(4):743–754
8. Hwang JI et al (2005) Molecular cloning and characterization of a novel phospholipase C, PLC-eta. Biochem J 389(Pt 1):181–186
9. Nakahara M et al (2005) A novel phospholipase C, PLC(eta)2, is a neuron-specific isozyme. J Biol Chem 280(32):29128–29134
10. Saunders CM et al (2002) PLC zeta: a sperm-specific trigger of Ca^{2+} oscillations in eggs and embryo development. Development 129(15):3533–3544

11. Albuquerque EX, Thesleff S (1967) Influence of phospholipase C on some electrical properties of the skeletal muscle membrane. J Physiol 190(1):123–137
12. Macchia V, Pastan I (1967) Action of phospholipase C on the thyroid. Abolition of the response to thyroid-stimulating hormone. J Biol Chem 242(8):1864–1869
13. Portela A et al (1966) Membrane response to phospholipase C and acetylcholine in cesium and potassium Ringer. Acta Physiol Lat Am 16(4):380–386
14. Trifaro JM et al (2002) Pathways that control cortical F-actin dynamics during secretion. Neurochem Res 27(11):1371–1385
15. Fukami K et al (2010) Phospholipase C is a key enzyme regulating intracellular calcium and modulating the phosphoinositide balance. Prog Lipid Res 49(4):429–437
16. Kadamur G, Ross EM (2013) Mammalian phospholipase C. Annu Rev Physiol 75:127–154
17. Sun MK, Alkon DL (2010) Pharmacology of protein kinase C activators: cognition-enhancing and antidementic therapeutics. Pharmacol Ther 127(1):66–77
18. Rosse C et al (2010) PKC and the control of localized signal dynamics. Nat Rev Mol Cell Biol 11(2):103–112
19. Akutagawa A et al (2006) Disruption of phospholipase Cdelta4 gene modulates the liver regeneration in cooperation with nuclear protein kinase C. J Biochem 140(5):619–625
20. Hashimoto A et al (2000) Cutting edge: essential role of phospholipase C-gamma 2 in B cell development and function. J Immunol 165(4):1738–1742
21. Hong J et al (2010) Bile acid reflux contributes to development of esophageal adenocarcinoma via activation of phosphatidylinositol-specific phospholipase Cgamma2 and NADPH oxidase NOX5-S. Cancer Res 70(3):1247–1255
22. Li M et al (2009) Phospholipase Cepsilon promotes intestinal tumorigenesis of Apc(Min/+) mice through augmentation of inflammation and angiogenesis. Carcinogenesis 30(8):1424–1432
23. Sun C et al (2009) Inhibition of phosphatidylcholine-specific phospholipase C prevents bone marrow stromal cell senescence in vitro. J Cell Biochem 108(2):519–528
24. Varela D et al (2007) Activation of H2O2-induced VSOR Cl⁻ currents in HTC cells require phospholipase Cgamma1 phosphorylation and Ca^{2+} mobilisation. Cell Physiol Biochem 20(6):773–780
25. Wahl MI et al (1989) Platelet-derived growth factor induces rapid and sustained tyrosine phosphorylation of phospholipase C-gamma in quiescent BALB/c 3T3 cells. Mol Cell Biol 9(7):2934–2943
26. Wang D et al (2000) Phospholipase Cgamma2 is essential in the functions of B cell and several Fc receptors. Immunity 13(1):25–35
27. Bunney TD, Katan M (2011) PLC regulation: emerging pictures for molecular mechanisms. Trends Biochem Sci 36(2):88–96
28. Suh PG et al (2008) Multiple roles of phosphoinositide-specific phospholipase C isozymes. BMB Rep 41(6):415–434
29. Rupwate SD, Rajasekharan R (2012) Plant phosphoinositide-specific phospholipase C: an insight. Plant Signal Behav 7(10):1281–1283
30. Harlan JE et al (1994) Pleckstrin homology domains bind to phosphatidylinositol-4,5-bisphosphate. Nature 371(6493):168–170
31. Ni T et al (2017) Structure and lipid-binding properties of the kindlin-3 pleckstrin homology domain. Biochem J 474(4):539–556
32. Kang JK et al (2017) Increased intracellular Ca^{2+} concentrations prevent membrane localization of PH domains through the formation of Ca^{2+}-phosphoinositides. Proc Natl Acad Sci U S A 114(45):11926–11931
33. Bilkova E et al (2017) Calcium directly regulates phosphatidylinositol 4,5-bisphosphate headgroup conformation and recognition. J Am Chem Soc 139(11):4019–4024
34. Jezyk MR et al (2006) Crystal structure of Rac1 bound to its effector phospholipase C-beta2. Nat Struct Mol Biol 13(12):1135–1140

35. Wen W, Yan J, Zhang M (2006) Structural characterization of the split pleckstrin homology domain in phospholipase C-gamma1 and its interaction with TRPC3. J Biol Chem 281(17):12060–12068
36. Touhara K et al (1994) Binding of G protein beta gamma-subunits to pleckstrin homology domains. J Biol Chem 269(14):10217–10220
37. Wang T et al (1999) Differential association of the pleckstrin homology domains of phospholipases C-beta 1, C-beta 2, and C-delta 1 with lipid bilayers and the beta gamma subunits of heterotrimeric G proteins. Biochemistry 38(5):1517–1524
38. Strazza M et al (2017) PLCepsilon1 regulates SDF-1alpha-induced lymphocyte adhesion and migration to sites of inflammation. Proc Natl Acad Sci U S A 114(10):2693–2698
39. Raaijmakers JH, Bos JL (2009) Specificity in Ras and Rap signaling. J Biol Chem 284(17):10995–10999
40. Kawasaki H, Kretsinger RH (1994) Calcium-binding proteins. 1: EF-hands. Protein Profile 1(4):343–517
41. Essen LO et al (1996) Crystal structure of a mammalian phosphoinositide-specific phospholipase C delta. Nature 380(6575):595–602
42. Rhee SG, Choi KD (1992) Regulation of inositol phospholipid-specific phospholipase C isozymes. J Biol Chem 267(18):12393–12396
43. Essen LO et al (1997) A ternary metal binding site in the C2 domain of phosphoinositide-specific phospholipase C-delta1. Biochemistry 36(10):2753–2762
44. Otterhag L, Sommarin M, Pical C (2001) N-terminal EF-hand-like domain is required for phosphoinositide-specific phospholipase C activity in Arabidopsis thaliana. FEBS Lett 497(2-3):165–170
45. Nomikos M et al (2015) Essential role of the EF-hand domain in targeting sperm phospholipase Czeta to membrane phosphatidylinositol 4,5-bisphosphate (PIP2). J Biol Chem 290(49):29519–29530
46. Theodoridou M et al (2013) Chimeras of sperm PLCzeta reveal disparate protein domain functions in the generation of intracellular Ca^{2+} oscillations in mammalian eggs at fertilization. Mol Hum Reprod 19(12):852–864
47. Ellis MV, S. U, Katan M (1995) Mutations within a highly conserved sequence present in the X region of phosphoinositide-specific phospholipase C-delta 1. Biochem J 307(Pt 1):69–75
48. Nagano N, Orengo CA, Thornton JM (2002) One fold with many functions: the evolutionary relationships between TIM barrel families based on their sequences, structures and functions. J Mol Biol 321(5):741–765
49. Williams RL (1999) Mammalian phosphoinositide-specific phospholipase C. Biochim Biophys Acta 1441(2-3):255–267
50. Ryu SH et al (1987) Bovine brain cytosol contains three immunologically distinct forms of inositolphospholipid-specific phospholipase C. Proc Natl Acad Sci U S A 84(19):6649–6653
51. Margolis B et al (1990) Effect of phospholipase C-gamma overexpression on PDGF-induced second messengers and mitogenesis. Science 248(4955):607–610
52. Meisenhelder J et al (1989) Phospholipase C-gamma is a substrate for the PDGF and EGF receptor protein-tyrosine kinases in vivo and in vitro. Cell 57(7):1109–1122
53. Wahl MI, Daniel TO, Carpenter G (1988) Antiphosphotyrosine recovery of phospholipase C activity after EGF treatment of A-431 cells. Science 241(4868):968–970
54. Ronnstrand L et al (1992) Identification of two C-terminal autophosphorylation sites in the PDGF beta-receptor: involvement in the interaction with phospholipase C-gamma. EMBO J 11(11):3911–3919
55. Kim HK et al (1991) PDGF stimulation of inositol phospholipid hydrolysis requires PLC-gamma 1 phosphorylation on tyrosine residues 783 and 1254. Cell 65(3):435–441
56. Gout I et al (1993) The GTPase dynamin binds to and is activated by a subset of SH3 domains. Cell 75(1):25–36
57. Bar-Sagi D et al (1993) SH3 domains direct cellular localization of signaling molecules. Cell 74(1):83–91

58. van Huizen R et al (1998) Two distantly positioned PDZ domains mediate multivalent INAD-phospholipase C interactions essential for G protein-coupled signaling. EMBO J 17(8):2285–2297

59. Yamaguchi T et al (1993) Two functionally different domains of rabphilin-3A, Rab3A p25/smg p25A-binding and phospholipid- and Ca^{2+}-binding domains. J Biol Chem 268(36):27164–27170

60. Luo JH, Weinstein IB (1993) Calcium-dependent activation of protein kinase C. The role of the C2 domain in divalent cation selectivity. J Biol Chem 268(31):23580–23584

61. Davletov BA, Sudhof TC (1993) A single C2 domain from synaptotagmin I is sufficient for high affinity Ca^{2+}/phospholipid binding. J Biol Chem 268(35):26386–26390

62. Kawasaki H, Nakayama S, Kretsinger RH (1998) Classification and evolution of EF-hand proteins. Biometals 11(4):277–295

63. Kim Y et al (2001) Chimeric HTH motifs based on EF-hands. J Biol Inorg Chem 6(2):173–181

64. Lomasney JW et al (1999) Activation of phospholipase C delta1 through C2 domain by a Ca^{2+}-enzyme-phosphatidylserine ternary complex. J Biol Chem 274(31):21995–22001

65. Montaville P et al (2007) The C2A-C2B linker defines the high affinity Ca^{2+} binding mode of rabphilin-3A. J Biol Chem 282(7):5015–5025

66. Busch E et al (2000) Calcium affinity, cooperativity, and domain interactions of extracellular EF-hands present in BM-40. J Biol Chem 275(33):25508–25515

67. Gifford JL, Walsh MP, Vogel HJ (2007) Structures and metal-ion-binding properties of the Ca^{2+}-binding helix-loop-helix EF-hand motifs. Biochem J 405(2):199–221

68. Linse S et al (1988) The role of protein surface charges in ion binding. Nature 335(6191):651–652

69. Fanning AS, Anderson JM (1996) Protein-protein interactions: PDZ domain networks. Curr Biol 6(11):1385–1388

70. Wang CK et al (2010) Extensions of PDZ domains as important structural and functional elements. Protein Cell 1(8):737–751

71. Kim JK et al (2011) Subtype-specific roles of phospholipase C-beta via differential interactions with PDZ domain proteins. Adv Enzym Regul 51(1):138–151

72. Paris L et al (2017) Phosphatidylcholine-specific phospholipase C inhibition reduces HER2-overexpression, cell proliferation and in vivo tumor growth in a highly tumorigenic ovarian cancer model. Oncotarget 8(33):55022–55038

73. Charnock-Jones DS, Day K, Smith SK (1996) Cloning, expression and genomic organization of human placental protein disulfide isomerase (previously identified as phospholipase C alpha). Int J Biochem Cell Biol 28(1):81–89

74. Bates RC et al (2014) Activation of Src and release of intracellular calcium by phosphatidic acid during Xenopus laevis fertilization. Dev Biol 386(1):165–180

75. Yelumalai S et al (2015) Total levels, localization patterns, and proportions of sperm exhibiting phospholipase C zeta are significantly correlated with fertilization rates after intracytoplasmic sperm injection. Fertil Steril 104(3):561–8.e4

76. Lagercrantz J et al (1995) Genomic organization and complete cDNA sequence of the human phosphoinositide-specific phospholipase C beta 3 gene (PLCB3). Genomics 26(3):467–472

77. Mao GF, Kunapuli SP, Koneti Rao A (2000) Evidence for two alternatively spliced forms of phospholipase C-beta2 in haematopoietic cells. Br J Haematol 110(2):402–408

78. Kim MJ et al (1998) A cytosolic, galphaq- and betagamma-insensitive splice variant of phospholipase C-beta4. J Biol Chem 273(6):3618–3624

79. Xiao W, Kawakami Y, Kawakami T (2013) Immune regulation by phospholipase C-beta isoforms. Immunol Res 56(1):9–19

80. Kawakami T, Xiao W (2013) Phospholipase C-beta in immune cells. Adv Biol Regul 53(3):249–257

81. Berstein G et al (1992) Phospholipase C-beta 1 is a GTPase-activating protein for Gq/11, its physiologic regulator. Cell 70(3):411–418

82. Runnels LW, Scarlata SF (1999) Determination of the affinities between heterotrimeric G protein subunits and their phospholipase C-beta effectors. Biochemistry 38(5):1488–1496
83. Hwang JI et al (2000) Regulation of phospholipase C-beta 3 activity by Na^+/H^+ exchanger regulatory factor 2. J Biol Chem 275(22):16632–16637
84. Camps M et al (1992) Isozyme-selective stimulation of phospholipase C-beta 2 by G protein beta gamma-subunits. Nature 360(6405):684–686
85. Lee SB et al (1993) Activation of phospholipase C-beta 2 mutants by G protein alpha q and beta gamma subunits. J Biol Chem 268(34):25952–25957
86. Wang T et al (1999) Selective interaction of the C2 domains of phospholipase C-beta1 and -beta2 with activated Galphaq subunits: an alternative function for C2-signaling modules. Proc Natl Acad Sci U S A 96(14):7843–7846
87. Scarlata S et al (2016) Phospholipase Cbeta connects G protein signaling with RNA interference. Adv Biol Regul 61:51–57
88. Martelli AM et al (1992) Nuclear localization and signalling activity of phosphoinositidase C beta in Swiss 3T3 cells. Nature 358(6383):242–245
89. Kim CG, Park D, Rhee SG (1996) The role of carboxyl-terminal basic amino acids in Gqalpha-dependent activation, particulate association, and nuclear localization of phospholipase C-beta1. J Biol Chem 271(35):21187–21192
90. Payrastre B et al (1992) A differential location of phosphoinositide kinases, diacylglycerol kinase, and phospholipase C in the nuclear matrix. J Biol Chem 267(8):5078–5084
91. Ratti S et al (2017) Nuclear inositide signaling via phospholipase C. J Cell Biochem 118(8):1969–1978
92. Poli A et al (2016) Nuclear phosphatidylinositol signaling: focus on phosphatidylinositol phosphate kinases and phospholipases C. J Cell Physiol 231(8):1645–1655
93. Piazzi M et al (2015) PI-PLCbeta1b affects Akt activation, cyclin E expression, and caspase cleavage, promoting cell survival in pro-B-lymphoblastic cells exposed to oxidative stress. FASEB J 29(4):1383–1394
94. Navaratnarajah P, Gershenson A, Ross EM (2017) The binding of activated Galphaq to phospholipase C-beta exhibits anomalous affinity. J Biol Chem 292(40):16787–16801
95. Wang HL (1997) Basic amino acids at the C-terminus of the third intracellular loop are required for the activation of phospholipase C by cholecystokinin-B receptors. J Neurochem 68(4):1728–1735
96. Adamski FM, Timms KM, Shieh BH (1999) A unique isoform of phospholipase Cbeta4 highly expressed in the cerebellum and eye. Biochim Biophys Acta 1444(1):55–60
97. Min DS et al (1993) Purification of a novel phospholipase C isozyme from bovine cerebellum. J Biol Chem 268(16):12207–12212
98. Alvarez RA et al (1995) cDNA sequence and gene locus of the human retinal phosphoinositide-specific phospholipase-C beta 4 (PLCB4). Genomics 29(1):53–61
99. Park D et al (1992) Cloning, sequencing, expression, and Gq-independent activation of phospholipase C-beta 2. J Biol Chem 267(23):16048–16055
100. Harden TK, Hicks SN, Sondek J (2009) Phospholipase C isozymes as effectors of Ras superfamily GTPases. J Lipid Res 50(Suppl):S243–S248
101. Snyder JT et al (2003) The pleckstrin homology domain of phospholipase C-beta2 as an effector site for Rac. J Biol Chem 278(23):21099–21104
102. Jhon DY et al (1993) Cloning, sequencing, purification, and Gq-dependent activation of phospholipase C-beta 3. J Biol Chem 268(9):6654–6661
103. Fukaya M et al (2008) Predominant expression of phospholipase Cbeta1 in telencephalic principal neurons and cerebellar interneurons, and its close association with related signaling molecules in somatodendritic neuronal elements. Eur J Neurosci 28(9):1744–1759
104. Watanabe M et al (1998) Patterns of expression for the mRNA corresponding to the four isoforms of phospholipase Cbeta in mouse brain. Eur J Neurosci 10(6):2016–2025
105. Yang YR et al (2016) Primary phospholipase C and brain disorders. Adv Biol Regul 61:80–85
106. Koh HY (2013) Phospholipase C-beta1 and schizophrenia-related behaviors. Adv Biol Regul 53(3):242–248

107. Schoonjans AS et al (2016) PLCB1 epileptic encephalopathies; Review and expansion of the phenotypic spectrum. Eur J Paediatr Neurol 20(3):474–479
108. Kim D et al (1997) Phospholipase C isozymes selectively couple to specific neurotransmitter receptors. Nature 389(6648):290–293
109. Cocco L et al (2016) Modulation of nuclear PI-PLCbeta1 during cell differentiation. Adv Biol Regul 60:1–5
110. Ramazzotti G et al (2017) PLC-beta1 and cell differentiation: An insight into myogenesis and osteogenesis. Adv Biol Regul 63:1–5
111. Brugnoli F et al (2017) Up-modulation of PLC-beta2 reduces the number and malignancy of triple-negative breast tumor cells with a CD133(+)/EpCAM(+) phenotype: a promising target for preventing progression of TNBC. BMC Cancer 17(1):617
112. Lu G et al (2016) Phospholipase C Beta 1: a candidate signature gene for proneural subtype high-grade glioma. Mol Neurobiol 53(9):6511–6525
113. Driscoll PC (2015) Exposed: the many and varied roles of phospholipase C gamma SH2 domains. J Mol Biol 427(17):2731–2733
114. Jang HJ et al (2013) Phospholipase C-gamma1 involved in brain disorders. Adv Biol Regul 53(1):51–62
115. Epple H et al (2008) Phospholipase Cgamma2 modulates integrin signaling in the osteoclast by affecting the localization and activation of Src kinase. Mol Cell Biol 28(11):3610–3622
116. Choi JH et al (2007) Phospholipase C-gamma1 potentiates integrin-dependent cell spreading and migration through Pyk2/paxillin activation. Cell Signal 19(8):1784–1796
117. Bunney TD et al (2012) Structural and functional integration of the PLCgamma interaction domains critical for regulatory mechanisms and signaling deregulation. Structure 20(12):2062–2075
118. Arkinstall S, Payton M, Maundrell K (1995) Activation of phospholipase C gamma in Schizosaccharomyces pombe by coexpression of receptor or nonreceptor tyrosine kinases. Mol Cell Biol 15(3):1431–1438
119. Phillippe M et al (2009) Role of nonreceptor protein tyrosine kinases during phospholipase C-gamma 1-related uterine contractions in the rat. Reprod Sci 16(3):265–273
120. Buitrago C, Gonzalez Pardo V, de Boland AR (2002) Nongenomic action of 1 alpha,25(OH)(2)-vitamin D3. Activation of muscle cell PLC gamma through the tyrosine kinase c-Src and PtdIns 3-kinase. Eur J Biochem 269(10):2506–2515
121. Kusuyama J et al (2018) Spleen tyrosine kinase influences the early stages of multilineage differentiation of bone marrow stromal cell lines by regulating phospholipase C gamma activities. J Cell Physiol 233(3):2549–2559
122. Rivas M, Santisteban P (2003) TSH-activated signaling pathways in thyroid tumorigenesis. Mol Cell Endocrinol 213(1):31–45
123. Kroczek C et al (2010) Swiprosin-1/EFhd2 controls B cell receptor signaling through the assembly of the B cell receptor, Syk, and phospholipase C gamma2 in membrane rafts. J Immunol 184(7):3665–3676
124. Sato K et al (2003) Reconstitution of Src-dependent phospholipase Cgamma phosphorylation and transient calcium release by using membrane rafts and cell-free extracts from Xenopus eggs. J Biol Chem 278(40):38413–38420
125. Kim MJ et al (2000) Direct interaction of SOS1 Ras exchange protein with the SH3 domain of phospholipase C-gamma1. Biochemistry 39(29):8674–8682
126. Braiman A et al (2006) Recruitment and activation of PLCgamma1 in T cells: a new insight into old domains. EMBO J 25(4):774–784
127. Barbosa CM et al (2014) PLCgamma2 and PKC are important to myeloid lineage commitment triggered by M-SCF and G-CSF. J Cell Biochem 115(1):42–51
128. Wen R et al (2002) Phospholipase C gamma 2 is essential for specific functions of Fc epsilon R and Fc gamma R. J Immunol 169(12):6743–6752
129. Todt JC, Hu B, Curtis JL (2004) The receptor tyrosine kinase MerTK activates phospholipase C gamma2 during recognition of apoptotic thymocytes by murine macrophages. J Leukoc Biol 75(4):705–713

130. Ting AT et al (1992) Fc gamma receptor activation induces the tyrosine phosphorylation of both phospholipase C (PLC)-gamma 1 and PLC-gamma 2 in natural killer cells. J Exp Med 176(6):1751–1755
131. Hiller G, Sundler R (2002) Regulation of phospholipase C-gamma 2 via phosphatidylinositol 3-kinase in macrophages. Cell Signal 14(2):169–173
132. Kagan JC, Medzhitov R (2006) Phosphoinositide-mediated adaptor recruitment controls Toll-like receptor signaling. Cell 125(5):943–955
133. Finco TS et al (1998) LAT is required for TCR-mediated activation of PLCgamma1 and the Ras pathway. Immunity 9(5):617–626
134. Stoica B et al (1998) The amino-terminal Src homology 2 domain of phospholipase C gamma 1 is essential for TCR-induced tyrosine phosphorylation of phospholipase C gamma 1. J Immunol 160(3):1059–1066
135. Zhang W et al (2000) Association of Grb2, Gads, and phospholipase C-gamma 1 with phosphorylated LAT tyrosine residues. Effect of LAT tyrosine mutations on T cell antigen receptor-mediated signaling. J Biol Chem 275(30):23355–23361
136. Dower NA et al (2000) RasGRP is essential for mouse thymocyte differentiation and TCR signaling. Nat Immunol 1(4):317–321
137. Ebinu JO et al (2000) RasGRP links T-cell receptor signaling to Ras. Blood 95(10):3199–3203
138. Shannon LA et al (2010) CCR7/CCL21 migration on fibronectin is mediated by phospholipase Cgamma1 and ERK1/2 in primary T lymphocytes. J Biol Chem 285(50):38781–38787
139. Ji QS et al (1997) Essential role of the tyrosine kinase substrate phospholipase C-gamma1 in mammalian growth and development. Proc Natl Acad Sci U S A 94(7):2999–3003
140. Wonerow P et al (2003) A critical role for phospholipase Cgamma2 in alphaIIbbeta3-mediated platelet spreading. J Biol Chem 278(39):37520–37529
141. Inoue O et al (2003) Integrin alpha2beta1 mediates outside-in regulation of platelet spreading on collagen through activation of Src kinases and PLCgamma2. J Cell Biol 160(5):769–780
142. Garcia-Diaz Barriga G et al (2017) 7,8-dihydroxyflavone ameliorates cognitive and motor deficits in a Huntington's disease mouse model through specific activation of the PLCgamma1 pathway. Hum Mol Genet 26(16):3144–3160
143. Sims R et al (2017) Rare coding variants in PLCG2, ABI3, and TREM2 implicate microglial-mediated innate immunity in Alzheimer's disease. Nat Genet 49(9):1373–1384
144. Jiang D et al (2017) Phospholipase Cgamma1 mediates intima formation through Akt-Notch1 signaling independent of the phospholipase activity. J Am Heart Assoc 6(7)
145. Zhu L et al (2016) PLC-gamma1 is involved in the inflammatory response induced by influenza A virus H1N1 infection. Virology 496:131–137
146. Jang HJ et al (2018) PLCgamma1: potential arbitrator of cancer progression. Adv Biol Regul 67:179–189
147. Cai S et al (2017) Expression of phospholipase C isozymes in human breast cancer and their clinical significance. Oncol Rep 37(3):1707–1715
148. Woyach JA et al (2014) Resistance mechanisms for the Bruton's tyrosine kinase inhibitor ibrutinib. N Engl J Med 370(24):2286–2294
149. Koss H et al (2014) Dysfunction of phospholipase Cgamma in immune disorders and cancer. Trends Biochem Sci 39(12):603–611
150. Meldrum E et al (1991) A second gene product of the inositol-phospholipid-specific phospholipase C delta subclass. Eur J Biochem 196(1):159–165
151. Allen V et al (1997) Regulation of inositol lipid-specific phospholipase cdelta by changes in Ca^{2+} ion concentrations. Biochem J 327(Pt 2):545–552
152. Kim YH et al (1999) Phospholipase C-delta1 is activated by capacitative calcium entry that follows phospholipase C-beta activation upon bradykinin stimulation. J Biol Chem 274(37):26127–26134
153. Yamaga M et al (1999) Phospholipase C-delta1 contains a functional nuclear export signal sequence. J Biol Chem 274(40):28537–28541

154. Kunrath-Lima M et al (2018) Phospholipase C delta 4 (PLCdelta4) is a nuclear protein involved in cell proliferation and senescence in mesenchymal stromal stem cells. Cell Signal 49:59–67

155. Stallings JD et al (2005) Nuclear translocation of phospholipase C-delta1 is linked to the cell cycle and nuclear phosphatidylinositol 4,5-bisphosphate. J Biol Chem 280(23):22060–22069

156. Yoko-o T et al (1993) The putative phosphoinositide-specific phospholipase C gene, PLC1, of the yeast Saccharomyces cerevisiae is important for cell growth. Proc Natl Acad Sci U S A 90(5):1804–1808

157. Murthy KS et al (2004) Activation of PLC-delta1 by Gi/o-coupled receptor agonists. Am J Phys Cell Phys 287(6):C1679–C1687

158. Kanemaru K et al (2017) Phospholipase Cdelta1 regulates p38 MAPK activity and skin barrier integrity. Cell Death Differ 24(6):1079–1090

159. Sakurai K et al (2011) Phospholipase Cdelta3 is a novel binding partner of myosin VI and functions as anchoring of myosin VI on plasma membrane. Adv Enzym Regul 51(1):171–181

160. Kouchi Z et al (2011) Phospholipase Cdelta3 regulates RhoA/Rho kinase signaling and neurite outgrowth. J Biol Chem 286(10):8459–8471

161. Nakamura Y et al (2005) Phospholipase C-delta1 and -delta3 are essential in the trophoblast for placental development. Mol Cell Biol 25(24):10979–10988

162. Shimohama S et al (1991) Aberrant accumulation of phospholipase C-delta in Alzheimer brains. Am J Pathol 139(4):737–742

163. Yagisawa H, Tanase H, Nojima H (1991) Phospholipase C-delta gene of the spontaneously hypertensive rat harbors point mutations causing amino acid substitutions in a catalytic domain. J Hypertens 9(11):997–1004

164. Vachel L et al (2015) The low PLC-delta1 expression in cystic fibrosis bronchial epithelial cells induces upregulation of TRPV6 channel activity. Cell Calcium 57(1):38–48

165. Nakamura Y et al (2008) Phospholipase C-delta1 is an essential molecule downstream of Foxn1, the gene responsible for the nude mutation, in normal hair development. FASEB J 22(3):841–849

166. Ichinohe M et al (2007) Lack of phospholipase C-delta1 induces skin inflammation. Biochem Biophys Res Commun 356(4):912–918

167. Fukami K et al (2003) Phospholipase Cdelta4 is required for Ca^{2+} mobilization essential for acrosome reaction in sperm. J Cell Biol 161(1):79–88

168. Fukami K et al (2001) Requirement of phospholipase Cdelta4 for the zona pellucida-induced acrosome reaction. Science 292(5518):920–923

169. Shao Q et al (2017) Phospholipase Cdelta1 suppresses cell migration and invasion of breast cancer cells by modulating KIF3A-mediated ERK1/2/beta- catenin/MMP7 signalling. Oncotarget 8(17):29056–29066

170. Satow R et al (2014) Phospholipase Cdelta1 induces E-cadherin expression and suppresses malignancy in colorectal cancer cells. Proc Natl Acad Sci U S A 111(37):13505–13510

171. Shimozawa M et al (2017) Phospholipase C delta1 negatively regulates autophagy in colorectal cancer cells. Biochem Biophys Res Commun 488(4):578–583

172. Shibatohge M et al (1998) Identification of PLC210, a Caenorhabditis elegans phospholipase C, as a putative effector of Ras. J Biol Chem 273(11):6218–6222

173. Sorli SC et al (2005) Signaling properties and expression in normal and tumor tissues of two phospholipase C epsilon splice variants. Oncogene 24(1):90–100

174. Lopez I et al (2001) A novel bifunctional phospholipase c that is regulated by Galpha 12 and stimulates the Ras/mitogen-activated protein kinase pathway. J Biol Chem 276(4):2758–2765

175. Song C et al (2001) Regulation of a novel human phospholipase C, PLCepsilon, through membrane targeting by Ras. J Biol Chem 276(4):2752–2757

176. Wing MR et al (2003) Direct activation of phospholipase C-epsilon by Rho. J Biol Chem 278(42):41253–41258

177. Wing MR, Bourdon DM, Harden TK (2003) PLC-epsilon: a shared effector protein in Ras-, Rho-, and G alpha beta gamma-mediated signaling. Mol Interv 3(5):273–280

178. Malik S et al (2015) G protein betagamma subunits regulate cardiomyocyte hypertrophy through a perinuclear Golgi phosphatidylinositol 4-phosphate hydrolysis pathway. Mol Biol Cell 26(6):1188–1198

179. Madukwe JC et al (2018) G protein betagamma subunits directly interact with and activate phospholipase C. J Biol Chem 293(17):6387–6397

180. Schmidt M et al (2001) A new phospholipase-C calcium signaling pathway mediated by cyclic AMP and a Rap GTPase. Nat Cell Biol 3(11):1020–1024

181. Evellin S et al (2002) Stimulation of phospholipase C-epsilon by the M3 muscarinic acetylcholine receptor mediated by cyclic AMP and the GTPase Rap2B. J Biol Chem 277(19):16805–16813

182. Jin TG et al (2001) Role of the CDC25 homology domain of phospholipase Cepsilon in amplification of Rap1-dependent signaling. J Biol Chem 276(32):30301–30307

183. Xiang SY et al (2013) PLCepsilon, PKD1, and SSH1L transduce RhoA signaling to protect mitochondria from oxidative stress in the heart. Sci Signal 6(306):ra108

184. Wang H et al (2005) Phospholipase C epsilon modulates beta-adrenergic receptor-dependent cardiac contraction and inhibits cardiac hypertrophy. Circ Res 97(12):1305–1313

185. Tadano M et al (2005) Congenital semilunar valvulogenesis defect in mice deficient in phospholipase C epsilon. Mol Cell Biol 25(6):2191–2199

186. Chan JJ, Katan M (2013) PLCvarepsilon and the RASSF family in tumour suppression and other functions. Adv Biol Regul 53(3):258–279

187. Tyutyunnykova A, Telegeev G, Dubrovska A (2017) The controversial role of phospholipase C epsilon (PLCepsilon) in cancer development and progression. J Cancer 8(5):716–729

188. Zhang RY et al (2016) PLCepsilon signaling in cancer. J Cancer Res Clin Oncol 142(4):715–722

189. Popovics P et al (2014) A canonical EF-loop directs Ca^{2+}-sensitivity in phospholipase C-eta2. J Cell Biochem 115(3):557–565

190. Smrcka AV, Brown JH, Holz GG (2012) Role of phospholipase Cepsilon in physiological phosphoinositide signaling networks. Cell Signal 24(6):1333–1343

191. Yang YR et al (2013) The physiological roles of primary phospholipase C. Adv Biol Regul 53(3):232–241

192. Stewart AJ et al (2005) Identification of a novel class of mammalian phosphoinositol-specific phospholipase C enzymes. Int J Mol Med 15(1):117–121

193. Zhou Y et al (2005) Molecular cloning and characterization of PLC-eta2. Biochem J 391(Pt 3):667–676

194. Kouchi Z et al (2005) The role of EF-hand domains and C2 domain in regulation of enzymatic activity of phospholipase Czeta. J Biol Chem 280(22):21015–21021

195. Lo Vasco VR (2011) Role of Phosphoinositide-Specific Phospholipase C eta2 in Isolated and Syndromic Mental Retardation. Eur Neurol 65(5):264–269

196. Popovics P et al (2013) Phospholipase C-eta2 is required for retinoic acid-stimulated neurite growth. J Neurochem 124(5):632–644

197. Popovics P, Stewart AJ (2012) Phospholipase C-eta activity may contribute to Alzheimer's disease-associated calciumopathy. J Alzheimers Dis 30(4):737–744

198. Cox LJ et al (2002) Sperm phospholipase Czeta from humans and cynomolgus monkeys triggers Ca^{2+} oscillations, activation and development of mouse oocytes. Reproduction 124(5):611–623

199. Fujimoto S et al (2004) Mammalian phospholipase Czeta induces oocyte activation from the sperm perinuclear matrix. Dev Biol 274(2):370–383

200. Jones KT et al (2000) Different Ca^{2+}-releasing abilities of sperm extracts compared with tissue extracts and phospholipase C isoforms in sea urchin egg homogenate and mouse eggs. Biochem J 346(Pt 3):743–749

201. Kashir J et al (2017) Antigen unmasking enhances visualization efficacy of the oocyte activation factor, phospholipase C zeta, in mammalian sperm. Mol Hum Reprod 23(1):54–67

202. Kashir J, Nomikos M, Lai FA (2018) Phospholipase C zeta and calcium oscillations at fertilisation: the evidence, applications, and further questions. Adv Biol Regul 67:148–162

203. Aarabi M et al (2014) Sperm-derived WW domain-binding protein, PAWP, elicits calcium oscillations and oocyte activation in humans and mice. FASEB J 28(10):4434–4440

204. Aarabi M et al (2010) Sperm-borne protein, PAWP, initiates zygotic development in Xenopus laevis by eliciting intracellular calcium release. Mol Reprod Dev 77(3):249–256

205. Wu AT et al (2007) PAWP, a sperm-specific WW domain-binding protein, promotes meiotic resumption and pronuclear development during fertilization. J Biol Chem 282(16):12164–12175

206. Escoffier J et al (2016) Homozygous mutation of PLCZ1 leads to defective human oocyte activation and infertility that is not rescued by the WW-binding protein PAWP. Hum Mol Genet 25(5):878–891

207. Kashir J et al (2015) PLCzeta or PAWP: revisiting the putative mammalian sperm factor that triggers egg activation and embryogenesis. Mol Hum Reprod 21(5):383–388

208. Nomikos M et al (2015) Functional disparity between human PAWP and PLCzeta in the generation of Ca^{2+} oscillations for oocyte activation. Mol Hum Reprod 21(9):702–710

209. Nomikos M et al (2014) Sperm-specific post-acrosomal WW-domain binding protein (PAWP) does not cause Ca^{2+} release in mouse oocytes. Mol Hum Reprod 20(10):938–947

210. Satouh Y, Nozawa K, Ikawa M (2015) Sperm postacrosomal WW domain-binding protein is not required for mouse egg activation. Biol Reprod 93(4):94

211. Grasa P et al (2008) The pattern of localization of the putative oocyte activation factor, phospholipase Czeta, in uncapacitated, capacitated, and ionophore-treated human spermatozoa. Hum Reprod 23(11):2513–2522

212. Kashir J et al (2014) Sperm-induced Ca^{2+} release during egg activation in mammals. Biochem Biophys Res Commun 450(3):1204–1211

213. Young C et al (2009) Phospholipase C zeta undergoes dynamic changes in its pattern of localization in sperm during capacitation and the acrosome reaction. Fertil Steril 91(5 Suppl):2230–2242

214. Heytens E et al (2009) Reduced amounts and abnormal forms of phospholipase C zeta (PLCzeta) in spermatozoa from infertile men. Hum Reprod 24(10):2417–2428

215. Kashir J et al (2013) Variance in total levels of phospholipase C zeta (PLC-zeta) in human sperm may limit the applicability of quantitative immunofluorescent analysis as a diagnostic indicator of oocyte activation capability. Fertil Steril 99(1):107–117

216. Yoon SY et al (2008) Human sperm devoid of PLC, zeta 1 fail to induce Ca^{2+} release and are unable to initiate the first step of embryo development. J Clin Invest 118(11):3671–3681

217. Swann K et al (2006) PLCzeta(zeta): a sperm protein that triggers Ca^{2+} oscillations and egg activation in mammals. Semin Cell Dev Biol 17(2):264–273

218. Nomikos M et al (2005) Role of phospholipase C-zeta domains in Ca^{2+}-dependent phosphatidylinositol 4,5-bisphosphate hydrolysis and cytoplasmic Ca^{2+} oscillations. J Biol Chem 280(35):31011–31018

219. Halet G et al (2003) Ca^{2+} oscillations at fertilization in mammals. Biochem Soc Trans 31(Pt 5):907–911

220. Marangos P, FitzHarris G, Carroll J (2003) Ca^{2+} oscillations at fertilization in mammals are regulated by the formation of pronuclei. Development 130(7):1461–1472

221. Amdani SN et al (2016) Phospholipase C zeta (PLCzeta) and male infertility: Clinical update and topical developments. Adv Biol Regul 61:58–67

222. Wang C et al (2005) Binding of PLCdelta1PH-GFP to PtdIns(4,5)P2 prevents inhibition of phospholipase C-mediated hydrolysis of PtdIns(4,5)P2 by neomycin. Acta Pharmacol Sin 26(12):1485–1491

223. Klein RR et al (2011) Direct activation of human phospholipase C by its well known inhibitor u73122. J Biol Chem 286(14):12407–12416

224. Dwyer L et al (2010) Phospholipase C-independent effects of 3M3FBS in murine colon. Eur J Pharmacol 628(1-3):187–194

225. Frei E, Hofmann F, Wegener JW (2009) Phospholipase C mediated Ca^{2+} signals in murine urinary bladder smooth muscle. Eur J Pharmacol 610(1-3):106–109
226. Xu S et al (2009) Phospholipase Cgamma2 is critical for Dectin-1-mediated Ca^{2+} flux and cytokine production in dendritic cells. J Biol Chem 284(11):7038–7046
227. Shi TJ et al (2008) Phospholipase C{beta}3 in mouse and human dorsal root ganglia and spinal cord is a possible target for treatment of neuropathic pain. Proc Natl Acad Sci U S A 105(50):20004–20008
228. Ibrahim S et al (2007) The transfer of VLDL-associated phospholipids to activated platelets depends upon cytosolic phospholipase A2 activity. J Lipid Res 48(7):1533–1538
229. Sickmann T et al (2008) Unexpected suppression of neuronal G protein-activated, inwardly rectifying K^+ current by common phospholipase C inhibitor. Neurosci Lett 436(2):102–106
230. Kim DD, Ramirez MM, Duran WN (2000) Platelet-activating factor modulates microvascular dynamics through phospholipase C in the hamster cheek pouch. Microvasc Res 59(1):7–13
231. Muto Y, Nagao T, Urushidani T (1997) The putative phospholipase C inhibitor U73122 and its negative control, U73343, elicit unexpected effects on the rabbit parietal cell. J Pharmacol Exp Ther 282(3):1379–1388
232. Amtmann E (1996) The antiviral, antitumoural xanthate D609 is a competitive inhibitor of phosphatidylcholine-specific phospholipase C. Drugs Exp Clin Res 22(6):287–294
233. Powis G et al (1992) Selective inhibition of phosphatidylinositol phospholipase C by cytotoxic ether lipid analogues. Cancer Res 52(10):2835–2840
234. Suzuki H et al (2002) Effects of RHC-80267, an inhibitor of diacylglycerol lipase, on excitation of circular smooth muscle of the guinea-pig gastric antrum. J Smooth Muscle Res 38(6):153–164
235. Bae YS et al (2003) Identification of a compound that directly stimulates phospholipase C activity. Mol Pharmacol 63(5):1043–1050
236. Bassett AR et al (2013) Highly efficient targeted mutagenesis of Drosophila with the CRISPR/Cas9 system. Cell Rep 4(1):220–228
237. Friedland AE et al (2013) Heritable genome editing in C. elegans via a CRISPR-Cas9 system. Nat Methods 10(8):741–743
238. Raimondi C et al (2016) A Small Molecule Inhibitor of PDK1/PLCgamma1 Interaction Blocks Breast and Melanoma Cancer Cell Invasion. Sci Rep 6:26142

Chapter 10
New Insights in the IP$_3$ Receptor and Its Regulation

Jan B. Parys and Tim Vervliet

Abstract The inositol 1,4,5-trisphosphate (IP$_3$) receptor (IP$_3$R) is a Ca^{2+}-release channel mainly located in the endoplasmic reticulum (ER). Three IP$_3$R isoforms are responsible for the generation of intracellular Ca^{2+} signals that may spread across the entire cell or occur locally in so-called microdomains. Because of their ubiquitous expression, these channels are involved in the regulation of a plethora of cellular processes, including cell survival and cell death. To exert their proper function a fine regulation of their activity is of paramount importance. In this review, we will highlight the recent advances in the structural analysis of the IP$_3$R and try to link these data with the newest information concerning IP$_3$R activation and regulation. A special focus of this review will be directed towards the regulation of the IP$_3$R by protein-protein interaction. Especially the protein family formed by calmodulin and related Ca^{2+}-binding proteins and the pro- and anti-apoptotic/autophagic Bcl-2-family members will be highlighted. Finally, recently identified and novel IP$_3$R regulatory proteins will be discussed. A number of these interactions are involved in cancer development, illustrating the potential importance of modulating IP$_3$R-mediated Ca^{2+} signaling in cancer treatment.

Keywords IP$_3$R · Ca^{2+} signaling · IP$_3$-induced Ca^{2+} release · Calmodulin · Bcl-2 · IRBIT · TESPA1 · PKM2 · BAP1 · Cancer

Abbreviations

a.a.	amino acids
BAP1	BRCA-associated protein 1
Bcl	B-cell lymphoma
BH	Bcl-2 homology

J. B. Parys (✉) · T. Vervliet
KU Leuven, Laboratory for Molecular and Cellular Signaling, Department of Cellular and Molecular Medicine & Leuven Kanker Instituut, Leuven, Belgium
e-mail: jan.parys@kuleuven.be

© Springer Nature Switzerland AG 2020
M. S. Islam (ed.), *Calcium Signaling*, Advances in Experimental Medicine and Biology 1131, https://doi.org/10.1007/978-3-030-12457-1_10

CaBP	neuronal Ca^{2+}-binding protein
CaM	calmodulin
CaM1234	calmodulin fully deficient in Ca^{2+} binding
cryo-EM	cryo-electron microscopy
DARPP-32	dopamine- and cAMP-regulated phosphoprotein of 32 kDa
ER	endoplasmic reticulum
IBC	IP_3-binding core
IICR	IP_3-induced Ca^{2+} release
IP_3	inositol 1,4,5-trisphosphate
IP_3R	IP_3 receptor
IRBIT	IP_3R-binding protein released by IP_3
MLCK	myosin light chain kinase
NCS-1	neuronal Ca^{2+} sensor-1
PK	pyruvate kinase
PKA	cAMP-dependent protein kinase
PKB	protein kinase B/Akt
PLC	phospholipase C
PTEN	phosphatase and tensin homolog
RyR	ryanodine receptor
TCR	T-cell receptor
TESPA1	thymocyte-expressed, positive selection-associated 1
TIRF	total internal reflection fluorescence
TKO	triple-knockout

10.1 Introduction

The inositol 1,4,5-trisphosphate (IP_3) receptor (IP_3R) is a ubiquitously expressed Ca^{2+}-release channel mainly located in the endoplasmic reticulum (ER) [1]. The IP_3R is activated by IP_3, produced by phospholipase C (PLC), following cell stimulation by for instance extracellular agonists, hormones, growth factors or neurotransmitters. The IP_3R is responsible for the initiation and propagation of complex spatio-temporal Ca^{2+} signals that control a multitude of cellular processes [2, 3]. Moreover, dysfunction of the IP_3R and deregulation of the subsequent Ca^{2+} signals is involved in many pathological situations [4–10].

There are at least three main reasons for the central role of the IP_3R in cellular signaling. First, IP_3R signaling is not only dependent on the production of IP_3, but is also heavily modulated by its local cellular environment, integrating multiple signaling pathways. Indeed, IP_3R activity is controlled by the cytosolic and the intraluminal Ca^{2+} concentrations, pH, ATP, Mg^{2+} and redox state, as well as by its phosphorylation state at multiple sites. Furthermore, a plethora of associated proteins can modulate localization and activity of the IP_3R [11–15]. Second, in higher organisms, three genes (ITPR1, ITPR2 and ITPR3) encode three isoforms (IP_3R1, IP_3R2, and IP_3R3). These isoforms have a homology of about 75% at the a.a. level, allowing for differences in sensitivity towards IP_3 ($IP_3R2 > IP_3R1 > IP_3R3$) as well as towards the various regulatory factors and proteins [12, 16–19]. Splice isoforms and the possibility to form both homo- and heterotetramers further increase IP_3R diversity. Third, the intracellular localization of the IP_3Rs determines their local effect [1]. Recently, an increased appreciation for the existence and functional importance of intracellular Ca^{2+} microdomains was obtained, e.g. between ER and mitochondria, lysosomes or plasma membrane where IP_3-induced Ca^{2+} release (IICR) occurs, allowing Ca^{2+} to control very local processes [20–24].

As a number of excellent reviews on various aspects of IP_3R structure and function have recently appeared [25–32], we will in present review highlight the most recent advances concerning the understanding of IP_3R structure and regulation, with special focus on recent insights obtained in relation to IP_3R modulation by associated proteins.

10.2 New Structural Information on the IP₃R

The IP_3Rs form large Ca^{2+}-release channels consisting of 4 subunits, each about 2700 a.a. long, that assemble to functional tetramers with a molecular mass of about 1.2 MDa. Each subunit consists of five distinct domains (Fig. 10.1a): the N-terminal coupling domain or suppressor domain (for IP_3R1: a.a. 1–225), the IP_3-binding core (IBC, a.a. 226–578), the central coupling domain or modulatory and transducing domain (a.a. 579–2275), the channel domain with 6 trans-membrane helices (a.a. 2276–2589) and the C-terminal tail or gatekeeper domain (a.a. 2590–2749) [33].

The crystal structure of the two N-terminal domains of the IP_3R1 were first resolved separately at a resolution of 2.2 Å (IBC with bound IP_3, [34]) and 1.8 Å (suppressor domain, [35]). Subsequent studies analyzed the crystal structure of the full ligand-binding domain, i.e. the suppressor domain and the IBC together, resolved in the presence and absence of bound IP_3 at a resolution between 3.0 and 3.8 Å [36, 37]. These studies indicated that the N-terminus of IP_3R1 consisted of two successive β-trefoil domains (β-TF) followed by an α-helical armadillo repeat domain. IP_3 binds in a cleft between the second β-trefoil domain and the α-helical armadillo repeat, leading to a closure of the IP_3-binding pocket and a conformational change of the domains involved [36–38]. Recently, Mikoshiba and co-workers succeeded to perform X-ray crystallography on the complete cytosolic part of the IP_3R [39]. This study was performed using truncated IP_3R1 proteins (IP_3R^{2217} and IP_3R^{1585}) in which additional point mutations (resp. $R^{937}G$ and $R^{922}G$) were incorporated in order to increase the quality of the obtained crystals. In addition to the three domains mentioned above (the two β-trefoil domains and the α-helical armadillo repeat domain), three large α-helical domains were described, i.e. HD1 (a.a. 605–1009), HD2 (a.a. 1026–1493) and HD3 (1593–2217) (Fig. 10.1b). Binding of IP_3 induces a conformation change that is transmitted from the IBC through HD1 and HD3, whereby a short, 21 a.a.-long domain (a.a. 2195–2215) called the leaflet domain is essential for IP_3R function.

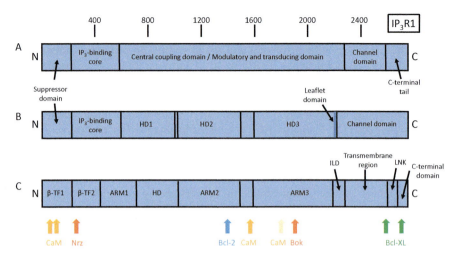

Fig. 10.1 Alignment of proposed IP$_3$R1 structures. (**a**) Linear representation of IP$_3$R1 [33]. (**b**) Linear representation of the IP$_3$R1 domains identified by X-ray crystallography [39]. (**c**) Linear representation of the IP$_3$R1 domains identified by cryo-EM [41]. For the various domains, the original nomenclature was used. Additionally, the interaction sites for calmodulin (CaM) and for the various Bcl-2 family members (Bcl-2, Bcl-XL, Nrz and Bok) are indicated with colored arrows at the bottom of the figure. Please note that the name of the interacting protein indicated at each arrow represents the protein for which binding was initially described. As discussed in the text, related proteins share in some cases common binding sites. The striped arrow indicates that this binding site is only present in a specific IP$_3$R1splice isoform. For further explanations, please see text

In parallel with the analysis of the IP$_3$R by X-ray crystallography, the structure of full-size IP$_3$R1 was investigated by several groups by cryo-electron microscopy (cryo-EM), obtaining increasingly better resolution [40]. The structure of the IP$_3$R1 at the highest resolution obtained by this method until now (4.7 Å) was published by Serysheva and co-workers and allowed modelling of the backbone topology of 2327 of the 2750 a.a [41]. As IP$_3$R1 was purified in the absence of IP$_3$ and as Ca^{2+} was depleted before vitrification, the obtained structure corresponds to the closed state of the channel (Fig. 10.2). In total, ten domains were identified: two contiguous β-trefoil domains (a.a. 1–436), followed by three armadillo solenoid folds (ARM1–ARM3, a.a. 437–2192) with an α-helical domain between ARM1 and 2, an intervening lateral domain (ILD, a.a. 2193–2272), the transmembrane region with six trans-membrane α-helices (TM1–6) (a.a. 2273–2600), a linker domain (LNK, a.a. 2601–2680) and the C-terminal domain containing an 80 Å α-helix (a.a. 2681–2731) (Fig. 10.1c). The latter domains of the four subunits form together with the four TM6 helices (~55 Å) a central core structure that is not found in other types of Ca^{2+} channels. The four transmembrane TM6 helices thereby line the Ca^{2+} conduction pathway and connect via their respective LNK domains with the cytosolic helices.

How binding of IP$_3$ is coupled to channel opening is still under investigation. An interesting aspect of the IP$_3$R structure thereby is the fact that either after

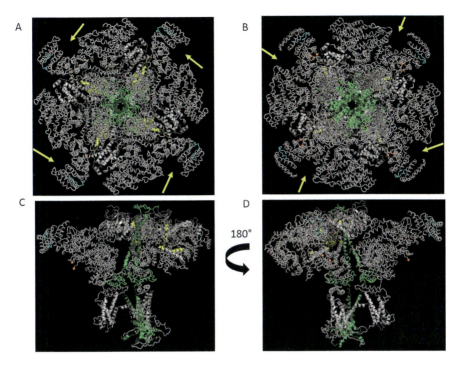

Fig. 10.2 Cryo-EM structure of IP$_3$R1. Structure of IP$_3$R1 fitted to the cryo-EM map (PDB 3JAV, [41]) showing (**a**) a cytosolic and (**b**) a luminal view of an IP$_3$R1 tetramer. (**c, d**) Side views of two neighboring IP$_3$R1 subunits as seen from the (**c**) inside or the (**d**) outside of the tetramer. The discontinuous CaM-binding site in the suppressor domain is indicated in yellow (a.a. 49–81 and a.a. 106–128). The yellow arrows in panels A and B indicate where the CaM-binding site in the central coupling domain should be located (a.a. 1564–1585). This could not be indicated on the structure itself because the part between a.a. 1488 and 1588 of the IP$_3$R is not resolved. The binding site for Bcl-2 and, to a lesser extent, Bcl-XL located in the central coupling domain is indicated in blue (a.a. 1389–1408). The C-terminal binding site for Bcl-2, Bcl-XL and Mcl-1 is shown in green (a.a. 2512–2749). The domains indicated in dark green (a.a. 2571–2606 and a.a. 2690–2732) thereby represent the BH3-like structures that were identified to bind to Bcl-XL. The region where Bok interacts with IP$_3$R1 (a.a. 1895–1903) was not resolved in this cryo-EM structure. The two orange spheres (a.a. 1883 and 1945) however show the boundaries of this non-characterized IP$_3$R1 region to which Bok binds. These images were obtained using PyMOL. For further explanations, please see text

mild trypsinisation of IP$_3$R1 [42] or after heterologous expression of the various IP$_3$R1 fragments corresponding to the domains obtained by trypsinisation [43], the resulting structure appeared both tetrameric and functional. This indicates that continuity of the polypeptide chain is not per se needed for signal transmission to the channel domain, although the resulting Ca^{2+} signals can differ, depending on the exact cleavage site and the IP$_3$R isoform under consideration [44, 45].

Meanwhile, various models for the transmission of the IP$_3$ signal to the channel region were proposed for IP$_3$R1, including a direct coupling between the N-terminus and the C-terminus [41, 46–48] and a long-range coupling mediated by the central coupling domain [48], via intra- and/or inter-subunit interactions [41]. Mechanisms for the latter can involve β-TF1 ➔ ARM3 ➔ ILD [41] or IBC ➔ HD1 ➔ HD3 ➔ leaflet [39].

In addition to the structural studies on IP$_3$R1 described above, the structure of human IP$_3$R3 was recently analyzed at high resolution (between 3.3 and 4.3 Å) under various conditions. Its apo state was compared to the structures obtained at saturating IP$_3$ and/or Ca^{2+} concentrations [49]. In the presence of IP$_3$, five different conformational states were resolved, suggesting a dynamic transition between intermediate states eventually leading to channel opening. Ca^{2+} binding appeared to eliminate the inter-subunit interactions present in the apo and the IP$_3$-bound states and provoke channel inhibition. Two Ca^{2+}-binding sites were identified, one just upstream of ARM2 and one just upstream of ARM3, though their relative function cannot be inferred from structural data alone.

Although IP$_3$R1 and IP$_3$R3 are structurally quite similar, they are differentially activated and regulated (see Sect. 10.1). Additional work, including performing a high-resolution cryo-EM analysis of IP$_3$-bound IP$_3$R1 and the further investigation of the effect of Ca^{2+} and other IP$_3$R modulators, including associated proteins, on IP$_3$R structure will therefore be needed to fully unravel the underlying mechanism of activation and to understand the functional differences between the various IP$_3$R isoforms.

10.3 Complexity of IP$_3$R Activation and Regulation

Concerning the mechanisms of activation and regulation of the IP$_3$R, progress has been made on several points recently.

10.3.1 IP$_3$ Binding Stoichiometry

First, a long-standing question in the field concerned the number of IP$_3$ molecules needed to evoke the opening of the IP$_3$R/Ca^{2+}-release channel. Some studies demonstrated a high cooperativity of IP$_3$ binding to its receptor, and suggested that minimally 3 IP$_3$ molecules should be bound to the IP$_3$R to evoke Ca^{2+} release [50, 51]. In contrast herewith, co-expression of an IP$_3$R apparently defective in IP$_3$ binding (R^{265}Q) and of a channel-dead IP$_3$R mutant (D^{2550}A) resulted in a partial IP$_3$-induced Ca^{2+} release, suggesting that one IP$_3$R subunit can gate another and that therefore not all subunits need to bind IP$_3$ to form an active channel [52]. Moreover, these results fit with the most recent cryo-EM data discussed above (see Sect. 10.2.; [41]).

Recently, a comprehensive study by Yule and co-workers demonstrated in triple-knockout (TKO) cells, devoid of endogenous IP$_3$R expression (DT-40 TKO and HEK TKO), that the activity of recomplemented IP$_3$Rs depends on the occupation of the 4 IP$_3$-binding sites by their ligand [53]. The strongest evidence for this was obtained by the expression of a concatenated IP$_3$R1 containing 3 wild-type subunits and 1 mutant subunit. The mutant subunit contained a triple mutation ($R^{265}Q/K^{508}Q/R^{511}Q$) in the ligand-binding domain precluding any IP$_3$ binding, as previously demonstrated [54], while the $R^{265}Q$ single mutant still retained 10% binding activity. Interestingly, the tetrameric IP$_3$R containing only 1 defective IP$_3$-binding site and expressed in cells fully devoid of endogenous IP$_3$Rs was completely inactive in Ca^{2+} imaging experiments, unidirectional Ca^{2+} flux experiments and in patch-clamp electrophysiological experiments [53]. Similar experiments were performed for IP$_3$R2, making use of its existing short splice isoform that lacks 33 a.a. in the suppressor domain rendering it non-functional [55]. These data strongly suggest that no opening of the IP$_3$R can occur, unless each subunit has bound IP$_3$. This characteristic would strongly limit the number of active IP$_3$Rs and protect the cell against unwanted Ca^{2+} release in conditions in which the IP$_3$ concentration is only slightly increased [50, 53]. However, in the case of IP$_3$R mutations affecting IP$_3$ binding / IP$_3$R activity it may explain why they are detrimental, even in heterozygous conditions [10].

10.3.2   Physiological Relevance of IP$_3$R Heterotetramer Formation

As already indicated above (see Sect. 10.1.), the high level of homology between the various IP$_3$R isoforms allows not only for the formation of homotetramers but also for that of heterotetramers [57–59]. The frequency of heterotetramer occurrence is however not completely clear. A study in COS-7 cells indicated that kinetic constrains affect the formation of heterotetramers and that therefore the level of heterotetramers composed of overexpressed IP$_3$R1 and of either endogenously expressed or overexpressed IP$_3$R3 was lower than what could be expected from a purely binomial distribution [60]. In contrast herewith, by using isoform-specific IP$_3$R antibodies for sequential depletion of the IP$_3$Rs, it was shown that in pancreas, over 90% of IP$_3$R3 is present in heterotetrameric complexes, generally with IP$_3$R2 [61]. This is significant as pancreas is a tissue in which IP$_3$R2 and IP$_3$R3 together constitute over 80% of the total amount of IP$_3$R [62, 63]. It is therefore meaningful to investigate whether the presence of IP$_3$R heterotetramers will contribute in increasing the diversity of the IP$_3$R Ca^{2+}-release channels, as is generally assumed. However, due to the fact that most cells express or can express various types of homo- and heterotetrameric IP$_3$Rs in unknown proportions, addressing this question is in most cell types not straightforward.

Overexpressing mutated IP$_3$R1 and IP$_3$R3 in COS-7 cells at least indicated that heterotetramers are functional [52]. The expression of concatenated dimeric IP$_3$R1-IP$_3$R2 (and IP$_3$R2-IP$_3$R1) in DT-40 TKO cells led to the formation of IP$_3$R heterotetramers with a defined composition (2:2) that could be compared with homotetrameric IP$_3$R1 or homotetrameric IP$_3$R2 that were similarly expressed [61]. Investigation of their electrophysiological properties via nuclear patch-clamp recordings indicated that in the IP$_3$R1-IP$_3$R2 2:2 heterotetramers the properties of the IP_3R2 dominated with respect to the induction of Ca^{2+} oscillations and their regulation by ATP [61]. A more recent study based on the same approach but now including combinations of all three IP_3R isoforms, demonstrated that 2:2 heterotetrameric IP_3Rs display an IP_3 sensitivity that is intermediate to that of their respective homotetramers [64] indicating that heterotetramerization successfully increases IP_3R diversity. In addition, the obtained results also demonstrate that IP_3R2 properties with respect to both the induction of Ca^{2+} oscillations and the regulation by ATP also dominated in IP_3R2-IP_3R3 2:2 heterotetramers. In contrast, when a tetrameric IP_3R containing 3 IP_3R1 and 1 IP_3R2 subunit was expressed, its properties were similar to that of a homotetrameric IP_3R1 [64]. Taken together, these experiments indicate that IP_3R heterotetramers increase the diversity of the IP_3Rs with respect to Ca^{2+} release and that further studies are needed to fully understand how IP_3R heterotetramers are regulated by other factors, including associated proteins.

10.3.3 Novel Crosstalk Mechanism Between cAMP and IICR

cAMP and Ca^{2+}, the two most important intracellular messengers, have numerous crosstalks between them [65]. At the level of the IP_3R, the most evident crosstalk is the sensitization of IP_3R1 by cAMP-dependent protein kinase (PKA) [66], while a similar regulatory role is highly probable for IP_3R2 but less likely for IP_3R3 [15, 65].

A novel line of regulation was discovered some time ago when it was shown that cAMP can, independently from PKA or cAMP-activated exchange proteins, potentiate the IP_3R [67–69]. In particular, it was shown in HEK cells that adenylate cyclase 6, which in those cells accounts for only a minor portion of the adenylate cyclase isoforms, is responsible for providing cAMP to a microdomain surrounding IP_3R2, increasing its activity [69]. Such mechanism would form a specific signaling complex in which locally a very high concentration of cAMP could be reached, without affecting its global concentration [65]. Recent work provided further evidence concerning the importance of cAMP for IP_3R functioning, showing that the presence of cAMP can uncover IP_3Rs that were insensitive to IP_3 alone [56]. Indeed, in HEK cells heterologously expressing the parathyroid hormone (PTH) receptor, it appears that PTH, via production of cAMP, evokes Ca^{2+} release after full depletion of the carbachol-sensitive Ca^{2+} stores. Although the identity of the Ca^{2+} stores could not yet be established, the obtained results are indicative that cAMP unmasks IP_3Rs with a high affinity for IP_3. This fits with the previous observation that IP_3R2, the IP_3R with the highest affinity for IP_3 (reviewed in [19]), is regulated by cAMP [69]. The molecular mechanism on how cAMP interacts with the IP_3R remains to be determined. At this moment no discrimination can be made between a low-affinity cAMP-binding site on the IP_3R itself or a similar binding site on an associated protein [65]. The possibility that the IP_3R-binding protein released by IP$_3$ (IRBIT), related to protein S-adenosylhomocysteine-hydrolase, known to bind cAMP, is involved was however already excluded by knockdown and overexpression experiments [56].

10.4 Complexity of Protein-Protein Interactions Affecting the IP$_3$R

In a comprehensive review published a few years ago, over 100 proteins that interact with the IP$_3$R have been listed [14]. For that reason, we will limit ourselves in the present review to either newly discovered interacting proteins or proteins for which new information about their interaction recently became available.

10.4.1 Calmodulin (CaM) and Related Ca^{2+}-Binding Proteins

CaM is the most ubiquitously expressed intracellular Ca^{2+} sensor. It is a relatively small protein (148 a.a.) with a typical dumbbell structure. A central, flexible linker region connects the globular N-terminal and C-terminal domains, each containing two Ca^{2+}-binding EF-hand motifs with a classical helix-loop-helix structure. The K$_d$ of CaM for Ca^{2+} ranges between 5×10^{-7} and 5×10^{-6} M, with the C-terminal Ca^{2+}-binding sites having a three to five-fold higher affinity than the N-terminal ones [70]. CaM therefore displays the correct Ca^{2+} affinity to sense changes in intracellular Ca^{2+} concentrations and serve as Ca^{2+} sensor. While apo-CaM has a rather compact structure, Ca^{2+}-CaM exposes in each domain a hydrophobic groove with acidic residues at its extremities that will allow interaction with their target [71]. A plethora of target proteins that are modulated by CaM exists, including various Ca^{2+}-transporting proteins [72]. These various proteins contain CaM-binding sites that can be categorized into various types of motifs [73].

Although the interaction of CaM with the IP$_3$R was already observed soon after the identification of the IP$_3$R as IP$_3$-sensitive Ca^{2+}-release channel [74] its exact mechanism of action is still not completely elucidated. Moreover, there are a number of interesting features related to the binding of CaM to the IP$_3$R: (i) the existence of multiple binding sites, (ii) the possibility for both Ca^{2+}-CaM and apo-CaM to affect IP$_3$R function and (iii) the use of some of the CaM-binding sites by other Ca^{2+}-binding proteins. The aim of this paragraph therefore is to present a comprehensive view on the relation between CaM (and some related Ca^{2+}-binding proteins) and the IP$_3$R.

On IP$_3$R1, three CaM-binding sites were described (Fig. 10.1). A high-affinity CaM-binding site (a.a. 1564–1585; Fig. 10.2a–b, indicated by the yellow arrows) was described in the central coupling domain [75], while a low-affinity one was found in the suppressor domain [76]. The latter site is discontinuous (a.a. 49–81 and a.a. 106–128; Fig. 10.2, indicated in yellow) and can bind to both apo-CaM and Ca^{2+}-CaM [77]. Finally, a third site was described on the S2(−) IP$_3$R1 splice isoform in which a.a. 1693–1732 are removed [78, 79]. CaM binding to this newly formed site is inhibited by PKA-mediated phosphorylation, probably on Ser^{1589} [79].

CaM interaction with the two other IP$_3$R isoforms was studied in less detail, but an IP$_3$R2 construct overlapping with the CaM-binding site in the central coupling domain interacted with CaM, supporting the conservation of this site [75]. However, no direct interaction between CaM and IP$_3$R3 could be measured [75, 80] though CaM can bind to IP$_3$R1-IP$_3$R3 heterotetramers [79].

Functional effects on the IP$_3$R have been described for both apo-CaM and Ca^{2+}-CaM. In fact, apo-CaM is equally potent in inhibiting IP$_3$ binding to full-length IP$_3$R1 as Ca^{2+}-CaM [81]. In agreement with the absence of CaM binding to IP$_3$R3, full-length IP$_3$R3 remained insensitive to regulation by CaM [80]. In contrast, a Ca^{2+}-independent inhibition of IP$_3$ binding was observed for the isolated ligand-binding domain of IP$_3$R1 [82] as well as for that of IP$_3$R2 and IP$_3$R3 [83].

Concerning IP$_3$-induced Ca^{2+} release, the situation is somewhat more complex. Ca^{2+} release by IP$_3$R1 is inhibited by CaM in a Ca^{2+}-dependent way [84, 85] while similar results were subsequently found for IP$_3$R2 and IP$_3$R3 [76, 86]. However, linking these functional effects molecularly to a CaM-binding site appeared more difficult, not only because of the apparent absence of a Ca^{2+}-dependent CaM-binding site on IP$_3$R3 but also because the mutation W^{1577}A that abolishes CaM binding to IP$_3$R1 [75], does not abolish the CaM-mediated inhibition of IICR [87].

Furthermore, other results suggested that the relation between CaM and the IP$_3$R was more complex than originally thought. A detailed analysis of the CaM-binding site located in the central coupling domain of IP$_3$R1 provided evidence that it consisted of a high-affinity Ca^{2+}-CaM and a lower affinity apo-CaM site [88]. Moreover, in the same study it was demonstrated that a CaM mutant deficient in Ca^{2+} binding (CaM1234) could inhibit IICR in a Ca^{2+}-dependent way with the same potency as CaM. In a separate study, it was demonstrated that a myosin light chain kinase (MLCK)-derived peptide, which binds to CaM with high affinity, fully inhibited the IP$_3$R [89]. This inhibition could be reversed by the addition of CaM but not of CaM1234 and the results were interpreted as evidence that endogenously bound CaM is needed for IP$_3$R activity. A follow-up study by another group [90] however proposed that the MLCK peptide is not removing endogenous CaM but is interacting with an endogenous CaM-like domain on IP$_3$R, thereby disrupting its interaction with a so-called 1–8-14 CaM-binding motif (a.a. 51–66) essential for IP$_3$R activity [91].

Meanwhile, the interaction of apo-CaM with the suppressor domain was studied via NMR analysis [92]. This study brought forward two main pieces of evidence. First, it was shown that the binding of apo-CaM to the suppressor domain induced an important, general conformational change to the latter. These changes further increased in the presence of Ca^{2+}. Secondly, analysis of the conformational change of CaM indicated that apo-CaM already binds with its C-lobe to the IP$_3$R1 suppressor domain, and that only after addition of Ca^{2+} also the N-lobe interacts with the suppressor domain. These results can therefore explain the importance of the CaM-binding sites in the suppressor domain in spite of their difficult accessibility ([92]; Fig. 10.2).

Finally, some Ca^{2+}-binding proteins related to CaM (e.g. neuronal Ca^{2+}-binding protein (CaBP) 1, calmyrin, also known as CIB1, and neuronal Ca^{2+} sensor-1 (NCS-1)) also regulate the IP$_3$R. Similarly to CaM, these proteins contain 4 EF-hand motifs but in contrast with CaM, not all of them bind Ca^{2+}. In CaBP1 and NCS-1 only 3 EF hands are functional (EF1, EF3, EF4 and EF2, EF3, EF4 resp.) and in calmyrin only 2 (EF3 and EF4). Moreover, some of the EF hands bind Mg^{2+} rather than Ca^{2+}. Furthermore, those proteins are myristoylated. Although early results suggested that CaBP1 and calmyrin could, in the absence of IP$_3$, activate the IP$_3$R under some circumstances [93, 94], there is presently a large consensus that they, similarly to CaM, generally inhibit the IP$_3$R [93, 95, 96].

CaBP1 was proposed to interact with the IP$_3$R1 with a higher affinity than CaM itself [94, 96], while in contrast to CaM it does not affect the ryanodine receptor (RyR), another family of intracellular Ca^{2+}-release channels. Additionally, the interaction with the IP$_3$R would be subject to regulation by caseine kinase 2, an enzyme that can phosphorylate CaBP1 on S^{120} [96]. Similarly to CaM, CaBP1 binds in a Ca^{2+}-independent way to the IP$_3$R1 suppressor domain, but in contrast to CaM, only to the first of the two non-contiguous binding sites described for CaM (Fig. 10.1). However, CaM and CaBP1 similarly antagonized the thimerosal-stimulated interaction between the suppressor domain and the IBC of IP$_3$R1, suggesting a common mechanism of action whereby they disrupt intramolecular interactions needed for channel activation [97]. More recent work confirmed the inhibitory effect of CaBP1 on IP$_3$R1, while expanding the knowledge concerning the CaBP1 binding site. In particular, NMR analysis indicated that CaBP1 interacts with its C lobe with the suppressor domain of the IP$_3$R and that even at saturating Ca^{2+} concentrations EF1 is bound to Mg^{2+}, precluding a conformational change of the N lobe [98]. The same study demonstrated that Ca^{2+}-bound CaBP1 bound with an 10-fold higher affinity than Mg^{2+}-bound CaBP1 and an at least 100-fold higher affinity than CaM itself. Functional analysis performed in DT-40 cells solely expressing IP$_3$R1 demonstrated that CaBP1 stabilized the closed conformation of the channel, probably by clamping inter-subunit interactions [99]. The interaction of specific hydrophobic a.a. in the C lobe of CaBP1 (V^{101}, L^{104}, V^{162}) that become more exposed in the presence of Ca^{2+} with hydrophobic a.a. in the IBC (L^{302}, I^{364}, L^{393}) appeared hereby essential.

The action of NCS-1 on the IP$_3$R forms a slightly different story. It co-immunoprecipitates with IP$_3$R1 and IP$_3$R2 in neuronal cells and in heart thereby stimulating IICR in a Ca^{2+}-dependent way [100, 101]. Interestingly, paclitaxel by binding to NCS-1 increases its interaction with IP$_3$R1 and so induces Ca^{2+} oscillations in various cell types [102, 103]. This Ca^{2+}-signaling pathway was proposed to lead to calpain activation and to underlie the origin of paclitaxel-induced peripheral neuropathy [104]. However, the interaction site of NCS-1 on the IP$_3$R, either direct or indirect, has not yet been identified.

Taken together these results confirm that Ca^{2+}-binding proteins interact in a complex way with the IP_3R and that the various Ca^{2+}-binding proteins have distinct, though sometimes overlapping, roles. The functional effect of CaM has been studied in detail and it appears to inhibit the IP_3R. The results described above support a view that the main action of CaM on the IP_3R is at the level of the suppressor domain. Indeed, apo-CaM can via its C lobe bind to the suppressor domain of all three IP_3R isoforms while a subsequent binding of the N lobe will depend on the Ca^{2+} concentration. The binding of CaM in that domain can disturb an intra-IP_3R interaction needed for IP_3R function and therefore inhibits IICR. This behavior can be particularly important in cells having high CaM expression levels, as for example Purkinje neurons that also demonstrate high levels of IP_3R1. In that case, CaM was proposed to be responsible for suppressing basal IP_3R activity [81]. Moreover, as the intracellular distribution of CaM can depend on intracellular Ca^{2+} dynamics, it was also hypothesized that it allows IP_3R regulation is a non-uniform way [84]. Additionally, it should be emphasized that CaM can act on other Ca^{2+}-transporting proteins in the cell, like the RyR [105], the plasma membrane Ca^{2+} ATPase [106] and various plasma membrane Ca^{2+} channels including voltage-operated Ca^{2+} channels and transient receptor potential channels [107, 108]. In all these cases CaM tends to inhibit Ca^{2+} influx into the cytosol (inhibition of IP_3Rs, RyRs and plasma membrane Ca^{2+} channels) while promoting Ca^{2+} efflux out of the cell (stimulation of plasma membrane Ca^{2+} ATPase).

An IP_3R-inhibiting behavior can similarly be expected for CaM-related Ca^{2+}-binding proteins, though their interaction sites are not strictly identical to that of CaM. The binding site for NCS-1, which rather stimulates the IP_3R, is even still unknown. In comparison to CaM, CaBP1 demonstrates a much higher affinity for the IP_3R [99] and a higher specificity, as it does not affect the RyR [96]. In cells expressing CaBP1, the major control of IICR will therefore depend on the interaction of the IP_3R with CaBP1, while RyR activity will depend on the presence and activation of CaM. Further work will be needed to completely unravel the exact role of these various proteins in the control of intracellular Ca^{2+} signaling. From the present results, it can already be expected that the relative role of the various Ca^{2+}-binding proteins in the control of IICR will strongly depend on the exact cell type in consideration.

10.4.2 The Bcl-2-Protein Family

The B-cell lymphoma (Bcl)-2 protein family has been extensively studied as critical regulator of apoptosis [109]. This family consists of both anti- and pro-apoptotic members. The anti-apoptotic family members inhibit apoptosis in at least two different manners. First, at the mitochondria anti-apoptotic Bcl-2 proteins such as Bcl-2, Bcl-XL and Mcl-1, bind to the pro-apoptotic Bcl-2-family members thereby inhibiting the permeabilization of the outer mitochondrial membrane by Bax and

Bak and subsequent release of cytochrome C [110, 111]. Second, the anti-apoptotic Bcl-2-family members also affect intracellular Ca^{2+} signaling. On the one hand they promote pro-survival Ca^{2+} oscillations while on the other hand they inhibit pro-apoptotic Ca^{2+} release from the ER that otherwise could lead to mitochondrial Ca^{2+} overload [112]. These combined actions mean that anti-apoptotic Bcl-2 proteins can, by modulating several protein families involved in intracellular Ca^{2+} signaling, both fine tune mitochondrial bio-energetics and inhibit Ca^{2+}-mediated mitochondrial outer membrane permeabilization [113–116]. Both the interaction between Bcl-2-family members and their ability to regulate intracellular Ca^{2+} signaling is critically dependent on the presence of so-called Bcl-2 homology (BH) domains. Anti-apoptotic Bcl-2 proteins contain four of these domains (BH1, 2, 3 and 4) [111]. The BH1 to 3 domains together form a hydrophobic cleft that inactivates the pro-apoptotic Bcl-2-family members via interaction with their BH3 domain. For regulating intracellular Ca^{2+} signaling events, anti-apoptotic Bcl-2 proteins rely to a great extent, however not exclusively, on their BH4 domain. In this review we will focus on how IP$_3$Rs are regulated by Bcl-2 proteins. For a more extensive revision of how Bcl-2-family members regulate the various members of the intracellular Ca^{2+} signaling machinery we would like to refer to our recent review on the subject [112].

The various IP$_3$R isoforms are important targets for several anti-apoptotic Bcl-2-family members [112]. To complicate matters, multiple binding sites on the IP$_3$R have been described for anti-apoptotic Bcl-2 proteins [117]. First, Bcl-2, Bcl-XL and Mcl-1 were shown to target the C-terminal part (a.a. 2512–2749) of IP$_3$R1 (Fig. 10.2, indicated in green) thereby stimulating pro-survival Ca^{2+} oscillations [114, 115, 118]. Additionally, Bcl-2, and Bcl-XL with lesser affinity, also target the central coupling domain (a.a. 1389–1408 of IP$_3$R1; Figs. 10.1 and 10.2, indicated in blue) of the IP$_3$R where binding of these proteins inhibits pro-apoptotic Ca^{2+}-release events [116, 118–120]. Finally, the zebrafish protein Nrz [121] and its mammalian homolog Bcl-2-like 10 [122] were shown to interact with the IBC and inhibit IICR.

The group of Kevin Foskett performed a more in-depth study into how the IP$_3$R is regulated by Bcl-XL and proposed a mechanism unifying the regulation at the C-terminal and at the central coupling domain of the IP$_3$R [123]. Two domains containing BH3-like structures (a.a. 2571–2606 and a.a. 2690–2732; Figs. 10.1 and 10.2, indicated in dark green) were identified in the C-terminal part of the IP$_3$R. When Bcl-XL is, via its hydrophobic cleft, bound to both BH3-like domains it sensitizes the IP$_3$R to low concentrations of IP$_3$, thereby stimulating Ca^{2+} oscillations. If Bcl-XL binds to only one of these BH3 like domains while also binding to the central coupling domain, it will inhibit IICR. Whether Bcl-XL occupies one or the two BH3-like domains at the C-terminus of the IP$_3$R was proposed to be dependent on Bcl-XL levels and on the intensity of IP$_3$R stimulation. Whether Bcl-2 operates in a similar manner is still unclear. As there is evidence that Bcl-2 shows a greater affinity than Bcl-XL for the inhibitory binding site in the central coupling domain it is likely that this site is the preferential target for Bcl-2 [118]. In addition, for Bcl-2 not its hydrophobic cleft but rather its transmembrane domain seems to play an important role for targeting and regulating the IP$_3$R via both its C-terminus and the site located in the central coupling domain [124].

Based on the recent cryo-EM structure of IP$_3$R1 [29, 41], this central site in the coupling domain resides in a relatively easily accessible area of IP$_3$R1 (Fig. 10.2, indicated in blue). The C-terminal transmembrane domain of Bcl-2 may thus serve to concentrate the protein at the ER near the IP$_3$R from where its N-terminal BH4 domain can more easily bind to the central coupling domain. In addition, sequestering Bcl-2 proteins at the ER membrane via their transmembrane domain may increase their ability to interact with the C-terminus of the IP$_3$R (Fig. 10.2, indicated in green). As this C-terminal binding site seems to be located more at the inside of the IP$_3$R1 tetramer one can expect a local high concentration of Bcl-2 proteins to be necessary for this interaction. Besides directly modulating IICR, Bcl-2 can serve as an anchor for targeting additional regulatory proteins to the IP$_3$R. It was shown that Bcl-2 attracts dopamine- and cAMP-regulated phosphoprotein of 32 kDa (DARPP-32) and calcineurin to the IP$_3$R thereby regulating the phosphorylation state of the latter and consequently its Ca^{2+}-release properties [125]. Finally, recent data indicate also for Bcl-2 an additional interaction site in the ligand-binding domain [126] highlighting the complexity of the interaction of the anti-apoptotic Bcl-2-family members with the IP$_3$R. Further research will be needed to unravel the precise function of each of these sites.

Another Bcl-2-family member that regulates the IP$_3$R is the zebrafish protein Nrz. The latter was shown to bind via its BH4 domain to the IBC of zebrafish IP$_3$R1, whereby E^{255} appeared essential for interaction (Fig. 10.1). Nrz prevents IP$_3$ binding to the IP$_3$R thereby inhibiting IICR [121]. Interestingly, although the Nrz BH4 domain is sufficient for interaction with the IP$_3$R, inhibition of IICR required the BH4-BH3-BH1 domains. Furthermore, phosphorylation of Nrz abolished its interaction with the IP$_3$R. Recently, Bcl-2-like 10, the human orthologue of Nrz, was shown, just like Nrz in zebrafish, to interact with the IBC, indicating a conserved function for this protein [122].

Besides anti-apoptotic Bcl-2-family members, also pro-apoptotic Bcl-2 proteins and other BH3 domain-containing proteins are known to target and regulate IP$_3$Rs. For instance, Bok, a pro-apoptotic Bcl-2-family member, interacts with the IP$_3$R (a.a. 1895–1903 of IP$_3$R1; Figs. 10.1 and 10.2) [127]. This interaction protects IP$_3$R1 and IP$_3$R2 from proteolytic cleavage by caspase 3 that results in a Ca^{2+} leak that may contribute to mitochondrial Ca^{2+} overload and thus apoptosis [128, 129]. Subsequent work demonstrated that the majority of all cellular Bok is bound to the IP$_3$R thereby stabilizing the Bok protein [130]. Unbound, newly synthesized Bok is rapidly turned over by the proteasome pathway. Both the association of mature Bok with the IP$_3$R and the rapid degradation of newly synthesized Bok by the proteasome restrict the pro-apoptotic functions of Bok thus preventing cell death induction.

From the above it is clear that the IP$_3$R is heavily regulated by both pro- and anti-apoptotic Bcl-2-family members. The occurrence of multiple binding sites for the same Bcl-2-family member further increases the complexity [112]. Furthermore, it should be stressed that the regulation of the IP$_3$R by Bcl-2 proteins is conserved during evolution. This is illustrated by the ability of the zebrafish Nrz protein to regulate IICR via its BH4 domain [121] and is further validated by the observation that the BH4 domains of Bcl-2 derived from different vertebrates are able to inhibit IICR with similar efficiency [131]. The large number of both pro-and anti-apoptotic Bcl-2 proteins that regulate the IP$_3$R, targeting it at multiple sites, suggests that throughout evolution regulating IICR became an important functional aspect of the Bcl-2-protein family.

Mcl-1, Bcl-2 and Bcl-XL all target the C-terminal region of the IP$_3$R stimulating the occurrence of pro-survival Ca^{2+} oscillations and thus Ca^{2+} transfer to the mitochondria [114, 115, 118]. These Ca^{2+} transfers into the mitochondria are important for normal cell functioning [113] but are also involved in cancer development and could potentially form a novel therapeutic target [132]. Mitochondrial Ca^{2+} contributes to maintaining proper ATP production. When Ca^{2+} transfer into the mitochondria is inhibited, ATP levels decrease, activating autophagy. At the same time the cell cycle progression is halted [113, 133]. In cancer cells, decreased Ca^{2+} transfer into the mitochondria, consecutive loss of ATP and the start of autophagy is not accompanied by a stop in the cell cycle. Continuing the cell cycle without sufficient building blocks and ATP results in necrotic cell death [132]. Cancer cells are therefore reliant on proper Ca^{2+} transfer to the mitochondria to maintain mitochondrial function, including the production of ATP and metabolites necessary for completing the cell cycle. It is therefore common for cancer cells to upregulate one or several anti-apoptotic Bcl-2 proteins. By interacting with the C-terminus of the IP$_3$R the Bcl-2 proteins may stimulate Ca^{2+} oscillations assuring proper mitochondrial Ca^{2+} uptake and an adequate mitochondrial metabolism. On the other hand, upregulation of Bcl-2 and/or Bcl-XL also protects the cells from excessive IP$_3$R-mediated Ca^{2+} release by binding to the central regulatory site [116, 118–120] and prevents apoptosis, even in the presence of cell death inducers [109, 134]. In healthy cells a similar regulation of IICR by Bcl-2 proteins occurs. However, when cell death is induced in the latter, the amount of anti-apoptotic Bcl-2 proteins declines [134] potentially decreasing the level of their association with the IP$_3$R. This alleviates the inhibitory actions on IICR allowing pro-death Ca^{2+} signals while also reducing the opportunities for the occurrence of pro-survival Ca^{2+} oscillations.

10.4.3   *Beclin 1*

Beclin 1 is a pro-autophagic BH3 domain-containing protein [135]. It interacts with various proteins involved in the regulation of autophagy, including Bcl-2 [136, 137]. The latter protein, by sequestering Beclin 1, prevents its pro-autophagic action. A first study presenting evidence that Beclin 1 also interacts with the IP$_3$R showed an interaction between Beclin 1 and the IP$_3$R that depended on Bcl-2 and which was disrupted by the IP$_3$R inhibitor xestospongin B [138]. The release of Beclin 1 from the Bcl-2/IP$_3$R complex resulted in the stimulation of autophagy which could be counteracted by overexpressing the IBC. This suggested that the IBC was able to sequester the xestospongin B-released Beclin 1 thus halting its pro-autophagic function. From subsequent work, it appeared that the role of Beclin 1 with respect to the IP_3R was more complex [139]. Indeed, the binding of Beclin 1 to the ligand-binding domain was confirmed, though it appeared that in IP_3R1 and to a lesser degree in IP_3R3 the suppressor domain (a.a. 1–225) played a more prominent role in the interaction than the IBC. Interestingly, during starvation-induced autophagy Beclin 1 binding to the IP_3R sensitized IICR that was shown to be essential for the autophagy process [139]. Using the $F^{123}A$ Beclin 1 mutant that does not interact with Bcl-2, it was shown that the sensitization of the IP_3R by Beclin 1 was not due to counteracting the inhibitory effect of Bcl-2, although, in agreement with the previous study [138] it appeared that Beclin 1 binding to Bcl-2 may be needed to target the protein in proximity of the IP_3R.

10.4.4   IRBIT

IRBIT regulates IICR by targeting the IP_3R ligand-binding domain thereby competing with IP_3. Moreover, this interaction is promoted by IRBIT phosphorylation [140]. Besides the IP_3R, IRBIT binds to several other targets regulating a wide range of cellular processes [141]. How IRBIT determines which target to interact with and modulate was recently described [142]. First, various forms of IRBIT exist: IRBIT, the long-IRBIT homologue and its splice variants, which were shown to have distinct expression patterns. Besides this, the N-terminal region of the various members of the IRBIT-protein family showed distinct differences. These differences, obtained by N-terminal splicing, are important in maintaining protein stability and in determining which target to interact with.

Recently, it was shown that Bcl-2-like 10, which binds to a distinct site in the ligand-binding domain (see Sect. 10.4.2), functionally and structurally interferes with the action of IRBIT on the IP_3R [122]. When both proteins are present, Bcl-2-like 10, via its BH4 domain, interacts with IRBIT, thereby mutually strengthening their interaction with the IP_3R and decreasing IICR in an additive way. Upon dephosphorylation of IRBIT, both IRBIT and Bcl-2-like 10 are released from the IP_3R, increasing pro-apoptotic Ca^{2+} transfer from the ER to the mitochondria. Interestingly, this study also showed that IRBIT is involved in regulating ER-mitochondrial contact sites as IRBIT knockout reduced the number of these contact sites [122].

10.4.5   Thymocyte-Expressed, Positive Selection-Associated 1 (TESPA1)

T-cell receptor (TCR) stimulation triggers a signaling cascade ultimately leading to the activation of PLC, production of IP_3 and IICR important for T-cell maturation [143]. TESPA1, a protein involved in the development/selection of T cells [144], has been shown to regulate these Ca^{2+} signals. TESPA1 has a significant homology with KRAS-induced actin-interacting protein [147], a protein that was already shown to interact with and control the IP$_3$R [145, 146]. TESPA1 similarly interacts with the various IP$_3$R isoforms and it appeared that the full ligand-binding domain was needed for this interaction. However, at first no functional effect was described for this interaction [147]. Recently this topic was revisited and it was shown that TESPA1 recruits IP$_3$R1 to the TCR where PLC signaling is initiated and IP$_3$ produced [143]. In this way, TESPA1 promotes IP$_3$R1 phosphorylation on Y^{353} by the tyrosine kinase Fyn, increasing the affinity of the IP$_3$R for IP$_3$. The combination of both these effects increases the efficiency by which Ca^{2+} signaling occurs after TCR stimulation, which is beneficial for T-cell selection and maturation [148]. Furthermore, in Jurkat cells TESPA1 interacts at the ER-mitochondria contact sites with GRP75 [149], a linker protein coupling IP$_3$R with the mitochondrial VDAC1 channel favoring Ca^{2+} transfer from ER to mitochondria [150]. Consequently, TESPA1 knockout diminished the TCR-evoked Ca^{2+} transfers to both mitochondria and cytosol and confirm the important role for TESPA1 in these processes.

10.4.6 Pyruvate Kinase (PK) M2

PKs catalyze the last step of glycolysis and convert phosphoenolpyruvate to pyruvate resulting in the production of ATP. Many cancer cells preferentially upregulate glycolysis over oxidative phosphorylation suggesting a potential role for the PK family in cancer development. Four distinct PK isoforms exists, having each a distinct tissue expression pattern but PKM2 has the peculiarity to be expressed at an elevated level in most tumoral cells where it has a growth-promoting function. Moreover, although PKM1 and PKM2 are nearly identical, differing in only 22 a.a., they are regulated differently and have non-redundant functions [151]. Besides its metabolic functions, PKM2 is also involved in several non-metabolic functions. The latter encompass a nuclear role in transcriptional regulation, protein kinase activity towards various proteins in different cellular organelles, and even an extracellular function as PKM2 is also present in exosomes [152, 153]. It is therefore interesting that also a role for PKM2 at the ER was described since a direct interaction was found between PKM2 and the central coupling domain of the IP$_3$R, inhibiting IICR in various cell types [154, 155]. Moreover, a recent study links the switch from oxidative phosphorylation to glycolysis in breast cancer cells with PKM2 methylation [156]. Methylated PKM2 promoted proliferation, migration and growth of various breast cancer cell lines. Strikingly, PKM2 methylation did not seem to alter its enzymatic activity but did however alter mitochondrial Ca^{2+} homeostasis by decreasing IP$_3$R levels. Finally, co-immunoprecipitation experiments showed an interaction between methylated PKM2 and IP$_3$R1 and IP$_3$R3, though in this study it was not investigated whether the interaction was direct or indirect [156]. As PKM2 is in a variety of cancers considered as a good prognostic marker with a strong potential as therapeutic target [152] these new data, linking directly a metabolic enzyme with an intracellular Ca^{2+}-release channel and ER-mitochondria Ca^{2+} transfer, provide new possibilities for therapeutic intervention.

10.4.7 BRCA-Associated Protein 1 (BAP1) and the F-Box Protein FBXL2

Prolonged stimulation of IP_3Rs leads to a downregulation of the IP_3R levels [157–159]. This downregulation is mainly due to IP_3R ubiquitination followed by their degradation via the proteasomal pathway [31, 160]. Ubiquitination is therefore an important IP_3R modification that may severely impact IICR signaling to for instance the mitochondria, thereby greatly affecting cell death and cell survival decisions. Recently a number of proto-oncogenes and tumor suppressors have been identified that critically control IP_3R3 ubiquitination.

BAP1 is a tumor suppressor with deubiquitinase activity that is known to have important roles in regulating gene expression, DNA stability, replication, and repair and in maintaining chromosome stability [161–164]. Besides this, BAP1 was also shown to influence cellular metabolism, suggesting potential roles for BAP1 outside the nucleus [165, 166]. Heterozygous loss of BAP1 results in decreased mitochondrial respiration while increasing glycolysis [167, 168]. These cells produced a distinct metabolite signature, indicative for the occurrence of the Warburg effect that is supporting cells towards malignant transformation. Heterozygous loss of BAP1 leads to a decreased ER-mitochondria Ca^{2+} transfer and altered mitochondrial metabolism [167]. BAP1 regulates this Ca^{2+} transfer by interacting with the N-terminal part (a.a. 1–800) of IP_3R3, a region which contains the complete ligand-binding domain and a small part of the central coupling domain. The deubiquitinase activity of BAP1 prevents degradation of IP_3R3 by the proteasome. Loss of BAP1 consequently results in excessive reduction of IP_3R3 levels thereby lowering mitochondrial Ca^{2+} uptake. This not only reduces the cell its responsiveness to Ca^{2+}-induced cell death but also promotes glycolysis over oxidative phosphorylation, both important aspects of malignant cell transformation. The nuclear function of BAP1 with respect to maintaining DNA integrity [161–164] together with its extra-nuclear role in regulating cell metabolism and sensitivity to Ca^{2+}-induced cell death [165–168] suggests that this protein may be an excellent target for cancer drug development.

F-box protein FBXL2 that forms a subunit of a ubiquitin ligase complex has the opposite effect of BAP1 on IP_3R3. FBXL2 interacts with a.a. 545–566 of IP_3R3, promoting its ubiquitination and its subsequent degradation. Reduced IP_3R3 leads to a decreased transfer of Ca^{2+} to the mitochondria and a reduced sensitivity towards apoptosis, thus promoting tumor growth [169]. The phosphatase and tensin homolog (PTEN) tumor suppressor could inhibit this pro-tumorigenic effect of FBXL2. PTEN not only promotes apoptosis by inhibiting protein kinase B/Akt (PKB) [170–

172] thereby counteracting PKB-mediated IP$_3$R3 phosphorylation [173, 174] but also by directly binding to IP$_3$R3 [169]. Binding of PTEN to IP$_3$R3 displaces FBXL2 from its binding site, reducing IP$_3$R3 ubiquitination, stabilizing IP$_3$R3 levels, and thus increasing pro-apoptotic Ca^{2+} signaling to the mitochondria [169]. In accordance with the fact that the FBXL2-binding site is only partially conserved in IP$_3$R1 and IP$_3$R2, the stability of the two latter isoforms appeared to be affected neither by FBXL2 nor by PTEN.

In several tumors, PTEN function is impaired which results in accelerated IP$_3$R3 degradation and impaired apoptosis induction. Treatment with drugs that stabilize IP$_3$R levels may therefore also be of interest for cancer therapy in cases where PTEN is affected.

10.5   Conclusions

Intracellular Ca^{2+} signaling is involved in a plethora of cellular processes. The ubiquitously expressed IP$_3$R Ca^{2+}-release channels play an important role in the generation of these signals and serve as signaling hubs for several regulatory factors and proteins/protein complexes. Since the first identification of the IP$_3$R [175], IP$_3$R-interacting proteins and their modulating roles on Ca^{2+} signaling and (patho)physiological processes have been the subject of many studies and well over 100 interaction partners were reported [14], though for many of them it is unclear how they exactly interact with the IP$_3$R and how they affect IP$_3$R function. Moreover, for many regulatory proteins, multiple binding sites were described of which the relative importance is not directly apparent. The recent (and future) advances in the elucidation of the IP$_3$R structure will pave the way for a better understanding how IP$_3$R gating exactly occurs and how different cellular factors and regulatory proteins influence IICR. As several of these proteins affect life and death decisions and/or play important roles in tumor development, the exact knowledge of their interaction site and their action of the IP$_3$R may lead to the development of new therapies for e.g. cancer treatment.

Acknowledgements   TV is recipient of a postdoctoral fellowship of the Research Foundation—Flanders (FWO). Work performed in the laboratory of the authors was supported by research grants of the FWO, the Research Council of the KU Leuven and the Interuniversity Attraction Poles Programme (Belgian Science Policy). JBP is member of the Transautophagy COST action CA15138.

References

1. Vermassen E, Parys JB, Mauger JP (2004) Subcellular distribution of the inositol 1,4,5-trisphosphate receptors: functional relevance and molecular determinants. Biol Cell 96:3–17
2. Berridge MJ, Bootman MD, Roderick HL (2003) Calcium signalling: dynamics, homeostasis and remodelling. Nat Rev Mol Cell Biol 4:517–529

3. Berridge MJ, Lipp P, Bootman MD (2000) The versatility and universality of calcium signalling. Nat Rev Mol Cell Biol 1:11–21
4. Berridge MJ (2016) The inositol trisphosphate/calcium signaling pathway in health and disease. Physiol Rev 96:1261–1296
5. Tada M, Nishizawa M, Onodera O (2016) Roles of inositol 1,4,5-trisphosphate receptors in spinocerebellar ataxias. Neurochem Int 94:1–8
6. Egorova PA, Bezprozvanny IB (2018) Inositol 1,4,5-trisphosphate receptors and neurodegenerative disorders. FEBS J 285:3547–3565
7. Hisatsune C, Mikoshiba K (2017) IP$_3$ receptor mutations and brain diseases in human and rodents. J Neurochem 141:790–807
8. Hisatsune C, Hamada K, Mikoshiba K (2018) Ca^{2+} signaling and spinocerebellar ataxia. Biochim Biophys Acta 1865:1733–1744
9. Kerkhofs M, Seitaj B, Ivanova H, Monaco G, Bultynck G, Parys JB (2018) Pathophysiological consequences of isoform-specific IP$_3$ receptor mutations. Biochim Biophys Acta 1865: 1707–1717
10. Terry LE, Alzayady KJ, Furati E, Yule DI (2018) Inositol 1,4,5-trisphosphate receptor mutations associated with human disease. Messenger 6:29–44
11. Fedorenko OA, Popugaeva E, Enomoto M, Stathopulos PB, Ikura M, Bezprozvanny I (2014) Intracellular calcium channels: Inositol-1,4,5-trisphosphate receptors. Eur J Pharmacol 739:39–48
12. Foskett JK, White C, Cheung KH, Mak DO (2007) Inositol trisphosphate receptor Ca^{2+} release channels. Physiol Rev 87:593–658
13. Parys JB, De Smedt H (2012) Inositol 1,4,5-trisphosphate and its receptors. Adv Exp Med Biol 740:255–279
14. Prole DL, Taylor CW (2016) Inositol 1,4,5-trisphosphate receptors and their protein partners as signalling hubs. J Physiol 594:2849–2866
15. Vanderheyden V, Devogelaere B, Missiaen L, De Smedt H, Bultynck G, Parys JB (2009) Regulation of inositol 1,4,5-trisphosphate-induced Ca^{2+} release by reversible phosphorylation and dephosphorylation. Biochim Biophys Acta 1793:959–970
16. Ivanova H, Vervliet T, Missiaen L, Parys JB, De Smedt H, Bultynck G (2014) Inositol 1,4,5-trisphosphate receptor-isoform diversity in cell death and survival. Biochim Biophys Acta 1843:2164–2183
17. Patel S, Joseph SK, Thomas AP (1999) Molecular properties of inositol 1,4,5-trisphosphate receptors. Cell Calcium 25:247–264
18. Taylor CW, Genazzani AA, Morris SA (1999) Expression of inositol trisphosphate receptors. Cell Calcium 26:237–251
19. Vervloessem T, Yule DI, Bultynck G, Parys JB (2015) The type 2 inositol 1,4,5-trisphosphate receptor, emerging functions for an intriguing Ca^{2+}-release channel. Biochim Biophys Acta 1853:1992–2005
20. Gutierrez T, Simmen T (2018) Endoplasmic reticulum chaperones tweak the mitochondrial calcium rheostat to control metabolism and cell death. Cell Calcium 70:64–75
21. La Rovere RM, Roest G, Bultynck G, Parys JB (2016) Intracellular Ca^{2+} signaling and Ca^{2+} microdomains in the control of cell survival, apoptosis and autophagy. Cell Calcium 60: 74–87
22. Marchi S, Bittremieux M, Missiroli S, Morganti C, Patergnani S, Sbano L et al (2017) Endoplasmic reticulum-mitochondria communication through Ca^{2+} signaling: the importance of mitochondria-associated membranes (MAMs). Adv Exp Med Biol 997:49–67
23. Marchi S, Patergnani S, Missiroli S, Morciano G, Rimessi A, Wieckowski MR et al (2018) Mitochondrial and endoplasmic reticulum calcium homeostasis and cell death. Cell Calcium 69:62–72
24. Raffaello A, Mammucari C, Gherardi G, Rizzuto R (2016) Calcium at the center of cell signaling: interplay between endoplasmic reticulum, mitochondria, and lysosomes. Trends Biochem Sci 41:1035–1049

25. Ando H, Kawaai K, Bonneau B, Mikoshiba K (2018) Remodeling of Ca^{2+} signaling in cancer: regulation of inositol 1,4,5-trisphosphate receptors through oncogenes and tumor suppressors. Adv Biol Regul 68:64–76

26. Garcia MI, Boehning D (2017) Cardiac inositol 1,4,5-trisphosphate receptors. Biochim Biophys Acta 1864:907–914

27. Kania E, Roest G, Vervliet T, Parys JB, Bultynck G (2017) IP$_3$ receptor-mediated calcium signaling and its role in autophagy in cancer. Front Oncol 7:140

28. Roest G, La Rovere RM, Bultynck G, Parys JB (2017) IP$_3$ receptor properties and function at membrane contact sites. Adv Exp Med Biol 981:149–178

29. Serysheva II, Baker MR, Fan G (2017) Structural insights into IP$_3$R function. Adv Exp Med Biol 981:121–147

30. Wang L, Alzayady KJ, Yule DI (2016) Proteolytic fragmentation of inositol 1,4,5-trisphosphate receptors: a novel mechanism regulating channel activity? J Physiol 594: 2867–2876

31. Wright FA, Wojcikiewicz RJ (2016) Chapter 4 – inositol 1,4,5-trisphosphate receptor ubiquitination. Prog Mol Biol Transl Sci 141:141–159

32. Eid AH, El-Yazbi AF, Zouein F, Arredouani A, Ouhtit A, Rahman MM et al (2018) Inositol 1,4,5-trisphosphate receptors in hypertension. Front Physiol 9:1018

33. Uchida K, Miyauchi H, Furuichi T, Michikawa T, Mikoshiba K (2003) Critical regions for activation gating of the inositol 1,4,5-trisphosphate receptor. J Biol Chem 278:16551–16560

34. Bosanac I, Alattia JR, Mal TK, Chan J, Talarico S, Tong FK et al (2002) Structure of the inositol 1,4,5-trisphosphate receptor binding core in complex with its ligand. Nature 420: 696–700

35. Bosanac I, Yamazaki H, Matsu-Ura T, Michikawa T, Mikoshiba K, Ikura M (2005) Crystal structure of the ligand binding suppressor domain of type 1 inositol 1,4,5-trisphosphate receptor. Mol Cell 17:193–203

36. Lin CC, Baek K, Lu Z (2011) Apo and InsP$_3$-bound crystal structures of the ligand-binding domain of an InsP$_3$ receptor. Nat Struct Mol Biol 18:1172–1174

37. Seo MD, Velamakanni S, Ishiyama N, Stathopulos PB, Rossi AM, Khan SA et al (2012) Structural and functional conservation of key domains in InsP$_3$ and ryanodine receptors. Nature 483:108–112

38. Bosanac I, Michikawa T, Mikoshiba K, Ikura M (2004) Structural insights into the regulatory mechanism of IP$_3$ receptor. Biochim Biophys Acta 1742:89–102

39. Hamada K, Miyatake H, Terauchi A, Mikoshiba K (2017) IP$_3$-mediated gating mechanism of the IP$_3$ receptor revealed by mutagenesis and X-ray crystallography. Proc Natl Acad Sci USA 114:4661–4666

40. Taylor CW, da Fonseca PC, Morris EP (2004) IP$_3$ receptors: the search for structure. Trends Biochem Sci 29:210–219

41. Fan G, Baker ML, Wang Z, Baker MR, Sinyagovskiy PA, Chiu W et al (2015) Gating machinery of InsP$_3$R channels revealed by electron cryomicroscopy. Nature 527:336–341

42. Yoshikawa F, Iwasaki H, Michikawa T, Furuichi T, Mikoshiba K (1999) Trypsinized cerebellar inositol 1,4,5-trisphosphate receptor. Structural and functional coupling of cleaved ligand binding and channel domains. J Biol Chem 274:316–327

43. Wang L, Wagner LE 2nd, Alzayady KJ, Yule DI (2017) Region-specific proteolysis differentially regulates type 1 inositol 1,4,5-trisphosphate receptor activity. J Biol Chem 292:11714–11726

44. Wang L, Yule DI (2018) Differential regulation of ion channels function by proteolysis. Biochim Biophys Acta 1865:1698–1706

45. Wang L, Wagner LE 2nd, Alzayady KJ, Yule DI (2018) Region-specific proteolysis differentially modulates type 2 and type 3 inositol 1,4,5-trisphosphate receptor activity in models of acute pancreatitis. J Biol Chem 293:13112–13124

46. Chan J, Yamazaki H, Ishiyama N, Seo MD, Mal TK, Michikawa T et al (2010) Structural studies of inositol 1,4,5-trisphosphate receptor: coupling ligand binding to channel gating. J Biol Chem 285:36092–36099

47. Schug ZT, Joseph SK (2006) The role of the S4-S5 linker and C-terminal tail in inositol 1,4,5-trisphosphate receptor function. J Biol Chem 281:24431–24440
48. Yamazaki H, Chan J, Ikura M, Michikawa T, Mikoshiba K (2010) Tyr-167/Trp-168 in type 1/3 inositol 1,4,5-trisphosphate receptor mediates functional coupling between ligand binding and channel opening. J Biol Chem 285:36081–36091
49. Paknejad N, Hite RK (2018) Structural basis for the regulation of inositol trisphosphate receptors by Ca^{2+} and IP_3. Nat Struct Mol Biol 25:660–668
50. Marchant JS, Taylor CW (1997) Cooperative activation of IP_3 receptors by sequential binding of IP_3 and Ca^{2+} safeguards against spontaneous activity. Curr Biol 7:510–518
51. Meyer T, Holowka D, Stryer L (1988) Highly cooperative opening of calcium channels by inositol 1,4,5-trisphosphate. Science 240:653–656
52. Boehning D, Joseph SK (2000) Direct association of ligand-binding and pore domains in homo- and heterotetrameric inositol 1,4,5-trisphosphate receptors. EMBO J 19:5450–5459
53. Alzayady KJ, Wang L, Chandrasekhar R, Wagner LE 2nd, Van Petegem F, Yule DI (2016) Defining the stoichiometry of inositol 1,4,5-trisphosphate binding required to initiate Ca^{2+} release. Sci Signal 9:ra35
54. Yoshikawa F, Morita M, Monkawa T, Michikawa T, Furuichi T, Mikoshiba K (1996) Mutational analysis of the ligand binding site of the inositol 1,4,5-trisphosphate receptor. J Biol Chem 271:18277–18284
55. Iwai M, Tateishi Y, Hattori M, Mizutani A, Nakamura T, Futatsugi A et al (2005) Molecular cloning of mouse type 2 and type 3 inositol 1,4,5-trisphosphate receptors and identification of a novel type 2 receptor splice variant. J Biol Chem 280:10305–10317
56. Konieczny V, Tovey SC, Mataragka S, Prole DL, Taylor CW (2017) Cyclic AMP recruits a discrete intracellular Ca^{2+} store by unmasking hypersensitive IP_3 receptors. Cell Rep 18:711–722
57. Joseph SK, Lin C, Pierson S, Thomas AP, Maranto AR (1995) Heteroligomers of type-I and type-III inositol trisphosphate receptors in WB rat liver epithelial cells. J Biol Chem 270:23310–23316
58. Monkawa T, Miyawaki A, Sugiyama T, Yoneshima H, Yamamoto-Hino M, Furuichi T et al (1995) Heterotetrameric complex formation of inositol 1,4,5-trisphosphate receptor subunits. J Biol Chem 270:14700–14704
59. Wojcikiewicz RJ, He Y (1995) Type I, II and III inositol 1,4,5-trisphosphate receptor co-immunoprecipitation as evidence for the existence of heterotetrameric receptor complexes. Biochem Biophys Res Commun 213:334–341
60. Joseph SK, Bokkala S, Boehning D, Zeigler S (2000) Factors determining the composition of inositol trisphosphate receptor hetero-oligomers expressed in COS cells. J Biol Chem 275:16084–16090
61. Alzayady KJ, Wagner LE 2nd, Chandrasekhar R, Monteagudo A, Godiska R, Tall GG et al (2013) Functional inositol 1,4,5-trisphosphate receptors assembled from concatenated homo- and heteromeric subunits. J Biol Chem 288:29772–29784
62. De Smedt H, Missiaen L, Parys JB, Henning RH, Sienaert I, Vanlingen S et al (1997) Isoform diversity of the inositol trisphosphate receptor in cell types of mouse origin. Biochem J 322:575–583
63. Wojcikiewicz RJ (1995) Type I, II, and III inositol 1,4,5-trisphosphate receptors are unequally susceptible to down-regulation and are expressed in markedly different proportions in different cell types. J Biol Chem 270:11678–11683
64. Chandrasekhar R, Alzayady KJ, Wagner LE 2nd, Yule DI (2016) Unique regulatory properties of heterotetrameric inositol 1,4,5-trisphosphate receptors revealed by studying concatenated receptor constructs. J Biol Chem 291:4846–4860
65. Taylor CW (2017) Regulation of IP_3 receptors by cyclic AMP. Cell Calcium 63:48–52
66. Wagner LE 2nd, Joseph SK, Yule DI (2008) Regulation of single inositol 1,4,5-trisphosphate receptor channel activity by protein kinase A phosphorylation. J Physiol 586:3577–3596
67. Meena A, Tovey SC, Taylor CW (2015) Sustained signalling by PTH modulates IP_3 accumulation and IP_3 receptors through cyclic AMP junctions. J Cell Sci 128:408–420

68. Tovey SC, Dedos SG, Rahman T, Taylor EJ, Pantazaka E, Taylor CW (2010) Regulation of inositol 1,4,5-trisphosphate receptors by cAMP independent of cAMP-dependent protein kinase. J Biol Chem 285:12979–12989

69. Tovey SC, Dedos SG, Taylor EJ, Church JE, Taylor CW (2008) Selective coupling of type 6 adenylyl cyclase with type 2 IP$_3$ receptors mediates direct sensitization of IP$_3$ receptors by cAMP. J Cell Biol 183:297–311

70. Chin D, Means AR (2000) Calmodulin: a prototypical calcium sensor. Trends Cell Biol 10:322–328

71. Villarroel A, Taglialatela M, Bernardo-Seisdedos G, Alaimo A, Agirre J, Alberdi A et al (2014) The ever changing moods of calmodulin: how structural plasticity entails transductional adaptability. J Mol Biol 426:2717–2735

72. Tidow H, Nissen P (2013) Structural diversity of calmodulin binding to its target sites. FEBS J 280:5551–5565

73. Yap KL, Kim J, Truong K, Sherman M, Yuan T, Ikura M (2000) Calmodulin target database. J Struct Funct Genom 1:8–14

74. Maeda N, Kawasaki T, Nakade S, Yokota N, Taguchi T, Kasai M et al (1991) Structural and functional characterization of inositol 1,4,5-trisphosphate receptor channel from mouse cerebellum. J Biol Chem 266:1109–1116

75. Yamada M, Miyawaki A, Saito K, Nakajima T, Yamamoto-Hino M, Ryo Y et al (1995) The calmodulin-binding domain in the mouse type 1 inositol 1,4,5-trisphosphate receptor. Biochem J 308:83–88

76. Adkins CE, Morris SA, De Smedt H, Sienaert I, Török K, Taylor CW (2000) Ca^{2+}-calmodulin inhibits Ca^{2+} release mediated by type-1, −2 and −3 inositol trisphosphate receptors. Biochem J 345:357–363

77. Sienaert I, Nadif Kasri N, Vanlingen S, Parys JB, Callewaert G, Missiaen L et al (2002) Localization and function of a calmodulin-apocalmodulin-binding domain in the N-terminal part of the type 1 inositol 1,4,5-trisphosphate receptor. Biochem J 365:269–277

78. Islam MO, Yoshida Y, Koga T, Kojima M, Kangawa K, Imai S (1996) Isolation and characterization of vascular smooth muscle inositol 1,4,5-trisphosphate receptor. Biochem J 316:295–302

79. Lin C, Widjaja J, Joseph SK (2000) The interaction of calmodulin with alternatively spliced isoforms of the type-I inositol trisphosphate receptor. J Biol Chem 275:2305–2311

80. Cardy TJ, Taylor CW (1998) A novel role for calmodulin: Ca^{2+}-independent inhibition of type-1 inositol trisphosphate receptors. Biochem J 334:447–455

81. Patel S, Morris SA, Adkins CE, O'Beirne G, Taylor CW (1997) Ca^{2+}-independent inhibition of inositol trisphosphate receptors by calmodulin: redistribution of calmodulin as a possible means of regulating Ca^{2+} mobilization. Proc Natl Acad Sci U S A 94:11627–11632

82. Sipma H, De Smet P, Sienaert I, Vanlingen S, Missiaen L, Parys JB et al (1999) Modulation of inositol 1,4,5-trisphosphate binding to the recombinant ligand-binding site of the type-1 inositol 1,4, 5-trisphosphate receptor by Ca^{2+} and calmodulin. J Biol Chem 274: 12157–12162

83. Vanlingen S, Sipma H, De Smet P, Callewaert G, Missiaen L, De Smedt H et al (2000) Ca^{2+} and calmodulin differentially modulate myo-inositol 1,4, 5-trisphosphate (IP$_3$)-binding to the recombinant ligand-binding domains of the various IP$_3$ receptor isoforms. Biochem J 346:275–280

84. Michikawa T, Hirota J, Kawano S, Hiraoka M, Yamada M, Furuichi T et al (1999) Calmodulin mediates calcium-dependent inactivation of the cerebellar type 1 inositol 1,4,5-trisphosphate receptor. Neuron 23:799–808

85. Missiaen L, Parys JB, Weidema AF, Sipma H, Vanlingen S, De Smet P et al (1999) The bell-shaped Ca^{2+} dependence of the inositol 1,4, 5-trisphosphate-induced Ca^{2+} release is modulated by Ca^{2+}/calmodulin. J Biol Chem 274:13748–13751

86. Missiaen L, DeSmedt H, Bultynck G, Vanlingen S, Desmet P, Callewaert G et al (2000) Calmodulin increases the sensitivity of type 3 inositol-1,4, 5-trisphosphate receptors to Ca^{2+} inhibition in human bronchial mucosal cells. Mol Pharmacol 57:564–567

87. Nosyreva E, Miyakawa T, Wang Z, Glouchankova L, Mizushima A, Iino M et al (2002) The high-affinity calcium-calmodulin-binding site does not play a role in the modulation of type 1 inositol 1,4,5-trisphosphate receptor function by calcium and calmodulin. Biochem J 365:659–367
88. Kasri NN, Bultynck G, Smyth J, Szlufcik K, Parys JB, Callewaert G et al (2004) The N-terminal Ca^{2+}-independent calmodulin-binding site on the inositol 1,4,5-trisphosphate receptor is responsible for calmodulin inhibition, even though this inhibition requires Ca^{2+}. Mol Pharmacol 66:276–284
89. Kasri NN, Török K, Galione A, Garnham C, Callewaert G, Missiaen L et al (2006) Endogenously bound calmodulin is essential for the function of the inositol 1,4,5-trisphosphate receptor. J Biol Chem 281:8332–8338
90. Sun Y, Taylor CW (2008) A calmodulin antagonist reveals a calmodulin-independent interdomain interaction essential for activation of inositol 1,4,5-trisphosphate receptors. Biochem J 416:243–253
91. Sun Y, Rossi AM, Rahman T, Taylor CW (2013) Activation of IP$_3$ receptors requires an endogenous 1-8-14 calmodulin-binding motif. Biochem J 449:39–49
92. Kang S, Kwon H, Wen H, Song Y, Frueh D, Ahn HC et al (2011) Global dynamic conformational changes in the suppressor domain of IP$_3$ receptor by stepwise binding of the two lobes of calmodulin. FASEB J 25:840–850
93. White C, Yang J, Monteiro MJ, Foskett JK (2006) CIB1, a ubiquitously expressed Ca^{2+}-binding protein ligand of the InsP$_3$ receptor Ca^{2+} release channel. J Biol Chem 281:20825–20833
94. Yang J, McBride S, Mak DO, Vardi N, Palczewski K, Haeseleer F et al (2002) Identification of a family of calcium sensors as protein ligands of inositol trisphosphate receptor Ca^{2+} release channels. Proc Natl Acad Sci U S A 99:7711–7716
95. Haynes LP, Tepikin AV, Burgoyne RD (2004) Calcium-binding protein 1 is an inhibitor of agonist-evoked, inositol 1,4,5-trisphosphate-mediated calcium signaling. J Biol Chem 279:547–555
96. Kasri NN, Holmes AM, Bultynck G, Parys JB, Bootman MD, Rietdorf K et al (2004) Regulation of InsP$_3$ receptor activity by neuronal Ca^{2+}-binding proteins. EMBO J 23:312–321
97. Bultynck G, Szlufcik K, Kasri NN, Assefa Z, Callewaert G, Missiaen L et al (2004) Thimerosal stimulates Ca^{2+} flux through inositol 1,4,5-trisphosphate receptor type 1, but not type 3, via modulation of an isoform-specific Ca^{2+}-dependent intramolecular interaction. Biochem J 381:87–96
98. Li C, Chan J, Haeseleer F, Mikoshiba K, Palczewski K, Ikura M et al (2009) Structural insights into Ca^{2+}-dependent regulation of inositol 1,4,5-trisphosphate receptors by CaBP1. J Biol Chem 284:2472–2481
99. Li C, Enomoto M, Rossi AM, Seo MD, Rahman T, Stathopulos PB et al (2013) CaBP1, a neuronal Ca^{2+} sensor protein, inhibits inositol trisphosphate receptors by clamping intersubunit interactions. Proc Natl Acad Sci U S A 110:8507–8512
100. Nakamura TY, Jeromin A, Mikoshiba K, Wakabayashi S (2011) Neuronal calcium sensor-1 promotes immature heart function and hypertrophy by enhancing Ca^{2+} signals. Circ Res 109:512–523
101. Schlecker C, Boehmerle W, Jeromin A, DeGray B, Varshney A, Sharma Y et al (2006) Neuronal calcium sensor-1 enhancement of InsP$_3$ receptor activity is inhibited by therapeutic levels of lithium. J Clin Invest 116:1668–1674
102. Zhang K, Heidrich FM, DeGray B, Boehmerle W, Ehrlich BE (2010) Paclitaxel accelerates spontaneous calcium oscillations in cardiomyocytes by interacting with NCS-1 and the InsP$_3$R. J Mol Cell Cardiol 49:829–835
103. Boehmerle W, Splittgerber U, Lazarus MB, McKenzie KM, Johnston DG, Austin DJ et al (2006) Paclitaxel induces calcium oscillations via an inositol 1,4,5-trisphosphate receptor and neuronal calcium sensor 1-dependent mechanism. Proc Natl Acad Sci USA 103:18356–18361

104. Boeckel GR, Ehrlich BE (2018) NCS-1 is a regulator of calcium signaling in health and disease. Biochim Biophys Acta 1865:1660–1667

105. Meissner G (2017) The structural basis of ryanodine receptor ion channel function. J Gen Physiol 149:1065–1089

106. Brini M, Cali T, Ottolini D, Carafoli E (2013) The plasma membrane calcium pump in health and disease. FEBS J 280:5385–5397

107. Hasan R, Zhang X (2018) Ca^{2+} regulation of TRP ion channels. Int J Mol Sci 19:1256

108. Saimi Y, Kung C (2002) Calmodulin as an ion channel subunit. Annu Rev Physiol 64:289–311

109. Letai AG (2008) Diagnosing and exploiting cancer's addiction to blocks in apoptosis. Nat Rev Cancer 8:121–132

110. Brunelle JK, Letai A (2009) Control of mitochondrial apoptosis by the Bcl-2 family. J Cell Sci 122:437–441

111. Davids MS, Letai A (2012) Targeting the B-cell lymphoma/leukemia 2 family in cancer. J Clin Oncol 30:3127–3135

112. Vervliet T, Parys JB, Bultynck G (2016) Bcl-2 proteins and calcium signaling: complexity beneath the surface. Oncogene 35:5079–5092

113. Cárdenas C, Miller RA, Smith I, Bui T, Molgó J, Müller M et al (2010) Essential regulation of cell bioenergetics by constitutive InsP$_3$ receptor Ca^{2+} transfer to mitochondria. Cell 142:270–283

114. Eckenrode EF, Yang J, Velmurugan GV, Foskett JK, White C (2010) Apoptosis protection by Mcl-1 and Bcl-2 modulation of inositol 1,4,5-trisphosphate receptor-dependent Ca^{2+} signaling. J Biol Chem 285:13678–13684

115. White C, Li C, Yang J, Petrenko NB, Madesh M, Thompson CB et al (2005) The endoplasmic reticulum gateway to apoptosis by Bcl-X$_L$ modulation of the InsP$_3$R. Nat Cell Biol 7:1021–1028

116. Rong YP, Bultynck G, Aromolaran AS, Zhong F, Parys JB, De Smedt H et al (2009) The BH4 domain of Bcl-2 inhibits ER calcium release and apoptosis by binding the regulatory and coupling domain of the IP$_3$ receptor. Proc Natl Acad Sci USA 106:14397–14402

117. Parys JB (2014) The IP$_3$ receptor as a hub for Bcl-2 family proteins in cell death control and beyond. Sci Signal 7:pe4

118. Monaco G, Beckers M, Ivanova H, Missiaen L, Parys JB, De Smedt H et al (2012) Profiling of the Bcl-2/Bcl-X$_L$-binding sites on type 1 IP$_3$ receptor. Biochem Biophys Res Commun 428:31–35

119. Monaco G, Decrock E, Akl H, Ponsaerts R, Vervliet T, Luyten T et al (2012) Selective regulation of IP$_3$-receptor-mediated Ca^{2+} signaling and apoptosis by the BH4 domain of Bcl-2 versus Bcl-xl. Cell Death Differ 19:295–309

120. Rong YP, Aromolaran AS, Bultynck G, Zhong F, Li X, McColl K et al (2008) Targeting Bcl-2-IP$_3$ receptor interaction to reverse Bcl-2's inhibition of apoptotic calcium signals. Mol Cell 31:255–265

121. Bonneau B, Nougarede A, Prudent J, Popgeorgiev N, Peyrieras N, Rimokh R et al (2014) The Bcl-2 homolog Nrz inhibits binding of IP$_3$ to its receptor to control calcium signaling during zebrafish epiboly. Sci Signal 7:ra14

122. Bonneau B, Ando H, Kawaai K, Hirose M, Takahashi-Iwanaga H, Mikoshiba K (2016) IRBIT controls apoptosis by interacting with the Bcl-2 homolog, Bcl2l10, and by promoting ER-mitochondria contact. elife 5:e19896

123. Yang J, Vais H, Gu W, Foskett JK (2016) Biphasic regulation of InsP$_3$ receptor gating by dual Ca^{2+} release channel BH3-like domains mediates Bcl-xL control of cell viability. Proc Natl Acad Sci USA 113:E1953–E1962

124. Ivanova H, Ritaine A, Wagner L, Luyten T, Shapovalov G, Welkenhuyzen K et al (2016) The trans-membrane domain of Bcl-2α, but not its hydrophobic cleft, is a critical determinant for efficient IP$_3$ receptor inhibition. Oncotarget 7:55704–55720

125. Chang MJ, Zhong F, Lavik AR, Parys JB, Berridge MJ, Distelhorst CW (2014) Feedback regulation mediated by Bcl-2 and DARPP-32 regulates inositol 1,4,5-trisphosphate receptor phosphorylation and promotes cell survival. Proc Natl Acad Sci U S A 111:1186–1191
126. Ivanova H, Wagner LE, 2nd, Tanimura A, Vandermarliere E, Luyten T, Welkenhuyzen K et al (2019) Bcl-2 and IP$_3$ compete for the ligand-binding domain of IP$_3$Rs modulating Ca^{2+} signaling output. Cell Mol Life Sci. In press
127. Schulman JJ, Wright FA, Kaufmann T, Wojcikiewicz RJ (2013) The Bcl-2 protein family member Bok binds to the coupling domain of inositol 1,4,5-trisphosphate receptors and protects them from proteolytic cleavage. J Biol Chem 288:25340–25349
128. Assefa Z, Bultynck G, Szlufcik K, Nadif Kasri N, Vermassen E, Goris J et al (2004) Caspase-3-induced truncation of type 1 inositol trisphosphate receptor accelerates apoptotic cell death and induces inositol trisphosphate-independent calcium release during apoptosis. J Biol Chem 279:43227–43236
129. Hirota J, Furuichi T, Mikoshiba K (1999) Inositol 1,4,5-trisphosphate receptor type 1 is a substrate for caspase-3 and is cleaved during apoptosis in a caspase-3-dependent manner. J Biol Chem 274:34433–34437
130. Schulman JJ, Wright FA, Han X, Zluhan EJ, Szczesniak LM, Wojcikiewicz RJ (2016) The stability and expression level of Bok are governed by binding to inositol 1,4,5-trisphosphate receptors. J Biol Chem 291:11820–11828
131. Ivanova H, Luyten T, Decrock E, Vervliet T, Leybaert L, Parys JB et al (2017) The BH4 domain of Bcl-2 orthologues from different classes of vertebrates can act as an evolutionary conserved inhibitor of IP$_3$ receptor channels. Cell Calcium 62:41–66
132. Cárdenas C, Müller M, McNeal A, Lovy A, Jana F, Bustos G et al (2016) Selective vulnerability of cancer cells by inhibition of Ca^{2+} transfer from endoplasmic reticulum to mitochondria. Cell Rep 14:2313–2324
133. Finkel T, Hwang PM (2009) The Krebs cycle meets the cell cycle: mitochondria and the G$_1$-S transition. Proc Natl Acad Sci USA 106:11825–11826
134. Distelhorst CW (2018) Targeting Bcl-2-IP$_3$ receptor interaction to treat cancer: a novel approach inspired by nearly a century treating cancer with adrenal corticosteroid hormones. Biochim Biophys Acta 1865:1795–1804
135. He C, Levine B (2010) The Beclin 1 interactome. Curr Opin Cell Biol 22:140–149
136. Decuypere JP, Parys JB, Bultynck G (2012) Regulation of the autophagic Bcl-2/Beclin 1 interaction. Cell 1:284–312
137. Erlich S, Mizrachy L, Segev O, Lindenboim L, Zmira O, Adi-Harel S et al (2007) Differential interactions between Beclin 1 and Bcl-2 family members. Autophagy 3:561–568
138. Vicencio JM, Ortiz C, Criollo A, Jones AW, Kepp O, Galluzzi L et al (2009) The inositol 1,4,5-trisphosphate receptor regulates autophagy through its interaction with Beclin 1. Cell Death Differ 16:1006–1017
139. Decuypere JP, Welkenhuyzen K, Luyten T, Ponsaerts R, Dewaele M, Molgo J et al (2011) Ins(1,4,5)P$_3$ receptor-mediated Ca^{2+} signaling and autophagy induction are interrelated. Autophagy 7:1472–1489
140. Ando H, Mizutani A, Matsu-ura T, Mikoshiba K (2003) IRBIT, a novel inositol 1,4,5-trisphosphate (IP$_3$) receptor-binding protein, is released from the IP$_3$ receptor upon IP$_3$ binding to the receptor. J Biol Chem 278:10602–10612
141. Ando H, Kawaai K, Mikoshiba K (2014) IRBIT: a regulator of ion channels and ion transporters. Biochim Biophys Acta 1843:2195–2204
142. Kawaai K, Ando H, Satoh N, Yamada H, Ogawa N, Hirose M et al (2017) Splicing variation of long-IRBIT determines the target selectivity of IRBIT family proteins. Proc Natl Acad Sci USA 114:3921–3926
143. Liang J, Lyu J, Zhao M, Li D, Zheng M, Fang Y et al (2017) Tespa1 regulates T cell receptor-induced calcium signals by recruiting inositol 1,4,5-trisphosphate receptors. Nat Commun 8:15732

144. Wang D, Zheng M, Lei L, Ji J, Yao Y, Qiu Y et al (2012) Tespa1 is involved in late thymocyte development through the regulation of TCR-mediated signaling. Nat Immunol 13:560–568

145. Dingli F, Parys JB, Loew D, Saule S, Mery L (2012) Vimentin and the K-Ras-induced actin-binding protein control inositol-(1,4,5)-trisphosphate receptor redistribution during MDCK cell differentiation. J Cell Sci 125:5428–5440

146. Fujimoto T, Machida T, Tanaka Y, Tsunoda T, Doi K, Ota T et al (2011) KRAS-induced actin-interacting protein is required for the proper localization of inositol 1,4,5-trisphosphate receptor in the epithelial cells. Biochem Biophys Res Commun 407:438–443

147. Matsuzaki H, Fujimoto T, Ota T, Ogawa M, Tsunoda T, Doi K et al (2012) Tespa1 is a novel inositol 1,4,5-trisphosphate receptor binding protein in T and B lymphocytes. FEBS Open Bio 2:255–259

148. Malissen B, Gregoire C, Malissen M, Roncagalli R (2014) Integrative biology of T cell activation. Nat Immunol 15:790–797

149. Matsuzaki H, Fujimoto T, Tanaka M, Shirasawa S (2013) Tespa1 is a novel component of mitochondria-associated endoplasmic reticulum membranes and affects mitochondrial calcium flux. Biochem Biophys Res Commun 433:322–326

150. Szabadkai G, Bianchi K, Varnai P, De Stefani D, Wieckowski MR, Cavagna D et al (2006) Chaperone-mediated coupling of endoplasmic reticulum and mitochondrial Ca^{2+} channels. J Cell Biol 175:901–911

151. Dayton TL, Jacks T, Vander Heiden MG (2016) PKM2, cancer metabolism, and the road ahead. EMBO Rep 17:1721–1730

152. Hsu MC, Hung WC (2018) Pyruvate kinase M2 fuels multiple aspects of cancer cells: from cellular metabolism, transcriptional regulation to extracellular signaling. Mol Cancer 17:35

153. Dong G, Mao Q, Xia W, Xu Y, Wang J, Xu L et al (2016) PKM2 and cancer: the function of PKM2 beyond glycolysis. Oncol Lett 11:1980–1986

154. Lavik AR (2016) The role of inositol 1,4,5-trisphosphate receptor-interacting proteins in regulating inositol 1,4,5-trisphosphate receptor-dependent calcium signals and cell survival. PhD thesis, Case Western Reserve University, USA. https://etd.ohiolink.edu/!etd.send_file?accession=case1448532307&disposition=inline

155. Lavik A, Harr M, Kerkhofs M, Parys JB, Bultynck G, Bird G et al (2018) IP$_3$Rs recruit the glycolytic enzyme PKM2 to the ER, promoting Ca^{2+} homeostasis and survival in hematologic malignancies. In: Abstract 66, 15th International meeting of the European Calcium Society. Hamburg, Germany

156. Liu F, Ma F, Wang Y, Hao L, Zeng H, Jia C et al (2017) PKM2 methylation by CARM1 activates aerobic glycolysis to promote tumorigenesis. Nat Cell Biol 19:1358–1370

157. Sipma H, Deelman L, Smedt HD, Missiaen L, Parys JB, Vanlingen S et al (1998) Agonist-induced down-regulation of type 1 and type 3 inositol 1,4,5-trisphosphate receptors in A7r5 and DDT1 MF-2 smooth muscle cells. Cell Calcium 23:11–21

158. Wojcikiewicz RJ, Furuichi T, Nakade S, Mikoshiba K, Nahorski SR (1994) Muscarinic receptor activation down-regulates the type I inositol 1,4,5-trisphosphate receptor by accelerating its degradation. J Biol Chem 269:7963–7969

159. Wojcikiewicz RJ, Nakade S, Mikoshiba K, Nahorski SR (1992) Inositol 1,4,5-trisphosphate receptor immunoreactivity in SH-SY5Y human neuroblastoma cells is reduced by chronic muscarinic receptor activation. J Neurochem 59:383–386

160. Oberdorf J, Webster JM, Zhu CC, Luo SG, Wojcikiewicz RJ (1999) Down-regulation of types I, II and III inositol 1,4,5-trisphosphate receptors is mediated by the ubiquitin/proteasome pathway. Biochem J 339:453–461

161. Lee HS, Lee SA, Hur SK, Seo JW, Kwon J (2014) Stabilization and targeting of INO80 to replication forks by BAP1 during normal DNA synthesis. Nat Commun 5:5128

162. Zarrizi R, Menard JA, Belting M, Massoumi R (2014) Deubiquitination of γ-tubulin by BAP1 prevents chromosome instability in breast cancer cells. Cancer Res 74:6499–6508

163. Yu H, Pak H, Hammond-Martel I, Ghram M, Rodrigue A, Daou S et al (2014) Tumor suppressor and deubiquitinase BAP1 promotes DNA double-strand break repair. Proc Natl Acad Sci U S A 111:285–290

164. Yu H, Mashtalir N, Daou S, Hammond-Martel I, Ross J, Sui G et al (2010) The ubiquitin carboxyl hydrolase BAP1 forms a ternary complex with YY1 and HCF-1 and is a critical regulator of gene expression. Mol Cell Biol 30:5071–5085

165. Baughman JM, Rose CM, Kolumam G, Webster JD, Wilkerson EM, Merrill AE et al (2016) NeuCode proteomics reveals Bap1 regulation of metabolism. Cell Rep 16:583–595

166. Ruan HB, Han X, Li MD, Singh JP, Qian K, Azarhoush S et al (2012) O-GlcNAc transferase/host cell factor C1 complex regulates gluconeogenesis by modulating PGC-1α stability. Cell Metab 16:226–237

167. Bononi A, Giorgi C, Patergnani S, Larson D, Verbruggen K, Tanji M et al (2017) BAP1 regulates IP$_3$R3-mediated Ca^{2+} flux to mitochondria suppressing cell transformation. Nature 546:549–553

168. Bononi A, Yang H, Giorgi C, Patergnani S, Pellegrini L, Su M et al (2017) Germline BAP1 mutations induce a Warburg effect. Cell Death Differ 24:1694–1704

169. Kuchay S, Giorgi C, Simoneschi D, Pagan J, Missiroli S, Saraf A et al (2017) PTEN counteracts FBXL2 to promote IP$_3$R3- and Ca^{2+}-mediated apoptosis limiting tumour growth. Nature 546:554–558

170. Worby CA, Dixon JE (2014) PTEN. Annu Rev Biochem 83:641–669

171. Carnero A, Paramio JM (2014) The PTEN/PI3K/AKT pathway in vivo, cancer mouse models. Front Oncol 4:252

172. Milella M, Falcone I, Conciatori F, Cesta Incani U, Del Curatolo A, Inzerilli N et al (2015) PTEN: multiple functions in human malignant tumors. Front Oncol 5:24

173. Bittremieux M, Parys JB, Pinton P, Bultynck G (2016) ER functions of oncogenes and tumor suppressors: modulators of intracellular Ca^{2+} signaling. Biochim Biophys Acta 1863: 1364–1378

174. Bononi A, Bonora M, Marchi S, Missiroli S, Poletti F, Giorgi C et al (2013) Identification of PTEN at the ER and MAMs and its regulation of Ca^{2+} signaling and apoptosis in a protein phosphatase-dependent manner. Cell Death Differ 20:1631–1643

175. Furuichi T, Yoshikawa S, Miyawaki A, Wada K, Maeda N, Mikoshiba K (1989) Primary structure and functional expression of the inositol 1,4,5-trisphosphate-binding protein P$_{400}$. Nature 342:32–38

Chapter 11
Expression of the Inositol 1,4,5-Trisphosphate Receptor and the Ryanodine Receptor Ca^{2+}-Release Channels in the Beta-Cells and Alpha-Cells of the Human Islets of Langerhans

Fabian Nordenskjöld, Björn Andersson, and Md. Shahidul Islam

Abstract Calcium signaling regulates secretion of hormones and many other cellular processes in the islets of Langerhans. The three subtypes of the inositol 1,4,5-trisphosphate receptors (IP3Rs), inositol 1,4,5-trisphosphate receptor type 1 (IP3R1), 1,4,5-trisphosphate receptor type 2 (IP3R2), 1,4,5-trisphosphate receptor type 3 (IP3R3), and the three subtypes of the ryanodine receptors (RyRs), ryanodine receptor 1 (RyR1), ryanodine receptor 2 (RyR2) and ryanodine receptor 3 (RyR3) are the main intracellular Ca^{2+}-release channels. The identity and the relative levels of expression of these channels in the alpha-cells, and the beta-cells of the human islets of Langerhans are unknown. We have analyzed the RNA sequencing data obtained from highly purified human alpha-cells and beta-cells for quantitatively identifying the mRNA of the intracellular Ca^{2+}-release channels in these cells. We found that among the three IP3Rs the IP3R3 is the most abundantly expressed one in the beta-cells, whereas IP3R1 is the most abundantly expressed one in the alpha-cells. In addition to the IP3R3, beta-cells also expressed the IP3R2, at a lower level. Among the RyRs, the RyR2 was the most abundantly expressed one in the beta-cells, whereas the RyR1 was the most abundantly expressed one in the alpha-cells. Information on the relative abundance of the different intracellular Ca^{2+}-release channels in the human alpha-cells and the beta-cells may help the understanding

F. Nordenskjöld · B. Andersson
Department of Cell and Molecular Biology, Karolinska Institutet, Stockholm, Sweden

M. S. Islam (✉)
Department of Clinical Science and Education, Södersjukhuset, Karolinska Institutet, Stockholm, Sweden

Department of Emergency Care and Internal Medicine, Uppsala University Hospital, Uppsala, Sweden
e-mail: Shahidul.Islam@ki.se

© Springer Nature Switzerland AG 2020
M. S. Islam (ed.), *Calcium Signaling*, Advances in Experimental Medicine and Biology 1131, https://doi.org/10.1007/978-3-030-12457-1_11

of their roles in the generation of Ca^{2+} signals and many other related cellular processes in these cells.

Keywords Human islets of Langerhans · Inositol 1,4,5-trisphosphate receptors in the beta-cells · Inositol 1,4,5-trisphosphate receptors in the alpha-cells · Ryanodine receptors in the alpha cells · Ryanodine receptors in the beta-cells · Ca^{2+} signaling in the islets · Human alpha-cells · Human beta-cells · RNA-sequencing

11.1   Introduction

Human islets of Langerhans are microorgans that contain the glucagon-secreting α-cells, the insulin-secreting β-cells, and the somatostatin-secreting δ-cells, dispersed throughout the islets [1, 2]. Islet research is important because of the roles of these microorgans in the secretion of insulin and glucagon, and the impairment of such secretions in diabetes mellitus, which is a global public health problem. The human islets contain 28–75% β-cell, 10–65% α-cells, and 1.2–22% δ-cells [1]. It is important to understand the molecular mechanisms of hormone secretion from the islet cells to understand their roles in the pathogenesis of different diseases including diabetes mellitus and pancreatogenous hyperinsulinemic hypoglycemia. A major obstacle in islet research is the difficulty in obtaining pure preparations of individual islet cells in sufficient numbers for experiments.

Insulin secretion from the β-cells is triggered by an increase in the cytoplasmic free Ca^{2+} concentration ($[Ca^{2+}]_c$) [3, 4]. The mechanism of increase in the $[Ca^{2+}]_c$ include Ca^{2+} entry through the plasma membrane Ca^{2+} channels, and Ca^{2+} release from the intracellular Ca^{2+} stores [3]. The intracellular Ca^{2+}-release channels are activated by inositol 1,4,5-trisphosphate and/or by Ca^{2+}, the latter process being called calcium-induced calcium release (CICR) [3, 5].

The two main families of the intracellular Ca^{2+}-release channels are the inositol 1,4,5-trisphosphate (IP3) receptor (IP3R) and the ryanodine receptors (RyR). They form tetrameric ion channels. In mammals and other higher organisms, three genes *ITPR1*, *ITPR2* and *ITPR3* encode the IP3R1, IP3R2, and IP3R3 respectively. The homology between the three isoforms is about 75% at amino acid level. They usually form homotetramers, but can also form heterotetramers, thereby altering the regulatory properties of the IP3Rs, and increasing the diversity of the spatial and temporal aspects of Ca^{2+} signaling mediated by the IP3Rs [6]. Ca^{2+}, at modest concentrations, acts as a co-activator of the IP3Rs, since in the absence of Ca^{2+}, IP3 alone cannot activate the channel [7]. At higher concentrations Ca^{2+} inhibits IP3-induced Ca^{2+} release [7].

Three genes *RYR1*, *RYR2*, and *RYR3* encode the ryanodine receptor 1 (RyR1), the ryanodine receptor 2 (RyR2), and the ryanodine receptor 3 (RyR3) respectively [8]. Previous studies have shown that ryanodine receptors participate in the generation of Ca^{2+} signals through CICR in the human β-cells [3, 5, 9, 10].

From numerous studies we know that the β-cells and the α-cells have both the IP3Rs and the RyRs [3, 11]. The functional properties and molecular regulation of the different types of the IP3Rs and the RyRs are different. In spite of over three decades of research, it is not known which types of the IP3Rs and the RyRs are present in the human islet cells. It is difficult to identify different types of IP3Rs and RyRs by immunohistochemistry because of lack of antibodies that selectively discriminate between the receptor subtypes. Difficulties in preparing highly purified α-cells and β-cells in sufficient amounts have further hampered the identification of the intracellular Ca^{2+}-release channels by conventional molecular techniques.

RNA sequencing or whole transcriptome shotgun sequencing of cDNA by using "next generation sequencing" is a powerful method for identifying the presence of, and measuring the quantity of different species of mRNA in the cells [12]. RNA sequencing is more reliable than hybridization-based microarrays for gene expression studies. We have used this approach for quantitatively identifying the transient receptor potential channels in the human β-cells [13]. By analyzing the RNA sequencing data obtained from highly purified α-cells and β-cells, we have now identified the level of expression of the different subtypes of the IP3Rs and RyRs in the human α-cells and the β-cells.

11.2 Methods and Materials

For identifying the intracellular Ca^{2+}-release channels, we used the transcriptomes of the purified human α-cells and β-cells reported by Blodgett et al. [14]. These investigators isolated islets from deceased human donors, dissociated those into single cell suspensions, fixed and permeabilized the cells to stain for the intracellular hormones, and sorted the cells by using fluorescence activated cell sorter (FACS). They sorted highly purified (> 97% pure) human α-cells and β-cells identified by anti-glucagon and anti-insulin antibodies respectively, under experimental conditions that minimized RNA degradation [14]. The δ-cells were stained by anti-somatostatin, and were gated out to obtain homogenous preparations of highly purified α-cells and β-cells [14]. The methods have been described in details by Blodgett et al. [14]. In short, they extracted RNA from the purified α-cells and β-cells, purified, quantified, and analyzed RNA for integrity. They constructed libraries by RNA fragmentation, first- and second-strand cDNA synthesis, ligation of adaptors, amplification, library validation, and ribosomal RNA removal [14]. They performed 91 base pair, paired-end sequencing on Illumina HiSeq 2000 [14].

The RNA sequencing data were made available for the public on the GEO database (https://www.ncbi.nlm.nih.gov/geo). We analyzed the data obtained from the α-cell and the β-cell samples from the adult donors of both sexes (5 males, 1 female, 1 undefined) of variable ages (4–60 years), and BMI (21.5–37 kg/m^2) [14]. Of the seven α-cell samples, RNA-sequencing data were produced from six samples. All of the seven β-cell samples yielded β- cell RNA-sequencing data.

We analyzed the RNA-sequencing data that consisted of 13 samples (6 α-cell and 7 β-cell) from 7 adult donors. We first filtered the data for the mitochondrial reads.

The data were mas mapped against the human mitochondrial genome (http://www.ncbi.nlm.nih.gov/nuccore/NC_012920.1) by using bowtie 2, and any mapped reads were removed from the data (option–un-conc-gz). On average 18% of the reads of each sample mapped to the mitochondrial genome. We analyzed gene expression by using RSEM software package [15], with bowtie 2 as mapping software (RSEM v1.2.25, bowtie 2 v2.2.6, standard RSEM in-parameters for bowtie 2). The filtered reads were mapped to the annotated human genome (version GRCh37.2). The resulting expression counts were normalized by RSEM to TPM-values (transcripts per million). We preferred TPM instead of FPKM (fragments per kilobase million) because TPM makes it convenient to compare the proportion of reads that maps to a gene in each sample. Gene level differential expression analysis was done by using the EBSeq R-package, >99% confidence [16].

11.3   Results

Of the six genes analyzed, four (*RYR2*, *ITPR1*, *ITPR2*, *ITPR3*) appeared to be differentially expressed in the two cell types. According to our differential expression analysis, *ITPR1* had a higher expression in the α-cells than in the β-cells, while *RYR2*, *ITPR2* and *ITPR3* had a higher expression in the β-cells than in the α-cells.

Figure 11.1 shows a heat map of the gene expression values of the IP3Rs and RyRs in the FACS purified human α-cell and β-cell preparations. IP3R1 was expressed more abundantly in the α-cells than in the β-cells, where it was almost absent. IP3R3 was more abundantly expressed in the β-cells than in the α-cells, where it was almost absent. In the β-cells the expression of the IP3R2 was lower than that of the IP3R3, but was higher than the expression of the IP3R2 in the α-cells, where it was almost absent. Highest expression of IP3R2 was observed in one of the β-cell preparations (β-4, white).

The expression of the RYR1 was higher in the α-cells than in the β-cells, where it was almost absent (Fig. 11.1). On the other hand, the expression of the RYR2 was higher in the β-cells than in the α-cells, where it was almost absent. The expression of the RYR3 was highly variable both in the α-cells and in the β-cells. There was very low expression of RYR3 in two preparations of the α-cells (α-5, α-7), and four preparations of β-cells (β-1, β5–7).

Figure 11.2 shows the level of expression of the three IP3Rs in six purified α-cell and seven purified β-cell preparations. In all α-cell preparations, the expressions of the IP3R1 were higher than those in any of the β-cell preparations. In six β-cell preparations (β-1, β-2, β-4 to 7), the expressions of the IP3R3 were higher than those in the α-cells. In one β-cell preparation (β-3) the expression of the IP3R3 was lower, similar to that in the α-cells.

The level of expressions of the three RyRs in the α-cells and the β-cells are shown in Fig. 11.3. In general, both the α-cells and the β-cells expressed RyRs at a much lower level than the IP3Rs. We found that *RYR1* was expressed, at low level, in five α-cell preparations (a1, a3, a4-a6), whereas it was almost absent in one α-cell

Fig. 11.1 Expression pattern of IP3Rs and the RyRs in the purified β-cells and α-cells of human islets of Langerhans. A heat map of the gene expression values is shown. The color indicates the relative expression of each gene (black, low; red, high; white, highest). For example, the expression of IP3R1 (*ITPR1*) is higher in the α-cells than in the β-cells (β-cells black); *RYR2* is more in the β-cells than in the α-cells (α-cells black)

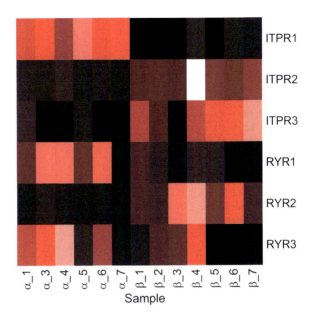

Fig. 11.2 Expression of the IP3Rs in the purified human β-cells and the α-cells of the islets of Langerhans. The scale is linear with all counts normalized to TPM. The figure shows the relative expression levels as bar plots. Green, IP3R1 (*ITPR1*); red, IP3R2 (*ITPR2*), blue, IP3R3 (*ITPR3*). TPM = transcripts per million. Results obtained from six α-cell preparations and seven β-cell preparations are shown

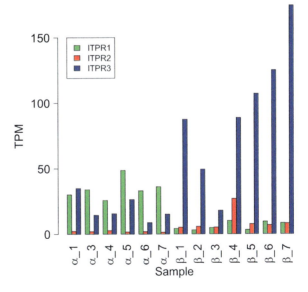

preparation (α-7). All of the seven samples of the β-cells expressed *RYR1* at very low level. On the other hand all of the samples of the β-cells expressed the *RYR2* at a level much higher than any of the α-cell preparations. The α-cells expressed *RYR2* only at very low level. We observed that both the α-cells and the β-cells expressed *RYR3* at low levels. Three of the β-cell preparations (β-2, β-3, β-4) expressed *RYR3* at low levels, whereas it was almost absent in the remaining four preparations (β-1, β-5, β-6, β-7). *RYR3* was expressed at a higher level in the α-cells than in the β-cells.

Fig. 11.3 Expression of the
RyRs in the purified human
β-cells and the α-cells of the
islets of Langerhans. The
figure shows the relative
expression levels as bar plots.
Green, RyR1 (*RYR1*); red,
RyR2 (*RYR2*); blue, RyR3
(*RYR3*). The scale is linear
with all counts normalized to
TPM. TPM = transcripts per
million. Note that the Y axis
of this figure is different from
that of Fig. 11.2. Results
obtained from six α-cell
preparations and seven β-cell
preparations are shown

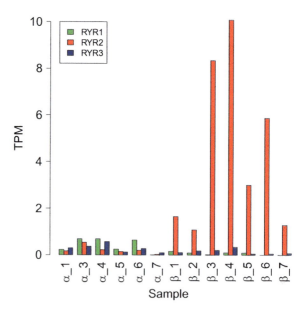

Four of the α-cell preparations expressed RYR3 at low level, whereas it was almost
absent in the remaining two (α-5 and α-7).

11.4   Discussion

By analyzing the next generation RNA-sequencing data, we have identified the
subtypes of the two major intracellular Ca^{2+}-release channels that are expressed
in the two major types of cells in the human islets of Langerhans. The transcriptome
data we used are reliable because these were obtained from α-cells and β-cells that
were highly purified by FACS after staining the cells for insulin, glucagon and
somatostatin, and after excluding the contaminating somatostatin-positive cells [14].

Our results showed that the most abundantly expressed IP3R in the β-cells was
the IP3R3, and the most abundantly expressed IP3R in the α-cells was the IP3R1.
The most abundantly expressed RyR in the β-cells was the RyR2 and the most
abundantly expressed RyR in the α-cells was the RyR1. In both of the cell types
the expression of IP3Rs was many fold higher than that of the RyRs.

A previous study reported that IP3R3 is readily detectable by immonoblotting in
the rat insulinoma RINm5F cells, and a hamster-derived insulinoma cell line HIT-
T15 cells [17, 18]. Wojcikiewicj et al. demonstrated that at protein level, 96% of the
IP3Rs in the RINm5F cells are IP3R3 and 4% IP3R1 [19]. It was unclear whether
the abundant expression of IP3R3 was a peculiarity limited to he transformed rodent
β-cells. Our study showed that IP3R3 was expressed abundantly even in the primary
human β-cells. Using RT-PCR, Blondel et al. could detect only IP3R3, but not IP3R2

and IP3R1 in the rat whole pancreatic islets [17]. Lee et al. showed that in the whole mouse islets, and rat islets the IP3R1 and the IP3R3 are the most abundant IP3Rs, respectively [20, 21]. In our study the primary human β-cells expressed the IP3R3 abundantly, and to a lesser extent also the IP3R2. This raises the possibility that in the β-cells IP3Rs may exist not only as homo-tetramers, but also as hetero-tetramers of IP3R3 and IP3R2. It may be noted that in pancreatic acinar cells, which express both IP3R3 and IP3R2, the majority of the IP3R3 form hetero-tetramer with IP3R2 [22]. IP3R hetero-tetramers composed of different IP3R subtypes show unique affinities for IP3 [6]. Chandrasekhar et al. have also shown that in IP3R hetero-tetramers, the properties of the IP3R2 dictates the regulatory properties and ATP sensitivity of the hetero-tetramer irrespective of the other subtypes present in the tetramer [6].

Even though the IP3Rs are primarily activated by inositol, 1,4,5-trisphosphate, and biphasically regulated by Ca^{2+}, there are many differences in the electrophysiological properties and molecular mechanisms that regulate the three subtypes of IP3Rs [6, 23, 24]. For instance, the three IP3Rs have different affinities for inositol, 1,4,5-trisphosphate (IP3R2 > IP3R1 > IP3R3).

The most abundant RyR in the β-cells was the RyR2. This is consistent with our previous report where we demonstrated by RNAse protection assay that mouse islets and βTC3 mouse insulinoma cells express RyR2 [25]. Mitchell et al. detected mRNA of both RyR1 and RyR2 but not of RyR3 in rat islets, mouse insulinoma MIN6 cells, and rat insulinoma INS-1 cells. Our results showed that primary human β-cells express RyR1 mRNA only at very low level. On the other hand, the most abundant RyR in the primary human α-cells was the RyR1, which was however expressed at a lower level compared to the RyR2 in the β-cells.

While all three RyRs can be activated by micromolar concentrations of Ca^{2+}, RyR2 is the one that is best known for physiological activation by Ca^{2+} entering through the voltage gated Ca^{2+} channels in the plasma membrane. Abundant expression of RyR2 in the β-cells as opposed to the α-cells, is consistent with the current views that CICR plays an important role in Ca^{2+} signaling in the β-cells [3, 5, 26]. Furthermore, in the β-cells, the RyR2-mediated CICR is facilitated by cAMP, and agonists like glucagon-like peptide 1, which increase cAMP [3, 25]. Glucagon-like peptide 1 and its analogues are frequently used in the treatment of type 2 diabetes.

We found that the expression of both the IP3Rs and the RyRs were highly variable in both the α-cell and the β-cell samples obtained from different donors. We speculate that the variability may partly be related to the procedures rather than the biology [14]. It should be noted that the human islets were collected after cold ischemia, and they were cultured prior to transport. These factors may reduce the yield of α-cells and β-cells after dissociation and FACS sorting, and may contribute to the variability of the RNA sequencing data from donor to donor.

It should be noted that we have estimated the relative abundance of the IP3R subtypes and the RyR subtypes in the human α-cells and β-cells only at the RNA level, and it remains a possibility that the expression of these channel subtypes at protein level may be different. At present, it is difficult to quantitatively estimate the relative abundance of these channel proteins because of difficulty in obtaining highly purified human α-cells and β-cells in sufficient amounts, and lack of sub-type specific antibodies against some of these channel proteins. Nevertheless, the comparison between the mRNA levels in the different cell types is still relevant and a strong indication of biological differences.

In summary, we have quantitatively identified the three subtypes of the IP3Rs and the RyRs in the human α-cells and β-cells, by analyzing the RNA sequencing data obtained from highly purified human α-cells and β-cells. Our results showed that the most abundant IP3R in the α-cells and the β-cells were the IP3R1 and the IP3R3 respectively; the most abundant RyRs in the α-cells and β-cells were the RyR1 and RyR2 respectively. Our results will be helpful in understanding the molecular mechanisms of generation of different types of Ca^{2+} signals and in regulating the Ca^{2+}-dependent processes in these two cell types of the human islets of Langerhans.

Acknowledgements Financial support was obtained from the Karolinska Institutet and the Uppsala County Council.

References

1. Brissova M, Fowler MJ, Nicholson WE, Chu A, Hirshberg B, Harlan DM et al (2005) Assessment of human pancreatic islet architecture and composition by laser scanning confocal microscopy. J Histochem Cytochem 53(9):1087–1097
2. Islam MS, Gustafsson AJ (2007) Islets of Langerhans: cellular structure and physiology. In: Ahsan N (ed) Chronic allograft failure: natural history, pathogenesis, diagnosis and management. Landes Bioscience, Austin, pp 229–232
3. Islam MS (2014) Calcium signaling in the islets. In: Islam MS (ed) Islets of Langerhans, 2nd edn. Springer, Dordrecht, pp 1–26
4. Islam MS (2010) Calcium signaling in the islets. Adv Exp Med Biol 654:235–259
5. Islam MS (2002) The ryanodine receptor calcium channel of beta-cells: molecular regulation and physiological significance. Diabetes 51(5):1299–1309
6. Chandrasekhar R, Alzayady KJ, Wagner LE 2nd, Yule DI (2016) Unique regulatory properties of heterotetrameric inositol 1,4,5-trisphosphate receptors revealed by studying concatenated receptor constructs. J Biol Chem 291(10):4846–4860
7. Taylor CW, Tovey SC (2010) IP(3) receptors: toward understanding their activation. Cold Spring Harb Perspect Biol 2(12):a004010
8. Santulli G, Lewis D, des Georges A, Marks AR, Frank J (2018) Ryanodine receptor structure and function in health and disease. Subcell Biochem 87:329–352
9. Holz GG, Leech CA, Heller RS, Castonguay M, Habener JF (1999) cAMP-dependent mobilization of intracellular Ca^{2+} stores by activation of ryanodine receptors in pancreatic beta-cells. A Ca^{2+} signaling system stimulated by the insulinotropic hormone glucagon-like peptide-1-(7-37). J Biol Chem 274(20):14147–14156
10. Gustafsson AJ, Islam MS (2005) Cellular calcium ion signalling–from basic research to benefits for patients. Lakartidningen 102(44):3214–3219
11. Hamilton A, Zhang Q, Salehi A, Willems M, Knudsen JG, Ringgaard AK et al (2018) Adrenaline stimulates glucagon secretion by Tpc2-dependent Ca^{2+} mobilization from acidic stores in pancreatic alpha-cells. Diabetes 67(6):1128–1139
12. Wang Z, Gerstein M, Snyder M (2009) RNA-Seq: a revolutionary tool for transcriptomics. Nat Rev Genet 10(1):57–63

13. Marabita F, Islam MS (2017) Expression of transient receptor potential channels in the purified human pancreatic beta-cells. Pancreas 46(1):97–101

14. Blodgett DM, Nowosielska A, Afik S, Pechhold S, Cura AJ, Kennedy NJ et al (2015) Novel observations from next-generation RNA sequencing of highly purified human adult and fetal islet cell subsets. Diabetes 64(9):3172–3181

15. Li B, Dewey CN (2011) RSEM: accurate transcript quantification from RNA-Seq data with or without a reference genome. BMC Bioinforma 12:323

16. Leng N, Dawson JA, Thomson JA, Ruotti V, Rissman AI, Smits BM et al (2013) EBSeq: an empirical Bayes hierarchical model for inference in RNA-seq experiments. Bioinformatics 29(8):1035–1043

17. Blondel O, Takeda J, Janssen H, Seino S, Bell GI (1993) Sequence and functional characterization of a third inositol trisphosphate receptor subtype, IP3R-3, expressed in pancreatic islets, kidney, gastrointestinal tract, and other tissues. J Biol Chem 268(15):11356–11363

18. De Smedt H, Missiaen L, Parys JB, Bootman MD, Mertens L, Van Den Bosch L et al (1994) Determination of relative amounts of inositol trisphosphate receptor mRNA isoforms by ratio polymerase chain reaction. J Biol Chem 269(34):21691–21698

19. Wojcikiewicz RJ (1995) Type I, II, and III inositol 1,4,5-trisphosphate receptors are unequally susceptible to down-regulation and are expressed in markedly different proportions in different cell types. J Biol Chem 270(19):11678–11683

20. Lee B, Bradford PG, Laychock SG (1998) Characterization of inositol 1,4,5-trisphosphate receptor isoform mRNA expression and regulation in rat pancreatic islets, RINm5F cells and betaHC9 cells. J Mol Endocrinol 21(1):31–39

21. Lee B, Laychock SG (2001) Inositol 1,4,5-trisphosphate receptor isoform expression in mouse pancreatic islets: effects of carbachol. Biochem Pharmacol 61(3):327–336

22. Alzayady KJ, Wagner LE 2nd, Chandrasekhar R, Monteagudo A, Godiska R, Tall GG et al (2013) Functional inositol 1,4,5-trisphosphate receptors assembled from concatenated homo- and heteromeric subunits. J Biol Chem 288(41):29772–29784

23. Vais H, Foskett JK, Mak DO (2010) Unitary Ca^{2+} current through recombinant type 3 InsP(3) receptor channels under physiological ionic conditions. J Gen Physiol 136(6):687–700

24. De Smet P, Parys JB, Vanlingen S, Bultynck G, Callewaert G, Galione A et al (1999) The relative order of IP3 sensitivity of types 1 and 3 IP3 receptors is pH dependent. Pflugers Arch 438(2):154–158

25. Islam MS, Leibiger I, Leibiger B, Rossi D, Sorrentino V, Ekstrom TJ et al (1998) In situ activation of the type 2 ryanodine receptor in pancreatic beta cells requires cAMP-dependent phosphorylation. Proc Natl Acad Sci U S A 95(11):6145–6150

26. Bruton JD, Lemmens R, Shi CL, Persson-Sjogren S, Westerblad H, Ahmed M et al (2003) Ryanodine receptors of pancreatic beta-cells mediate a distinct context-dependent signal for insulin secretion. FASEB J 17(2):301–303

Chapter 12
Evolution of Excitation-Contraction Coupling

John James Mackrill and Holly Alice Shiels

Abstract In mammalian cardiomyocytes, Ca^{2+} influx through L-type voltage-gated Ca^{2+} channels (VGCCs) is amplified by release of Ca^{2+} via type 2 ryanodine receptors (RyR2) in the sarcoplasmic reticulum (SR): a process termed Ca^{2+}-induced Ca^{2+}-release (CICR). In mammalian skeletal muscles, VGCCs play a distinct role as voltage-sensors, physically interacting with RyR1 channels to initiate Ca^{2+} release in a mechanism termed depolarisation-induced Ca^{2+}-release (DICR). In the current study, we surveyed the genomes of animals and their close relatives, to explore the evolutionary history of genes encoding three proteins pivotal for ECC: L-type VGCCs; RyRs; and a protein family that anchors intracellular organelles to plasma membranes, namely junctophilins (JPHs). In agreement with earlier studies, we find that non-vertebrate eukaryotes either lack VGCCs, RyRs and JPHs; or contain a single homologue of each protein. Furthermore, the molecular features of these proteins thought to be essential for DICR are only detectable within vertebrates and not in any other taxonomic group. Consistent with earlier physiological and ultrastructural observations, this suggests that CICR is the most basal form of ECC and that DICR is a vertebrate innovation. This development was accompanied by the appearance of multiple homologues of RyRs, VGCCs and junctophilins in vertebrates, thought to have arisen by 'whole genome replication' mechanisms. Subsequent gene duplications and losses have resulted in distinct assemblies of ECC components in different vertebrate clades, with striking examples being the apparent absence of RyR2 from amphibians, and additional duplication events for all three ECC proteins in teleost fish. This is consistent with teleosts possessing the most derived mode of DICR, with their $Ca_v1.1$ VGCCs completely lacking in Ca^{2+} channel activity.

J. J. Mackrill (✉)
Department of Physiology, School of Medicine, University College Cork, Cork, Ireland
e-mail: J.Mackrill@ucc.ie

H. A. Shiels
Division of Cardiovascular Sciences, Faculty of Biology, Medicine and Health, University of Manchester, Manchester, UK
e-mail: Holly.Shiels@manchester.ac.uk

© Springer Nature Switzerland AG 2020
M. S. Islam (ed.), *Calcium Signaling*, Advances in Experimental Medicine and Biology 1131, https://doi.org/10.1007/978-3-030-12457-1_12

Keywords Excitation-contraction coupling · Ryanodine receptor · Voltage-gated Ca^{2+} channel · Junctophilin · Evolution · Whole-genome duplication

12.1 Introduction

It is estimated that animals (Metazoa) evolved and diversified into major extant groups before the end of the Ediacaran Period, approximately 542 million years ago (Ma) [1]. Metazoa are taxonomically organised based on their degree of symmetry, Fig. 12.1. The body plan of radially symmetrical animals (Radiata) is circular and includes animals in the phyla Cnidaria and Ctenophora. Bilaterally symmetrical

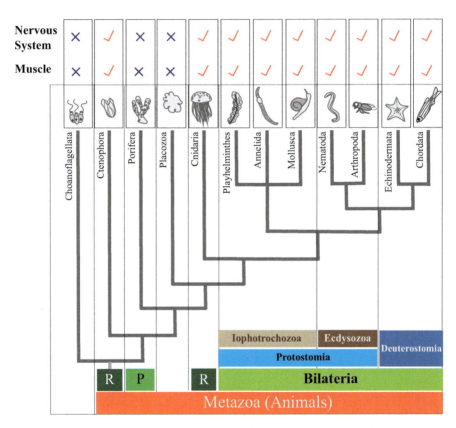

Fig. 12.1 Relationships between metazoan taxa in the context of excitation-contraction coupling (ECC). Diagrammatic representation of the relationships between major metazoan taxa and their sister group, Choanoflagellata. Also indicated is the presence of absence of nervous systems and muscles in these organisms: key players in ECC. Choanoflagellates are eukaryotes that are closely related to animals and can form colonies of cells. The relationships between basal metazoans, the Radiata ('R', ctenophores and cnidarians) and Parazoa ('P', poriferans) are unclear: shown is one consensus based on phylogenetic analysis of genomic data from these groups [36]

animals (Bilateria) have two symmetrical halves and are divided into three main groups that are differentiated by molecular and morphological characteristics: (1) the Ecdysozoa, which includes Arthropoda and Nematoda and have a three-layered cuticle that moults; (2) the Iophotrochozoa, a diverse group that includes the phyla Brachiopoda, Phoronida, Ectoprocta, Platyhelmintes, Nemertea, Mollusca and Annelida, who are characterised by the presence of either a lophophore, a crown of ciliated tentacles, or a distinct trocohophore larval stage (hence the name of this group: lopho + trocho); (3) the Deuterostoma, that include the phyla Echinodermata and Chordata who are distinguished by their embryonic development, during which the blastopore (the first opening) becomes the anus (as opposed to the mouth forming from the blastopore in the ecdysozoans and lophotrochozoans).

Molecular phylogenetic analyses indicate that certain metazoan lineages, the Poriferans (sponges) and Cnidarians (jellyfish, anemones and corals), emerged 750–800 Ma, during the Cryogenian Period [2]. Living cnidarians possess the characteristics that define animal life: multicellularity and the specialisation of particular tissue types. One innovation resulting from these developments was the evolution of specialised contractile myoepithelial and muscle cells, which generate force by interactions between ATP-dependent motors (myosins) and cytoskeletal components (actin) [2]. The concurrent development of the nervous system permitted the rapid regulation of force production by contractile tissues in response to intrinsic and extrinsic stimuli, such as neuronal inputs [3]. Regulation of force production in contractile cells initiated by such cues is termed excitation-contraction coupling (ECC). Prior to the development of actin-myosin dependent contractile machinery, eukaryotic cells moved exclusively using slower and weaker systems dependent on ciliary or flagellar beating. For example choanoflagellates, a closely related outgroup to metazoa, use the beating of their single flagellum to move and to ingest water and food. The development of ECC produced several selective advantages for metazoans, both in terms of capturing prey and of eluding predators [4].

12.1.1 Excitation-Contraction Coupling (ECC) in Various Muscle Types

Some features of ECC are conserved throughout animal evolution and in distinct types of muscle. In vertebrates, there are two main types of muscle: smooth and striated. Smooth muscle cells line the walls of hollow organs and are characterised by relatively slow contractions. They are defined by the net-like organisation of the contractile apparatus, which is anchored to structures called dense bodies. In contrast, cardiac and skeletal muscles are striated, due to an arrangement of the actin-myosin complexes within repeated units called sarcomeres, delimited by transverse structures called z-discs (also called z-lines).

As a second messenger, Ca^{2+} plays a key role in ECC in all muscle types. The rise of intracellular Ca^{2+} serves to activate the contractile apparatus in smooth muscle and in the cardiac and skeletal forms of striated muscle. The pathways subsequently diverge, with Ca^{2+} binding to the actin filament-associated troponin complex in striated muscles; but in smooth muscle, to calmodulin (CaM) forming an active complex which activates the myosin filament via myosin light chain kinase (MLCK). Ca^{2+} interactions with troponin, or activation of MLCK following binding of Ca^{2+}-CaM, initiates the cross-bridge cycling machinery and causes cell shortening, in line with the sliding filament theory of muscle contraction.

The excitatory mechanisms that raise intracellular Ca^{2+} and the effectors that transmit the chemical signal to mechanical force differ between muscle types. Cardiac- and skeletal muscle-type myosin heavy chain genes are thought to have arisen early during vertebrate evolution, prior to the divergence of Actinopterygian (ray-finned) and Sarcopterygian (lobe-finned) fish [5]. Furthermore, phylogenomic analyses indicate that the contractile machinery of striated muscles evolved independently in bilaterian and non-bilaterian animals. Radiata (cnidarians and ctenophores) lack key proteins of bilaterian striated muscle contractile machinery, such as components of the troponin complex. However, poriferans and non-metazoan eukaryotes possess homologues of other sarcomeric contractile proteins, such as muscle-type actins and myosins, even though they do not have muscle cells. Indeed, Ca^{2+}-stimulated, actin-myosin dependent contraction occurs in several single-celled eukaryote taxa, including amoebozoans (eg. *Amoeba, Dictyostelium*) and basal plants (eg. *Volvox*) [6]. This suggests that the metazoan contractile apparatus was derived by addition of new proteins to pre-existing actin-myosin force-producing machinery and that this occurred independently, in at least two animal lineages [7].

A key step in ECC is the delivery of sufficient quantities of Ca^{2+} to the vicinity of the contractile apparatus, to initiate and maintain cross-bridge cycling. Regardless of whether this Ca^{2+} is extracellular in origin or is released from intracellular stores, a major consideration is that Ca^{2+} is heavily buffered within the myoplasm. Buffering occurs through interactions of Ca^{2+} with anionic proteins and other large, negatively charged biomolecules. As a result, Ca^{2+} diffuses through cytoplasm about ten times more slowly than it does through water [8]. This places constraints on the distances within cells over which Ca^{2+} is able to operate effectively as a second messenger. This limitation has been circumvented in several ways during the evolution of metazoans, including the employment of muscles of small diameter (limiting distances over which Ca^{2+} has to diffuse), or of invaginations of the surface membrane called T-tubules. Slow movement of Ca^{2+} within the myoplasm is also a strength of this second messenger system, as it lends itself toward compartmentalisation, thus providing the scope for one signal to initiate multiple and distinct reactions separated in space and time [9].

12.1.2   Two Modes of Excitation-Contraction Coupling in Mammalian Striated Muscles

Mammalian cardiomyocytes use a mechanism of ECC considered to have developed early during animal evolution [10, 11]. Propagating action potentials, initiated in nodal cells, are detected by a constituent of L-type voltage-gated Ca^{2+} channel complexes (VGCCs) known as $Ca_v1.2$, or the $\alpha1C$ subunit of the cardiac dihydropyridine receptor [12]. This sarcolemmal protein acts as both a voltage-sensor and a Ca^{2+} channel, allowing this second messenger to enter the myoplasm [13]. Other components of L-type VGCC complexes are involved in modifying the electrophysiological properties and trafficking of these channels; and include the β, $\alpha2\delta$ and γ subunits [14], see Sect. 12.2.

In mammalian cardiomyocytes, the quantity of Ca^{2+} influx via Ca_v1x channels (where 'x' means an undefined member of the Ca_v1 family) is insufficient to initiate contraction, and so it must be amplified by additional Ca^{2+} release from the sarcoplasmic reticulum (SR). This release occurs via an SR cation channel called the type 2 ryanodine receptor (RyR2), that is activated by Ca_v1x-dependent entry of this ion in a process termed Ca^{2+}-induced Ca^{2+}-release (CICR) [15]. Communication between $Ca_v1.2$ (plus $Ca_v1.3$) and RyR2 occurs at junctions between SR and sarcolemmal membranes, forming Ca^{2+} Release Units (CRUs) [16]. CRUs are found at the cell surface where the sarcolemmal and SR membrane systems interact in regions known as peripheral couplings. CRUs also exist at "dyadic" junctions between the SR and infoldings of the sarcolemma membrane called transverse- (T-) tubules. These invaginations allow membrane depolarisation to propagate deep into the myoplasm, circumventing the limitations on signalling imposed by Ca^{2+} buffering and permitting the development of cardiomyocytes of high cross-sectional area, capable of generating large forces [17]. However, not all cardiomyocytes possess well-developed T-tubular systems. For example, in mammals T-tubules can be completely absent or variable in density in adult atrial myocytes [18], and are lacking in all cardiomyocyte types of embryonic and newborn animals [19]. Moreover, T-tubules are absent in both the atria and the ventricle of non-mammalian vertebrates, namely aves, non-avian reptiles, the amphibians and fish [20, 21]. In these cases, ECC occurs at the peripheral couplings where VGCC and RyR2 channels form CRUs at the surface membrane. Here, and in mammalian myocytes without T-tubules, CICR-dependent propagation of Ca^{2+} waves travelling through the myoplasm are amplified by RyRs located within non-junctional, or "corbular" SR. These extra-junctional RyRs effectively act as "relay stations", propagating Ca^{2+} waves deeper into the myoplasm than would be practical by diffusion alone [20, 166].

In mammalian skeletal muscle fibres, ECC is initiated by nerve impulses at the end-plates of motor neurons. These action potentials propagate into myofibres via T-tubules, where a distinct member of the L-type VGCC family, based on the $Ca_v1.1$ protein, is enriched [21]. Instead of acting as Ca^{2+}-influx channels, myoplasmic regions of $Ca_v1.1$ communicate directly with RyR1 channels located in the terminal cisternae of the SR. This allosteric, voltage-induced, or depolarisation-induced Ca^{2+} release (DICR) mechanism involves long-range protein-protein interactions, spanning the 10–12 nm gap between the T-tubule and apposed terminal cisternae SR [22]. Within myofibres, T-tubules are usually associated with paired termini of the SR, forming CRUs at "triad" junctions, see Fig. 12.2. Based on ultrastructural observations, DICR in skeletal muscle is an innovation that occurred either in pre-vertebrates or early during vertebrate evolution [10]. This mechanism displays a number of distinctions from the CICR-based ECC employed in cardiomyocytes: (1) no direct requirement for extracellular Ca^{2+} and influx of this ion (a voltage-dependent conformational change in the VGCC is communicated to the SR Ca^{2+} release channels); (2) faster kinetics, due to the lack of a requirement of triggering Ca^{2+} to diffuse through the junctional myoplasm; and (3) use of different members of the VGCC ($Ca_v 1.1$ versus $Ca_v 1.2$) and Ca^{2+} release channel (RyR1 versus RyR2) families.

In vertebrate skeletal muscle, not all RyR1 channels are directly coupled to $Ca_v 1.1$ voltage-sensors. Ultrastructural data from toadfish skeletal myocytes indicate that alternate RyRs (which appear as "feet structures" by transmission electron microscopy) are directly coupled to "tetrads", likely to represent groups of four VGCC complexes [22]. Consequently, non-coupled RyR channels would have to be activated by an additional mechanism, with CICR triggered by Ca^{2+} release from neighbouring coupled RyR channels being the strongest candidate. Although present at low abundance in most mammalian muscle tissues, a third RyR type (RyR3) is thought to play a role in amplifying RyR1-mediated Ca^{2+} signals in some skeletal muscle types, via CICR at extra-junctional sites. This amplification of DICR occurs in neonatal skeletal myocytes [23] and in a subset of muscles of cephalic or somitomeric origin in adult mice [24]. RyR3 amplification of the Ca^{2+} signal is also present the heart; not in the working atrial and ventricular myocytes, but in cells of the conduction system [25].

In addition to the molecular communication between RyR and $Ca_v 1x$ channels, multiple other protein-protein interactions modulate the process of ECC in the striated muscles of mammals and other vertebrates [11]. These accessory proteins are associated with the sarcolemma (β-subunits of VGCCs), the myoplasm (calmodulin, FK506-binding proteins, protein kinases and phosphatases), the SR membrane (triadins, junctophilins, kinase anchor proteins, selenoprotein N) or the SR lumen (calsequestrin, histidine rich Ca^{2+} binding protein) [26]. Recently, it has been demonstrated that five proteins are essential for *de novo* reconstitution of the DICR-mode of ECC in a non-muscle mammalian cell-line: $Ca_v 1.1$, RyR1, the β_{1a}-subunit of the VGCC, SH3 and cysteine rich domain 3 (STAC3) protein and junctophilin 2 (JPH2) [27]. Junctophilins are SR/ER integral membrane proteins that are essential for the formation of the junctions (peripheral, dyadic or triadic) between the SR and sarcolemmal membranes [28]. They hold the CRUs in register and thus together with the Ca^{2+} cycling proteins, are important for CICR and DICR, in facilitating ECC.

12.1.3   Aim of This Study

In order to simplify reconstruction of the evolutionary history of ECC, we have focused on three proteins that are essential for the formation of CRUs in striated muscles: the Ca_v1x subunit of L-type VGCCs; the RyR Ca^{2+} release channels; and the junctophilins, Fig. 12.2. Many of the observations in this study come from reviewing the literature. However, we also include significant novel *in silico* findings (Figs. 12.3, 12.4, 12.5, 12.6, and 12.7), where we test the hypotheses on the evolution of ECC gleaned from published work. The mode of ECC in vertebrate and invertebrate smooth muscle differs significantly from that in striated muscle and

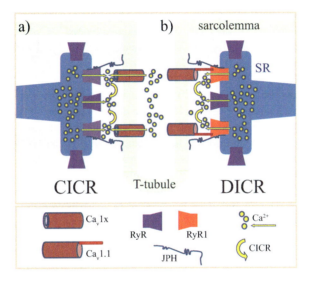

Fig. 12.2   Distinct modes of excitation-contraction coupling in mammalian striated muscles. Cartoon depicting the two modes of ECC employed within mammalian striated muscles: (**a**) Ca^{2+}-induced Ca^{2+} release (CICR) involves amplification of a small influx of Ca^{2+} via Ca_v1x VGCCs by calcium release via RyR channels in the SR [13, 15]. In mammals, this archetypally occurs at the dyad junctions of cardiomyocytes, but operates in other membranes and cell types, including the subplasmalemmal junctions of neurons [161] and of smooth muscle cells [29]. In the mammalian heart, junctophilin-2 (JPH2) plays critical roles in forming membrane junctions and in regulating communication between $Ca_v1.2$ and RyR2 [140, 141, 147]. This mode of ECC is evolutionarily most basal, with all of the necessary signalling protein components being present in choanoflagellates, a sister group to animals; (**b**) Depolarisation-induced Ca^{2+} release (DICR) in mammalian skeletal muscle involves direct, physical communication between $Ca_v1.1$ voltage-sensors in the T-tubule and RyR1 in the junctional SR. These interactions take place at triad junctions, whose formation is facilitated by both JPH1 and JPH2. This mode of signalling might operate in other mammalian cell types, including neurons [161]. In some cases, Ca^{2+} signals produced by DICR are amplified by CICR from RyRs that are not physically coupled to VGCCs [20, 22–24]. In terms of evolution, DICR appears to have arisen most recently and is probably a vertebrate innovation [10]

thus is not included in the subsequent analysis. For a detailed review of the smooth muscle literature *please see* [29].

12.2 The Evolutionary History of L-Type Voltage-Gated Ca^{2+} Channels (VGCC)

VGCC α subunits are part of a four-domain voltage-gated cation channel (FVCC) superfamily: proteins that share a common topology of four repeats, each containing six transmembrane helices, connected by extracellular and cytoplasmic loops [30]. The FVCC superfamily is comprised of five families: voltage-gated sodium channels (Na_v); Ca^{2+} channels that are activated by large changes in membrane potential ("high voltage activated" (HVA)); "low-voltage activated" (LVA) Ca^{2+} channels; sodium leak channels; and fungal Ca^{2+} channels [31]. Mammalian VGCCs consist of ten members grouped into three families, with Ca_v3x being LVA; and Ca_v1x and Ca_v2x being HVA. Of these, Ca_v1x forms L-type Ca^{2+} channels, characterised by their long open durations ("L-type") and sensitivity to dihydropyridine drugs [32]. There are four members of the Ca_v1x family in mammals, with $Ca_v1.1$ underlying ECC in skeletal muscle and $Ca_v1.2$ playing the predominant role in this process in cardiomyocytes. In addition to their channel-forming and voltage-sensing $\alpha1$ subunit, L-type VGCCs complexes also contain β and $\alpha2\delta$ subunits, along with a γ subunit in the case of $Ca_v1.1$ assemblies. These accessory proteins serve roles in channel complex trafficking and/or can modify VGCC electrophysiological properties [32]. The evolutionary histories of these accessory proteins have been described in detail elsewhere and will not be addressed further in the current work [33–35].

Published phylogenetic surveys have indicated that members of the FVCC superfamily were present in the common ancestor of all extant eukaryotes and subsequently diversified into five families, including the HVA (Ca_v1x and Ca_v2x) Ca^{2+} channels [31, 36]. Reconstruction of the evolution of the Ca_v1x family has been problematic, particularly because of controversies surrounding the relationships between early emerging animal groups [34, 36, 37]. Choanoflagellates are a sister group of metazoans, that morphologically resemble the choanocytes (flagellated "collar cells") of sponges. The genome of one member of this group, *Salpingoeca rossetta*, encodes one HVA and one LVA Ca_v homologue [33].

12.2.1 VGCC in Metazoan Invertebrates

Trichoplax adhaerens is the sole known living representative of the phylum Placozoa. Anatomically, these animals are the simplest that have been studied, comprising of only six cell types; lacking in muscle cells and in electrical or chemical synapses [38]. Despite this apparent minimalism, the *T.adhaerens* genome encodes one member of each VGCC subfamily (Ca_v1x, Ca_v2x, Ca_v3x) and multiple homologues of proteins involved in neurotransmission [39]. Possession of Ca_v1x channels without the necessary cell-types to perform ECC might be resolved using similar rationales employed to explain paradoxes in the evolution of the actin-myosin contractile apparatus: such proteins were present prior to the development of these processes, but served distinct roles [7]. Recent investigation of *T.adhaerans* behaviours lends support to this hypothesis: this placozoan can alter its ciliary-based motility in response to gravity, or to neuropeptides, despite having no nervous system. These sensory responses utilise specialised crystal cells [40], or neurosecretory cells [41], respectively. In the latter case, the Ca_v3x homologue localises to secretory cells, suggesting that it plays a positive feedback role in secretion of neuropeptides: these neurotransmitter-like molecules could activate VGCCs, stimulating Ca^{2+} influx, which would elicit further exocytosis [42]. Elucidation of the roles of Ca_v1x channels in this organism could bring new insights into the evolution of ECC.

The sponge *Amphimedon queenslandica* possesses one gene encoding a HVA VGCC, which is intermediate between Ca_v1x and Ca_v2x in terms of its primary structure [33]. The ctenphore *Mnemiopsis leidyi* (the warty comb jelly) also has one HVA homologue, which is most closely related to vertebrate Ca_v2x proteins [33]. Some cnidarians, such as the starlet anemone *Nematostella vectensis* and the coral *Acropora millepora*, possess one Ca_v1x homologue, three members of the Ca_v2x family and two genes encoding Ca_v3x channels. In contrast, the hydra *Hydra magnipapillata*, another cnidarian, has only a single member of each family. This indicates expansion of Ca_v families in some cnidarian groups but not in others, presumably by lineage-specific gene duplication. Using *in situ* hybridisation microscopy, the mRNA encoding the Ca_v1x homologue was detected at greatest abundance in the muscle and/or motor neurons of *N.vectensis*, in keeping with a candidate role in ECC [33].

12.2.1.1 Protostomes

Among bilaterians (animals with bilateral symmetry), all protostomes (including arthropods, molluscs and worms) investigated have one representative of each of the three Ca_v subfamilies (Ca_v1x, Ca_v2x and Ca_v3x) [33, 35]. Nematode worms lack a heart and cardiomyocytes. They use either obliquely striated muscles in their body walls, or smooth muscles in other contractile organs. The contractile apparatus of obliquely striated muscles is anchored to staggered dense bodies, rather than to transverse z-discs such as those used by vertebrates [43]. The surface membrane of myofibres from these worms lacks T-tubules, but interacts with subsarcolemmal junctional SR to form candidate CRUs [44]. The model nematode *Caenorhabditis elegans* has a single Ca_v1x protein, encoded by the *egl-19* gene. Complete loss-of-function mutations in this gene cause embryonic lethality; partial loss-of-function mutants display hypotonia (flaccidity); and gain-of-function mutant worms are myotonic, with impaired relaxation of the body [45, 46]. These observations strongly support an essential role for Ca_v1x VGCCs in nematode ECC.

All arthropod muscles are striated, with the contractile apparatus anchored to transverse z-discs [43]. Observations using transmission and freeze-fracture electron microscopy indicate that flight muscles from dragonflies, pedipalp muscles from a scorpion, or leg muscle from a fly, contain T-tubules. The SR-facing regions of these membranes bear large proteinaceous particles, resembling those which are accepted to be L-type VGCCs in vertebrate T-tubules [47]. In the fruit fly *Drosophila melanogaster*, a Ca_v1x protein is encoded by the *Dmca1D* gene and forms dihydropyridine-sensitive Ca^{2+} channel, required for contraction of the body wall muscles [48]. The *D.melanogaster* circulatory system is powered by a heart tube, which has both anterior and posterior pacemaker regions. Action potentials propagate from these sites in a manner that is dependent on Ca^{2+} influx through dihydropyridine-sensitive channels [49]. Biochemical, pharmacological and electrophysiological investigations indicate that crayfish striated muscle [50], isopod abdominal extensor muscles [51], and locust leg muscle [52] all possess dihydropyridine-sensitive Ca^{2+}-conducting L-type VGCCs. This suggests that ECC dependent on Ca^{2+}-influx through Ca_v1x channels, augmented to varying degrees by CICR from the SR, is a common feature of arthropod muscles. This is analogous to the type of ECC that occurs in vertebrate cardiomyocytes.

In contrast, the superficial abdominal flexor muscles of the shrimp *Atya lanipes* are electrically inexcitable and during contraction, Ca^{2+} influx is too small to be measured electrophysiologically. However, this contraction is strictly dependent on extracellular Ca^{2+} and could be modulated by agonistic (BayK 8644) or antagonistic (nifedipine) dihydropyridines [53, 54]. The proposed solution to these paradoxical observations is that "silent" L-type VGCCs allow an undetectable influx of Ca^{2+} into myocytes, which is amplified by CICR from the SR by a mechanism with extremely high gain. Consistent with this proposal, *A.lanipes* skeletal muscle has T-tubules, abundant SR and contracts in response to the RyR agonist caffeine [54]. Alternatively, this might represent development of the DICR mode of ECC in a relatively early branching group of bilaterians, with its dependency on extracellular Ca^{2+} being due to a requirement to replenish depleted intracellular Ca^{2+} stores, rather than due to a direct dependence on Ca^{2+} influx.

Annelids and molluscs use transversely striated, obliquely striated or smooth myocytes in different organ systems [43]. L-type Ca^{2+} channels have been reported in atrial cells from the Pacific oyster, *Crassostrea gigas* [55]. Ventricular myocytes of the pond snail *Lymnaea stagnalis* are myogenic, displaying rhythmic depolarization and contractions independently of neural inputs. These mollusc cardiomyocytes contain HVA and LVA VGCCs, which contribute to both pace-making and to ECC [56]. Heterologous expression and electrophysiological analysis of the Ca_v1x protein from *L.stagnalis* indicate that it has very similar electrophysiological properties to mammalian $Ca_v1.2$, with the most notable difference being a reduced sensitivity to dihydropyridine drugs [57]. Contraction of smooth muscle cells from the sea-slug *Aplysia kurodai* can be activated by acetylcholine or high extracellular K^+, via a mechanism that could be inhibited by nifedipine [58]. Similarly, contraction of the obliquely striated muscle of *Octopus vulgaris* arms is dependent on rapidly activating and slowly inactivating currents carried by L-type VGCCs [59]. These findings demonstrate that Ca_v1x-dependent ECC, resembling that in vertebrate cardiomyocytes, is a common feature of molluscan muscles.

12.2.1.2 Deuterstomes

Deuterstomes are a major group of bilaterian animals that includes echinoderms (sea urchins, starfish, sea cucumbers and their relatives), hemichordates (such as acorn worms) and chordates (cephalochordates, tunicates and craniates). Of these, echinoderms use a distinctive mode of locomotion: walking on "tube-feet" (podia) dependent on hydrostatic forces powered by the contraction of smooth muscle-like cells of myoepithelial origin [60]. Although Ca^{2+} plays a key role in the contraction of these myocytes, there are no available data on the presence and function of L-type VGCCs in echinoderm contractile tissues [61]. However, Ca_v1x and Ca_v2x have been detected in sea urchin sperm, at the level of both mRNA and protein [62]. We were unable to find any published data on the role of L-type VGCC in the ECC of hemichordate muscles.

Chordates are bilaterian animals that possess a notochord: a flexible rod that runs along the long-axis of the body. Cephalochordates are a free-living, aquatic subphylum of chordates that superficially resemble fish or worms. They are exemplified by *Branchiostoma sp.*, also known as lancelets or amphioxus. Like vertebrates, the body wall muscles of lancelets are striated and arranged in repeated segments called somites [63]. In the current study, we identified a single Ca_v1x homologue in the genome of *B. blecheri*. Lancelet somites are thin fibres, meaning that Ca^{2+} influx through VGCCs might be adequate for ECC [64]. However, electrophysiological and biochemical investigations indicate that *B. lanceolatum* trunk muscles possess dyad junctions and under certain conditions, can contract in the absence of extracellular Ca^{2+} [65, 66]. This suggests that the DICR mode of ECC could have evolved early in the chordate lineage [10].

Tunicates, or sea quirts, are a sister group to vertebrates and to cephalochordates, with a free-swimming larval stage and a sessile adult form. They possess skeletal, cardiac and smooth muscle types. Tunicate genomes encode a single Ca_v1x homologue, which displays low sensitivity to dihydropyridines [67]. There appears to be considerable diversity in the modes of ECC employed by tunicates. For example, muscle fibres from the body wall of adult *Doliolum denticulatum* lack T-tubules and an SR system, with their contraction being ablated by the dihydropyridine, nifedipine [68]. In contrast, caudal muscles from *Botryllus schlosseri* (the golden star tunicate) larvae possess a highly developed T-tubule system that forms dyad junctions with the SR [69].

12.2.2   VGCC in Metazoan Vertebrates

The two-rounds of whole genome duplication (2R-WGD) process, originally postulated by Susumu Ohno, describes a mechanism that facilitated the rapid evolution
and expansion of vertebrates [70]. Gene duplication by polyploidy, and the resulting
potential for functional diversification of duplicated genes, permitted rapid increases
in the complexity and adaptability of this group. Phylogenetic and cytogenetic
data support the concept of these rounds of genomic duplication occurring prior to
the divergence of agnathan (jawless fish, including the cyclostomes) and gnathos-
tome (jawed) vertebrates [71]. The 2R-WGD mechanism probably permitted the
innovation of the DICR-mode of ECC in skeletal muscle, either prior to, or early
during, the evolution of vertebrates. Cyclostomes are the most basal of extant
vertebrates, comprising of hagfish and lampreys. Depolarisation-evoked contraction
of caudal hearts (i.e. accessory hearts, analogous to those of the lymphatic system),
slow-twitch and fast-twitch skeletal muscle from the inshore hagfish, *Eptatretus
burgeri*, persists in nominally Ca^{2+}-free seawater, indicating a DICR mode of
ECC [72]. Electrically-stimulated twitches in skeletal muscle fibres from the river
lamprey, *Lampetra planeri*, were maintained in the presence of levels of colbalt ions
that blocked Ca^{2+} influx via dihydropyridine-sensitive channels, again providing
evidence for a DICR mechanism [73]. Freeze-fracture electron microscopy of
skeletal muscle from the Pacific hagfish *E. stoutii* and the river lamprey, *L. planeri*,
indicates that membrane particles corresponding to L-type VGCC complexes are
arranged in tetrads [10], a feature thought to depend on physical interactions with
underlying RyRs at CRUs [74]. Despite such physiological evidence of a DICR
mode of ECC in cyclostome skeletal myocytes, we were only able to detect partial
homologues of a Ca_V1 VGCC encoded in the genome of the sea lamprey *Petromyzon
marinus*. Furthermore, this homologue displays greatest identity with $Ca_V1.2$, rather
than $Ca_V1.1$. However, failure to detect homologues of skeletal muscle-type $Ca_V1.1$
in *P.marinus* might be a consequence of incomplete sequencing or poor annotation
of its genome.

   Amphibians, non-avian reptiles, birds and mammals generally have four Ca_V1x
homologues ($Ca_V1.1$ to $Ca_V1.4$), probably generated by the 2R-WGD mechanism.
This could have permitted functional specialisation of the $Ca_V1.1$ type in skeletal
muscles, allowing it to mechanically couple to RyR1 in the SR. In mammalian
skeletal muscle, Ca^{2+} influx via VGCCs is not required for ECC, since mice
engineered with a non-conducting mutant of $Ca_V1.1$ (N617D) do not show any
overt changes in skeletal myocyte function [75]. Teleosts are the largest group
within the class Actinopterygii (ray-finned fish), that underwent an additional third
round of WGD between 320 and 350 Ma [76]. This facilitated the development
of new innovations, such as two non-conducting homologues of $Ca_V1.1$ in the
skeletal muscles of teleost fish, including the zebrafish (*Danio rerio*), pike characin
(*Ctenolucius hujeta*), and rice medaka (*Oryzias latipes*) [77]. Salmonids may be
of particular interest in relation to functional specialisation of VGCCs, because of

an additional and relatively recent WGD event (the salmonid-specific 4R-WGD or Ss4R) that has been dated 25–100 Ma [78].

The sterlet *Acipenser ruthenus* is a basal ray-finned sturgeon, whose $Ca_v1.1$ homologue displays a Ca^{2+} conductance intermediate between that of mammals and teleosts. Together with phylogenetic evidence, this suggests that teleost non-conducting $Ca_v1.1$ voltage-sensors are the most derived forms, with Ca^{2+}-conducting $Ca_v1.1$ channels of mammals and other sarcopterygian (lobe-finned fish) descendants being the most basal [79].

12.3   Evolution of Ryanodine Receptor (RyR) Ca^{2+} Release Channels

Along with inositol 1,4,5-trisphosphate receptors (ITPRs), the ryanodine receptors (RyRs) form a superfamily of tetrameric, ligand-gated cation channels that release Ca^{2+} from intracellular storage organelles including the ER and SR [26]. ITPR homologues are encoded in the genomes of most eukaryotes, but have apparently been lost by some groups, such as late branching fungi and green plants [80]. The Ca^{2+} release channel superfamily is subdivided between the ITPR-A and ITPR-B families, the latter of which is probably ancestral to the RyRs [81]. Proteins related to RyRs, both in terms of sequence identity and protein domain architecture, first appeared over 950 Ma, prior to the divergence of metazoa and their sister groups choanoflagellata and filasterea [80].

12.3.1   RyRs of Invertebrates

Among animals, RyR homologues have been detected in the genomes of sponges such as *Amphimedon queenslandica*, *Oscarella carmela* and *Sycon ciliatum* [81], though not in ctenophores or cnidaria, Fig. 12.3. If sponges are basal metazoans [33, 82], this implies that RyRs were subsequently lost from ctenophore and cnidarian lineages. However, the striated swimming muscles of the hydrozoan jellyfish *Aglantha digitale* [83] and of the giant smooth muscles of the ctenophore *Beroe ovata* [84] are reported to contain an SR system. This indicates that either RyRs have not been detected in cnidarian and ctenophore genomes because of poor sequence quality or incomplete annotation; or that Ca^{2+} might be released from the SR of these organisms by a distinct mechanism, such as CICR via ITPRs. Of these possibilities, we consider the latter to be more likely, given the increasing numbers of cnidarian and ctenophore species whose genomes have been deduced. Furthermore, contraction of radial muscles from the jellyfish *Polyorchis penicillatus* is abolished by low extracellular Ca^{2+} or the Ca_v1x antagonist nitrendipine, but is also reduced by 25% in the presence of 10 mM caffeine [85]. In addition to its use as

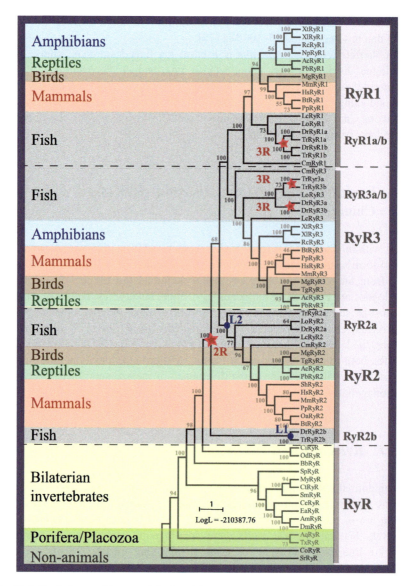

Fig. 12.3 Evolutionary history of RyR calcium release channels. The evolutionary history of proteins related to RyR2 from *Homo sapiens* (Accession Number Q92736.3) was inferred using the Maximum Parsimony method, conducted using MEGA5 [162]. The percentage of trees (from 500 replicates) in which taxa are clustered was determined by using the bootstrap method, with 50% being taken as significant [163]. Bootstrap values are indicated next to branches. Also shown is a scale-bar of the number of amino acid substitutions per position and the logarithmic likelihood value for this consensus tree. Red stars ('2R' and '3R') indicate potential whole genome duplications; blue circles ('L1' and 'L2') candidate gene losses. This analysis involved 68 proteins (full details available from the corresponding author, upon request). Species analysed were mammals: *Homo sapiens* ('Hs', human) *Mus musculus* ('Mm', house-mouse), *Bos taurus* ('Bt', cattle) and *Panthera pardus* ('Pp', leopard); birds: *Meleagris gallopavo* ('Mg', turkey) and

an RyR agonist, caffeine is also an antagonist of ITPRs [86]. Since this drug did not stimulate contraction of *P.penicillatus* striated muscles in its own right, its effects were probably mediated by ITPR antagonism, rather than depletion of intracellular stores by sustained activation of RyR channels.

12.3.1.1 Protostomes

Most bilaterian animal groups possess a single RyR homologue, most likely operating via CICR [80]. The extent of the contribution of Ca^{2+} released in this manner to contraction varies between organisms, tissues and muscle types. Platyhelminthes, or flatworms, are considered to be basal bilaterians: they lack a body cavity, do not possess a specialised cardiovascular system and contain two layers of smooth muscle-like contractile tissue in their body walls. The plant alkaloid ryanodine is specific, high-affinity and irreversible ligand for RyRs [87], that can only bind to these channels when they are in an open state. Consequently, [^{3}H]ryanodine binding has been employed to gauge the open state of RyR channels in response to physiological and pharmacological stimuli [26, 88]. Membrane vesicles prepared from the trematode flatworm *Schistosoma mansoni* contain binding sites for ryanodine, whose abundance could be increased by the RyR agonist caffeine and decreased by the antagonist dantrolene [88]. Both release of $^{45}Ca^{2+}$ from microsomal membranes and contraction of the body wall muscle of *S.mansoni* were enhanced by ryanodine [89]. Caffeine could stimulate body wall contraction in *S.mansoni* and in the turbellarian flatworms *Dugesia tigrina* and *Procerodes littoralis*, in a manner that was inhibited by high concentrations of ryanodine, although not by other characteristic antagonists of mammalian RyRs, neomycin and ruthenium red [90]. Along with phylogenomic data (Fig. 12.3), these observations suggest that a single RyR homologue contributes to muscle contraction in flatworms, and that their Ca^{2+} release channels are pharmacologically distinct from their vertebrate counterparts.

Fig. 12.3 (continued) *Taeniopygia guttata* ('Tg', zebrafinch); reptiles: *Anolis carolinensis* ('Ac', anole) and *Python bivittatus* ('Pb', python); amphibians: *Xenopus tropicalis* ('Xt', Western clawed frog), *Xenopus laevis* ('Xl', African clawed frog), *Rana catesbeiana* ('Rc', American bullfrog) and *Nanorana parkeri* ('Np', High Himalayan frog); fish: *Latimeria chalumnae* ('Lc', coelacanth), *Lepisosteus oculatus* ('Lc', spotted gar), *Danio rerio* ('Dr', zebrafish), *Takifugu rubripes* ('Tr', Japanese pufferfish) and *Callorhinchus milii* ('Cm', elephant shark); tunicates: *Ciona intestinalis* ('Ci', a sea squirt) and *Oikopleura dioica* ('Od'); cephalochordate: *Branchiostoma belcheri* ('Bb', Belcher's lancelet); echinoderm: *Strongylocentrotus purpuratus* ('Sp', Atlantic purple sea urchin); mollusc: *Mizuhopecten yessoensis* ('My', Yesso scallop); annelid: *Capitella teleta* ('Ct'); arthropods: *Eurytemora affinis* ('Ea', marine copepod), *Apis mellifera* ('Am', European honeybee) and *Drosophila melanogaster* ('Dm', fruit-fly); nematode: *Caenorhabditis elegans* ('Ce'); platyhelminth: *Schistosoma mansoni* ('Sm'); porifera: *Amphimedon queenslandica* ('Aq', sponge); placozoa: *Trichoplax adhaerens* ('Tr'); choanoflagellata: *Salpingoeca rosetta* ('Sr'); and filasterea *Capsaspora owczarzaki* ('Co')

Nematode worms are coelomate (possess a body cavity), but lack a specialised circulatory system. Deletion of the single RyR homologue (*unc-68*) in the model nematode worm *C.elegans* results in incomplete flaccid paralysis, indicating that this channel is not essential for, but is supportive of, body muscle ECC [91]. Contraction of body wall muscle from *Ascaris suum* (large roundworm of the pig), stimulated by the cholinergic ant-helminthic drug levamisole, was abolished in Ca^{2+}-free extracellular medium and inhibited by 44% by 100 nM ryanodine [92]. This might suggest that RyRs are non-essential for nematode ECC, but might augment this process under some circumstances. Alternatively, the *A.suum* RyRs might display a very low open probability under these experimental conditions, such that only a small proportion of them would be open and available to bind to ryanodine.

RyR-mediated Ca^{2+} release appears to play a critical role in arthropod ECC. One of the earliest reports of a physiological effect of ryanodine was an increase in oxygen consumption in the cockroach *Periplaneta americana*, probably due to enhanced muscle contraction and ATP utilisation [93]. An insertional mutagenesis strategy was used to demonstrate that a single RyR homologue is essential for ECC in the body wall, visceral and circulatory muscles of *D.melanogaster* [94]. In this fruit fly, high concentrations of ryanodine exerted a negative chronotropic effect on the cardiovascular system, slowing heart rate by about 60% [95]. Tritiated ryanodine binding to membranes prepared from thoracic tissues of the moth *Heliothis virescens* is biphasically dependent on Ca^{2+}, stimulated by ATP, but unaffected by caffeine. Single channel electrophysiology of RyR complexes isolated from the same tissue demonstrated a cation channel that was activated by Ca^{2+} and ATP; inhibited by ruthenium red; and modified by ryanodine [96]. In the honeybee, *Apis melifera*, both caffeine and 4-chloro-*meta*-cresol (another RyR agonist) could stimulate skeletal muscle contraction [97]. Overall, these studies support the presence of a single RyR homologue in insect muscles, which displays some characteristics that are similar to their vertebrate relatives, as well as some that are distinct. Such pharmacological differences have prompted the commercial development of insecticides selectively targeting insect RyRs, as a novel strategy for pest control [98].

Photolysis of a caged Ca^{2+} molecule, nitr-5, was used to directly demonstrate CICR in muscle fibres of the giant barnacle *Balanus nubilus* [99]. In skeletal muscle fibres from the crayfish *Procambarus clarkia*, the RyR antagonists procaine and tetracaine inhibit Ca^{2+} transients triggered by depolarisation, without decreasing Ca^{2+} currents across the sarcolemma [100]. Unlike in vertebrates, muscle from the intestine of this crustacean is striated and also uses a CICR-mode of ECC, as indicated by its dependence on extracellular Ca^{2+} and inhibition by ryanodine [101]. A RyR protein isolated from abdominal muscles the lobster *Homarus americanus*, displays single-channel properties that are similar to their vertebrate counterparts, with the exception of relative insensitivity to activating Ca^{2+}, or to the agonist caffeine [102, 103]. These studies demonstrate that arthropod muscles use Ca^{2+} influx to activate contraction, augmented to varying degrees by the release of Ca^{2+} from the SR via RyRs. This is further supported by ultrastructural observations of insect and arachnid muscles, which display organised "feet structures" (presumed to be RyRs) at junctions with the sarcolemma, which lack overt tetrad organisation of the overlying membrane particles (thought to be Ca_V1 channel complexes) [47].

The obliquely striated abductor muscle from the Noble scallop, *Chlamys nobilis*, does not contain T-tubules, but bears feet structures at presumptive subsarcolemmal CRUs. A protein of similar biochemical properties to vertebrate RyRs was isolated from these muscles [104]. Ryanodine, caffeine and the second messenger cyclic ADP ribose all stimulated Ca^{2+} release from SR vesicles prepared from the abductor muscle of *Pecten jacobaeus*, the Mediterranean scallop [105]. In smooth muscle from the body wall of the wedge sea-hare *Dolabella auricularia*, extracellular K^+-induced depolarisation triggered contraction that persisted in the presence of the Ca^{2+} chelator ethylene glycol-bis(β-aminoethyl ether)-N,N,N′,N′-tetraacetic acid, but which was inhibited by the RyR antagonist procaine. Ultrastructural examination of these muscles indicated the presence of T-tubules coupled to SR, spanned by bridging feet structures [106]. This suggests that a DICR-like form of ECC operates in the muscles of at least some molluscs. The longitudinal muscle of the body wall (LMBW) of several echinoderms possesses subsarcolemmal SR, which can release Ca^{2+} in response to the RyR agonist caffeine, *for review see* [61].

12.3.1.2 Deuterstomes

Ryanodine increases the amplitude and decreases the frequency of spontaneous contractions of the LMBW of the sea-cucumber *Sclerodactyla briareus* [107]. This suggests that RyRs play a role in ECC in echinoderms. Indeed, there is strong similarity between the RyR from sea urchin eggs (suRyR) and mammalian RyRs in crucial regions such as the selectivity filter, the pore helix and helix bundle crossing region [108]. Interestingly, the activation of suRyR differs substantially from the mammalian RyR isoforms [109]; cyclic adenosine diphosphoribose can trigger release but 'crude sea urchin egg homogenate' was required to substantially activate the channels, suggesting the requirement of an unknown cellular activator and/or binding protein.

Among the chordates, the genome of the cephalochordate *B.belcheri* contains a single RyR gene, Fig. 12.3. This contrasts with an earlier suggestion that *B.floridae* contains three RyR orthologues [80]. The most likely cause of this discrepancy is that the original observation was based on conceptual translations of partial DNA sequences, the products of which were not definitively grouped with one RyR type or another. As discussed in Sect. 12.2.1, the trunk muscles of *B.lanceolatum* can contract in the absence of extracellular Ca^{2+} under some circumstances, including the presence of caffeine [64, 66]. This suggests that cephalochordates can use the DICR-mode of ECC under some conditions, but this conclusion conflicts with ultra-structural studies, which show that lancelet muscles lack a tetradic organisation of dihydropyridine receptor/VGCC complexes [10]. Sarcolemmal tetrads are thought to a hallmark of the molecular interaction between Ca_V1x and RyR proteins [74]. However, it is possible that several different mechanisms of DICR have evolved in distinct metazoan lineages, not all of which are dependent on the arrangement of $Ca_v 1x$ in tetrads.

The muscles of some, though not all, tunicates contain RyR-dependent SR Ca^{2+} stores. For example, myocytes generated by differentiation of growth-arrested blastocysts from the ascidian *Halocynthia roretzi* (the sea pineapple) display functional coupling between VGCCs and caffeine-/ryanodine-sensitive intracellular Ca^{2+} stores [110]. BLAST searches of two tunicate genomes suggest that both *Oikopleura dioica* and *Ciona intestinalis* encode a single RyR homologue, Fig. 12.3.

12.3.2 RyRs of Vertebrates

Most vertebrates possess three RyR paralogues, indicative of two rounds of duplication during the 2R-WGD process, followed by loss of one member. These paralogues are employed for different physiological roles (DICR versus CICR) and display distinct but partially overlapping tissue distributions [111, 112]. For example, in mammals RyR1 is most abundant in skeletal muscle, whereas RyR2 is prevalent in cardiomyocytes. In addition, the teleost specific third round of WGD [76] has generated a greater number of these paralogues, with six (RyR1a, RyR1b, RyR2a, RyR3a and RyR3b) being reported in the genomes of *D.rerio* (zebrafish), *Fundulus heteroclitus* (Atlantic killifish), *Gasterosteus aculeatus* (three-spined stickleback) and *Takifugu rubripes* (Japanese pufferfish) [113]. The basal ray-finned fish *Polypterus ornatipinnis* (the ornate bichir) possesses duplicated RyR1 genes, but single copies of the two other vertebrate RyR types. The authors of this work proposed that this resulted from generation of multiple RyR paralogues during 3R-WGD, followed by losses of some of the duplicated genes [114]. However, other workers have suggested that the bichir RyR1a and RyR1b proteins are generated by alternate splicing of transcripts from a single gene, rather than being the products of separate genes. The current and earlier studies have demonstrated that another basal actinopterygian, *Lepisosteus oculatus* (the spotted gar), encodes just three RyR paralogues in its genome, Fig. 12.3 [113].

A novel observation in the current work is that in all four of the amphibian genomes investigated, the only RyR homologues that could be detected correspond to RyR1 and RyR3, i.e. this class of vertebrate apparently lacks RyR2. This is unexpected, because previous studies, based on use of anti-RyR2 antibodies of incompletely defined subtype selectivity, demonstrated that this channel protein is present in SR-enriched microsomal membranes prepared from the hearts of *Rana pipiens* (the Northern leopard frog) [115] and of *Xenopus laevis* (African clawed frog) [116]. It is unlikely that this disparity is due to poor quality or curation of genomes, since those of *X.laevis* and *X.tropicalis* (Western clawed frog) are particularly well annotated. Contraction of amphibian cardiomyocytes is largely dependent on extracellular Ca^{2+} influx, with release of this ion from the SR playing little part in ventricular myocyte ECC under normal conditions [117]. A lack of involvement of Ca^{2+} release in heart contraction has been reported in other ectothermic vertebrates, such as fish, in which RyRs play an incompletely defined auxiliary role [11, 118], only contributing to this process under situations demanding increased cardiac output [119]. An investigation using immunofluorescent microscopy with a non-subtype selective antibody, indicate that RyRs are present in junctional SR of the atria of *Rana temporaria* (the European Common frog) [120]. Furthermore, mRNAs encoding RyR1 and RyR3 have been detected in the heart of *R.esculenta* (the Edible frog) [121].

It is hypothesised that a WGD event in a sterile, diploid, hybrid ancestor of *Xenopus sp.* generated a fertile, tetraploid species. As a consequence, *X.laevis* possesses two non-identical copies of each chromosome, termed the long ("L") and short ("S") forms. In cases where losses had not occurred (>56% of genes), this also resulted in the availability of four copies of each gene (two from each ancestor). Often, these genes evolved asymmetrically, with one ancestral pair being relatively preserved and the other being subject to deletion, divergence, loss or reduced transcription [122]. In the current study, searching of a boutique *Xenopus* genomic and transcriptomic database (Xenbase: http://www.xenbase.org/ [123]) revealed that the two copies of RyR1, termed RyR1L and RyR1S, are encoded by chromosomes 8L and 8S. In terms of transcription in adult animals, *X.laevis* RyR1L is expressed at highest levels in skeletal muscle and is also present in the eye; whereas RyR1S mRNA is not detected at high abundance in any tissue examined. Like RyR1, *X.laevis* RyR3 genes are located in chromosomes 8L and 8S. In adult African clawed frogs, the mRNA encoding RyR3L is most abundant in the heart, followed by skeletal muscle, brain and then the eye. RyR3S transcripts are also abundant in skeletal muscle, but were only present at low levels in the heart and in other tissues. These findings suggest a potential role for RyR3L in *Xenopus* cardiomyocyte ECC. This also implies redundancy among RyRs in this process, in that RyR3 might be able to substitute for RyR2 under some circumstances.

12.4   Evolution of Depolarisation-Induced Ca^{2+} Release

Although related processes might exist in some arthropods, molluscs, echinoderms and cephalochordates, the DICR mode of ECC, dependent on the organisation of VGCC into tetrads, is probably a vertebrate innovation [10, 74]. DICR is characteristic of skeletal muscle, but also plays roles in other tissues, such as the brain [124]. This form of ECC is dependent on physical interactions between $Ca_v1.1$-containing VGCCs in the sarcolemma and RyR1 Ca^{2+} release channels in the terminal cisternae of the SR. Additional accessory proteins, such as the β_{1a}-subunits of L-type VGCCs [27, 125], modulate DICR but are not critical for it. The three-dimensional architecture of interactions between RyR1 and L-type VGCCs are currently being unveiled using cryo-electron microscopy approaches [26, 126]. Although these models are currently of low resolution, it is likely that future advances in such techniques will reveal greater molecular detail of the RyR1-$Ca_v1.1$ interaction.

An investigation comparing rates of evolution between different RyRs and $Ca_v1.1$, as an indicator of co-evolution, generated some unexpected findings: that the only pairs of genes displaying significantly positive correlations were RyR3-RyR2 and RyR3-$Ca_v1.1$. It was anticipated that the evolution rates of RyR1 and $Ca_v1.1$ genes would be correlated, given that they encode a pair of interacting proteins. The authors explained the unexpected association between the evolutionary rates of RyR3 and $Ca_v1.1$ by suggesting an indirect effect: that DICR via RyR1 could trigger further release of Ca^{2+} via RyR3, which not directly coupled to VGCCs [127]. However, in this scenario, it would be predicted that RyR1 and RyR3 should also display significant co-evolution, which was not observed. An alternative explanation offered was that there could be a yet undiscovered direct interaction between RyR3 and $Ca_v1.1$ [127].

Clearer insights into the molecular mechanisms of DICR have been obtained by the generation of mutated and chimeric forms of the interacting partners involved. The spontaneously arising mouse mutant *dysgenic* lacks functional $Ca_v1.1$ gene and in its homozygous state, dies perinatally due to asphyxia. Myotubes cultured from the skeletal muscles of these mice provided an excellent system for testing which components of VGCC α1-subunits (Ca_v1) are critical for skeletal muscle-type DICR [128]. In a ground-breaking study, a panel of chimeras between cDNAs encoding $Ca_v1.1$ and $Ca_v1.2$ was generated and expressed in dysgenic myotubes, in order to determine which region(s) are required to restore DICR. This work established a key role for the cytoplasmic loop between transmembrane repeats II and III of $Ca_v1.1$ [129]. This "skeletal" II-III loop contains the information required not only for activation of RyR1 (orthograde signalling), but also for receiving a signal from the Ca^{2+} release channel that enhances VGCC currents. Chimeras containing $Ca_v1.1$ loops that could restore DICR to dysgenic myotubes also resulted in the formation of VGCC tetrads, as analysed by freeze-fracture electron microscopy. Constructs containing the mouse $Ca_v1.2$ ("cardiac") II-III loop did not form tetrads, whereas those bearing the II-III loop from the Ca_v1 homologue of *Musca domestica* (the housefly) did, despite being unable to support the DICR mode of ECC [130]. These observations suggest that the formation of VGCC tetrads is necessary, though not sufficient for DICR.

To further map the molecular features of the $Ca_v1.1$ II-III loop that are critical for DICR, constructs containing large deletions of this region were expressed in dyspedic myotubes. Using this strategy, it was found that deletion of residues 720–765 or 724–743 abolished depolarisation induced Ca^{2+} transients [131]. Such experiments have also revealed that additional regions of $Ca_v1.1$ such as the III-IV loop [132], or other components of the L-type VGCC such as the β_{1a}-subunit, contribute to DICR [125]. Introduction of a series of single amino acid substitutions into the II-III loop of *M.domestica*, revealed which features are critical for bidirectional communication between mammalian $Ca_v1.1$ and RyR1. These are: i) four conserved amino acid residues within a DICR motif; ii) a cluster of acid amino acid residues at the centre of this loop; and iii) the secondary structure of this negatively charged region [133]. Identification of these critical features has enabled us to search for them within all of the available Ca_v1 channel sequences from

Ca$_v$1.1

DICR motif

Species		Position
Homo	LKIDEFESNVNEVKDPYPSADFPGDDEEDEPEIPLSPRPRPLAELQ	720-765
Mus	LKIDEFESNVNEVKDPYPSADFPGDDEEDEPEIPVSPRPRPLAELQ	720-765
Mondelphis	LKVDEFESNVNEVKDPYPSADFPGDDEEEEPEIPLSPRPRPLAELQ	721-766
Birds/Reptiles	LKVDEFESNVNEIKDPYPSADFPGDDEEDEPEIPLSPRPRPLAELQ	Conserved
Xenopus	LKIDEFESNVNEIKDPYPSADFPGDDEEEEPEIPLSPRPRPLAELQ	716-761
Nanorana	LKIDEFESNVNEIKDPYPSADFPGDDEEEEPEIPISPRPRPLAELQ	719-764
Danio 1a	LKVDEFESNVNEIKDPFPPADFPGDDEEEEPEIPLSPRPRPMADLQ	732-777
Danio 1b	LKIDEFESNVNEVKDPFPPADFPGDDEEEEPEIPISPRPRPMADLQ	740-785
Acipenser	LRVDEFESNVNEIKDPYPSADFPGDDEEEEPDIPLSPRPRPMAELQ	728-773
Lepisosteus	LKVDEFESNVNEIKDPYPSADFPGDDEEEEPEIPLSPRPRPLAELQ	737-782
Latimeria	LKIDEFESNVNEIKDPYPSADFPGDDEEEEPEIPLSPRPRPLAELQ	727-772
Callorhinchus	LKVDEFEANVNEVKDPYPSDDFPGDDEEEEPEIPLSPRPRPLAELQ	199-244
	*.:.****.****.***.* ********.**.**.******.*.**	

Ca$_v$1.2

Species		Position
Homo	-INMDDLQPNENEDKSPYPNPETTGEE-DEEEPEMPVGPRPRPLSELH	670-715
Mus	-INMDDLQPSENEDKSPHSNPDTAGEE-DEEEPEMPVGPRPRPLSELH	879-924
Mondelphis	-INVDDLQPNENEKGPYPNPEAGGEE-EEEEPEMPVGPRPRPLSELH	862-907
Gallus	-INMDDYQPNENEEKSPYPTTEAPAEE-DEEEPEMPVGPRPRPMSELH	874-919
Taeniopygia	-INVDDYQPNENEEKSPYPTTEAPAEE-DEEEPEMPVGPRPRPMSELH	822-867
Anolis	KSNIDEYQPNENEEKNPYPTTEAPGEEEEEEEPEMPVGPRPRPMSELH	850-897
Python	KSNIDEYQPNENEEKNPYPTAETPGEEE-EEEPEMPVGPRPRPMSELH	796-842
Xenopus	-ISVEEFSSNENDEKSPYPSSNETPAEEEEEEPEMPVGPRPRPLSELH	820-866
Nanorana	-INVEDYPSNENDEKSPYPSSNENPGEEEEEEEPEMPVGPRPRPLSELH	793-839
Danio	-INIDEYTGEDNEEKNPYPVNDFPGED-DEEEPEMPVGPRPRPLSDIQ	855-900
Latimeria	-IAVEDYQLDGNEEKSSYPVNDFPGEE-EEEEPEMPVGPRPRPLSELQ	817-862
Callorhinchus	KIGLDEYQVEENEEKNPYPTNEFPGEE-EEDEPEMPVGPRPRPLSDLQ	855-901
	::: . *::*. : : *.************.*.:::	

Negatively charged region

Fig. 12.4 Features of Ca$_v$1.1 required for depolarisation-induced calcium release (DICR). Multiple Sequence Alignment (MSA) of the region of the II-III loop of Ca$_v$1.1 that is critical for the DICR-mode of ECC. This region from *Homo sapiens* Ca$_v$1.1 (L720-Q765) was aligned to corresponding homologues from the following species, using Clustal Omega software (https://www.ebi.ac.uk/Tools/msa/clustalo/, [164]): *Mus musculus* (mouse), *Monodelphis domestica* (grey short-tailed opossum), *Gallus gallus* (chicken), *Taeniopygia guttata* (zebra-finch) *Python bivittatus* (Burmese python), *Anolis carolinensis* (anole, a lizard), *Xenopus laevis* (African clawed frog), *Nanorana parkeri* (High Himalayan frog), *Danio rerio* (zebrafish), *Acipenser ruthenus* (sterlet), *Lepisosteus oculatus* (spotted gar), *Latimeria chalumnae* (the coelacanth, a lobe-finned fish) and *Callorhinchus milii* (Elephant shark). Positions of these sequences are shown to the right; for Ca$_v$1.1 those from bird and reptile species are pooled, since they are 100% identical. The corresponding region from *C.milii* is derived from a partial sequence, explaining its distinct position in this protein. Also shown is an MSA from the corresponding region of Ca$_v$1.2 homologues. Below each alignment, asterisks (*) indicate amino acid identity, colons (:) homology and full-stops (.) similarity. Although these Ca$_v$1.2 II-III loops contain a negatively charged central region, they lack the DICR motif that is highly conserved among Ca$_v$1.1 homologues

metazoans, Fig. 12.4. Although limited by the availability or curation of genomes from basal vertebrates and non-vertebrate chordates, these searches support the notion that DICR, or skeletal muscle-type ECC, is a vertebrate innovation [10]. All vertebrate Ca$_v$1.1 proteins share the consensus four residue DICR motif followed by a negatively charged region. The only exception to this is an alanine to aspartic acid substitution in the first residue of the motif, in the $Ca_v1.1$ II-III loop of *Callorhinchus milii* (the Elephant shark), a basal vertebrate. In contrast, the motif absent from the II-III loops of all vertebrate $Ca_v1.2$ proteins and from Ca_v1 proteins from non-vertebrates.

Mapping of the regions of RyR1 that are crucial for bidirectional communication with $Ca_v1.1$ has progressed using a similar combination of biochemical, molecular, electrophysiological and Ca^{2+}-imaging approaches. The availability of skeletal muscle myotubes from a transgenic RyR1 knockout mouse, termed *dyspedic* (lacking in feet structures), has facilitated determination of the functional effects of expressing RyR chimeras, deletions and point mutations. By monitoring in vitro interactions between fusion proteins, it has been demonstrated that a 37 amino acid region of RyR1, Arg1076-Asp1112, interacts with both the II-III and III-IV cytoplasmic loops of $Ca_v1.1$ [132, 134]. Analyses of RyR/RyR2 chimeras revealed that residues 1837–2168 of RyR1 contribute to DICR, but are not essential for it [135]. Assessment of additional RyR1/RyR2 chimeras for their ability to restore DICR and to promote assembly of L-type VGCC complexes into tetrads in dyspedic myotubes, revealed that multiple regions of RyR1 were important for these processes. In particular, chimeras containing residues 1635–3720 or 1635–2559 of RyR1 effectively restored DICR, whereas residues 1635–3720 or 2659–3720 strongly promoted tetrad formation [74].

Three regions within vertebrate RyRs, termed the divergent regions (DR1, DR2, DR3), display the lowest sequence identity between paralogues [136]. Deletion of DR2 (residues 1303-1406) of RyR1 abolishes DICR, without altering caffeine-mediated Ca^{2+} release in dyspedic myotubes. Additional regions of RyR1 must contribute to skeletal muscle-type ECC, since replacement of its DR2 with that from RyR2 did not alter reciprocal communication with $Ca_v1.1$ [137]. Similarly, replacement of residues 1-1680 in RyR3 with the corresponding region of RyR1 could restore DICR. Deletion of residues 1272-1455, containing DR2, from RyR1 abolished DICR. However, this region proved to be necessary but not sufficient for this mode of ECC, since it did not restore DICR to chimeras with an RyR3 backbone [138]. An *N*-ethyl-*N*-nitrosourea mutagenesis approach revealed that an E4242G mutation of RyR1 inhibited retrograde (RyR1 to $Ca_v1.1$), though not orthograde ($Ca_v1.1$ to RyR1) communication with VGCCs [139].

Overall, these investigations indicate that multiple regions of RyR1 participate in communication with $Ca_v1.1$. In the current work, we performed BLAST searches for the 37 residue RyR1 (R1076-N1112) sequence reported to interact with the II-III and III-V loops of $Ca_v1.1$ [132, 134]. As was the case for $Ca_v1.1$, this analysis of RyR1 lends support to the earlier preposition that DICR arose in vertebrates [10]. In addition, a partial RyR1 sequence from the lamprey *P.marinus* contains a sequence that is closely related to the 37 residue $Ca_v1.1$ binding region, Fig. 12.5. Multiple sequence alignments of this $Ca_v1.1$-interacting region of RyR1 homologues, and comparison with the corresponding region of RyR2 proteins, has uncovered novel detail of this molecular interaction. The alanine residue at position 1105 and the arginine at residue 1110 are found in all vertebrate RyR1 homologues examined,

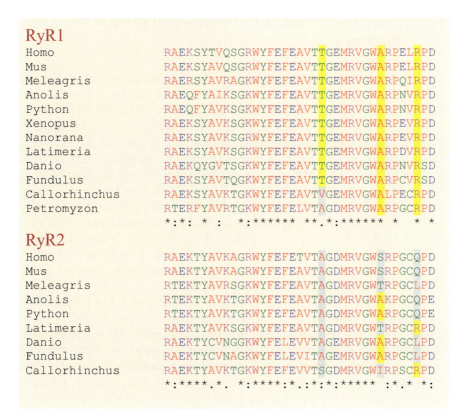

Fig. 12.5 Conservation of Ca_v1.1-interacting region from RyR1. Multiple Sequence Alignment (MSA) of the 37 amino acid Ca$_v$1.1-interacting region of RyR1. This region from *Homo sapiens* RyR1 (R1075-N1111) was aligned to corresponding homologues from the species listed in Fig. 12.4 and the following additional species, using Clustal Omega software [164]: *Meleagris gallopavo* (turkey), *Fundulus heteroclitus* (Atlantic killifish) and *Petromyzon marinus* (the sea lamprey). An MSA for the corresponding regions of RyR2 is also shown. Below each alignment, asterisks (*) indicate amino acid identity, colons (:) homology and full-stops (.) similarity. Yellow-shaded residues indicate those which are conserved in RyR1, though not in RyR2, suggesting that they could have a specific role in determining DICR

but are not conserved in RyR2. The threonine at position 1097 is not found in any RyR2 homologue, but is conserved in all RyR1 proteins, with the exception of those from the basal vertebrates *C.milii* and *P.marinus*, Fig. 12.5. The candidate role of these conserved residues in DICR is a compelling target for future mutagenesis, expression and functional reconstitution studies.

12.5   Junctophilins and ECC: Proteins that Bring It All Together

Junctophilins (JPHs) are a family of proteins that link intracellular membranes (ER and SR) to the plasma membrane, thereby promoting formation of junctions [28], Fig. 12.2. As such, they are critical for formation of CRUs and for supporting both CICR- and DICR-modes of ECC. This anchoring role is facilitated by a single ER/SR transmembrane region close to the C-terminus, an intervening helical domain and eight repeats of a lipid binding "membrane occupation and recognition nexus" (MORN) repeat motif, distributed between the middle and the N-terminus of these proteins [140, 141]. JPHs were discovered using an antibody-based biochemical approach to identify proteins that were highly enriched in rabbit skeletal muscle triads. Initially, three genes encoding JPHs were identified in mice: JPH1 is predominantly expressed in skeletal muscle; JPH2 in both skeletal and cardiac muscle; and JPH3 in the nervous system [28]. A fourth JPH paralogue, JPH4, was subsequently identified and found to be predominantly expressed in brain [142], but is also present in T-lymphocytes [143].

Genetic ablation of JPH2 in transgenic mice is lethal prior to mid-gestation, with the SR of heart tissues becoming enlarged, vacuolated and distributed at a greater distance from the sarcolemma than in wild-type animals [28]. Cardiomyocytes from the JPH2 knockout mice display abnormal spontaneous Ca^{2+}-transients, which are resistant to inhibition by removal of extracellular Ca^{2+}. Overexpression of JPH2, by means of adenovirus-delivered gene therapy, rescued deficits in ventricular contractility and T-tubule architecture in a mouse aortic constriction model of heart failure [144]. Conditional knockdown of JPH2 in mice, using RNA interference, decreased contractility; increased heart failure and death; and reduced gain during ECC. This impaired gain is due to enhanced spontaneous activity of RyR2 in JPH2 knockdown mice, leading to chronic depletion of SR Ca^{2+} stores. This indicates that there is an inhibitory interaction between the two proteins, supported by the observation that JPH2 co-immunoprecipitates with RyR2 [145]. In the dog left ventricular cardiomyocytes, JPH2 forms complexes with a range of Ca^{2+}-handling proteins, including RyR2, SERCA2a, calsequestrin and $Ca_v1.2$, increasing the amplitude of Ca^{2+} currents via L-type VGCCs [146]. Imaging of Ca^{2+} and single protein molecules in cardiomyocytes from transgenic mice overexpressing JPH2 showed an enlargement of the physical size of dyadic CRUs, without an anticipated increase in spontaneous Ca^{2+} release events. This was accompanied by an increase in the ratio between JPH2 and RyR2 at these CRUs, suggesting that these anchoring proteins inhibit gating of the Ca^{2+} release channels [147].

JPH1 knockout mice are deficient in milk suckling behaviour and die within a day of birth. Skeletal muscle from JPH1 deficient mice produces less force in response to low-frequency electrical stimulation than controls and displays reduced abundance of triadic junctions [148]. Both JPH1 and JPH2 can interact with a 12-residue motif, close to the C-terminus of $Ca_v1.1$ (residues 1595-1606, EGIFRRTGGLFG). This motif is highly conserved amongst vertebrate $Ca_v1.1$ and $Ca_v1.2$ proteins, but is not found in mammalian $Ca_v2.1$, nor within invertebrate Ca_v1 homologues. When expressed in mouse C2C12 myotubes, a dominant negative form of JPH1, lacking the SR-anchoring transmembrane domain, inhibits DICR without modifying Ca^{2+} store loading [149]. Again, this implies that JPHs not only act as anchors in the formation of CRUs, but also regulate the Ca^{2+} signalling that occurs at these junctions.

Although the evolutionary history of JPH family has been investigated previously [150], its deeper relationships with other protein families have not been reported. In the current study, we have examined the phylogeny of JPHs and related families, throughout the tree of life. We have found that JPHs form part of a MORN repeat-containing superfamily of proteins, which contains members in bacteria, archaea and eukaryotes, suggesting that they arose early during the evolution of cellular life, Fig. 12.6. In addition to the MORN domain proteins previously reported to be related to JPHs, namely phosphatidylinositol-4-phosphate 5-kinases (PIP5Ks) and histone-lysine N-methyltransferases [150], we have found four families of MORN-repeat proteins (MORN1-4), 2-isopropylmalate synthases, ankyrin repeat and MYND domain-containing (ANKMY) proteins, radial spoke-head (RSPH) proteins and alsin-like (ALS) proteins. We have also identified candidate MORN superfamily members in the genomes of the viruses *Pandoravirus dulcis* and of *Bodo saltans* virus.

It is expected that all of these protein families are able to interact with lipids, in particular phospholipids, via their MORN repeats. As is the case for JPHs, certain MORN domain proteins play additional roles in determining the subcellular distribution and stability of protein complexes. For example, MORN4 family members act as adaptors, tethering class III myosin motor proteins to membranes [151]. In eukaryotes, RSPH proteins are located in the central pair of microtubules of cilia/flagellae and regulate force production by them, via interactions with the motor protein dynein [152]. Of all members of the MORN repeat superfamily, ALS proteins are most closely related to JPHs, Fig. 12.6. In humans, alsin is mutated in amyotrophic lateral sclerosis-2 [153], a disease characterised by progressive loss of motor neurons, leading to paralysis. ALS regulates endocytosis by activation of the small G-protein Rab5, again implying that the MORN family has conserved roles in cellular motility and trafficking [154].

A key difference between JPHs and all other MORN repeat proteins examined is that the former possesses a transmembrane segment, crucial for localising it to intracellular organelles. In the current analysis, the most basal organism possessing a JPH homologue was *Salpingoeca rossetta*, belonging to the phylum choanoflagellata, a sister group to metazoans, Fig. 12.6. This suggests that the *S.rossetta*

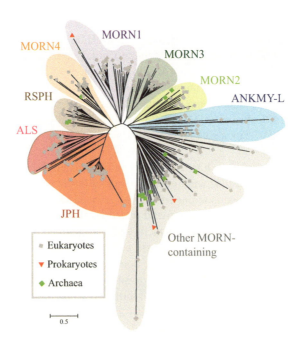

Fig. 12.6 Phylogenetic tree of the MORN repeat containing protein superfamily. An extensive evolutionary history of proteins related to JPH1 from *Homo sapiens* (Accession Number NP_065698.1), inferred using the Maximum Parsimony method, conducted using MEGA5 [162]. The tree is to scale, with branch lengths corresponding to amino acid substitutions per site, calculated using the average pathway method [165]. This analysis involved 238 amino acid sequences, from a range of eukaryotes, archaea and bacteria, the details of which are available upon request. The clustering of this tree has enabled us to envisage eight distinct families of protein containing membrane occupation and recognition nexus (MORN) repeat-containing proteins: the junctophilins (JPH), alsins (ALS), radial spokehead homology (RSPH) proteins, four MORN-containing families (MORN1 to MORN4) and an ankyrin repeat and MYND domain-containing protein 1-like (ANKMY-L) family; along with another less clearly defined group of proteins

genome encodes at least three of the proteins required for DICR (RyR, a HVA Ca_V1, JPH), but at present there is no experimental evidence to either reject or to support the hypothesis that this form of Ca^{2+}-signalling operates in a choanoflagellate. Furthermore, both RyR [80, 81, 155] and JPH homologues were undetectable in another choanoflagellates species, *Monsiga brevicolis*, suggesting that *S.rossetta* might represent a transitional stage in the evolution of Ca^{2+} signalling. JPH homologues were not found in filasterea, another sister group to animals represented by the organism *Capsaspora owczarzaki* [88].

In agreement with an earlier study, single JPH homologues were detected in placozoans (*T.adhaerens*), poriferans (*A.queenslandica*) and in cnidarians (*N.vectensis* and *H.vulgaris*) [150]. However, in the previous publication, the cnidarian JPH homologues clustered with bilaterian animals such as echinoderms and vertebrates, whereas in the current work, they formed a separate group clustered with poriferans. This is likely to reflect the use of a distinct subset of proteins, with a greater range of JPH homologues from non-bilaterians; and the use of a more robust algorithm for reconstruction of phylogenetic trees (Maximum-Parsimony instead of Neighbour-Joining) [156]. Prediction of the transmembrane topology of these JPH homolgues, using both Phobius (http://phobius.sbc.su.se/ [157]) and THHM (http://www.cbs.dtu.dk/services/TMHMM-2.0/ [158]) points toward a lack of a transmembrane spanning segment in these proteins from *T.adhaerens*, *N.vectensis* and *H.vulgaris*. One explanation for this observation could be poor sequence quality or annotation, leading to artefactually truncated hypothetical proteins. Alternatively, these basal metazoan JPH homologues could play roles in the regulation of VGCC channels, as described for the engineered C-terminally truncated form of mammalian JPH1, that inhibits gating of $Ca_v1.1$ [149].

All other invertebrate genomes analysed are predicted to encode a single JPH homologue. For only a few organisms, the presence of a JPH homologue has been verified experimentally. In the nematode worm *C.elegans*, JPH is transcribed in muscle cells and knockdown of its mRNA leads to hypolocomotion, supporting a role in ECC [44]. Tissue-specific knockdown or overexpression of JPH in the insect *D.melanogaster* generates similar phenotypes to those that that arise through genetic modification of its mouse paralogues: skeletal myopathies for JPH1; alterations in T-tubule formation for JPH2; and neuronal defects in the case of JPH3 and/or JPH4 [159]. This implies that the single JPH protein in an arthropod can be functionally equivalent to its four mammalian paralogues.

Amongst vertebrates, amphibians, non-avian reptiles and mammals contain four JPH paralogues (JPH1, JPH2, JPH3 and JPH4), consistent with a 2R-WGD mechanism of evolution [70]. In contrast, elasmobranch (shark, ray and skate) and sarcopterygian (lobe-finned fish) genomes encode three JPH paralogues and lack JPH4, Figs. 12.7 and 12.8. This implies that an early vertebrate ancestor possessed four JPH genes and that one of these was lost in certain lineages, although not in others. Our preliminary analyses of teleost fish genomes show that they encode four JPH homologues, but in contrast to tetrapods (amphibians, reptiles, birds and mammals), these represent JPH2, JPH3 and a duplicated pair of JPH1 genes. A mechanism for the evolution of these genes could be duplication of the three ancestral JPH paralogues during the teleost-specific third round of WGD [76], followed by losses of one of the copies of JPH2 and of JPH3. In the tetrapod lineage, birds alone appear to have lost the JPH4 gene. Similar avian-specific gene losses have been reported for other protein families, such as components of the endocrine system that are highly conserved among other vertebrates [160].

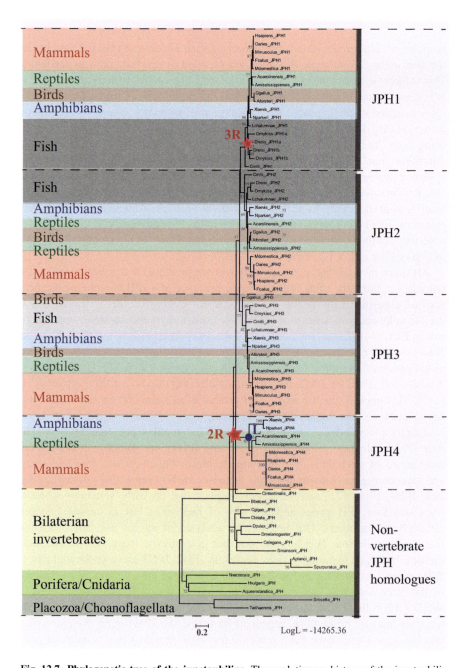

Fig. 12.7 Phylogenetic tree of the junctophilins. The evolutionary history of the junctophilin (JPH) family of MORN-repeat containing proteins was inferred using the methodology described in Fig. 12.5. This analysis involved 71 amino acid sequences, with full details being available upon request. Red stars ('2R' and '3R') indicate putative gene duplications; the blue circle ('L') candidate gene losses. This phylogenetic tree indicates that invertebrates possess a single JPH homologue and that this was duplicated twice early during vertebrate evolution to generate four paralogues. JPH4 was subsequently lost from fish and birds. In contrast, JPH1 was duplicated again in the teleost-specific 3R-WGD. Presumably, other JPH types were either not replicated in this event, or were duplicated and then one of the copies was subsequently lost

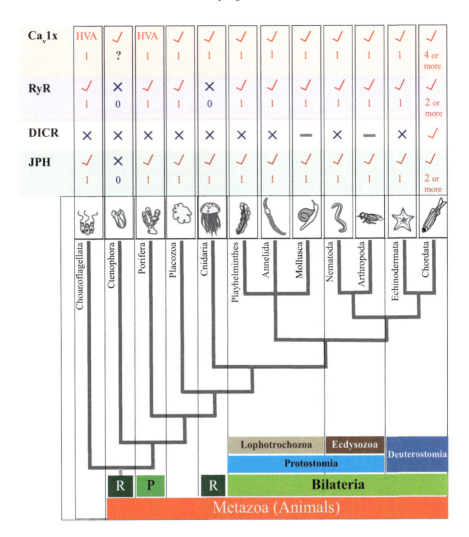

	Choanoflagellata	Ctenophora	Porifera	Placozoa	Cnidaria	Playhelminthes	Annelida	Mollusca	Nematoda	Arthropoda	Echinodermata	Chordata
Ca_v1x	HVA 1	✓ ?	HVA 1	✓ 1	✓ 1	✓ 1	✓ 1	✓ 1	✓ 1	✓ 1	✓ 1	✓ 4 or more
RyR	✓ 1	✗ 0	✓ 1	✓ 1	✗ 0	✓ 1	✓ 1	✓ 1	✓ 1	✓ 1	✓ 1	✓ 2 or more
DICR	✗	✗	✗	✗	✗	✗	✗	–	✗	–	✗	✓
JPH	✓ 1	✗ 0	✓ 1	✓ 1	✓ 1	✓ 1	✓ 1	✓ 1	✓ 1	✓ 1	✓ 1	✓ 2 or more

Fig. 12.8 Summary of the Presence of ECC components in metazoa and choanoflagellates. The presence of Ca_v1x (or HVA, high voltage-activated) voltage-gated calcium channels, ryanodine receptor (RyR) Ca^{2+} release channels, depolarisation-induced Ca^{2+}-release (DICR) and junctophilin (JPH) homologues in metazoa and choanoflagellata is indicated. The '?' indicates that Ca_v1x member used is unknown; the '-'indicates uncertainty of the mode of ECC employed. Where homologues of these ECC proteins have been found, the number detected is shown. Further details are given in Fig. 12.1

12.6 Conclusions and Future Directions

Our phylogenetic analyses and literature review of the evolution of ECC has generated several new insights into this group of signal transduction mechanisms. Concordant with the physiological and ultrastructural investigations of other researchers, we have found that the DICR mode of ECC is likely to be a vertebrate innovation [10]. The CICR form of ECC is employed in the vertebrate heart and in different muscle types from the majority of invertebrates. The time at which this Ca^{2+} signalling process first appeared is less certain. Choanoflagellates, a sister group to animals, possess proteins related to those required for CICR and possibly, even for DICR [27]: RyRs, HVA Cav1 channels, a JPH homologue, VGCC β-subunits [33] and STAC3 paralogues (*JJM, preliminary analyses*). This could represent operation of ECC-like mechanisms in choanoflagellates, or by analogy to the development of the actin-myosin contractile apparatus [7], the possession of all components required for this process, prior to its development in the metazoan lineage.

Distinct modes of ECC are employed by different metazoan groups. RyR homologues could not be detected in cnidarians or ctenophores, hinting that ECC is largely dependent on Ca^{2+}-influx via VGCC in these animals. However, experimental evidence suggests that CICR occurs in some cnidarians, with a likely scenario being amplification of Ca^{2+}-influx by Ca^{2+} release via ITPRs. However, ctenophores also lack detectable ITPR homologues [80].

Some invertebrates, including arthropods, molluscs and lancelets, display forms of ECC that resemble DICR. This manifests mainly in the apparent lack of dependency on Ca^{2+}-influx. However, where data are available, RyR and Ca_v1 proteins of these organisms lack the molecular features that are required for DICR in vertebrates. Furthermore, their VGCCs are not ordered into tetrads in junctional membranes, another feature thought to be essential for DICR. Possible reasons for this discrepancy are either operation of CICR with very high gain, as proposed for ECC in the abdominal flexor muscles of the shrimp *A.lanipes*, or a yet uncharacterised mode of DICR that is distinct from that in vertebrate skeletal muscle.

The genes encoding proteins involved in ECC have been subject to expansions and losses in different vertebrate classes. Such events have underpinned the evolution of vertebrate innovations, such as the DICR-mode of ECC. This is exemplified in teleost fish, whereby the 3R-WGD process has allowed the development of duplicated non-conducting VGCCs ($Ca_v1.1a$ and $Ca_v1.1b$) and RyRs (RyR1a and RyR1b), which operate within distinct skeletal muscle fibre types. A remarkable example of gene loss appears to have occurred in amphibians, which apparently lack the gene encoding RyR2. This loss appears to have been compensated for by the expression of RyR3 and RyR1 in the heart, implying that there might be some redundancy between these Ca^{2+} release channels, at least within the CICR-mode of ECC.

Many of the observations in the current work are based on limited genomic and experimental data, making some of them tenuous. For improved understanding of the evolution of ECC, more information needs to be gathered. For example, the number of genomes sequenced and the quality of the annotation of these data needs to increase, in particular for some vertebrate species in which key events may have occurred, such as agnathans and amphibians. Although Ca^{2+} signalling has been explored experimentally in some unicellular eukaryotes, such *C.owczarzaki* and *Paramecium tetraurelia* [155, 81], the roles of ECC proteins in choanoflagellates, a sister group of animals which lacks muscles, could be particularly informative. Contemporary technologies, such as genome-editing and determination of three-dimensional protein structures by cryo-electron microscopy, are likely to reveal the exquisite interactions that underlie ECC in greater detail, in a wider range of animal species and their relatives.

References

1. Cunningham JA, Liu AG, Bengtson S, Donoghue PC (2017) The origin of animals: can molecular clocks and the fossil record be reconciled? BioEssays 39(1):1–12. https://doi.org/10.1002/bies.201600120
2. Erwin DH (2015) Early metazoan life: divergence, environment and ecology. Philos Trans R Soc Lond Ser B Biol Sci 370(1684):20150036. https://doi.org/10.1098/rstb.2015.0036
3. Bosch TCG, Klimovich A, Domazet-Loso T, Grunder S, Holstein TW, Jekely G, Miller DJ, Murillo-Rincon AP, Rentzsch F, Richards GS, Schroder K, Technau U, Yuste R (2017) Back to the basics: cnidarians start to fire. Trends Neurosci 40(2):92–105. https://doi.org/10.1016/j.tins.2016.11.005
4. Seipel K, Schmid V (2005) Evolution of striated muscle: jellyfish and the origin of triploblasty. Dev Biol 282(1):14–26. https://doi.org/10.1016/j.ydbio.2005.03.032
5. McGuigan K, Phillips PC, Postlethwait JH (2004) Evolution of sarcomeric myosin heavy chain genes: evidence from fish. Mol Biol Evol 21(6):1042–1056. https://doi.org/10.1093/molbev/msh103
6. Brunet T, Arendt D (2016) From damage response to action potentials: early evolution of neural and contractile modules in stem eukaryotes. Philos Trans R Soc Lond Ser B Biol Sci 371(1685):20150043. https://doi.org/10.1098/rstb.2015.0043
7. Steinmetz PR, Kraus JE, Larroux C, Hammel JU, Amon-Hassenzahl A, Houliston E, Worheide G, Nickel M, Degnan BM, Technau U (2012) Independent evolution of striated muscles in cnidarians and bilaterians. Nature 487(7406):231–234. https://doi.org/10.1038/nature11180
8. Allbritton NL, Meyer T, Stryer L (1992) Range of messenger action of calcium ion and inositol 1,4,5-trisphosphate. Science 258(5089):1812–1815
9. Parekh AB (2011) Decoding cytosolic Ca^{2+} oscillations. Trends Biochem Sci 36(2):78–87. https://doi.org/10.1016/j.tibs.2010.07.013
10. Di Biase V, Franzini-Armstrong C (2005) Evolution of skeletal type e-c coupling: a novel means of controlling calcium delivery. J Cell Biol 171(4):695–704. https://doi.org/10.1083/jcb.200503077
11. Shiels HA, Galli GL (2014) The sarcoplasmic reticulum and the evolution of the vertebrate heart. Physiology 29(6):456–469. https://doi.org/10.1152/physiol.00015.2014
12. Catterall WA, Striessnig J, Snutch TP, Perez-Reyes E, International Union of P (2003) International Union of Pharmacology. XL. Compendium of voltage-gated ion channels: calcium channels. Pharmacol Rev 55(4):579–581. https://doi.org/10.1124/pr.55.4.8
13. Bers DM, Perez-Reyes E (1999) Ca channels in cardiac myocytes: structure and function in Ca influx and intracellular Ca release. Cardiovasc Res 42(2):339–360
14. Campiglio M, Flucher BE (2015) The role of auxiliary subunits for the functional diversity of voltage-gated calcium channels. J Cell Physiol 230(9):2019–2031. https://doi.org/10.1002/jcp.24998
15. Endo M (1977) Calcium release from the sarcoplasmic reticulum. Physiol Rev 57(1):71–108. https://doi.org/10.1152/physrev.1977.57.1.71

16. Franzini-Armstrong C, Protasi F, Ramesh V (1999) Shape, size, and distribution of Ca^{2+} release units and couplons in skeletal and cardiac muscles. Biophys J 77(3):1528–1539. https://doi.org/10.1016/S0006-3495(99)77000-1
17. Hong T, Shaw RM (2017) Cardiac T-tubule microanatomy and function. Physiol Rev 97(1):227–252. https://doi.org/10.1152/physrev.00037.2015
18. Dibb KM, Clarke JD, Eisner DA, Richards MA, Trafford AW (2013) A functional role for transverse (t-) tubules in the atria. J Mol Cell Cardiol 58:84–91. https://doi.org/10.1016/j.yjmcc.2012.11.001
19. Orchard C, Brette F (2008) t-Tubules and sarcoplasmic reticulum function in cardiac ventricular myocytes. Cardiovasc Res 77(2):237–244. https://doi.org/10.1093/cvr/cvm002
20. Perni S, Iyer VR, Franzini-Armstrong C (2012) Ultrastructure of cardiac muscle in reptiles and birds: optimizing and/or reducing the probability of transmission between calcium release units. J Muscle Res Cell Motil 33(2):145–152. https://doi.org/10.1007/s10974-012-9297-6
21. Brandt NR, Kawamoto RM, Caswell AH (1985) Dihydropyridine binding sites on transverse tubules isolated from triads of rabbit skeletal muscle. J Recept Res 5(2–3):155–170
22. Block BA, Imagawa T, Campbell KP, Franzini-Armstrong C (1988) Structural evidence for direct interaction between the molecular components of the transverse tubule/sarcoplasmic reticulum junction in skeletal muscle. J Cell Biol 107(6 Pt 2):2587–2600
23. Bertocchini F, Ovitt CE, Conti A, Barone V, Scholer HR, Bottinelli R, Reggiani C, Sorrentino V (1997) Requirement for the ryanodine receptor type 3 for efficient contraction in neonatal skeletal muscles. EMBO J 16(23):6956–6963. https://doi.org/10.1093/emboj/16.23.6956
24. Conti A, Reggiani C, Sorrentino V (2005) Selective expression of the type 3 isoform of ryanodine receptor Ca^{2+} release channel (RyR3) in a subset of slow fibers in diaphragm and cephalic muscles of adult rabbits. Biochem Biophys Res Commun 337(1):195–200. https://doi.org/10.1016/j.bbrc.2005.09.027
25. Daniels RE, Haq KT, Miller LS, Chia EW, Miura M, Sorrentino V, McGuire JJ, Stuyvers BD (2017) Cardiac expression of ryanodine receptor subtype 3; a strategic component in the intracellular Ca^{2+} release system of Purkinje fibers in large mammalian heart. J Mol Cell Cardiol 104:31–42. https://doi.org/10.1016/j.yjmcc.2017.01.011
26. Meissner G (2017) The structural basis of ryanodine receptor ion channel function. J Gen Physiol 149(12):1065–1089. https://doi.org/10.1085/jgp.201711878
27. Perni S, Lavorato M, Beam KG (2017) De novo reconstitution reveals the proteins required for skeletal muscle voltage-induced Ca^{2+} release. Proc Natl Acad Sci U S A 114(52):13822–13827. https://doi.org/10.1073/pnas.1716461115
28. Takeshima H, Komazaki S, Nishi M, Iino M, Kangawa K (2000) Junctophilins: a novel family of junctional membrane complex proteins. Mol Cell 6(1):11–22
29. Somlyo AV, Siegman MJ (2012) Smooth muscle myocyte ultrastructure and contractility. In: Hill JA, Olson EN (eds) Muscle. Boston/Waltham, Academic, pp 1117–1132. https://doi.org/10.1016/B978-0-12-381510-1.00083-1
30. Catterall WA (2010) Signaling complexes of voltage-gated sodium and calcium channels. Neurosci Lett 486(2):107–116. https://doi.org/10.1016/j.neulet.2010.08.085
31. Pozdnyakov I, Matantseva O, Skarlato S (2018) Diversity and evolution of four-domain voltage-gated cation channels of eukaryotes and their ancestral functional determinants. Sci Rep 8(1):3539. https://doi.org/10.1038/s41598-018-21897-7
32. Dolphin AC (2016) Voltage-gated calcium channels and their auxiliary subunits: physiology and pathophysiology and pharmacology. J Physiol 594(19):5369–5390. https://doi.org/10.1113/JP272262
33. Moran Y, Zakon HH (2014) The evolution of the four subunits of voltage-gated calcium channels: ancient roots, increasing complexity, and multiple losses. Genome Biol Evol 6(9):2210–2217. https://doi.org/10.1093/gbe/evu177
34. Moran Y, Barzilai MG, Liebeskind BJ, Zakon HH (2015) Evolution of voltage-gated ion channels at the emergence of Metazoa. J Exp Biol 218(Pt 4):515–525. https://doi.org/10.1242/jeb.110270

35. Jeziorski MC, Greenberg RM, Anderson PA (2000) The molecular biology of invertebrate voltage-gated Ca^{2+} channels. J Exp Biol 203(Pt 5):841–856

36. Senatore A, Raiss H, Le P (2016) Physiology and evolution of voltage-gated calcium channels in early diverging animal phyla: cnidaria, placozoa, porifera and ctenophora. Front Physiol 7:481. https://doi.org/10.3389/fphys.2016.00481

37. Whelan NV, Kocot KM, Moroz LL, Halanych KM (2015) Error, signal, and the placement of Ctenophora sister to all other animals. Proc Natl Acad Sci U S A 112(18):5773–5778. https://doi.org/10.1073/pnas.1503453112

38. Schierwater B, de Jong D, Desalle R (2009) Placozoa and the evolution of Metazoa and intrasomatic cell differentiation. Int J Biochem Cell Biol 41(2):370–379. https://doi.org/10.1016/j.biocel.2008.09.023

39. Smith CL, Varoqueaux F, Kittelmann M, Azzam RN, Cooper B, Winters CA, Eitel M, Fasshauer D, Reese TS (2014) Novel cell types, neurosecretory cells, and body plan of the early-diverging metazoan Trichoplax adhaerens. Curr Biol 24(14):1565–1572. https://doi.org/10.1016/j.cub.2014.05.046

40. Mayorova TD, Smith CL, Hammar K, Winters CA, Pivovarova NB, Aronova MA, Leapman RD, Reese TS (2018) Cells containing aragonite crystals mediate responses to gravity in Trichoplax adhaerens (Placozoa), an animal lacking neurons and synapses. PLoS One 13(1):e0190905. https://doi.org/10.1371/journal.pone.0190905

41. Senatore A, Reese TS, Smith CL (2017) Neuropeptidergic integration of behavior in Trichoplax adhaerens, an animal without synapses. J Exp Biol 220(Pt 18):3381–3390. https://doi.org/10.1242/jeb.162396

42. Smith CL, Abdallah S, Wong YY, Le P, Harracksingh AN, Artinian L, Tamvacakis AN, Rehder V, Reese TS, Senatore A (2017) Evolutionary insights into T-type Ca^{2+} channel structure, function, and ion selectivity from the Trichoplax adhaerens homologue. J Gen Physiol 149(4):483–510. https://doi.org/10.1085/jgp.201611683

43. Paniagua R, Royuela M, Garcia-Anchuelo RM, Fraile B (1996) Ultrastructure of invertebrate muscle cell types. Histol Histopathol 11(1):181–201

44. Yoshida M, Sugimoto A, Ohshima Y, Takeshima H (2001) Important role of junctophilin in nematode motor function. Biochem Biophys Res Commun 289(1):234–239. https://doi.org/10.1006/bbrc.2001.5951

45. Jospin M, Jacquemond V, Mariol MC, Segalat L, Allard B (2002) The L-type voltage-dependent Ca^{2+} channel EGL-19 controls body wall muscle function in Caenorhabditis elegans. J Cell Biol 159(2):337–348. https://doi.org/10.1083/jcb.200203055

46. Lee RY, Lobel L, Hengartner M, Horvitz HR, Avery L (1997) Mutations in the alpha1 subunit of an L-type voltage-activated Ca^{2+} channel cause myotonia in Caenorhabditis elegans. EMBO J 16(20):6066–6076. https://doi.org/10.1093/emboj/16.20.6066

47. Takekura H, Franzini-Armstrong C (2002) The structure of Ca^{2+} release units in arthropod body muscle indicates an indirect mechanism for excitation-contraction coupling. Biophys J 83(5):2742–2753. https://doi.org/10.1016/S0006-3495(02)75284-3

48. Hara Y, Koganezawa M, Yamamoto D (2015) The Dmca1D channel mediates Ca^{2+} inward currents in Drosophila embryonic muscles. J Neurogenet 29(2–3):117–123. https://doi.org/10.3109/01677063.2015.1054991

49. Lin N, Badie N, Yu L, Abraham D, Cheng H, Bursac N, Rockman HA, Wolf MJ (2011) A method to measure myocardial calcium handling in adult Drosophila. Circ Res 108(11):1306–1315. https://doi.org/10.1161/CIRCRESAHA.110.238105

50. Krizanova O, Novotova M, Zachar J (1990) Characterization of DHP binding protein in crayfish striated muscle. FEBS Lett 267(2):311–315

51. Erxleben C, Rathmayer W (1997) A dihydropyridine-sensitive voltage-dependent calcium channel in the sarcolemmal membrane of crustacean muscle. J Gen Physiol 109(3):313–326

52. Findsen A, Overgaard J, Pedersen TH (2016) Reduced L-type Ca^{2+} current and compromised excitability induce loss of skeletal muscle function during acute cooling in locust. J Exp Biol 219(Pt 15):2340–2348. https://doi.org/10.1242/jeb.137604

53. Monterrubio J, Lizardi L, Zuazaga C (2000) Silent calcium channels in skeletal muscle fibers of the crustacean Atya lanipes. J Membr Biol 173(1):9–17
54. Bonilla M, Garcia MC, Orkand PM, Zuazaga C (1992) Ultrastructural and mechanical properties of electrically inexcitable skeletal muscle fibers of the crustacean Atya lanipes. Tissue Cell 24(4):525–535
55. Pennec JP, Talarmin H, Droguet M, Giroux-Metges MA, Gioux M, Dorange G (2004) Characterization of the voltage-activated currents in cultured atrial myocytes isolated from the heart of the common oyster Crassostrea gigas. J Exp Biol 207(Pt 22):3935–3944. https://doi.org/10.1242/jeb.01221
56. Yeoman MS, Brezden BL, Benjamin PR (1999) LVA and HVA Ca^{2+} currents in ventricular muscle cells of the Lymnaea heart. J Neurophysiol 82(5):2428–2440. https://doi.org/10.1152/jn.1999.82.5.2428
57. Senatore A, Boone A, Lam S, Dawson TF, Zhorov B, Spafford JD (2011) Mapping of dihydropyridine binding residues in a less sensitive invertebrate L-type calcium channel (LCa v 1). Channels (Austin) 5(2):173–187. https://doi.org/10.4161/chan.5.2.15141
58. Huang Z, Ishii Y, Watari T, Liu H, Miyake S, Suzaki T, Tsuchiya T (2005) Sources of activator calcium ions in the contraction of smooth muscles in Aplysia kurodai. Zool Sci 22(8):923–932. https://doi.org/10.2108/zsj.22.923
59. Rokni D, Hochner B (2002) Ionic currents underlying fast action potentials in the obliquely striated muscle cells of the octopus arm. J Neurophysiol 88(6):3386–3397. https://doi.org/10.1152/jn.00383.2002
60. Cavey MJ, Wood RL (1981) Specializations for excitation-contraction coupling in the podial retractor cells of the starfish Stylasterias forreri. Cell Tissue Res 218(3):475–485
61. Hill RB (2001) Role of Ca^{2+} in excitation-contraction coupling in echinoderm muscle: comparison with role in other tissues. J Exp Biol 204(Pt 5):897–908
62. Granados-Gonzalez G, Mendoza-Lujambio I, Rodriguez E, Galindo BE, Beltran C, Darszon A (2005) Identification of voltage-dependent Ca^{2+} channels in sea urchin sperm. FEBS Lett 579(29):6667–6672. https://doi.org/10.1016/j.febslet.2005.10.035
63. Peachey LD (1961) Structure of the longitudinal body muscles of amphioxus. J Biophys Biochem Cytol 10(4 Suppl):159–176
64. Hagiwara S, Henkart MP, Kidokoro Y (1971) Excitation-contraction coupling in amphioxus muscle cells. J Physiol 219(1):233–251
65. Benterbusch R, Herberg FW, Melzer W, Thieleczek R (1992) Excitation-contraction coupling in a pre-vertebrate twitch muscle: the myotomes of Branchiostoma lanceolatum. J Membr Biol 129(3):237–252
66. Melzer W (1982) Twitch activation in Ca^{2+} -free solutions in the myotomes of the lancelet (Branchiostoma lanceolatum). Eur J Cell Biol 28(2):219–225
67. Okamura Y, Izumi-Nakaseko H, Nakajo K, Ohtsuka Y, Ebihara T (2003) The ascidian dihydropyridine-resistant calcium channel as the prototype of chordate L-type calcium channel. Neurosignals 12(3):142–158. https://doi.org/10.1159/000072161
68. Inoue I, Tsutsui I, Bone Q (2002) Excitation-contraction coupling in isolated loco-motor muscle fibres from the pelagic tunicate Doliolum which lack both sarcoplas-mic reticulum and transverse tubular system. J Comp Physiol B 172(6):541–546. https://doi.org/10.1007/s00360-002-0280-1
69. Schiaffino S, Nunzi MG, Burighel P (1976) T system in ascidian muscle: organization of the sarcotubular system in the caudal muscle cells of Botryllus schlosseri tadpole larvae. Tissue Cell 8(1):101–110
70. Ohno S (1993) Patterns in genome evolution. Curr Opin Genet Dev 3(6):911–914
71. Caputo Barucchi V, Giovannotti M, Nisi Cerioni P, Splendiani A (2013) Genome duplication in early vertebrates: insights from agnathan cytogenetics. Cytogenet Genome Res 141(2–3):80–89. https://doi.org/10.1159/000354098
72. Inoue I, Tsutsui I, Bone Q (2002) Excitation-contraction coupling in skeletal and caudal heart muscle of the hagfish Eptatretus burgeri Girard. J Exp Biol 205(Pt 22):3535–3541

73. Inoue I, Tsutsui I, Bone Q, Brown ER (1994) Evolution of skeletal muscle excitation-contraction coupling and the appearance of dihydropyridine-sensitive intramembrane charge movement. Proc R Soc London Ser B Biol Sci 255(1343):181–187

74. Protasi F, Paolini C, Nakai J, Beam KG, Franzini-Armstrong C, Allen PD (2002) Multiple regions of RyR1 mediate functional and structural interactions with alpha(1S)-dihydropyridine receptors in skeletal muscle. Biophys J 83(6):3230–3244. https://doi.org/10.1016/S0006-3495(02)75325-3

75. Dayal A, Schrotter K, Pan Y, Fohr K, Melzer W, Grabner M (2017) The Ca^{2+} influx through the mammalian skeletal muscle dihydropyridine receptor is irrelevant for muscle performance. Nat Commun 8(1):475. https://doi.org/10.1038/s41467-017-00629-x

76. Glasauer SM, Neuhauss SC (2014) Whole-genome duplication in teleost fishes and its evolutionary consequences. Mol Gen Genomics 289(6):1045–1060. https://doi.org/10.1007/s00438-014-0889-2

77. Schredelseker J, Shrivastav M, Dayal A, Grabner M (2010) Non-Ca^{2+}-conducting Ca^{2+} channels in fish skeletal muscle excitation-contraction coupling. Proc Natl Acad Sci U S A 107(12):5658–5663. https://doi.org/10.1073/pnas.0912153107

78. Lien S, Koop BF, Sandve SR, Miller JR, Kent MP, Nome T, Hvidsten TR, Leong JS, Minkley DR, Zimin A, Grammes F, Grove H, Gjuvsland A, Walenz B, Hermansen RA, von Schalburg K, Rondeau EB, Di Genova A, Samy JK, Olav Vik J, Vigeland MD, Caler L, Grimholt U, Jentoft S, Vage DI, de Jong P, Moen T, Baranski M, Palti Y, Smith DR, Yorke JA, Nederbragt AJ, Tooming-Klunderud A, Jakobsen KS, Jiang X, Fan D, Hu Y, Liberles DA, Vidal R, Iturra P, Jones SJ, Jonassen I, Maass A, Omholt SW, Davidson WS (2016) The Atlantic salmon genome provides insights into rediploidization. Nature 533(7602):200–205. https://doi.org/10.1038/nature17164

79. Schrotter K, Dayal A, Grabner M (2017) The mammalian skeletal muscle DHPR has larger Ca^{2+} conductance and is phylogenetically ancient to the early ray-finned fish sterlet (Acipenser ruthenus). Cell Calcium 61:22–31. https://doi.org/10.1016/j.ceca.2016.10.002

80. Mackrill JJ (2012) Ryanodine receptor calcium release channels: an evolutionary perspective. Adv Exp Med Biol 740:159–182. https://doi.org/10.1007/978-94-007-2888-2_7

81. Alzayady KJ, Sebe-Pedros A, Chandrasekhar R, Wang L, Ruiz-Trillo I, Yule DI (2015) Tracing the evolutionary history of inositol, 1, 4, 5-trisphosphate receptor: insights from analyses of Capsaspora owczarzaki Ca^{2+} Release Channel Orthologs. Mol Biol Evol 32(9):2236–2253. https://doi.org/10.1093/molbev/msv098

82. Simion P, Philippe H, Baurain D, Jager M, Richter DJ, Di Franco A, Roure B, Satoh N, Queinnec E, Ereskovsky A, Lapebie P, Corre E, Delsuc F, King N, Worheide G, Manuel M (2017) A large and consistent Phylogenomic dataset supports sponges as the sister group to all other animals. Curr Biol 27(7):958–967. https://doi.org/10.1016/j.cub.2017.02.031

83. Singla CL (1978) Locomotion and neuromuscular system of Aglantha digitale. Cell Tissue Res 188(2):317–327

84. Cario C, Malaval L, Hernandez-Nicaise ML (1995) Two distinct distribution patterns of sarcoplasmic reticulum in two functionally different giant smooth muscle cells of Beroe ovata. Cell Tissue Res 282(3):435–443

85. Lin YC, Grigoriev NG, Spencer AN (2000) Wound healing in jellyfish striated muscle involves rapid switching between two modes of cell motility and a change in the source of regulatory calcium. Dev Biol 225(1):87–100. https://doi.org/10.1006/dbio.2000.9807

86. Missiaen L, Parys JB, De Smedt H, Himpens B, Casteels R (1994) Inhibition of inositol trisphosphate-induced calcium release by caffeine is prevented by ATP. Biochem J 300(Pt 1):81–84

87. Sutko JL, Ito K, Kenyon JL (1985) Ryanodine: a modifier of sarcoplasmic reticulum calcium release in striated muscle. Fed Proc 44(15):2984–2988

88. Mackrill JJ (2010) Ryanodine receptor calcium channels and their partners as drug targets. Biochem Pharmacol 79(11):1535–1543. https://doi.org/10.1016/j.bcp.2010.01.014

89. Silva CL, Cunha VM, Mendonca-Silva DL, Noel F (1998) Evidence for ryanodine receptors in Schistosoma mansoni. Biochem Pharmacol 56(8):997–1003

90. Day TA, Haithcock J, Kimber M, Maule AG (2000) Functional ryanodine receptor channels in flatworm muscle fibres. Parasitology 120(Pt 4):417–422

91. Maryon EB, Coronado R, Anderson P (1996) unc-68 encodes a ryanodine receptor involved in regulating C. elegans body-wall muscle contraction. J Cell Biol 134(4):885–893

92. Robertson AP, Clark CL, Martin RJ (2010) Levamisole and ryanodine receptors. I: a contraction study in Ascaris suum. Mol Biochem Parasitol 171(1):1–7. https://doi.org/10.1016/j.molbiopara.2009.12.007

93. Hassett CC (1948) Effect of ryanodine on the oxygen consumption of Periplaneta americana. Science 108(2797):138–139. https://doi.org/10.1126/science.108.2797.138

94. Sullivan KM, Scott K, Zuker CS, Rubin GM (2000) The ryanodine receptor is essential for larval development in Drosophila melanogaster. Proc Natl Acad Sci U S A 97(11):5942–5947. https://doi.org/10.1073/pnas.110145997

95. Frolov RV, Singh S (2012) Inhibition of ion channels and heart beat in Drosophila by selective COX-2 inhibitor SC-791. PLoS One 7(6):e38759. https://doi.org/10.1371/journal.pone.0038759

96. Scott-Ward TS, Dunbar SJ, Windass JD, Williams AJ (2001) Characterization of the ryanodine receptor-Ca^{2+} release channel from the thoracic tissues of the lepidopteran insect Heliothis virescens. J Membr Biol 179(2):127–141

97. Collet C (2009) Excitation-contraction coupling in skeletal muscle fibers from adult domestic honeybee. Pflugers Arch 458(3):601–612. https://doi.org/10.1007/s00424-009-0642-6

98. Lahm GP, Cordova D, Barry JD (2009) New and selective ryanodine receptor activators for insect control. Bioorg Med Chem 17(12):4127–4133. https://doi.org/10.1016/j.bmc.2009.01.018

99. Lea TJ, Ashley CC (1990) Ca^{2+} release from the sarcoplasmic reticulum of barnacle myofibrillar bundles initiated by photolysis of caged Ca^{2+}. J Physiol 427:435–453

100. Gyorke S, Palade P (1992) Calcium-induced calcium release in crayfish skeletal muscle. J Physiol 457:195–210

101. Brenner TL, Wilkens JL (2001) Physiology and excitation-contraction coupling in the intestinal muscle of the crayfish Procambarus clarkii. J Comp Physiol B 171(7):613–621

102. Xiong H, Feng X, Gao L, Xu L, Pasek DA, Seok JH, Meissner G (1998) Identification of a two EF-hand Ca^{2+} binding domain in lobster skeletal muscle ryanodine receptor/Ca^{2+} release channel. Biochemistry 37(14):4804–4814. https://doi.org/10.1021/bi971198b

103. Zhang JJ, Williams AJ, Sitsapesan R (1999) Evidence for novel caffeine and Ca^{2+} binding sites on the lobster skeletal ryanodine receptor. Br J Pharmacol 126(4):1066–1074. https://doi.org/10.1038/sj.bjp.0702400

104. Abe T, Ishida H, Matsuno A (1997) Foot structure and foot protein in the cross striated muscle of a pecten. Cell Struct Funct 22(1):21–26

105. Panfoli I, Burlando B, Viarengo A (1999) Cyclic ADP-ribose-dependent Ca^{2+} release is modulated by free $[Ca^{2+}]$ in the scallop sarcoplasmic reticulum. Biochem Biophys Res Commun 257(1):57–62. https://doi.org/10.1006/bbrc.1999.0405

106. Sugi H, Suzuki S (1978) Ultrastructural and physiological studies on the longitudinal body wall muscle of Dolabella auricularia. I. Mechanical response and ultrastructure. J Cell Biol 79(2 Pt 1):454–466

107. Devlin CL, Amole W, Anderson S, Shea K (2003) Muscarinic acetylcholine receptor compounds alter net Ca^{2+} flux and contractility in an invertebrate smooth muscle. Invertebr Neurosci 5(1):9–17. https://doi.org/10.1007/s10158-003-0023-3

108. Shiwa M, Murayama T, Ogawa Y (2002) Molecular cloning and characterization of ryanodine receptor from unfertilized sea urchin eggs. Am J Physiol Regul Integr Comp Physiol 282(3):R727–R737. https://doi.org/10.1152/ajpregu.00519.2001

109. Lokuta AJ, Darszon A, Beltran C, Valdivia HH (1998) Detection and functional characterization of ryanodine receptors from sea urchin eggs. J Physiol 510(Pt 1):155–164

110. Nakajo K, Chen L, Okamura Y (1999) Cross-coupling between voltage-dependent Ca^{2+} channels and ryanodine receptors in developing ascidian muscle blastomeres. J Physiol 515(Pt 3):695–710

111. Giannini G, Conti A, Mammarella S, Scrobogna M, Sorrentino V (1995) The ryanodine receptor/calcium channel genes are widely and differentially expressed in murine brain and peripheral tissues. J Cell Biol 128(5):893–904
112. Mackrill JJ, Challiss RA, O'Connell DA, Lai FA, Nahorski SR (1997) Differential expression and regulation of ryanodine receptor and myo-inositol 1,4,5-trisphosphate receptor Ca^{2+} release channels in mammalian tissues and cell lines. Biochem J 327(Pt 1):251–258
113. Holland EB, Goldstone JV, Pessah IN, Whitehead A, Reid NM, Karchner SI, Hahn ME, Nacci DE, Clark BW, Stegeman JJ (2017) Ryanodine receptor and FK506 binding protein 1 in the Atlantic killifish (Fundulus heteroclitus): a phylogenetic and population-based comparison. Aquat Toxicol 192:105–115. https://doi.org/10.1016/j.aquatox.2017.09.002
114. Darbandi S, Franck JP (2009) A comparative study of ryanodine receptor (RyR) gene expression levels in a basal ray-finned fish, bichir (Polypterus ornatipinnis) and the derived euteleost zebrafish (Danio rerio). Comp Biochem Physiol B Biochem Mol Biol 154(4):443–448. https://doi.org/10.1016/j.cbpb.2009.09.003
115. Lai FA, Liu QY, Xu L, el-Hashem A, Kramarcy NR, Sealock R, Meissner G (1992) Amphibian ryanodine receptor isoforms are related to those of mammalian skeletal or cardiac muscle. Am J Phys 263(2 Pt 1):C365–C372. https://doi.org/10.1152/ajpcell.1992.263.2.C365
116. Jeyakumar LH, Ballester L, Cheng DS, McIntyre JO, Chang P, Olivey HE, Rollins-Smith L, Barnett JV, Murray K, Xin HB, Fleischer S (2001) FKBP binding characteristics of cardiac microsomes from diverse vertebrates. Biochem Biophys Res Commun 281(4):979–986. https://doi.org/10.1006/bbrc.2001.4444
117. Klitzner T, Morad M (1983) Excitation-contraction coupling in frog ventricle. Possible Ca^{2+} transport mechanisms. Pflugers Arch 398(4):274–283
118. Shiels HA, Sitsapesan R (2015) Is there something fishy about the regulation of the ryanodine receptor in the fish heart? Exp Physiol 100(12):1412–1420. https://doi.org/10.1113/EP085136
119. Cros C, Salle L, Warren DE, Shiels HA, Brette F (2014) The calcium stored in the sarcoplasmic reticulum acts as a safety mechanism in rainbow trout heart. Am J Physiol Regul Integr Comp Physiol 307(12):R1493–R1501. https://doi.org/10.1152/ajpregu.00127.2014
120. Tijskens P, Meissner G, Franzini-Armstrong C (2003) Location of ryanodine and dihydropyridine receptors in frog myocardium. Biophys J 84(2 Pt 1):1079–1092. https://doi.org/10.1016/S0006-3495(03)74924-8
121. Perin P, Botta L, Tritto S, Laforenza U (2012) Expression and localization of ryanodine receptors in the frog semicircular canal. J Biomed Biotechnol 2012:398398. https://doi.org/10.1155/2012/398398
122. Session AM, Uno Y, Kwon T, Chapman JA, Toyoda A, Takahashi S, Fukui A, Hikosaka A, Suzuki A, Kondo M, van Heeringen SJ, Quigley I, Heinz S, Ogino H, Ochi H, Hellsten U, Lyons JB, Simakov O, Putnam N, Stites J, Kuroki Y, Tanaka T, Michiue T, Watanabe M, Bogdanovic O, Lister R, Georgiou G, Paranjpe SS, van Kruijsbergen I, Shu S, Carlson J, Kinoshita T, Ohta Y, Mawaribuchi S, Jenkins J, Grimwood J, Schmutz J, Mitros T, Mozaffari SV, Suzuki Y, Haramoto Y, Yamamoto TS, Takagi C, Heald R, Miller K, Haudenschild C, Kitzman J, Nakayama T, Izutsu Y, Robert J, Fortriede J, Burns K, Lotay V, Karimi K, Yasuoka Y, Dichmann DS, Flajnik MF, Houston DW, Shendure J, DuPasquier L, Vize PD, Zorn AM, Ito M, Marcotte EM, Wallingford JB, Ito Y, Asashima M, Ueno N, Matsuda Y, Veenstra GJ, Fujiyama A, Harland RM, Taira M, Rokhsar DS (2016) Genome evolution in the allotetraploid frog Xenopus laevis. Nature 538(7625):336–343. https://doi.org/10.1038/nature19840
123. Bowes JB, Snyder KA, Segerdell E, Gibb R, Jarabek C, Noumen E, Pollet N, Vize PD (2008) Xenbase: a Xenopus biology and genomics resource. Nucleic Acids Res 36(Database issue):D761–D767. https://doi.org/10.1093/nar/gkm826
124. Mouton J, Marty I, Villaz M, Feltz A, Maulet Y (2001) Molecular interaction of dihydropyridine receptors with type-1 ryanodine receptors in rat brain. Biochem J 354(Pt 3):597–603
125. Cheng W, Altafaj X, Ronjat M, Coronado R (2005) Interaction between the dihydropyridine receptor Ca^{2+} channel beta-subunit and ryanodine receptor type 1 strength-

ens excitation-contraction coupling. Proc Natl Acad Sci U S A 102(52):19225–19230. https://doi.org/10.1073/pnas.0504334102

126. Samso M (2015) 3D structure of the Dihydropyridine receptor of skeletal muscle. Eur J Transl Myol 25(1):4840. https://doi.org/10.4081/ejtm.2015.4840

127. McKay PB, Griswold CK (2014) A comparative study indicates both positive and purifying selection within ryanodine receptor (RyR) genes, as well as correlated evolution. J Exp Zool A Ecol Genet Physiol 321(3):151–163. https://doi.org/10.1002/jez.1845

128. Tanabe T, Beam KG, Powell JA, Numa S (1988) Restoration of excitation-contraction coupling and slow calcium current in dysgenic muscle by dihydropyridine receptor complementary DNA. Nature 336(6195):134–139. https://doi.org/10.1038/336134a0

129. Tanabe T, Beam KG, Adams BA, Niidome T, Numa S (1990) Regions of the skeletal muscle dihydropyridine receptor critical for excitation-contraction coupling. Nature 346(6284):567–569. https://doi.org/10.1038/346567a0

130. Takekura H, Paolini C, Franzini-Armstrong C, Kugler G, Grabner M, Flucher BE (2004) Differential contribution of skeletal and cardiac II-III loop sequences to the assembly of dihydropyridine-receptor arrays in skeletal muscle. Mol Biol Cell 15(12):5408–5419. https://doi.org/10.1091/mbc.E04-05-0414

131. Ahern CA, Bhattacharya D, Mortenson L, Coronado R (2001) A component of excitation-contraction coupling triggered in the absence of the T671-L690 and L720-Q765 regions of the II-III loop of the dihydropyridine receptor alpha(1s) pore subunit. Biophys J 81(6):3294–3307. https://doi.org/10.1016/S0006-3495(01)75963-2

132. Leong P, MacLennan DH (1998) The cytoplasmic loops between domains II and III and domains III and IV in the skeletal muscle dihydropyridine receptor bind to a contiguous site in the skeletal muscle ryanodine receptor. J Biol Chem 273(45):29958–29964

133. Kugler G, Weiss RG, Flucher BE, Grabner M (2004) Structural requirements of the dihydropyridine receptor alpha1S II-III loop for skeletal-type excitation-contraction coupling. J Biol Chem 279(6):4721–4728. https://doi.org/10.1074/jbc.M307538200

134. Leong P, MacLennan DH (1998) A 37-amino acid sequence in the skeletal muscle ryanodine receptor interacts with the cytoplasmic loop between domains II and III in the skeletal muscle dihydropyridine receptor. J Biol Chem 273(14):7791–7794

135. Proenza C, O'Brien J, Nakai J, Mukherjee S, Allen PD, Beam KG (2002) Identification of a region of RyR1 that participates in allosteric coupling with the alpha(1S) (Ca(V)1.1) II-III loop. J Biol Chem 277(8):6530–6535. https://doi.org/10.1074/jbc.M106471200

136. Sorrentino V, Volpe P (1993) Ryanodine receptors: how many, where and why? Trends Pharmacol Sci 14(3):98–103

137. Yamazawa T, Takeshima H, Shimuta M, Iino M (1997) A region of the ryanodine receptor critical for excitation-contraction coupling in skeletal muscle. J Biol Chem 272(13):8161–8164

138. Perez CF, Mukherjee S, Allen PD (2003) Amino acids 1-1,680 of ryanodine receptor type 1 hold critical determinants of skeletal type for excitation-contraction coupling. Role of divergence domain D2. J Biol Chem 278(41):39644–39652. https://doi.org/10.1074/jbc.M305160200

139. Bannister RA, Sheridan DC, Beam KG (2016) Distinct components of retrograde Ca(V)1.1-RyR1 coupling revealed by a lethal mutation in RyR1. Biophys J 110(4):912–921. https://doi.org/10.1016/j.bpj.2015.12.031

140. Takeshima H, Hoshijima M, Song LS (2015) Ca(2)(+) microdomains organized by junctophilins. Cell Calcium 58(4):349–356. https://doi.org/10.1016/j.ceca.2015.01.007

141. Landstrom AP, Beavers DL, Wehrens XH (2014) The junctophilin family of proteins: from bench to bedside. Trends Mol Med 20(6):353–362. https://doi.org/10.1016/j.molmed.2014.02.004

142. Nishi M, Sakagami H, Komazaki S, Kondo H, Takeshima H (2003) Coexpression of junctophilin type 3 and type 4 in brain. Brain Res Mol Brain Res 118(1–2):102–110

143. Woo JS, Srikanth S, Nishi M, Ping P, Takeshima H, Gwack Y (2016) Junctophilin-4, a component of the endoplasmic reticulum-plasma membrane junctions, regu-

lates Ca^{2+} dynamics in T cells. Proc Natl Acad Sci U S A 113(10):2762–2767. https://doi.org/10.1073/pnas.1524229113

144. Reynolds JO, Quick AP, Wang Q, Beavers DL, Philippen LE, Showell J, Barreto-Torres G, Thuerauf DJ, Doroudgar S, Glembotski CC, Wehrens XH (2016) Junctophilin-2 gene therapy rescues heart failure by normalizing RyR2-mediated Ca^{2+} release. Int J Cardiol 225:371–380. https://doi.org/10.1016/j.ijcard.2016.10.021

145. van Oort RJ, Garbino A, Wang W, Dixit SS, Landstrom AP, Gaur N, De Almeida AC, Skapura DG, Rudy Y, Burns AR, Ackerman MJ, Wehrens XH (2011) Disrupted junctional membrane complexes and hyperactive ryanodine receptors after acute junctophilin knockdown in mice. Circulation 123(9):979–988. https://doi.org/10.1161/CIRCULATIONAHA.110.006437

146. Jiang M, Zhang M, Howren M, Wang Y, Tan A, Balijepalli RC, Huizar JF, Tseng GN (2016) JPH-2 interacts with Cai-handling proteins and ion channels in dyads: contribution to premature ventricular contraction-induced cardiomyopathy. Heart Rhythm 13(3):743–752. https://doi.org/10.1016/j.hrthm.2015.10.037

147. Munro ML, Jayasinghe ID, Wang Q, Quick A, Wang W, Baddeley D, Wehrens XH, Soeller C (2016) Junctophilin-2 in the nanoscale organisation and functional signalling of ryanodine receptor clusters in cardiomyocytes. J Cell Sci 129(23):4388–4398. https://doi.org/10.1242/jcs.196873

148. Ito K, Komazaki S, Sasamoto K, Yoshida M, Nishi M, Kitamura K, Takeshima H (2001) Deficiency of triad junction and contraction in mutant skeletal muscle lacking junctophilin type 1. J Cell Biol 154(5):1059–1067. https://doi.org/10.1083/jcb.200105040

149. Nakada T, Kashihara T, Komatsu M, Kojima K, Takeshita T, Yamada M (2018) Physical interaction of junctophilin and the CaV1.1 C terminus is crucial for skeletal muscle contraction. Proc Natl Acad Sci U S A 115(17):4507–4512. https://doi.org/10.1073/pnas.1716649115

150. Garbino A, van Oort RJ, Dixit SS, Landstrom AP, Ackerman MJ, Wehrens XH (2009) Molecular evolution of the junctophilin gene family. Physiol Genomics 37(3):175–186. https://doi.org/10.1152/physiolgenomics.00017.2009

151. Mecklenburg KL, Freed SA, Raval M, Quintero OA, Yengo CM, O'Tousa JE (2015) Invertebrate and vertebrate class III myosins interact with MORN repeat-containing adaptor proteins. PLoS One 10(3):e0122502. https://doi.org/10.1371/journal.pone.0122502

152. Kohno T, Wakabayashi K, Diener DR, Rosenbaum JL, Kamiya R (2011) Subunit interactions within the Chlamydomonas flagellar spokehead. Cytoskeleton (Hoboken) 68(4):237–246. https://doi.org/10.1002/cm.20507

153. Sheerin UM, Schneider SA, Carr L, Deuschl G, Hopfner F, Stamelou M, Wood NW, Bhatia KP (2014) ALS2 mutations: juvenile amyotrophic lateral sclerosis and generalized dystonia. Neurology 82(12):1065–1067. https://doi.org/10.1212/WNL.0000000000000254

154. Topp JD, Gray NW, Gerard RD, Horazdovsky BF (2004) Alsin is a Rab5 and Rac1 guanine nucleotide exchange factor. J Biol Chem 279(23):24612–24623. https://doi.org/10.1074/jbc.M313504200

155. Plattner H (2015) Molecular aspects of calcium signalling at the crossroads of unikont and bikont eukaryote evolution – the ciliated protozoan Paramecium in focus. Cell Calcium 57(3):174–185. https://doi.org/10.1016/j.ceca.2014.12.002

156. Hasegawa M, Fujiwara M (1993) Relative efficiencies of the maximum likelihood, maximum parsimony, and neighbor-joining methods for estimating protein phylogeny. Mol Phylogenet Evol 2(1):1–5. https://doi.org/10.1006/mpev.1993.1001

157. Kall L, Krogh A, Sonnhammer EL (2004) A combined transmembrane topology and signal peptide prediction method. J Mol Biol 338(5):1027–1036. https://doi.org/10.1016/j.jmb.2004.03.016

158. Sonnhammer EL, von Heijne G, Krogh A (1998) A hidden Markov model for predicting transmembrane helices in protein sequences. Proc Int Conf Intell Syst Mol Biol 6:175–182

159. Calpena E, Lopez Del Amo V, Chakraborty M, Llamusi B, Artero R, Espinos C, Galindo MI (2018) The Drosophila junctophilin gene is functionally equivalent to its four mammalian counterparts and is a modifier of a Huntingtin poly-Q expansion and the Notch pathway. Dis Model Mech 11(1):dmm029082. https://doi.org/10.1242/dmm.029082

160. Mello CV, Lovell PV (2018) Avian genomics lends insights into endocrine function in birds. Gen Comp Endocrinol 256:123–129. https://doi.org/10.1016/j.ygcen.2017.05.023
161. Berridge MJ (1998) Neuronal calcium signaling. Neuron 21(1):13–26
162. Tamura K, Peterson D, Peterson N, Stecher G, Nei M, Kumar S (2011) MEGA5: molecular evolutionary genetics analysis using maximum likelihood, evolutionary distance, and maximum parsimony methods. Mol Biol Evol 28(10):2731–2739. https://doi.org/10.1093/molbev/msr121
163. Felsenstein J (1985) Confidence-limits on phylogenies – an approach using the bootstrap. Evolution 39(4):783–791
164. Sievers F, Wilm A, Dineen D, Gibson TJ, Karplus K, Li W, Lopez R, McWilliam H, Remmert M, Soding J, Thompson JD, Higgins DG (2011) Fast, scalable generation of high-quality protein multiple sequence alignments using Clustal omega. Mol Syst Biol 7:539. https://doi.org/10.1038/msb.2011.75
165. Nei M, Kumar S (2000) Molecular evolution and phylogenetics. Oxford University Press, New York
166. Sheard TM, Kharche SR, Pinali C, Shiels HA. 3D ultrastructural organisation of calcium release units in the avian sarcoplasmic reticulum. Journal of Experimental Biology. 2019 Jan 1:jeb-197640.

Chapter 13
Molecular Insights into Calcium Dependent Regulation of Ryanodine Receptor Calcium Release Channels

Naohiro Yamaguchi

Abstract Ryanodine receptor calcium release channels (RyRs) play central roles in controlling intracellular calcium concentrations in excitable and non-excitable cells. RyRs are located in the sarcoplasmic or endoplasmic reticulum, intracellular Ca^{2+} storage compartment, and release Ca^{2+} during cellular action potentials or in response to other cellular stimuli. Mammalian cells express three structurally related isoforms of RyR. RyR1 and RyR2 are the major RyR isoforms in skeletal and cardiac muscle, respectively, and RyR3 is expressed in various tissues along with the other two isoforms. A prominent feature of RyRs is that the Ca^{2+} release channel activities of RyRs are regulated by calcium ions; therefore, intracellular Ca^{2+} release controls positive- and negative-feedback phenomena through the RyRs. RyR channel activities are also regulated by Ca^{2+} indirectly, i.e. through Ca^{2+} binding proteins at both cytosolic and sarco/endoplasmic reticulum luminal sides. Here, I summarize Ca^{2+}-dependent feedback regulation of RyRs including recent progress in the structure/function aspects.

Keywords Ryanodine receptor · Excitation-contraction coupling · Calcium release channel · Intracellular calcium · Calmodulin

Transient increase of intracellular Ca^{2+} concentration plays a pivotal role in numerous cell functions, including muscle contraction, neuronal plasticity, and immune responses. Multiple sources of Ca^{2+} are involved in this signaling, including Ca^{2+} influx from the extracellular spaces and Ca^{2+} release from intracellular Ca^{2+} stores: the endo/sarcoplasmic reticulum (ER/SR), nuclear envelope, and mitochondria [1–3]. Ryanodine receptors (RyRs) are Ca^{2+} release channels located in the ER/SR

N. Yamaguchi (✉)
Cardiac Signaling Center, University of South Carolina, Medical University of South Carolina and Clemson University, Charleston, SC, USA

Department of Regenerative Medicine and Cell Biology, Medical University of South Carolina, Charleston, SC, USA
e-mail: yamaguch@musc.edu

© Springer Nature Switzerland AG 2020
M. S. Islam (ed.), *Calcium Signaling*, Advances in Experimental Medicine and Biology 1131, https://doi.org/10.1007/978-3-030-12457-1_13

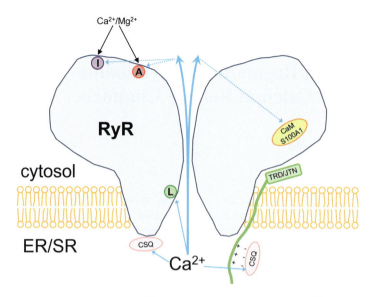

Fig. 13.1 A schematic of the RyR Ca^{2+} release channel regulation by Ca^{2+}. RyRs have been suggested to possess at least three Ca^{2+} regulatory sites; cytoplasmic high affinity activation (A) and low affinity inhibition (I) sites, and luminal regulatory (L) site. Mg^{2+} inhibits RyRs by binding to A or I sites. Calcium ions passing through the RyR from ER/SR lumen to the cytoplasm are considered to bind to RyR-A and I sites and cytoplasmic accessory proteins, CaM and S100A1. The low affinity luminal Ca^{2+} binding protein, CSQ, forms a macromolecular complex via its interaction with RyR accessory proteins, triadin (TRD) and junctin (JTN). CSQ may also bind to RyR directly

and play a primary role in Ca^{2+} release from SR during skeletal and cardiac muscle contraction [4–6]. Additionally, RyRs play an important role in smooth muscle, neurons, and other cell types by co-existing with another family of Ca^{2+} release channels called inositol-trisphosphate receptors [3]. RyRs release Ca^{2+} from the ER/SR and are regulated by Ca^{2+} [4–6], suggesting that RyRs have a self-regulatory mechanism controlled by Ca^{2+}, i.e. RyR-released Ca^{2+} regulates the same or neighboring RyRs. This chapter focuses on the structure-function aspect of RyR regulation by Ca^{2+} and Ca^{2+} binding proteins such as calmodulin and calsequestrin (Fig. 13.1).

13.1 Molecular Structure of RyRs

In striated muscle, RyRs are localized in the junctional SR membrane in close proximity to transverse (T)-tubule membranes, invaginations of the plasma membrane into the myofibrils. In skeletal muscle the SR is typically on both sides of the T-tubule (triad), while in cardiomyocytes, the SR is only on one side of the

T-tubule (dyad). In both triadic and dyadic junctions, electron microscopy shows foot structures spanning between the SR and T-tubule [4]. Molecular identification of RyRs was first performed with rabbit fast twitch skeletal muscle using ryanodine, a specific ligand of RyRs. Isolated RyRs are homotetramers of a ~500 kDa polypeptide. Morphological analysis of the reconstituted purified proteins identified RyRs as the foot structures [7–10]. Molecular cloning of RyRs showed that mammals express three different RyR isoforms [11–16]. Skeletal muscle expresses primarily RyR1. The dominant RyR isoform in cardiac muscle is RyR2. RyR3 was initially identified in the brain; however, the brain expresses all three RyR isoforms. Although expression patterns depend on the locations in the brain, in general, RyR2 is widely dispersed over the whole brain [17]. RyR3 is also expressed together with RyR1 in the diaphragm and slow twitch muscle [18, 19], thus functional characterization of RyR1 is mainly performed with fast twitch muscle. In amphibians and avian skeletal muscle, two RyR isoforms, αRyR and βRyR, are recognized and correspond to the mammalian RyR1 and RyR3, respectively [20–22].

All three isoforms of RyR have a large cytoplasmic domain, which possesses multiple regulatory sites for channel activity. The carboxyl-terminal end of the RyR spans the SR membrane six times, in which a pore helix and the transmembrane segment form the channel pore [23–27]. RyRs are activated by micromolar Ca^{2+} and adenine nucleotide, and are inhibited by millimolar Ca^{2+} and Mg^{2+} [4–6]. A number of proteins have been found to interact with RyRs and regulate their channel activity. These include triadin, junctin, FK506-binding proteins, protein kinases and phosphatases, and Ca^{2+} binding proteins such as calmodulin and S100A1. Recently, cryo-electron microscopy and 3D image reconstruction of the purified full-length RyRs and crystal structural analysis of truncated recombinant RyRs have detailed the structures of RyR1 and RyR2 at near atomic resolution [24–30].

13.2   Activation by Cytoplasmic Ca^{2+}

Skeletal and cardiac muscle contractions are triggered by SR Ca^{2+} release mediated by RyR1 and RyR2. Two different mechanisms are now recognized to open RyRs. In skeletal muscle, direct interaction between RyR1 and the T-tubule voltage sensor, also recognized as the DHP receptor L-type Ca^{2+} channel (Cav1.1), opens RyR1 during skeletal action potential (voltage-induced Ca^{2+} release) [31]. Alternatively, in cardiac muscle, small Ca^{2+} influx through the cardiac L-type Ca^{2+} channel (Cav1.2) increases intracellular Ca^{2+}, and at ~micromolar concentrations opens RyR2 by means of Ca^{2+}-induced Ca^{2+} release (CICR) [32]. The CICR mechanism was initially recognized in skeletal muscle contraction [33, 34]; however, elimination of Ca^{2+} from the extracellular space or blocking Ca^{2+} influx through Cav1.1 did not abolish voltage-dependent intracellular Ca^{2+} transients [35, 36]. Thus, CICR in skeletal muscle (RyR1) is not considered a trigger for muscle contraction.

Furthermore, slower kinetics of CICR in contrast to the rapid Ca^{2+} release in skeletal muscle also supported the idea that CICR is not a physiological trigger for skeletal muscle contraction [37]. However, CICR may play a role in amplifying Ca^{2+} signaling by activating RyR1s which do not couple with DHP receptors or the small population of RyR3s [38]. Calcium-dependent activation of RyR1 can be altered by RyR1 missense mutations associated with skeletal myopathies such as malignant hyperthermia, thus, CICR may impact these pathologies [37, 39]. Ca^{2+}-dependent activation of RyRs has been well characterized using isolated membrane fractions, intact cells and muscle fibers, purified RyR proteins, and recombinant RyRs by several different methods including muscle tension measurements, Ca^{2+} flux measurements using Ca^{2+} indicator dyes or radioactive $^{45}Ca^{2+}$, single channel recordings, and specific ligand ($[^3H]$ryanodine) binding assays [37, 40, 41]. All three mammalian RyR isoforms are activated by $\sim$0.5–5 μM Ca^{2+} depending on assay conditions. Several potential Ca^{2+} binding sites were initially identified using truncated RyR1 proteins and $^{45}Ca^{2+}$ overlays [42–44]. Subsequently, site-directed mutagenesis showed that E3987 in RyR2 (E4032 in RyR1) was critical for Ca^{2+}-dependent activation of RyRs [45, 46]. The mutant RyRs showed impaired Ca^{2+} dependent activation in single channel recordings and $[^3H]$ryanodine binding assay. E4032A-RyR1 expressing myotubes were impaired in caffeine-induced Ca^{2+} release, but the aberrant function was restored in the presence of ryanodine [47]. Recently, near-atomic level cryo-electron microscopy analysis of RyR1 ($\sim$4 Å resolution) determined the open and closed state conformations of RyR1 [30]. The structure of RyR1 with 30 μM Ca^{2+}, which is optimal for RyR activation, identified a new Ca^{2+} binding site in RyR1. The Ca^{2+} binding site is formed by 3 essential amino acids, E3893, E3967, and T5001, together with two auxiliary amino acids, H3895 and Q3970, for secondary coordination of the Ca^{2+} sphere [30]. In this structural model, E4032 is distal from the bound calcium ion, but forms an interface with carboxyl terminal tail where T5001 locates. This suggests that E4032 contribute to stabilize the conformation of Ca^{2+} bound RyR1 [30]. Murayama and colleagues introduced point mutations on RyR2 amino acids corresponding to the RyR1 E3893, E3967, and Q3970, and found that the mutations altered Ca^{2+}-dependent activation of RyR2 [48]. These functional results support the idea that the identified Ca^{2+} binding site serves as a functional Ca^{2+} regulatory site. Further detail analysis including assessments of other amino acids combined in the presence of other channel agonists and antagonists will further advance structure and function relationship of Ca^{2+}-dependent activation of RyRs. Another $\sim$6 Å resolution cryo-electron micrograph of RyR1 suggested that 10 mM Ca^{2+} changed the conformation of the EF hand-type Ca^{2+} binding domain of RyR1; therefore, it was proposed as a Ca^{2+} activation site [24]. However, studies with recombinant proteins including the EF hand domain showed Ca^{2+} affinity was >60 μM [49, 50], which is much higher than the RyR-activating Ca^{2+} concentration. Also, functional study scrambling of the EF hand sequence in RyR1 and deletion of the entire EF hand in RyR2 did not affect the Ca^{2+} activation of RyRs [51, 52]. Considering that the structural analysis was determined with 10 mM Ca^{2+}, the EF hand site is likely to be a Ca^{2+}

inactivation site [53]. We also found that the EF hand domain contributes to the isoform-specific regulation of RyRs by calmodulin (see below) [54].

Ca^{2+}-dependent activation of RyR1 and RyR2 are similar in single channel recordings and flux measurements in the SR vesicles; however, Ogawa and colleagues pointed out that RyR2 in rat ventricular SR or as a recombinant form exhibited a suppressed activity at 10–100 μM Ca^{2+} using [^{3}H]ryanodine equilibrium binding assay [55]. Similar suppressed RyR2 activities were observed in our own study with rabbit recombinant RyR2 using the same technique [56]. Surprisingly, this suppressed activity was restored by $\sim$1 mM Mg^{2+} [55], which is usually considered to be an inhibitor of RyR channel activity by competing off Ca^{2+} at the Ca^{2+} activation site or binding to the Ca^{2+} inhibitory site [40, 57]. RyR2 in the rabbit ventricular SR showed this suppression only when AMP or caffeine, RyR activators, were added, suggesting that the suppressed effects depend on the type of RyR2 sample. One possibility for this mechanism is therefore that regulatory factors were removed during the sample preparations. Another possible explanation is that the RyR2 conformation is not very stable under long time (>8 h) equilibrium conditions in the [^{3}H]ryanodine binding assays. We found that replacement of the RyR1-EF hand domain with corresponding RyR2 sequence or the introduction of point mutations in the cytoplasmic loop between the second and the third transmembrane segments (S2-S3 loop) of RyR1 resulted in suppressed activity at 10–100 μM Ca^{2+} [53, 58]. The results suggested that the EF hand and S2-S3 cytoplasmic loop of RyRs are involved in the conformational stability and Ca^{2+}-dependent regulation (activation/inhibition) of RyR channels.

13.3 Inhibition by Cytoplasmic Ca^{2+}

While RyRs are activated by micromolar cytosolic Ca^{2+}, higher concentrations of Ca^{2+} (>1 mM) inhibit RyR channel activities. Thus, RyRs have a high affinity Ca^{2+} activation site and a low affinity Ca^{2+} inactivation site (A and I sites, respectively in Fig. 13.1). These sites are also implicated in Mg^{2+} inhibition, namely submillimolar Mg^{2+} competes with activating Ca^{2+} at A site and millimolar Mg^{2+} binds to the I site for inhibitory effect [57, 59]. Although the physiological significance of RyR inactivation by millimolar levels of Ca^{2+} has been questioned, local rise of cytosolic Ca^{2+} around the RyRs may be sufficient to inhibit RyR channel activity. Single channel recording showed that Ca^{2+} flux from the lumen to the cytosolic side resulted in a decrease of open probability of both the RyR1 and RyR2 channel, supporting Ca^{2+}-dependent inactivation of RyRs by the released Ca^{2+} in intact tissues [60, 61]. All three mammalian isoforms of RyR are inhibited by high concentrations of Ca^{2+}; however, affinity for inhibitory Ca^{2+} in RyR1 is 5–10 times higher than those in RyR2 and RyR3 [6, 53, 62]. Deletion of 52 amino acids including a large cluster (42 amino acids) of negatively charged amino acids in RyR1 resulted in a threefold decrease in Ca^{2+} inactivation affinity [63]; yet, this change in the local electrostatics property may have caused a large

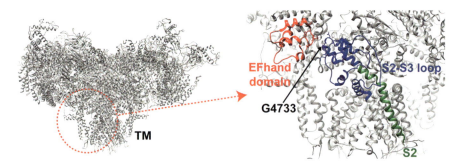

Fig. 13.2 High resolution cryo-electron microscopy structure of RyR1. The closed state of RyR1 (Protein Data Bank Accession 5TB0 [30]) is presented by UCSF Chimera program (https://www.cgl.ucsf.edu/chimera/) [121]. (*Left panel*) Structure of tetrameric RyR1. TM denotes transmembrane region. (*Right panel*) Enlargement of region marked with red circle in *left panel*. The EF hand domain (*red*) is shown to be adjacent to the S2-S3 loop (*blue*) in the neighboring subunit [25]. In this structure, Gly4733 and neighboring amino acids are located in close proximity to the EF hand domain, and point mutations on these amino acid residues altered Ca^{2+}-dependent inactivation of RyR1 [58]. Thus, the S2-S3 loop of RyR may transduce its Ca^{2+}-dependent inhibitory signal through the EF hand domain. Note that the S2 transmembrane (*green*) has also been shown to be involved in Ca^{2+}-dependent inactivation of RyRs [53]

conformational change. Construction and characterization of RyR1/RyR2 chimeras highlighted differences of Ca^{2+}-inactivation affinity between the two RyR isoforms. Chimeric RyRs showed that RyR isoform specific Ca^{2+} inactivation depends on the sequence of the carboxyl-terminal quarter (~1000 amino acids) [62, 64, 65]. Further characterization suggested that the second transmembrane segment (S2) and EF hand type Ca^{2+} binding motifs are involved in the isoform-specific Ca^{2+}-dependent inactivation of RyRs [53]. In agreement with these observations, scrambling one EF hand sequence (EF1) in RyR1 resulted in a twofold reduction in the affinity of Ca^{2+}-dependent inhibition [51]. In near-atomic level cryo-electron microscopy, the EF hand domain and S2-S3 cytoplasmic loop are in close proximity [25]. In another structural model, 10 mM Ca^{2+} changed the conformation of the EF hand domain [24]. Site-directed mutagenesis of the S2-S3 loop of RyR1 impaired the affinity for Ca^{2+}-dependent inactivation, and resulted in RyR2-type Ca^{2+}-dependent activity profiles [58]. Considering the Ca^{2+} affinity of the recombinant EF hand domain (60 μM-4 mM) [49, 50], the Ca^{2+} inactivation site of the RyR is the EF hand motif. One possible mechanism is that the S2-S3 loop transduces the signal of Ca^{2+} binding to the EF hand domain to the channel pore region including S2 [58]. It should be noted that a point mutation in G4733 of RyR1, which is in close proximity to the EF hand domain (Fig. 13.2), significantly suppressed Ca^{2+}-dependent inactivation [58].

13.4 Regulation by Luminal Ca^{2+}

RyRs could also be regulated by SR luminal Ca^{2+}, as during Ca^{2+} release the junctional SR Ca^{2+} concentration drastically drops. This suggests that RyR channel gating can be regulated directly by luminal Ca^{2+}; e.g. SR Ca^{2+} filling status regulates RyR channel opening and closing. It is known that the SR Ca^{2+} store with a certain level of Ca^{2+} exhibits spontaneous Ca^{2+} release in mammalian cardiac muscle cells [66, 67]. Chen and colleagues found that the store overload-induced Ca^{2+} release (SOICR) was observed in heterologous cells expressing recombinant RyR channels; therefore, it is likely an intrinsic property of RyRs. [68]. SOICR mechanisms were implicated in the aberrant Ca^{2+} signaling found in RyR mutation-related skeletal and cardiac muscle diseases [68–70]. The muscular disease-associated RyR mutations reduce the threshold for SOICR; therefore, spontaneous Ca^{2+} release (Ca^{2+} spills) occurs when the SR Ca^{2+} store loading is increased by the triggers of pathologies such as catecholamine release. The luminal Ca^{2+}-sensing gate of RyRs was investigated by site-directed mutagenesis, revealing that E4872 on the inner pore helix (S6 transmembrane segment) of RyR2 is essential for luminal Ca^{2+} activation of RyR2 and SOICR [71]. Knock-in mice harboring the E4872Q-RyR2 mutation were resistant to Ca^{2+}-dependent ventricular tachycardia, suggesting that SOICR is a critical mechanism for arrhythmogenesis [71].

It also should be noted that luminal Ca^{2+} can also access cytosolic Ca^{2+} activation and inactivation sites [60, 61] (Fig. 13.1). In single channel measurements of RyR1 and RyR2, luminal Ca^{2+} passed through RyRs to the cytosolic side in conjunction with potassium ions under a voltage gradient, and activated and inhibited the same RyR channels depending on luminal Ca^{2+} concentration [60, 61], which suggests that during excitation-contraction coupling, local cytoplasmic Ca^{2+} concentrations can reach millimolar levels and are sufficient for Ca^{2+}-dependent inactivation of RyRs.

13.5 Regulation by Calmodulin and S100A1

Calmodulin (CaM) is a 16.7-kDa protein that possesses 2 EF hand-type Ca^{2+} binding sites on both the amino and carboxyl-terminal. Thus, CaM works as a Ca^{2+} sensing subunit of multiple ion channels [72]. CaM modifies RyR channel function independently from regulation by Ca^{2+}; therefore, RyRs have "dual" cytosolic Ca^{2+} dependent regulatory mechanisms (direct and indirect). RyRs are regulated by not only the Ca^{2+} bound form of CaM, but also by CaM at cellular resting Ca^{2+} concentrations ($\sim$0.1 μM). Ca^{2+} bound CaM inhibits all three mammalian isoforms of RyR, while CaM activates RyR1 and RyR3 and inhibits RyR2 at submicromolar Ca^{2+} concentrations [73–77], suggesting that CaM constitutively binds to RyRs to regulate their channel activities by sensing cytoplasmic Ca^{2+} concentrations. In vitro experiments also showed that CaM regulation of the RyR depends on redox state. Affinities for CaM regulation of RyR channel activity at the oxidized condition are 2–20 fold lower than at the reduced condition [77, 78]. The results are consistent with observations that CaM is dissociated from RyR2, resulting in a Ca^{2+} leak from SR in failing hearts [79], in which the redox balance possibly shifts to the more oxidized condition [80, 81].

Purified RyR1 and RyR2 as well as the recombinant RyR3 bind 4 CaM per tetrameric RyR, i.e. one RyR subunit binds one CaM [56, 77, 78, 82]. The CaM binding and regulatory domain was identified by trypsin digestion, binding of synthetic RyR1 peptides, and site directed mutagenesis of RyR1 amino acids 3614–3643 [82–84]. This domain was confirmed to be conserved in RyR2 and RyR3 by site-directed mutagenesis [56, 78]. Crystal structure analysis of a synthetic RyR1 peptide (amino acids 3614–3643) and CaM complex revealed that the carboxyl-terminal lobe of CaM binds to the peptide, while the amino-terminal lobe binds with low affinity or is free [85], which may explain that multiple RyR domain peptides or fusion proteins can bind to CaM [44, 86–88]. Point mutations in RyR1 3614–3643 or the corresponding RyR2 and RyR3 domains eliminated CaM binding and regulation of channel activities [56, 78, 82, 89]; thus, this conserved domain likely plays a primary role for CaM-dependent regulation. Although the primary CaM regulatory domain is well conserved, RyR isoform-specific CaM regulation at submicromolar (cellular resting level) Ca^{2+} concentrations, namely activation of RyR1 and RyR3 versus inhibition of RyR2, was investigated using RyR1/RyR2 chimera channels. Replacing the flanking regions of the RyR2 CaM binding domain with the RyR1 sequence abolished CaM regulation of RyR2 at submicromolar Ca^{2+} concentrations [90]. More recently, the EF hand domain and large N-terminal region were shown to be important for isoform-specific CaM regulation of RyRs [54]. These domains possibly mediate long-range interaction between the CaM binding domain and the functional effects on the channel, as the CaM binding domain is $\sim$10 nm apart from the RyR channel pore region in cryo-electron micrographs [91].

In vivo significance of CaM regulation of RyR1 and RyR2 was studied with genetically modified mice. Knock-in mice carrying point mutations in the RyR2 CaM regulatory domain (W3587A/L3591D/F3603A: ADA mutations) were impaired in CaM binding and regulation of cardiac RyR2 [89]. The mice showed rapidly developing cardiac hypertrophy and died 2–3 weeks after birth. Cardiomyocytes isolated from the mutant mouse hearts exhibited long durations of the spontaneous Ca^{2+} transients or Ca^{2+} sparks, indicating that CaM inhibition of RyR2 contributes to the termination of SR Ca^{2+} release, which is important for heart physiology and growth [89, 92]. The knock-in ADA mice were impaired in CaM regulation of RyR2 at both diastolic (submicromolar) and systolic (micromolar) Ca^{2+} concentrations, while knock-in mice with a single mutation (L3591D), were only impaired in CaM regulation of RyR2 during diastole and showed more modest levels of cardiac hypertrophy, suggesting that CaM regulation of RyR2 at systolic Ca^{2+} levels plays a major role in vivo [93]. The corresponding RyR1 mutation (RyR1-L3624D) attenuated both CaM activation and inhibition at submicromolar and micromolar Ca^{2+} concentrations [82]. However, knock-in mice carrying RyR1-L3624D showed only modest effects on skeletal muscle excitation-contraction coupling without lethality, suggesting that CaM regulation of RyR1 plays a minor role in skeletal physiology [94]. More recently, missense mutations in calmodulin genes were identified in patients with catecholaminergic polymorphic ventricular tachycardia (CPVT) [95], in which CaM mutations likely alter RyR2 regulation [96–98]. Thus, the CaM-RyR2 interaction can be a good therapeutic target for the cardiac pathologies [99].

S100A1, another EF hand type Ca^{2+} binding protein, is also expressed in skeletal and cardiac muscle and regulates intracellular Ca^{2+} signaling by interacting with multiple Ca^{2+} handling proteins including RyRs [100–103]. Competitive binding experiments showed that S100A1 shares a common binding site on RyR1 with CaM [104, 105]. Consistently, the L3624D-RyR1 mutation impaired both CaM and S100A1 activation of RyR1 at submicromolar Ca^{2+} concentrations in single channel recordings [94]. On the other hand, the corresponding L3591D-RyR2 mutation abolished CaM regulation only at submicromolar Ca^{2+} level, while S100A1 regulation of the mutant RyR2 was impaired at both submicromolar and micromollar Ca^{2+} concentrations [93]. The mutations and functional experiments suggested that S100A1 and CaM do not share exactly the same binding site in RyR2. Recent FRET experiments also showed that S100A1 interacts allosterically with the CaM binding site in RyR1 and RyR2 rather than through direct binding [106].

13.6 Regulation by Calsequestrin

On the SR luminal side, the low affinity but high capacity Ca^{2+} binding protein, calsequestrin (CSQ), localizes to the junctional SR [107, 108]. RyR-associated proteins, triadin and junctin appear to anchor CSQ to the junctional SR through charge interactions [109–111] (Fig. 13.1). In addition, it was recently shown that cardiac CSQ could also directly bind to the luminal side of RyR2 [112]. Two isoforms, CSQ1 and CSQ2, are dominantly expressed in skeletal and cardiac muscle, respectively. Direct regulation of RyR channel activities by CSQ have been investigated by planar bilayer single channel recordings, where luminal conditions can be controlled. CSQ regulates the RyR channel in a luminal Ca^{2+} concentration dependent manner. With high luminal Ca^{2+} concentrations, CSQ was dissociated from the RyR accessory proteins, triadin and junctin, while CSQ inhibited the RyR channel through the accessory proteins at the intermediate Ca^{2+} concentration [113–115].

Gene knockout of the CSQ in mice demonstrated both its physiological and pathological significance. CSQ1 (*Casq1*) knockout mice were viable and fertile; however, modest structural and functional changes were observed in the fast twitch skeletal muscle. Ablation of CSQ1 resulted in slightly slower force development and relaxation of the fast twitch muscle. Structural analysis showed that CSQ1 knockout muscle exhibited low SR volume and high mitochondria density, suggesting that CSQ1 is important for muscle development [116]. CSQ2 has been implicated in cardiac pathology. Missense mutations in human *CASQ2* gene, resulting in gene knockout or single amino acid substitutions, were found in patients afflicted with catecholaminergic polymorphic ventricular tachycardia [117, 118]. Both mouse models exhibited arrhythmogenesis under the stress conditions of exercise or catecholamine infusion [119, 120]. Consistently, intracellular Ca^{2+} handling was altered by catecholamine in mutant cardiomyocytes isolated from mouse hearts. These results indicate that CSQ2 regulation of SR Ca^{2+} and RyR2 channels is pathologically important.

13.7   Closing Remarks

Almost 50 years have passed since Ca^{2+}-induced Ca^{2+} release, that we now know is associated with Ca^{2+}-dependent activation of RyR, was first reported [33, 34]. In the last 20 years molecular biology and genetic techniques greatly advanced our understanding of the structure/function relationship of RyR channel regulation by small molecules and proteins. More recently, high resolution three dimensional structural analyses have revealed the detailed protein conformations of the RyR channel complexes under different conditions corresponding to the open/closed channels [24–27, 30]. Combining these approaches and using computational modeling will provide more detailed molecular insights into RyR regulation by Ca^{2+} and Ca^{2+} binding proteins at near atomic levels.

Acknowledgements I would like to appreciate Dr. Gerhard Meissner for his mentorship on RyR structure/function analysis when I was working in his laboratory and for his continuous advices and encouragements on my current studies. I am very thankful to Dr. Martin Morad, the director of Cardiac Signaling Center, for providing me with wonderful environment to study Ca^{2+} signaling, and valuable suggestions and discussions on my studies. I am also grateful to Angela C. Gomez and Jordan S. Carter for their contributions to the RyR mutagenesis studies in my laboratory and for the comments on this manuscript. This study was supported by National Institutes of Health Grants R03AR061030, P20GM103499, and UL1TR001450.

References

1. Clapham DE (2007) Calcium signaling. Cell 131(6):1047–1058
2. Bootman MD (2012) Calcium signaling. Cold Spring Harb Perspect Biol 4(7):a011171
3. Berridge MJ (2016) The inositol trisphosphate/calcium signaling pathway in health and disease. Physiol Rev 96(4):1261–1296
4. Franzini-Armstrong C, Protasi F (1997) Ryanodine receptors of striated muscles: a complex channel capable of multiple interactions. Physiol Rev 77(3):699–729
5. Lanner JT, Georgiou DK, Joshi AD, Hamilton SL (2010) Ryanodine receptors: structure, expression, molecular details, and function in calcium release. Cold Spring Harb Perspect Biol 2(11):a003996
6. Meissner G (2017) The structural basis of ryanodine receptor ion channel function. J Gen Physiol 149(12):1065–1089
7. Inui M, Saito A, Fleischer S (1987) Purification of the ryanodine receptor and identity with feet structures of junctional terminal cisternae of sarcoplasmic reticulum from fast skeletal muscle. J Biol Chem 262(4):1740–1747

8. Imagawa T, Smith JS, Coronado R, Campbell KP (1987) Purified ryanodine receptor from skeletal muscle sarcoplasmic reticulum is the Ca^{2+}−permeable pore of the calcium release channel. J Biol Chem 262(34):16636–16643

9. Lai FA, Erickson HP, Rousseau E, Liu QY, Meissner G (1988) Purification and reconstitution of the calcium release channel from skeletal muscle. Nature 331(6154):315–319

10. Anderson K, Lai FA, Liu QY, Rousseau E, Erickson HP, Meissner G (1989) Structural and functional characterization of the purified cardiac ryanodine receptor-Ca^{2+} release channel complex. J Biol Chem 264(2):1329–1335

11. Takeshima H, Nishimura S, Matsumoto T, Ishida H, Kangawa K, Minamino N et al (1989) Primary structure and expression from complementary DNA of skeletal muscle ryanodine receptor. Nature 339(6224):439–445

12. Zorzato F, Fujii J, Otsu K, Phillips M, Green NM, Lai FA et al (1990) Molecular cloning of cDNA encoding human and rabbit forms of the Ca^{2+} release channel (ryanodine receptor) of skeletal muscle sarcoplasmic reticulum. J Biol Chem 265(4):2244–2256

13. Otsu K, Willard HF, Khanna VK, Zorzato F, Green NM, MacLennan DH (1990) Molecular cloning of cDNA encoding the Ca^{2+} release channel (ryanodine receptor) of rabbit cardiac muscle sarcoplasmic reticulum. J Biol Chem 265(23):13472–13483

14. Nakai J, Imagawa T, Hakamata Y, Shigekawa M, Takeshima H, Numa S (1990) Primary structure and functional expression from cDNA of the cardiac ryanodine receptor/calcium release channel. FEBS Lett 271(1–2):169–177

15. Hakamata Y, Nakai J, Takeshima H, Imoto K (1992) Primary structure and distribution of a novel ryanodine receptor/calcium release channel from rabbit brain. FEBS Lett 312(2–3):229–235

16. Takeshima H (1993) Primary structure and expression from cDNAs of the ryanodine receptor. Ann N Y Acad Sci 707:165–177

17. Furuichi T, Furutama D, Hakamata Y, Nakai J, Takeshima H, Mikoshiba K (1994) Multiple types of ryanodine receptor/Ca^{2+} release channels are differentially expressed in rabbit brain. J Neurosci 14(8):4794–4805

18. Conti A, Gorza L, Sorrentino V (1996) Differential distribution of ryanodine receptor type 3 (RyR3) gene product in mammalian skeletal muscles. Biochem J 316(Pt 1):19–23

19. Murayama T, Ogawa Y (1997) Characterization of type 3 ryanodine receptor (RyR3) of sarcoplasmic reticulum from rabbit skeletal muscles. J Biol Chem 272(38):24030–24037

20. Oyamada H, Murayama T, Takagi T, Iino M, Iwabe N, Miyata T et al (1994) Primary structure and distribution of ryanodine-binding protein isoforms of the bullfrog skeletal muscle. J Biol Chem 269(25):17206–17214

21. Percival AL, Williams AJ, Kenyon JL, Grinsell MM, Airey JA, Sutko JL (1994) Chicken skeletal muscle ryanodine receptor isoforms: ion channel properties. Biophys J 67(5):1834–1850

22. Ottini L, Marziali G, Conti A, Charlesworth A, Sorrentino V (1996) Alpha and beta isoforms of ryanodine receptor from chicken skeletal muscle are the homologues of mammalian RyR1 and RyR3. Biochem J 315(Pt 1):207–216

23. Du GG, Sandhu B, Khanna VK, Guo XH, MacLennan DH (2002) Topology of the Ca^{2+} release channel of skeletal muscle sarcoplasmic reticulum (RyR1). Proc Natl Acad Sci U S A 99(26):16725–16730

24. Efremov RG, Leitner A, Aebersold R, Raunser S (2015) Architecture and conformational switch mechanism of the ryanodine receptor. Nature 517(7532):39–43

25. Zalk R, Clarke OB, des Georges A, Grassucci RA, Reiken S, Mancia F et al (2015) Structure of a mammalian ryanodine receptor. Nature 517(7532):44–49

26. Yan Z, Bai X, Yan C, Wu J, Li Z, Xie T et al (2015) Structure of the rabbit ryanodine receptor RyR1 at near-atomic resolution. Nature 517(7532):50–55

27. Peng W, Shen H, Wu J, Guo W, Pan X, Wang R et al (2016) Structural basis for the gating mechanism of the type 2 ryanodine receptor RyR2. Science 354(6310):aah5324

28. Zhong X, Liu Y, Zhu L, Meng X, Wang R, Van Petegem F et al (2013) Conformational dynamics inside amino-terminal disease hotspot of ryanodine receptor. Structure 21(11):2051–2060

29. Yuchi Z, Yuen SM, Lau K, Underhill AQ, Cornea RL, Fessenden JD et al (2015) Crystal structures of ryanodine receptor SPRY1 and tandem-repeat domains reveal a critical FKBP12 binding determinant. Nat Commun 6:7947

30. des Georges A, Clarke OB, Zalk R, Yuan Q, Condon KJ, Grassucci RA et al (2016) Structural basis for gating and activation of RyR1. Cell 167(1):145–157. e17

31. Schneider MF, Chandler WK (1973) Voltage dependent charge movement of skeletal muscle: a possible step in excitation-contraction coupling. Nature 242(5395):244–246

32. Nabauer M, Callewaert G, Cleemann L, Morad M (1989) Regulation of calcium release is gated by calcium current, not gating charge, in cardiac myocytes. Science 244(4906):800–803

33. Ford LE, Podolsky RJ (1970) Regenerative calcium release within muscle cells. Science 167(3914):58–59

34. Endo M, Tanaka M, Ogawa Y (1970) Calcium induced release of calcium from the sarcoplasmic reticulum of skinned skeletal muscle fibres. Nature 228(5266):34–36

35. Brum G, Rios E, Stefani E (1988) Effects of extracellular calcium on calcium movements of excitation-contraction coupling in frog skeletal muscle fibres. J Physiol 398:441–473

36. Nakai J, Dirksen RT, Nguyen HT, Pessah IN, Beam KG, Allen PD (1996) Enhanced dihydropyridine receptor channel activity in the presence of ryanodine receptor. Nature 380(6569):72–75

37. Endo M (2009) Calcium-induced calcium release in skeletal muscle. Physiol Rev 89(4):1153–1176

38. Rios E (2018) Calcium-induced release of calcium in muscle: 50 years of work and the emerging consensus. J Gen Physiol 150(4):521–537

39. Murayama T, Kurebayashi N, Ogawa H, Yamazawa T, Oyamada H, Suzuki J et al (2016) Genotype-phenotype correlations of malignant hyperthermia and central Core disease mutations in the central region of the RYR1 channel. Hum Mutat 37(11):1231–1241

40. Meissner G (1994) Ryanodine receptor/Ca^{2+} release channels and their regulation by endogenous effectors. Annu Rev Physiol 56:485–508

41. Coronado R, Morrissette J, Sukhareva M, Vaughan DM (1994) Structure and function of ryanodine receptors. Am J Physiol 266(6 Pt 1):C1485–C1504

42. Chen SR, Zhang L, MacLennan DH (1992) Characterization of a Ca^{2+} binding and regulatory site in the Ca^{2+} release channel (ryanodine receptor) of rabbit skeletal muscle sarcoplasmic reticulum. J Biol Chem 267(32):23318–23326

43. Chen SR, Zhang L, MacLennan DH (1993) Antibodies as probes for Ca^{2+} activation sites in the Ca^{2+} release channel (ryanodine receptor) of rabbit skeletal muscle sarcoplasmic reticulum. J Biol Chem 268(18):13414–13421

44. Chen SR, MacLennan DH (1994) Identification of calmodulin-, Ca^{2+}-, and ruthenium red-binding domains in the Ca^{2+} release channel (ryanodine receptor) of rabbit skeletal muscle sarcoplasmic reticulum. J Biol Chem 269(36):22698–22704

45. Chen SR, Ebisawa K, Li X, Zhang L (1998) Molecular identification of the ryanodine receptor Ca^{2+} sensor. J Biol Chem 273(24):14675–14678

46. Li P, Chen SR (2001) Molecular basis of Ca^{2+} activation of the mouse cardiac Ca^{2+} release channel (ryanodine receptor). J Gen Physiol 118(1):33–44

47. Fessenden JD, Chen L, Wang Y, Paolini C, Franzini-Armstrong C, Allen PD et al (2001) Ryanodine receptor point mutant E4032A reveals an allosteric interaction with ryanodine. Proc Natl Acad Sci U S A 98(5):2865–2870

48. Murayama T, Ogawa H, Kurebayashi N, Ohno S, Horie M, Sakurai T (2018) A tryptophan residue in the caffeine-binding site of the ryanodine receptor regulates Ca^{2+} sensitivity. Commun Biol 1:98

49. Xiong H, Feng X, Gao L, Xu L, Pasek DA, Seok JH et al (1998) Identification of a two EF-hand Ca^{2+} binding domain in lobster skeletal muscle ryanodine receptor/Ca^{2+} release channel. Biochemistry 37(14):4804–4814

50. Xiong L, Zhang JZ, He R, Hamilton SL (2006) A Ca^{2+}-binding domain in RyR1 that interacts with the calmodulin binding site and modulates channel activity. Biophys J 90(1):173–182

51. Fessenden JD, Feng W, Pessah IN, Allen PD (2004) Mutational analysis of putative calcium binding motifs within the skeletal ryanodine receptor isoform, RyR1. J Biol Chem 279(51):53028–53035

52. Guo W, Sun B, Xiao Z, Liu Y, Wang Y, Zhang L et al (2016) The EF-hand Ca^{2+} binding domain is not required for cytosolic Ca^{2+} activation of the cardiac ryanodine receptor. J Biol Chem 291(5):2150–2160

53. Gomez AC, Yamaguchi N (2014) Two regions of the ryanodine receptor calcium channel are involved in Ca^{2+}-dependent inactivation. Biochemistry 53(8):1373–1379

54. Xu L, Gomez AC, Pasek DA, Meissner G, Yamaguchi N (2017) Two EF-hand motifs in ryanodine receptor calcium release channels contribute to isoform-specific regulation by calmodulin. Cell Calcium 66:62–70

55. Chugun A, Sato O, Takeshima H, Ogawa Y (2007) Mg^{2+} activates the ryanodine receptor type 2 (RyR2) at intermediate Ca^{2+} concentrations. Am J Physiol Cell Physiol 292(1):C535–C544

56. Yamaguchi N, Xu L, Pasek DA, Evans KE, Meissner G (2003) Molecular basis of calmodulin binding to cardiac muscle Ca^{2+} release channel (ryanodine receptor). J Biol Chem 278(26):23480–23486

57. Laver DR, Baynes TM, Dulhunty AF (1997) Magnesium inhibition of ryanodine-receptor calcium channels: evidence for two independent mechanisms. J Membr Biol 156(3):213–229

58. Gomez AC, Holford TW, Yamaguchi N (2016) Malignant hyperthermia-associated mutations in the S2-S3 cytoplasmic loop of type 1 ryanodine receptor calcium channel impair calcium-dependent inactivation. Am J Physiol Cell Physiol 311(5):C749–C757

59. Murayama T, Kurebayashi N, Ogawa Y (2000) Role of Mg^{2+} in Ca^{2+}-induced Ca^{2+} release through ryanodine receptors of frog skeletal muscle: modulations by adenine nucleotides and caffeine. Biophys J 78(4):1810–1824

60. Tripathy A, Meissner G (1996) Sarcoplasmic reticulum lumenal Ca^{2+} has access to cytosolic activation and inactivation sites of skeletal muscle Ca^{2+} release channel. Biophys J 70(6):2600–2615

61. Xu L, Meissner G (1998) Regulation of cardiac muscle Ca^{2+} release channel by sarcoplasmic reticulum lumenal Ca^{2+}. Biophys J 75(5):2302–2312

62. Du GG, MacLennan DH (1999) Ca^{2+} inactivation sites are located in the COOH-terminal quarter of recombinant rabbit skeletal muscle Ca^{2+} release channels (ryanodine receptors). J Biol Chem 274(37):26120–26126

63. Hayek SM, Zhu X, Bhat MB, Zhao J, Takeshima H, Valdivia HH et al (2000) Characterization of a calcium-regulation domain of the skeletal-muscle ryanodine receptor. Biochem J 351(Pt 1):57–65

64. Nakai J, Gao L, Xu L, Xin C, Pasek DA, Meissner G (1999) Evidence for a role of C-terminus in Ca^{2+} inactivation of skeletal muscle Ca^{2+} release channel (ryanodine receptor). FEBS Lett 459(2):154–158

65. Du GG, Khanna VK, MacLennan DH (2000) Mutation of divergent region 1 alters caffeine and Ca^{2+} sensitivity of the skeletal muscle Ca^{2+} release channel (ryanodine receptor). J Biol Chem 275(16):11778–11783

66. Fabiato A, Fabiato F (1979) Calcium and cardiac excitation-contraction coupling. Annu Rev Physiol 41:473–484

67. Orchard CH, Eisner DA, Allen DG (1983) Oscillations of intracellular Ca^{2+} in mammalian cardiac muscle. Nature 304(5928):735–738

68. Jiang D, Xiao B, Yang D, Wang R, Choi P, Zhang L et al (2004) RyR2 mutations linked to ventricular tachycardia and sudden death reduce the threshold for store-overload-induced Ca^{2+} release (SOICR). Proc Natl Acad Sci U S A 101(35):13062–13067

69. Jiang D, Wang R, Xiao B, Kong H, Hunt DJ, Choi P et al (2005) Enhanced store overload-induced Ca^{2+} release and channel sensitivity to luminal Ca^{2+} activation are common defects of RyR2 mutations linked to ventricular tachycardia and sudden death. Circ Res 97(11):1173–1181

70. Jiang D, Chen W, Xiao J, Wang R, Kong H, Jones PP et al (2008) Reduced threshold for luminal Ca^{2+} activation of RyR1 underlies a causal mechanism of porcine malignant hyperthermia. J Biol Chem 283(30):20813–20820
71. Chen W, Wang R, Chen B, Zhong X, Kong H, Bai Y et al (2014) The ryanodine receptor store-sensing gate controls Ca^{2+} waves and Ca^{2+}-triggered arrhythmias. Nat Med 20(2):184–192
72. Saimi Y, Kung C (2002) Calmodulin as an ion channel subunit. Annu Rev Physiol 64:289–311
73. Tripathy A, Xu L, Mann G, Meissner G (1995) Calmodulin activation and inhibition of skeletal muscle Ca^{2+} release channel (ryanodine receptor). Biophys J 69(1):106–119
74. Buratti R, Prestipino G, Menegazzi P, Treves S, Zorzato F (1995) Calcium dependent activation of skeletal muscle Ca^{2+} release channel (ryanodine receptor) by calmodulin. Biochem Biophys Res Commun 213(3):1082–1090
75. Chen SR, Li X, Ebisawa K, Zhang L (1997) Functional characterization of the recombinant type 3 Ca^{2+} release channel (ryanodine receptor) expressed in HEK293 cells. J Biol Chem 272(39):24234–24246
76. Fruen BR, Bardy JM, Byrem TM, Strasburg GM, Louis CF (2000) Differential Ca^{2+} sensitivity of skeletal and cardiac muscle ryanodine receptors in the presence of calmodulin. Am J Physiol Cell Physiol 279(3):C724–C733
77. Balshaw DM, Xu L, Yamaguchi N, Pasek DA, Meissner G (2001) Calmodulin binding and inhibition of cardiac muscle calcium release channel (ryanodine receptor). J Biol Chem 276(23):20144–20153
78. Yamaguchi N, Xu L, Pasek DA, Evans KE, Chen SR, Meissner G (2005) Calmodulin regulation and identification of calmodulin binding region of type-3 ryanodine receptor calcium release channel. Biochemistry 44(45):15074–15081
79. Ono M, Yano M, Hino A, Suetomi T, Xu X, Susa T et al (2010) Dissociation of calmodulin from cardiac ryanodine receptor causes aberrant Ca^{2+} release in heart failure. Cardiovasc Res 87(4):609–617
80. Terentyev D, Gyorke I, Belevych AE, Terentyeva R, Sridhar A, Nishijima Y et al (2008) Redox modification of ryanodine receptors contributes to sarcoplasmic reticulum Ca^{2+} leak in chronic heart failure. Circ Res 103(12):1466–1472
81. Oda T, Yang Y, Uchinoumi H, Thomas DD, Chen-Izu Y, Kato T et al (2015) Oxidation of ryanodine receptor (RyR) and calmodulin enhance Ca release and pathologically alter, RyR structure and calmodulin affinity. J Mol Cell Cardiol 85:240–248
82. Yamaguchi N, Xin C, Meissner G (2001) Identification of apocalmodulin and Ca^{2+}-calmodulin regulatory domain in skeletal muscle Ca^{2+} release channel, ryanodine receptor. J Biol Chem 276(25):22579–22585
83. Moore CP, Rodney G, Zhang JZ, Santacruz-Toloza L, Strasburg G, Hamilton SL (1999) Apocalmodulin and Ca^{2+} calmodulin bind to the same region on the skeletal muscle Ca^{2+} release channel. Biochemistry 38(26):8532–8537
84. Rodney GG, Moore CP, Williams BY, Zhang JZ, Krol J, Pedersen SE et al (2001) Calcium binding to calmodulin leads to an N-terminal shift in its binding site on the ryanodine receptor. J Biol Chem 276(3):2069–2074
85. Maximciuc AA, Putkey JA, Shamoo Y, Mackenzie KR (2006) Complex of calmodulin with a ryanodine receptor target reveals a novel, flexible binding mode. Structure 14(10):1547–1556
86. Menegazzi P, Larini F, Treves S, Guerrini R, Quadroni M, Zorzato F (1994) Identification and characterization of three calmodulin binding sites of the skeletal muscle ryanodine receptor. Biochemistry 33(31):9078–9084
87. Guerrini R, Menegazzi P, Anacardio R, Marastoni M, Tomatis R, Zorzato F et al (1995) Calmodulin binding sites of the skeletal, cardiac, and brain ryanodine receptor Ca^{2+} channels: modulation by the catalytic subunit of cAMP-dependent protein kinase? Biochemistry 34(15):5120–5129
88. Lau K, Chan MM, Van Petegem F (2014) Lobe-specific calmodulin binding to different ryanodine receptor isoforms. Biochemistry 53(5):932–946

89. Yamaguchi N, Takahashi N, Xu L, Smithies O, Meissner G (2007) Early cardiac hypertrophy in mice with impaired calmodulin regulation of cardiac muscle Ca^{2+} release channel. J Clin Invest 117(5):1344–1353

90. Yamaguchi N, Xu L, Evans KE, Pasek DA, Meissner G (2004) Different regions in skeletal and cardiac muscle ryanodine receptors are involved in transducing the functional effects of calmodulin. J Biol Chem 279(35):36433–36439

91. Samso M, Wagenknecht T (2002) Apocalmodulin and Ca^{2+}-calmodulin bind to neighboring locations on the ryanodine receptor. J Biol Chem 277(2):1349–1353

92. Arnaiz-Cot JJ, Damon BJ, Zhang XH, Cleemann L, Yamaguchi N, Meissner G et al (2013) Cardiac calcium signalling pathologies associated with defective calmodulin regulation of type 2 ryanodine receptor. J Physiol 591(17):4287–4299

93. Yamaguchi N, Chakraborty A, Huang TQ, Xu L, Gomez AC, Pasek DA et al (2013) Cardiac hypertrophy associated with impaired regulation of cardiac ryanodine receptor by calmodulin and S100A1. Am J Physiol Heart Circ Physiol 305(1):H86–H94

94. Yamaguchi N, Prosser BL, Ghassemi F, Xu L, Pasek DA, Eu JP et al (2011) Modulation of sarcoplasmic reticulum Ca^{2+} release in skeletal muscle expressing ryanodine receptor impaired in regulation by calmodulin and S100A1. Am J Physiol Cell Physiol 300(5):C998–C1012

95. Nyegaard M, Overgaard MT, Sondergaard MT, Vranas M, Behr ER, Hildebrandt LL et al (2012) Mutations in calmodulin cause ventricular tachycardia and sudden cardiac death. Am J Hum Genet 91(4):703–712

96. Hwang HS, Nitu FR, Yang Y, Walweel K, Pereira L, Johnson CN et al (2014) Divergent regulation of ryanodine receptor 2 calcium release channels by arrhythmogenic human calmodulin missense mutants. Circ Res 114(7):1114–1124

97. Sondergaard MT, Tian X, Liu Y, Wang R, Chazin WJ, Chen SR et al (2015) Arrhythmogenic calmodulin mutations affect the activation and termination of cardiac ryanodine receptor-mediated Ca^{2+} release. J Biol Chem 290(43):26151–26162

98. Vassilakopoulou V, Calver BL, Thanassoulas A, Beck K, Hu H, Buntwal L et al (2015) Distinctive malfunctions of calmodulin mutations associated with heart RyR2-mediated arrhythmic disease. Biochim Biophys Acta 1850(11):2168–2176

99. Liu B, Walton SD, Ho HT, Belevych AE, Tikunova SB, Bonilla I et al (2018) Gene transfer of engineered calmodulin alleviates ventricular arrhythmias in a calsequestrin-associated mouse model of catecholaminergic polymorphic ventricular tachycardia. J Am Heart Assoc 7(10):e008155

100. Treves S, Scutari E, Robert M, Groh S, Ottolia M, Prestipino G et al (1997) Interaction of S100A1 with the Ca^{2+} release channel (ryanodine receptor) of skeletal muscle. Biochemistry 36(38):11496–11503

101. Most P, Remppis A, Pleger ST, Loffler E, Ehlermann P, Bernotat J et al (2003) Transgenic overexpression of the Ca^{2+}-binding protein S100A1 in the heart leads to increased in vivo myocardial contractile performance. J Biol Chem 278(36):33809–33817

102. Volkers M, Rohde D, Goodman C, Most P (2010) S100A1: a regulator of striated muscle sarcoplasmic reticulum Ca^{2+} handling, sarcomeric, and mitochondrial function. J Biomed Biotechnol 2010:178614

103. Prosser BL, Hernandez-Ochoa EO, Schneider MF (2011) S100A1 and calmodulin regulation of ryanodine receptor in striated muscle. Cell Calcium 50(4):323–331

104. Prosser BL, Wright NT, Hernandez-Ochoa EO, Varney KM, Liu Y, Olojo RO et al (2008) S100A1 binds to the calmodulin-binding site of ryanodine receptor and modulates skeletal muscle excitation-contraction coupling. J Biol Chem 283(8):5046–5057

105. Wright NT, Prosser BL, Varney KM, Zimmer DB, Schneider MF, Weber DJ (2008) S100A1 and calmodulin compete for the same binding site on ryanodine receptor. J Biol Chem 283(39):26676–26683

106. Rebbeck RT, Nitu FR, Rohde D, Most P, Bers DM, Thomas DD et al (2016) S100A1 protein does not compete with calmodulin for ryanodine receptor binding but structurally alters the ryanodine receptor.Calmodulin complex. J Biol Chem 291(30):15896–15907

107. Saito A, Seiler S, Chu A, Fleischer S (1984) Preparation and morphology of sarcoplasmic reticulum terminal cisternae from rabbit skeletal muscle. J Cell Biol 99(3):875–885

108. Franzini-Armstrong C, Kenney LJ, Varriano-Marston E (1987) The structure of calsequestrin in triads of vertebrate skeletal muscle: a deep-etch study. J Cell Biol 105(1):49–56

109. Guo W, Campbell KP (1995) Association of triadin with the ryanodine receptor and calsequestrin in the lumen of the sarcoplasmic reticulum. J Biol Chem 270(16):9027–9030

110. Zhang L, Kelley J, Schmeisser G, Kobayashi YM, Jones LR (1997) Complex formation between junctin, triadin, calsequestrin, and the ryanodine receptor. Proteins of the cardiac junctional sarcoplasmic reticulum membrane. J Biol Chem 272(37):23389–23397

111. Kobayashi YM, Alseikhan BA, Jones LR (2000) Localization and characterization of the calsequestrin-binding domain of triadin 1. Evidence for a charged beta-strand in mediating the protein-protein interaction. J Biol Chem 275(23):17639–17646

112. Handhle A, Ormonde CE, Thomas NL, Bralesford C, Williams AJ, Lai FA et al (2016) Calsequestrin interacts directly with the cardiac ryanodine receptor luminal domain. J Cell Sci 129(21):3983–3988

113. Beard NA, Sakowska MM, Dulhunty AF, Laver DR (2002) Calsequestrin is an inhibitor of skeletal muscle ryanodine receptor calcium release channels. Biophys J 82(1 Pt 1):310–320

114. Gyorke I, Hester N, Jones LR, Gyorke S (2004) The role of calsequestrin, triadin, and junctin in conferring cardiac ryanodine receptor responsiveness to luminal calcium. Biophys J 86(4):2121–2128

115. Beard NA, Casarotto MG, Wei L, Varsanyi M, Laver DR, Dulhunty AF (2005) Regulation of ryanodine receptors by calsequestrin: effect of high luminal Ca^{2+} and phosphorylation. Biophys J 88(5):3444–3454

116. Paolini C, Quarta M, Nori A, Boncompagni S, Canato M, Volpe P et al (2007) Reorganized stores and impaired calcium handling in skeletal muscle of mice lacking calsequestrin-1. J Physiol 583(Pt 2):767–784

117. Lahat H, Pras E, Olender T, Avidan N, Ben-Asher E, Man O et al (2001) A missense mutation in a highly conserved region of CASQ2 is associated with autosomal recessive catecholamine-induced polymorphic ventricular tachycardia in bedouin families from Israel. Am J Hum Genet 69(6):1378–1384

118. Postma AV, Denjoy I, Hoorntje TM, Lupoglazoff JM, Da Costa A, Sebillon P et al (2002) Absence of calsequestrin 2 causes severe forms of catecholaminergic polymorphic ventricular tachycardia. Circ Res 91(8):e21–e26

119. Knollmann BC, Chopra N, Hlaing T, Akin B, Yang T, Ettensohn K et al (2006) Casq2 deletion causes sarcoplasmic reticulum volume increase, premature Ca^{2+} release, and catecholaminergic polymorphic ventricular tachycardia. J Clin Invest 116(9):2510–2520

120. Song L, Alcalai R, Arad M, Wolf CM, Toka O, Conner DA et al (2007) Calsequestrin 2 (CASQ2) mutations increase expression of calreticulin and ryanodine receptors, causing catecholaminergic polymorphic ventricular tachycardia. J Clin Invest 117(7):1814–1823

121. Pettersen EF, Goddard TD, Huang CC, Couch GS, Greenblatt DM, Meng EC et al (2004) UCSF chimera–a visualization system for exploratory research and analysis. J Comput Chem 25(13):1605–1612

Chapter 14
Sarco-Endoplasmic Reticulum Calcium Release Model Based on Changes in the Luminal Calcium Content

Agustín Guerrero-Hernández, Víctor Hugo Sánchez-Vázquez, Ericka Martínez-Martínez, Lizeth Sandoval-Vázquez, Norma C. Perez-Rosas, Rodrigo Lopez-Farias, and Adan Dagnino-Acosta

Abstract The sarcoplasmic/endoplasmic reticulum (SR/ER) is the main intracellular calcium (Ca^{2+}) pool in muscle and non-muscle eukaryotic cells, respectively. The reticulum accumulates Ca^{2+} against its electrochemical gradient by the action of sarco/endoplasmic reticulum calcium ATPases (SERCA pumps), and the capacity of this Ca^{2+} store is increased by the presence of Ca^{2+} binding proteins in the lumen of the reticulum. A diversity of physical and chemical signals, activate the main Ca^{2+} release channels, i.e. ryanodine receptors (RyRs) and inositol (1, 4, 5) trisphosphate receptors (IP$_3$Rs), to produce transient elevations of the cytoplasmic calcium concentration ($[Ca^{2+}]_i$) while the reticulum is being depleted of Ca^{2+}. This picture is incomplete because it implies that the elements involved in the Ca^{2+} release process are acting alone and independently of each other. However, it appears that the Ca^{2+} released by RyRs and IP$_3$Rs is trapped in luminal Ca^{2+} binding proteins (Ca^{2+} lattice), which are associated with these release channels, and the activation of these channels appears to facilitate that the trapped Ca^{2+} ions become available for release. This situation makes the initial stage of the Ca^{2+} release process a highly efficient one; accordingly, there is a large increase in the $[Ca^{2+}]_i$ with minimal reductions in the bulk of the free luminal SR/ER $[Ca^{2+}]$ ($[Ca^{2+}]_{SR/ER}$). Additionally, it has been shown that active SERCA pumps

A. Guerrero-Hernández (✉) · V. H. Sánchez-Vázquez · E. Martínez-Martínez
L. Sandoval-Vázquez
Department of Biochemistry, Cinvestav, Mexico city, Mexico
e-mail: aguerrero@cinvestav.mx

N. C. Perez-Rosas
Bioquant, University of Heidelberg, Heidelberg, Germany

R. Lopez-Farias
CONACYT-Consorcio CENTROMET, Querétaro, Mexico

A. Dagnino-Acosta
CONACYT-Universidad de Colima (Centro Universitario de Investigaciones Biomédicas), Colima, Mexico

© Springer Nature Switzerland AG 2020
M. S. Islam (ed.), *Calcium Signaling*, Advances in Experimental Medicine and Biology 1131, https://doi.org/10.1007/978-3-030-12457-1_14

are required for attaining this highly efficient Ca^{2+} release process. All these data indicate that Ca^{2+} release by the SR/ER is a highly regulated event and not just Ca^{2+} coming down its electrochemical gradient via the open release channels. One obvious advantage of this sophisticated Ca^{2+} release process is to avoid depletion of the ER Ca^{2+} store and accordingly, to prevent the activation of ER stress during each Ca^{2+} release event.

Keywords Endoplasmic reticulum (ER) · Sarcoplasmic reticulum (SR) · Ryanodine receptors (RyRs) · IP$_3$ receptors (IP$_3$Rs) · Sarco-endoplasmic reticulum calcium ATPase (SERCA pump) · Free luminal ER [Ca^{2+}] ([Ca^{2+}]$_{ER}$) · Cytoplasmic [Ca^{2+}] ([Ca^{2+}]$_i$) · Calcium buffer capacity · Kinetics on demand (KonD)

14.1 Elements of the SR/ER Involved in Ca^{2+} Release

14.1.1 Calcium Ion as Second Messenger

A transient elevation of the cytoplasmic calcium concentration ([Ca^{2+}]$_i$) leads to changes in a large array of cellular functions [1]. These are muscle contraction, gland secretion, neurotransmission, respiration, cell movement, cell proliferation, gene transcription, cell death, among others. There are two main sources of calcium ions, the external milieu and the intracellular calcium stores [2]. The latter, in turn, are formed by two main Ca^{2+} pools, the sarco-endoplasmic reticulum and the acidic Ca^{2+} stores. This review will focus on the former rather on the latter. However, before reviewing what we know on how the sarco-endoplasmic reticulum provides Ca^{2+} for different cellular events, we will discuss some principles associated with Ca^{2+} ion as one of the many different second messengers that cells use to respond to changes in the environment.

Calcium ions are toxic in principle [3], since a sustained elevation of the [Ca^{2+}]$_i$ leads to cell death. For this reason cells invest a considerable amount of energy, coming directly or indirectly from ATP hydrolysis, to keep cytoplasmic [Ca^{2+}] in the 100 nM range by actively transporting this ion to either outside the cell or inside intracellular compartments [2]. The latter are known as intracellular Ca^{2+} stores. Generally, second messengers are molecules synthesized and degraded by enzymes, which are essential in the initiation and termination of second messenger activities. However, the situation of Ca^{2+} being a second messenger is different, in this case Ca^{2+} is moved from one cell compartment to another and those concentration changes are recognized by proteins that lead to changes in cell behavior [2]. Therefore, the plasma membrane and the different intracellular membranes are endowed with a large variety of Ca^{2+} permeable ion channels that respond to different stimuli. Accordingly, the diffusion of Ca^{2+} through the open pore of these proteins occurs in response to different signals that are physical (voltage, heat, pressure) or chemical (hormones, neurotransmitters, etc.) in nature. Once the

$[Ca^{2+}]_i$ has been elevated by the increased activity of Ca^{2+} permeable ion channels, this divalent cation is bound by two different types of proteins, buffers and effectors, the former limit the increase of the $[Ca^{2+}]_i$ and this gives time to Ca^{2+} pumps to expel this ion out of the cytoplasm, while the latter are characterized by the ability of forming a Ca^{2+}-protein complex that modifies the activity of different enzymes (kinases, phosphatases, proteases, etc.) that change the cell behavior allowing cells to adapt and respond to different stimuli [2, 4].

14.1.2 Intracellular Calcium Stores

It has become evident then that intracellular Ca^{2+} pools are, at the same time, sources of Ca^{2+} in response to certain stimuli [1, 5] and also Ca^{2+} buffering compartments that help cells survive the cytotoxic effect of increased $[Ca^{2+}]_i$ [6–8]. This paradoxical situation has been the motor behind the evolution of very interesting solutions that will be reviewed below. There are basically two different types of internal Ca^{2+} stores, one represented by the sarcoplasmic/endoplasmic reticulum (SR/ER) and the other referred to as acidic Ca^{2+} store. The former is a single organelle while the latter encompasses a variety of organelles that have a luminal pH below 7.0, these are the Golgi apparatus [9, 10], lysosomes [11] and secretory granules [12], among other organelles. These two different types of stores can be observed using electron microscopy to detect intracellular sites with elevated $[Ca^{2+}]$ [13, 14]. Alternatively, these two stores can be observed by depleting the SR/ER followed by maneuvers to alkalinize and open Ca^{2+} permeable channels in the acidic organelles [10, 15].

The SR/ER is a membrane organelle that traverses all the cytoplasm. Actually, it forms the nuclear envelope and for this reason this organelle goes from the nucleus all the way to the plasma membrane at the periphery of the cell [16–18]. The reticulum is formed by tubules and saccules that are interconnected and it appears that their lumen do not have any diffusion barriers [7, 19–21]. At the same time these reticular structures are dynamic because they can move, to a certain degree, within the cytoplasm; using the microtubules as their rail-roads [17, 22, 23]. Reticular membranes also have the characteristic of being highly fusogenic, resulting in a constant reshaping of the reticulum [22–24].

The endoplasmic reticulum of muscle cells is the sarcoplasmic reticulum and, particularly in striated muscle cells, it has specialized in moving large quantities of Ca^{2+} for both muscle contraction and relaxation [25, 26]. In non-muscle cells, the endoplasmic reticulum is the main intracellular Ca^{2+} store, although this organelle also carries out many more functions such as protein and phospholipid synthesis, drug detoxification, synthesis of cholesterol, etc. [17]. The function of the sarcoplasmic/endoplasmic reticulum as Ca^{2+} store involves the following essential elements: (a) Ca^{2+} pumps to accumulate this ion in the store, against its electrochemical gradient, (b) release Ca^{2+} permeable channels and (c) luminal Ca^{2+} binding proteins that increase the capacity of the lumen to accumulate Ca^{2+} in the store. The characteristics of each one of these elements will be reviewed briefly.

14.1.3 SERCA Pumps

SERCA stands for Sarco-Endoplasmic Reticulum Calcium ATPase, is an SR/ER integral membrane protein codified by three different genes, ATP2A1, ATP2A2 and ATP2A3. All three messenger RNAs from these genes show alternative splicing in the $3'$ end region, generating a large assortment of SERCA pumps. SERCA1a is the fastest pump and it is expressed in fast-twitch skeletal muscle, SERCA2a is expressed mainly in heart cells and SERCA2b is called house-keeping pump because it is present in smooth muscle cells and non-muscle cells having an ER [27].

Catalytic cycle of SERCA pumps consists of binding two Ca^{2+} ions from the cytoplasm, this in turn promotes ATP binding followed by SERCA own phosphorylation in a highly conserved aspartic residue, which results in conformational changes that close the access of cytoplasmic Ca^{2+} to the protein, and these would allow access to the lumen of the endoplasmic reticulum with a reduced affinity for Ca^{2+}. In the end, cytoplasmic Ca^{2+} is delivered into the lumen of the SR/ER in exchange for protons. This final event promotes dephosphorylation of the enzyme to reinitiate another catalytic cycle [28]. All these steps in the catalytic cycle involve large conformational changes whose structures have been determined by X-ray diffraction studies [29]. This series of conformational changes has very interesting implications for cell physiology. The slow rate of 100–1000 ions/second showed by SERCA pump [30] does not counteract the much higher Ca^{2+} rate of one million ions/sec in the RyR [31]. The leak activity of release channels then could be critical because it would impose an energy burden on the cell that could easily drain all its energy resources [32]. Accordingly, SERCA pump is one of the thermogenic sources in the body [33], to the extent that increased Ca^{2+} leakage from the SR, that in turn accelerates SERCA pump activity, can lead to malignant hyperthermia, a fatal side effect of anesthetics, like halothane, this complication can be overcome by using dantrolene, an inhibitor of Ca^{2+} release channel, by reducing the SR Ca^{2+} leak [34]. Furthermore, deletion of sarcolipin (a protein proposed to switch SERCA pump from Ca^{2+} translocation to thermogenesis) generates obese mice with the development of insulin resistance [35]. Thus, this catalytic cycle is essential for SERCA pump to produce heat and to accumulate Ca^{2+} in the lumen of the SR/ER.

It is easy to see then that any interference with the SERCA pump catalytic cycle will result in inhibition of its activity. It has been described a large variety of SERCA pump inhibitors that are either codified by the same cell or are chemicals from other sources. The former involve a series of peptides that are small, single span integral membrane proteins, that by binding to SERCA they inhibit its pump activity; these are phospholamban and sarcolipin [36], myoregulin, endorgulin, and another-regulin [37]. The second group of SERCA pump inhibitors comprises chemicals that bind to SERCA pump and inhibit the catalytic cycle, these are thapsigargin, cyclopiazonic acid and *tert*-butyl hydroquinone [38, 39]. Additionally, high levels of cholesterol and saturated fatty acids can also inhibit the activity of SERCA pump, allegedly by increasing the rigidity of the ER membrane [40]. Interestingly, it has been found a peptide, codified by a long-noncoding RNA, named DWORF, which is able to increase the activity of SERCA pump by displacing the inhibitory peptides from the SERCA pump [41]. It is becoming clear that the activity of SERCA pump not only respond to an increase in the $[Ca^{2+}]_i$ but also to the presence of these regulatory peptides.

14.1.4 Calcium Release Channels

There are two main Ca^{2+} release channels in the SR/ER. The SR expresses the Ryanodine Receptor (RyR) while the ER is endowed with the IP_3R [42, 43]. These release channels are tetramers that form Ca^{2+} permeable, non-selective cation channels [44]. Both types of channels are activated by low cytoplasmic $[Ca^{2+}]$ and inhibited by higher concentrations of this divalent cation [45, 46]. The majority of the protein is facing in the cytoplasm and only a small fraction, in the carboxy terminal region of this protein, is inserted in the sarco-endoplasmic reticulum membrane to form the ion channel pore. Release channels are tetrameric, Ca^{2+} permeable non selective cation channels and each subunit of the RYR is more than 5000 amino acids long (the whole channel weighs around 2.2 MDa) while each subunit of the IP_3R is around 2500 amino acids long (the whole channel weight close to 1 MDa). There are three different genes for each one of these release channels and each gene produces alternative splicing isoforms. RyR1 is expressed mainly in fast-twitch skeletal muscle and is activated by what is known as mechanical coupling [47]. The membrane depolarization of the T tubules generates conformational changes in the voltage-gated calcium channels (VGCC, specifically the dihydropiridine receptor) that are directly transmitted to the RyR1 resulting in one of the fastest Ca^{2+} release events (1–2 msec time to peak). The key issue here is that RyRs in the SR require making contact with VGCC in the T tubules of the plasma membrane. Typically, each one of the subunits of RyRs connects with one VGCC, so one RyRs is connected with four VGCC. The RyRs alternate between those connected to the VGCC and those that are not connected, and the idea is that calcium-induced calcium release (CICR) would be activating those RyRs that are not connected [42]. The isoform 2 of RyR is expressed in heart cells and in this case the association with VGCCs is not that clear, RyR2s are not connected physically to VGCC, but these two proteins are located very close to each other [48]. The Ca^{2+} entering via VGCCs triggers the activation of RyR2s to produce Ca^{2+} release that results in heart cell contraction. It has been calculated that for all the Ca^{2+} involved in contraction, as much as 90% can be provided by RyR2s and only a small fraction by VGCCs, although these figures vary according to the species studied [49]. Finally, RyR3 was the last one to be cloned and is present in different type of cells, for instance in the diaphragm, brain cells and smooth muscle cells [47]. Actually, it is very interesting that smooth muscle cells express all three RyRs with different localization within the SR [50]. RyRs in smooth muscle cells appear to be involved more in relaxation than contraction [51], the idea here is that localized Ca^{2+} release events involving the activation of RyRs, which are known as Ca^{2+} sparks, would activate Ca^{2+}-dependent, high conductance potassium channels that are heavily expressed in smooth muscle cells, this would result in membrane hyperpolarization and concomitant deactivation of VGCC, which provide most of the Ca^{2+} involved in smooth muscle contraction. This is the case because CICR is rather inefficient in smooth muscle cells [52, 53].

The IP_3R is activated by the combination of Ca^{2+} and IP_3 in a very complex manner, to the extent that IP_3Rs are activated in a very small window of IP_3 concentration [46]. This is very important because there is no correlation between the amount of IP_3 produced and the amplitude of the $[Ca^{2+}]_i$ response, actually, the main difference is time, because if an agonist produces a large amount of IP_3, this will reach the threshold for Ca^{2+} release, before the other agonist that has a much lower rate of IP_3 production [54, 55]. The IP_3 binding site is located in the amino terminus of the protein, while the ion channel is formed by the carboxy end of the IP_3R. The structure of this channel revealed by cryo-electron microscopy has shown that the very end of this protein, the carboxy terminal domain (CTD), is an alpha helix that goes all the way from the channel domain to contact the IP_3 binding region of the next subunit. This peculiar conformation of CTD might explain the allosterism displayed by IP_3 and Ca^{2+} to activate this release channel [56]. In addition to IP_3 and Ca^{2+} there are a number of proteins that can modulate the activity of IP_3Rs, examples are RACK, IRBIT, Homer, BCL2, Presenilin, Huntingtin, among others [57]. Thus, the activity of release channels is under the control of a large series of chemicals and protein interactions.

14.1.5 Luminal Calcium Binding Proteins

The $[Ca^{2+}]$ in the cytoplasm is in the submicromolar range and accordingly, the Ca^{2+} binding proteins that are present in the cytoplasm have high affinity and selectivity over Mg^{2+} to be able to bind Ca^{2+} and carry out their functions. This situation implies that the cytoplasmic Ca^{2+} binding proteins, participating in signal transduction, have evolved to display high affinity, but low capacity for Ca^{2+} ions [4].

This picture is completely different inside the lumen of the SR/ER where the free luminal $[Ca^{2+}]$ is in the submillimolar range and total $[Ca^{2+}]$ could be in the tens of millimolar range. There are two main luminal calcium binding proteins in the lumen of the reticulum, these are calsequestrin [58–60] and calreticulin [61]; the former appears to be the main Ca^{2+} binding protein in the lumen of the sarcoplasmic reticulum of striated muscles [62], while the latter is the main one in some smooth muscle cells and non-muscle cells [63]. These proteins are characterized by having low affinity but high capacity. This means that these proteins have several Ca^{2+} binding sites but their affinity for Ca^{2+} is low when compared with those proteins in the cytoplasm. The underlying idea is that these proteins are responsible for increasing the Ca^{2+} buffering power of the sarco-endoplasmic reticulum. That is, they increase the capacity of internal stores to accumulate Ca^{2+}, which allow internal Ca^{2+} stores to be able to provide Ca^{2+} to the cytoplasm and to trigger cellular events that are driven by an increase of the $[Ca^{2+}]_i$.

Additionally, if the stores have a role in buffering the elevation of the $[Ca^{2+}]_i$, then these proteins increase the capacity of internal stores to accumulate Ca^{2+}, which in turn, avoid the cytotoxic effects of high levels of the $[Ca^{2+}]_i$. This situation has been clearly shown in neurons, where membrane depolarization activates VGCC that allow Ca^{2+} entry to the cytoplasm, and then SERCA pumps direct cytoplasmic Ca^{2+} to the lumen of the ER. Once Ca^{2+} is inside the ER, there are specific regions that accumulate this ion to concentrations as high as 40 mM [64]. These regions are adjacent to other regions of the ER that do not accumulate Ca^{2+} at all; this is not easy to explain since both regions are located within the same lumen of the ER [65]. The idea here is that the distribution of the Ca^{2+} binding proteins is not homogeneous [66], and that these proteins are trapping Ca^{2+} to reduce the activity of this ion (i.e. Ca^{2+} ions are no longer free). In the same neurons, it was observed that the inhibition of SERCA pumps with thapsigargin, leads to mitochondria Ca^{2+} accumulation due to the inability of the ER to buffer the incoming Ca^{2+} [64, 65]. Although it was not shown in this study, the increased mitochondrial matrix $[Ca^{2+}]$ would trigger mitochondria dysfunction and quite possible apoptosis [8].

Calsequestrin does not appear to be freely distributed inside the SR but it is associated with RyRs either directly [67] or indirectly via the association with triadin and junctin [47, 68]. However, it has been shown that the role of calsequestrin is more complex than the mere increase of the SR Ca^{2+} buffering capacity since it also regulates RyRs-mediated Ca^{2+} release events [58]. A very interesting example has been observed studying a single point mutation of $CASQ2^{R33Q}$, which was discovered in a patient afflicted with CPVT (catecholaminergic polymorphic ventricular tachycardia), an inherited arrhythmogenic condition that is life threatening [60]. Interestingly, the presence of CASQ2 confers RyR2s sensitivity to luminal $[Ca^{2+}]$ to the extent that there is a peak ion channel activity at 1 mM luminal $[Ca^{2+}]$ [69]. Remarkably, the mutant $CASQ2^{R33Q}$ increases the RyR2 ion channel activity by luminal Ca^{2+} at concentrations as low as 10 μM [69]. Then the presence of $CASQ2^{R33Q}$ makes RyR2 a leaky channel to the point that this results in a very low free luminal SR $[Ca^{2+}]$ in heart cells, even lower than the levels attained during normal Ca^{2+} release process [60]. Unexpectedly, this extremely low free luminal SR $[Ca^{2+}]$ with $CASQ2^{R33Q}$ results in Ca^{2+} waves and Ca^{2+} sparks of larger amplitude than those recorded in normal heart cells that have a much higher free luminal SR $[Ca^{2+}]$ [60]. Moreover, the Ca^{2+} systolic-induced contraction was only slightly smaller in myocytes expressing the mutant $CASQ^{R33Q}$ than in those expressing wild type CASQ [70]. Deletion of CASQ2 greatly decreased the amplitude of Ca^{2+} sparks despite having a normal reduction in the luminal SR $[Ca^{2+}]$ suggesting that CASQ2 is indeed the Ca^{2+} buffering protein in the SR of heart cells [60]. However, it is also clear that CASQ2 beyond being a Ca^{2+} buffering protein also has other roles in the Ca^{2+} release event since it shows a complex control of the activity of RyR2s. Assuming that the Kd for Ca^{2+} binding of $CASQ^{R33Q}$ is the same as the one for CASQ wild type, and because the free luminal SR $[Ca^{2+}]$ is much smaller than normal, then the saturation of CASQ^{R33Q} should also be much smaller, these two conditions (low free luminal SR [Ca^{2+}] and smaller saturation of CASQ2) should result in a [Ca^{2+}]$_i$ response of smaller amplitude in the presence of CASQ2^{R33Q}. Since this is not the case [60], then it appears that CSQ2^{R33Q} mutant clearly exemplifies the idea that the amount of Ca^{2+} that is released by activation of RyR2 cannot be predicted from the observed changes in the free luminal SR [Ca^{2+}].

14.1.6   The Biophysical Vision of the SR/ER Ca^{2+} Stores

It is clear then that an intracellular Ca^{2+} store is a membrane compartment that forms a closed entity and involves the participation of several components, which are Ca^{2+} pumps, luminal Ca^{2+} binding proteins and Ca^{2+} release channels. The prevailing vision, particularly for mathematical models (Fig. 14.1), is that all these

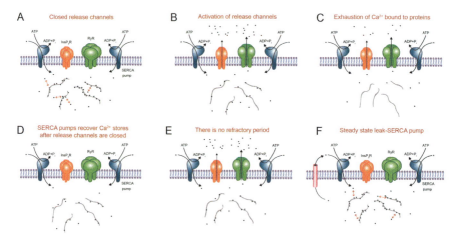

Fig. 14.1 Graphical model of Ca^{2+} release event where the free luminal SR/ER [Ca^{2+}] is in equilibrium with the Ca^{2+} bound to the luminal proteins. This cartoon depicts the critical elements involved in a Ca^{2+} release event. (**a**) The Ca^{2+} bound to luminal proteins is in equilibrium with the free luminal SR/ER [Ca^{2+}] (as represented by the double-headed arrows). (**b**) The activation of release channels would decrease the free luminal SR/ER [Ca^{2+}] and this results in Ca^{2+} unbinding from luminal proteins, as long as the release channels stay open. This scenario cannot produce an increase of the free luminal SR/ER [Ca^{2+}] in response to the activation of the release channels. (**c**) The Ca^{2+} release process terminates when all the Ca^{2+} bound to luminal proteins has been released. (**d**) SERCA pump needs that release channels to be close in order to recover the free luminal SR/ER [Ca^{2+}]. (**e**) This scenario cannot produce a refractory period for RyRs because the recovery of the free luminal SR/ER [Ca^{2+}] implies also complete restauration of the Ca^{2+} bound to proteins. (**f**) The steady state implies that Ca^{2+} leak activity is counterbalanced by SERCA pump activity to keep constant the free luminal SR/ER [Ca^{2+}] and also the total amount of Ca^{2+}. Any reduction in the free luminal SR/ER [Ca^{2+}] implies a corresponding reduction in the total amount of Ca^{2+} stored in the SR/ER

three components work independently of each other and that it is only via changes in the free luminal [Ca^{2+}] that these three components communicate among each other [71, 72]. Accordingly, these models have explored the role played by free luminal [Ca^{2+}] in modulating the activity of both release channels and Ca^{2+} pumps. However, there are several experimental studies indicating that these three components are not working independently of each other. This situation was clearly unexpected particularly for SERCA pump and release channels, because they are supposed to be antagonistic elements of the Ca^{2+} pools since one increases while the other reduces the Ca^{2+} content of the intracellular store. Nevertheless, there is evidence, which will be discussed in the rest of the review, for a communication between Ca^{2+} release channels and SERCA pump to achieve an efficient Ca^{2+} release event (i.e. an increase in the [Ca^{2+}]$_i$ with minimal or no reduction in the free luminal SR/ER [Ca^{2+}]).

Mathematical models reflect our incomplete knowledge of the Ca^{2+} release event. One example can be found in a study investigating the role played by Ca^{2+} diffusion inside the SR in the termination of Ca^{2+} sparks [73, 74]. Interestingly, this model uses a previously reported Ca^{2+} diffusion coefficient in the non-junctional SR of 0.6×10^{-10} m^2/s [75], but to have an efficient Ca^{2+} release process, this coefficient has to be five-fold larger of 3.5×10^{-10} m^2/s in junctional SR [73, 74]. The problem with this difference is that the model uses a previously estimated concentration of calsequestrin of 6 mM [76], while the model needed to increase the concentration by five-fold (30 mM) in the junctional SR [73, 74]. These conditions generate a paradox because the idea is that calsequestrin is the main Ca^{2+} buffering protein in this compartment and since calsequestrin behaves as a non-mobile protein within the SR, because it is associated with the RyRs, therefore increasing the concentration of calsequestrin in the SR should lower the diffusion of Ca^{2+} ions. However, this is the opposite of what the model predicts to be happening in the junctional SR because it was increased five-fold. Thus, it appears then that our current picture of how calsequestrin is working would lead to a strong competition for the free luminal Ca^{2+} between RyRs and calsequestrin and since this does not seem to be happening in the Ca^{2+} release event, then there is still missing information on how these proteins produce an efficient Ca^{2+} release process.

14.2 Contributions of Noriaki Ikemoto to the SR Ca^{2+} Release Model

14.2.1 The Activation of RyR1 Produces an Initial Increase in the Free Luminal SR [Ca^{2+}] Before the Opening of this Release Channel

A very interesting observation made by Ikemoto's group consists in finding that stimulation of RyRs in skeletal muscle terminal cisterna SR vesicles produces an elevation of the free luminal [Ca^{2+}] that occurs before any Ca^{2+} has come out from the store to the cytoplasm, in other words, the free luminal [Ca^{2+}] is increased in response to the activation of RyRs but before the opening of these channels [77]. The implications of this observation are so profound and paradigm changing that we are still in the process of incorporating them in our current models on how Ca^{2+} pools are working. The importance of this observation resides in the fact that current mathematical models of Ca^{2+} release cannot generate an increase in the free luminal [Ca^{2+}] in response to the activation of RyRs. The current idea on how Ca^{2+} release occurs requires the opening of RyRs to decrease the free luminal [Ca^{2+}], as an initial event, and this in turn would drive the dissociation of Ca^{2+} from calsequestrin to further amplify the amount of Ca^{2+} that is released from the SR. This model means that Ca^{2+} dissociation from calsequestrin is driven by a reduction of the free luminal [Ca^{2+}], which is the result of the opening of RyRs. This picture seems to coincide with the results obtained by Ikemoto using high concentrations of polylysine to activate RyRs [77]. However, using lower concentrations of polylysine resulted in a clear increase of the free luminal [Ca^{2+}], as measured with tetramethylmurexide, before any reduction in the total amount of Ca^{2+} (as measured with $^{45}Ca^{2+}$) stored in the SR vesicles [77]. This work suggests that activation of RyRs, before the opening of RyRs, results in the increase of the free luminal SR [Ca^{2+}] as an initial step followed then by the actual opening of these release channels.

14.2.2   Role of Calsequestrin in Producing a Transient Increase of the Luminal SR [Ca^{2+}]

Another crucial observation done by Ikemoto's laboratory concerns the role played by calsequestrin in generating an efficient Ca^{2+} release process [77]. They found that activation of RyRs with polylysine did not produce any increase in the free luminal [Ca^{2+}] when the SR vesicles were devoid of calsequestrin. This indicates that the Ca^{2+} bound to this protein is the one that is released by RyRs. However, the rather radical observation is that calsequestrin inside the SR vesicles is not enough to reconstitute the Ca^{2+} release event, but that calsequestrin needs to be associated with the RyRs of the SR vesicle to recover the increase in the free luminal [Ca^{2+}] response [77]. This signifies that the Ca^{2+} buffering power provided by calsequestrin is not enough to have a correct Ca^{2+} release event, but that RyRs and calsequestrin have to interact, either directly or indirectly, in order to reconstitute an efficient Ca^{2+} release event. Again the implications of this observation are so ample that it has been difficult to incorporate in our current thinking. The same conclusion was reached by a different group and their data was summarized in a model where there were two different RyRs, whose main difference was to have or not attached calsequestrin [78]. Those RyRs having attached calsequestrin respond very fast and produced more than 50% of the Ca^{2+} release event, while those without calsequestrin have a slower activation and provide less Ca^{2+} to the response [78]. It is important to bear in mind that the rate of Ca^{2+} release is a critical characteristic to produce an increase in the $[Ca^{2+}]_i$ because, there are Ca^{2+} buffering proteins in the cytoplasm and Ca^{2+} pumps at the plasma membrane, that would reduce the effect of release channels on the $[Ca^{2+}]_i$. Accordingly, this means that those RyRs with calsequestrin bound would have a much higher effect on the $[Ca^{2+}]_i$ than those RyRs without calsequestrin, despite releasing similar amounts of Ca^{2+} from the SR but, the main difference is that one is faster than the other. These data are also suggesting that calsequestrin appears to be trapping Ca^{2+} next to the RyRs, meaning that this entrapped Ca^{2+} is not in equilibrium with the free luminal SR $[Ca^{2+}]$; if this is the case, then it is easy to explain why the activation of RyRs, before the opening of RyRs, produces an initial increase of the free luminal SR $[Ca^{2+}]$. Additionally, this situation also implies that the free luminal $[Ca^{2+}]$ is not a good predictor of the amount of Ca^{2+} released by activation of RyRs because the Ca^{2+} trapped by calsequestrin does not appear to be in equilibrium with the free luminal SR $[Ca^{2+}]$, which is the one reported by the fluorescent Ca^{2+} indicators. Yet another important consequence that can be derived from these works is that there will be a refractory period in the Ca^{2+} release event that would depend on the time taken by RyRs and calsequestrin to reconstitute the trapped Ca^{2+} complex. It appears that only closed conformation of RyRs allows the reconstitution of this complex, a conclusion that was derived from experiments using ryanodine in smooth muscle cells [79]. High concentrations of ryanodine lock RyRs in the open conformation but the Ca^{2+} flow is blocked because ryanodine binds to the open pore [45, 47]. In the cell, this effect of ryanodine allows SERCA pump to recover the luminal SR $[Ca^{2+}]$ because ryanodine has blocked the RyRs [79]. The most interesting observation is that in the presence of ryanodine and with normal SR luminal $[Ca^{2+}]$, the application of IP_3-producing agonists results in a much smaller increase of the $[Ca^{2+}]_i$ [79, 80]. Collectively, these data suggest that the SR Ca^{2+} store is still empty despite the presence of a high luminal SR $[Ca^{2+}]$. The interpretation for these data could be that the complex (release channel-Ca^{2+} binding protein) cannot be reassembled when the RyRs have been locked, by ryanodine, in the open conformation and this condition reduces the ability of nearby IP_3Rs to produce an efficient $[Ca^{2+}]_i$ response.

14.2.3   Evidence that There Is a Communication Between SERCA Pump and RyR1

At the end of the Ca^{2+} release event, the combined reduction in the free luminal SR/ER $[Ca^{2+}]$ and the associated increase of the $[Ca^{2+}]_i$ should lead to SERCA pump activation. At least that is the idea because high luminal $[Ca^{2+}]$ would make harder for SERCA pump the exchange of H^+ for Ca^{2+} in the lumen of the SR [81]. However, work done by Ikemoto's group has demonstrated that activation of RyR1s affects the activity of SERCA pump suggesting some kind of communication between these two proteins [82–86]. The most interesting part is that this communication occurs right after the activation of RyRs and before any reduction in the free luminal SR $[Ca^{2+}]$. Even more difficult to explain is the observation that this activation of the SERCA pump occurs during the RyR-triggered increase of the free luminal $[Ca^{2+}]$ [87]. This is quite paradoxical since it is expected that the increase in the free luminal $[Ca^{2+}]$ should lead to further inhibition of the SERCA pump activity [81]. It is still unknown how the activation of RyRs, before their opening, results in SERCA pump activation; particularly because SERCA pump and RyRs are not necessarily localized next to each other in the SR; however, this communication could indirectly occur via luminal Ca^{2+} binding proteins [83]. Regardless how these two proteins communicate, it is easy to argue that this communication between release channels and SERCA pump is an important mechanism to avoid life-threatening situations such as ER-stress due to depletion of the ER Ca^{2+} store.

14.3 Simultaneous Recording of the Redution in the Free Luminal SR/ER $[Ca^{2+}]$ and the Associated Changes in the $[Ca^{2+}]_i$

Since the 80's, when the organic fluorescent Ca^{2+} indicators were synthesized, it became customary to place cells in the absence of external $[Ca^{2+}]$ to look at the $[Ca^{2+}]_i$ response induced by different stimuli, and to take amplitude of this response as an indirect measure of the amount of Ca^{2+} present in the SR/ER Ca^{2+} pools. However, this type of approach can be deceiving because there are different factors that shape the $[Ca^{2+}]_i$ response. One example of this problem was the observation in smooth muscle cells that two applications of saturating concentrations of caffeine, which are separated by only 30 s, results in a second $[Ca^{2+}]_i$ response that is 80% smaller than the first response [88]. The initial and incorrect interpretation of these data was that 30 s was not enough time to refill the SR Ca^{2+} pool, yet when using Mag-fura-2 to see the free luminal SR $[Ca^{2+}]$, it became evident that the SR $[Ca^{2+}]$ is fully recovered from caffeine-induced depletion after a time of only 30 s [89]. Therefore it is clear the importance of recording simultaneously the changes in the $[Ca^{2+}]_i$ and the luminal SR/ER $[Ca^{2+}]$ with sufficiently high time resolution to be able to define an efficient Ca^{2+} release event. It is this approach what has made evident that the changes in the luminal SR/ER $[Ca^{2+}]$ does not present a simple correlation with the corresponding changes in the $[Ca^{2+}]_i$.

Initial attempts to look at the luminal ER $[Ca^{2+}]$ were carried out using Mag-fura-2, a low affinity Ca^{2+} indicator that does not go easily into the ER, and for this reason much of the work done with this indicator was carried out in permeabilized cells [90]. It turned out that Mag-fluo-4, a tricarboxylic low affinity Ca^{2+} indicator, goes into the ER more easily [91]. However, since Mag-fluo-4 lacks ER retention mechanism, not all dye stays in the ER, some of it goes into the Golgi apparatus and from there goes into the vesicles of the secretory pathway [10]. The advantage of organic dyes is that they are easy to be loaded and to calibrate, their stoichiometry with Ca^{2+} is 1:1 and they are relatively insensitive to pH changes. Nevertheless, the main limitation is that they lack a targeting mechanism and this issue has been solved developing Genetically Encoded Ca Indicators (GECIs) that are targeted to different organelles, the ER/SR among others [10, 92]. These GECIs use EF-hand proteins as Ca^{2+} sensor moiety and the signal could be chemo-luminescent or fluorescent in nature. The disadvantage of GECIs is that they are not homogeneously expressed in cells, their stoichiometry is not 1:1, and aequorin is irreversibly oxidized on Ca^{2+} binding, so it is a single shot Ca^{2+} indicator while GFP fluorescence is sensitive to changes in the pH [10].

Initial studies on the luminal ER $[Ca^{2+}]$ using aequorin in HeLa cells, found that the agonist-induced reduction in the free luminal ER $[Ca^{2+}]$ and the associated $[Ca^{2+}]_i$ response, were both of the same amplitude, whether the ER Ca^{2+} pool was fully loaded or practically depleted [93]. These data suggest that the Ca^{2+} released by the agonist is not in equilibrium with the free luminal ER $[Ca^{2+}]$. In pancreatic acinar cells it has been shown that a small concentration of acetylcholine (200 nM) increased the activity of Ca^{2+}-dependent Cl^- channels in the absence of any reduction in the luminal ER $[Ca^{2+}]$, despite the activation of IP_3R by acetylcholine [19]. These data indicate that activation of a small number of IP_3Rs produces no reduction in the free luminal ER $[Ca^{2+}]$ despite Ca^{2+} has been released. Again, these data suggest that the free luminal ER $[Ca^{2+}]$ is not in equilibrium with the Ca^{2+} that is released to the cytoplasm by IP_3Rs. A new cameleon targeted to the ER known as D1ER, showed that agonist-induced elevation of the $[Ca^{2+}]_i$ precedes any significant reduction in the free luminal $[Ca^{2+}]$. Actually, the majority of the reduction in the free luminal ER $[Ca^{2+}]$ occurs during the reduction of the agonist-induced $[Ca^{2+}]_i$ response [94]. These data suggest that the Ca^{2+} buffering capacity of the ER is extremely high, during the initial stages of Ca^{2+} release; later on, it switches to a form of low capacity [95–97]. This lack of correlation between the reduction in the free luminal ER $[Ca^{2+}]$ and the increase in the $[Ca^{2+}]_i$ cannot be explained by saturation of the luminal Ca^{2+} indicator. This is the case because in single uterine smooth muscle cells that were loaded with both fura-2 and Mag-fluo-4, the activation of SERCA pump increased the Mag-fluo-4 fluorescence implying that Mag-fluo-4 cannot be saturated inside the SR [91, 98]. This lack of correlation can be seen in different cell types and with activation of either RyRs or IP_3Rs. For instance, the addition of 1 mM ATP, to activate purinergic receptors in HeLa cells, results in the activation of IP_3Rs which produced a delayed reduction in the free luminal ER $[Ca^{2+}]$ with respect to the time course of the $[Ca^{2+}]_i$ response [99]. Another group studying HeLa cells that had been loaded with Indo-1 and an ER-targeted cameleon, observed that the histamine-induced increase of the $[Ca^{2+}]_i$ was associated with a rather small reduction of the free luminal ER $[Ca^{2+}]$ [100]. This was the case either in the presence or in the absence of external $[Ca^{2+}]$; indicating that the explanation for this lack of correlation is not due to the presence of Ca^{2+} entry at the plasma membrane. In heart cells, localized, fast, transient increases of the $[Ca^{2+}]_i$ known as Ca^{2+} sparks, which are due to the activation of a cluster of RyR2s, are associated with a transient decreased of the free luminal SR [Ca^{2+}], known as blinks [101]. Remarkably, the large majority of these blinks show a slower Ca^{2+} decrease rate than the rate of Ca^{2+} increase seen with the corresponding spark [101, 102]. In skeletal muscle, the reduction in the free luminal SR [Ca^{2+}] that are associated with a Ca^{2+} spark are known as skraps [97], which also have a much slower time course when compared with their corresponding sparks [96, 97]. These data indicate that even at the subcellular level the time course of the reduction in the luminal SR [Ca^{2+}] does not coincide with the corresponding increase in the [Ca^{2+}]$_i$ [96]. To complicate matters even further, it has been shown in heart cells that activation of SERCA pump with isoproterenol produces an increase of the luminal SR [Ca^{2+}] while Ca^{2+} is simultaneously released to the cytoplasm [103, 104]. Moreover, an efficient Ca^{2+} wave propagation requires active SERCA pumps [105]. All these data indicate that there is no correlation between the changes in the free luminal SR/ER [Ca^{2+}] and the increase in the [Ca^{2+}]$_i$ despite the fact that the SR/ER is the source for the Ca^{2+} that appears in the cytoplasm.

A more detailed analysis of the Ca^{2+} release process carried out by either RyRs or IP$_3$Rs, in smooth muscle cells from the urinary bladder, reveals the existence of four different phases for a Ca^{2+} release event [95]. The first one involves the largest increase in the [Ca^{2+}]$_i$ with just a small reduction in the luminal SR [Ca^{2+}] (we call this an efficient Ca^{2+} release event); phase 2 is characterized by the largest reduction in the luminal SR [Ca^{2+}], without any effect on the [Ca^{2+}]$_i$; while phase three is defined by the reduction in the [Ca^{2+}]$_i$ but the luminal SR is still depleted because the release channels stay open and the fourth phase, is when release channels have been closed and the free luminal SR [Ca^{2+}] recovers to their normally high levels by the action of SERCA pumps [95]. Additionally, in the same work, it was used a low concentration of heparin, to partially inhibit IP$_3$Rs, and the application of carbachol, to activate IP$_3$Rs, results in both a transient elevation of the free luminal SR [Ca^{2+}] and a transient increase of the [Ca^{2+}]$_i$ [95]. The elevation in the [Ca^{2+}]$_{SR}$ occurred at the same time that the increase in the [Ca^{2+}]$_i$. The implication of these data is that activation of IP$_3$Rs unleashes Ca^{2+} trapped in luminal proteins that under normal conditions (fully activated IP$_3$Rs) will gain immediate access to the cytoplasm. However, with the use of heparin, to partially reduce the activity of IP$_3$Rs, the just liberated Ca^{2+} from luminal Ca^{2+} binding proteins, diffuses back to the bulk of the SR and it is seen as an increase of the [Ca^{2+}]$_{SR}$. This lack of correlation between the changes in the [Ca^{2+}]$_i$ and the SR/ER [Ca^{2+}] can also be detected using GECIs in cells, that are either isolated or in the tissue [92]. A newly developed GECI that uses apoaequorin to sense Ca^{2+} and GFP to have a fluorescence signal has been used together with Fura-2. Simultaneous recording of the luminal ER [Ca^{2+}] and the [Ca^{2+}]$_i$ show that the [Ca^{2+}]$_i$ response precedes the agonist-induced reduction in the luminal ER [Ca^{2+}] both in HeLa cells and neurons in hippocampal slices [92, 106]. Using a different set of GECIs, it was found that the application of both bradykinin and CPA results in a nuclear [Ca^{2+}] elevation, detected with H2B-D3cpv probe, that peaks well before the nadir of the free luminal ER [Ca^{2+}] reduction [107]. All these data undermine the idea that SR/ER is a Ca^{2+}

pool where the physical state of Ca^{2+} ions is switching from free to protein bound, but that this ion appears to be also trapped in protein complexes that release Ca^{2+} ions in response to the activation of RyRs or IP_3Rs, a picture that was portrayed by Ikemoto [77] and others [78].

The idea that Ca^{2+} does not seem to be in equilibrium inside the SR/ER has been shown using microanalysis with electron microscopes. The activation of VGCC in neurons, loads the ER with Ca^{2+} due to the activity of SERCA pump. However, the increase in Ca^{2+} was not homogenous inside the ER. The same cisterna, can have regions with high Ca^{2+} next to regions where the total amount of Ca^{2+} was not changed at all by the activity of SERCA pumps [64, 65]. Again, these data indicate that there are regions of the SR/ER where Ca^{2+} activity is reduced (meaning that Ca^{2+} has been trapped) without changing the Ca^{2+} activity of contiguous regions despite the absence of evident diffusion barriers. The easiest way to explain these observations is that Ca^{2+} is trapped by proteins and that these ions are not in equilibrium with the free luminal ER $[Ca^{2+}]$. If this is the case on how the SR/ER accumulates Ca^{2+} it might explain why there is a lack of correlation between the Ca^{2+} supplied to the cytoplasm by the SR/ER and the associated changes in the free luminal SR/ER $[Ca^{2+}]$.

A question that is still open is how luminal Ca^{2+} binding proteins are able to trap Ca^{2+} inside the SR/ER. Nevertheless, there are some hints on how this might be happening. It has been shown that calsequestrin is able to increase the number of Ca^{2+} binding sites as the Ca^{2+} concentration is increased. The underlying mechanism for the increase in Ca^{2+} binding sites depends on the ability of calsequestrin to polymerize [62]. However, since calsequestrin is not expressed by all cells and since an efficient Ca^{2+} release is a generalized cellular event, then there should be other proteins doing the same function as calsequestrin. Actually, knockout studies of both CSQ1 and CSQ2 have suggested that there must be other proteins in the lumen of the SR/ER because RyRs were still able to release Ca^{2+} [58]. Thus either calreticulin, or maybe other Ca^{2+} binding proteins in the lumen of the SR/ER, have also the property of increasing the number of Ca^{2+} binding sites in response to an increase of the luminal $[Ca^{2+}]$.

Another very interesting situation to study involves the phenomenon known as the refractory period of Ca^{2+} release that might reflect how the SR/ER is trapping releasable Ca^{2+} [88, 89, 101, 102]. If it is true that the free SR $[Ca^{2+}]$ is in equilibrium with the Ca^{2+} released by RyRs, then this time should be more than enough to fully recover the caffeine-induced $[Ca^{2+}]_i$ response. So the question remains on how to explain the presence of a refractory period for caffeine even when the free luminal SR $[Ca^{2+}]$ has reached normal levels. The same situation is observed in heart cells, where the refractory period between Ca^{2+} sparks from the same site is significantly longer than the time required for recovery of the free luminal SR $[Ca^{2+}]$ [101, 102]. Moreover, keep in mind that the inactivation of the RyRs cannot be the explanation for the presence of a refractory period because; it has been shown that the reduction in the free luminal SR $[Ca^{2+}]$ is basically the same for both Ca^{2+} release events. This observation implies that the same number of RyRs was activated during the two Ca^{2+} release events although the second application of caffeine produced a five times smaller $[Ca^{2+}]_i$ response. Thus, all these data imply that the changes in the free luminal SR $[Ca^{2+}]$ does not reflect the total amount of Ca^{2+} that is released from the SR to the cytoplasm. If this is the case, then we think that the most likely explanation for the presence of a refractory period is that an efficient Ca^{2+} release event involves the liberation by RyRs of Ca^{2+} trapped by proteins, which are localized contiguous to the release channel [108]; however, it appears that this complex (RyRs-Ca^{2+} trapping proteins) takes a longer time to assemble than the time taken to recover the free luminal SR $[Ca^{2+}]$ (see Sect. 14.4.2).

14.4 Current View on How the SR/ER Ca^{2+} Store Is Working to Produce an Efficient Ca^{2+} Release Event

14.4.1 The Luminal Ca^{2+} Binding Proteins Compete for Ca^{2+} with the Open Release Channel

The picture depicted by Ikemoto in the '90s, on how the SR releases Ca^{2+}, appears to be a generalized situation not only for the SR but also for the ER, that is, for the IP$_3$Rs. However, this picture is still sketchy because we do not have enough spatial and temporal resolution to have all the elements involved in an efficient Ca^{2+} release event. For this reason, we have resorted on the use of mathematical models to find the conditions necessary to generate phase one during Ca^{2+} release from the ER/SR, as a reminder, phase one is characterized by an increase of the $[Ca^{2+}]_i$ without any reduction in the free luminal ER/SR $[Ca^{2+}]$ [95, 96]. The simplest solution we were able to figure out was reducing the number of luminal Ca^{2+} binding sites in response to a reduction of the luminal SR $[Ca^{2+}]$ [109]. We called this situation Kinetics on Demand (KonD) because the number of Ca^{2+} binding sites increases as the result of an elevation of the $[Ca^{2+}]$ while they are diminished by a reduction in the $[Ca^{2+}]$. This type of kinetics is completely different to the traditional one, where the number of Ca^{2+} binding sites is fixed and the reduction in the free $[Ca^{2+}]$ increases the number of Ca^{2+} binding sites that are unoccupied by Ca^{2+}. The mathematical model reveals that these unoccupied Ca^{2+}-binding sites, which are increasing in number during the Ca^{2+} release process, become a strong competitor for Ca^{2+} with the open release channels and naturally, this competition would slow the Ca^{2+} release process. This kind of problem would not be presented by KonD model because the number of Ca^{2+} binding sties is being reduced during the Ca^{2+} release event. We want to stress that KonD model was proposed based on the observation that the number of Ca^{2+} binding sites in calsequestrin increases as a function of its polymerization [62]. Therefore, the changes in the paradigm were based on understanding how calsequestrin is working and the main difference consisted on assuming that the total number of Ca^{2+} binding sites are not fixed but varies depending on the presence of Ca^{2+} [62].

Phase one reflects a mechanism for an efficient Ca^{2+} release process and we think that involves Ca^{2+} trapped in luminal Ca^{2+}-binding proteins, next to release channels. This situation implies that the Ca^{2+} that is released by RyRs or IP$_3$Rs is not in equilibrium with the free luminal SR/ER $[Ca^{2+}]$. However, there could be alternative explanations for phase one, the idea that phase one is due to saturation of the luminal Ca^{2+} indicator has been discarded because there are different experimental conditions that produce an increase of the free luminal $[Ca^{2+}]$, which argue against the idea that the Ca^{2+} indicator inside the SR/ER is saturated [91, 95, 97, 110]. In skeletal muscle it has been suggested that high cooperativity of Ca^{2+} binding to calsequestrin might explain phase one [58, 97, 111]. The main limitation with this explanation is that the high cooperativity means a smaller range of $[Ca^{2+}]$ before reaching saturation and the ability of SR/ER to buffer Ca^{2+} is quite the opposite since it covers several orders of magnitude [64]. Mathematical models have suggested that luminal Ca^{2+} binding proteins diffuse away from the RyRs while Ca^{2+} would be diffusing towards the open release channels during the release process [73, 74]. We think this seems to be unlikely because we do not see what could be the driving force for Ca^{2+}-binding proteins to diffuse away from the RyRs. Another scenario might be a restricted diffusion of Mag-Fluo-4, which is kept in the bulk of the SR away from the RyRs, so this dye will not be in rapid equilibrium with calsequestrin, which is known to be associated with RyRs [112] and if Ca^{2+} slowly diffuses between these two compartments, then it is expected to see recovery of the free luminal SR Ca^{2+} level (provided that SERCA pumps are located where Mag-Fluo-4 is) even when the RyRs are still open. However, this scenario has not been observed. Release channels need to be in the closed conformation for SERCA pump being able to recover the free luminal SR/ER $[Ca^{2+}]$. This observation implies that there is no diffusion barrier for the Ca^{2+} indicator. The absence of diffusion barriers for Ca^{2+} and Ca^{2+} indicators within the lumen of the SR/ER should make it difficult to see the transition between phase one and phase two. Indeed, the fusion of these two phases in a single phase has been observed but only in particular conditions; for instance when SERCA pumps had been inhibited with thapsigargin [95] or when it has not given enough time for stores to recover from a previous release event [88, 109, 113]. These data add a new element of complexity since it argues for the importance of having an active SERCA pump to be able to see phase one separated from phase two. Additionally, our mathematical model has stressed the importance that Ca^{2+} binding proteins rapidly switch from high to low Ca^{2+} buffering capacity during the Ca^{2+} release process, a scenario that we have called Kinetics on Demand [109]. The importance of this switch is that the empty Ca^{2+}-binding sites of the luminal Ca^{2+}-binding proteins will not compete with release channels for free luminal Ca^{2+}. The absence of this competition is reflected in release channels having enough free luminal Ca^{2+} for an efficient release event even when the free luminal SR/ER $[Ca^{2+}]$ is not decreasing. Obviously, it is important to know the molecular nature of the proteins responsible for sequestering Ca^{2+} next to release channels and the molecular mechanism involved to switch from sequestered Ca^{2+} to free Ca^{2+} inside the SR/ER and how the activation of release channels gain access to this trapped Ca^{2+}. Based mainly, but not exclusively, on our own work and the work done by Ikemoto's group, we have developed a graphical model that might explain the differences observed for the time courses of the $[Ca^{2+}]_i$ and the $[Ca^{2+}]_{SR/ER}$ during a Ca^{2+} release event (Fig. 14.2).

14.4.2 Proposed Graphical Model on How Release Channels in the SR/ER Produce an Efficient $[Ca^{2+}]_i$ Response

The graphical model shown in Fig. 14.2 is a cartoon on how the activation of Ca^{2+} release channels leads to an efficient increase of the $[Ca^{2+}]_i$. Initially (Fig. 14.2a), release channels are closed and the Ca^{2+} lattice represents the luminal Ca^{2+} binding proteins that are present in the form of polymers attached to the RyRs [114]. These have trapped Ca^{2+} inside the lattice and accordingly, this Ca^{2+} will not be

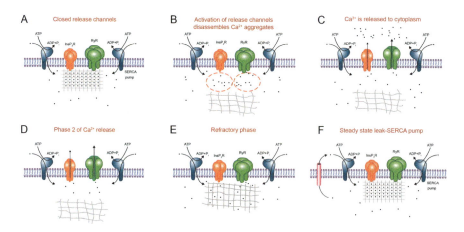

Fig. 14.2 Graphical model of an efficient Ca^{2+} release event by either RyRs or IP$_3$Rs in the SR/ER Ca^{2+} stores. This cartoon depicts those steps that are considered to be critical for generating an efficient Ca^{2+} release event that is reflected in a transient increase of the $[Ca^{2+}]_i$. (**a**) Ca^{2+} source is in the form of a lattice that is attached to the release channels but disconnected from the bulk of the free luminal SR/ER $[Ca^{2+}]$. (**b**) The activation of the release channels would liberate the Ca^{2+} trapped in the lattice. (**c**) The open conformation of release channels allow the movement of Ca^{2+}, down its electrochemical gradient, to the cytoplasm, without the need to reduce the bulk luminal SR/ER $[Ca^{2+}]$ (phase one). (**d**) Phase two of Ca^{2+} release, which is the inefficient part of the Ca^{2+} release event, mainly because the Ca^{2+} flux from the bulk of the SR/ER is rather slow when compared with the cytoplasmic Ca^{2+} removal mechanisms. (**e**) Once the release channels have been closed there will be a refractory period, because although the bulk of the free luminal SR/ER $[Ca^{2+}]$ has returned to normal levels, probably the Ca^{2+} lattice has not recovered just yet. (**f**) Once the Ca^{2+} lattice has reformed and reattached to the release channels, they are ready for another efficient Ca^{2+} release event

in equilibrium with the free luminal Ca^{2+} that is in the bulk of the SR/ER. This scenario might explain why there is an elevation in the free luminal SR/ER $[Ca^{2+}]$ in response to the activation of release channels; if this is the case then it is easy to see why the free luminal SR/ER $[Ca^{2+}]$ is an inadequate predictor of the amount of Ca^{2+} that the internal stores deliver to the cytoplasm. The activation of release channels (Fig. 14.2b) leads to conformational changes that disassemble the Ca^{2+} lattice producing a local elevation of the free luminal $[Ca^{2+}]$ next to the release channels. This elevation in the free luminal $[Ca^{2+}]$ might help release channels to reach the open conformation. If the release channel is not open or is blocked then this Ca^{2+} would start to diffuse to the bulk of the SR/ER, and this can be observed as a transient elevation of the free luminal SR/ER $[Ca^{2+}]$. This scenario might be the explanation behind the ability of polylysine, a substance that is both an activator and partial blocker of RyRs [115, 116], to reveal a transient elevation of the free luminal $[Ca^{2+}]$ during the activation of RyRs. Following this reasoning, we have used a low concentration of heparin to partially block IP_3Rs that had been previously activated in response to the application of carbachol in smooth muscle cells [117]. Actually, the rate of rising of the $[Ca^{2+}]_i$ when IP_3Rs had been inhibited by heparin, was strongly augmented [95]. The idea here is that Ca^{2+} would accumulate next to those partially blocked release channels and this will increase the driving force resulting in a higher Ca^{2+} exit via those few channels that were not blocked by heparin. Under normal circumstances, the release channels will be fully open (Fig. 14.2c) and this will allow free diffusion of Ca^{2+} from the SR/ER to the cytoplasm producing a significant, transient $[Ca^{2+}]_i$ response but with one very important characteristic, there will be basically no reduction of the bulk free luminal SR/ER $[Ca^{2+}]$. Since most of the Ca^{2+} indicator is in the bulk of the SR/ER, this dye will not see any reduction in the surrounding $[Ca^{2+}]$ because the Ca^{2+} next to the release channel is in a higher concentration than in the bulk of the SR/ER, at least initially in the release event. This is what we have called phase one of the Ca^{2+} release event [95], which is an efficient Ca^{2+} release process because there is an increase in the $[Ca^{2+}]_i$ with minimal or no reduction in the free luminal SR/ER $[Ca^{2+}]$. Phase one of the Ca^{2+} release process has been seen in all kind of cells, during slow or fast Ca^{2+} release events, for RyRs or IP_3Rs and using synthetic or genetically encoded Ca^{2+} indicators [96]. Phase two is observed when the Ca^{2+} supply that comes from the Ca^{2+} lattice has been exhausted and the free luminal Ca^{2+} in the bulk of the SR/ER starts to be drained via the still open release channels (Fig. 14.2d); however, in this case the Ca^{2+} movement is so slow that there is a clear reduction in the bulk of the SR/ER $[Ca^{2+}]$ but without any effect on the $[Ca^{2+}]_i$ (this is the unproductive part of the Ca^{2+} release event). Once the release channels are either closed or inactivated (Fig. 14.2e), then SERCA pump can start accumulating Ca^{2+} by hydrolyzing ATP until the high resting luminal SR/ER $[Ca^{2+}]$ has been reached, at least in the bulk of the SR/ER. However, there would be a time period larger than the time it will take to reach the normal free luminal $[Ca^{2+}]$, where it can be considered that the Ca^{2+} store has been refilled (high free luminal $[Ca^{2+}]$) but if the release channels were to be exposed to an activator in this time, the result would be an unproductive

Ca^{2+} release event [88, 95, 101, 102, 109]. We do not know the reason for this situation, but it is feasible that even when the free luminal SR/ER [Ca^{2+}] is high, close to the normal levels, the Ca^{2+} lattice appears that it has neither reassembled nor connected to the release channels yet. Thus, the characteristic of this stage is a small time window where the free luminal [Ca^{2+}] is basically normal but the activation of the release channels produces a much smaller [Ca^{2+}]$_i$ response [95, 101]. The implication of this situation is that Ca^{2+} buffering in the lumen of the SR/ER does not appear to be a reversible process. Actually, this process shows hysteresis [95, 96, 109], a situation that is not expected for proteins with a low affinity Ca^{2+} binding sites.

Activation of all RyRs, for instance with high concentrations of caffeine, displays this refractory phase in the order of tens of seconds [88, 95]. This implies that this is the time taken by RyRs to recover an efficient Ca^{2+} release event. Nevertheless, the frequency of Ca^{2+} release events in heart and skeletal muscle is much higher than the time taken by RyRs to recover an efficient Ca^{2+} release event. Then we think that muscles cope with this limitation by using only a fraction of all the RyRs present in the SR. In summary, a picture is emerging where the Ca^{2+} released by RyRs or IP$_3$Rs is trapped in a lattice that is not in equilibrium with the bulk of the SR/ER and it is essential for having an efficient [Ca^{2+}]$_i$ response. Additionally, we propose that the time it takes for this Ca^{2+} lattice to reconstitute itself and to attach to release channels determines the refractory period. We think that a reduction in the SERCA pump activity or the number of functional RyRs with an assembled Ca^{2+} lattice, or even worse, the presence of both conditions, lead to heart failure [118]. Under this condition, Ca^{2+} release via RyRs cannot be efficient since the Ca^{2+} flux is not fast enough to produce a large increase of the [Ca^{2+}]$_i$. One important issue here is that the formation of the Ca^{2+} lattice (Fig. 14.2f) requires both active SERCA pumps and release channels being in the close conformation. These two conditions are not met during heart failure and this might explain why the Ca^{2+} lattice would not be formed equally well and the Ca^{2+} release process is not efficient enough in this pathological condition. It is known that high concentrations of ryanodine lock RyRs in the open conformation, but at the same time, their ion pore is blocked by ryanodine [47]. This situation facilitates recovery of the high luminal [Ca^{2+}]$_{SR/ER}$ by the activity of SERCA pump after the application of caffeine, but it results in the lack of IP$_3$R-mediated [Ca^{2+}]$_i$ response despite the normal free luminal SR [Ca^{2+}] [79, 89, 117]. Thus, the idea is that the Ca^{2+} lattice to be reformed and attach to the release channels requires these channels to be in the closed conformation.

14.4.3   The Relevance of the Model Where Release Channels Used Ca^{2+} Trapped in the Ca^{2+} Lattice

The importance of this proposed mechanism used by the SR/ER to release Ca^{2+} is that these organelles can function as Ca^{2+} pools for cytoplasmic Ca^{2+} transients without running the risk of triggering ER stress response due to depletion of the ER luminal [Ca^{2+}] [40, 119–121]. There would be no Ca^{2+} depletion of the SR/ER store when release channels stay open only for the period of time that encompasses phase one of the Ca^{2+} release process. The reason would be that release channels would not reach phase two of the release process and no depletion of the SR/ER would occur. The time required to reach phase two of Ca^{2+} release is much longer than the time involved in a Ca^{2+} spark or a Ca^{2+} puff. Therefore, this model of Ca^{2+} release has the compelling advantage that it can release large amounts of Ca^{2+} to the cytoplasm without eliciting depletion of the SR/ER and accordingly, without triggering the ER stress response that might limit the role of the SR/ER as Ca^{2+} pool. The price to pay for this type of Ca^{2+} release process is the presence of a refractory period because of the time needed for reassembling the Ca^{2+} lattice. However, this limitation can be overcome relatively easy by having a reserve of release channels with a Ca^{2+} lattice. The dynamic of this reserve might be compromised during the fatigue process [122].

This model of SR/ER Ca^{2+} release mechanism combines data obtained with different cell types, different Ca^{2+} indicators and with activation of either RyRs or IP$_3$Rs, and the observation by Ikemoto's group that activation of RyRs results in an elevation of the free luminal [Ca^{2+}]$_{SR}$ and that this effect requires calsequestrin attached to RyRs. Additionally, this model also considers that calsequestrin increases its Ca^{2+}-binding capacity as a function of its Ca^{2+}-induced polymerization [62, 109]. Despite all these data that support this model; it is still incomplete because the lack of information with sufficiently high temporal and spatial resolution to see the disassembling of the Ca^{2+} lattice induced by the activation of the release channels. To achieve these resolutions is not trivial since the distance is too short, a few nanometers, and the rate of this process is in the order of milliseconds. Additionally, it is quite feasible that calsequestrin might be one of many different proteins that can generate a Ca^{2+} lattice in association with RyRs and IP$_3$Rs because the double knockout of this protein was not lethal [58] and even worse, it resulted in a much smaller reduction than expected of skeletal muscle luminal Ca^{2+} buffer capacity [58]. The few attempts that have been done to see the disassembling of the Ca^{2+} lattice in real time have failed because it was tried by reducing the free luminal SR [Ca^{2+}] with ionomycin, instead of directly activating RyRs with caffeine [60]. Interestingly, the time constant for calsequestrin depolymerization was extremely slow when driven by a reduction in the luminal SR [Ca^{2+}], which made authors think that this mechanism cannot be operating in cardiac cells because systole has a much higher frequency. However, our own interpretation is that the reduction in the free luminal [Ca^{2+}] is not the signal that drives the fast disassemble of the Ca^{2+} lattice, it is the opening of the RyRs. This scenario might explain why in the absence of external Ca^{2+}, that reduces the free luminal ER [Ca^{2+}], IP$_3$-inducing agonists produce an [Ca^{2+}]$_i$ response that was barely decreased in amplitude by the absence of external Ca^{2+} [93]. In another study, it became clear that the spatial resolution was not good enough to see the subtle changes associated with modifications in the degree of calsequestrin polymerization during the reduction of the free luminal [Ca^{2+}] [123]. So these few attempts suggest that it will be more difficult than previously thought to show the dynamics of the Ca^{2+} lattice associated with release channels during the Ca^{2+} release process.

14.5   Other Factors Involved in Having an Efficient Ca^{2+} Release Event

All these studies reviewed here clearly show that Ca^{2+} release from intracellular Ca^{2+} pools, particularly the SR/ER driven by release channels, is more complex than a simple analogy with a water tank (Fig. 14.1). The main difference appears to be that Ca^{2+} is trapped by luminal proteins and accordingly, it is not in equilibrium with the bulk of the SR/ER $[Ca^{2+}]$. It is only the activation of release channels (RyRs or IPRs) that liberates this Ca^{2+} from the lattice (Fig. 14.2). This situation implies that the change in the free luminal $[Ca^{2+}]$ is not a good indicator of how much Ca^{2+} has been released to the cytoplasm. Additionally, the activity of release channels is under the control of both the luminal $[Ca^{2+}]$ and the interaction with luminal proteins, as it is the case for RyRs and calsequestrin. Moreover, the distribution of release channels and Ca^{2+} binding proteins is not homogenous in the SR/ER and even when their lumen lack diffusion barriers, the tortuosity of these internal Ca^{2+} stores contribute with geometrical issues that also need to be considered. If all these factors are not adding enough complexity, there is still another layer of complexity due to the fact that all the elements involved in the Ca^{2+} release process work interconnected. In this regard, there is data showing that an efficient Ca^{2+} release process requires an active SERCA pump [89]. This was unexpected because these two activities, SERCA pump and release channels, should be antagonistic between each other and instead it turns out that an active SERCA pump is a critical element to achieve an efficient Ca^{2+} release event.

One example of this communication between SERCA pump and Ca^{2+} release channels is observed in the effect of activation of β adrenergic receptor in heart cells. In this case the phosphorylation of phospholamban by PKA dissociates this protein from SERCA pump resulting in a higher pump activity. However, this effect is reflected in a larger $[Ca^{2+}]_i$ response that paradoxically, cannot be explained by a higher free luminal SR $[Ca^{2+}]$ [20]. Moreover, the high speed of Ca^{2+} wave propagation requires active SERCA pumps in heart cells [105], a condition that is unexpected. It has been shown that induction of Ca^{2+} release due to activation of RyR1 activates SERCA pump even during the initial phase of Ca^{2+} release when the free luminal SR $[Ca^{2+}]$ is the highest [87], arguing against the idea that changes in the free luminal SR $[Ca^{2+}]$ is the way that RyRs control the activity of SERCA pump [81]. Additionally, rapid inhibition of SERCA pump with thapsigargin decreases both the amplitude and the rate of rising of the cytoplasmic Ca^{2+} response to caffeine or to acetylcholine in smooth muscle cells. This is happening even when thapsigargin does not reduce immediately the free luminal $[Ca^{2+}]$ [89]. All these data argue again for the need for active SERCA pumps to have an efficient Ca^{2+} release event. Efficient histamine-induced Ca^{2+} release in HeLa cells also requires that SERCA pump is active [124]. Pancreatic acinar cells from RGS2 knockout mice express a compensatory twofold-increase in the expression of SERCA pump and this is associated with a larger and more sensitive $[Ca^{2+}]_i$ response to acetylcholine that cannot be explained by a larger production of IP_3 [125]. Although this type of experiment does not explain why the elimination of RGS2 has resulted in a compensatory two-fold increase in the expression of SERCA pump, these results are in line with the idea that higher activity of SERCA pump produces larger agonist-induced Ca^{2+} release events.

14.6 Ca^{2+} Signaling Between the SR/ER and Other Organelles

14.6.1 The Connection Between the ER and Mitochondria

It has been shown that close association between the ER and mitochondria plays an important role in ATP production, lipids synthesis, mitochondria dynamics and Ca^{2+} signaling [126, 127]. One of the first studies about this association showed that the majority of mitochondria, around 80%, was forming contacts sites with the ER [128]. However, using a high resolution microscope in the 90s, it was found that the mitochondrial surface in contact with the ER was estimated to be around 10% of the total mitochondrial network [129]. Subsequent works have found that the percentage of mitochondria in contact with the ER was close to 20%, and that these values can vary when cells are challenged with different experimental conditions, although this interaction has never reached 80% [130–133]. Interestingly, Scorrano's group has observed that the fraction of mitochondria interacting with the ER is near 80% [134], as it was initially described by Montisano et al. [128]. Recently, using super-resolution microscopes, it has been shown that mitochondria are the main organelle in association with the ER, although these studies did not quantify the degree of this interaction [23, 24, 135]. Collectively, these data show an uncertain scenario regarding the fraction of the mitochondrial network that is associated with the ER, since these numbers vary so widely. At the present time it is not clear the reason behind this variability among the different studies.

14.6.2 The Association Between Mitochondria and the ER Is Stabilized by a Large Assortment of Protein Tethers

At the biochemical level, it has been shown that the ER-mitochondria interaction is involved in phospholipids synthesis, this process involves the bidirectional exchange of phospholipids between the ER and mitochondria [136]. It has been shown that this exchange occurs in specific contact sites called Mitochondria-associated ER membranes (MAMs). These structures facilitate different signaling transduction processes [137] and probably, Ca^{2+} transfer from the ER to mitochondria. Actually, it has been described several different protein tethers that modulate Ca^{2+} transfer from the ER to mitochondria [138].

It has been reported that the Voltage Dependent Anion Channel (VDAC), a resident protein in the outer mitochondrial membrane (OMM) connects with the IP_3R in the ER by interacting with the cytosolic chaperone glucose-regulated protein 75 (Grp75). This tether enhances the Ca^{2+} uptake by mitochondria [139]. Additionally, Cyclophilin D, a mitochondrial protein that modulates the opening of the permeability transition pore, interacts with the VDAC-Grp75-IP_3R complex; the loss of this protein decreases both ER-mitochondria interactions and Ca^{2+} transfer to mitochondria. This situation leads to insulin resistance in hepatic cells, suggesting that this tether might be controlling glucose homeostasis [140]. Moreover, the overexpression of α-synuclein (a central protein in Parkinson disease) in HeLa cells has promoted an increase in mitochondrial Ca^{2+} uptake by enhancing ER-mitochondria interactions. These data suggest that α-synuclein has an important role in mitochondrial Ca^{2+} homeostasis [141]. It is clear then, from these examples, that protein tethers are involved in Ca^{2+} signaling. Indeed, mitofusin 2 (Mfn-2) a GTPase that is located both in the OMM and the ER membrane, is involved in the formation of ER-mitochondria tethers. Mfn-2 ablation increases the distance between the ER and mitochondria, and this results in a reduced Ca^{2+} influx into mitochondria in response to IP_3 in both HeLa cells and mouse embryonic fibroblasts [134, 142]. Moreover, ablation of Mfn-2 leads to metabolic disorders such as glucose intolerance and impaired insulin signaling in both liver and muscle [143]. Conversely, Pozzan's group has demonstrated that silencing of Mfn-2 increases contact sites between the ER and mitochondria and augments mitochondria Ca^{2+} uptake. These data suggest that Mfn-2 appears to have a protective role because it forms a tether that prevents Ca^{2+} overload into the mitochondria [130]. Although both groups have shown that Mfn-2 is a molecular tether in the ER-mitochondria contact site, silencing this protein produces the opposite effects on Ca^{2+} signaling. Interestingly, a picture is emerging where Mfn-2 is able to increase the number of contact sites between the ER and mitochondria when they are low and it does the opposite when these sites are high. Further work is needed to understand the role played by Mfn-2 in Ca^{2+} transfer from the ER to mitochondria.

14.6.3 Mitocondrial Calcium Uniporter (MCU) Is a Finely Regulated Inner Mitochondrial Membrane Ion Channel

Ca^{2+} transfer from the ER to mitochondria is an essential event that regulates cell bioenergetics by increasing the activity of different Krebs cycle dehydrogenase; such as pyruvate dehydrogenase, isocitrate dehydrogenase and α-ketoglutarate dehydrogenase [144]. This Ca^{2+} entry in the mitochondria matrix involves the activation of Ca^{2+} release channels from the ER, mainly the IP_3R [129]. Although the closeness between the ER and mitochondria should lead to mitochondria Ca^{2+} transient before that in the cytoplasm, it turns out that this is not the case. Indeed, GECIs targeted to mitochondria and the ER in combination with fura-2 to detect changes in cytoplasmic $[Ca^{2+}]$ have shown that despite mitochondria being close to the ER, the cytosolic Ca^{2+} increases well before it does in mitochondria [110], suggesting that closeness between these two organelles is not enough to elevate Ca^{2+} in mitochondria. Alternatively, mitochondria might have mechanisms that prevent an immediate Ca^{2+} entry in the mitochondrial matrix. The molecular identification of the Mitochondrial Ca^{2+} Uniporter or MCU [145, 146] as the mitochondria Ca^{2+} permeable channel for Ca^{2+} entry to the mitochondria matrix, has demonstrated that this channel is indeed a protein complex formed by different types of proteins, such as EMRE, MCUb and members of the MICU family [147]. MICU1, MICU2 and MICU3 are EF-hand proteins that inhibit MCU activity at low $[Ca^{2+}]_i$ (around 500 nM), thus acting as a channel gatekeeper that prevents mitochondria Ca^{2+} overload. Remarkably, MICU1 silencing leads to neurologic and muscular problems during development [148]. That is, these proteins act as natural inhibitors of MCU that block Ca^{2+} entry into the mitochondria. This might explain the mitochondria delayed Ca^{2+} increase after activation of IP_3R and increase of cytoplasm $[Ca^{2+}]$. In this regard, it has been shown that histamine, which produces a sustained Ca^{2+} release event, leads to a robust Ca^{2+} entry in mitochondria; however, glutamate, which produces a transient Ca^{2+} release event due to deactivation of IP_3Rs, produces a much smaller Ca^{2+} increase in mitochondria matrix [149]. These data suggest that Ca^{2+} entry in mitochondria requires more a sustained Ca^{2+} release from the ER than a localized elevation of cytoplasmic $[Ca^{2+}]$. Collectively, a picture is emerging showing that closeness between the ER and mitochondria is necessary but not sufficient for an elevation of the mitochondria $[Ca^{2+}]$. Moreover, an elevation of the mitochondrial $[Ca^{2+}]$ can trigger apoptosis [150]. In conclusion, the Ca^{2+} transfer from the ER to the mitochondria is important for cell respiration and for tuning ATP production while at the same time could lead to an apoptotic event. It is extremely complex how the same signal, i.e. Ca^{2+} ions, results in so divergent cell responses. Thus, essential pieces of the puzzle are still missing to fully unravel the role of Ca^{2+} in the interaction between the ER and mitochondria.

14.6.4   The Connection Between the ER and Lysosomes

The lysosome is another organelle that associates with the ER [23]. This organelle is vesicular in nature, it is filled with hydrolytic enzymes and characterized by having an extremely acidic luminal pH, around 5.0 [151]. This acidic pH is generated by the activity of a V-type proton ATPase [152, 153] and it has been considered that the activity of this pump is essential for the Ca^{2+} accumulating activity of lysosomes [154, 155]. However, studies using Ca^{2+} indicators targeted directly to lysosomes and agonists for TRPML1 channel in lysosomes, have found that the acidic luminal pH is not essential for lysosomes to accumulate Ca^{2+} but that a still undefined mechanism allows lysosomes to accumulate Ca^{2+} that has been released by IP_3Rs from the ER [11, 156]. The dynamics of the ER and lysosomes has been recently observed with high spatial and temporal resolutions and it appears that lysosomes are able to reshape the ER [23]. However, it is still a long way to understand the regulation of Ca^{2+} transfer from the ER to lysosomes in the autophagic process [157]; although the presence of high spatial and temporal superresolution microscopes, as GI-SIM [23], would make easier to unravel the role played by Ca^{2+} both in the lysosomes and in the ER in the activity of these acidic organelles.

14.7 Concluding Remarks

All these studies reviewed here allow us to depict a picture where an efficient Ca^{2+} release event from the SR/ER requires the activity not only of release channels, but also of SERCA pumps and the luminal Ca^{2+} binding proteins. Additionally, it is clear that the free luminal SR/ER $[Ca^{2+}]$ cannot predict the amount of Ca^{2+} that will be released to the cytoplasm, the most likely explanation is that the Ca^{2+} released during the activation of release channels involve the participation of luminal Ca^{2+} binding proteins that trap Ca^{2+} next to the release channels and that the formation of this complex requires active SERCA pumps. This scenario could explain several situations of the Ca^{2+} release event; for instance, why the amplitude of the Ca^{2+} release event in the cytoplasm does not show any correlation with the reduction in the free luminal SR/ER $[Ca^{2+}]$? Why the increase in the $[Ca^{2+}]_i$ during Ca^{2+} release is associated with a minimal reduction in the bulk free luminal SR/ER $[Ca^{2+}]$? Why is the refractory period for Ca^{2+} release longer than the recovery of the free luminal SR/ER $[Ca^{2+}]$? It is clear that we are still far from understanding how release channels produce an efficient Ca^{2+} release event but the development of the GI-SIM superresolution microscope should help [23]. This microscope has both enhanced spatial and temporal resolutions and can be used with the current Ca^{2+} indicators, so it should be easier to follow changes in the $[Ca^{2+}]$ of both the lumen of the ER and the cytoplasm to gather a better picture on how the Ca^{2+} lattice produces an efficient Ca^{2+} release event. Additionally, it is clear that the SR/ER is the main Ca^{2+} source, not only for the cytoplasm, but also for other organelles as mitochondria and lysosomes. In conclusion, it appears that an efficient Ca^{2+} release event occurs only during the initial Ca^{2+} release process and that this requires the participation of release channels, luminal Ca^{2+} binding proteins and the SERCA pump. This means that the communication among all these elements makes the Ca^{2+} release event more complex than previously envisioned but very robust because it can successfully fulfill the role of SR/ER as Ca^{2+} source without interfering with the need of having a high luminal SR/ER $[Ca^{2+}]$.

References

1. Berridge MJ, Bootman MD, Roderick HL (2003) Calcium signalling: dynamics, homeostasis and remodelling. Nat Rev Mol Cell Biol 4(7):517–529
2. Clapham DE (2007) Calcium signaling. Cell 131(6):1047–1058
3. Zhivotovsky B, Orrenius S (2011) Calcium and cell death mechanisms: a perspective from the cell death community. Cell Calcium 50(3):211–221
4. Yanez M, Gil-Longo J, Campos-Toimil M (2012) Calcium binding proteins. In: Calcium signaling. Springer, Dordrecht, pp 461–482
5. Berridge MJ, Lipp P, Bootman MD (2000) The versatility and universality of calcium signalling. Nat Rev Mol Cell Biol 1(1):11–21
6. Tombal B, Denmeade SR, Gillis JM, Isaacs JT (2002) A supramicromolar elevation of intracellular free calcium ($[Ca^{2+}]i$) is consistently required to induce the execution phase of apoptosis. Cell Death Differ 9(5):561–573
7. Verkhratsky A (2005) Physiology and pathophysiology of the calcium store in the endoplasmic reticulum of neurons. Physiol Rev 85(1):201–279
8. Rizzuto R, Marchi S, Bonora M, Aguiari P, Bononi A, De Stefani D et al (2009) Ca^{2+} transfer from the ER to mitochondria: when, how and why. Biochim Biophys Acta Bioenerg 1787(11):1342–1351
9. Llopis J, McCaffery JM, Miyawaki A, Farquhar MG, Tsien RY (1998) Measurement of cytosolic, mitochondrial, and Golgi pH in single living cells with green fluorescent proteins. Proc Natl Acad Sci U S A 95(12):6803–6808
10. Gallegos-Gómez M-L, Greotti E, López-Méndez M-C, Sánchez-Vázquez V-H, Arias J-M, Guerrero-Hernández A (2018) The trans golgi region is a labile intracellular Ca^{2+} store sensitive to emetine. Sci Rep 8(1):17143
11. Garrity AG, Wang W, Collier CMD, Levey SA, Gao Q, Xu H (2016) The endoplasmic reticulum, not the pH gradient, drives calcium refilling of lysosomes. elife 5:e15887
12. Dickson EJ, Duman JG, Moody MW, Chen L, Hille B (2012) Orai-STIM-mediated Ca^{2+} release from secretory granules revealed by a targeted Ca^{2+} and pH probe. Proc Natl Acad Sci U S A 109(51):E3539–E3548
13. Pezzati R, Bossi M, Podini P, Meldolesi J, Grohovaz F (1997) High-resolution calcium mapping of the endoplasmic reticulum-Golgi-exocytic membrane system. Electron energy loss imaging analysis of quick frozen-freeze dried PC12 cells. Mol Biol Cell 8(8):1501–1512
14. Yagodin S, Pivovarova NB, Andrews SB, Sattelle DB (1999) Functional characterization of thapsigargin and agonist-insensitive acidic Ca^{2+} stores in Drosophila melanogaster S2 cell lines. Cell Calcium 25(6):429–438
15. Fasolato C, Zottinis M, Clementis E, Zacchettis D, Meldolesi J, Pozzan T (1991) Intracellular Ca^{2+} pools in PC12 cells. J Biol Chem 266(30):20159–20167
16. Phillips MJ, Voeltz GK (2016) Structure and function of ER membrane contact sites with other organelles. Nat Rev Mol Cell Biol 17(2):69–82
17. Voeltz GK, Rolls MM, Rapoport TA (2002) Structural organization of the endoplasmic reticulum. EMBO Rep 3(10):944–950
18. Shibata Y, Voeltz GK, Rapoport TA (2006) Rough sheets and smooth tubules. Cell 126(3):435–439
19. Park MK, Petersen OH, Tepikin AV (2000) The endoplasmic reticulum as one continuous Ca^{2+} pool: visualization of rapid Ca^{2+} movements and equilibration. EMBO J 19(21):5729–5739

20. Bers DM, Shannon TR (2013) Calcium movements inside the sarcoplasmic reticulum of cardiac myocytes. J Mol Cell Cardiol 58(1):59–66
21. Jones VC, Rodríguez JJ, Verkhratsky A, Jones OT (2009) A lentivirally delivered photoactivatable GFP to assess continuity in the endoplasmic reticulum of neurones and glia. Pflugers Arch Eur J Physiol 458(4):809–818
22. Friedman JR, Webster BM, Mastronarde DN, Verhey KJ, Voeltz GK (2010) ER sliding dynamics and ER-mitochondrial contacts occur on acetylated microtubules. J Cell Biol 190(3):363–375
23. Guo Y, Li D, Zhang S, Lippincott-schwartz J, Betzig E, Guo Y et al (2018) Visualizing intracellular organelle and cytoskeletal interactions at nanoscale resolution on millisecond resource visualizing intracellular organelle and cytoskeletal interactions at nanoscale resolution on millisecond timescales. Cell 175:1430–1442
24. Shim S-H, Xia C, Zhong G, Babcock HP, Vaughan JC, Huang B et al (2012) Super-resolution fluorescence imaging of organelles in live cells with photoswitchable membrane probes. Proc Natl Acad Sci 109(35):13978–13983
25. Hernández-Ochoa EO, Schneider MF (2018) Voltage sensing mechanism in skeletal muscle excitation-contraction coupling: coming of age or midlife crisis? Skelet Muscle 8(1):22
26. Sweeney HL, Hammers DW (2018) Muscle contraction. Cold Spring Harb Perspect Biol 10(2):a023200
27. Zarain-Herzberg A, García-Rivas G, Estrada-Avilés R (2014) Regulation of SERCA pumps expression in diabetes. Cell Calcium 56(5):302–310
28. Toyoshima C, Inesi G (2004) Structural basis of ion pumping by Ca^{2+}-ATPase of the sarcoplasmic reticulum. Annu Rev Biochem 73:269–292
29. Toyoshima C, Nomura H, Tsuda T (2004) Lumenal gating mechanism revealed in calcium pump crystal structures with phosphate analogues. Nature 432(7015):361–368
30. Dyla M, Terry DS, Kjaergaard M, Sørensen TLM, Andersen JL, Andersen JP et al (2017) Dynamics of P-type ATPase transport revealed by single-molecule FRET. Nature 551(7680):346–351
31. Mejía-Alvarez R, Kettlun C, Ríos E, Stern M, Fill M (1999) Unitary Ca^{2+} current through cardiac ryanodine receptor channels under quasi-physiological ionic conditions. J Gen Physiol 113(2):177–186
32. Ikeda K, Kang Q, Yoneshiro T, Camporez JP, Maki H, Homma M et al (2017) UCP1-independent signaling involving SERCA2b mediated calcium cycling regulates beige fat thermogenesis and systemic glucose homeostasis. Nat Med 23(12):1454–1465
33. De Meis L, Arruda AP, Carvalho DP (2005) Role of sarco/endoplasmic reticulum Ca^{2+}-ATPase in thermogenesis. Biosci Rep 25(3–4):181–190
34. Fruen BR, Mickelson JR, Louis CF (1997) Dantrolene inhibition of sarcoplasmic reticulum Ca^{2+} release by direct and specific action at skeletal muscle ryanodine receptors. J Biol Chem 272(43):26965–26971
35. Bal NC, Maurya SK, Sopariwala DH, Sahoo SK, Gupta SC, Shaikh SA et al (2012) Sarcolipin is a newly identified regulator of muscle-based thermogenesis in mammals. Nat Med 18(10):1575–1579
36. Bhupathy P, Babu GJ, Periasamy M (2007) Sarcolipin and phospholamban as regulatos of cardiac sarcoplasmic reticulum Ca^{2+} ATPase. J Mol Cell Cardiol 42(5):903–911
37. Anderson DM, Makarewich CA, Anderson KM, Shelton JM, Bezprozvannaya S, Bassel-Duby R et al (2016) Widespread control of calcium signaling by a family of SERCA-inhibiting micropeptides. Sci Signal 9(457):ra119
38. Dettbarn C, Palade P (1998) Effects of three sarcoplasmic/endoplasmic reticulum Ca++ pump inhibitors on release channels of intracellular stores. J Pharmacol Exp Ther 285(2):739–745
39. Chen J, De Raeymaecker J, Hovgaard JB, Smaardijk S, Vandecaetsbeek I, Wuytack F et al (2017) Structure/activity relationship of thapsigargin inhibition on the purified golgi/secretory pathway Ca^{2+}/Mn2+–transport ATPase (SPCA1a). J Biol Chem 292(17):6938–6951

40. Gustavo Vazquez-Jimenez J, Chavez-Reyes J, Romero-Garcia T, Zarain-Herzberg A, Valdes-Flores J, Manuel Galindo-Rosales J et al (2016) Palmitic acid but not palmitoleic acid induces insulin resistance in a human endothelial cell line by decreasing SERCA pump expression. Cell Signal 28(1):53–59
41. Nelson BR, Makarewich CA, Anderson DM, Winders BR, Troupes CD, Wu F et al (2016) Muscle physiology: a peptide encoded by a transcript annotated as long noncoding RNA enhances SERCA activity in muscle. Science 351(6270):271–275
42. Fill M, Copello JA (2002) Ryanodine receptor calcium release channels. Physiol Rev 82(4):893–922
43. Mikoshiba K (2015) Role of IP$_3$ receptor signaling in cell functions and diseases. Adv Biol Regul 57:217–227
44. Seo M-D, Velamakanni S, Ishiyama N, Stathopulos PB, Rossi AM, Khan SA et al (2012) Structural and functional conservation of key domains in InsP$_3$ and ryanodine receptors. Nature 483(7387):108–112
45. Meissner G (2004) Molecular regulation of cardiac ryanodine receptor ion channel. Cell Calcium 35(6):621–628
46. Foskett JK, White C, Cheung K, Mak DD (2007) Inositol trisphosphate receptor Ca^{2+} release channels. Am Physiol Soc 87:593–658
47. Meissner G (2017) The structural basis of ryanodine receptor ion channel function. J Gen Physiol 149(12):1065–1089
48. Scriven DRL, Dan P, Moore EDW (2000) Distribution of proteins implicated in excitation-contraction coupling in rat ventricular myocytes. Biophys J 79(5):2682–2691
49. Bers DM (2002) Cardiac excitation–contraction coupling. Nature 415(6868):198–205
50. Clark JH, Kinnear NP, Kalujnaia S, Cramb G, Fleischer S, Jeyakumar LH et al (2010) Identification of functionally segregated sarcoplasmic reticulum calcium stores in pulmonary arterial smooth muscle. J Biol Chem 285(18):13542–13549
51. Nelson MT, Cheng H, Rubart M, Santana LF, Bonev AD, Knot HJ et al (1995) Relaxation of arterial smooth muscle by calcium sparks. Science 270(5236):633–637
52. Kirber MT, Guerrero-Hernández A, Bowman DS, Fogarty KE, Tuft RA, Singer JJ et al (2000) Multiple pathways responsible for the stretch-induced increase in Ca^{2+} concentration in toad stomach smooth muscle cells. J Physiol 524(1):3–17
53. Kotlikoff MI (2003) Calcium-induced calcium release in smooth muscle: the case for loose coupling. Prog Biophys Mol Biol 83(3):171–191
54. van der Wal J, Habets R, Várnai P, Balla T, Jalink K, Varnai P et al (2001) Monitoring agonist-induced phospholipase C activation in live cells by fluorescence resonance energy transfer. J Biol Chem 276(18):15337–15344
55. Dickson EJ, Falkenburger BH, Hille B (2013) Quantitative properties and receptor reserve of the IP3 and calcium branch of G(q)-coupled receptor signaling. J Gen Physiol 141(5):521–535
56. Baker MR, Fan G, Serysheva II (2017) Structure of IP$_3$R channel: high-resolution insights from cryo-EM. Curr Opin Struct Biol 46:38–47
57. Choe CU, Ehrlich BE (2006) The inositol 1,4,5-trisphosphate receptor (IP3R) and its regulators: sometimes good and sometimes bad teamwork. Sci STKE 2006(363):re15
58. Royer L, Ríos E (2009) Deconstructing calsequestrin. Complex buffering in the calcium store of skeletal muscle. J Physiol 587(13):3101–3111
59. Ikemoto N, Ronjat M, Meszaros LG, Koshita M (1989) Postulated role of calsequestrin in the regulation of calcium release from sarcoplasmic reticulum. Biochemistry 28(16):6764–6771
60. Terentyev D, Kubalova Z, Valle G, Nori A, Vedamoorthyrao S, Terentyeva R et al (2008) Modulation of SR Ca release by luminal Ca and calsequestrin in cardiac myocytes: effects of CASQ2 mutations linked to sudden cardiac death. Biophys J 95(4):2037–2048
61. Michalak M, Groenendyk J, Szabo E, Gold LI, Opas M (2009) Calreticulin, a multi-process calcium-buffering chaperone of the endoplasmic reticulum. Biochem J 417(3):651–666
62. Park H, Park IY, Kim E, Youn B, Fields K, Dunker AK et al (2004) Comparing skeletal and cardiac calsequestrin structures and their calcium binding: a proposed mechanism for coupled calcium binding and protein polymerization. J Biol Chem 279(17):18026–18033

63. Arnaudeau S, Frieden M, Nakamura K, Castelbou C, Michalak M, Demaurex N (2002) Calreticulin differentially modulates calcium uptake and release in the endoplasmic reticulum and mitochondria. J Biol Chem 277(48):46696–46705
64. Pozzo-Miller LD, Connor JA, Andrews B (2000) Microheterogeneity of calcium signalling in dendrites. J Physiol 525(1):53–61
65. Pozzo-Miller LD, Pivovarova NB, Connor JA, Reese TS, Andrews SB (1999) Correlated measurements of free and total intracellular calcium concentration in central nervous system neurons. Microsc Res Tech 46(6):370–379
66. Papp S, Dziak E, Michalak M, Opas M (2003) Is all of the endoplasmic reticulum created equal? The effects of the heterogeneous distribution of endoplasmic reticulum Ca^{2+}-handling proteins. J Cell Biol 160(4):475–479
67. Handhle A, Ormonde CE, Thomas NL, Bralesford C, Williams AJ, Lai FA et al (2016) Calsequestrin interacts directly with the cardiac ryanodine receptor luminal domain. J Cell Sci 129:3983–3988
68. Wang Y, Li X, Duan H, Fulton TR, Eu JP, Meissner G (2009) Altered stored calcium release in skeletal myotubes deficient of triadin and junctin. Cell Calcium 45(1):29–37
69. Qin J, Valle G, Nani A, Nori A, Rizzi N, Priori SG et al (2008) Luminal Ca^{2+} regulation of single cardiac ryanodine receptors: insights provided by Calsequestrin and its mutants. J Gen Physiol 131(4):325–334
70. Terentyev D, Viatchenko-Karpinski S, Gyorke I, Volpe P, Williams SC, Gyorke S (2003) Calsequestrin determines the functional size and stability of cardiac intracellular calcium stores: mechanism for hereditary arrhythmia. Proc Natl Acad Sci 100(20):11759–11764
71. Wray S, Burdyga T (2010) Sarcoplasmic reticulum function in smooth muscle. Physiol Rev 90(1):113–178
72. Laver DR (2007) Ca^{2+} stores regulate ryanodine receptor Ca^{2+} release channels via luminal and cytosolic Ca^{2+} sites. Biophys J 92(10):3541–3555
73. Laver DR, Kong CHT, Imtiaz MS, Cannell MB (2013) Termination of calcium-induced calcium release by induction decay: an emergent property of stochastic channel gating and molecular scale architecture. J Mol Cell Cardiol 54(1):98–100
74. Cannell MB, Kong CHT, Imtiaz MS, Laver DR (2013) Control of sarcoplasmic reticulum Ca^{2+} release by stochastic RyR gating within a 3D model of the cardiac dyad and importance of induction decay for CICR termination. Biophys J 104(10):2149–2159
75. Wu X, Bers DM (2006) Sarcoplasmic reticulum and nuclear envelope are one highly interconnected Ca^{2+} store throughout cardiac myocyte. Circ Res 99(3):283–291
76. Murphy RM, Mollica JP, Beard NA, Knollmann BC, Lamb GD (2011) Quantification of calsequestrin 2 (CSQ2) in sheep cardiac muscle and Ca^{2+}-binding protein changes in CSQ2 knockout mice. Am J Physiol Heart Circ Physiol 300(2):H595–H604
77. Ikemoto N, Antoniu B, Kang JJ, Mészáros LG, Ronjat M (1991) Intravesicular calcium transient during calcium release from sarcoplasmic reticulum. Biochemistry 30(21):5230–5237
78. Yamaguchi N, Igami K, Kasai M (1997) Kinetics of depolarization-induced calcium release from skeletal muscle triads in vitro. J Biochem 121(3):432–439
79. Gomez-Viquez L, Rueda A, Garcia U, Guerrero-Hernandez A (2005) Complex effects of ryanodine on the sarcoplasmic reticulum Ca^{2+} levels in smooth muscle cells. Cell Calcium 38:121–130
80. Rueda A, García L, Guerrero-Hernández A (2002) Luminal Ca^{2+} and the activity of sarcoplasmic reticulum Ca^{2+} pumps modulate histamine-induced all-or-none Ca^{2+} release in smooth muscle cells. Cell Signal 14(6):517–527
81. Inesi G, Tadini-Buoninsegni F (2014) Ca^{2+}/H^+ exchange, lumenal Ca^{2+} release and Ca^{2+}/ATP coupling ratios in the sarcoplasmic reticulum ATPase. J Cell Commun Signal 8(1):5–11
82. Saiki Y, Ikemoto N (1999) Coordination between Ca^{2+} release and subsequent re-uptake in the sarcoplasmic reticulum. Biochemistry 38(10):3112–3119

83. Saiki Y, Ikemoto N (1997) Fluorescence probe study of the lumenal Ca^{2+} of the sarcoplasmic reticulum vesicles during Ca^{2+} uptake and Ca^{2+} release. Biochem Biophys Res Commun 241(1):181–186

84. Mészáros LG, Ikemoto N (1989) Non-identical behavior of the Ca^{2+}-ATPase in the terminal cisternae and the longitudinal tubules fractions of sarcoplasmic reticulum. Eur J Biochem 186(3):677–681

85. Yano M, Yamamoto T, Ikemoto N, Matsuzaki M (2005) Abnormal ryanodine receptor function in heart failure. Pharmacol Ther 107(3):377–391

86. Mészáros LG, Ikemoto N (1985) Conformational changes of the Ca^{2+}-ATPase as early events of Ca^{2+} release from sarcoplasmic reticulum. J Biol Chem 260(30):16076–16079

87. Ikemoto N, Yamamoto T (2000) The luminal Ca^{2+} transient controls Ca^{2+} release/re-uptake of sarcoplasmic reticulum. Biochem Biophys Res Commun 279(3):858–863

88. Guerrero A, Singer JJ, Fay FS (1994) Simultaneous measurement of Ca^{2+} release and influx into smooth muscle cells in response to caffeine. A novel approach for calculating the fraction of current carried by calcium. J Gen Physiol 104(2):395–422

89. Gómez-Viquez L, Guerrero-Serna G, García U, Guerrero-Hernández A (2003) SERCA pump optimizes Ca^{2+} release by a mechanism independent of store filling in smooth muscle cells. Biophys J 85(1):370–380

90. Hofer AM, Machen TE (1993) Technique for in situ measurement of calcium in intracellular inositol 1,4,5-trisphosphate-sensitive stores using the fluorescent indicator mag-fura-2 (gastric glands/thapsigargin/heparin). Proc Natl Acad Sci 90(April):2598–2602

91. Shmigol AV, Eisner DA, Wray S (2001) Simultaneous measurements of changes in sarcoplasmic reticulum and cytosolic $[Ca^{2+}]$ in rat uterine smooth muscle cells. J Physiol 531(1):707–713

92. Navas-Navarro P, Rojo-Ruiz J, Rodriguez-Prados M, Ganfornina MD, Looger LL, Alonso MT et al (2016) GFP-aequorin protein sensor for ex vivo and in vivo imaging of Ca^{2+} dynamics in high-Ca^{2+} organelles. Cell Chem Biol 23(6):738–745

93. Barrero MJ, Montero M, Alvarez J (1997) Dynamics of $[Ca^{2+}]$ in the endoplasmic reticulum and cytoplasm of intact HeLa cells: a comparative study. J Biol Chem 272(44):27694–27699

94. Palmer AE, Jin C, Reed JC, Tsien RY (2004) Bcl-2-mediated alterations in endoplasmic reticulum Ca^{2+} analyzed with an improved genetically encoded fluorescent sensor. Proc Natl Acad Sci 101(50):17404–17409

95. Dagnino-Acosta A, Guerrero-Hernández A (2009) Variable luminal sarcoplasmic reticulum Ca^{2+} buffer capacity in smooth muscle cells. Cell Calcium 46(3):188–196

96. Guerrero-Hernandez A, Dagnino-Acosta A, Verkhratsky A (2010) An intelligent sarco-endoplasmic reticulum Ca^{2+} store: release and leak channels have differential access to a concealed Ca^{2+} pool. Cell Calcium 48(2–3):143–149

97. Launikonis BS, Zhou J, Royer L, Shannon TR, Brum G, Ríos E (2006) Depletion "skraps" and dynamic buffering inside the cellular calcium store. Proc Natl Acad Sci U S A 103(8):2982–2987

98. Shmigol A, Wray S (2005) Modulation of agonist-induced Ca^{2+} release by SR Ca^{2+} load: direct SR and cytosolic Ca^{2+} measurements in rat uterine myocytes. Cell Calcium 37(3):215–223

99. Missiaen L, Van Acker K, Van Baelen K, Raeymaekers L, Wuytack F, Parys JB et al (2004) Calcium release from the Golgi apparatus and the endoplasmic reticulum in HeLa cells stably expressing targeted aequorin to these compartments. Cell Calcium 36(6):479–487

100. Ishii K, Hirose K, Iino M (2006) Ca^{2+} shuttling between endoplasmic reticulum and mitochondria underlying Ca^{2+} oscillations. EMBO Rep 7(4):390–396

101. Brochet DXP, Yang D, Di Maio A, Lederer WJ, Franzini-Armstrong C, Cheng H (2005) Ca^{2+} blinks: rapid nanoscopic store calcium signaling. Proc Natl Acad Sci U S A 102(8):3099–3104

102. Zima AV, Picht E, Bers DM, Blatter LA (2008) Termination of cardiac Ca^{2+} sparks: role of intra-SR $[Ca^{2+}]$, release flux, and intra-SR Ca^{2+} diffusion. Circ Res 103(8):e105–e115

103. Maxwell JT, Blatter LA (2012) Facilitation of cytosolic calcium wave propagation by local calcium uptake into the sarcoplasmic reticulum in cardiac myocytes. J Physiol 590:6037–6045
104. Maxwell JT, Blatter LA (2017) A novel mechanism of tandem activation of ryanodine receptors by cytosolic and SR luminal Ca^{2+} during excitation–contraction coupling in atrial myocytes. J Physiol 595(12):3835–3845
105. Keller M, Kao JPY, Egger M, Niggli E (2007) Calcium waves driven by "sensitization" wavefronts. Cardiovasc Res 74(1):39–45
106. Rodriguez-Garcia A, Rojo-Ruiz J, Navas-Navarro P, Aulestia FJ, Gallego-Sandin S, Garcia-Sancho J et al (2014) GAP, an aequorin-based fluorescent indicator for imaging Ca^{2+} in organelles. Proc Natl Acad Sci U S A 111(7):2584–2589
107. Greotti E, Wong A, Pozzan T, Pendin D, Pizzo P (2016) Characterization of the ER-targeted low affinity Ca^{2+} probe D4ER. Sensors 16(9):1–13
108. Baddeley D, Crossman D, Rossberger S, Cheyne JE, Montgomery JM, Jayasinghe ID et al (2011) 4D super-resolution microscopy with conventional fluorophores and single wavelength excitation in optically thick cells and tissues. PLoS One 6(5):e20645
109. Perez-Rosas NC, Gomez-Viquez NL, Dagnino-Acosta A, Santillan M, Guerrero-Hernandez A (2015) Kinetics on demand is a simple mathematical solution that fits recorded caffeine-induced luminal SR Ca^{2+} changes in smooth muscle cells. PLoS One 10(9):e0138195
110. Suzuki J, Kanemaru K, Ishii K, Ohkura M, Okubo Y, Iino M (2014) Imaging intraorganellar Ca^{2+} at subcellular resolution using CEPIA. Nat Commun 5:4153
111. Manno C, Ríos E (2015) A better method to measure total calcium in biological samples yields immediate payoffs. J Gen Physiol 145(3):167–171
112. Györke I, Hester N, Jones LR, Györke S (2004) The role of calsequestrin, triadin, and junctin conferring cardiac ryanodine receptor responsiveness to luminal calcium. Biophys J 86(4):2121–2128
113. Gómez-Viquez NL, Guerrero-Serna G, Arvizu F, García U, Guerrero-Hernández A (2010) Inhibition of SERCA pumps induces desynchronized RyR activation in overloaded internal Ca^{2+} stores in smooth muscle cells. Am J Physiol Cell Physiol 298(5):C1038–C1046
114. Song XW, Tang Y, Lei CH, Cao M, Shen YF, Yang YJ (2016) In situ visualizing T-tubule/SR junction reveals the ultra-structures of calcium storage and release machinery. Int J Biol Macromol 82:7–12
115. Cifuentes ME, Ronjat M, Ikemoto N (1989) Polylysine induces a rapid Ca^{2+} release from sarcoplasmic reticulum vesicles by mediation of its binding to the foot protein. Arch Biochem Biophys 273(2):554–561
116. El-Hayekt R, Yano M, Ikemoto N (1995) A conformational change in the junctional foot protein is involved in the regulation of Ca^{2+} release from sarcoplasmic reticulum: studies on polylysine-induced Ca^{2+} release. J Biol Chem 270(26):15634–15638
117. Rueda A, García L, Soria-Jasso LE, Arias-Montaño JA, Guerrero-Hernández A (2002) The initial inositol 1,4,5-trisphosphate response induced by histamine is strongly amplified by Ca^{2+} release from internal stores in smooth muscle. Cell Calcium 31(4):161–173
118. Bers DM, Eisner DA, Valdivia HH (2003) Sarcoplasmic reticulum Ca^{2+} and heart failure. Circ Res 93(6):487–490
119. Park SW, Zhou Y, Lee J, Lee J, Ozcan U (2010) Sarco(endo)plasmic reticulum Ca^{2+}-ATPase 2b is a major regulator of endoplasmic reticulum stress and glucose homeostasis in obesity. Proc Natl Acad Sci U S A 107(45):19320–19325
120. Sammels E, Parys JB, Missiaen L, De Smedt H, Bultynck G (2010) Intracellular Ca^{2+} storage in health and disease: a dynamic equilibrium. Cell Calcium 47(4):297–314
121. Fu S, Watkins SM, Hotamisligil GS (2012) The role of endoplasmic reticulum in hepatic lipid homeostasis and stress signaling. Cell Metab 15(5):623–634
122. Allen DG, Lamb GD, Westerblad H (2007) Impaired calcium release during fatigue. J Appl Physiol 104(1):296–305

123. Manno C, Figueroa LC, Gillespie D, Fitts R, Kang C, Franzini-Armstrong C et al (2017) Calsequestrin depolymerizes when calcium is depleted in the sarcoplasmic reticulum of working muscle. Proc Natl Acad Sci 114(4):E638–E647

124. Aguilar-Maldonado B, Gómez-Viquez L, García L, Del Angel RM, Arias-Montaño JA, Guerrero-Hernández A (2003) Histamine potentiates IP3-mediated Ca^{2+} release via thapsigargin-sensitive Ca^{2+} pumps. Cell Signal 15(7):689–697

125. Wang X, Huang G, Luo X, Penninger JM, Muallem S (2004) Role of regulator of G protein signaling 2 (RGS2) in Ca^{2+} oscillations and adaptation of Ca^{2+} signaling to reduce excitability of $RGS2^{-/-}$ cells. J Biol Chem 279(40):41642–41649

126. Rowland AA, Voeltz GK (2012) Endoplasmic reticulum–mitochondria contacts: function of the junction. Nat Rev Mol Cell Biol 13(10):607–615

127. Csordás G, Renken C, Várnai P, Walter L, Weaver D, Buttle KF et al (2006) Structural and functional features and significance of the physical linkage between ER and mitochondria. J Cell Biol 174(7):915–921

128. Montisano D, Cascarano J, Pickett C, James T (1982) Association between mitochondria and rough endoplasmic reticulum in rat liver. Anat Rec 203:441–450

129. Rizzuto R (1998) Close contacts with the endoplasmic reticulum as determinants of mitochondrial Ca^{2+} responses. Science 280(5370):1763–1766

130. Filadi R, Greotti E, Turacchio G, Luini A, Pozzan T, Pizzo P (2015) Mitofusin 2 ablation increases endoplasmic reticulum–mitochondria coupling. Proc Natl Acad Sci 112(17):E2174–E2181

131. Bravo R, Vicencio JM, Parra V, Troncoso R, Munoz JP, Bui M et al (2011) Increased ER-mitochondrial coupling promotes mitochondrial respiration and bioenergetics during early phases of ER stress. J Cell Sci 124(14):2511–2511

132. Bravo-sagua R, López-crisosto C, Parra V, Rodriguez M, Rothermel BA, Quest AFG et al (2016) mTORC1 inhibitor rapamycin and ER stressor tunicamycin induce differential patterns of ER- mitochondria coupling. Sci Rep 6:36394

133. Harmon M, Larkman P, Hardingham G, Jackson M, Skehel P (2017) A Bi-fluorescence complementation system to detect associations between the endoplasmic reticulum and mitochondria. Sci Rep 7:1–12

134. de Brito OM, Scorrano L (2008) Mitofusin 2 tethers endoplasmic reticulum to mitochondria. Nature 456(7222):605–610

135. Valm AM, Cohen S, Legant WR, Melunis J, Hershberg U, Wait E et al (2017) Applying systems-level spectral imaging and analysis to reveal the organelle interactome. Nature 546(7656):162–167

136. Vance E (1990) Phospholipid synthesis in a membrane fraction associated with mitochondria. J Biol Chem 265(13):7248–7256

137. Filadi R, Theurey P, Pizzo P (2017) The endoplasmic reticulum-mitochondria coupling in health and disease: molecules, functions and significance. Cell Calcium 62:1–15

138. Marchi S, Patergnani S, Missiroli S, Morciano G, Rimessi A, Wieckowski MR et al (2018) Mitochondrial and endoplasmic reticulum calcium homeostasis and cell death. Cell Calcium 69:62–72

139. Szabadkai G, Bianchi K, Várnai P, De Stefani D, Wieckowski MR, Cavagna D et al (2006) Chaperone-mediated coupling of endoplasmic reticulum and mitochondrial Ca^{2+} channels. J Cell Biol 175(6):901–911

140. Rieusset J, Fauconnier J, Paillard M, Belaidi E, Tubbs E, Chauvin M et al (2016) Disruption of calcium transfer from ER to mitochondria links alterations of mitochondria-associated ER membrane integrity to hepatic insulin resistance. Diabetologia 59:614–623

141. Cali T, Ottolini D, Negro A, Brini M (2012) α-Synuclein controls mitochondrial calcium homeostasis by enhancing endoplasmic reticulum-mitochondria interactions. J Biol Chem 287(22):17914–17929

142. Naon D, Zaninello M, Giacomello M, Varanita T, Grespi F, Lakshminaranayan S et al (2016) Critical reappraisal confirms that Mitofusin 2 is an endoplasmic reticulum–mitochondria tether. Proc Natl Acad Sci 113(40):11249–11254

143. Sebastián D, Hernández-alvarez MI, Segalés J, Sorianello E (2012) Mitofusin 2 (Mfn2) links mitochondrial and endoplasmic reticulum function with insulin signaling and is essential for normal glucose homeostasis. Proc Natl Acad Sci 109(14):5523–5528

144. Rossi A, Pizzo P, Filadi R (2018) Calcium, mitochondria and cell metabolism: a functional triangle in bioenergetics. BBA Mol Cell Res 1865(11):1–33

145. De Stefani D, Raffaello A, Teardo E, Szabò I, Rizzuto R (2011) A forty-kilodalton protein of the inner membrane is the mitochondrial calcium uniporter. Nature 476(7360):336–340

146. Baughman JM, Perocchi F, Girgis HS, Plovanich M, Belcher-Timme CA, Sancak Y et al (2011) Integrative genomics identifies MCU as an essential component of the mitochondrial calcium uniporter. Nature 476(7360):341–345

147. Mammucari C, Gherardi G, Rizzuto R (2017) Structure, activity regulation, and role of the mitochondrial calcium uniporter in health and disease. Front Oncol 7:139

148. Liu JC, Liu J, Holmstrom KM, Menazza S, Parks RJ, Fergusson MM et al (2016) MICU1 serves as a molecular gatekeeper to prevent in vivo mitochondrial calcium overload. Cell Rep 16(6):1561–1573

149. Szabadkai G, Simoni AM, Rizzuto R (2003) Mitochondrial Ca^{2+} uptake requires sustained Ca^{2+} release from the endoplasmic reticulum. J Biol Chem 278(17):15153–15161

150. Pinton P, Giorgi C, Siviero R, Zecchini E, Rizzuto R (2008) Calcium and apoptosis: ER-mitochondria Ca^{2+} transfer in the control of apoptosis. Oncogene 27:6407–6418

151. Christensen KA, Myers JT, Swanson JA (2002) pH-dependent regulation of lysosomal calcium in macrophages. J Cell Sci 115(3):599–607

152. Bowman EJ, Siebers A, Altendorf K (1988) Bafilomycins: a class of inhibitors of membrane ATPases from microorganisms, animal cells, and plant cells. Proc Natl Acad Sci U S A 85(21):7972–7976

153. Yoshimori T, Yamamoto A, Moriyama Y, Futai M, Tashiro Y (1991) Bafilomycin A1, a specific inhibitor of vacuolar-type H^+-ATPase, inhibits acidification and protein degradation in lysosomes of cultured cells. J Biol Chem 266(26):17707–17712

154. Sanjurjo CIL, Tovey SC, Prole DL, Taylor CW (2012 Jan) Lysosomes shape IP$_3$-evoked Ca^{2+} signals by selectively sequestering Ca^{2+} released from the endoplasmic reticulum. J Cell Sci 126(Pt 1):289–300

155. Sanjurjo CIL, Tovey SC, Taylor CW (2014) Rapid recycling of Ca^{2+} between IP$_3$-sensitive stores and lysosomes. PLoS One 9(10):e111275

156. Ronco V, Potenza DM, Denti F, Vullo S, Gagliano G, Tognolina M et al (2015) A novel Ca^{2+}-mediated cross-talk between endoplasmic reticulum and acidic organelles: implications for NAADP-dependent Ca^{2+} signalling. Cell Calcium 57(2):89–100

157. Augustine MK, Choi MD, Ryter SW, Beth Levine MD (2014) Autophagy in human health and disease. N Engl J Med 368:651–662

Chapter 15
Pyridine Nucleotide Metabolites and Calcium Release from Intracellular Stores

Antony Galione and Kai-Ting Chuang

Abstract Ca^{2+} signals are probably the most common intracellular signaling cellular events, controlling an extensive range of responses in virtually all cells. Many cellular stimuli, often acting at cell surface receptors, evoke Ca^{2+} signals by mobilizing Ca^{2+} from intracellular stores. Inositol trisphosphate (IP_3) was the first messenger shown to link events at the plasma membrane to release Ca^{2+} from the endoplasmic reticulum (ER), through the activation of IP_3-gated Ca^{2+} release channels (IP_3 receptors). Subsequently, two additional Ca^{2+} mobilizing messengers were discovered, cADPR and NAADP. Both are metabolites of pyridine nucleotides, and may be produced by the same class of enzymes, ADP-ribosyl cyclases, such as CD38. Whilst cADPR mobilizes Ca^{2+} from the ER by activation of ryanodine receptors (RyRs), NAADP releases Ca^{2+} from acidic stores by a mechanism involving the activation of two pore channels (TPCs). In addition, other pyridine nucleotides have emerged as intracellular messengers. ADP-ribose and $2'$-deoxy-ADPR both activate TRPM2 channels which are expressed at the plasma membrane and in lysosomes.

Keywords Calcium · Cyclic ADP-ribose · NAADP · CD38 · Ryanodine · Two-pore channels · Inositol trisphosphate · ADP-ribose · TRPM2 channel · Lysosome · Endoplasmic reticulum · Calcium microdomain

15.1 Introduction

Studies of cardiac contractility at the close of the nineteenth century by Sidney Ringer showed a requirement for Ca^{2+} ions in perfusion solutions [1]. Use of jellyfish photoproteins, such as aequorin, provided the first measurements of

A. Galione (✉) · K.-T. Chuang
Department of Pharmacology, University of Oxford, Oxford, UK
e-mail: antony.galione@pharm.ox.ac.uk; kai-ting_chuang@hms.harvard.edu

© Springer Nature Switzerland AG 2020

371

M. S. Islam (ed.), *Calcium Signaling*, Advances in Experimental Medicine and Biology 1131, https://doi.org/10.1007/978-3-030-12457-1_15

cytosolic Ca^{2+} in muscle cells. Importantly, Ca^{2+} transients were found to precede contractions and this realization was important in generating the concept of a messenger role for Ca^{2+} ions in cell contractility [2].

Work studying transmitter release from neurons and hormone secretion [3], also led to a growing appreciation of the role of Ca^{2+} ions in stimulus-response coupling. An important source of Ca^{2+} was that which could be mobilized from internal stores in response to hormones and neurotransmitters acting at cell surface receptors [4]. In the mid-1970s it was hypothesized that receptors could stimulate cellular Ca^{2+} signals by stimulating the hydrolysis of inositol lipids [5]. Importantly, the initial inositol lipid hydrolysed was found to be phosphatidylinositol 4,5 bisphosphate [6]. The enzyme involved, phospholipase C thus generated diacylglycerol which activates protein kinase C, and inositol 1,4,5 trisphosphate (IP_3). A pivotal finding was that IP_3 added to permeabilized pancreatic acinar cells released Ca^{2+} from a non-mitochondrial store in a way that was mimicked by activating plasma membrane muscarinic acetylcholine receptors [7]. Thus IP_3 was proposed as a Ca^{2+} mobilizing messenger linking the activation of cell surface receptors to mobilization of Ca^{2+} from intracellular stores. Biochemical purification studies [8, 9] and molecular cloning experiments [10, 11] defined the principal targets for IP_3 on intracellular stores as homo/heterotetrameric Ca^{2+} release channels termed IP_3 receptors (IP_3Rs). An understanding of how IP_3 regulates IP_3Rs has emerged from recent IP_3R structural studies [12]. The IP_3 signaling pathway is now well established, ubiquitous, and plays key roles in mediating many of the actions of a variety of cellular stimuli [13]. The first intact cell in which IP_3 was shown to evoke a cellular response was the sea urchin egg [14]. IP_3 microinjection induced exocytosis of cortical granules resulting in the raising of the fertilization envelope which acts as a barrier to polyspermy. At around the same time, sea urchin egg homogenates containing Ca^{2+} sequestering vesicles were found to be sensitive to IP_3 which discharged Ca^{2+} from non-mitochondrial stores [15] . Following the establishment of egg homogenates to study Ca^{2+} release mechanisms, Lee and colleagues found that in addition to IP_3, the pyridine nucleotides, NAD^+ and NADP, at micromolar concentrations, were also found to release Ca^{2+} by mechanisms independent from those regulated by IP_3 [16]. NAD^+ released Ca^{2+} from a subcellular fraction which was also sensitive to IP_3, but after a delay of several seconds. In contrast, NADP rapidly released Ca^{2+} from a denser fraction of vesicles. Subsequent analysis revealed that the Ca^{2+} mobilizing properties of NAD^+ was due to an enzyme-produced metabolite identified as cyclic ADP-ribose (cADPR) [17], and Ca^{2+} release evoked by NADP was due to a contaminant, nicotinic acid adenine dinucleotide phosphate (NAADP) [18]. An abbreviated summary of our current understanding of Ca^{2+} homeostasis in animal cells is shown in Fig. 15.1.

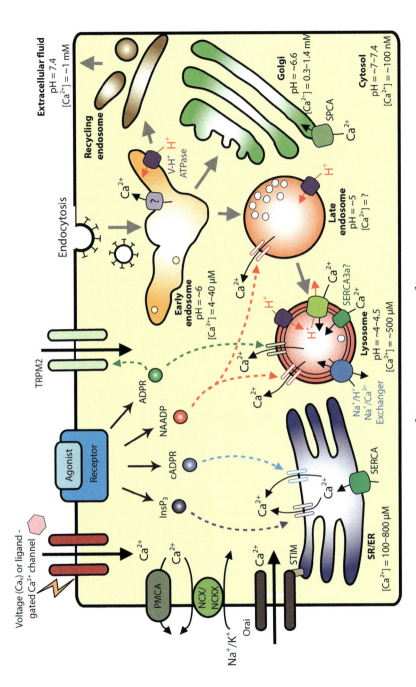

Fig. 15.1 Schematic representation of non-mitochondrial Ca^{2+} pools in animal cells. Ca^{2+} can enter the cytosol from the extracellular space via plasma membrane ion channels, or from intracellular Ca^{2+} sequestering stores such as the ER or the acidic Ca^{2+} stores in response to second messengers: IP_3, cADPR and NAADP. Ca^{2+} released to the cytosol is then exchanged or actively transported back to the Ca^{2+} stores or the extracellular space to restore a low cytosolic concentration of Ca^{2+}

15.2   Enzymology of cADPR and NAADP Synthesis and Metabolism

A family of multifunctional enzymes, termed ADP-ribosyl cyclases, has been characterized that are capable of both the synthesis and metabolism of both cADPR and NAADP. An enzyme activity responsible for the synthesis of cADPR was first indicated by the finding that NAD^+ mobilized Ca^{2+} with a delay from sea urchin egg homogenates but not from purified microsomes, indicating that egg homogenate supernatant contained an activity responsible for the conversion of NAD^+ to an active metabolite [16], later identified as cADPR [17]. This enzyme activity was also widespread in rat tissues and shown to be an enzyme showing stereo-specificity for substrate, pH and temperature-dependence, as well as protease-sensitivity [19]. The first ADP-ribosyl cyclase that was purified and characterized at the molecular level was that from *Aplysia* ovotestis [20–22]. The rationale for this was that during the study of ADP-ribosylation of G proteins by endotoxins in this tissue, a protein factor from ovotestis was uncovered that inhibited this reaction by competing for NAD^+ as a substrate. This protein factor, which was localized to ovotestis granules, was subsequently purified and cloned and found to catalyse the cyclization of NAD^+ to cADPR (Fig. 15.2). *Aplysia* ADP-ribosyl cyclase was the founding member of a class of enzymes that by sequence homology was found to include the mammalian proteins CD38 and CD157 [23]. In contrast to *Aplysia* ADP-ribosyl cyclase, CD38 is a multi- functional enzyme. It is an NAD glycohydrolase (NADase), producing ADP-ribose from NAD^+ [24]. But in addition, not only does it cyclize NAD to

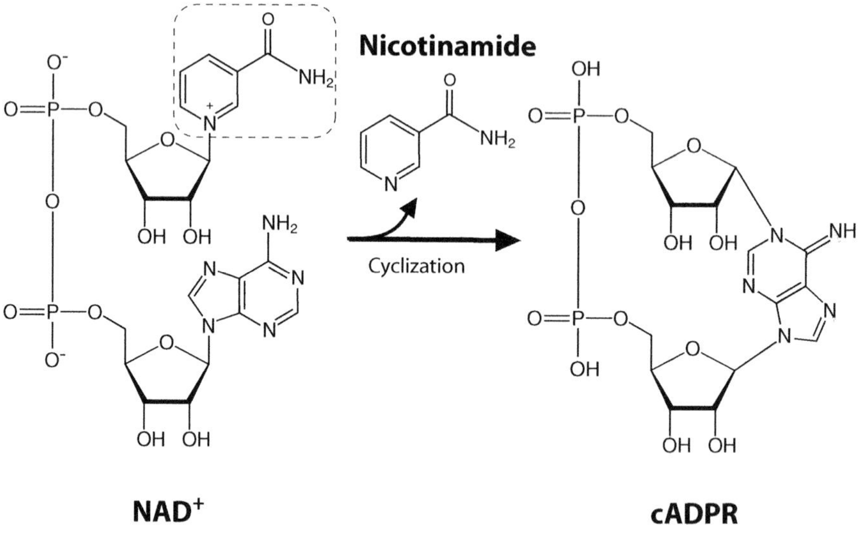

Fig. 15.2 cADPR synthesis. Synthesis of cADPR by cyclization of NAD^+ catalyzed by ADP-ribosyl cyclases

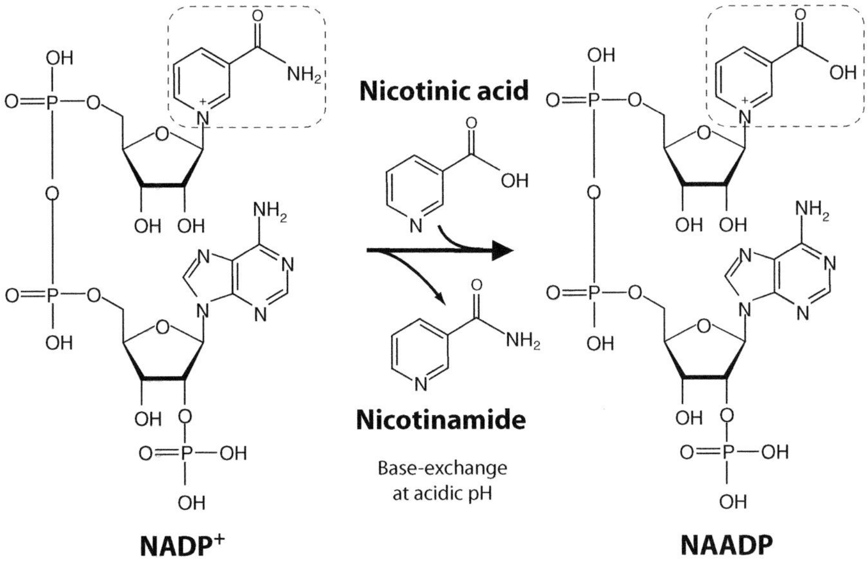

Fig. 15.3 NAADP synthesis. Synthesis of NAADP from NADP by base-exchange of the nicoti-namide moiety for nicotinic acid at acidic pH catalyzed by ADP-ribosyl cyclases

cADPR, it also has an appreciable hydrolase activity that converts cADPR to ADP-ribose [25]. Furthermore, CD38 may also use the alternate substrate NADP, and in the presence of nicotinic acid may catalyse a base- exchange reaction generating NAADP too (Fig. 15.3) [26]. Evidence has also emerged that CD38 may also hydrolyse NAADP to ADP-ribose 2-phosphate [27], although cellular phosphatases may also convert NAADP to inactive NAAD [28]. Thus CD38 is responsible for both the synthesis of a number of Ca^{2+} signaling regulators, and may also catalyse their metabolism.

Detailed mechanistic studies following the crystallization of both *Aplysia* ADP-ribosyl cyclase and CD38 have emerged in recent years to explain the various activities of these proteins (reviewed in [29]). As also mentioned below, an expanded family of ADP-ribosyl cyclases have been cloned from sea urchins, and various forms are localized inside intracellular vesicles or expressed as ecto-enzymes at the cell surface [30–33].

15.3 cADPR-Mediated Ca^{2+} Release

Initial studies showed that IP_3, cADPR and NAADP likely released Ca^{2+} by activating distinct mechanisms in sea urchin eggs, since all three compounds independently self-desensitized their respective Ca^{2+} release mechanisms to a second challenge [18]. Pharmacological analysis of cADPR-evoked Ca^{2+} release in sea urchin egg homogenates and intact eggs showed that the pathway for cADPR-evoked Ca^{2+} release was likely ryanodine receptors (RyRs) [34]. This is also the case in mammalian cells, where cADPR is now recognized as a widespread Ca^{2+} mobilizing messenger [35]. RyRs along with IP_3Rs, with which they share degrees of homology in both primary sequence and structure, were discovered as the principal Ca^{2+} release channels of the sarcoplasmic reticulum of striated muscle [36]. However, in common with IP_3Rs, they are also widely expressed in most cell types, including sea urchin eggs [37–39], where they are often both present together in the membranes of the endoplasmic reticulum (ER).

A key property of RyRs and indeed IP_3Rs are that they are also regulated by Ca^{2+} itself in a complex manner [40]. This may lead to Ca^{2+}-induced Ca^{2+} release (CICR), a phenomenon responsible for the globalization of local Ca^{2+} signals as propagating Ca^{2+} waves or repetitive Ca^{2+} spikes, hallmarks of Ca^{2+} signaling in all cells [41]. Increases in cytoplasmic Ca^{2+} were found to potentiate cADPR-evoked Ca^{2+} release both in cell free systems, and also in intact cells [42]. Thus it has been proposed that cADPR sensitizes RyRs to activation by Ca^{2+}. This hypothesis has a pleasing symmetry with the way in which IP_3 is thought to regulate IP_3R gating by also by modulating Ca^{2+} sensitivity of a Ca^{2+} release channel [43].

The exact mechanism by which cADPR regulates RyRs is currently unclear. However, pivotal roles for additional accessory proteins which interact with the large cytoplasmic domains of RyR subunits have been suggested. The radiolabelled photoaffinity cADPR derivative, $[^{32}P]$8-azido-cADPR, labels a 100 kDa and a140 kDa protein in sea urchin egg extracts, too small for RyRs, but these have not been identified. A key finding was that in the sea urchin egg microsomal system a soluble protein factor was required to confer cADPR-sensitivity of RyRs [44]. This was found to be calmodulin, a well-known component of RyR macromolecular complexes [45]. Furthermore, it was found that cycles of dissociation and re-association of calmodulin could account for the desensitization and resensitization of cADPR- evoked Ca^{2+} release from sea urchin egg homogenates [46]. In mammalian systems a role for FKBP12.6, an immunophilin with prolyl isomerase activity, has been proposed as important for RyR cADPR sensitivity [47–52]. cADPR has been suggested to induce the dissociation of FKBP12.6 from RyRs which destabilizes the channel causing an increased probability in their openings. The dependence of accessory proteins on the cADPR-sensitivity of RyRs may explain in part the variations in cADPR sensitivity of purified RyRs reconstituted in planar lipid bilayers [53]. More recently GADPH has been proposed to be a cADPR binding protein with a micromolar affinity for cADPR, and apparently required for cADPR-evoked Ca^{2+} release [54]. The development of 8-substituted analogues of cADPR as highly selective cADPR antagonists have been invaluable in dissecting cADPR-dependent signaling pathways [55, 56]. There is also some evidence that cADPR may activate SERCA pumps in some cells and thus increasing the Ca^{2+} loading of stores which could lead to increased Ca^{2+} release by a Ca^{2+} overload mechanism [57, 58].

15.4   NAADP-Mediated Ca^{2+} Release

Of the three major Ca^{2+} releasing messengers, NAADP is the most potent, often effective in cells at concentrations as low as 1–10 nM [59, 60]. Its mode of action intrigued researchers from its very discovery, since it appeared not to target the two principal Ca^{2+} release channels, IP$_3$Rs/RyRs, but rather a novel Ca^{2+} release channel. In sea urchin eggs and homogenates, NAADP-evoked Ca^{2+} release, which is rapid and likely mediated by a channel [61], is not affected by IP$_3$Rs/RyRs or cADPR inhibitors such as heparin, ryanodine or 8-amino-cADPR [18], but is selectively antagonized by voltage-gated cation channel blockers such as certain dihydropyridines [62]. In addition, in contrast to IP$_3$Rs or RyRs, the NAADP-sensitive Ca^{2+} release mechanism is not potentiated by divalent cations leading to the proposal that it does not function as a CICR channel [63, 64]. Furthermore, the NAADP- sensitive channel appeared not to reside on the ER [64, 65]. Fractionation of sea urchin egg homogenates showed that the NAADP-sensitive store was generally denser than IP$_3$ or cADPR-sensitive microsomes [16]. Treatment of homogenates with the SERCA pump inhibitor, thapsigargin, whilst completely abolishing Ca^{2+} release to either IP$_3$ or cADPR, did not prevent NAADP-evoked Ca^{2+} release. In studies in stratified eggs where ER accumulates near the nucleus, IP$_3$ and cADPR were found to mobilize Ca^{2+} from this region, whilst NAADP released Ca^{2+} from structures at the opposite pole. In detailed studies of NAADP-evoked Ca^{2+} release from sea urchin eggs, NAADP was found to induce an initial Ca^{2+} release which was followed by a series of further Ca^{2+} spikes [66, 67]. The initial Ca^{2+} release was insensitive to thapsigargin, whereas subsequent Ca^{2+} spikes were abolished by thapsigargin or IP$_3$ and RyR inhibitors [67]. It was therefore proposed that NAADP was initially releasing Ca^{2+} from a distinct organelle which then triggered further rounds of Ca^{2+} signals by stimulating Ca^{2+} release from the ER [67]. Further purification of the sea urchin egg NAADP- sensitive stores revealed them as rich in lysosomal markers and acidic in nature, since they stained with lysotracker red [68]. Furthermore, Ca^{2+} uptake whilst insensitive to thapsigargin, was dependent on proton gradients created by the action of bafilomycin-sensitive vacuolar proton pumps. In intact eggs, the lysomolytic agent, glycyl-L- phenylalanine 2-naphthylamide (GPN) was found to lyse lysotracker stained vesicles, which also caused bursts of localized Ca^{2+} release. Treatment with GPN also selectively abolished NAADP-evoked Ca^{2+} release whilst having no effect on either IP$_3$ or cADPR-induced Ca^{2+} signals. From this study it was proposed that NAADP selectively targets lysosome-like organelles in the sea urchin egg [68–70].

Building on these results from sea urchin eggs, the action of NAADP as a Ca^{2+} mobilizing molecule was investigated in a variety of mammalian cells. In the first study, pancreatic acinar cells were found to be exquisitely sensitive to NAADP which produced effects at considerably lower concentrations than either

IP$_3$ or cADPR [71]. Several important principles for mammalian NAADP signaling were proposed from this study. First the concentration-response curve for NAADP-evoked Ca^{2+} release (assessed by activation of Ca^{2+}-activated currents) is bell-shaped, with high concentrations of NAADP causing no effect on account of induction of rapid desensitization of NAADP receptors. Secondly, the response to NAADP required functional IP$_3$ Rs/RyRs, and thirdly, Ca^{2+} release by the secretagogue, cholecystokinin (CCK) required functional NAADP receptors. Until the development of selective NAADP antagonists such as Ned-19 [72], use of high, desensitizing, NAADP concentrations was the major way in which to implicate NAADP in Ca^{2+} signaling processes such as CCK signal transduction here [71]. The finding that NAADP required functional IP$_3$Rs/RyRs indicated that as in the sea urchin egg, one major action of NAADP-evoked Ca^{2+} release in mammalian cells is to trigger further Ca^{2+} release by recruiting ER-based CICR channels [73]. Similarly, activation of the lysosomal channel, TRPML1 causes similar recruitment of ER Ca^{2+} stores [74].

NAADP has now been shown to have a widespread if not universal action in cells as a mobilizer of Ca^{2+} from acidic stores such as lysosomes [60, 75] and endosomes [76]. Although these organelles contain considerably smaller amounts of Ca^{2+} than the ER, they nevertheless may play an important role in Ca^{2+} signaling by locally targeting Ca^{2+} to specific effectors. Questions still remain about the precise way in which Ca^{2+} is sequestered into lysosomes and related organelles. The proton gradient across organellar membranes is required and direct or indirect Ca^{2+}/H$^+$ exchange has been proposed [68]. However, such exchangers are only expressed in lower vertebrates and invertebrates so additional mechanisms may operate [77]. In addition, SERCA3 has been proposed to mediate Ca^{2+} uptake in part, in NAADP-sensitive acidic Ca^{2+} stores of platelets [78]. Others have suggested that filling of Ca^{2+} occurs by privileged communication with the ER by a pH-independent mechanism [79]. Interestingly, in cells from patients with the lysosomal storage disease, Niemann Pick C, lysosomes have defects in Ca^{2+} sequestration, have a low intralysosomal Ca^{2+} concentration, and show a much reduced response to NAADP [80]. Whilst reduced calcium uptake was suggested to account for reduced lysosomal calcium and NAADP action, subsequent studies suggested that this may due to enhanced calcium release [81]

In sea urchin egg membranes, [^{32}P]NAADP binds with very high affinity [66, 82, 83]. Intriguingly, NAADP binding becomes essentially irreversible and it has been proposed that it becomes occluded due to a conformational change of the receptor, a phenomenon that requires the presence of K$^+$ ions [84] and phospholipid [85]. Binding may stabilize protein complexes of solubilized proteins [86, 87]. These unusual binding phenomena have been linked with the equally unusual desensitization properties of the NAADP-sensitive Ca^{2+} release mechanism. Prior exposure with concentrations of NAADP which are subthreshold for evoking Ca^{2+} release can desensitize the mechanism to subsequent suprathreshold NAADP concentrations in a concentration and time-dependent manner [66, 88, 89].

15.5   Two Pore Channels as NAADP Targets

Two principles for NAADP-mediated Ca^{2+} release have emerged in recent years. The first was that NAADP-gated channels have distinct properties from known Ca^{2+} release channels such as IP$_3$Rs and RyRs, and their pharmacology more closely resembled that of voltage-gated cation and TRP channels [62]. Secondly, the NAADP-sensitive release mechanism principally resides on acidic stores such as lysosomes and lysosome-related organelles [68].

Inspection of genomic sequences emerging from a variety of organisms including that of sea urchins, pointed to two families of channels as possible targets. The first was mucolipin-1, a lysosomal TRP channel whose mutations may lead to the lysosomal storage disease, mucolipidosis IV [90–92], and second, a poorly characterized family of channels termed Two-pore channels (TPCs) [93]. TPCs are members of the superfamily of voltage-gated channels which comprise of around 150 members with predicted molecular weights ranging between 80 and 100 kDa. TPCs are predicted to have two domains each containing six transmembrane segments and a single pore loop for each domain. As such they represent a proposed evolutionary intermediate between single domain a subunits which tetramerise to form shaker-like K^+ channels, and the single pore-forming four homologous domain a subunits of voltage-gated Ca^{2+} and Na^+ channels. These channels are thought to have evolved by successive rounds of gene duplication [94–96].

A two pore channel (TPC1) had first been identified from sequences homologous to voltage-gated ion Ca^{2+} channels from rat kidney cDNA [97]. This was followed by the identification of a TPC1 from the genome of the plant *Arabidopsis* [98]. Thus it was the plant channel that was most intensively investigated initially. Importantly, it was shown to be localized to the plant vacuole, the principal acidic organelle in plants, and to act as a Ca^{2+} release channel [99]. On account of a pair of EF hands in the region between the two 6 transmembrane domain (TMD) repeats, not seen in mammalian TPCs, it was also proposed to function as a CICR channel. Electrophysiological analysis of AtTPC1 showed that it likely accounts for the slow vacuolar current and likely to play a key role in plant physiology [100].

At this time Michael Zhu cloned a novel mammalian TPC sequence termed TPC2, and heterologous expression showed that it largely localized to lysosomes, and thus TPCs emerged as plausible candidates as NAADP-gated channels. These data were finally reported in 2009 [101], as described below. Both mucoplin-1 and TPCs have now been proposed as NAADP-gated channels. Although there has been some evidence presented for mucoplin-1 as an NAADP-gated channel [102], this has not been seen by others [103, 104]. In contrast, a number of papers have emerged over the last few years firmly implicating TPCs as central components of NAADP-sensitive Ca^{2+} release channels [105, 106]. Heterologous expression of HsTPC2 in HEK293 cells greatly increased the responsiveness of these cells to NAADP so that now NAADP evoked biphasic Ca^{2+} signals [101]. Pharmacological analysis revealed that the initial Ca^{2+} release is due to Ca^{2+} release from acidic stores whilst the second larger release is mediated by activation of IP3Rs. This coupling between lysosomes and ER nicely mirrors previous studies of NAADP mediated Ca^{2+} release through organellar cross-talk with NAADP acting in a triggering role [67, 71]. Indeed in pulmonary arteriolar smooth muscle cells, both endothelin-1 and NAADP mediated Ca^{2+} signals initiate in a subcellular region where lysosomes and ER are closely apposed [107, 108]. In contrast, expression of TPC1, which localizes to endosomal vesicles, when activated by NAADP, mediates localized Ca^{2+} signals apparently uncoupled from ER-based Ca^{2+} release mechanisms [101, 109]}. Importantly, sea urchins also express TPC isoforms, and expression of both TPC1 and TPC2 also enhance the responsiveness of cells to NAADP generating characteristic biphasic Ca^{2+} signals [110, 111]. Sea urchins, in common with many animals, express three isoforms, although TPC3 is not expressed in man, mouse or rats. In one report, TPC3 appeared to act as a dominant negative suppressing the effects of NAADP on both small endogenous Ca^{2+} release or enhanced release due to TPC2 overexpression [110]. Another important finding is that immunopurified endogenous TPCs bind [^{32}P]NAADP with nanomolar affinity and recapitulates key properties of NAADP binding to native egg membrane fractions [110]. Electrophysiological studies either from isolated lysosomes [112, 113], immunopurified TPC2 reconstituted into lipid bilayers [114] or channels redirected to the plasma membrane by mutating lysosomal targeting sequences [115] have shown that TPCs are indeed NAADP-gated cation channels which can pass Ca^{2+} ions.

Interestingly, TPC2 channel activity is modulated by luminal pH, and increased luminal Ca^{2+} greatly increases their sensitivity to activation by NAADP [114].

Evidence from cells derived from TPC2 knockout mice also supports a key role for TPC2 in mediating NAADP-evoked Ca^{2+} release. In pancreatic beta cells, NAADP evokes Ca^{2+} activated plasma membrane currents which are absent in those from *Tpcn2−/−* mice [101] and may be important in regulating insulin secretion [116]. Furthermore, application of exogenous NAADP to diabetic mice restores insulin secretion [117]. In bladder smooth muscle, whilst NAADP contracts permeabilised myocytes, it fails to do so in cells from *Tpcn2$^{−/−}$* mice, and now agonist- mediated contractions are due entirely to SR-mediated Ca^{2+} release since agonist-coupling to Ca^{2+} release from acidic stores is now abolished [118]. RNA interference approaches are now emerging. For example, knockdown of TPC2 with siRNA has revealed important specific roles for the NAADP/TPC signaling pathway in striated muscle differentiation [119] Importantly, this effect is phenocopied by use of the membrane-permeant NAADP antagonist, Ned-19 or disruption of Ca^{2+} storage by lysosomes/acidic stores by bafilomycin [119]. Importantly, cells from *Tpcn1$^{−/−}$/Tpcn2$^{−/−}$* mice do not mobilize Ca^{2+} in response to NAADP, but do so on expression of either TPC1 or TPC2 proteins, but not TRPML1 [120].

Taken together, there is now compelling evidence that TPCs are key components of the NAADP-sensitive Ca^{2+} release mechanism. A study comparing cADPR and NAADP actions shows clearly that in HEK293 cells, that expression of RyRs confers cADPR-sensitivity, but TPC expression is needed for NAADP-evoked Ca^{2+} release [109] crystallizing the hypotheses that cADPR targets RyRs whilst NAADP targets TPCs. Direct patch-clamping of lysosomes has given further insight into the biophysical properties of TPCs. TPCs are likely non-selective cation channels with substantial Na^+ conductances [121]. However, even a minimal Ca^{2+} conductance may be sufficient to generate physiologically important Ca^{2+} microdomains particularly at lysosomal-ER contact sites [120].

Molecular structures of TPCs have recently been reported by X-ray crystallography or cryo-EM [122–126]. The role of TPCs is disease is a new and growing field [127, 128].

RyRs themselves have been proposed as direct targets for NAADP in some cell types [129, 130] although indirect activation of RyRs by local Ca^{2+} release from acidic stores is not trivial to discount. However, since it is likely that NAADP may bind to a channel accessory protein rather than a channel itself [131, 132], it is possible that NAADP may regulate multiple Ca^{2+} release channels under certain conditions [133, 134]. The requirement for an elusive NAADP-binding protein may explain why there is variability in responsiveness of TPC channel currents recorded under conditions of whole-lysosome patch clamp [112, 113, 121]. A similar case may operate for cADPR modulation of RyRs [54].

NAADP-evoked Ca^{2+} release from acidic stores has been proposed to work in three major ways to regulate cellular processes (Fig. 15.4) [135]. The first is to

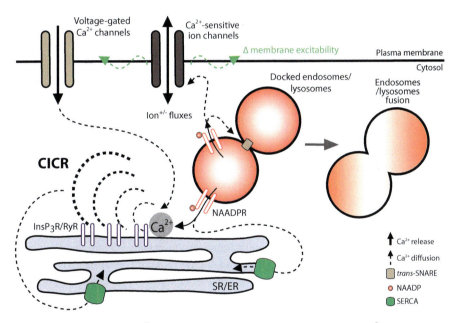

Fig. 15.4 Local to global Ca^{2+} responses mediated by NAADP. Locally, the Ca^{2+} release from acidic stores is likely to be important for normal functions of the endo-lysosomal system such as vesicle fusion or fission in lysosomal biogenesis. Small local Ca^{2+} release evoked by NAADP from lysosomes may also act as a trigger to initiate CICR from the SR/ER and generate global signals. NAADP-mediated Ca^{2+} release near the plasma membrane modulates membrane excitability (excitable cells), or ion fluxes (non-excitable cells) by opening Ca^{2+} activated channels. Changes in the membrane potential could further activate, for example, voltage-gated Ca^{2+} channels, to allow Ca^{2+} influx and initiate a global response via CICR

coordinate Ca^{2+} release by organelle cross-talk at junctions between lysosomes and the ER [136–138]. The second is to produce local Ca^{2+} signals in the sub-plasma membrane space to control Ca^{2+}-activated plasma membrane channels and thus regulate processes such as fluid secretion in non-excitable cells such as exocrine gland cells [71], or membrane excitability in excitable cells. Examples of the latter are the polyspermy blocking cortical flash in sea urchin eggs [139] or activation of membrane currents in starfish oocytes [140], and depolarization of pancreatic beta cells [101] or neurones [141]. Thirdly, local Ca^{2+} release via TPCs may regulate endolysosomal trafficking and organelle biogenesis by regulating membrane fission and fusion processes. Overexpression of TPCs or their block by the NAADP antagonist Ned-19 induces deregulation of endocytosis, lysosome biogenesis and trafficking mimicking features of lysosomal storage diseases [110, 142, 143] which may be relevant to disease [60, 128].

15.6 ADP-Ribose and Other Metabolites

As mentioned above, cADPR is metabolized to ADPR by CD38's intrinsic hydrolase activity. Other NADases may directly catalyse the formation of ADPR. ADPR although inactive at releasing Ca^{2+} from the ER, was found to stimulate Ca^{2+} influx. This was first observed in ascidian oocytes where low ADPR concentrations may activate plasma membrane currents [144]. An important target for ADPR is TRPM2. TRPM2 channels are polymodal-gated channels responding to not only pyridine nucleotides but also to calcium, oxidants and raised temperature [145]. They have been proposed as chanzymes in that they express an intrinsic enzyme activity, ADPR pyrophosphate conferred by a nudix box containing domain at their C-terminus [146], although this has been challenged [147]. Both cADPR and NAADP have also been reported to activate TRPM2, but at high micromolar concentrations, substantially higher than reported concentrations of these molecules in cells and tissues [148], although this has also been questioned [149]. In a T cell line, a second messenger role for ADP-ribose in mediating concanavalin A-activated Ca^{2+} influx via TRPM2 channels has been proposed [150]. Interestingly, although TRPM2 channels are mainly found in the plasma membrane, they have also been reported in lysosomes [151]. In addition to ADPR, ADP-ribosyl cyclases may also generate additional nucleotides including cADPR-phosphate, and adenine dinucleotides [152] which may also be active in some cells at generating Ca^{2+} signals. A recent report has suggested that 2'-deoxyadenosine 5'-diphosphoribose is an endogenous TRPM2 superagonist [153]. The recent description of the molecular structure of TRPM2 from cryo-EM studies will shed more light on the molecular architecture of the pyridine nucleotide binding site [154].

15.7 Receptor-Mediated NAADP and cADPR-Mediated Ca^{2+} Signaling

Both cADPR and NAADP have satisfied all the criteria originally mandated by Sutherland, to be unambiguously assigned as second or intracellular messengers [155].

Endogenous levels of both cADPR and NAADP are found in a wide range of tissues and cells from across the phyla. Similarly, increases in levels of cADPR [35] and NAADP [75] have been reported in response to cell activation by a variety of stimuli and cell surface receptor families. There have been two principal ways in which cADPR and NAADP levels have been measured [156]. The first are radioreceptor assays based on the high affinity binding sites for these molecules on sea urchin egg membranes [157, 158], and the second is a cycling assay using ADP-ribosyl cyclase in reverse to generate NAD which is then coupled to the generation of a fluorescent product [159, 160]. The latter requires initial treatment with enzymes to completely remove endogenous NAD before proceeding. An additional important development was the use of NGD as an alternative substrate to NAD. True ADP- ribosyl cyclase as opposed to most ADP glycohydrolases tend to cyclise NGD to cGDPR which is fluorescent [161]], and this assay has been extensively used to demonstrate ADP-ribosyl cylase activities in many preparations, and in some cases regulation of activities by various stimuli.

The first indication that cADPR levels could be regulated by phosphorylation processes was the finding that cGMP via G kinase stimulates Ca^{2+} release via cADPR synthesis in sea urchin eggs [162, 163]. cAMP on the other hand may selectively enhance NAADP forming base-exchange activity by PKA or EPAC [164], thus differential phosphorylation or regulation by cGMP or cAMP may dictate which messenger is generated on enzyme stimulation [165].

As a general principle, NAADP increases rapidly and transiently on cell stimulation, whilst cADPR remains elevated for many minutes [166, 167]. From this it has been proposed that NAADP plays a major role in triggering Ca^{2+} signals, whilst cADPR may have a longer term role in producing a long lasting increase in sensitivity of RyRs to CICR. Indeed prolonged elevations of cADPR have been associated with the circadian clock in plants as well as increased proliferative state of cells in culture [168].

CD38 knockout mice are providing important insights into the role of CD38 in agonist-mediated Ca^{2+} signaling mediated by both cADPR and NAADP. Several studies have indicated that various tissues from $Cd38^{-/-}$ mice have substantially reduced endogenous cADPR levels. In addition, the ability of a number of stimuli to elevate cADPR levels is impaired. Concomitantly, a number of important physiological processes are abrogated [23]. Defects include reduced insulin secretion from pancreatic beta cells [169], abolition of Ca^{2+} mobilization in pancreatic acinar cells [170, 171], cardiac hypertrophy [172], changes in airway [173] and vascular smooth muscle pharmaco-mechanical coupling [174], defects in neutrophil chemotaxis with increased susceptibility to bacterial infection [175], impaired oxytocin signaling with associated behavioral correlates [176].

A recent report has suggested a novel way in which cADPR signaling is regulated in sperm. These cells with minimal signaling machineries apparently acquire CD38 and RyRs from secreted prostasomes from prostate gland cells and competence in cADPR signaling is required for sperm motility and thus ability to fertilize ova [177].

The finding that, at least in vitro, CD38 can synthesise NAADP [27], has also led to the investigation of this enzyme in receptor-mediated NAADP production. Using cells from *Cd38−/−* mice it was found that in pancreatic acinar cells, NAADP production stimulated by physiological concentrations of CCK was abolished [171], as was that produced by angiotensin II in hepatic stellate cells [178], and by IL8 in lymphokine-activated killer cells [179]. Recent evidence has been presented to show that NAADP synthesis is also linked to beta- adrenergic signalling mechanisms in cardiac myocytes [180]. However, agonist-induced NAADP production has been reported in tissues from *Cd38−/−* mouse tissues, raising the possibility of alternate synthetic pathways [181]. CD157, also known as BST1, shows ADP-ribosyl cyclase activity, but apparently is poor at catalysing NAADP synthesis by base exchange [182].

The great plasticity of Ca^{2+} signaling pathways is exemplified by the ability of high affinity CCK receptors in Cd38−/− pancreatic acinar cells now to switch from NAADP production to IP -mediated Ca^{2+} signaling [171]. Conversely, CCK evoked Ca^{2+} signals in the ADP-ribosyl cyclase-deficient AR42J rat pancreatic cell line switch from IP -mediated Ca^{2+} signaling to NAADP signaling and Ca^{2+} release from acidic stores on transfection with CD38 cDNA [171].

A major question concerning the notion that CD38 is the major synthetic enzyme for cADPR and NAADP synthesis is the membrane topology of this protein [183]. It was originally noted as a plasma membrane ectoenzyme with its active site facing the extracellular space. De Flora and colleagues presented evidence that NAD may leak from the cell via connexins which is acted upon by extracellular CD38 to yield extracellular cADPR. cADPR may be transported back into the cell via nucleoside transporters [183]. However, an appreciable amount of CD38 is intracellular and associated with organelles such as secretory granules or endosomes, and this may increase on cell stimulation. This appears to be the case in sea urchin eggs, and evidence has been provided that NAD may be transported into organelles, converted to cADPR, which is transported into the cytoplasm to its site of action [30, 31]. This also raises the possibility that cADPR and NAADP may also be stored in intracellular compartments, and cell stimuli may act to regulate the egress of these molecules into the cytoplasm as for Ca^{2+} mobilization from intracellular stores. Another possibility, recently proposed, is that CD38 may exist in two topologies with the active site of one form, termed type III, actually facing the cytoplasm, [184, 185, 186]. The cytosolic facing form has been shown to interact with the Ca^{2+} binding CIB1 protein which may enhance its activity [186]. Whatever the situation, more clarification is needed to understand how cellular stimuli are coupled to increases in ADP-ribosyl cyclase activities.

15.8   Conclusions

The emergence of cADPR and NAADP as Ca^{2+} mobilizing messengers and the elucidation of their place in signaling pathways and the identification of their molecular targets over the last two decades has provided many surprises and advances in our understanding of cellular regulatory processes. The field has moved on from the role of these molecules at fertilization of invertebrate eggs to central players in mammalian Ca^{2+} homeostasis and signaling. We can expect more surprises and answers in the years to come, but the next key development will be to identify cADPR and NAADP binding proteins and ascertain how they modulate their respective targets.

References

1. Ringer S (1882) Concerning the influence exerted by each of the constituents of the blood on the contraction of the ventricle. J Physiol 3:380–393
2. Ashley CC, Ridgway EB (1968) Simultaneous recording of membrane potential, calcium transient and tension in single muscle fibers. Nature 219:1168–1169
3. Douglas WW, Poisner AM (1964) Stimulus-secretion coupling in a neurosecretory organ: the role of calcium in the release of vasopressin from the neurohypophysis. J Physiol 172:1–18
4. Nielsen SP, Petersen OH (1972) Transport of calcium in the perfused submandibular gland of the cat. J Physiol 223:685–697
5. Michell RH (1975) Inositol phospholipids and cell surface receptor function. Biochim Biophys Acta 415:81–47
6. Berridge MJ (1983) Rapid accumulation of inositol trisphosphate reveals that agonists hydrolyse polyphosphoinositides instead of phosphatidylinositol. Biochem J 212:849–858
7. Streb H, Irvine RF, Berridge MJ, Schulz I (1983) Release of Ca^{2+} from a nonmitochondrial intracellular store in pancreatic acinar cells by inositol-1,4,5-trisphosphate. Nature 306:67–69
8. Supattapone S, Worley PF, Baraban JM, Snyder SH (1988) Solubilization, purification, and characterization of an inositol trisphosphate receptor. J Biol Chem 263:1530–1534
9. Maeda N, Niinobe M, Mikoshiba K (1990) A cerebellar Purkinje cell marker P400 protein is an inositol 1,4,5-trisphosphate (InsP$_3$) receptor protein. Purification and characterization of InsP3 receptor complex. Embo J 9:61–67
10. Furuichi T, Yoshikawa S, Miyawaki A, Wada K, Maeda N, Mikoshiba K (1989) Primary structure and functional expression of the inositol 1,4,5-trisphosphate-binding protein P400. Nature 342:32–38
11. Mignery GA, Sudhof TC, Takei K, De Camilli P (1989) Putative receptor for inositol 1,4,5-trisphosphate similar to ryanodine receptor. Nature 342:192–195
12. Serysheva II, Baker MR, Fan G (2017) Structural Insights into IP^{3}R Function. Adv Exp Med Biol 981:121–147
13. Berridge MJ (1993) Inositol trisphosphate and calcium signalling. Nature 361: 315–325
14. Whitaker M, Irvine RF (1984) Inositol 1,4,5 trisphosphate microinjection activates sea urchin eggs. Nature 312:636–639
15. Clapper DL, Lee HC (1985) Inositol trisphosphate induces calcium release from nonmitochondrial stores i sea urchin egg homogenates. J Biol Chem 260:13947–13954

16. Clapper DL, Walseth TF, Dargie PJ, Lee HC (1987) Pyridine nucleotide metabolites stimulate calcium release from sea urchin egg microsomes desensitized to inositol trisphosphate. J Biol Chem 262:9561–9568

17. Lee HC, Walseth TF, Bratt GT, Hayes RN, Clapper DL (1989) Structural determination of a cyclic metabolite of NAD$^+$ with intracellular Ca^{2+}-mobilizing activity. J Biol Chem 264:1608–1615

18. Lee HC, Aarhus R (1995) A derivative of NADP mobilizes calcium stores insensitive to inositol trisphosphate and cyclic ADP-ribose. J Biol Chem 270:2152–2157

19. Rusinko N, Lee HC (1989) Widespread occurrence in animal tissues of an enzyme catalyzing the conversion of NAD$^+$ into a cyclic metabolite with intracellular Ca^{2+}-mobilizing activity. J Biol Chem 264:11725–11731

20. Hellmich MR, Strumwasser F (1991) Purification and characterization of a molluscan egg-specific NADase, a second-messenger enzyme. Cell Regul 2:193–202

21. Glick DL, Hellmich MR, Beushausen S, Tempst P, Bayley H, Strumwasser F (1991) Primary structure of a molluscan egg-specific NADase, a second-messenger enzyme. Cell Regul 2:211–218

22. Lee HC, Aarhus R (1991) ADP-ribosyl cyclase: an enzyme that cyclizes NAD$^+$ into a calcium-mobilizing metabolite. Cell Regul 2:203–209

23. Malavasi F, Deaglio S, Funaro A, Ferrero E, Horenstein AL, Ortolan E et al (2008) Evolution and function of the ADP ribosyl cyclase/CD38 gene family in physiology and pathology. Physiol Rev 88:841–886

24. Chini EN, Chini CCS, Espindola Netto JM, de Oliveira GC, van Schooten W (2018) The pharmacology of CD38/NADase: an emerging target in cancer and diseases of aging. Trends Pharmacol Sci 39:424–436

25. Howard M, Grimaldi JC, Bazan JF, Lund FE, Santos-Argumedo L, Parkhouse RM et al (1993) Formation and hydrolysis of cyclic ADP-ribose catalyzed by lymphocyte antigen CD38. Science 262:1056–1059

26. Aarhus R, Graeff RM, Dickey DM, Walseth TF, Lee HC (1995) ADP-ribosyl cyclase and CD38 catalyze the synthesis of a calcium-mobilizing metabolite from NADP. J Biol Chem 270:30327–30333

27. Graeff R, Liu Q, Kriksunov IA, Hao Q, Lee HC (2006) Acidic residues at the active sites of CD38 and ADP-ribosyl cyclase determine nicotinic acid adenine dinucleotide phosphate (NAADP) synthesis and hydrolysis activities. J Biol Chem 281:28951–28957

28. Berridge G, Cramer R, Galione A, Patel S (2002) Metabolism of the novel Ca^{2+}-mobilizing messenger nicotinic acid-adenine dinucleotide phosphate via a 2′-specific Ca2$^+$-dependent phosphatase. Biochem J 365:295–301

29. Lee HC (2000) Enzymatic functions and structures of CD38 and homologs. Chem Immunol 75:39–59

30. Churamani D, Boulware MJ, Geach TJ, Martin AC, Moy GW, Su YH et al (2007) Molecular characterization of a novel intracellular ADP-ribosyl cyclase. PLoS One 2:e797

31. Davis LC, Morgan AJ, Ruas M, Wong JL, Graeff RM, Poustka AJ et al (2008) Ca^{2+} signaling occurs via second messenger release from intraorganelle synthesis sites. Curr Biol 18:1612–1618

32. Churamani D, Boulware MJ, Ramakrishnan L, Geach TJ, Martin AC, Vacquier VD et al (2008) Molecular characterization of a novel cell surface ADP-ribosyl cyclase from the sea urchin. Cell Signal 20:2347–2355

33. Ramakrishnan L, Muller-Steffner H, Bosc C, Vacquier VD, Schuber F, Moutin MJ et al (2010) A single residue in a novel ADP-ribosyl cyclase controls production of the calcium-mobilizing messengers cyclic ADP-ribose and nicotinic acid adenine dinucleotide phosphate. J Biol Chem 285:19900–19909

34. Galione A, Lee HC, Busa WB (1991) Ca^{2+}-induced Ca^{2+} release in sea urchin egg homogenates: modulation by cyclic ADP-ribose. Science 253:1143–1146

35. Galione A, Churchill GC (2000) Cyclic ADP ribose as a calcium-mobilizing messenger. Sci STKE 2000:pe1

36. Fill M, Copello JA (2002) Ryanodine receptor calcium release channels. Physiol Rev 82:893–922
37. McPherson SM, McPherson PS, Mathews L, Campbell KP, Longo FJ (1992) Cortical localization of a calcium release channel in sea urchin eggs. J Cell Biol 116:1111–1121
38. Lokuta AJ, Darszon A, Beltran C, Valdivia HH (1998) Detection and functional characterization of ryanodine receptors from sea urchin eggs. J Physiol 510:155–164
39. Shiwa M, Murayama T, Ogawa Y (2002) Molecular cloning and characterization of ryanodine receptor from unfertilized sea urchin eggs. Am J Physiol Regul Integr Comp Physiol 282:R727–R737
40. Taylor CW (1998) Inositol trisphosphate receptors: Ca^{2+}-modulated intracellular Ca^{2+} channels. Biochim Biophys Acta 1436:19–33
41. Roderick HL, Berridge MJ, Bootman MD (2003) Calcium-induced calcium release. Curr Biol 13:R425
42. Lee HC (1993) Potentiation of calcium- and caffeine-induced calcium release by cyclic ADP-ribose. J Biol Chem 268:293–299
43. Parys JB, De Smedt H (2012) Inositol 1,4,5-trisphosphate and its receptors. Adv Exp Med Biol 740:255–279
44. Lee HC, Aarhus R, Graeff RM (1995) Sensitization of calcium-induced calcium release by cyclic ADP-ribose and calmodulin. J Biol Chem 270:9060–9066
45. Zhu X, Ghanta J, Walker JW, Allen PD, Valdivia HH (2004) The calmodulin binding region of the skeletal ryanodine receptor acts as a self-modulatory domain. Cell Calcium 35:165–177
46. Thomas JM, Summerhill RJ, Fruen BR, Churchill GC, Galione A (2002) Calmodulin dissociation mediates desensitization of the cADPR-induced Ca^{2+} release mechanism. Curr Biol 12:2018–2022
47. Morita K, Kitayama T, Kitayama S, Dohi T (2006) Cyclic ADP-ribose requires FK506-binding protein to regulate intracellular Ca^{2+} dynamics and catecholamine release in acetylcholine-stimulated bovine adrenal chromaffin cells. J Pharmacol Sci 101:40–51
48. Noguchi N, Takasawa S, Nata K, Tohgo A, Kato I, Ikehata F et al (1997) Cyclic ADP-ribose binds to FK506-binding protein 12.6 to release Ca^{2+} from islet microsomes. J Biol Chem 272:3133–3136
49. Tang WX, Chen YF, Zou AP, Campbell WB, Li PL (2002) Role of FKBP12.6 in cADPR-induced activation of reconstituted ryanodine receptors from arterial smooth muscle. Am J Physiol Heart Circ Physiol 282:H1304–H1310
50. Wang YX, Zheng YM, Mei QB, Wang QS, Collier ML, Fleischer S et al (2004) FKBP12.6 and cADPR regulation of Ca^{2+} release in smooth muscle cells. Am J Physiol Cell Physiol 286:C538–C546
51. Zhang X, Tallini YN, Chen Z, Gan L, Wei B, Doran R et al (2009) Dissociation of FKBP12.6 from ryanodine receptor type 2 is regulated by cyclic ADP-ribose but not beta-adrenergic stimulation in mouse cardiomyocytes. Cardiovasc Res 84:253–262
52. Zheng J, Wenzhi B, Miao L, Hao Y, Zhang X, Yin W et al (2010) Ca^{2+} release induced by cADP-ribose is mediated by FKBP12.6 proteins in mouse bladder smooth muscle. Cell Calcium 47:449–457
53. Copello JA, Qi Y, Jeyakumar LH, Ogunbunmi E, Fleischer S (2001) Lack of effect of cADP-ribose and NAADP on the activity of skeletal muscle and heart ryanodine receptors. Cell Calcium 30:269–284
54. Zhang K, Sun W, Huang L, Zhu K, Pei F, Zhu L et al (2017) Identifying glyceraldehyde 3-phosphate dehydrogenase as a cyclic adenosine diphosphoribose binding protein by photoaffinity protein-ligand labeling approach. J Am Chem Soc 139:156–170
55. Walseth TF, Lee HC (1993) Synthesis and characterization of antagonists of cyclic-ADP-ribose-induced Ca^{2+} release. Biochim Biophys Acta 1178:235–242
56. Sethi JK, Empson RM, Bailey VC, Potter BV, Galione A (1997) 7-Deaza-8-bromo-cyclic ADP-ribose, the first membrane-permeant, hydrolysis-resistant cyclic ADP-ribose antagonist. J Biol Chem 272:16358–16363

57. Lukyanenko V, Gyorke I, Wiesner TF, Gyorke S (2001) Potentiation of Ca^{2+} release by cADP-ribose in the heart is mediated by enhanced SR Ca^{2+} uptake into the sarcoplasmic reticulum. Circ Res 89:614–622

58. Yamasaki-Mann M, Demuro A, Parker I (2009) cADPR stimulates SERCA activity in Xenopus oocytes. Cell Calcium 45:293–299

59. Guse AH, Lee HC (2008) NAADP: a universal Ca^{2+} trigger. Sci Signal 1:re10

60. Galione A (2015) A primer of NAADP-mediated Ca^{2+} signalling: from sea urchin eggs to mammalian cells. Cell Calcium 58:27–47

61. Genazzani AA, Mezna M, Summerhill RJ, Galione A, Michelangeli F (1997) Kinetic properties of nicotinic acid adenine dinucleotide phosphate-induced Ca^{2+} release. J Biol Chem 272:7669–7675

62. Genazzani AA, Mezna M, Dickey DM, Michelangeli F, Walseth TF, Galione A (1997) Pharmacological properties of the Ca^{2+}-release mechanism sensitive to NAADP in the sea urchin egg. Br J Pharmacol 121:1489–1495

63. Chini EN, Dousa TP (1996) Nicotinate-adenine dinucleotide phosphate-induced Ca^{2+}-release does not behave as a Ca^{2+}-induced Ca^{2+}-release system. Biochem J 316:709–711

64. Genazzani AA, Galione A (1996) Nicotinic acid-adenine dinucleotide phosphate mobilizes Ca^{2+} from a thapsigargin-insensitive pool. Biochem J 315:721–725

65. Lee HC, Aarhus R (2000) Functional visualization of the separate but interacting calcium stores sensitive to NAADP and cyclic ADP-ribose. J Cell Sci 113:4413–4420

66. Aarhus R, Dickey DM, Graeff RM, Gee KR, Walseth TF, Lee HC (1996) Activation and inactivation of Ca^{2+} release by NAADP+. J Biol Chem 271:8513–8516

67. Churchill GC, Galione A (2001) NAADP induces Ca^{2+} oscillations via a two-pool mechanism by priming IP3- and cADPR-sensitive Ca^{2+} stores. EMBO J 20:2666–2671

68. Churchill GC, Okada Y, Thomas JM, Genazzani AA, Patel S, Galione A (2002) NAADP mobilizes Ca^{2+} from reserve granules, lysosome-related organelles, in sea urchin eggs. Cell 111:703–708

69. Morgan AJ, Platt FM, Lloyd-Evans E, Galione A (2011) Molecular mechanisms of endolysosomal Ca^{2+} signalling in health and disease. Biochem J 439:349–374

70. Lee HC (2012) Cyclic ADP-ribose and nicotinic acid adenine dinucleotide phosphate (NAADP) as messengers for calcium mobilization. J Biol Chem 287:31633–31640

71. Cancela JM, Churchill GC, Galione A (1999) Coordination of agonist-induced Ca^{2+}-signalling patterns by NAADP in pancreatic acinar cells. Nature 398:74–76

72. Naylor E, Arredouani A, Vasudevan SR, Lewis AM, Parkesh R, Mizote A et al (2009) Identification of a chemical probe for NAADP by virtual screening. Nat Chem Biol 5:220–226

73. Patel S, Churchill GC, Galione A (2001) Coordination of Ca^{2+} signalling by NAADP. Trends Biochem Sci 26:482–489

74. Kilpatrick BS, Yates E, Grimm C, Schapira AH, Patel S (2016) Endo-lysosomal TRP mucolipin-1 channels trigger global ER Ca^{2+} release and Ca^{2+} influx. J Cell Sci 129:3859–3867

75. Galione A, Morgan AJ, Arredouani A, Davis LC, Rietdorf K, Ruas M et al (2010) NAADP as an intracellular messenger regulating lysosomal calcium-release channels. Biochem Soc Trans 38:1424–1431

76. Menteyne A, Burdakov A, Charpentier G, Petersen OH, Cancela JM (2006) Generation of specific Ca^{2+} signals from Ca^{2+} stores and endocytosis by differential coupling to messengers. Curr Biol 16:1931–1937

77. Melchionda M, Pittman JK, Mayor R, Patel S (2016) Ca^{2+}/H^+ exchange by acidic organelles regulates cell migration in vivo. J Cell Biol 212:803–813

78. Jardin I, Lopez JJ, Pariente JA, Salido GM, Rosado JA (2008) Intracellular calcium release from human platelets: different messengers for multiple stores. Trends Cardiovasc Med 18:57–61

79. Garrity AG, Wang W, Collier CM, Levey SA, Gao Q, Xu H (2016) The endoplasmic reticulum, not the pH gradient, drives calcium refilling of lysosomes. Elife 5

80. Lloyd-Evans E, Morgan AJ, He X, Smith DA, Elliot-Smith E, Sillence DJ et al (2008) Niemann-Pick disease type C1 is a sphingosine storage disease that causes deregulation of lysosomal calcium. Nat Med 14:1247–1255

81. Hoglinger D, Haberkant P, Aguilera-Romero A, Riezman H, Porter FD, Platt FM et al (2015) Intracellular sphingosine releases calcium from lysosomes. Elife 4

82. Billington RA, Genazzani AA (2000) Characterization of NAADP$^+$ binding in sea urchin eggs. Biochem Biophys Res Commun 276:112–116

83. Patel S, Churchill GC, Galione A (2000) Unique kinetics of nicotinic acid-adenine dinucleotide phosphate (NAADP) binding enhance the sensitivity of NAADP receptors for their ligand. Biochem J 352:725–729

84. Dickinson GD, Patel S (2003) Modulation of NAADP (nicotinic acid-adenine dinucleotide phosphate) receptors by K^{2+} ions: evidence for multiple NAADP receptor conformations. Biochem J 375:805–812

85. Churamani D, Dickinson GD, Patel S (2005) NAADP binding to its target protein in sea urchin eggs requires phospholipids. Biochem J 386:497–504

86. Berridge G, Dickinson G, Parrington J, Galione A, Patel S (2002) Solubilization of receptors for the novel Ca^{2+}-mobilizing messenger, nicotinic acid adenine dinucleotide phosphate. J Biol Chem 277:43717–43723

87. Churamani D, Dickinson GD, Ziegler M, Patel S (2006) Time sensing by NAADP receptors. Biochem J 397:313–320

88. Genazzani AA, Empson RM, Galione A (1996) Unique inactivation properties of NAADP-sensitive Ca^{2+} release. J Biol Chem 271:11599–11602

89. Churchill GC, Galione A (2001) Prolonged inactivation of nicotinic acid adenine dinucleotide phosphate-induced Ca^{2+} release mediates a spatiotemporal Ca^{2+} memory. J Biol Chem 276:11223–11225

90. Bargal R, Avidan N, Ben-Asher E, Olender Z, Zeigler M, Frumkin A et al (2000) Identification of the gene causing mucolipidosis type IV. Nat Genet 26:118–123

91. Sun M, Goldin E, Stahl S, Falardeau JL, Kennedy JC, Acierno JS Jr et al (2000) Mucolipidosis type IV is caused by mutations in a gene encoding a novel transient receptor potential channel. Hum Mol Genet 9:2471–2478

92. Bach G (2001) Mucolipidosis type IV. Mol Genet Metab 73:197–203

93. Galione A, Evans AM, Ma J, Parrington J, Arredouani A, Cheng X et al (2009) The acid test: the discovery of two-pore channels (TPCs) as NAADP-gated endolysosomal Ca^{2+} release channels. Pflugers Arch 458:869–876

94. Zhu MX, Ma J, Parrington J, Calcraft PJ, Galione A, Evans AM (2010) Calcium signaling via two-pore channels: local or global, that is the question. Am J Physiol Cell Physiol 298:C430–C441

95. Cai X, Patel S (2010) Degeneration of an intracellular ion channel in the primate lineage by relaxation of selective constraints. Mol Biol Evol 27:2352–2359

96. Rahman T, Cai X, Brailoiu GC, Abood ME, Brailoiu E, Patel S (2014) Two-pore channels provide insight into the evolution of voltage-gated Ca^{2+} and Na^+ channels. Sci Signal 7:ra109

97. Ishibashi K, Suzuki M, Imai M (2000) Molecular cloning of a novel form (two-repeat) protein related to voltage-gated sodium and calcium channels. Biochem Biophys Res Commun 270:370–376

98. Furuichi T, Cunningham KW, Muto S (2001) A putative two pore channel AtTPC1 mediates Ca^{2+} flux in Arabidopsis leaf cells. Plant Cell Physiol 42:900–905

99. Peiter E, Maathuis FJ, Mills LN, Knight H, Pelloux J, Hetherington AM et al (2005) The vacuolar Ca^{2+}-activated channel TPC1 regulates germination and stomatal movement. Nature 434:404–408

100. Hedrich R, Marten I (2011) TPC1-SV channels gain shape. Mol Plant 4:428–441

101. Calcraft PJ, Ruas M, Pan Z, Cheng X, Arredouani A, Hao X et al (2009) NAADP mobilizes calcium from acidic organelles through two-pore channels. Nature 459:596–600

102. Zhang F, Li PL (2007) Reconstitution and characterization of a nicotinic acid adenine dinucleotide phosphate (NAADP)-sensitive Ca^{2+} release channel from liver lysosomes of rats. J Biol Chem 282:25259–25269

103. Pryor PR, Reimann F, Gribble FM, Luzio JP (2006) Mucolipin-1 is a lysosomal membrane protein required for intracellular lactosylceramide traffic. Traffic 7:1388–1398

104. Yamaguchi S, Jha A, Li Q, Soyombo AA, Dickinson GD, Churamani D et al (2011) Transient receptor potential mucolipin 1 (TRPML1) and two-pore channels are functionally independent organellar ion channels. J Biol Chem 286:22934–22942

105. Zong X, Schieder M, Cuny H, Fenske S, Gruner C, Rotzer K et al (2009) The two-pore channel TPCN2 mediates NAADP-dependent Ca^{2+}-release from lysosomal stores. Pflugers Arch 458:891–899

106. Brailoiu E, Churamani D, Cai X, Schrlau MG, Brailoiu GC, Gao X et al (2009) Essential requirement for two-pore channel 1 in NAADP-mediated calcium signaling. J Cell Biol 186:201–209

107. Kinnear NP, Boittin FX, Thomas JM, Galione A, Evans AM (2004) Lysosome-sarcoplasmic reticulum junctions. A trigger zone for calcium signaling by nicotinic acid adenine dinucleotide phosphate and endothelin-1. J Biol Chem 279:54319–54326

108. Kinnear NP, Wyatt CN, Clark JH, Calcraft PJ, Fleischer S, Jeyakumar LH et al (2008) Lysosomes co-localize with ryanodine receptor subtype 3 to form a trigger zone for calcium signalling by NAADP in rat pulmonary arterial smooth muscle. Cell Calcium 44:190–201

109. Ogunbayo OA, Zhu Y, Rossi D, Sorrentino V, Ma J, Zhu MX et al (2011) Cyclic adenosine diphosphate ribose activates ryanodine receptors, whereas NAADP activates two-pore domain channels. J Biol Chem 286:9136–9140

110. Ruas M, Rietdorf K, Arredouani A, Davis LC, Lloyd-Evans E, Koegel H et al (2010) Purified TPC isoforms form NAADP receptors with distinct roles for Ca^{2+} signaling and endolysosomal trafficking. Curr Biol 20:703–709

111. Brailoiu E, Hooper R, Cai X, Brailoiu GC, Keebler MV, Dun NJ et al (2010) An ancestral deuterostome family of two-pore channels mediates nicotinic acid adenine dinucleotide phosphate-dependent calcium release from acidic organelles. J Biol Chem 285:2897–2901

112. Schieder M, Rotzer K, Bruggemann A, Biel M, Wahl-Schott CA (2010) Characterization of two-pore channel 2 (TPCN2)-mediated Ca^{2+} currents in isolated lysosomes. J Biol Chem 285:21219–21222

113. Jha A, Ahuja M, Patel S, Brailoiu E, Muallem S (2014) Convergent regulation of the lysosomal two-pore channel-2 by Mg^{2+}, NAADP, $PI(3,5)P_2$ and multiple protein kinases. Embo J 33:501–511

114. Pitt SJ, Funnell TM, Sitsapesan M, Venturi E, Rietdorf K, Ruas M et al (2010) TPC2 is a novel NAADP-sensitive Ca^{2+} release channel, operating as a dual sensor of luminal pH and Ca^{2+}. J Biol Chem 285:35039–35046

115. Brailoiu E, Rahman T, Churamani D, Prole DL, Brailoiu GC, Hooper R et al (2010) An NAADP-gated two-pore channel targeted to the plasma membrane uncouples triggering from amplifying Ca^{2+} signals. J Biol Chem 285:38511–38516

116. Arredouani A, Ruas M, Collins SC, Parkesh R, Clough F, Pillinger T et al (2015) Nicotinic Acid Adenine Dinucleotide Phosphate (NAADP) and endolysosomal two-pore channels modulate membrane excitability and stimulus-secretion coupling in mouse pancreatic beta cells. J Biol Chem 290:21376–21392

117. Park KH, Kim BJ, Shawl AI, Han MK, Lee HC, Kim UH (2013) Autocrine/paracrine function of nicotinic acid adenine dinucleotide phosphate (NAADP) for glucose homeostasis in pancreatic beta-cells and adipocytes. J Biol Chem 288:35548–35558

118. Tugba Durlu-Kandilci N, Ruas M, Chuang KT, Brading A, Parrington J, Galione A (2010) TPC2 proteins mediate nicotinic acid adenine dinucleotide phosphate (NAADP)- and agonist-evoked contractions of smooth muscle. J Biol Chem 285:24925–24932

119. Aley PK, Mikolajczyk AM, Munz B, Churchill GC, Galione A, Berger F (2010) Nicotinic acid adenine dinucleotide phosphate regulates skeletal muscle differentiation via action at two-pore channels. Proc Natl Acad Sci USA 107:19927–19932

120. Ruas M, Davis LC, Chen CC, Morgan AJ, Chuang KT, Walseth TF et al (2015) Expression of Ca^{2+}-permeable two-pore channels rescues NAADP signalling in TPC-deficient cells. Embo J 34:1743–1758

121. Wang X, Zhang X, Dong XP, Samie M, Li X, Cheng X et al (2012) TPC proteins are phosphoinositide- activated sodium-selective ion channels in endosomes and lysosomes. Cell 151:372–383

122. Guo J, Zeng W, Chen Q, Lee C, Chen L, Yang Y et al (2016) Structure of the voltage-gated two-pore channel TPC1 from Arabidopsis thaliana. Nature 531:196–201

123. Hedrich R, Mueller TD, Becker D, Marten I (2018) Structure and function of TPC1 vacuole SV channel gains shape. Mol Plant 11:764–775

124. Kintzer AF, Stroud RM (2016) Structure, inhibition and regulation of two-pore channel TPC1 from Arabidopsis thaliana. Nature 531:258–262

125. Kintzer AF, Stroud RM (2018) On the structure and mechanism of two-pore channels. Febs J 285:233–243

126. Patel S, Penny CJ, Rahman T (2016) Two-pore channels enter the atomic era: structure of plant TPC revealed. Trends Biochem Sci 41:475–477

127. Grimm C, Butz E, Chen CC, Wahl-Schott C, Biel M (2017) From mucolipidosis type IV to Ebola: TRPML and two-pore channels at the crossroads of endo-lysosomal trafficking and disease. Cell Calcium 67:148–155

128. Patel S, Kilpatrick BS (2018) Two-pore channels and disease. Biochim Biophys Acta Mol Cell Res 1865:1678–1686

129. Gerasimenko JV, Maruyama Y, Yano K, Dolman NJ, Tepikin AV, Petersen OH et al (2003) NAADP mobilizes Ca^{2+} from a thapsigargin-sensitive store in the nuclear envelope by activating ryanodine receptors. J Cell Biol 163:271–282

130. Dammermann W, Guse AH (2005) Functional ryanodine receptor expression is required for NAADP-mediated local Ca^{2+} signaling in T-lymphocytes. J Biol Chem 280:21394–21399

131. Lin-Moshier Y, Walseth TF, Churamani D, Davidson SM, Slama JT, Hooper R et al (2012) Photoaffinity labeling of nicotinic acid adenine dinucleotide phosphate (NAADP) targets in mammalian cells. J Biol Chem 287:2296–2307

132. Walseth TF, Lin-Moshier Y, Jain P, Ruas M, Parrington J, Galione A et al (2012) Photoaffinity labeling of high affinity nicotinic acid adenine dinucleotide phosphate (NAADP)-binding proteins in sea urchin egg. J Biol Chem 287:2308–2315

133. Galione A, Petersen OH (2005) The NAADP receptor: new receptors or new regulation? Mol Interv 5:73–79

134. Guse AH, Diercks BP (2018) Integration of nicotinic acid adenine dinucleotide phosphate (NAADP)-dependent calcium signalling. J Physiol 596:2735–2743

135. Galione A (2011) NAADP receptors. Cold Spring Harb Perspect Biol 3:a004036

136. Morgan AJ, Davis LC, Wagner SK, Lewis AM, Parrington J, Churchill GC et al (2013) Bidirectional Ca^{2+} signaling occurs between the endoplasmic reticulum and acidic organelles. J Cell Biol 200:789–805

137. Kilpatrick BS, Eden ER, Schapira AH, Futter CE, Patel S (2013) Direct mobilisation of lysosomal Ca^{2+} triggers complex Ca^{2+} signals. J Cell Sci 126:60–66

138. Kilpatrick BS, Eden ER, Hockey LN, Yates E, Futter CE, Patel S (2017) An endosomal NAADP-sensitive two-pore Ca^{2+} channel regulates ER-endosome membrane contact sites to control growth factor signaling. Cell Rep 18:1636–1645

139. Churchill GC, O'Neill JS, Masgrau R, Patel S, Thomas JM, Genazzani AA et al (2003) Sperm deliver a new second messenger: NAADP. Curr Biol 13:125–128

140. Moccia F, Lim D, Kyozuka K, Santella L (2004) NAADP triggers the fertilization potential in starfish oocytes. Cell Calcium 36:515–524

141. Brailoiu GC, Brailoiu E, Parkesh R, Galione A, Churchill GC, Patel S et al (2009) NAADP-mediated channel 'chatter' in neurons of the rat medulla oblongata. Biochem J 419:91–97.

142. Hockey LN, Kilpatrick BS, Eden ER, Lin-Moshier Y, Brailoiu GC, Brailoiu E et al (2015) Dysregulation of lysosomal morphology by pathogenic LRRK2 is corrected by TPC2 inhibition. J Cell Sci 128:232–238

143. Lin-Moshier Y, Keebler MV, Hooper R, Boulware MJ, Liu X, Churamani D et al (2014) The Two-Pore Channel (TPC) interactome unmasks isoform-specific roles for TPCs in endolysosomal morphology and cell pigmentation. Proc Natl Acad Sci USA 111:13087–13092

144. Wilding M, Russo GL, Galione A, Marino M, Dale B (1998) ADP-ribose gates the fertilization channel in ascidian oocytes. Am J Physiol 275:C1277–C1283

145. Sumoza-Toledo A, Penner R (2011) TRPM2: a multifunctional ion channel for calcium signalling. J Physiol 589:1515–1525

146. Perraud AL, Fleig A, Dunn CA, Bagley LA, Launay P, Schmitz C et al (2001) ADP-ribose gating of the calcium-permeable LTRPC2 channel revealed by Nudix motif homology. Nature 411:595–599

147. Iordanov I, Mihalyi C, Toth B, Csanady L (2016) The proposed channel-enzyme transient receptor potential melastatin 2 does not possess ADP ribose hydrolase activity. Elife 5: e17600

148. Beck A, Kolisek M, Bagley LA, Fleig A, Penner R (2006) Nicotinic acid adenine dinucleotide phosphate and cyclic ADP-ribose regulate TRPM2 channels in T lymphocytes. FASEB J 20:962–964

149. Toth B, Iordanov I, Csanady L (2015) Ruling out pyridine dinucleotides as true TRPM2 channel activators reveals novel direct agonist ADP-ribose-2′-phosphate. J Gen Physiol 145:419–430

150. Gasser A, Glassmeier G, Fliegert R, Langhorst MF, Meinke S, Hein D et al (2006) Activation of T cell calcium influx by the second messenger ADP-ribose. J Biol Chem 281:2489–2496

151. Lange I, Yamamoto S, Partida-Sanchez S, Mori Y, Fleig A, Penner R (2009) TRPM2 functions as a lysosomal Ca^{2+}-release channel in beta cells. Sci Signal 2:ra23

152. Basile G, Taglialatela-Scafati O, Damonte G, Armirotti A, Bruzzone S, Guida L et al (2005) ADP-ribosyl cyclases generate two unusual adenine homodinucleotides with cytotoxic activity on mammalian cells. Proc Natl Acad Sci USA 102:14509–14514

153. Fliegert R, Bauche A, Wolf Perez AM, Watt JM, Rozewitz MD, Winzer R et al (2017) 2′-Deoxyadenosine 5′-diphosphoribose is an endogenous TRPM2 superagonist. Nat Chem Biol 13:1036–1044

154. Huang Y, Winkler PA, Sun W, Lu W, Du J (2018) Architecture of the TRPM2 channel and its activation mechanism by ADP-ribose and calcium. Nature 562:145–149

155. Sutherland EW (1972) Studies on the mechanism of hormone action. Science 177:401–408

156. Morgan AJ, Galione A (2008) Investigating cADPR and NAADP in intact and broken cell preparations. Methods 46:194–203

157. Churamani D, Carrey EA, Dickinson GD, Patel S (2004) Determination of cellular nicotinic acid-adenine dinucleotide phosphate (NAADP) levels. Biochem J 380:449–454

158. Lewis AM, Masgrau R, Vasudevan SR, Yamasaki M, O'Neill JS, Garnham C et al (2007) Refinement of a radioreceptor binding assay for nicotinic acid adenine dinucleotide phosphate. Anal Biochem 371:26–36

159. Graeff R, Lee HC (2002) A novel cycling assay for cellular cADP-ribose with nanomolar sensitivity. Biochem J 361:379–384

160. Graeff R, Lee HC (2002) A novel cycling assay for nicotinic acid-adenine dinucleotide phosphate with nanomolar sensitivity. Biochem J 367:163–168

161. Graeff RM, Walseth TF, Fryxell K, Branton WD, Lee HC (1994) Enzymatic synthesis and characterizations of cyclic GDP-ribose. A procedure for distinguishing enzymes with ADP-ribosyl cyclase activity. J Biol Chem 269:30260–30267

162. Galione A, White A, Willmott N, Turner M, Potter BV, Watson SP (1993) cGMP mobilizes intracellular Ca^{2+} in sea urchin eggs by stimulating cyclic ADP-ribose synthesis. Nature 365:456–459

163. Graeff RM, Franco L, De Flora A, Lee HC (1998) Cyclic GMP-dependent and -independent effects on the synthesis of the calcium messengers cyclic ADP-ribose and nicotinic acid adenine dinucleotide phosphate. J Biol Chem 273:118–125

164. Kim BJ, Park KH, Yim CY, Takasawa S, Okamoto H, Im MJ et al (2008) Generation of nicotinic acid adenine dinucleotide phosphate and cyclic ADP-ribose by glucagon-like peptide-1 evokes Ca^{2+} signal that is essential for insulin secretion in mouse pancreatic islets. Diabetes 57:868–878

165. Wilson HL, Galione A (1998) Differential regulation of nicotinic acid-adenine dinucleotide phosphate and cADP-ribose production by cAMP and cGMP. Biochem J 331(Pt 3):837–843

166. Yamasaki M, Thomas JM, Churchill GC, Garnham C, Lewis AM, Cancela JM et al (2005) Role of NAADP and cADPR in the induction and maintenance of agonist-evoked Ca^{2+} spiking in mouse pancreatic acinar cells. Curr Biol 15:874–878

167. Gasser A, Bruhn S, Guse AH (2006) Second messenger function of nicotinic acid adenine dinucleotide phosphate revealed by an improved enzymatic cycling assay. J Biol Chem 281:16906–16913

168. Dodd AN, Gardner MJ, Hotta CT, Hubbard KE, Dalchau N, Love J et al (2007) The Arabidopsis circadian clock incorporates a cADPR-based feedback loop. Science 318:1789–1792

169. Kato I, Yamamoto Y, Fujimura M, Noguchi N, Takasawa S, Okamoto H (1999) CD38 disruption impairs glucose-induced increases in cyclic ADP-ribose, $[Ca^{2+}]i$, and insulin secretion. J Biol Chem 274:1869–1872

170. Fukushi Y, Kato I, Takasawa S, Sasaki T, Ong BH, Sato M et al (2001) Identification of cyclic ADP-ribose-dependent mechanisms in pancreatic muscarinic Ca^{2+} signaling using CD38 knockout mice. J Biol Chem 276:649–655

171. Cosker F, Cheviron N, Yamasaki M, Menteyne A, Lund FE, Moutin MJ et al (2010) The ecto-enzyme CD38 is a nicotinic acid adenine dinucleotide phosphate (NAADP) synthase that couples receptor activation to Ca^{2+} mobilization from lysosomes in pancreatic acinar cells. J Biol Chem 285:38251–38259

172. Takahashi J, Kagaya Y, Kato I, Ohta J, Isoyama S, Miura M et al (2003) Deficit of CD38/cyclic ADP-ribose is differentially compensated in hearts by gender. Biochem Biophys Res Commun 312:434–440

173. Deshpande DA, White TA, Guedes AG, Milla C, Walseth TF, Lund FE et al (2005) Altered airway responsiveness in CD38-deficient mice. Am J Respir Cell Mol Biol 32:149–156

174. Mitsui-Saito M, Kato I, Takasawa S, Okamoto H, Yanagisawa T (2003) CD38 gene disruption inhibits the contraction induced by alpha-adrenoceptor stimulation in mouse aorta. J Vet Med Sci 65:1325–1330

175. Partida-Sanchez S, Cockayne DA, Monard S, Jacobson EL, Oppenheimer N, Garvy B et al (2001) Cyclic ADP-ribose production by CD38 regulates intracellular calcium release, extracellular calcium influx and chemotaxis in neutrophils and is required for bacterial clearance in vivo. Nat Med 7:1209–1216

176. Jin D, Liu HX, Hirai H, Torashima T, Nagai T, Lopatina O et al (2007) CD38 is critical for social behaviour by regulating oxytocin secretion. Nature 446:41–45

177. Park KH, Kim BJ, Kang J, Nam TS, Lim JM, Kim HT et al (2011) Ca^{2+} signaling tools acquired from prostasomes are required for progesterone-induced sperm motility. Sci Signal 4:ra31

178. Kim SY, Cho BH, Kim UH (2010) CD38-mediated Ca^{2+} signaling contributes to angiotensin II-induced activation of hepatic stellate cells: attenuation of hepatic fibrosis by CD38 ablation. J Biol Chem 285:576–582

179. Rah SY, Mushtaq M, Nam TS, Kim SH, Kim UH (2010) Generation of cyclic ADP-ribose and nicotinic acid adenine dinucleotide phosphate by CD38 for Ca^{2+} signaling in interleukin-8-treated lymphokine-activated killer cells. J Biol Chem 285:21877–21887

180. Lin WK, Bolton EL, Cortopassi WA, Wang Y, O'Brien F, Maciejewska M et al (2017) Synthesis of the Ca^{2+}-mobilizing messengers NAADP and cADPR by intracellular CD38 enzyme in the mouse heart: role in beta-adrenoceptor signaling. J Biol Chem 292:13243–13257

181. Soares S, Thompson M, White T, Isbell A, Yamasaki M, Prakash Y et al (2007) NAADP as a second messenger: neither CD38 nor base-exchange reaction are necessary for in vivo generation of NAADP in myometrial cells. Am J Physiol Cell Physiol 292:C227–C239
182. Higashida H, Liang M, Yoshihara T, Akther S, Fakhrul A, Stanislav C et al (2017) An immunohistochemical, enzymatic, and behavioral study of CD157/BST-1 as a neuroregulator. BMC Neurosci 18:35
183. De Flora A, Guida L, Franco L, Zocchi E (1997) The CD38/cyclic ADP-ribose system: a topological paradox. Int J Biochem Cell Biol 29:1149–1166
184. Lee HC (2011) Cyclic ADP-ribose and NAADP: fraternal twin messengers for calcium signaling. Sci China Life Sci 54(8):699–711
185. Zhao YJ, Lam CM, Lee HC (2012) The membrane-bound enzyme CD38 exists in two opposing orientations. Sci Signal 5:ra67
186. Liu J, Zhao YJ, Li WH, Hou YN, Li T, Zhao ZY et al (2017) Cytosolic interaction of type III human CD38 with CIB1 modulates cellular cyclic ADP-ribose levels. Proc Natl Acad Sci USA

Chapter 16
Calcium Signaling in the Heart

Derek A. Terrar

Abstract The aim of this chapter is to discuss evidence concerning the many roles of calcium ions, Ca^{2+}, in cell signaling pathways that control heart function. Before considering details of these signaling pathways, the control of contraction in ventricular muscle by Ca^{2+} transients accompanying cardiac action potentials is first summarized, together with a discussion of how myocytes from the atrial and pacemaker regions of the heart diverge from this basic scheme. Cell signaling pathways regulate the size and timing of the Ca^{2+} transients in the different heart regions to influence function. The simplest Ca^{2+} signaling elements involve enzymes that are regulated by cytosolic Ca^{2+}. Particularly important examples to be discussed are those that are stimulated by Ca^{2+}, including Ca^{2+}-calmodulin-dependent kinase (CaMKII), Ca^{2+} stimulated adenylyl cyclases, Ca^{2+} stimulated phosphatase and NO synthases. Another major aspect of Ca^{2+} signaling in the heart concerns actions of the Ca^{2+} mobilizing agents, inositol trisphosphate (IP$_3$), cADP-ribose (cADPR) and nicotinic acid adenine dinucleotide phosphate, (NAADP). Evidence concerning roles of these Ca^{2+} mobilizing agents in different regions of the heart is discussed in detail. The focus of the review will be on short term regulation of Ca^{2+} transients and contractile function, although it is recognized that Ca^{2+} regulation of gene expression has important long term functional consequences which will also be briefly discussed.

Keywords Heart · Cardiac · Calcium · Signaling · Calcium-stimulated enzymes · CaMKII · AC1 · AC8 · IP$_3$ · cADPR · NAADP

D. A. Terrar (✉)
Department of Pharmacology, University of Oxford, Oxford, UK
e-mail: derek.terrar@pharm.ox.ac.uk

© Springer Nature Switzerland AG 2020
M. S. Islam (ed.), *Calcium Signaling*, Advances in Experimental Medicine and Biology 1131, https://doi.org/10.1007/978-3-030-12457-1_16

16.1   Introduction

Readers of this volume will be well aware of the importance of calcium in all plant and animal species arising from the many roles of calcium ions in all biological processes. The statement that calcium is everything (generally attributed to Loewi) might be regarded as inspiring awe or exasperation in relation to the design of experiments to explore the functions of calcium. The focus of this chapter concerns the many roles of calcium ions, Ca^{2+}, in the heart.

The primary function of the heart is to provide a co-ordinated muscle contraction system to pump blood to the lungs and body, enabling the body tissues to receive oxygen and nutrients. The elements of the process that couples electrical activity to contraction of cardiac muscle will be outlined briefly to set the scene for how this relatively simple process might be regulated by many calcium-dependent systems. These calcium-dependent regulatory pathways will make up the bulk of the chapter. They include calcium-dependent enzymes (including the Ca^{2+}-calmodulin-dependent protein kinase, CaMKII, and other enzymes influenced by Ca^{2+}, particularly those that regulate cAMP and cGMP signaling systems), as well as the pathways involving the Ca^{2+} mobilizing agents inositol trisphosphate (IP$_3$), cADP-ribose (cADPR) and nicotinic acid adenine-dinucleotide phosphate (NAADP). In the case of cADPR and NAADP, evidence concerning the actions of these substances will be reviewed separately, but a brief summary will discuss the possibility of synergistic actions of these two substance in the heart, since their actions appear to be complementary and it has recently been suggested that the synthesis of both is catalyzed by the same enzyme, CD38, in the heart [1].

Although the headlines of the process coupling the electric signal of the action potential to contraction (EC coupling) are simple and generally agreed (see reviews [2]; [3]), there are details of the underlying mechanisms that remain open to debate. The outstanding questions are difficult to address experimentally precisely because of the many overlapping roles of calcium. These experimental problems are further compounded by the nature of different compartments inside cardiac cells, not just different membrane delimited compartments (such as the sarcoplasmic reticulum, mitochondria and lysosomes) but also the presence of microdomains in the cytosol where the concentration of Ca^{2+} might differ for functionally important periods of time from that in the bulk cytosol, for example in the subsarcolemmal space beneath the plasma membrane, or spaces between intracellular organelles. Although the poorly understood aspects of EC coupling are functionally important, most of the discussion of the calcium-dependent signaling pathways that form the bulk of this chapter can be followed with just a knowledge of the agreed simple framework.

In addition to the rapid processes controlling contraction, many of the roles of Ca^{2+} in the heart concern long term processes involving turning genes on or off to change protein expression, and are important not only in development but in adaptive processes such as hypertrophy that might be either beneficial or harmful. These roles of Ca^{2+} over periods of days or longer will be covered here only in summary, and such long-term calcium signaling mechanisms are described in detail in: [4]; [5].

Before addressing our knowledge of calcium signaling in the heart, it should be emphasized that although all of the points raised above relate to cardiac muscle in general, there are important differences between cell types in different regions of the heart with particular functions. Much of the discussion will concern ventricular muscle that has generally received most extensive experimental study, but it is clear that cells from the upper atrial chambers of the heart show important variations in the control of their Ca^{2+} signals that are related to their different functions, and an even greater divergence from the ventricular pattern is shown in the pacemaker cells from the SA node. These differences are addressed in particular sections below, after first outlining a basic scheme for ventricular myocytes. The atrioventricular node and conduction systems also show particular specialisations but will not be discussed in detail here (see [6]).

16.2 Simple Scheme of EC Coupling in Ventricular Myocytes

Excitation-contraction (EC) coupling in the heart has been extensively reviewed, for example in [2], [7], as well as in an excellent recent paper from [3], from which Fig. 16.1a is taken. EC coupling links electrical activity in the form of a cardiac action potential, generally lasting hundreds of ms, to a rise in free Ca^{2+} in the cytosol, which in turn activates the contractile machinery of the actin and myosin filaments. Ventricular myocytes show a striated appearance arising from overlap of the thick myosin and thin actin filaments, but also reflecting the division of the length of the myocyte into sarcomeres, which are slightly less than 2 μm long and act as functional units. Each end of a sarcomere in ventricular muscle is bounded by a transverse tubule, which is an invagination of the surface membrane running deep into the myocyte, as shown schematically in Fig. 16.1a (from [3]), together with a confocal image in Fig. 16.1b illustrating an example of this arrangement in a rabbit ventricular myocyte (from [8]). An important membrane system within the myocyte is the endoplasmic reticulum (termed sarcoplasmic reticulum, SR, in myocytes), and this functions as a Ca^{2+} store in which the luminal Ca^{2+} concentration far exceeds that in the cytoplasm. The regions where the SR approaches the transverse tubules are referred to as terminal cisternae, and these two membrane components, together with the restricted space between them, form an important unit for EC coupling, as shown in Figs. 16.1a and 16.1b. This unit of transverse tubule and SR in cardiac muscle is generally referred to as a dyad. The headline of EC coupling is that Ca^{2+} enters the cytosol from the extracellular space to trigger further release of Ca^{2+} ions from the SR by a process termed Ca^{2+}-induced-Ca^{2+}-release (CICR). The Ca^{2+} enters via voltage-gated L-type Ca^{2+} channels located predominantly in the transverse tubules, and the release of Ca^{2+} is via ryanodine receptors (RyR2) in the terminal cisternae of the SR. At body temperature the L-type Ca^{2+} channels are rapidly activated by the depolarization of the action potential, reaching a peak

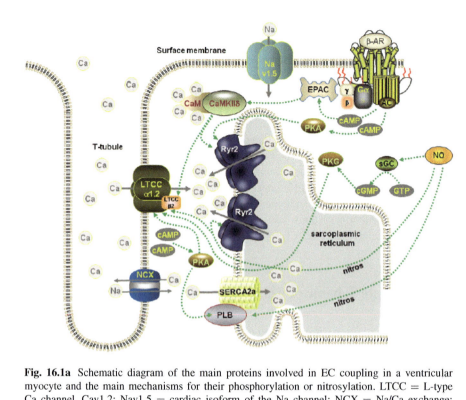

Fig. 16.1a Schematic diagram of the main proteins involved in EC coupling in a ventricular myocyte and the main mechanisms for their phosphorylation or nitrosylation. LTCC = L-type Ca channel, Cav1.2; Nav1.5 = cardiac isoform of the Na channel; NCX = Na/Ca exchange; RyR2 = ryanodine receptor; PLB = phospholamban; AC = adenylyl cyclase; sGC = soluble guanylyl cyclase; EPAC = exchange protein activated by cAMP; PKA = protein kinase A; PKG = protein kinase G. The transverse tubule (T tubule) is shown as an invagination of the surface membrane forming a dyad with the adjacent terminal cisterna of the SR. CICR occurs in the microdomain of the dyad between these two structures. (Reproduced from [3])

of activation in about 3 ms, and this leads to CICR via RyR2 within the dyad followed by a rise in bulk cytosolic Ca^{2+} concentration to activate the myofilaments. The RyR2 is key to this process and is a large protein with multiple regulatory sites (see: [9–12]). The brief rise in Ca^{2+} in the bulk cytosol associated with this process of CICR is termed the Ca^{2+} transient, CaT. Restoration of cytosolic Ca^{2+} concentration back to the resting level (resulting in the decline of the CaT) is largely achieved by pumping of Ca^{2+} back into the SR by a sarcoplasmic-endoplasmic reticulum Ca^{2+} ATP-ase, SERCA, and by extrusion of Ca^{2+} back into the extracellular space, predominantly by the secondarily active and electrogenic Na^+/Ca^{2+} exchange, NCX. There is a minor role in most circumstances of a Ca^{2+} ATP-ase in the surface membrane.

It is worth briefly mentioning mitochondria which occupy about a third of the cellular volume of ventricular myocytes and are essential for supplying ATP to meet the energy demands associated with ion transport and contraction. In the context of

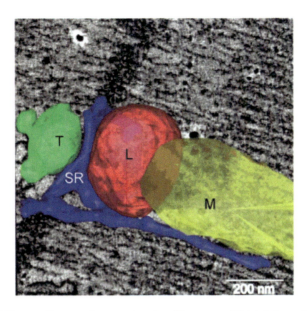

Fig. 16.1b 3D Electron Tomography reconstruction of lysosomes, SR, t-tubules and mitochondria. Representative electron tomography (ET) image of lysosomes near calcium signalling organelles, showing reconstructed organelles in 3D of sarcoplasmic reticulum, mitochondria and t-tubule in rabbit left ventricular tissue. Dual-axis ET and IMOD software were used to image, reconstruct and model lysosomes (L, red), sarcoplasmic reticulum (SR, blue), mitochondria (M, yellow), and t-tubules (T, green) in 3D. Isovolumetric voxel size = 1.206 nm, Z-depth = 275 nm. Scale bar = 200 nm. See original paper for Supplementary Video 1 for 3D animation of tomographic reconstruction of cell. The structural element formed by the t-tubule and SR is important for CICR, while the microdomains between the lysosome and the SR, and between the lysosome and the mitochondria, will be discussed in the section concerning cardiac actions of the Ca^{2+} mobilising agent, NAADP. (Reproduced from [8])

Ca^{2+} signaling, mitochondria may influence changes in cytosolic Ca^{2+} during EC coupling [13], and in turn Ca^{2+} within the mitochondria regulates ATP production [14]. Mitochondria may play a particular role in shaping cytosolic Ca^{2+} signals in atrial myocytes [15].

The potentially explosive process of CICR is subject to local control, and clusters of RyR2 act as release units, giving rise to Ca^{2+} 'sparks' [16]. These are transient localized increases in cytosolic Ca^{2+} arising from Ca^{2+} release from the SR, and it is generally believed the CaT is made of many Ca^{2+} spark-like events; see: [17–19]. These Ca^{2+} sparks are approximately elliptical when viewed in 2D and cover a distance comparable to the sarcomere spacing. The occurrence of sparks is influenced by Ca^{2+} close to RyR2 both in the cytosol and in the lumen of the SR. Recent discussions of the influences of cytosolic and luminal SR Ca^{2+} on RyR2 function can be found in: [20]; [21]; and [19] (and see [22]).

A key aspect of the EC coupling process described above concerns the timing of the CaT relative to the time course of the action potential. It is important to note that all the processes of CICR are very temperature dependent. Temperature is particularly influential for Ca^{2+} entry via L-type Ca^{2+} channels [23] and for Ca^{2+} release from the SR via RyR2 [24]. The transport processes that are responsible for Ca^{2+} extrusion from the cell (principally Na^+/Ca^{2+} exchange) and for Ca^{2+} uptake into the SR (by the sarcoplasmic/endoplasmic reticulum ATP-ase, SERCA) are also very temperature dependent. In the ventricular muscle of most mammalian species, including human (although not adult rat or mouse), the action potential that controls contraction has a prolonged plateau at a potential that is much more depolarized than the resting level of approximately -85 mV. This plateau lasts hundreds of ms and is often close to 0 mV. At body temperature the peak and a substantial part of the decline of the CaT occurs while the plateau is still elevated. This is illustrated in Fig. 16.2 (from [25]; see also [26]).

One aspect of the importance of temperature and the timing of CaT relative to the AP plateau concerns the balance between Ca^{2+} extrusion from the cytosol to the extracellular space (predominantly by NCX) and uptake of Ca^{2+} from the cytosol into the SR. These two Ca^{2+} removal processes are essentially in competition. Extrusion of Ca^{2+} via NCX depends on the cytosolic concentration of Ca^{2+} ions (which changes rapidly in the subsarcolemmal space from the resting level of 100 nM to a peak concentration at least 100 times greater, though still much smaller than the free extracellular Ca^{2+} concentration of 1–2 mM), and on the concentrations of sodium ions on either side of the membrane (which are likely to be relatively stable during a single action potential at over 100 mM on the extracellular side and approximately 5 mM in the cytosol, since the influx of Na^+ is too small on this time scale to have any substantial effect on these relatively large Na^+ concentrations). The other determinant of NCX is the membrane potential since Ca^{2+} extrusion normally depends primarily on the driving force for Na^+ entry (though the Ca^{2+} gradient is also important). The precise reversal potential

Fig. 16.2 Simultaneous recording of AP and CaT measured using optical methods in whole guinea-pig heart at body temperature. The AP waveform is shown as optical signal in which depolarization is represented as a downward deflection, inverting the normal AP appearance. The CaT is shown below the AP. It is clear that the peak and a major component of the decay of CaT occurred during the AP plateau (see text). (Reproduced from [25])

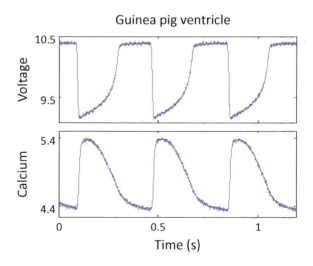

(in other words the potential for zero net current) for NCX is controversial, and will change as the subsarcolemmal Ca^{2+} changes, but it is clear that the extrusion of Ca^{2+} at a plateau potential close to 0 mV will be much less than that at the resting potential of approximately -85 mV because the driving force for Na^+ entry will be much greater at the resting potential. It could be argued that one factor underlying the evolution of action potentials with a prolonged plateau results from the above arguments concerning Ca^{2+} balance: while the action potential remains at the plateau potential the balance between Ca^{2+} extrusion via NCX and Ca^{2+} uptake into the SR by SERCA greatly favours Ca^{2+} uptake. This might be seen as an efficient mechanism so that the large fraction of Ca^{2+} release from the SR (contributing perhaps 75% of the amplitude of the CaT, depending on species) can be taken back into the SR with minimal opposition from NCX at plateau potentials.

Note that at room temperature (which is often used even for experiments on mammalian ventricular myocytes) the time course of the CaT is greatly prolonged relative to the timing of the action potential so that even the peak but also generally all of the decay of the CaT occur at or close to the resting potential, which favours Ca^{2+} extrusion from the cytosol via NCX over uptake of Ca^{2+} back into the SR by SERCA. This situation departs from the physiological and can give rise to difficulties in the interpretation of some experimental observations.

One area where these difficulties are particularly important in the context of Ca^{2+} signaling in the heart concerns the effect of an agent that increases the Ca^{2+} sensitivity of RyR2 to cytosolic Ca^{2+}, as appears to be the case for caffeine. Some have argued that both on theoretical and experimental grounds such an agent will have only a transient effect since although there is an initial increase in CaT amplitude, the increase is not sustained because the amount of Ca^{2+} loaded in the SR declines until there is an exact compensation of SR Ca^{2+} load to restore the original CaT amplitude. When caffeine or similar agent is removed there is an initial decline in CaT amplitude followed by restoration of CaT amplitude as the SR Ca^{2+} load re-adjusts [27]. It is stressed that in the steady state Ca^{2+} influx equals Ca^{2+} efflux during every beat. The experiments to investigate this compensation have generally been done under conditions in which the peak of the CaT occurs after repolarization of the initiating AP or voltage-clamp pulse (mostly at room temperature). We have found that in guinea-pig ventricular myocytes maintained close to body temperature (36 °C), a low concentration of caffeine (250 μM) caused an increase in CaT amplitude, and although there was some decline after the initial effect presumably reflecting partial compensation, the CaT remained elevated for the 30 s of caffeine exposure (unpublished observations of Rakovic & Terrar, presented to the 2006 meeting of Biophysical Society). However, at room temperature (after the heater for the cell superfusion system was turned off) the full compensation described by Eisner et al. was observed. One interpretation of these apparently conflicting observations is that full compensation occurs when there is a clear separation between the timing of the initiating Ca^{2+} influx via L-type Ca^{2+} channels and Ca^{2+} removal from the cytosol by SERCA and NCX (particularly when NCX dominates over SERCA at potentials close to the resting level), as is the case when the timing of the CaT is clearly slower than the action potential at room temperature.

However, at body temperature when the bulk of the CaT occurs at plateau potentials, additional Ca^{2+} released from the SR can be taken back up by SERCA with little or no opposition from NCX, and therefore without contravening the need to maintain equal Ca^{2+} influx and efflux during every beat at the steady state.

Measuring the amount and timing of Ca^{2+} entry via voltage-activated L-type Ca^{2+} channels during an action potential with a long positive plateau is also difficult, if not impossible, to measure with currently available methods. It might be thought that the availability of very selective blockers for L-type Ca^{2+} channels, such as nisoldipine, would allow estimation of this current by difference methods. However, if the action potential waveform is applied under voltage clamp conditions, and nisoldipine is added this will block not only the current through L-type Ca^{2+} channels but also any effects of the Ca^{2+} entering the cell on other processes that could affect other currents (such as Ca^{2+}-dependent changes in K^+, or Ca^{2+}-activated Cl^- currents), but in particular the current associated with extrusion of Ca^{2+} by electrogenic NCX. The best experiments (e.g [28]) have tried to take account of NCX with the use of cytosolic Ca^{2+} buffers such as BAPTA, but the difficulty here is that such buffers may also have effects in addition to suppressing NCX (such as a reduced inactivation of L-type Ca^{2+} channels). Often modeling of ventricular APs includes a significant amount of L-type Ca^{2+} current towards the end of the plateau (e.g [29]), although this does not seem to fit with the substantial inactivation of L-type Ca^{2+} channels towards the end of the plateau, as reported by [30] and [31].

It should be also noted that many aspects of the structure and function of ventricular myocytes are altered in the failing heart, as summarized in [3].

16.3 Atria

The purpose of this section is to outline major differences between atrial and ventricular myocytes that are relevant for understanding of calcium-signaling pathways described below. Recent reviews of the mechanism of EC coupling in atrial myocytes can be found in [32].

The most important feature of atrial myocytes is that transverse tubules are either absent, or present at a much lower density, than is the case in ventricular cells, with the consequence that CICR operates with major differences from the scheme outlined above. Fig. 16.3a-e shows a confocal image of an atrial myocyte stained with an antibody recognizing RyR2. It can be seen that there is a line of RyR2 at the periphery, consistent with a location just beneath the sarcolemma. This is referred to as the junctional SR. There are additional bands of SR with a separation consistent with sarcomere spacing of slightly less than 2 μm and this is referred to as non-junctional SR. In terms of EC coupling, the simplest scheme is that Ca^{2+} enters through L-type Ca^{2+} channels in the surface membrane and activates Ca^{2+} release via RyR in the junctional SR.

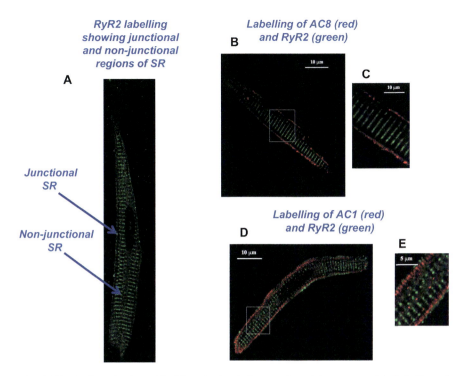

Fig. 16.3(a-e) Organisation of RyR2 and related elements in atrial myocytes. (**a**). Labelling of RyR2 (green) to show junctional and non-junctional SR (observation by T.P. Collins in the Terrar lab). (**b–e**) (Reproduced from [33]). Immunolocalisation of RyRs and Ca^{2+}-stimulated adenylyl cyclases B, RyR2 (green) and AC1 (red). C, enlarged section of panel A identified by white box. D, RyR2 (green) and AC8 (red). (**e**), enlarged section of panel C identified by white box

The way in which the rise in Ca^{2+} at the surface propagates to regions deeper within the atrial myocyte is the subject of intense recent research. It is clear that the spread of Ca^{2+} from periphery to centre is normally slow relative to that in a ventricular myocytes (in which transverse tubules greatly speed the spread of Ca^{2+} release). To gain insight into the underlying mechanisms in atrial myocytes, confocal microscopy has been used to make simultaneous measurements of Ca^{2+} in the cytosol (using rhod-2) and SR (using fluo-5N) [34]. The magnitude of the cytosolic CaT close to junctional SR was higher than that in the vicinity of non-junctional SR. As might be expected the fall in junctional SR Ca^{2+} was well synchronised with the local rise in cytosolic Ca^{2+}, but in the non-junctional SR there was a small rise in SR Ca^{2+} as the Ca^{2+} wavefront progressed deeper into the cell, presumably resulting from Ca^{2+} uptake by SERCA, with the additional luminal Ca^{2+} leading to sensitization of RyR2 to the Ca^{2+} signal in the cytosol. The rise of cytosolic Ca^{2+} at individual release sites of the non-junctional SR thus preceded the depletion of SR Ca^{2+} accompanying Ca^{2+} release in this region. It was proposed that Ca^{2+} release from non-junctional SR is activated by both cytosolic and luminal

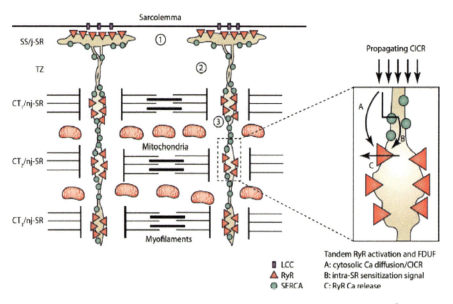

Fig. 16.3f Atrial ECC: tandem RyR activation and FDUF mechanism. AP-induced Ca^{2+} release from junctional-SR (jSR) by L-type Ca^{2+} channel activation [1], followed by propagation through mitochondria-free transition zone (TZ; 2) and activation of centripetal propagating CICR [3] from central (CT) non-junctional (nj-SR) CRUs (CT1 → CT2 → CT3 → ...). Inset: FDUF mechanism. Tandem RyR activation by cytosolic CICR (**a**) and luminal RyR sensitization (**b**) by elevated [Ca]SR (SR Ca^{2+} sensitization signal) resulting in Ca^{2+} release (**c**). SR Ca^{2+} sensitization signal is generated by Ca uptake at the activation front by SERCA. (Reproduced from [32])

Ca^{2+}. This was described as tandem RyR activation and was thought to occur via a novel 'fire-diffuse-uptake-fire' or FDUF mechanism, in which Ca^{2+} uptake by SERCA at the propagation front elevates local SR Ca^{2+}, which in turn causes RyR sensitization, lowering the threshold for activation of CICR by cytosolic Ca^{2+} [34]. This is shown schematically in Fig. 16.3f. Recent discussion of these mechanisms can be found in: [20] and [19].

The above evidence is both intriguing and convincing, but there may also be some functions of sparse transverse tubules in atrial myocytes, as described in an interesting rapid signaling mechanism involving axial tubule junctions [35].

Particular roles of Ca^{2+} stimulated adenylyl cyclases and of CaMKII in atrial myocytes are discussed below.

16.4   SA Node

The mechanisms underlying pacemaker activity in the SA node remain poorly understood and are the subject of extensive discussion. It could be argued that since pacemaker activity initiates the heartbeat that is essential for life, evolution has led to the development of a system of several potentially redundant mechanisms that combine to create a secure system in which if one component should fail another can step forward to ensure that the heart never stops beating. Two major lines of argument in this field in recent years concern whether the primary mechanism for controlling the timing of pacemaker activity in the SA node is a 'membrane' clock or a 'calcium' clock, but it seems most likely that under normal physiological conditions both these timing mechanisms play important roles in an integrated system, as reviewed in [36] and [37]. It will be argued that both membrane and calcium clock components are subject to modification by Ca^{2+} signaling processes.

For the membrane clock the timing is controlled by the ebb and flow of ionic currents, determined predominantly by the sequential opening and closing of different ion channels (see [38] and [39] for recent work in this area).

For the calcium clock the cyclic nature of uptake and release of Ca^{2+} by the SR provides another timing mechanism. Some proponents of the calcium clock give prominence to 'local Ca^{2+} release events' which are similar to Ca^{2+} sparks in SA node cells [36], although others argue that these events are not essential for pacemaker activity [37]. Although there is an excellent correlation between the timing of local release events and the rate of firing of spontaneous action potentials in the SA node, this relationship need not be causal. The timing of the repetitive cyclic activity of uptake and release of Ca^{2+} from the SR might perhaps depend on a mechanism to sense the level of Ca^{2+} in the SR (as appears to be the case in ventricular myocytes under 'overload' conditions [40]).

Some proponents of a membrane clock give prominence to the 'funny' current (I_f, or in neurons I_h) which is activated by hyperpolarization, or repolarization of the action potential under physiological conditions. This current usually exerts a depolarizing influence, and is sometimes referred to as 'the' pacemaker current, but while it often exerts a modulatory influence on heart rate, some authors argue that it is not essential for pacemaker activity. The cardiac channels supporting this current are located primarily in the SA node, and are referred to as HCN since they are activated by hyperpolarization and regulated by cyclic nucleotides. Another very important specific characteristic of the SA node concerns the lack of I_{K1} channels that set the 'resting' potential in atria and ventricular muscle. In the absence of I_{K1} channels, de-activation of voltage-activated K channels (such as I_{Kr} and I_{Ks}) can make an important contribution to pacemaker depolarization. Two other channel types contributing to this slow depolarization provide 'background' current and sustained inward current, both carried at least in part by Na^+. Recent evidence shows that the sustained inward current channels are related to Ca_V 1.3, a voltage-gated Ca^{2+} channel [41]. Ca_V 1.3 channels can in their own right also exert a slow depolarizing influence, in addition to supplementing opening of Ca_V 1.2 channels during the upstroke of the action potential [38]. Other ion channels must also be taken into account, including the I_{KACh} channel that is activated by ACh, but can be open to conduct ions even in the absence of this neurotransmitter. (See [37] for detailed discussion).

In addition to currents through ion channels, ionic currents through the Na^+/Ca^{2+} exchange mechanism (NCX) contribute to both membrane and Ca^{2+}

clock mechanisms (and may provide a link between the two). Some argue that NCX plays a role throughout the cardiac cycle [42]. Interesting new insights concerning the role of NCX during pacemaker activity can be found in [43].

It is interesting to consider whether the wave like spread of Ca^{2+} described in atrial myocytes based on the work of [34] might have a parallel in SA node cells. Labelling of RyR2 in SA node cells shows what appear to be junctional and non-junctional components (see Fig. 1 in [44]), though these possible compartments were not described in this way at the time of publication. This topic deserves further experimental investigation.

Clinical observations and results of genetic manipulation show that heart rate is disturbed when there are abnormalities in components of either the membrane or Ca^{2+} clock mechanisms. It seems likely that pacemaker activity is a complex summation of a variety of the above mechanisms, although it seems clear that Ca^{2+} ions play essential and diverse roles in the initiation and regulation of the pacemaker activity controlling the heartbeat.

16.5   CaMKII

Perhaps the most important Ca^{2+}-stimulated enzyme in cardiac myocytes is the Ca^{2+} and calmodulin dependent protein kinase, CaMKII, which is central to Ca^{2+}-dependent regulation of cardiac function and has been the subject of many reviews (e.g [45–54]. Since this enzyme has been so extensively discussed, only a summary of its contributions will be presented here.

Two targets of CaMKII, L-type Ca^{2+} channels and RyR2 are shown in Fig. 16.1. One function of CaMKII concerns facilitation of I_{CaL} during repetitive stimulation. This arises because during repetitive electrical stimulation, extra Ca^{2+} enters through L-type Ca^{2+} channel as the heart rate increases, and this additional Ca^{2+} stimulates CaMKII leading to phosphorylation of the same channels to enhance their activity [55]; [56]. A second function of CaMKII concerns phosphorylation of RyR2 to modulate its sensitivity to Ca^{2+}, particularly following β-adrenoceptor stimulation [57]. There is also a frequency-dependent acceleration of relaxation, which is thought to arise from an effect of CaMKII to increase SERCA activity, although this was not thought to require phosphorylation of phospholamban [58]. In addition to the link between PKA and CaMKII, a parallel activation pathway involving EPAC. exchange protein activated by cAMP, needs to be taken into account [59].

In addition to effects of CaMKII on L-type Ca^{2+} channels and RyR2, there are also regulatory effects on voltage-gated K^+ and Na^+ channels (e.g [60–63]).

CaMKII makes particularly important contributions to the pacemaker function of the SA node [49]. Additional discussion of CaMKII in atria is included below.

16.6   AC1 and AC8

Work in neurons first established a role for Ca^{2+}-stimulated adenylyl cyclases to control neural activity, predominantly via HCN ion channels that are activated by hyperpolarization and directly regulated by cAMP (e.g [64]; and see [65]; [66]. The first work in heart on this topic showed the presence of AC1 and AC8 in SA node and atrial myocytes but not ventricular myocytes [67]. One function of these Ca^{2+}-stimulated adenylyl cyclases appeared to be regulation of HCN in guinea pig SA node cells. Activation of HCN currents by hyperpolarization was reduced (shifted in the hyperpolarizing direction) following application of membrane permeant BAPTA-AM to load the cells with Ca^{2+} chelator or by the adenylyl cyclase inhibitor, MDL12330A, but the effects of MDL12330 were reduced or abolished in cells pre-treated with BAPTA-AM. Immunohistochemistry showed an intracellular location of AC1 and AC8 in atrial and SA node cells primarily close to the cell surface [67]. Additional work on spontaneous CaT accompanying action potentials in SA node also supports the functional importance of AC1 and AC8 in pacemaker cells, as well as a location of these enzymes at or close to the surface membrane [68]. Interestingly in dogs that have had pacemaker function suppressed by blockade of the atrio-ventricular node, introduction of AC1/HCN2 constructs into left bundle branches provided a highly efficient biological pacing mechanism [69], demonstrating the ability of this Ca^{2+} stimulated enzyme pathway to determine pacemaker function under these conditions.

The functional importance of Ca^{2+}-stimulated adenylyl cyclases has also been demonstrated in atrial myocytes. The presence and activity of these enzymes in the absence of hormonal or neurotransmitter stimulation, probably accounts for the ability of muscarinic acetylcholine antagonists or NO to reduce CaT and L-type Ca^{2+} currents in atrial myocytes, in contrast to ventricular myocytes which lack resting adenylyl cyclase activity, with the major adenylyl cyclases being AC5 and AC6 [70]. Ventricular myocytes do, however, respond to agents that inhibit adenylyl cyclase once this has been activated for example by isoproterenol. In guinea pig atrial myocytes, MDL12,330A reduced the amplitudes of CaT and L-type Ca^{2+} currents, as did application of BAPTA-AM. However, BAPTA-AM was without effect on the amplitude of L-type Ca^{2+} currents in cells loaded with a high concentration of cAMP (100 micro M) from a patch pipette. Immunohistochemistry showed that AC8 appeared to be localized together with RyR2 in the junctional SR just beneath the membrane (although it is difficult to exclude some location in nearby caveolae of the plasmalemma), while AC1 appeared to be in the space between junctional and non-junctional SR [33], and see Fig. 16.3b–e.

An interesting additional conclusion from the above experiments was that the location and function of CaMKII appeared to differ in atrial and ventricular myocytes. In guinea-pig atrial myocytes, the CaMKII appeared to be present with RyR2 in the non-junctional SR, but the junctional SR showed RyR with little or no CaMKII. In other words the CaMKII seemed to be quite distant from L-type Ca^{2+} channels, though close to non-junctional RyR in atrial myocytes. In contrast, ventricular cells showed the usual striated pattern of RyR2 presumably close to L-type Ca^{2+} channels in dyads formed by SR and transverse tubules (which are rare or absent in guinea-pig atrial myocytes), and there was a reasonable amount of colocalization with CaMKII. It therefore appears that in ventricular myocytes, CaMKII is located close to L-type Ca^{2+} channels in transverse tubules as expected from its important role in determining the rate-dependence of this current.

L-type Ca^{2+} currents in guinea pig ventricular myocytes were reduced by the CaMKII kinase inhibitor, KN93, with little effect of the inactive analogue KN92, while BAPTA-AM was without effect in the presence of KN93. In other words all of the effects of BAPTA-AM in ventricular myocytes can be attributed to inhibition of CaMKII rather than adenylyl cyclase. In the case of guinea pig atrial myocytes, KN93 again reduced L-type Ca^{2+} currents, and there was no effect of KN92, but addition of BAPTA-AM in the presence of KN93 caused a further reduction of L-type Ca^{2+} current amplitude. Nevertheless, as mentioned above, BAPTA-AM was without effect in atrial myocytes under conditions in which the cytosolic cAMP concentrations were maintained at a high level from a patch pipette. If the actions of BAPTA-AM in atrial myocytes result entirely from inhibition of Ca^{2+}-stimulated adenylyl cyclase, how might the effects of KN-93 on L-type Ca^{2+} current arise? It was suggested that in atrial myocytes, the primary effect of inhibition of CaMKII is on SR proteins leading to a reduction in the Ca^{2+} released from the SR, as a consequence of effects on uptake and or release, leading to partial inhibition of adenylyl cyclase activity. In support of this view, inhibition of SR function with a combination of ryanodine and thapsigargin reduced L-type Ca^{2+} current, but under these conditions inhibition of CaMKII with KN-93 had no further effect. In summary, CaMKII influences SR proteins and L-type Ca^{2+} channels in ventricular myocytes, while in atrial myocytes CaMKII only affects SR proteins. The effects of Ca^{2+} chelation in atrial myocytes can be entirely accounted for by suppression of Ca^{2+}-stimulated adenylyl cyclase activity, but some activation of these adenylyl cyclases results from Ca^{2+} release from the SR [33].

Possible activation of AC8 or AC1 by Ca^{2+} release from the SR via the action of the IP_3-dependent Ca^{2+} signaling system in atrial myocytes will be considered later.

16.7   Other Calcium-Sensitive Enzymes

An interesting early discussion of Ca^{2+} regulation of enzymes can be found in [71]. This includes an account of how Ca^{2+} stimulation of phosphodiesterase was discovered by Cheung in 1970. In more recent experiments on cardiac muscle, expression of the Ca^{2+} stimulated phosphodiesterase, PDE1A, was found to be fivefold greater in SA node tissues than ventricle. Interestingly nimodipine was thought to act as a selective PDE1 inhibitor, and this compound suppressed total PDE activity by approximately 40% in SA node cell lysates but by only 4% in ventricular cell lysates. In SA node cells that had been permeabilised to avoid effects of nimodipine on L-type Ca^{2+} channels, the PDE1 inhibitor increased the activity of local Ca^{2+} release events. In spontaneously beating HL-1 cells, suppression of PDE1 protein with a selective siRNA increased spontaneous beating frequency. Immunohistochemistry in rabbit SA node cells showed that PDE1A was located on the cytosolic side of the HCN protein in the surface membrane, and separated by approximately 150 nm [72].

It is interesting to note here that Ca^{2+} signaling pathways have been implicated in the cardiac hypertrophy that arises as a maladaptive response to increased workload and excessive cardiac stimulation by hormones including Angiotensin II and β-adrenoceptor agonists; calcineurin and CaMKII are among calcium signaling molecules that have been proposed to play key roles in the development of cardiac hypertrophy [73]. This is a topic that will be revisited in discussion below of the actions of IP_3, cADPR and NAADP. In the case of PDE1C, this enzyme has also been proposed to be involved in cardiac hypertrophy [74]. When hypertrophy was induced by aortic constriction, PDE1C expression was increased. In addition mice that were deficient in PDE1C showed reduced hypertrophy compared to WT in response to aortic constriction. Ventricular myocytes grown in culture for 72 h after isolation from mice that were deficient in PDE1C showed a reduced hypertrophic response to Angiotensin II compared with myocytes from WT mice [74].

NO synthase enzymes have calmodulin containing domains allowing regulation by cytosolic Ca^{2+} [75]. The roles of endothelial NO synthase (eNOS) and neuronal NO synthase (nNOS) have been reviewed by [76] [10] and [77]. Despite the names of these enzymes, which are based on tissues in which they were first discovered, eNOS and nNOS are present in cardiac muscle and eNOS is located predominantly in caveolae, which are invaginations of the surface membrane, while nNOS is associated with SR, at least under normal physiological conditions in healthy heart. Regulation of eNOS by Ca^{2+} released from the SR by IP_3 will be discussed later.

The above summary of Ca^{2+}-regulated enzymes is not an exhaustive list, and it seems likely that there will be many additions to this family.

16.8 IP_3 Signaling

Effects of inositol-(1,4,5)- trisphosphate (IP_3) on CICR in the heart can be traced back at least to observations from Fabiato [78]. However, interest in IP_3-dependent Ca^{2+} signaling as a physiological mechanism for short term effects on CaT associated with action potentials was established by the work of Lipp et al. [79]. These observations showed that the relative importance of IP_3 dependent signaling for short term effects on CaT was much greater in atrial than ventricular muscle, and atrial actions of IP_3 will be discussed first, though in ventricular myocytes there are functionally import long term changes resulting from modification of gene expression following activation of IP_3 receptors in perinuclear membranes [4], and perhaps other effects of IP_3 receptors in the cardiac cell periphery. The functional importance of junctional SR beneath the plasmalemma in atrial myocytes was discussed above, and it was noted that Ca^{2+} entry through L-type Ca^{2+} channels triggers CICR directly in these membranes. Type II IP_3 receptors (IP_3R) are present in atrial myocytes, and interestingly show a precise location in the juntional SR, but not the non-junctional SR that runs deep within the cell with the usual sarcomere spacing. The location of IP_3R in junctional but not non-junctional SR (in contrast to RyR2 which were present in all SR membranes) was shown in both 2D and 3D images from the immunohistochemical observations. To test for functional effects of IP_3, Lipp et al. [79] applied a membrane permeant ester, IP_3-BM, to atrial myocytes leading to cytosolic release of IP_3 following the action of intracellular esterases. Using these methods, cytosolic IP_3 was shown to increase the amplitude of CaT accompanying action potentials in electrically stimulated cells. Application of IP_3-BM also increased Ca^{2+} spark frequency in atrial myocytes. Long exposures to IP_3-BM, presumably achieving higher concentrations of cytosolic IP_3 led to the development of arrhythmogenic Ca^{2+} waves [79].

One cell surface receptor in cat atrial myocytes that appears to be linked to the IP_3-dependent Ca^{2+} signaling pathway is that activated by endothelin-1, which caused an increase in the amplitude of CaT accompanying action potentials initiated by electrical stimulation. Endothelin-1 also caused the appearance of spontaneous arrhythmogenic Ca^{2+} events. In atrial cells that were not electrically stimulated, endothelin-1 increased the frequency and amplitude of Ca^{2+} sparks. All these effects of endothelin-1 were blocked by 2-aminoethoxydiphenylborane, 2-APB, at 2 μM, a concentration which appears to be reasonably selective for IP_3R. In permeabilised atrial myocytes, direct application of IP_3 increased Ca^{2+} spark frequency, as did the IP_3R activator, adenophostin. The effects of IP_3 in permeabilized cells were blocked by the IP_3 antagonists, heparin and 2-APB. These effects are all consistent with the importance of the IP_3 signaling pathway and atrial myocytes and the effectiveness of endothelin-1 in activating this pathway [80].

Ca^{2+} release events via RyRs are generally referred to as Ca^{2+} sparks, while those through IP_3Rs are referred to as Ca^{2+} 'puffs' [81]. These Ca^{2+} puffs are smaller in amplitude and slower to rise and decay than Ca^{2+} sparks, and in the presence of tetracaine to block Ca^{2+} sparks via RyR, both IP_3 and adenophostin were observed to cause the appearance of Ca^{2+} puffs with appropriate amplitude and time characteristics [80].

Further evidence in support of the importance of an IP_3-dependent signaling pathway for effects of endothelin-1 in atria was provided in experiments using transgenic mice, in which atrial myocytes from mice lacking Type II IP_3R failed to show an increase in CaT amplitude following endothelin-1 application, while myocytes from WT mice showed the usual enhancement, and this was blocked by 2-APB [82].

Another surface receptor pathway that appears to be linked to IP_3-dependent Ca^{2+} signaling is the α_1-adrenoceptor. Phenylephrine caused an approximate doubling of L-type Ca^{2+} current in cat atrial myocytes, and these effects were blocked by the α_1-adrenoceptor antagonist, prazosin. The effects of phenylephrine on Ca^{2+} currents were blocked by cytosolic heparin, 2-APB and the IP_3 inhibitor xestospongin, all of which would be expected to reduce IP_3 cellular actions. Interestingly, the effects were also suppressed by L-N^5-1-iminoethylornithine, an inhibitor of NO synthase, by oxadiazolo [4,3-alpha] quinoxaline-1-one (ODQ), an inhibitor of guanylate cyclase, and by H89, a PKA inhibitor. Taken together these observations are consistent with an α-adrenoceptor mediated increase in IP_3 causing Ca^{2+} release from junctional SR where the IP_3 receptors are located which activates NO synthase (eNOS) in nearby caveolae in the plasmalemma, and the resultant NO activates guanylyl cyclase leading in turn to cGMP-dependent inhibition of phosphodiesterase thus causing an increase in cAMP and PKA dependent enhancement of L-type Ca^{2+} channel activity [83]. Further evidence in support of this mechanism was provided by direct measurements of NO production in atrial myocytes in response to phenylephrine, and the observations that NO production was suppressed by 2-APB and by ryanodine [83].

The evidence for the dominance of the NO mediated pathway involving IP_3-dependent Ca^{2+} release is convincing for cat atrial myocytes, although another simpler pathway deserves consideration, perhaps more important in other species. In guinea pig atrial myocytes, the Type II IP_3R again seem to be preferentially located in junctional SR, and phenylephrine caused an increase in the amplitude of CaT accompanying action potential. These effects were suppressed by prazosin and by 2-APB, consistent with α-adrenoceptor activation of the IP_3-dependent pathway to bring about the increase in CaT [84]. The importance of Ca^{2+}-stimulated adenylyl cyclases, AC8 and AC1, in atrial myocytes was reported above, and these enzymes were thought to be located either in junctional SR (for AC8) or in the space between junctional and non-junctional SR (for AC1). IP_3-mediated Ca^{2+} release could therefore directly activate nearby AC8, and perhaps even AC1, without the intervention of the NO-dependent pathway. Recent observations show that photoreleased IP_3 increased CaT and that these effects were reduced by the adenylyl cyclase inhibitor, MDL-12,330, and by 2-APB [85], but the possible involvement of eNOS and cGMP-dependent pathways in these effects has yet to be tested.

Additional detailed reviews of IP_3-dependent Ca^{2+} signaling in atrial myocytes can be found in [86–89]; [32].

It was mentioned above that in ventricular myocytes IP_3 receptors in the perinuclear membrane may be important for local Ca^{2+} signals that may function to regulate gene expression [4].

Although IP_3 receptors are thought to play less of a role in controlling CaT in ventricular myocytes, effects of stimulation of enothelin-1 receptors have been shown to increase the amplitude of CaT accompanying action potentials in rabbit ventricular myocytes, and the effects were suppressed by low concentrations of 2-APB, supporting a role for the IP_3-dependent signaling mechanism [90]. Recent evidence also supports a role for IP_3 and stimulation of CaMKII in the hypertrophic effects of both endothelin and the α-adrenoceptor agonist phenylephrine in rat ventricular myocytes [91].

Effects of phenylephrine on Ca^{2+} spark-like events with increased width, perhaps involving IP_3 actions at Type I IP_3R in the perinuclear region have been described by [92]. Evidence concerning IP_3-dependent nuclear Ca^{2+} signaling in the mammalian heart has been presented by [93]

In view of the major effects of IP_3 in atrial myocyes described above, it is perhaps unsurprising that IP_3 also appears to play a prominent role in SA node. Application of the membrane permeant form of IP_3 (IP_3-BM) was shown to exert a positive chronotropic effect [94] in mouse SA node cells. In these experiments, Type II IP_3R appeared to be located close to the surface membrane, but also deeper in the cells perhaps associated with both junctional and non-junctional SR, although IP_3R were said to be predominantly close to the plasmalemma. In these cells, application of IP_3-BM increased the frequency of Ca^{2+} sparks particularly close to the cell surface. IP_3-BM also increased both the spontaneous frequency and the amplitude of CaT accompanying action potentials in mouse SA node cells. In mice lacking Type II IP_3R, the effect of endothelin-1 on the amplitude and frequency of CaT was prevented as expected, while the effects of isoproterenol acting through β-adrenoceptor pathways appeared to be unchanged [94]. A summary of these effects, with new observations, was presented in a review again showing that IP_3-BM increased Ca^{2+} spark activity, particularly at the periphery of SA node cells, and that endothelin-1 increased the frequency and amplitude of whole cell CaT [95].

Interesting new observations, broadly supporting the above hypotheses have recently been presented comparing effects of 2-APB (a membrane permeant IP_3 receptor antagonist) in SA node myocytes from WT mice and KO mice lacking NCX. SA node myocytes from WT mice showed spontaneous whole cell CaT that were synchronized across the cell by the accompanying action potentials, while those from mice lacking NCX showed spontaneous Ca^{2+} transients that were 'uncoupled' from the surface membrane in that these events were not able to initiate action potentials to accompany the Ca^{2+} transients [43]. These spontaneous Ca^{2+} transients in cells lacking NCX were nevertheless surprisingly regular in their frequency, but very irregular in their spatial characteristics, presumably because of the lack of synchronizing influence of the action potential. 2-APB at a concentration of 2 μM had a surprisingly large effect to reduce the frequency of CaT in SA node from WT mice, and it was argued that the effects might result from blockade of IP_3R, but perhaps also from non-specific effects, for example on L-type Ca^{2+} currents. In the case of the spontaneous activity of SA node cells from mice lacking NCX, 2-APB also greatly reduced the frequency of spontaneous activity, which could not be attributed to effects of 2-APB on surface membrane channels, but was interpreted again to reflect the importance of IP_3-signaling to influence spontaneous rate [43]. In addition, blockade of phospholipase C (which is essential for IP_3 production from PIP2) abolished spontaneous activity in both WT SA node cells and those lacking NCX, while an inactive structural analogue was without effect. Phenylephrine increased the spontaneous frequency of whole cell Ca^{2+} transients in both WT and pacemaker cells lacking NCX, and in both types of cell the effects were greatly suppressed by 2-APB. Interestingly, a high concentration of ryanodine (100 μM) completely suppressed spontaneous CaT activity in WT SA node cells, and under these conditions phenylephrine was unable to restore activity, while there still seemed to be Ca^{2+} in the SR that could be released with a high concentration of caffeine [43].

Overall the observations provide a convincing case for a contribution of IP_3-dependent Ca^{2+} signaling both at rest and after stimulation by endothelin-1 or the alpha-adrenoceptor agonist, phenylephrine.

Before moving on to other Ca^{2+} mobilization agents, two other points should be mentioned. First, IP_3-dependent Ca^{2+} signaling may make an increased contribution under conditions of heart failure, and this has been shown to be functionally important in atrial myocytes [96]. In addition to activation of IP_3-dependent mechanisms by coupling to hormones and neurotransmitter receptors, the possible influence of mechanical stimuli must also be considered, since another interesting activator of IP_3-dependent mechanisms in atrial myocytes is shear stress [97], [98].

16.9 cADPR/NAADP Signaling

The Ca^{2+} mobilizing agents cADP-ribose and NAADP are now widely recognized as playing major roles in a wide variety of plant and animal cells, and influence the function in processes as diverse as egg fertilization, neuronal processing and the closing of stomata on leaves. General reviews of the actions of these substances can be found in [99–103].

These two signaling molecules are included under a single heading here since it is becoming clearer that in the heart they play a co-operative and perhaps synergistic role. Cardiac actions of cADP-ribose (reviewed in [104] and [9]) will be discussed first, before going on to NAADP actions, which have been reviewed more recently [105], and finally the possible parallel combined actions of these two signaling molecules will be considered. The general view presented here is that cADPR actions primarily increase Ca^{2+} release from the SR, while NAADP actions primarily increase Ca^{2+} uptake into the SR, though qualifications of this broad statement will be necessary. An important recent observation is that the CD38 enzyme (an ADP-ribosyl cyclase) in the heart appears to be located at the SR and can catalyse the synthesis of both cADPR and NAADP [1], as discussed in more detail below. A scheme to represent current hypotheses is shown in Fig. 16.6.

16.10 cADP-Ribose

The first observations concerning cADPR in cardiac tissue showed increased Ca^{2+} release from cardiac microsomes, and enhanced opening of Ca^{2+} release channels (RyR2) incorporated into artificial membranes [106]. Some of the these early observed effects on RyR2 channel opening were difficult to reconcile with observations that cADPR competed with ATP for sites on RyR2 [107]. More recently, evidence concerning how cADP might increase Ca^{2+} release from SR via RyR2 in cardiac myocytes has been reviewed in [9], and this question will be considered in more detail below.

16.10.1   Detection of cADPR in Cardiac Muscle and Possible Synthetic Mechanisms

Synthesis of cADPR by cardiac tissue was first shown by [108]. A variety of tissue extracts, including heart, were shown to support cADPR synthesis when incubated with NAD. Endogenous cADPR was detected by Walseth et al. [109], at approximately 1 pmol/mg protein, and the cytosolic concentration was estimated to be about 200 nM. The kinetics of cADPR synthesis from NAD by heart muscle were investigated by [110].

An important observation concerning the regulation of cADPR synthesis showed that this was enhanced by exposure of rat cardiac myocytes to a beta-agonist [111]. This was studied either as [^{3}H]cADPR from [^{3}HJ]NAD, or in a fluorescent assay in which the βNGD, a fluorescent analogue of NAD, was converted to cGDPR. Isoproterenol increased the synthesis in a concentration dependent manner, and the effects were inhibited by the β-adrenoceptor antagonist, propranolol. The ability of isoproterenol to stimulate cADPR synthesis was blocked by cholera toxin, supporting a role for the stimulatory G protein, Gs, in the pathway coupling the β-adrenoceptor to enhancement of the activity of the synthetic enzyme [111].

Using the NGD assay, synthesis of cADPR by guinea-pig cardiac ventricular membranes was shown to be enhanced following application of the catalytic subunit of PKA. In whole guinea pig hearts, exposure to isoprenaline increased cADPR levels, and these effects were blocked by both propranolol and H89, consistent with involvement of PKA in these effects [104]. More recent observations supporting an increase in cADPR levels following exposure to isoproterenol is provided by [113].

Another signalling pathway that leads to increased cADPR synthesis is that linked to Angiotensin II receptors [114].

16.10.2   Role of CD38 in Cardiac cADPR Synthesis

Several ADP-ribosyl cyclase enzymes have been identified that can synthesise cADPR from betaNAD [115].

Gul et al. [116], [117] have presented evidence for such a synthetic enzyme in the heart with activity increased by Angiotensin II, but this enzyme was not thought to be CD38 since synthetic activity was still observed in cardiac muscle from $CD38^{-/-}$ mice lacking the ability to express CD38.

However, CD38 in endosomes was thought to be responsible for an increase in cADPR synthesis following beta-adrenoceptor stimulation by isoproterenol, since this increase was no longer observed in cardiac tissue from $CD38^{-/-}$ mice [118]. As expected from information above concerning PKA, the effects of isoproterenol on cADPR synthesis were blocked by H89 [118].

In our experiments on mixed membrane preparations from mouse heart, a resting synthesis of cADPR was present in WT but not $CD38^{-/-}$ mice [1], consistent with

CD38 as the major synthesizing enzyme under these conditions. A sheep heart SR preparation that is commonly used for the study of cardiac SR proteins also supported cADPR synthesis [1].

SAN4825 is a drug developed to inhibit cardiac ADP-ribosyl cyclases, and was shown to suppress the ability of a rat SR membrane preparation to synthesize fluorescent cGDPR from NGD [119]. SAN4825 also inhibited cGDPR synthesis in the mixed membrane preparation from mouse heart in which CD38 was thought to be the major synthetic enzyme [1].

16.10.3 cADPR Actions on Ca^{2+} Transients Associated with Contraction

The first evidence that cADPR exerts a functional effect in intact cardiac myocytes came from the use of the 8-amino-cADPR, which had been shown to suppress cADPR-induced Ca^{2+} release through RyR in sea urchin egg preparations [120]. Ca^{2+} transients (CaT) accompanying action potentials were recorded in guinea pig ventricular myocytes. These were investigated both using the fluorescent probe, fura-2 and by a less direct method involving measurement of 'Ca^{2+}-activated' currents predominantly carried by NCX. Both methods of measurement showed a reduction in the amplitude of CaT following cytosolic application of 8-amino-cADPR. The less direct method based on measurements 'Ca^{2+}-activated' currents was thought to provide a more accurate indication of the timing of the CaT since it avoids the inevitable buffering caused by fluorescent probes (which increases as the affinity of the probe for Ca^{2+} increases). Myocyte contractions accompanying action potentials were recorded in the same myocytes. All experiments were carried out close to body temperature (36 °C). In these experiments, 8-amino-cADPR caused an approximately 40% reduction of the amplitude of both CaT and contraction [121]. Under voltage-clamp conditions, 8-amino-cADPR did not cause any reduction of L-type Ca^{2+} currents. In additional experiments using fura-2, 8-amino-cADPR was again found to reduce the amplitude of CaT. When SR function was inhibited with ryanodine, the CaT constructed from 'Ca^{2+}-activated' currents was greatly reduced (particularly the early peak that occurred approximately 50 ms after the upstroke of the action potential that appears to be associated with CICR), as was the accompanying contraction, and under these conditions 8-amino-cADPR did not cause any further reductions in CaT or contraction [121]. The observations were thought to be consistent with a role for cADPR to promote Ca^{2+} release from the SR, as proposed for sea urchin egg and many other tissues, in response to Ca^{2+} entry via L-type Ca^{2+} currents during the action potential.

In the above experiments, the amount of Ca^{2+} stored in the SR under these conditions was estimated from the size of the contraction associated with a rapid application of a high concentration of caffeine. Under the conditions in which 8-amino-cADPR reduced the amplitudes of CaT and contractions, in the same cells

8-amino-cADPR did not reduce the amount of Ca^{2+} stored in the SR, and indeed there appeared to be a small but significant increase in SR Ca^{2+} load under these conditions [121].

The first experiments to detect an effect cADPR applied from a patch pipette in intact myocytes showed an increase in the amplitude of contractions associated with action potentials (1 Hz), and the increase in contraction showed a concentration-dependence with an EC_{50} of 2–3 μM. CaT amplitude measured with fura-2 was also increased by patch-applied cADPR. These enhancing effects of exogenous cADPR were blocked by 8-amino-cADPR. Pretreatment of these guinea pig ventricular myocytes with ryanodine and thapsigargin to suppress SR function also prevented the effects of exogenous cADPR [122].

Another convincing demonstration of cADPR actions to increase the amplitude of CaT in intact guinea-pig ventricular myocytes was provided by experiments in which cADPR was photoreleased from a caged compound. In these experiments CaT was measured with fluo-3 (a probe with lower Ca^{2+} affinity than fura-2). The cells were stimulated to fire action potentials once every 2 s, and following a single photorelease of cADPR there was a progressive increase in the amplitudes of CaT over a period of about 15 s. Accompanying the change in amplitude of CaT, there was a change in its time course, so that the peak appeared slightly earlier in the action potential [123]. Further observations supported an increase in the amplitude of CaT and contractions accompanying action potentials following cADPR applied either by photorelease or directly from a patch pipette, and in both cases there was no change in SR Ca^{2+} load [124]. As above SR load was judged from the response to rapid application of a high concentration of caffeine, but in these experiments two different methods were used to assess SR Ca^{2+}. In one SR Ca^{2+} was estimated from the amplitude of the cytosolic Ca^{2+} change measured using fluo4 following rapid caffeine exposure, and in the other the amount of SR Ca^{2+} was assessed from the time integral of caffeine-induced current under voltage-clamp conditions (predominantly NCX while extruding the Ca^{2+} released from the SR). With both assessment methods, SR Ca^{2+} appeared unchanged while the CaT or contraction accompanying action potentials was clearly enhanced by the cADPR application [124]. In another series of experiments, CaT were evoked by repeating voltage-clamp pulses (from -40 to 0 mV for 200 ms; 0.33 Hz) rather than action potentials, and photoreleased cADPR increased the amplitude of CaT without changing the decay time, and without changing the amplitude of the L-type Ca^{2+} current that evoked the CaT, again consistent with an increase in the gain of CICR [124].

16.10.4   cADPR Actions on the Frequency and Amplitude of Ca^{2+} Sparks

Guinea pig ventricular myocytes are unusual in showing few if any Ca^{2+} sparks, while rat ventricular myocytes regularly show Ca^{2+} spark activity. Before measuring this Ca^{2+} spark activity, photorelease of cADPR in rat myocytes was again shown to increase the amplitude of CaT measured with fluo-3. In rat myocytes at rest (not stimulated to fire action potentials) photorelease of cADPR caused an increase in Ca^{2+} spark frequency by approximately two to threefold. Prior application of 8-amino-cADPR prevented the effect of photoreleased cADPR on Ca^{2+} spark frequency. It is important to note that spark characteristics, in particular spark amplitude were not changed, as would have been the case if SR Ca^{2+} load had been increased [123]. Further observations in support of this hypothesis, with effects of cADPR to increase CaT amplitude, and at high doses to provoke Ca^{2+} waves have been made in rat ventricular myocytes [125].

Taken together the observations so far described above are consistent with the hypothesis that cytosolic cADPR in some way increases the gain of CICR via RyR2, though this should not be taken to imply that a direct effect of cADPR on RyR2 is necessarily involved.

It is important to recognize that not all reports show observations consistent with the above hypothesis, and some have failed to detect effects of cADPR and 8-amino-cADPR on CaT in rat ventricular myocytes [126]. However, these negative observations were made at room temperature, and it was later shown that the effects of cADPR, and of the antagonists, 8-amino-cADPR and 8-Br-cADPR, were all temperature dependent, showing clear agonist or antagonist effects at 36 °C, but not having detectable effects on CaT when the cell superfusion heater was switched off [122].

The above observations on Ca^{2+} sparks in rat ventricular myocytes were from intact cells. Different conclusions were made on the basis of later observations on saponin permeabilised rat ventricular myocytes, and on rat heart microsomes, all at room temperature, and it has been claimed that the primary target for cADPR in the heart is SERCA-dependent Ca^{2+} uptake into the SR [127]. Under the conditions of these experiments in permeabilised myocytes, unlike all the observations reported above, there was a change in the amount of Ca^{2+} loaded into the SR following exposure to cADPR, tested by rapid application of caffeine. The measurements of SR content were made 5 min after application of cADPR in permeabilized myocytes. The observations also showed that when rat heart microsomes had been exposed to cADPR (in the presence of ruthenium red to block Ca^{2+} release via RyR2), the Ca^{2+} uptake by the microsomes was faster in the presence than in the absence of cADPR, consistent with an enhanced activity of SERCA [127]. These observations provoked a re-investigation of cADPR actions, and it was found that in rat saponin permeabilised ventricular myocytes at 36 °C application of 10 μM cADPR caused an approximate doubling in Ca^{2+} spark frequency after 30 s, while the SR Ca^{2+} load tested with rapid caffeine application was not increased at this time [124]. The increase in spark frequency caused by cADPR after 30 s was prevented by prior exposure to the antagonist, 8-amino-cADPR. However, if the exposure to cADPR was prolonged to 10 min, the increase in Ca^{2+} spark frequency was maintained but under these conditions there was indeed an increase in SR Ca^{2+} load as assessed from the response to rapid caffeine application [124].

Furthermore, while the increase in Ca^{2+} spark frequency at 30s occurred without a change in spark amplitude or decay time (consistent with no change in SR load or uptake by SERCA), in contrast after 10 min exposure to cADPR under these conditions Ca^{2+} spark amplitude was increased and Ca^{2+} spark decay time was quickened [124]. These changes in spark characteristics after 10 min exposure to cADPR are consistent with an increase in SERCA activity, since it has been shown that even at room temperature SERCA contributes significantly to spark decay, and isoproterenol was observed to shorten Ca^{2+} spark decay time [128]. In these experiments the protocol was designed to maintain a comparable SR Ca^{2+} load in the presence and absence of isoproterenol, and therefore to avoid an increase in Ca^{2+} spark amplitude that might otherwise have occurred.

The main conclusion from the above experiments is that there is a clear effect of cADPR on Ca^{2+} release from the SR at short exposure times that does not depend on an increased loading of the SR with Ca^{2+}. The effect of cADPR on Ca^{2+} uptake by SERCA at very long exposure times has been observed in particular conditions (permeabilization of the cell membrane with saponin, or isolated microsomal preparations). It is not yet clear how the slow increase in SERCA activity arises, and whether it might be a secondary consequence of the cytosolic Ca^{2+} changes associated with the rapid effects of cADPR. The mechanism involving an increase in SERCA activity may represent an additional slow pathway for effects of cADPR on CaT (perhaps involving activation of CaMKII by the additional Ca^{2+} associated with a prolonged increase in Ca^{2+} spark frequency), and it remains for future study whether this pathway also operates under physiological conditions in intact myocytes. However, it is clear that cADPR has important effects apparently to increase the gain of CICR under conditions in which there is little or no change in SR Ca^{2+} load, and that these effects are associated with an increase in the CaT accompanying action potentials.

16.10.5 Possible Mechanisms for the Increase in Release of Ca^{2+} from the SR Mediated by cADPR

Although the observations summarized above provide strong support for an effect of an increase in cytosolic cADPR (whether from patch applied or photoreleased cADPR) to increase CICR without a change in SR Ca^{2+} load, the underlying mechanism of action of cADPR remains unclear. It should first be acknowledged that these methods of application of cADPR to the bulk cytosol are unlikely to simulate the precise consequences of synthesis of cADPR, perhaps by CD38 in a signaling microdomain close to ryanodine receptors. It has already been mentioned that a direct effect of cADPR on RyR seems difficult to reconcile with observations on RyR2 activity under well controlled conditions [9]. It remains possible that a binding protein for cADPR (or another intermediary step) may be interposed between synthesis of cADPR and the consequential change in CICR in intact myocytes. It was mentioned in the EC coupling section that some have argued that an agent acting to increase the amount of Ca^{2+} released via RyRs cannot have a lasting effect, since the SR Ca^{2+} load will adjust to compensate [27]. It was, however, argued above that at body temperature when a substantial fraction of the CaT occurs while the AP is at elevated plateau potentials, additional Ca^{2+} can be released from the SR and taken back up with little or no opposition from NCX and therefore without contravening the need for balance between Ca^{2+} influx and efflux for a single beat. Whatever the merits of the theoretical arguments, repeated observations described in this section show increases in CaT amplitude which follow cADPR application to the cytosol by different methods in both atrial and ventricular myocytes without a change in the measured SR Ca^{2+} load. Additional information relating to possible mechanisms is listed below.

16.10.6 Possible Influence of Calmodulin (and CaMKII)

In sea urchin egg homogenates, it has been shown that calmodulin is a necessary requirement for cADPR mediated Ca^{2+} release [129–131], and it appears that CaMKII is required for cADPR effects in pancreatic islets [132]. More recently calmodulin dissociation from RyR has been shown to underlie the specific 'densensitization' of the cADPR-induced Ca^{2+} release mechanism in sea urchin homogenates, so that binding of calmodulin to RyR appears to be essential for the cADPR mediated Ca^{2+} release [133]. In guinea pig ventricular myocytes, effects of patch applied cADPR on the amplitude of CaT accompanying action potentials were prevented by the calmodulin antagonists, calmidazolium and W7, while calmidazolium also prevented the effect of photoreleased cADPR on the amplitude of CaT. Effects of photoreleased cADPR on CaT were also prevented by the CaMKII antagonist KN93 [104]. The underlying mechanisms deserve further study.

16.10.7 Possible Influence of FKBP

FKB12.6 is a protein that associates with RyR2, although its functions are controversial [9]. Evidence for an involvement of this protein in the actions of cADPR has been provided by experiments in which cADPR increased Ca^{2+} spark frequency in permeabilized myocytes from WT mice, but not those lacking FKB12.6 [134]. The possible significance of these observations is discussed in more detail in [9].

16.10.8 Contributions of cADPR Signaling to Cardiac Arrhythmia and Hypertrophy

The ability of beta-adrenoceptor agonists to increase cADPR synthesis in heart muscle has been described in detail above. High concentrations of beta-adrenoceptor agonist are known to provoke arrhythmias, and it is interesting to investigate whether cADPR-dependent mechanisms might contribute to the arrhythmogenic effects in addition to the well known involvement of other proteins including L-type Ca^{2+} channels, phospholamban in conjunction with SERCA, and perhaps RyR2 (as reviewed in [7]). High concentrations of isoproterenol (50 nM) gave rise to spontaneous action potentials and Ca^{2+} waves associated with SR Ca^{2+} overload in guinea pig ventricular myocytes, and these spontaneous events were suppressed by application of the cADPR antagonist, 8-amino-cADPR [135].

Gul et al. have shown that the hypertrophy that is normally seen following chronic exposure to isoproterenol is reduced in mice lacking CD38 as compared to WT, and propose that this results at least in part from reduction in cADPR synthesis [118]. It has been shown that the acute arrhythmogenic response of isolated hearts to high concentrations of isoproterenol is reduced in hearts from $CD38^{-/-}$ mice as compared to WT, and that the ADP-ribosyl cyclase inhibitor, SAN4825 also reduced isoproterenol-induced arrhythmias in WT hearts, consistent with a contribution of the cADPR signaling pathway to arrhythmogenic mechanisms [1], although the possible combined actions of cADPR and NAADP acting in concert on different aspects of the control of CaT amplitude in heart will be discussed later.

It was mentioned above that Angiotensin II can increase cADPR formation in neonatal rat cardiac myocytes [114], and chronic stimulation of this pathway is thought to lead to cardiac hypertrophy since the cADPR antagonist, 8-Br-cADPR, suppressed cardiac hypertrophy caused by AngII [117]. More recently it has been shown that ROS and AngII mechanisms co-operate to bring about cardiac hypertrophy [136].

16.11 NAADP

The first observations of actions of NAADP in heart showed the presence of binding sites for NAADP and the ability of NAADP to release Ca^{2+} from rabbit heart microsomes [137].

16.11.1 Detection of NAADP in Cardiac Muscle and Possible Synthetic Mechanisms

The ability of rat heart tissue to synthesise NAADP was shown by [138], and endogenous levels of NAADP in mammalian hearts were reported by [139]. It is important to note that the levels of NAADP that are necessary for activation of this signaling system in a variety of cell types seem to be in the submicromolar range (generally less than 100 nM), and indeed higher levels of NAADP can cause a self-inactivation or desensitization of the pathway [103]. The physiological levels are therefore likely to be very low and consequently difficult to measure. Synthetic mechanisms for NAADP remain controversial. Evidence was presented in the preceding section that ADP-ribosyl cyclases can catalyse cADPR formation, and that CD38 is the most likely candidate enzyme in the heart [1]. At least under in vitro conditions, CD38 can also catalyse the formation of NAADP from NADP and nicotinic acid by a base exchange mechanism in a variety of cell systems [103]. Support for the action of CD38 to underlie NAADP synthesis in mouse heart is provided by the observation that a cardiac membrane fraction from WT mice was able to catalyse the formation of NAADP from NADP and nicotinic acid, but this synthesis was not observed with cardiac membrane from $CD38^{-/-}$ mice. The base exchange reaction for formation of NAADP occurs most readily at acidic pH [140] [103], but NAADP synthesis from NADP and nicotinic acid was observed at a cytosolic pH of 7.2 in both in mouse cardiac mixed membranes and in a sheep cardiac SR preparation [1] that has been used to study SR proteins including RyR2 [141]. When mouse cardiac myocytes were permeabilised with saponin to allow access of NADP and nicotinic acid to the cytoplasm, NAADP synthesis was detected, but little or no synthesis was detected in the absence of saponin. This substance primarily permeabilises the cholesterol-containing membranes such as the plasmalemma, while Triton-X-100 permeabilises all cell membranes including the SR, and might potentially expose additional enzymes (including those with lumen-facing active sites in various organelles). However, there was little or no difference between saponin and Triton-X-100, applied alone or in combination, to permit NAADP synthesis, consistent with NAADP synthesis by an enzyme with an active site facing the cytosol [1].

It should be noted that another paper takes a different view, agreeing that CD38 promotes cADPR synthesis in the heart, but proposing that an additional enzyme is responsible for NAADP synthesis [118]. In these experiments CaTs accompanying action potentials were not recorded (see [1] for more detailed discussion).

An enzyme inhibitor, SAN4825, was developed to suppress cADPR synthesis [119], and in our experiments mentioned above, SAN4825 was found to inhibit cADPR production by cardiac membrane preparations. In addition, NAADP production by cardiac membrane preparations was also inhibited by SAN4825, and in silico modeling studies supported binding of SAN4825 at a single site on CD38 to inhibit both cADPR and NAADP synthesis [1].

Evidence was presented above that the amounts of cADPR synthesised in cardiac preparations, and the functional effects of cADPR in intact myocytes and whole hearts were increased following beta-adrenoceptor stimulation. Exposure of guinea-pig hearts to beta-adrenoceptor stimulation also increased NAADP synthesis, as detected by two different methods [142], [113]. This seems unsurprising if the same enzyme, CD38, is responsible for synthesis of cADPR and NAADP, although more work is needed to show how modification of CD38, perhaps by PKA dependent mechanisms, increases synthesis of cADPR and NAADP by two very different chemical reactions.

16.11.2 Location of CD38 in Cardiac Myocytes

The location of CD38 in heart is of particular interest in view of its possible functions supporting synthesis of both cADPR and NAADP. Observations in mouse WT cardiac myocytes showed that an antibody against CD38 appeared to be predominantly localized at the SR (although there was also a small subfraction associated with the plasmalemma). Labelling showed a banded appearance corresponding to the striations of the SR (separation slightly less than 2 μm). However, cardiac myocytes lacking CD38 showed minimal labeling without striations, supporting the selectivity of the antibody for CD38 in the mouse WT myocytes. Experiments were also performed on rabbit ventricular and atrial myocytes, and again the pattern and spacing of immunolabeling of CD38 was consistent with a location of this enzyme at or close to the SR. When antibodies to RyR2 and CD38 were applied to the same rabbit ventricular myocytes, there was a close correspondence between the two labeling patterns, again supporting an SR location for CD38 [1]. The location of CD38 in cardiac myocytes is illustrated in Fig. 16.4.

16.11.3 Orientation of CD38

CD38 is generally thought of as a type II transmembrane protein with its catalytic domain located on the outside of the cell, or pointing away from the cytosol in intracellular organelles, but recent evidence shows that this is not always the case [143–145] and the possible cytosolic synthesis of cADPR and NAADP by CD38 associated with SR in cardiac myocytes is discussed in detail in [1]. Future research will be necessary to provide clear information on the orientation of CD38 in cardiac membranes, but the suggestion that NAADP production can occur in the cytosol is consistent with the observations mentioned above that NAADP production occurs in media at pH7.2 after permeabilisation of cardiac myocytes with saponin, and is similar when Triton X100 is used as the permeabilisation agent.

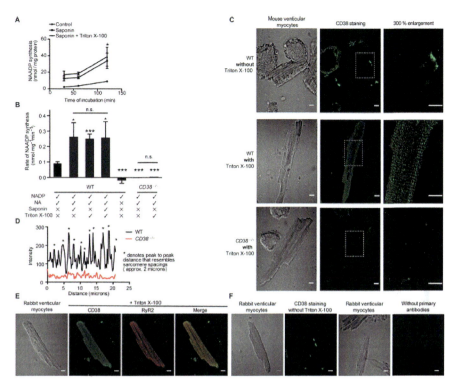

Fig. 16.4 Permeabilization of cardiac myocytes with Triton X-100 and/or saponin enhanced NAADP production and permitted immunolabeling of CD38. (**a**), the rate of NAADP synthesis was higher after permeabilization of the cell membranes with saponin alone, and permeabilization with Triton X-100 in addition to saponin did not further increase the rate of NAADP synthesis (n = 3 in each group). (**b**), this point is further illustrated in the bar graph, which also shows that the ability to synthesize NAADP was lost in myocytes from CD38$^{-/-}$ mice (n = 4). Omission of NA also abolished synthesis of NAADP in intact cells (n = 5). (**c**), immunolabeling of CD38 using rabbit anti-human CD38 antibody without Triton X-100 showed little staining, although there were surface patches with higher fluorescence intensity (top panels). Following membrane permeabilization with Triton X-100 to allow access of the antibody to the cell interior, there was clear staining with a striated pattern in WT (center panels) but not in permeabilized CD38$^{-/-}$ cardiac myocytes (bottom panels). (**d**), the representative intensity-distance plot (bottom panel) shows that, in permeabilized cells, the staining observed in the WT myocyte had a much higher intensity than in the CD38$^{-/-}$ myocyte and showed multiple peaks with a repeating distance interval that resembled the sarcomere spacing. (**e**), similar observations in rabbit ventricular myocytes. The fluorescent images of myocytes permeabilized with Triton X-100 showed clear labeling with CD38 antibody. There was a striated pattern with a similar spacing as that shown by immunolabeling of RyR2. (**f**), no labeling with a striated pattern was observed when Triton X-100 or primary antibodies were omitted. The images show representative staining of the major observation (> 75%) in each group (n > 20). Scale bars 10 mm, n = number of cells. Data are expressed as mean % S.E. *, p < 0.05; ***, p < 0.001; n.s., not significant. (Reproduced from [1])

16.11.4 Importance of Lysosomes for Cardiac Actions of NAADP

In a variety of cell types the actions of NAADP appear to depend on the presence and function of acidic organelles, such as lysosomes or similar structures [146], [103]. An important experimental tool to investigate these pathways is bafilomycin A which blocks a V-type H^+ ATP-ase that sets up the proton gradient that is necessary for the functions of acidic organelles, including uptake of Ca^{2+} by a Ca^{2+}/H^+ exchange mechanism. Ca^{2+} release from acidic organelles has been shown to be a key component of the actions of NAADP in many cell types, and bafilomycin A is well established as useful experimental tool to suppress these NAADP effects [103]. The term lysosome will be used here to describe the acidic organelle involved in NAADP actions in the heart, although it is recognized that it is difficult to distinguish between lysosomes and other related organelles such as endosomes [147]. In our experiments described below on the functional effects of NAADP on Ca^{2+} transients and contraction of cardiac myocytes, bafilomycin A consistently suppressed NAADP actions, supporting the key role of acidic lysosomes for NAADP actions in the heart.

Given the proposed location of CD38 at or close to SR membranes and the proposed role of CD38 for NAADP synthesis, it is of special interest that lysosomes in the heart are localized close to the SR. This was shown by immunohistochemistry with conventional microscopy, labeling microsomes with lysotracker, and with EM methods [8]; see Figs. 16.1b and 16.5. Conventional 2D EM studies showed a close association of approximately 20 nm between lysosomes and SR. In the case of studies using 3D electron tomography the separation between SR and lysosomes appeared to be 3.3 nm. These separations are clearly consistent with microdomains between lysosomes and SR that may be important for proposed actions of NAADP to release Ca^{2+} from lysosomes, with actions of the released Ca^{2+} in the interposed microdomains to influence SR function. It is interesting to note that in addition to the positioning of lysosomes close to the SR it was found that lysosomes also form close associations with mitochondria. In this case 3D electron tomography showed a median separation of 6.2 nm [8]. This additional possible signaling microdomain will be important for some of the functional effects of NAADP that involve mitochondria and these are described in more detail below.

16.11.5 Importance of TPC2 and TPC1 Proteins for Cardiac Actions of NAADP

Generally actions of NAADP in heart are consistent with reported actions of this Ca^{2+} signaling molecule in many other tissues. The target of NAADP action, whether directly or after the intervention of an associated binding protein, is thought to be two-pore cation channels (TPCs) in a variety of tissues [148], [149]. There are

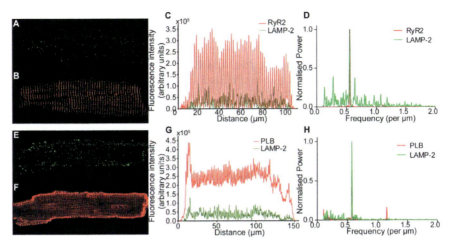

Fig. 16.5 Lysosomes are localised in a periodic pattern with sarcomeric distribution, similar to RyR and PLB, in fixed, isolated rabbit ventricular myocytes. (Top line) Representative example of LAMP-2 and RyR2 co-labelling experiment. Immunolabelling of LAMP-2 (**a**) and RyR2 (**b**) co-stained in the same, fixed, myocyte. (**c**) Intensity plot illustrating total pixel intensity for each column of pixels, for the two stains. Results of Fourier analyses (**d**) indicate a dominant frequency of 0.56 striations μm $-$ 1, or a signal peak every 1.79 μm, for RyR2 and LAMP-2. (Middle line) Representative example of LAMP-2 and PLB co-labelling experiment. Confocal images of LAMP-2 (**e**) and PLB (**f**) with merged column intensity plot of both stains (**g**). Fourier analysis (**h**) of each signal indicate a dominant frequency of 0.57 signal peaks μm $-$ 1 for both PLB and LAMP-2, equivalent to a peak every 1.74 μm. A correlation plot of dominant frequencies for LAMP-2 against either RyR2 or PLB in co-labelling experiments (not shown; see original paper) revealed strong correlation (Pearson product-moment correlation co-efficient (r) = 0.72, n = 22 cells (n = 18 for RyR2 vs LAMP-2 and n = 4 for PLB vs LAMP-2). (Reproduced from [8])

at least three types of these channels, although TPC1 and TPC2 are thought to be most important in mammalian tissues. There is some controversy over the major cations that flow to support the functional effects of NAADP, and some propose that TPCs are channels activated by PI(3,5)P2 and that the primary cation flowing through the channels is Na$^+$ [150], although at least under some conditions NAADP activates Ca^{2+} flux through lysosomal TPC2 [149], [151]. Our evidence in the heart also supports the hypothesis of NAADP activated Ca^{2+} flux, and experiments comparing NAADP actions in cardiac muscle from WT mice or those lacking TPC2 proteins show that NAADP effects require TPC2. This was true both for isolated ventricular myocytes and for intact hearts [152], and the involvement of TPC2 proteins in the effects of NAADP on calcium transients, contraction and related processes are discussed in more detail below.

A useful chemical tool for the study of NAADP mechanisms is Ned-19, which acts as an antagonist of NAADP [153]. When used in conjunction with genetic methods to suppress the expression of TPC2, Ned-19 would be predicted to have effects in WT mice but to have little or no effect KO mice lacking TPC2, as was found to be the case in [152], for NAADP actions concerning effects on Ca^{2+} transients, hypertrophy and arrhythmias that are described in detail below.

It appears that TPC1 proteins may be involved in a quite separate set of actions of NAADP in heart. When the heart is subject to a restricted blood flow, and then reperfused with oxygenated solution, the cardiac myocytes show substantial damage and sometimes cell death. The damage caused by such exposures to ischaemia and reperfusion is associated with Ca^{2+} oscillations. Cell death associated with ischaemia and subsequent perfusion was reduced by an antagonist of NAADP termed Ned-K (a structural analogue of the Ned-19 described above), supporting a role for NAADP in the underlying mechanisms of these effects [154]. It seems likely that mitochondria are involved in these actions of NAADP, and the observation described above that lysosomes are located close to mitochondria [8] is consistent with a signaling microdomain between these two organelles for actions of Ca^{2+} released from the lysosomes. Interestingly, transgenic mice lacking TPC1 proteins showed reduced injury following ischaemia and reperfusion, consistent with the importance of TPC1 proteins in this particular aspect of NAADP dependent Ca^{2+} signaling [154].

Most of the actions of NAADP described in this chapter concern involvement of cytosolic NAADP on intracellular signaling pathways, but there are some additional reported extracellular effects that appear to be related to the observations concerning ischaemia and reperfusion. Extracellular NAADP can increase intracellular levels of cADPR and NAADP by activating a signaling pathway that is thought to be mediated by a P2Y11 GPCR receptor on the surface membrane that is antagonised by NF157 [155]. Production of cADPR and NAADP in neonatal rat cardiomyocyte cultures was also enhanced by the selective P2Y11 agonist, NF456 [155]. The possible link with the observations above on TPC1 proteins involved in ischaemia-reperfusion injury is that 1 μM extracellular NAADP was observed to have a significant protective effect, and P2Y11 antagonist, NF157 suppressed this protection [156].

16.11.6   NAADP Actions on Ca^{2+} Transients Associated with Contraction

The discussion above alludes to a key signaling role of NAADP in cardiac muscle involving modulation of the Ca^{2+} transients, CaT, and contractions associated with action potentials that are essential for the normal physiology of the heart. Effects of cytosolic NAADP on CaT associated with action potentials were first shown in experiments in which NAADP was photoreleased from a caged compound that was applied to the cytosol of guinea-pig ventricular myocytes via a patch pipette [142]. CaTs were measured in cells in which action potentials were stimulated every 2 s, and a single pulse of UV light to photorelease NAADP caused a progressive increase in CaT amplitude accompanying successive action potentials. The effects on CaT amplitude continued to develop for at least a minute after a single photorelease, and were dependent on concentration of caged compound over the range 0.5–5 μM. It is difficult to be sure of the cytosolic concentration of free NAADP corresponding to these concentrations of caged compound, but this is likely to have been less than 100 nM, perhaps between 10 and 100 nM if the efficiency of release from the caged compound at the above concentrations were approximately 2%.

It is difficult to record similar effects of NAADP applied directly to the cytosol from a patch pipette because it seems likely that there will be overlap between the possibly enhancing effects of low concentrations NAADP and the self-inactivating or desensitizing effects of higher concentrations of this messenger [157], as mentioned at the start of this section on NAADP. However, high concentrations of 0.1 or 1 mM NAADP predominantly cause self-inactivation of the NAADP signaling mechanism, and as expected prior application of NAADP from a patch pipette reduced (0.1 mM) or abolished (1 mM) the effects of photoreleased NAADP on the amplitude of CaT [142].

The effects of photoreleased NAADP on CaT amplitude were also suppressed by prior application of bafilomycin A, consistent with the involvement of lysosomes [142].

There were no detectable effects of photoreleased NAADP on the amplitude of L-type Ca^{2+} currents and it therefore seems likely that the actions of NAADP on CaT amplitude involve the SR, and perhaps the signaling microdomain proposed above between lysosomes and SR. As expected pretreatment of myocytes with thapsigargin and ryanodine (to inhibit both uptake and release of Ca^{2+} by the SR) prevented the effects of photoreleased NAADP on CaT. The effects of photoreleased NAADP were accompanied by an increased amount of Ca^{2+} loaded into the SR (as tested by a rapid caffeine application to release SR Ca^{2+}), and taken together the observations are consistent with an effect of NAADP to release Ca^{2+} from the lysosome leading to an increase in SR Ca^{2+} and therefore the amplitude of the CaT [142]. The possibility of an amplification mechanism so that the amount of extra Ca^{2+} loaded into the SR may be greater than that released from the lysosome will be described later.

Another method for application of NAADP is to apply a membrane permeant ester (NAADP-AM) to the extracellular solution, allowing intracellular accumulation of NAADP-AM followed by release of free NAADP resulting from the action of cytosolic esterases [158]. Applying this approach to guinea pig cardiac ventricular cells, rapid application of 60 nM NAADP-AM (to favour the enhancing effects of low concentrations of NAADP before the development of high self-inactivating cytosolic concentrations of NAADP) increased the amplitude of contractions in cells stimulated to fire action potentials. In rat ventricular cells, which show Ca^{2+} sparks, application of 60 nM NAADP increased both the frequency and amplitude of Ca^{2+} sparks (effects which are both consistent with the proposed increase in amount of Ca^{2+} loaded into the SR). The effects NAADP-AM on contractions in guinea pig ventricular myocytes and on Ca^{2+} spark characteristics in rat myocytes were suppressed by bafilomycin A, again consistent with the involvement of the proposed lysosomal pathway in these effects [142].

Some have argued that the predominant effect of NAADP is to have a direct influence on sensitivity of RyR2 to increase Ca^{2+} release (e.g [159], [160]). This would be consistent with the increase in Ca^{2+} spark frequency mentioned above, but not the increase in amplitude. Interesting effects of patch applied NAADP have been observed in mouse ventricular myocyte, which unlike those from guinea pig, retain Ca^{2+} in the SR even when the myocytes are not electrically stimulated to fire action potentials. Under these conditions, NAADP applied from a patch pipette cause a slow wave of Ca^{2+} release from the SR. The concentration dependence of this effect showed a bell shaped curve, in which the response to 375 nM NAADP was substantially smaller than that to 37.5 nM, consistent with the desensitizing or self-inactivating effects of NAADP described above. The response to 37.5 nM NAADP was suppressed by a selective NAADP antagonist (BZ194, 10 microM), as well as by ruthenium red (which prevents Ca^{2+} release from SR) and bafilomycin A (consistent with the involvement of lysosomes as described above).

Observations described above in this section have all been made in ventricular myocytes, though it seems likely that the NAADP signaling system may be present throughout the heart. In atrial myocytes, effects of NAADP on CaT accompanying action potentials have been found to be broadly similar to those in ventricular cells. This was shown again using photorelease of NAADP from a caged compound, with a single release of cytosolic NAADP causing a progressive increase in the amplitude of successive CaT over a 2 min period [161]. As with the observations in ventricular cells there was no effect of NAADP on L-type Ca^{2+} currents, but the actions of NAADP were prevented when SR release of Ca^{2+} was suppressed with a combination of thapsigargin and ryanodine. Once again, the effects appeared to be dependent on lysosomes since the effects of photoreleased NAADP were prevented by prior exposure to bafilomycin A [161].

Guinea pig atrial myocytes (surprisingly unlike ventricular myocytes from the same species) show Ca^{2+} sparks in perhaps 30–50% of cells, and it is interesting that, as had been observed in rat ventricular myocytes, membrane-permeant NAADP-AM caused an increase in Ca^{2+} spark amplitude. This was consistent with an increase in the amount of Ca^{2+} loaded into the SR following the intracellular actions of NAADP, and this interpretation was further reinforced by the observation that Ca^{2+} release from the SR caused by rapid application of a high concentration of caffeine was also increased following NAADP-AM application. Broadly the increase in SR Ca^{2+} in atrial myocytes following the increased levels of cytosolic NAADP under these conditions appeared to be approximately 30% [161].

16.11.7 An Amplification Mechanism Linking Lysosomal Ca^{2+} Release and Increased Ca^{2+} Uptake by the SR in the Heart

Given the relatively small size of lysosomes compared with the extensive SR system, it seems unlikely that lysosomes could release sufficient Ca^{2+} to account directly for the substantial observed increase in SR Ca^{2+} content. One obvious potential amplification mechanism is CaMKII, particularly taking into account the evidence for a signaling microdomain between lysosomes and SR [8] that was described above.

This was tested in a series of experiments using two inhibitors of CaMKII, KN93 and the peptide AIP (autocamtide inhibitor peptide). The observations were made in guinea pig ventricular myocytes, mouse ventricular myocytes and guinea pig atrial myocytes. In all cases inhibition of CaMKII suppressed the ability of photoreleased NAADP to increase the amplitude of CaT accompanying action potentials. The interpretation that the observed effects of KN93 arose from inhibition of CaMKII was further supported by the failure of KN92, a structural analogue of KN93 without effects on CaMKII, to prevent actions of NAADP on CaT [152].

16.11.8 Extent of Ongoing Activity in NAADP Signaling Pathway Under Resting Conditions in Cardiac Muscle

The evidence presented above using photorelease of NAADP from a caged compound or application of the membrane permeant NAADP-AM make a convincing case for cytosolic actions of NAADP, predominantly to increase Ca^{2+} content of the SR and therefore the amplitude of CaT. It seems reasonable to ask whether there might be ongoing activity of this signaling system at rest and whether resting activity might be enhanced, for example following actions of hormones or neurotransmitters. Early evidence for a resting role for NAADP in guinea-pig ventricular cells came from the observation that the amplitude of CaT accompanying action potentials was reduced both by a high concentration of NAADP (presumably by the desensitizing or self inactivation mechanism) and by bafilomycin A (to suppress lysosomal function), and this reduction was approximately 20–25% [142].

An ongoing activity of the NAADP signaling system also seemed to operate in atrial myocytes, since bafilomycin A again caused a reduction in the amplitude of CaT accompanying action potentials. Interestingly the resting activity of the NAADP system seemed to be greater in atrial than in ventricular myocytes, since the reduction in CaT amplitude following bafilomycin was approximately 50% [161]. Whether this apparent difference between ventricular and atrial myocytes is real deserves further study in other species, but it is interesting to note in this context that atrial cells show an ongoing activity of adenyl cyclases (thought to be AC1 and AC8) as described above in the section on Ca^{2+} stimulated adenylyl cyclases, and ongoing activity of adenylyl cyclases is not observed in resting ventricular myocytes. One possibility is that the ongoing activity of adenylyl cyclases in atrial myocytes acts via a cAMP dependent pathway to increase NAADP synthesis by CD38, which in turn increases the contribution of the NAADP signaling pathway to the control of the amplitude of the CaT. The ability of beta-adrenoceptor stimulation to increase levels of NAADP in myocytes and whole hearts was described above, and the importance of cAMP dependent mechanisms to regulate functionally important aspects of NAADP signaling will be discussed in more detail in the next section.

16.11.9   Enhanced NAADP Effects Following
         Beta-Adrenoceptor Stimulation in Heart

Early evidence showing the influence of β-adrenoceptor stimulation on the function of NAADP signaling pathway in controlling CaT amplitude was provided in experiments taking advantage of the desensitizing or self inactivating effect of high concentrations of NAADP. It was found that high concentrations (1 mM) of NAADP applied via a patch pipette in guinea pig ventricular myocytes caused a larger reduction in CaT amplitude in the presence than in the absence of β-adrenoceptor stimulation, consistent with a greater role of this pathway when cAMP levels are elevated. If NAADP concentrations are substantially increased following beta-adrenoceptor signalling, it might be expected that further addition of NAADP would have less effect than would be the case when NAADP levels are at their resting level, and consistent with this line of argument photoreleased NAADP had a smaller effect on CaT amplitude after beta-adrenoceptor stimulation than under resting conditions [142].

   In guinea pig atrial myocytes, the effects of beta-adrenoceptor stimulation to increase the amplitude of CaT were less when lysosomal function was inhibited by bafilomycin A. Another experimental procedure to suppress lysosomal function is to apply GPN to disrupt these organelles. As expected, pretreatment with GPN also reduced the increase in CaT caused by beta-adrenoceptor stimulation [161]. It should be noted even if there were a functionally important contribution of the NAADP signaling pathway to the effects of beta-adrenoceptor stimulation, this would be expected to be additional to the well known effects of β-adrenoceptor stimulation to increase the amplitude of Ca^{2+} currents (e.g [23]. and review by [7]) and to increase Ca^{2+} uptake by SERCA into the SR following phosphorylation of phospholamban [7]. Based on the observations with bafilomycin and GPN it appears that in guinea-pig atrial myocytes the NAADP Ca^{2+} signaling mechanism involving lysosomes contributes over 30% to the total increase in CaT caused by beta-adrenoceptor stimulation.

   A separate piece of evidence supporting an involvement of lysosomes in the effects of β-adrenoceptor stimulation comes from observations in which the pH of the normally acidic lysosomes was measured. Isoproterenol caused an alkalinisation of lysosomes in atrial myocytes similar to that caused by NAADP-AM [161]. The maintenance of a relatively high concentration of Ca^{2+} within the lysosome is thought to depend on Ca^{2+}/H^+ exchange, and alkalinisation may occur following Ca^{2+} release from the lysosome, potentially by a change in the lysosomal buffering capacity or in the direction of Ca^{2+}/H^+ exchange.

   Two further tests of the ability of β-adrenoceptor stimulation to increase the functional effects of NAADP signaling are to use the NAADP-antagonist, Ned-19 or to use myocytes lacking TPC2 receptors. In guinea pig ventricular myocytes Ned-19 pretreatment reduced the effects of β-adrenoceptor stimulation to a similar extent as that observed in separate experiments in which lysosomal function was suppressed by bafilomycin A. As expected, the effects of beta-adrenoceptor stimulation on the amplitude of L-type Ca^{2+} currents was unaffected by Ned-19. Mouse ventricular myocytes from WT and those lacking TPC2 proteins were compared, and the effects of β-adrenoceptor stimulation on CaT and contraction were less in myocytes lacking TPC2 proteins. However, the effects of β-adrenoceptor stimulation on the amplitude of L-type Ca^{2+} currents were similar in WT myocytes and those lacking TPC2 proteins [152].

Another important piece of evidence concerns the synthetic pathway for NAADP, and observations described above support the key role of CD38. While bafilomycin A reduced the increase in CaT in response to β-adrenoceptor stimulation in WT mice, there was no such effect of bafilomycin A in mice lacking CD38 [1].

Taken together, the observations above are consistent with the hypothesis that β-adrenoceptor stimulation increases NAADP synthesis, and that NAADP causes Ca^{2+} release from lysosomes via TPC2 proteins into a signaling domain between lysosomes and SR, and this leads to increased Ca^{2+} uptake by the SR by a CaMKII dependent mechanism. The contribution of the NAADP signaling pathway to the effects of β-adrenoceptor stimulation in atrial and ventricular myocytes appeared to be about 30%.

16.11.10   Involvement of NAADP Signaling in Cardiac Arrhythmia and Hypertrophy

Cardiac arrhythmias associated with NAADP actions following high concentrations of beta-adrenoceptor agonist have been described by Guse and colleagues. In these experiments on mouse ventricular myocytes, high concentrations of beta adrenoceptor agonist increased the amplitude of CaT associated action potentials, but also led to additional spontaneous Ca^{2+} transients that were taken to indicate arrhythmias. The spontaneous events were suppressed by bafilomycin A and by an NAADP antagonist, BZ194 [159]. In a subsequent paper, measurements of cytosolic Ca^{2+} were supplemented by additional detection of changes in SR luminal Ca^{2+} using the low affinity Ca^{2+} probe, mag-fura-2. High concentrations of beta-adrenoceptor agonist again caused spontaneous arrhythmogenic activity, and in the spontaneous activity was suppressed by SAN4825, which as described above inhibits synthesis of both cADPR and NAADP [160].

Cardiac arrhythmias measured in an in vivo mouse model when exposed to high concentrations of β-adrenoceptor agonist were suppressed by the NAADP antagonist, BZ194, supporting a role for the NAADP pathway in arrhythmias at the whole animal level [159].

Another approach supporting a role for NAADP in arrhythmogenic properties is to compare the effects of a high concentration of beta-adrenoceptor agonist in WT mice and those lacking TPC2 proteins that were shown to be essential for the arrhythmogenic response, as described above. Isolated whole hearts from

$Tpcn2^{-/-}$ mice were resistant to arrhythmias caused by high concentrations of beta-adrenoceptor agonist as compared to hearts from WT animals [152].

Hearts from $CD38^{-/-}$ mice lacking CD38, the key enzyme for synthesis of NAADP, were also resistant to arrhythmias caused by high concentrations of beta adrenoceptor agonist as compared to WT hearts. A similar suppression of the arrhythmogenic effects of beta-adrenoceptor stimulation was seen following exposure of WT hearts to SAN4825, the inhibitor of cADPR and NAADP synthesis [1].

Prolonged exposure of animals to beta-agonist also causes cardiac hypertrophy. The involvement of the NAADP signaling mechanism in the stress pathway linking β-adrenoceptor stimulation to cardiac hypertrophy was demonstrated by the observation that following chronic exposure to β-adrenoceptor agonist hearts from $Tpcn2^{-/-}$ mice showed less cardiac hypertrophy and an increased threshold for arrhythmogenesis compared to WT controls [152]. In addition, hearts from $CD38^{-/-}$ mice were also resistant to cardiac hypertrophy and associated arrhythmias following chronic β-adrenoceptor stimulation compared to WT [118].

16.11.11   Trigger Hypothesis

The trigger hypothesis in which a transient change in NAADP concentration leads to important functional effects is well supported by very convincing evidence in a wide variety of cell types [162], [163], but this may not necessarily be the case in heart. As discussed above the evidence in heart seems to support an ongoing activity of the NAADP pathway under resting conditions (in the sense that cells were not exposed to hormones or neurotransmitters), so that suppression of the pathway, for example by bafilomycin A, reduces CaT accompanying action potentials by 20–25% in ventricular myocytes, and by approximately 50% in atrial myocytes. Although in whole hearts following beta-adrenoceptor stimulation with isoproterenol the rise in NAADP levels appeared to be transient in comparison with the changes in cADPR concentrations, the effects of isoproterenol on contraction were also transient [113]. The reason for the transience of the effect of isoproterenol on contraction are unknown, but might perhaps result from the difficulty of maintaining oxygen supply and cytosolic pH when contraction is substantially increased in hearts perfused with physiological solution rather than blood. Oxygen supply and pH can be more easily maintained in isolated myocytes, but changes in NAADP levels have not yet been measured in myocytes exposed to isoproterenol in the superfusing solution. However, the component of the effects of isoproterenol on CaT and contraction that involves NAADP (blocked by bafilomycin A and absent in myocytes lacking TPC2) appears to be well maintained at least over the several minutes that effects were measured [152]. Further experiments are required to shed more light on the question of whether NAADP acts as a trigger or an ongoing influence in cardiac muscle, both at rest and following β-adrenoceptor stimulation.

16.12 Combined Actions of cADPR and NAADP

It is argued above that there are maintained concentrations of both cADPR and NAADP for their effects in heart muscle on CaT and associated contractions accompanying action potentials under physiological conditions. Both agents are produced by CD38, which in mouse heart appeared to be the major synthetic enzyme for these two Ca^{2+} mobilizing messengers. Evidence was presented that CD38 in heart is predominantly associated with SR, and there appears to be an important Ca^{2+} signaling domain between lysosomes and the SR. TPC2 proteins appear to be essential for NAADP actions on CaT and contractions (while TPC1 proteins are important for ischaemia-reperfusion effects, perhaps involving a separate signaling microdomain between lysosomes and mitochondria). β-adrenoceptor stimulation increased the levels of both substances, and in ventricular myocytes both cADPR and NAADP pathways seemed to contribute to the β-adrenoceptor mediated effects on the amplitudes of CaT and contractions accompanying action potentials.

The evidence also supports the contention that at least for short periods (perhaps less than 3 min) the predominant effect of cADPR alone is on Ca^{2+} release from the SR (though this should not be taken to indicate that this necessarily results from a direct effect on RyR2 without a change in the amount of Ca^{2+} loaded into the SR, while NAADP actions primarily result from increased Ca^{2+} loading into the SR. It is the author's belief that any additional effects of cADPR alone on Ca^{2+} uptake into the SR after prolonged exposure to cADPR in permeabilised myocytes are less important than those on the Ca^{2+} release process (and have yet to be established under more physiological conditions), but this seems a fruitless argument if cADPR and NAADP normally act in concert as a result of their synthesis by the same enzyme, CD38, associated with the SR, and their complementary actions on CICR. In a similar way, the argument discussed above about whether cADPR could have a sustained effect on the amplitude of CaT and contractions if its only action were to influence Ca^{2+} release via RyR seems unlikely to be productive for a physiological situation in which cADPR and NAADP act together. The dual effects of NAADP on Ca^{2+} load and cADPR on Ca^{2+} release are expected to combine to create a powerful effect on CaT and contractions accompanying action potentials. One reservation arises from the presumed location of SERCA and RyR2 in various schemes of EC coupling in which these two proteins are normally shown far apart with the RyR2 at the terminal cisternae, and the SERCA closer to the middle of the sarcomere, as might be expected if Ca^{2+} is taken up after contraction in one part of the SR before transport of Ca^{2+} to the terminal cisternae for release via RyR2. However, recent evidence supports the presence of both SERCA and RyR2 in the terminal cisternae [112] (See also [19] for discussion of JSR and NSR)

A scheme summarizing the actions of cADPR and NAADP is shown in Fig. 16.6.

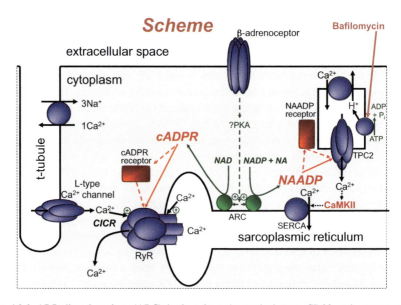

Fig. 16.6 ADP-ribosyl cyclase (ARC) is thought to be equivalent to CD38 and appears to be located on (or close to) the SR (see text). Cardiac ARC catalyses formation of both cADPR (with NAD as substrate) and NAADP (from NADP and nicotinic acid (NA)). cADPR appears to act primarily on RyRs, although there may be an intermediary step possibly involving a binding protein, and additional slow effects of cADPR on Ca^{2+} uptake by the SR cannot be excluded. NAADP causes Ca^{2+} to be released from endolysosomes via TPC2 proteins (perhaps with mediation of an additional NAADP receptor protein). The Ca^{2+} released from the endolysosome leads to additional Ca^{2+} uptake by the SR (an effect that appears to involve CaMKII, perhaps by phosphorylation of phospholamban to increase SERCA activity). Additional Ca^{2+} loaded into the SR will increase CICR both by providing a greater gradient for Ca^{2+} release and by increasing opening probability of RyRs. Additional effects of NAADP on RyRs cannot be excluded. cADPR and NAADP may function together in a cooperative or synergistic way in ventricular myocytes (perhaps also interacting with IP3-dependent Ca^{2+} signaling in atrial myocytes). Stimulation of beta-adrenoceptors increases synthesis of both cADPR and NAADP, by mechanism that remains to be fully determined, but may involve an action of PKA on ARC

16.13   Summary

It was recognized at the outset of this article that Ca^{2+} is involved in all aspects of biology, and evidence presented here shows that Ca^{2+} in the heart not only controls its primary function to provide rhythmic contractions to pump blood to the lungs and tissues, but also that diverse Ca^{2+} signaling pathways regulate this basic function. The simplest regulatory elements are Ca^{2+} stimulated enzymes. In addition Ca^{2+} mobilizing agents play important roles in stimulating Ca^{2+} release from the intracellular organelles of the SR and lysosomes. These Ca^{2+} regulated processes form an interconnecting network that determines function, and not surprisingly there are variations in the contributions and relative importance of components of this network in different regions of the heart with different functions.

IP$_3$ is particularly important for the short term control of Ca^{2+} transients in atria and pacemaker tissue of the SA node, although long term effects on gene expression occur throughout the heart. It is suggested that cADPR and NAADP are synthesized in heart by the same enzyme, CD38, located at or close to the SR, where these two substances act in a complementary way to regulate Ca^{2+} release by increasing the gain of CICR and increasing the amount of Ca^{2+} stored in the SR. Although the relationship between cADPR and NAADP actions is argued to be particularly close, it is recognized that overall function is determined by complex interactions with IP$_3$ and Ca^{2+} stimulated enzymes. Actions of hormones and neurotransmitters provide a regulatory influence on these complex pathways. The contributions of Ca^{2+} signaling pathways to the overall effect of receptor stimulation on Ca^{2+} transients are often surprisingly large. Although disentangling the complex interacting Ca^{2+}-dependent processes provides a challenge for the design of experiments, it is clear that Ca^{2+} signaling in the heart is a major determinant of function.

References

1. Lin WK, Bolton EL, Cortopassi WA, Wang Y, O'Brien F, Maciejewska M et al (2017) Synthesis of the Ca^{2+}-mobilizing messengers NAADP and cADPR by intracellular CD38 enzyme in the mouse heart: role in beta-adrenoceptor signaling. J Biol Chem 292(32):13243–13257
2. Bers DM (2002) Cardiac excitation-contraction coupling. Nature 415(6868):198–205
3. MacLeod KT (2016) Recent advances in understanding cardiac contractility in health and disease. F1000 Res 5:1770
4. Dewenter M, von der Lieth A, Katus HA, Backs J (2017) Calcium signaling and transcriptional regulation in cardiomyocytes. Circ Res 121(8):1000–1020
5. Harada M, Luo X, Murohara T, Yang B, Dobrev D, Nattel S (2014) MicroRNA regulation and cardiac calcium signaling: role in cardiac disease and therapeutic potential. Circ Res 114(4):689–705
6. Li J, Greener ID, Inada S, Nikolski VP, Yamamoto M, Hancox JC et al (2008) Computer three-dimensional reconstruction of the atrioventricular node. Circ Res 102(8):975–985
7. Bers DM (2008) Calcium cycling and signaling in cardiac myocytes. Annu Rev Physiol 70:23–49
8. Aston D, Capel RA, Ford KL, Christian HC, Mirams GR, Rog-Zielinska EA et al (2017) High resolution structural evidence suggests the sarcoplasmic reticulum forms microdomains with acidic stores (lysosomes) in the heart. Sci Rep 7:40620
9. Venturi E, Pitt S, Galfre E, Sitsapesan R (2012) From eggs to hearts: what is the link between cyclic ADP-ribose and ryanodine receptors? Cardiovasc Ther 30(2):109–116
10. Lim G, Venetucci L, Eisner DA, Casadei B (2008) Does nitric oxide modulate cardiac ryanodine receptor function? Implications for excitation-contraction coupling. Cardiovasc Res 77(2):256–264
11. O'Brien F, Venturi E, Sitsapesan R (2015) The ryanodine receptor provides high throughput Ca^{2+}−release but is precisely regulated by networks of associated proteins: a focus on proteins relevant to phosphorylation. Biochem Soc Trans 43(3):426–433
12. Camors E, Valdivia HH (2014) CaMKII regulation of cardiac ryanodine receptors and inositol triphosphate receptors. Front Pharmacol 5:101
13. Lukyanenko V, Chikando A, Lederer WJ (2009) Mitochondria in cardiomyocyte Ca^{2+} signaling. Int J Biochem Cell Biol 41(10):1957–1971

14. Williams GS, Boyman L, Lederer WJ (2015) Mitochondrial calcium and the regulation of metabolism in the heart. J Mol Cell Cardiol 78:35–45

15. Hohendanner F, Maxwell JT, Blatter LA (2015) Cytosolic and nuclear calcium signaling in atrial myocytes: IP3-mediated calcium release and the role of mitochondria. Channels 9(3):129–138

16. Cheng H, Lederer WJ, Cannell MB (1993) Calcium sparks: elementary events underlying excitation-contraction coupling in heart muscle. Science 262(5134):740–744

17. Cannell MB, Kong CH, Imtiaz MS, Laver DR (2013) Control of sarcoplasmic reticulum Ca^{2+} release by stochastic RyR gating within a 3D model of the cardiac dyad and importance of induction decay for CICR termination. Biophys J 104(10):2149–2159

18. Wescott AP, Jafri MS, Lederer WJ, Williams GS (2016) Ryanodine receptor sensitivity governs the stability and synchrony of local calcium release during cardiac excitation-contraction coupling. J Mol Cell Cardiol 92:82–92

19. Sobie EA, Williams GSB, Lederer WJ (2017) Ambiguous interactions between diastolic and SR Ca^{2+} in the regulation of cardiac Ca^{2+} release. J Gen Physiol 149(9):847–855

20. Rios E (2017) Perspectives on "control of Ca release from within the cardiac sarcoplasmic reticulum". J Gen Physiol 149(9):833–836

21. Jones PP, Guo W, Chen SRW (2017) Control of cardiac ryanodine receptor by sarcoplasmic reticulum luminal Ca(2). J Gen Physiol 149(9):867–875

22. Ching LL, Williams AJ, Sitsapesan R (2000) Evidence for Ca^{2+} activation and inactivation sites on the luminal side of the cardiac ryanodine receptor complex. Circ Res 87(3):201–206

23. Mitchell MR, Powell T, Terrar DA, Twist VW (1983) Characteristics of the second inward current in cells isolated from rat ventricular muscle. Proc R Soc Lond Ser B Biol Sci 219(1217):447–469

24. Sitsapesan R, Montgomery RA, MacLeod KT, Williams AJ (1991) Sheep cardiac sarcoplasmic reticulum calcium-release channels: modification of conductance and gating by temperature. J Physiol 434:469–488

25. Lee P, Wang K, Woods CE, Yan P, Kohl P, Ewart P et al (2012) Cardiac electrophysiological imaging systems scalable for high-throughput drug testing. Pflugers Arch: Eur J Physiol 464(6):645–656

26. Winter J, Bishop MJ, Wilder CDE, O'Shea C, Pavlovic D, Shattock MJ (2018) Sympathetic nervous regulation of calcium and action potential alternans in the intact heart. Front Physiol 9:16

27. Eisner DA, Caldwell JL, Kistamas K, Trafford AW (2017) Calcium and excitation-contraction coupling in the heart. Circ Res 121(2):181–195

28. Janvier NC, Harrison SM, Boyett MR (1997) The role of inward Na(+)-Ca^{2+} exchange current in the ferret ventricular action potential. J Physiol 498(Pt 3):611–625

29. O'Hara T, Virag L, Varro A, Rudy Y (2011) Simulation of the undiseased human cardiac ventricular action potential: model formulation and experimental validation. PLoS Comput Biol 7(5):e1002061

30. White E, Terrar DA (1992) Inactivation of Ca current during the action potential in guinea-pig ventricular myocytes. Exp Physiol 77(1):153–164

31. Linz KW, Meyer R (1998) Control of L-type calcium current during the action potential of guinea-pig ventricular myocytes. J Physiol 513(Pt 2):425–442

32. Blatter LA (2017) The intricacies of atrial calcium cycling during excitation-contraction coupling. J Gen Physiol 149(9):857–865

33. Collins TP, Terrar DA (2012) Ca^{2+}-stimulated adenylyl cyclases regulate the L-type Ca^{2+} current in guinea-pig atrial myocytes. J Physiol 590(8):1881–1893

34. Maxwell JT, Blatter LA (2017) A novel mechanism of tandem activation of ryanodine receptors by cytosolic and SR luminal Ca^{2+} during excitation-contraction coupling in atrial myocytes. J Physiol 595(12):3835–3845

35. Brandenburg S, Kohl T, Williams GS, Gusev K, Wagner E, Rog-Zielinska EA et al (2016) Axial tubule junctions control rapid calcium signaling in atria. J Clin Invest 126(10):3999–4015

36. Lakatta EG, Maltsev VA, Vinogradova TM (2010) A coupled system of intracellular Ca^{2+} clocks and surface membrane voltage clocks controls the timekeeping mechanism of the heart's pacemaker. Circ Res 106(4):659–673
37. Capel RA, Terrar DA (2015) The importance of Ca^{2+}-dependent mechanisms for the initiation of the heartbeat. Front Physiol 6:80
38. Mesirca P, Torrente AG, Mangoni ME (2015) Functional role of voltage gated Ca^{2+} channels in heart automaticity. Front Physiol 6:19
39. DiFrancesco D (2015) HCN4, sinus bradycardia and atrial fibrillation. Arrhythmia Electrophysiol Rev 4(1):9–13
40. Chen W, Wang R, Chen B, Zhong X, Kong H, Bai Y et al (2014) The ryanodine receptor store-sensing gate controls Ca^{2+} waves and Ca^{2+}−triggered arrhythmias. Nat Med 20(2):184–192
41. Toyoda F, Mesirca P, Dubel S, Ding WG, Striessnig J, Mangoni ME et al (2017) CaV1.3 L-type Ca^{2+} channel contributes to the heartbeat by generating a dihydropyridine-sensitive persistent Na(+) current. Sci Rep 7(1):7869
42. Sanders L, Rakovic S, Lowe M, Mattick PA, Terrar DA (2006) Fundamental importance of Na^{+}-Ca^{2+} exchange for the pacemaking mechanism in guinea-pig sino-atrial node. J Physiol 571(Pt 3):639–649
43. Kapoor N, Tran A, Kang J, Zhang R, Philipson KD, Goldhaber JI (2015) Regulation of calcium clock-mediated pacemaking by inositol-1,4,5-trisphosphate receptors in mouse sinoatrial nodal cells. J Physiol 593(12):2649–2663
44. Rigg L, Heath BM, Cui Y, Terrar DA (2000) Localisation and functional significance of ryanodine receptors during beta-adrenoceptor stimulation in the guinea-pig sino-atrial node. Cardiovasc Res 48(2):254–264
45. Maier LS (2012) Ca^{2+}/calmodulin-dependent protein kinase II (CaMKII) in the heart. Adv Exp Med Biol 740:685–702
46. Swaminathan PD, Purohit A, Hund TJ, Anderson ME (2012) Calmodulin-dependent protein kinase II: linking heart failure and arrhythmias. Circ Res 110(12):1661–1677
47. Gray CB, Heller BJ (2014) CaMKIIdelta subtypes: localization and function. Front Pharmacol 5:15
48. Mustroph J, Maier LS, Wagner S (2014) CaMKII regulation of cardiac K channels. Front Pharmacol 5:20
49. Wu Y, Anderson ME (2014) CaMKII in sinoatrial node physiology and dysfunction. Front Pharmacol 5:48
50. Hund TJ, Mohler PJ (2015) Role of CaMKII in cardiac arrhythmias. Trends Cardiovasc Med 25(5):392–397
51. Mattiazzi A, Bassani RA, Escobar AL, Palomeque J, Valverde CA, Vila Petroff M et al (2015) Chasing cardiac physiology and pathology down the CaMKII cascade. Am J Physiol Heart Circ Physiol 308(10):H1177–H1191
52. Johnston AS, Lehnart SE, Burgoyne JR (2015) Ca^{2+} signaling in the myocardium by (redox) regulation of PKA/CaMKII. Front Pharmacol 6:166
53. Anderson ME (2015) Oxidant stress promotes disease by activating CaMKII. J Mol Cell Cardiol 89(Pt B):160–167
54. Feng N, Anderson ME (2017) CaMKII is a nodal signal for multiple programmed cell death pathways in heart. J Mol Cell Cardiol 103:102–109
55. Richard S, Perrier E, Fauconnier J, Perrier R, Pereira L, Gomez AM et al (2006) 'Ca^{2+}-induced Ca^{2+} entry' or how the L-type Ca^{2+} channel remodels its own signalling pathway in cardiac cells. Prog Biophys Mol Biol 90(1–3):118–135
56. Xu L, Lai D, Cheng J, Lim HJ, Keskanokwong T, Backs J et al (2010) Alterations of L-type calcium current and cardiac function in CaMKII{delta} knockout mice. Circ Res 107(3):398–407
57. Ferrero P, Said M, Sanchez G, Vittone L, Valverde C, Donoso P et al (2007) Ca^{2+}/calmodulin kinase II increases ryanodine binding and Ca^{2+}−induced sarcoplasmic reticulum Ca^{2+} release kinetics during beta-adrenergic stimulation. J Mol Cell Cardiol 43(3):281–291

58. DeSantiago J, Maier LS, Bers DM (2002) Frequency-dependent acceleration of relaxation in the heart depends on CaMKII, but not phospholamban. J Mol Cell Cardiol 34(8):975–984

59. Pereira L, Metrich M, Fernandez-Velasco M, Lucas A, Leroy J, Perrier R et al (2007) The cAMP binding protein Epac modulates Ca^{2+} sparks by a Ca^{2+}/calmodulin kinase signalling pathway in rat cardiac myocytes. J Physiol 583(Pt 2):685–694

60. Bers DM, Grandi E (2009) Calcium/calmodulin-dependent kinase II regulation of cardiac ion channels. J Cardiovasc Pharmacol 54(3):180–187

61. Nerbonne JM (2011) Repolarizing cardiac potassium channels: multiple sites and mechanisms for CaMKII-mediated regulation. Heart Rhythm 8(6):938–941

62. Grandi E, Herren AW (2014) CaMKII-dependent regulation of cardiac Na(+) homeostasis. Front Pharmacol 5:41

63. Belardinelli L, Giles WR, Rajamani S, Karagueuzian HS, Shryock JC (2015) Cardiac late Na(+) current: proarrhythmic effects, roles in long QT syndromes, and pathological relationship to CaMKII and oxidative stress. Heart Rhythm 12(2):440–448

64. Luthi A, McCormick DA (1999) Modulation of a pacemaker current through Ca^{2+}-induced stimulation of cAMP production. Nat Neurosci 2(7):634–641

65. Halls ML, Cooper DM (2011) Regulation by Ca^{2+}−signaling pathways of adenylyl cyclases. Cold Spring Harb Perspect Biol 3(1):a004143

66. Wang H, Zhang M (2012) The role of Ca(2)(+)-stimulated adenylyl cyclases in bidirectional synaptic plasticity and brain function. Rev Neurosci 23(1):67–78

67. Mattick P, Parrington J, Odia E, Simpson A, Collins T, Terrar D (2007) Ca^{2+}−stimulated adenylyl cyclase isoform AC1 is preferentially expressed in guinea-pig sino-atrial node cells and modulates the I(f) pacemaker current. J Physiol 582(Pt 3):1195–1203

68. Younes A, Lyashkov AE, Graham D, Sheydina A, Volkova MV, Mitsak M et al (2008) Ca^{2+}-stimulated basal adenylyl cyclase activity localization in membrane lipid microdomains of cardiac sinoatrial nodal pacemaker cells. J Biol Chem 283(21):14461–14468

69. Boink GJ, Nearing BD, Shlapakova IN, Duan L, Kryukova Y, Bobkov Y et al (2012) Ca^{2+}-stimulated adenylyl cyclase AC1 generates efficient biological pacing as single gene therapy and in combination with HCN2. Circulation 126(5):528–536

70. Timofeyev V, Myers RE, Kim HJ, Woltz RL, Sirish P, Heiserman JP et al (2013) Adenylyl cyclase subtype-specific compartmentalization: differential regulation of L-type Ca^{2+} current in ventricular myocytes. Circ Res 112(12):1567–1576

71. Cheung WY (1980) Calmodulin plays a pivotal role in cellular regulation. Science 207(4426):19–27

72. Lukyanenko YO, Younes A, Lyashkov AE, Tarasov KV, Riordon DR, Lee J et al (2016) Ca^{2+}/calmodulin-activated phosphodiesterase 1A is highly expressed in rabbit cardiac sinoatrial nodal cells and regulates pacemaker function. J Mol Cell Cardiol 98:73–82

73. Tham YK, Bernardo BC, Ooi JY, Weeks KL, McMullen JR (2015) Pathophysiology of cardiac hypertrophy and heart failure: signaling pathways and novel therapeutic targets. Arch Toxicol 89(9):1401–1438

74. Knight WE, Chen S, Zhang Y, Oikawa M, Wu M, Zhou Q et al (2016) PDE1C deficiency antagonizes pathological cardiac remodeling and dysfunction. Proc Natl Acad Sci U S A 113(45):E7116–E7E25

75. Forstermann U, Sessa WC (2012) Nitric oxide synthases: regulation and function. Eur Heart J 33(7):829–837, 37a–37d

76. Casadei B (2006) The emerging role of neuronal nitric oxide synthase in the regulation of myocardial function. Exp Physiol 91(6):943–955

77. Zhang YH (2017) Nitric oxide signalling and neuronal nitric oxide synthase in the heart under stress. F1000 Res 6:742

78. Fabiato A (1992) Two kinds of calcium-induced release of calcium from the sarcoplasmic reticulum of skinned cardiac cells. Adv Exp Med Biol 311:245–262

79. Lipp P, Laine M, Tovey SC, Burrell KM, Berridge MJ, Li W et al (2000) Functional InsP3 receptors that may modulate excitation-contraction coupling in the heart. Curr Biol: CB 10(15):939–942

80. Zima AV, Blatter LA (2004) Inositol-1,4,5-trisphosphate-dependent Ca^{2+} signalling in cat atrial excitation-contraction coupling and arrhythmias. J Physiol 555(Pt 3):607–615
81. Yao Y, Choi J, Parker I (1995) Quantal puffs of intracellular Ca^{2+} evoked by inositol trisphosphate in xenopus oocytes. J Physiol 482(Pt 3):533–553
82. Li X, Zima AV, Sheikh F, Blatter LA, Chen J (2005) Endothelin-1-induced arrhythmogenic Ca^{2+} signaling is abolished in atrial myocytes of inositol-1,4,5-trisphosphate(IP3)-receptor type 2-deficient mice. Circ Res 96(12):1274–1281
83. Wang YG, Dedkova EN, Ji X, Blatter LA, Lipsius SL (2005) Phenylephrine acts via IP3-dependent intracellular NO release to stimulate L-type Ca^{2+} current in cat atrial myocytes. J Physiol 567(Pt 1):143–157
84. Collins TP, Mikoshiba K, Terrar DA (2007) Possible novel mechanism for the positive inotropic action of alpha-1 adrenoceptor agonists in guinea-pig isolated atrial myocytes. Biophys J:433A–433A
85. Terrar DA, Capel RA, Collins TP, Rajasumdaram S, Ayagamar T, Burton RAB (2018) Cross talk between IP3 and adenylyl cyclase signaling pathways in cardiac atrial Myocytes. Biophys J 114(3):466A–466A
86. Kockskamper J, Zima AV, Roderick HL, Pieske B, Blatter LA, Bootman MD (2008) Emerging roles of inositol 1,4,5-trisphosphate signaling in cardiac myocytes. J Mol Cell Cardiol 45(2):128–147
87. Berridge MJ (2009) Inositol trisphosphate and calcium signalling mechanisms. Biochim Biophys Acta 1793(6):933–940
88. Hohendanner F, McCulloch AD, Blatter LA, Michailova AP (2014) Calcium and IP3 dynamics in cardiac myocytes: experimental and computational perspectives and approaches. Front Pharmacol 5:35
89. Garcia MI, Boehning D (2017) Cardiac inositol 1,4,5-trisphosphate receptors. Biochim Biophys Acta 1864(6):907–914
90. Domeier TL, Zima AV, Maxwell JT, Huke S, Mignery GA, Blatter LA (2008) IP3 receptor-dependent Ca^{2+} release modulates excitation-contraction coupling in rabbit ventricular myocytes. Am J Physiol Heart Circ Physiol 294(2):H596–H604
91. Subedi KP, Son MJ, Chidipi B, Kim SW, Wang J, Kim KH et al (2017) Signaling pathway for endothelin-1- and phenylephrine-induced cAMP response element binding protein activation in rat ventricular Myocytes: role of inositol 1,4,5-Trisphosphate receptors and CaMKII. Cell Physiol Biochem 41(1):399–412
92. Hirose M, Stuyvers B, Dun W, Ter Keurs H, Boyden PA (2008) Wide long lasting perinuclear Ca^{2+} release events generated by an interaction between ryanodine and IP3 receptors in canine Purkinje cells. J Mol Cell Cardiol 45(2):176–184
93. Zima AV, Bare DJ, Mignery GA, Blatter LA (2007) IP3-dependent nuclear Ca^{2+} signalling in the mammalian heart. J Physiol 584(Pt 2):601–611
94. Ju YK, Liu J, Lee BH, Lai D, Woodcock EA, Lei M et al (2011) Distribution and functional role of inositol 1,4,5-trisphosphate receptors in mouse sinoatrial node. Circ Res 109(8):848–857
95. Ju YK, Woodcock EA, Allen DG, Cannell MB (2012) Inositol 1,4,5-trisphosphate receptors and pacemaker rhythms. J Mol Cell Cardiol 53(3):375–381
96. Hohendanner F, Walther S, Maxwell JT, Kettlewell S, Awad S, Smith GL et al (2015) Inositol-1,4,5-trisphosphate induced Ca^{2+} release and excitation-contraction coupling in atrial myocytes from normal and failing hearts. J Physiol 593(6):1459–1477
97. Kim JC, Woo SH (2015) Shear stress induces a longitudinal Ca^{2+} wave via autocrine activation of P2Y1 purinergic signalling in rat atrial myocytes. J Physiol 593(23):5091–5109
98. Son MJ, Kim JC, Kim SW, Chidipi B, Muniyandi J, Singh TD et al (2016) Shear stress activates monovalent cation channel transient receptor potential melastatin subfamily 4 in rat atrial myocytes via type 2 inositol 1,4,5-trisphosphate receptors and Ca^{2+} release. J Physiol 594(11):2985–3004
99. Lee HC (2001) Physiological functions of cyclic ADP-ribose and NAADP as calcium messengers. Annu Rev Pharmacol Toxicol 41:317–345

100. Higashida H, Salmina AB, Olovyannikova RY, Hashii M, Yokoyama S, Koizumi K et al (2007) Cyclic ADP-ribose as a universal calcium signal molecule in the nervous system. Neurochem Int 51(2–4):192–199

101. Lee HC (2012) Cyclic ADP-ribose and nicotinic acid adenine dinucleotide phosphate (NAADP) as messengers for calcium mobilization. J Biol Chem 287(38):31633–31640

102. Wei W, Graeff R, Yue J (2014) Roles and mechanisms of the CD38/cyclic adenosine diphosphate ribose/Ca^{2+} signaling pathway. World J Biol Chem 5(1):58–67

103. Galione A (2015) A primer of NAADP-mediated Ca^{2+} signalling: from sea urchin eggs to mammalian cells. Cell Calcium 58(1):27–47

104. Rakovic S, Terrar DA (2002) Calcium signaling by cADPR in cardiac myocytes. In: Lee HC (ed) Cyclic ADP-ribose and NAADP. Structures, metabolism and functions. Kluwer, Dordecht, pp 319–341

105. Terrar DA (2015) The roles of NAADP, two-pore channels and lysosomes in Ca^{2+} signaling in cardiac muscle. Messenger 4:23–33

106. Meszaros LG, Bak J, Chu A (1993) Cyclic ADP-ribose as an endogenous regulator of the non-skeletal type ryanodine receptor Ca^{2+} channel. Nature 364(6432):76–79

107. Sitsapesan R, McGarry SJ, Williams AJ (1994) Cyclic ADP-ribose competes with ATP for the adenine nucleotide binding site on the cardiac ryanodine receptor Ca^{2+}-release channel. Circ Res 75(3):596–600

108. Rusinko N, Lee HC (1989) Widespread occurrence in animal tissues of an enzyme catalyzing the conversion of NAD+ into a cyclic metabolite with intracellular Ca^{2+}−mobilizing activity. J Biol Chem 264(20):11725–11731

109. Walseth TF, Aarhus R, Zeleznikar RJ Jr, Lee HC (1991) Determination of endogenous levels of cyclic ADP-ribose in rat tissues. Biochim Biophys Acta 1094(1):113–120

110. Meszaros V, Socci R, Meszaros LG (1995) The kinetics of cyclic ADP-ribose formation in heart muscle. Biochem Biophys Res Commun 210(2):452–456

111. Higashida H, Egorova A, Higashida C, Zhong ZG, Yokoyama S, Noda M et al (1999) Sympathetic potentiation of cyclic ADP-ribose formation in rat cardiac myocytes. J Biol Chem 274(47):33348–33354

112. Kolstad TR, Stokke MK, Stang E, Brorson SH, William LE, Sejersted OM (2015) Serca located in the junctional SR shapes calcium release in cardiac myocytes. Biophys J 108(2):503A–503A

113. Lewis AM, Aley PK, Roomi A, Thomas JM, Masgrau R, Garnham C et al (2012) Beta-adrenergic receptor signaling increases NAADP and cADPR levels in the heart. Biochem Biophys Res Commun 427(2):326–329

114. Higashida H, Zhang J, Hashii M, Shintaku M, Higashida C, Takeda Y (2000) Angiotensin II stimulates cyclic ADP-ribose formation in neonatal rat cardiac myocytes. Biochem J 352(Pt 1):197–202

115. Ferrero E, Lo Buono N, Horenstein AL, Funaro A, Malavasi F (2014) The ADP-ribosyl cyclases–the current evolutionary state of the ARCs. Front Biosci 19:986–1002

116. Gul R, Kim SY, Park KH, Kim BJ, Kim SJ, Im MJ et al (2008) A novel signaling pathway of ADP-ribosyl cyclase activation by angiotensin II in adult rat cardiomyocytes. Am J Physiol Heart Circ Physiol 295(1):H77–H88

117. Gul R, Park JH, Kim SY, Jang KY, Chae JK, Ko JK et al (2009) Inhibition of ADP-ribosyl cyclase attenuates angiotensin II-induced cardiac hypertrophy. Cardiovasc Res 81(3):582–591

118. Gul R, Park DR, Shawl AI, Im SY, Nam TS, Lee SH et al (2016) Nicotinic Acid Adenine Dinucleotide Phosphate (NAADP) and Cyclic ADP-Ribose (cADPR) mediate Ca^{2+} signaling in cardiac hypertrophy induced by beta-adrenergic stimulation. PLoS One 11(3):e0149125

119. Kannt A, Sicka K, Kroll K, Kadereit D, Gogelein H (2012) Selective inhibitors of cardiac ADPR cyclase as novel anti-arrhythmic compounds. Naunyn Schmiedeberg's Arch Pharmacol 385(7):717–727

120. Walseth TF, Lee HC (1993) Synthesis and characterization of antagonists of cyclic-ADP-ribose-induced Ca^{2+} release. Biochim Biophys Acta 1178(3):235–242

121. Rakovic S, Galione A, Ashamu GA, Potter BV, Terrar DA (1996) A specific cyclic ADP-ribose antagonist inhibits cardiac excitation-contraction coupling. Curr Biol: CB 6(8):989–996

122. Iino S, Cui Y, Galione A, Terrar DA (1997) Actions of cADP-ribose and its antagonists on contraction in guinea pig isolated ventricular myocytes. Influence Temp Circ Res 81(5):879–884

123. Cui Y, Galione A, Terrar DA (1999) Effects of photoreleased cADP-ribose on calcium transients and calcium sparks in myocytes isolated from guinea-pig and rat ventricle. Biochem J 342(Pt 2):269–273

124. Macgregor AT, Rakovic S, Galione A, Terrar DA (2007) Dual effects of cyclic ADP-ribose on sarcoplasmic reticulum Ca^{2+} release and storage in cardiac myocytes isolated from guinea-pig and rat ventricle. Cell Calcium 41(6):537–546

125. Prakash YS, Kannan MS, Walseth TF, Sieck GC (2000) cADP ribose and $[Ca^{2+}](i)$ regulation in rat cardiac myocytes. Am J Physiol Heart Circ Physiol 279(4):H1482–H1489

126. Guo X, Laflamme MA, Becker PL (1996) Cyclic ADP-ribose does not regulate sarcoplasmic reticulum Ca^{2+} release in intact cardiac myocytes. Circ Res 79(1):147–151

127. Lukyanenko V, Gyorke I, Wiesner TF, Gyorke S (2001) Potentiation of Ca^{2+} release by cADP-ribose in the heart is mediated by enhanced SR Ca^{2+} uptake into the sarcoplasmic reticulum. Circ Res 89(7):614–622

128. Gomez AM, Cheng H, Lederer WJ, Bers DM (1996) Ca^{2+} diffusion and sarcoplasmic reticulum transport both contribute to $[Ca^{2+}]i$ decline during Ca^{2+} sparks in rat ventricular myocytes. J Physiol 496(Pt 2):575–581

129. Lee HC, Aarhus R, Graeff R, Gurnack ME, Walseth TF (1994) Cyclic ADP ribose activation of the ryanodine receptor is mediated by calmodulin. Nature 370(6487):307–309

130. Lee HC, Aarhus R, Graeff RM (1995) Sensitization of calcium-induced calcium release by cyclic ADP-ribose and calmodulin. J Biol Chem 270(16):9060–9066

131. Tanaka Y, Tashjian AH Jr (1995) Calmodulin is a selective mediator of Ca^{2+}-induced Ca^{2+} release via the ryanodine receptor-like Ca^{2+} channel triggered by cyclic ADP-ribose. Proc Natl Acad Sci U S A 92(8):3244–3248

132. Takasawa S, Ishida A, Nata K, Nakagawa K, Noguchi N, Tohgo A et al (1995) Requirement of calmodulin-dependent protein kinase II in cyclic ADP-ribose-mediated intracellular Ca^{2+} mobilization. J Biol Chem 270(51):30257–30259

133. Thomas JM, Summerhill RJ, Fruen BR, Churchill GC, Galione A (2002) Calmodulin dissociation mediates desensitization of the cADPR-induced Ca^{2+} release mechanism. Curr Biol: CB 12(23):2018–2022

134. Zhang X, Tallini YN, Chen Z, Gan L, Wei B, Doran R et al (2009) Dissociation of FKBP12.6 from ryanodine receptor type 2 is regulated by cyclic ADP-ribose but not beta-adrenergic stimulation in mouse cardiomyocytes. Cardiovasc Res 84(2):253–262

135. Rakovic S, Cui Y, Iino S, Galione A, Ashamu GA, Potter BV et al (1999) An antagonist of cADP-ribose inhibits arrhythmogenic oscillations of intracellular Ca^{2+} in heart cells. J Biol Chem 274(25):17820–17827

136. Gul R, Shawl AI, Kim SH, Kim UH (2012) Cooperative interaction between reactive oxygen species and Ca^{2+} signals contributes to angiotensin II-induced hypertrophy in adult rat cardiomyocytes. Am J Physiol Heart Circ Physiol 302(4):H901–H909

137. Bak J, Billington RA, Timar G, Dutton AC, Genazzani AA (2001) NAADP receptors are present and functional in the heart. Curr Biol: CB 11(12):987–990

138. Chini EN, Dousa TP (1995) Enzymatic synthesis and degradation of nicotinate adenine dinucleotide phosphate (NAADP), a Ca^{2+}-releasing agonist, in rat tissues. Biochem Biophys Res Commun 209(1):167–174

139. Chini EN, Chini CC, Kato I, Takasawa S, Okamoto H (2002) CD38 is the major enzyme responsible for synthesis of nicotinic acid-adenine dinucleotide phosphate in mammalian tissues. Biochem J 362(Pt 1):125–130

140. Vasudevan SR, Galione A, Churchill GC (2008) Sperm express a Ca^{2+}-regulated NAADP synthase. Biochem J 411(1):63–70

141. Galfre E, Pitt SJ, Venturi E, Sitsapesan M, Zaccai NR, Tsaneva-Atanasova K et al (2012) FKBP12 activates the cardiac ryanodine receptor Ca^{2+}−release channel and is antagonised by FKBP12.6. PLoS One 7(2):e31956

142. Macgregor A, Yamasaki M, Rakovic S, Sanders L, Parkesh R, Churchill GC et al (2007) NAADP controls cross-talk between distinct Ca^{2+} stores in the heart. J Biol Chem 282(20):15302–15311

143. Zhao YJ, Zhang HM, Lam CM, Hao Q, Lee HC (2011) Cytosolic CD38 protein forms intact disulfides and is active in elevating intracellular cyclic ADP-ribose. J Biol Chem 286(25):22170–22177

144. Zhao YJ, Lam CM, Lee HC (2012) The membrane-bound enzyme CD38 exists in two opposing orientations. Sci Signal 5(241):ra67

145. Zhao YJ, Zhu WJ, Wang XW, Zhang LH, Lee HC (2015) Determinants of the membrane orientation of a calcium signaling enzyme CD38. Biochim Biophys Acta 1853(9):2095–2103

146. Churchill GC, Okada Y, Thomas JM, Genazzani AA, Patel S, Galione A (2002) NAADP mobilizes Ca^{2+} from reserve granules, lysosome-related organelles, in sea urchin eggs. Cell 111(5):703–708

147. Repnik U, Cesen MH, Turk B (2013) The endolysosomal system in cell death and survival. Cold Spring Harb Perspect Biol 5(1):a008755

148. Ruas M, Rietdorf K, Arredouani A, Davis LC, Lloyd-Evans E, Koegel H et al (2010) Purified TPC isoforms form NAADP receptors with distinct roles for Ca^{2+} signaling and endolysosomal trafficking. Curr Biol: CB 20(8):703–709

149. Ruas M, Davis LC, Chen CC, Morgan AJ, Chuang KT, Walseth TF et al (2015) Expression of Ca(2)(+)-permeable two-pore channels rescues NAADP signalling in TPC-deficient cells. EMBO J 34(13):1743–1758

150. Wang X, Zhang X, Dong XP, Samie M, Li X, Cheng X et al (2012) TPC proteins are phosphoinositide- activated sodium-selective ion channels in endosomes and lysosomes. Cell 151(2):372–383

151. Jentsch TJ, Hoegg-Beiler MB, Vogt J (2015) Departure gate of acidic Ca(2)(+) confirmed. EMBO J 34(13):1737–1739

152. Capel RA, Bolton EL, Lin WK, Aston D, Wang Y, Liu W et al (2015) Two-pore channels (TPC2s) and nicotinic acid adenine dinucleotide phosphate (NAADP) at Lysosomal-sarcoplasmic reticular junctions contribute to acute and chronic beta-adrenoceptor signaling in the heart. J Biol Chem 290(50):30087–30098

153. Naylor E, Arredouani A, Vasudevan SR, Lewis AM, Parkesh R, Mizote A et al (2009) Identification of a chemical probe for NAADP by virtual screening. Nat Chem Biol 5(4):220–226

154. Davidson SM, Foote K, Kunuthur S, Gosain R, Tan N, Tyser R et al (2015) Inhibition of NAADP signalling on reperfusion protects the heart by preventing lethal calcium oscillations via two-pore channel 1 and opening of the mitochondrial permeability transition pore. Cardiovasc Res 108(3):357–366

155. Djerada Z, Millart H (2013) Intracellular NAADP increase induced by extracellular NAADP via the P2Y11-like receptor. Biochem Biophys Res Commun 436(2):199–203

156. Djerada Z, Peyret H, Dukic S, Millart H (2013) Extracellular NAADP affords cardioprotection against ischemia and reperfusion injury and involves the P2Y11-like receptor. Biochem Biophys Res Commun 434(3):428–433

157. Cancela JM, Churchill GC, Galione A (1999) Coordination of agonist-induced Ca^{2+}−signalling patterns by NAADP in pancreatic acinar cells. Nature 398(6722):74–76

158. Parkesh R, Lewis AM, Aley PK, Arredouani A, Rossi S, Tavares R et al (2008) Cell-permeant NAADP: a novel chemical tool enabling the study of Ca^{2+} signalling in intact cells. Cell Calcium 43(6):531–538

159. Nebel M, Schwoerer AP, Warszta D, Siebrands CC, Limbrock AC, Swarbrick JM et al (2013) Nicotinic acid adenine dinucleotide phosphate (NAADP)-mediated calcium signaling and arrhythmias in the heart evoked by beta-adrenergic stimulation. J Biol Chem 288(22):16017–16030

160. Warszta D, Nebel M, Fliegert R, Guse AH (2014) NAD derived second messengers: role in spontaneous diastolic Ca^{2+} transients in murine cardiac myocytes. DNA Repair 23:69–78
161. Collins TP, Bayliss R, Churchill GC, Galione A, Terrar DA (2011) NAADP influences excitation-contraction coupling by releasing calcium from lysosomes in atrial myocytes. Cell Calcium 50(5):449–458
162. Zhu MX, Ma J, Parrington J, Calcraft PJ, Galione A, Evans AM (2010) Calcium signaling via two-pore channels: local or global, that is the question. Am J Physiol Cell Physiol 298(3):C430–C441
163. Kinnear NP, Boittin FX, Thomas JM, Galione A, Evans AM (2004) Lysosome-sarcoplasmic reticulum junctions. A trigger zone for calcium signaling by nicotinic acid adenine dinucleotide phosphate and endothelin-1. J Biol Chem 279(52):54319–54326

Chapter 17
Molecular Basis and Regulation of Store-Operated Calcium Entry

Jose J. Lopez, Isaac Jardin, Letizia Albarrán, Jose Sanchez-Collado, Carlos Cantonero, Gines M. Salido, Tarik Smani, and Juan A. Rosado

Abstract Store-operated Ca^{2+} entry (SOCE) is a ubiquitous mechanism for Ca^{2+} influx in mammalian cells with important physiological implications. Since the discovery of SOCE more than three decades ago, the mechanism that communicates the information about the amount of Ca^{2+} accumulated in the intracellular Ca^{2+} stores to the plasma membrane channels and the nature of these channels have been matters of intense investigation and debate. The stromal interaction molecule-1 (STIM1) has been identified as the Ca^{2+} sensor of the intracellular Ca^{2+} compartments that activates the store-operated channels. STIM1 regulates two types of store-dependent channels: the Ca^{2+} release-activated Ca^{2+} (CRAC) channels, formed by Orai1 subunits, that conduct the highly Ca^{2+} selective current I_{CRAC} and the cation permeable store-operated Ca^{2+} (SOC) channels, which consist of Orai1 and TRPC1 proteins and conduct the non-selective current I_{SOC}. While the crystal structure of *Drosophila* CRAC channel has already been solved, the architecture of the SOC channels still remains unclear. The dynamic interaction of STIM1 with the store-operated channels is modulated by a number of proteins that either support the formation of the functional STIM1-channel complex or protect the cell against Ca^{2+} overload.

Keywords SOCE · Orai · STIM · CRAC · Transient receptor potential (TRP) channels

J. J. Lopez · I. Jardin (✉) · L. Albarrán · J. Sanchez-Collado · C. Cantonero · G. M. Salido
· J. A. Rosado
Department of Physiology, Cell Physiology Research Group and Institute of Molecular Pathology
Biomarkers, University of Extremadura, Cáceres, Spain
e-mail: ijp@unex.es

T. Smani
Department of Medical Physiology and Biophysics and Group of Cardiovascular
Pathophysiology, Institute of Biomedicine of Sevilla (IBiS), Hospital Universitario Virgen del
Rocío/CSIC/University of Sevilla, Sevilla, Spain

© Springer Nature Switzerland AG 2020
M. S. Islam (ed.), *Calcium Signaling*, Advances in Experimental Medicine
and Biology 1131, https://doi.org/10.1007/978-3-030-12457-1_17

Abbreviations

$[Ca^{2+}]_c$	cytosolic free Ca^{2+} concentration
AMPA	alpha-amino-3-hydroxy-5-methylisoxazole-4-propionate
CRAC channels	Ca^{2+}-release activated Ca^{2+} channels
CTID	C-terminal inhibitory domain
ER	endoplasmic reticulum
IP_3	inositol 1,4,5-trisphosphate
NAADP	nicotinic acid adenine dinucleotide phosphate
NMDA	N-methyl-D-aspartate
OAG	1-oleoyl-2-acetyl-sn-glycerol
PM	plasma membrane
ROC	receptor-operated channels
ROS	reactive oxygen species
SERCA	sarco/endoplasmic reticulum Ca^{2+} ATPase
SMOC	second messenger-operated channels
SOAP	STIM1-Orai1 Association Pocket
SOC channels	store-operated Ca^{2+} channels
SOCE	store-operated Ca^{2+} entry

17.1 Receptor-Operated Calcium Entry

Among the second messengers that mediate the function of cellular agonists, cytosolic free-Ca^{2+} stands out both for its versatility and ubiquity. In contrast to other intracellular messengers, Ca^{2+} signals do not depend on the reversible generation or covalent modification of signalling molecules but on the transport of Ca^{2+} across membranes. Cellular agonists increase cytosolic free-Ca^{2+} concentration ($[Ca^{2+}]_c$) either by releasing compartmentalized Ca^{2+} and/or by promoting Ca^{2+} influx from the extracellular medium through plasma membrane (PM) channels [1]. Both mechanisms occur in favour of an electrochemical gradient and, therefore, do not consume energy. Once cellular stimulation is terminated a number of mechanisms, involved in the recovery and maintenance of low resting $[Ca^{2+}]_c$, activate in order to allow further Ca^{2+} signals and to prevent cytosolic Ca^{2+} overload. Ca^{2+} removal from the cytosol is mediated by different mechanisms that includes Ca^{2+} reuptake by the sarco/endoplasmic reticulum Ca^{2+}-ATPase (SERCA) into intracellular organelles, especially the sarcoplasmic/endoplasmic reticulum [2, 3] and a variety of acidic, lysosome-related, organelles [4, 5]. SERCA pumps are also responsible to maintain a low resting $[Ca^{2+}]_c$ by opposing Ca^{2+} leak into the cytosol from the intracellular stores [6]. Further mechanisms for cytosolic Ca^{2+} clearance includes Ca^{2+} extrusion across the PM through the collaborative actions of the plasma membrane Ca^{2+}-ATPase and the Na^+/Ca^{2+} exchangers [7] (Fig. 17.1), as well as mitochondrial Ca^{2+} uptake, which also shapes the responses to agonists.

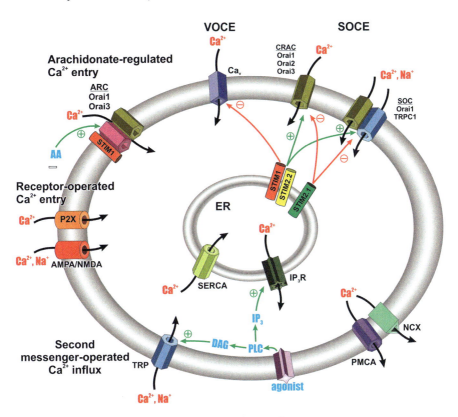

Fig. 17.1 Schematic representation of the main cellular Ca^{2+} entry mechanims. Cell stimulation with physiological agonists results in the activation of phospholipase C (PLC), which, in turn, leads to the generation of inositol 1,4,5 trisphosphate (IP$_3$) and diacylglycerol (DAG). IP$_3$ releases Ca^{2+} from intracellular stores, which results in the activation of store-operated Ca^{2+} entry (SOCE)via CRAC and SOC channels through the participation of STIM1 and STIM2.2 isoforms and the modulator STIM2.1. On the other hand, DAG is the endogenous activator of certain TRP channels that participate in second messenger-operated Ca^{2+} influx. PLA$_2$-generated arachidonate activates a Ca^{2+} selective channel formed by Orai1 and Orai3 subunits named arachidonate-regulated channels (ARC). Agonists might also induce direct receptor-operated Ca^{2+} entry gating ionotropic receptors such as purinergic P2X and AMPA o NMDA glutamate receptors. Finally, voltage-operated Ca^{2+} channels play a major role in Ca^{2+} entry in electrically excitable cells. Calcium removal from the cytosol is mediated by ATPases and transporters, such as the sarco/endoplasmic reticulum Ca^{2+} ATPase (SERCA), the plasma membrane Ca^{2+} ATPase (PMCA) and the Na$^+$/Ca^{2+} exchangers

Physiological agonists induce Ca^{2+} release from intracellular compartments through the generation of a variety of signalling molecules that gate Ca^{2+} channels located in the membrane of the internal Ca^{2+} stores, i.e. the ER, secretory granules and lysosome-related organelles. Ca^{2+}-releasing molecules include the inositol 1,4,5-trisphosphate (IP$_3$), nicotinic acid adenine dinucleotide phosphate (NAADP), cyclic ADP ribose or sphingosine 1-phosphate [8, 9]. Given the negligible electrical potential across the ER membrane [10], Ca^{2+} efflux from this organelles is only driven by a chemical gradient generated by SERCA pumps. This is most likely the mechanism involved in Ca^{2+} efflux from secretory vesicles and lysosome-related organelles, where a slightly positive membrane potential (lumen positive) has been reported [11, 12].

Ca^{2+} efflux from finite intracellular compartments is transient and sometimes insufficient for the activation of certain cellular processes that require higher or more sustained rises in $[Ca^{2+}]_c$. In this context Ca^{2+} entry through PM Ca^{2+}-permeable channels plays an essential functional role. Ca^{2+} influx leads to more sustained and/or greater increase in $[Ca^{2+}]_c$, which is required for intracellular Ca^{2+} store refilling and full activation of certain cellular processes. Extracellular Ca^{2+} might enter the cell by different mechanisms including voltage-operated Ca^{2+} entry and receptor-operated Ca^{2+} entry.

Voltage-operated Ca^{2+} entry takes place in electrically excitable cells where membrane depolarization leads to the opening of Ca^{2+} channels, which, in turn, results in Ca^{2+} transients that initiate different physiological events (Fig.17.1). Voltage-operated Ca^{2+} channels are heteromeric complexes where the pore-forming $\alpha 1$ subunits consists of four repeated domains, each containing six transmembrane domains (S1–S6) and a loop between transmembrane segments S5 and S6. The channel also contains an intracellular β subunit, a disulphide-linked $\alpha 2\delta$ dimer and a transmembrane γ subunit [13]. Voltage-operated Ca^{2+} channels are named according to the principal permeation ion (Ca) followed by the principal physiological modulator, i.e. voltage (v) as a subscript (Ca_v), then the numerical order corresponds to the nomenclature of the $\alpha 1$ subunit [14]. According to this nomenclature, three subfamilies have been described: $Ca_v 1$, which mediate L-type Ca^{2+} currents, $Ca_v 2$, responsible of the P/Q-type, N-type and R-type currents, and $Ca_v 3$, which mediate T-type Ca^{2+} currents [15].

Despite a functional expression of voltage-gated Ca^{2+} channels has been reported in a variety of electrically nonexcitable cells, such as T and B lymphocytes and mast cells [16–19], their role in cell biology is not as relevant as in excitable cells. In non-excitable cells, Ca^{2+} influx is mostly mediated by receptor-operated pathways. Heterologous regulation of voltage-operated channels upon activation of certain receptor-operated pathways has been described, where STIM1, the ER Ca^{2+} sensor for the activation of SOCE, induces inactivation of voltage-gated Ca^{2+} channels [20, 21] (Fig. 17.1). Receptor-operated Ca^{2+} entry groups the pathways for Ca^{2+} influx activated by membrane receptor occupation. This includes the Ca^{2+} currents that occurs by gating the receptor-channel itself, the properly named receptor-operated Ca^{2+} channels (ROC) and the currents via second messenger-operated channels (SMOC) or store-operated channels [22].

Receptor-operated Ca^{2+} entry occurs through a number of Ca^{2+} permeable ionotropic receptors activated by different ligands, such as the purinergic P2X receptors [23] or the N-methyl-D-aspartate (NMDA) and alpha-amino-3-hydroxy-5-methylisoxazole-4-propionate (AMPA) receptors activated by glutamate [24–26] (Fig. 17.1). On the other hand, second messenger-operated Ca^{2+} entry is mediated by Ca^{2+}-permeable SMOCs activated by diffusible messengers, which are generated by receptor occupation. Among the second messengers that have been reported to gate SMOCs are the phospholipase C product diacylglycerol, or its membrane permeant analogue 1-oleoyl-2-acetyl-sn-glycerol (OAG) [27–30], cyclic ADP ribose [31] or arachidonate [32–35]. The function of different SMOCs is modulated by phosphoinositides, such as phosphatidylinositol 4,5-bisphosphate [36, 37]. The third major pathway for Ca^{2+} influx activated by receptor occupation is store-operated Ca^{2+} entry (SOCE), a route regulated by the Ca^{2+} concentration into the agonist-sensitive stores [38].

17.2 Store-Operated Ca^{2+} Entry

Probably the most ubiquitous mechanism for receptor-operated Ca^{2+} influx is store-operated Ca^{2+} entry (SOCE), a mechanism for the entry of Ca^{2+} across PM-resident channels that is activated when the Ca^{2+} concentration decreases in the intracellular Ca^{2+} stores [39]. SOCE has been demonstrated upon discharge of different agonist-sensitive intracellular Ca^{2+} pools, including the ER [40] and bafilomycin A1-sensitive acidic compartments [41]. Store-operated or "capacitative" Ca^{2+} entry was initially proposed by Putney in 1986 [40] as a mechanism for Ca^{2+} influx regulated by the filling state of the intracellular Ca^{2+} stores. A previous study by Casteels and Droogmans reported similar observations in vascular smooth muscle cells, suggesting two pathways for the transport of Ca^{2+} from the extracellular medium to the noradrenalin-sensitive intracellular store including a direct route into the store [42]. This hypothesis, which limited the role of SOCE to the refilling of the intracellular Ca^{2+} compartments during or after agonist stimulation, was later on demonstrated to be imprecise as reported by Kwan and co-workers in 1990 [43]. In this study the authors used lanthanum to prevent Ca^{2+} entry and extrusion and demonstrated that Ca^{2+} released into the cytoplasm could replenish the stores. These findings provide initial evidence that the refilling process does not involve a direct route into the intracellular stores but rather is the result of a sequential Ca^{2+} transport into the cytoplasm and subsequent uptake from the stores by SERCA pumps [43, 44]. Therefore, SOCE is a relevant mechanism for Ca^{2+} store refilling and is also essential to sustain a number of cellular functions, including cell proliferation, differentiation, platelet aggregation and secretion, contraction or migration [45–52]. The study by Kwan and co-workers also provided a key conceptual characteristic of SOCE, by demonstrating that SOCE does not require the synthesis of IP$_3$ since Ca^{2+} influx mediated by methacholine and thapsigargin, an inhibitor of SERCA that is unable to increase the cellular level of IP$_3$, was mediated by the same mechanism activated by the discharge of the intracellular Ca^{2+} stores [43].

Since the identification of SOCE both the nature of the channels in the PM and the mechanism of activation of Ca^{2+} influx have been widely investigated. Early studies on the activation mechanism led to a variety of mutually exclusive hypotheses that can be grouped into (1) those suggesting a constitutive conformational coupling between elements in the ER and the PM, (2) the indirect coupling between the ER and the PM via diffusible messengers, (3) the insertion in the PM of preformed channels located in internal vesicles and (4) a reversible *de novo* conformational coupling between a Ca^{2+} sensor in the ER and the Ca^{2+} channels in the PM [53]. In 2005, the *ST*romal *I*nteraction *M*olecule-1 (STIM1) was identified as the ER Ca^{2+} sensor, using an RNA interference-based screen to identify genes involved in thapsigargin-evoked Ca^{2+} entry [54]. Furthermore, it was found that STIM1 communicates the information about the filling state of the ER to the SOC channels, as demonstrated by immunofluorescence, electron microscopy and EF-hand mutants of STIM1 that result in store-independent SOCE activation [55] (Fig. 17.1).

17.3 STIM1

Human STIM1 is a type-1 transmembrane protein of 685 amino acids that was initially reported to be involved in cell-to-cell interaction [56] and cell growth suppression with a relevant role in tumoral progression [57, 58]. STIM1 has been presented as a 90-kDa phosphoprotein ubiquitously expressed in the PM as well as in intracellular membranes of a number of human primary and tumoral cells [59]. The C-terminal region of STIM1 is cytosolic whatever its location (ER or PM), while the N-terminal region is located either in the ER lumen (for STIM1 located in the ER membrane) or the extracellular medium (for PM-resident STIM1) [60–62]. Structural studies revealed several functional domains in the STIM1 N-terminal region, including an ER signal peptide (aa 1–22), a canonical EF-hand Ca^{2+}-binding motif (aa 63–96) as well as a hidden EF-hand (aa 97–128), and a sterile alpha motif (SAM; aa 132–200) that plays an important role in protein-protein interaction [63–65]. STIM1 transmembrane domain (aa 214–234) is followed by the cytosolic region, which contains an ezrin/radixin/moesin domain, highly conserved among the STIM proteins. This region includes three coiled-coil domains, named CC1 (aa 238–342), CC2 (aa 364–389) and CC3 (aa 399–423), a CRAC-modulatory domain (aa 470–491), a proline/serine-rich region (aa 600–629), and a polybasic lysine-rich region (aa 671–685) [66–70].

The key region of STIM1 for the interaction with the store-operated channels was identified almost at the same time by four research groups. This region overlaps with the coiled-coil domains present in the cytosolic region of STIM1 and received different names: SOAR (STIM1 Orai-activating region, aa 344–442) [68], OASF (Orai-activating small fragment, 233–450; an extended version of OASF encompasses aa 234–491) [71, 72], CAD (CRAC-activating domain, aa 342–448) [73] and CCb9 (for coiled-coil domain containing region b9 encompassing aa 339–444) [74]. The crystal structure of the SOAR domain has revealed that SOAR appears as a dimer. Each subunit comprises four α-helical regions (α1, α2, α3, α4) [75], where the proposed strong Orai1 activation site is located between helixes α1 and $\alpha 3$, with the F^{394} residue playing a predominant role, and the $\alpha 4$ helix plays a relevant role in the maintenance of the SOAR dimer [76].

Overlapping with the CRAC-modulatory domain, a C-terminal inhibitory domain (CTID) has been described to modulate STIM1 function through the interaction with the modulator SARAF (see below). Deletion of the CTID region has been reported to induce spontaneous clustering of STIM1 and activation of Orai1 independently of the filling state of the Ca^{2+} stores [77]. STIM1 is subjected to a variety of post-translational modifications, including phosphorylation on serine [78] and tyrosine residues [79], and N-linked glycosylation at N^{131} and N^{171} located within in the SAM domain [80].

An alternative spliced long variant, STIM1L, has been found to be expressed in human skeletal muscle [81, 82], neonatal rat cardiomyocytes [83] and differentiated myotubes [82]. STIM1L is generated by alternative splicing and extension of exon 11 leading to a variant that contains an extra 106 amino acids (aa 515-620) insert in the cytosolic region. This region has been reported to provide STIM1 the ability to interact with Orai1 channels leading to the formation of permanent STIM1-Orai1 clusters that is associated with the rapid (< 1 s) activation of SOCE in skeletal muscle cells in comparison with other cells [84].

17.4   STIM2

The STIM1 homologue, STIM2, was identified in 2001 as a ubiquitously expressed protein that in vivo forms oligomers with STIM1 indicating a possible functional interaction between both proteins [78]. STIM2 is a type-1 transmembrane protein with similar structure to STIM1. The N-terminal region contains the functional Ca2+-binding EF-hand motif (aa 67–100) and SAM domain (aa 136–204). The cytosolic C-terminal region contains the ezrin/radixin/moesin domain with three coiled-coil domains [85]. The SOAR overlaps with this domain [86], and, as for STIM1, is involved in the activation of Orai [87]. Adjacent to the SOAR domain there is a proline- and histidine-rich region whose function is still unclear [78, 88]. Close to the end of the C-terminal region there is a calmodulin-binding region and a polybasic lysine-rich region [88, 89].

STIM2 is expressed in human and mouse tissues [78, 90], and it is the dominant STIM isoform in mouse brain, pancreas, placenta and heart; however, STIM2 is almost absent in skeletal muscle, kidney, liver and lung. It has been found co-expressed together with STIM1 in many human cell lines [78] and cell types [91–93], which indicates that both STIM isoforms co-exist and suggesting that they might interact functionally. STIM2 has been observed located in the ER membrane and in acidic intracellular stores [41, 85, 94], but, at present, no description of its location in the plasma membrane has been reported, in contrast to STIM1 [61, 67, 95, 96].

Three STIM2 splice variants have been described to date STIM2.1 (STIM2β), STIM2.2 (STIM2α) and STIM2.3. The best known and characterized variant is

STIM2.2 [97] and most of the previously published studies on STIM2 refer to STIM2.2 [98]. The *STIM2* gene comprises 13 exons, but the STIM2.2 mRNA is encoded by 12 exons (excluding exon 9) leading to a protein of 833 amino acids [87]. The STIM2.1 variant contains an eight-residue insert (383-VAASYLIQ-392) within the SOAR domain encoded by exon 9 that was initially found to impair the interaction with Orai1 and its activation [87]. However, more recent studies have revealed that the heterodimer consisting of the SOAR regions of STIM1 and STIM2.1 is able to induce full activation of Orai1 while preventing cross-linking and clustering of this channel [99]. STIM2.1 is ubiquitously expressed and might heterodimerize with STIM1 and STIM2.2 leading to the attenuation of Ca^{2+} influx through SOCE [87]. The third STIM2 variant, STIM2.3, expresses an alternative exon 13 that results in an upstream end of translation leading to a protein approximately 17 kDa smaller [97]. The expression of STIM2.3 is limited and its function is still uncertain [97].

The EF-hand motif of STIM1 and STIM2 binds Ca^{2+} and provide STIM proteins the ability to sense the ER Ca^{2+} concentration ($[Ca^{2+}]_{ER}$). STIM2 EF-hand exhibits a greater affinity for Ca^{2+} than that of STIM1 (STIM2 EF-hand motif Kd~0.5 mM, STIM1 EF-hand motif Kd~0.6 mM) [64, 100]. As a result, STIM2 is expected to show a greater sensitivity to minor changes in $[Ca^{2+}]_{ER}$ than STIM1. Consequently, STIM2 has been reported to be partially active at resting $[Ca^{2+}]_{ER}$ and is further activated by small reductions in $[Ca^{2+}]_{ER}$; on the other hand, STIM1 requires larger reductions in $[Ca^{2+}]_{ER}$ to become active [101]. The greater sensitivity for free-Ca^{2+} confers STIM2 the ability to sense $[Ca^{2+}]_{ER}$ fluctuations, stabilizing both basal cytosolic and endoplasmic reticulum Ca^{2+} levels, as well as to activate earlier than STIM1 upon agonist stimulation [101]. Furthermore, STIM2 has been reported to trigger a conformational remodeling of STIM1 C-terminus, leading to STIM1-Orai1 interaction at relatively high $[Ca^{2+}]_{ER}$ [102].

The characterization of STIM1 as the Ca^{2+} sensor facilitated the identification of the channels involved in SOCE. After intense research and debate two types of STIM1-modulated SOC channels have been described so far: the Ca^{2+}-release activated Ca^{2+} (CRAC) channels, which involves Orai1 subunits, and store-operated Ca^{2+} (SOC) channels, involving subunits of Orai1 and the canonical transient receptor potential (TRPC) family member TRPC1 [103–106].

17.5   STIM and Orai: CRAC Channels

Soon after the identification of SOCE, a highly Ca^{2+}-selective current activated by discharge of the intracellular Ca^{2+} stores was identified in mast cells and named Calcium-Release Activated Current (I_{CRAC}). This current was described as a voltage-insensitive current with a characteristic inward rectification [107]. More than a decade after the identification of I_{CRAC}, in 2006, the channel conducting this current was identified, through whole-genome screening of *Drosophila* S2 cells and gene mapping in patients with the hereditary severe combined immune deficiency

(SCID) syndrome induced by I_{CRAC} deficiency, as the protein Orai1 [108–110]. In Greek mythology the Orai is the keeper of the heaven´s gate [111]. Orai1 is a protein that shares no homology with other known ion channels. It is a protein of 301 amino acids that shows four transmembrane domains (TM1-TM4) and cytosolic N- and C-terminal tails [112–115], which are both required for STIM1 interaction and regulation [68, 71, 73, 116–118].

Despite the initial studies pointed out to a tetramer as the most likely Orai1 subunit stoichiometry of the mammalian CRAC channels [119–121], crystallization of *Drosophila* Orai has revealed that the channel consists of an hexameric assembly of Orai subunits [122]. Although the crystal structure of mammalian Orai1 is not available the analysis of the biophysical properties of concatenated hexameric and tetrameric human Orai1 channels has shown that a tetrameric architecture displays the highly Ca^{2+}-selective conductance characteristic of I_{CRAC} and endogenous CRAC channels, while the hexameric structure forms a non-selective cation channel [123].

Concerning the hexameric CRAC channel, the pore is located amid the hexamer formed by the six TM1 domains and also including the residues 74–90 (named ETON region) within the N-terminus, which is essential for the interaction with STIM1 [116]. The channel pore has been reported to acts as a funnel formed by an external vestibule, which consist of negatively charged residues (D110, D112 and D114) that attract Ca^{2+} to the immediacies to the pore, followed by the selectivity filter (aa E106), which is required for the CRAC channel high Ca^{2+} selectivity [124, 125], a hydrophobic region (aa V102, F99 and L95) and a basic region (aa R91, K87 and R83). The channel pore is surrounded by three rings comprised by TM2, TM3 and TM4 [122]. A number of residues within TM2-TM4 have been reported to be key regulators of CRAC channel function. The Orai1 mutations L138F and P245L, located in TM2 and TM4, have been shown to trigger constitutive currents that underlie the pathogenesis of different disorders, such as tubular aggregate myopathy (TAM) [126] and Stormorken disease [127], respectively.

The Orai family includes three human homologs, Orai1, Orai2 and Orai3. All three Orai isoforms are activated by Ca^{2+} store depletion when co-expressed with STIM1 [128, 129]; however, the Orai isoforms show different sensitivity to 2-aminoethoxydiphenyl borate (2-APB), an IP_3 receptor antagonist and Ca^{2+} channel blocker [130, 131]. While 2-APB has been reported to act as a direct Orai3 activator, its effect on Orai1 and Orai2 function is concentration-dependent. Low 2-APB concentrations (5–10 μM) activate Orai1 and Orai2 currents when co-expressed with STIM1; by contrast, at high concentrations (50 μM), 2-APB is a potent inhibitor of Orai1 currents [124, 132, 133]. In addition to the inhibitory role on Orai1 channels, 2-APB can also directly inhibit STIM1 function by promoting intramolecular interaction between the CC1 and SOAR regions of STIM1 [134]. Orai3 expression has recently been reported to be regulated by the protein stanniocalcin-2 in mouse platelets, so that platelets from mice lacking stanniocalcin-2 exhibit a significantly greater Orai3 expression, leading to enhanced agonist-induced Ca^{2+} mobilization and platelet aggregation [135].

The mechanism of activation of CRAC channels by STIM1 has long been investigated. In its resting state the canonical EF-hand domain is occupied by Ca^{2+}, STIM1 appears as a dimer and the SOAR region is hidden from the channel due to an autoinhibitory electrostatic interaction between an acidic segment within the CC1 and a polybasic region (aa 382–386) within the SOAR domain [136, 137]. Discharge of the ER upon agonist stimulation leads to Ca^{2+} dissociation from the EF-hand, which, in turn, changes the EF-hand-SAM conformation facilitating the formation of STIM1 oligomers [100, 138]. This conformational change lessens the angle of interaction of the TM region within the ER membrane bringing the C-termini together [139] and inducing the release of the SOAR domain from the autoinhibitory interaction with the CC1 region [116, 140–142]. STIM1 interacts with the N- and C-termini of Orai1. The interaction between STIM1 and Orai1 C-terminus was solved by NMR and involves the positively charged residues that interact with the CC1 autoinhibitory domain (K382, K384, K385 and K386), two aromatic residues Y361 and Y362 and four hydrophobic amino acids (L347, L351, L373 and A376) within STIM1 as well as Orai1 residues L273, L276, R281, L286 and R289 from its C-terminus, forming altogether the so called STIM1-Orai1 Association Pocket (SOAP) [141]. It has been reported that STIM binding to the Orai C-terminus orients STIM for effective interaction with the N-terminus [143]. The interaction between STIM1 and Orai1 N-terminus is essential for channel gating [144]; however, the sequence in STIM1 and in Orai1 N-terminus is not completely solved yet, although it has been proposed that Orai1 N-terminus might interact with a sequence adjacent to the SOAP previously described [145]. Furthermore, the STIM1-Orai1 stoichiometry in an efficient interaction has not been determined yet and further studies are still required to define these aspects.

17.6   STIM1, Orai1 and TRPC1: SOC Channels

In addition to I_{CRAC}, a number of store-operated currents with different biophysical properties have been described in a variety of cell types, including A431 epidermal cells, messangial cells, endothelial cells as well as aortic, portal vein and pulmonary artery myocytes, which have been named I_{SOC}. I_{SOC} currents are not selective for Ca^{2+} and exhibit a greater conductance than I_{CRAC} [146]. The activation of I_{SOC} involves the interaction among STIM1, Orai1 and TRPC1 [103, 104, 106, 147–150].

The TRP proteins form non-selective cation channels identified in the trp mutant of *Drosophila*. In 1969, Cosens and Manning reported a *Drosophila* mutant with altered electroretinogram [151]. In *Drosophila* photoreceptors, the sustained light-sensitive ionic current due to Na^+ and Ca^{2+} influx is conducted by two Ca^{2+}-permeable channels encoded by the *trp* and *trpl* genes [152, 153]. By contrast, the trp mutant exhibited transient, rather than sustained, light-sensitive receptor potential, which gave the name to transient receptor potential (TRP) channels [154]. Soon after the identification of *Drosophila* TRP proteins, the first mammalian TRP was identified, TRPC1, both in human [155, 156] and in mouse [157]. Since the identification of TRPC1 several TRP proteins have been identified in vertebrates, which have been grouped into seven subfamilies: four are closely related to *Drosophila* TRP (TRPC, TRPV, TRPA and TRPM), two are more distantly related subfamilies (TRPP and TRPML), and, finally, the TRPN group, which is expressed in fish, flies and worms [158]. Each subfamily includes different members, i.e. the TRPC subfamily comprises seven members (TRPC1-TRPC7), the vanilloid TRP subfamily (TRPV) includes six members (TRPV1–TRPV6), the TRPA (ankyrin) subfamily includes only one protein, TRPA1, the melastatin TRP subfamily (TRPM) comprises eight members (TRPM1-TRPM8) and both the TRPP (polycystin) and the TRPML (mucolipin) subfamilies include three members each [159, 160].

TRP channels are permeable to monovalent and divalent cations and exhibit a Ca^{2+}/Na^+ permeability ratio <10 [156], with some exceptions such as TRPM4 and TRPM5, which are selective for monovalent cations, and the highly Ca^{2+} selective TRPV5 and TRPV6 [161, 162]. The lack of Ca^{2+} selectivity for TRP channels raises the possibility that these channels are involved in the SOC channels [104]. TRP proteins share a common architecture, which resembles other ionic channels, including six transmembrane domains (TM1-TM6), with cytosolic N- and C-termini and a pore loop region between TM5 and TM6 [156, 163]. The C-terminal region includes a characteristic TRP signature motif (EWKFAR), involved in allosteric channel activation [164] and a CIRB (calmodulin/IP$_3$ receptor-binding) region, involved in the regulation of TRP channel gating [165, 166]. The N- and C-termini of TRP channels included coiled-coil domains that play a relevant role in the interaction with STIM1 [167].

Despite *Drosophila* TRP channels were described as receptor-operated channels activated by second messengers [168], the involvement of TRP channels in SOCE has been widely investigated and debated. Particular attention has been focused on the TRPC subfamily members, which have been found to be gated by Ca^{2+} store depletion in a number of cell types by using different approaches, from overexpression of specific TRPC proteins to knockdown of endogenous TRPs and knockout models [169–172]. There is now a consensus that TRPC1 forms a complex with STIM1 and Orai1, which is responsible of the less selective I_{SOC} current. There are two hypotheses concerning the molecular basis of I_{SOC}. One such hypothesis supports that the I_{SOC} current represents the sum of I_{CRAC} and a less selective current through TRPC channels [173], whereas another model suggests that I_{SOC} involves channels, of still unknown composition, but including both TRPC1 and Orai1 subunits [174].

STIM1 has been reported to interact with TRPC channels by direct association of the STIM1 SOAR domain with the N- and C-terminal coiled-coil domains of TRPC, as single mutations in these domains has been shown to reduce the interaction and activation of TRPC1 by STIM1 [167]. Apparently, STIM1 directly associates with certain TRPC proteins, such as TRPC1, TRPC4 and TRPC5, but the regulation of other TRPC channels, such as TRPC3 or TRPC6, is indirect and depends on TRPC1 and TRPC4, respectively. According to this model, at the resting state, TRPC3 N- and C-terminal coiled-coil domains interact with each other to prevent the interaction of the C-terminal coiled-coil domain with STIM1. Upon store depletion, the TRPC1 C-terminal coiled-coil domain dissociates this interaction to allow the association between the SOAR region within STIM1 and the TRPC3 C-terminal coiled-coil domain, and, thus, the regulation of TRPC3 channel by STIM1 [167]. In addition to the interaction between STIM1 and TRPCs described above, the activation of these channels by STIM1 require the electrostatic interaction between two positively charged residues located in the polybasic lysine-rich domain of STIM1 (K^{684} and K^{685}) and two negatively charged aspartates in TRPC1 (D^{639} and D^{640}), with a similar sequence described in TRPC3 [175].

17.7   Regulation of SOCE

Given the functional relevance of SOCE in cell biology, this mechanism is finely regulated by a number of biochemical interactions involving signalling molecules, physico-chemical conditions or Ca^{2+} itself. Soon after the identification of I_{CRAC} it was reported that an elevated $[Ca^{2+}]_c$ was able to inactivate the channel function in two different ways termed fast Ca^{2+}-dependent inactivation (FCDI), which occurs within milliseconds, and slow Ca^{2+}-dependent inactivation (SCDI) that initiates tens of seconds after CRAC channel activation [176–179].The rapid activation of FCDI, together with the fact that it does not depend on the I_{CRAC} amplitude led to the proposal that the inactivation is driven by Ca^{2+} binding to nearby cytosolic inactivating sites in close proximity to the channel pore [176, 180]. Different regions of Orai1 have been analysed and the two most prominent residues were identified as W76 and Y80 [181], located in the proposed Orai1 calmodulin-binding domain. However, the more recent publication of the crystal architecture of the pore-forming region of Orai1 revealed that residues W76 and Y80 face the pore lumen and, therefore, they are not freely accessible to calmodulin, as predicted [122]. Further studies using calmodulin dominant negative mutants have confirmed that FCDI might occur independently of calmodulin and suggest that W76 and Y80 might participate in a conformational change within the pore region leading to FCDI [182].

In contrast to FCDI, SCDI depends on global rises in $[Ca^{2+}]_c$ [177, 179]. Mitochondria have been reported to play a role in the modulation of SCDI. Ca^{2+} uptake by mitochondria upon rises in $[Ca^{2+}]_c$ due to the opening of CRAC channels slows down store refilling and attenuates the rate and extent of SCDI [183]. Current evidence supports a role of the protein SARAF (store-operated Ca^{2+} entry associated regulatory factor) in the activation of SCDI. SARAF is a 339-amino acid type I transmembrane protein that has been found to be located in the ER [184], as well as in the PM [35], location where its expression is regulated by STIM1 [185]. SARAF has been reported to interact with STIM1 modulating the Ca^{2+} fluxes and protecting the cell against Ca^{2+} overload [184]. Despite no functional domains have been identified yet, studies using SARAF constructs have revealed that the N-terminal region is essential for SARAF activity, while the C-terminal region is responsible for the interaction with STIM1 and the migration to ER-PM junctions in a STIM1-dependent manner [184]. SARAF interacts with the STIM1 SOAR region at rest to prevent spontaneous activation of STIM1 [184]. The interaction of SARAF with the SOAR region is modulated by an adjacent region termed CTID (aa 448–530), so that, in the resting state, the STIM1 SOAR-CTID region facilitates access of SARAF to SOAR to prevent spontaneous activation of SOCE, while store depletion, leading to a conformational change in STIM1, induces the dissociation of SARAF from the SOAR domain that facilitates the activation of SOCE. This process is followed by a more sustained reinteraction of SARAF with STIM1 (Fig. 17.2), which has been proposed as the mechanism underlying SCDI [77]. Analysis of the SARAF-STIM1 interaction has revealed that this association reaches a minimum 30s after store depletion by thapsigargin and then rises [186], a time-course that is compatible with the activation of SCDI [177, 179]. In addition to STIM1, we have recently reported that SARAF interacts with and modulates the function of Orai1 and TRPC1 channels. Studies in cells with low STIM1 expression have revealed that SARAF enhances Orai1-mediated Ca^{2+} entry and, conversely, attenuates Ca^{2+} influx mediated by TRPC1 channels [186, 187]. Altogether, the role of SARAF in the regulation of Orai1 and TRPC1 function strongly depends on STIM1 expression. In STIM1-expressing cells, SARAF negatively regulates Ca^{2+} influx through the

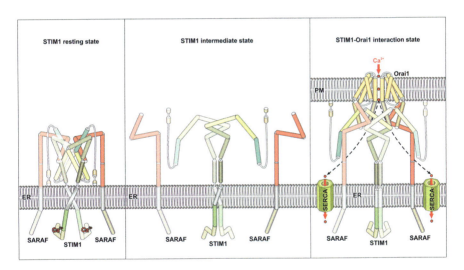

Fig. 17.2 Interaction of SARAF with STIM1. In the resting state, the canonical EF-hand motif is occupied by Ca^{2+} and STIM1 appears as a dimer in a autoinhibited configuration mediated by electrostatic interaction between an acidic segment within the CC1 and a polybasic region within the SOAR domain [136, 137]. In this configuration, SARAF is associated with the SOAR region. Discharge of the intracellular Ca^{2+} stores is followed by Ca^{2+} dissociation from the EF-hand motif, which leads to the beginning of the conformational change of STIM1 towards an extended state and the dissociation of SARAF. STIM1, in the extended state, interacts with Orai1 channels in the plasma membrane and a few seconds after the discharge of the Ca^{2+} stores SARAF re-associates with STIM1, which might lead to SCDI [77]. The Ca^{2+} that enters the cytoplasm through the Orai1 channels is pumped into the ER by the sarco/endoplasmic reticulum Ca^{2+} ATPase (SERCA)

channels modulated by STIM1, including the store-operated CRAC and SOC channels [77, 184, 186], and the store-independent arachidonate-regulated Ca^{2+} (ARC) channel [35], probably as a mechanism to prevent Ca^{2+} overload. On the other hand, in the absence of STIM1, SARAF enhances Orai1 function while attenuating TRPC1 activation [186, 187], probably as a mechanism to prevent cation influx through a channel with greater conductance and facilitate Ca^{2+} entry through a more selective low-conductance channel. The role of SARAF in SCDI is observed only when the STIM1-Orai1 complex is located within $PI(4,5)P_2$-rich microdomain [188], which indicates that plasma membrane nanodomains play an important role in STIM1-Orai1 function [145].

In addition to SARAF, other molecular modulators have been proposed to regulate SOCE to prevent excessive Ca^{2+} influx. Among them, Golli, a member of the family of myelin basic proteins, has been reported to attenuate SOCE in Jurkat T cells, while T cells from Golli-deficient mice exhibits enhanced Ca^{2+} influx evoked by T cell receptor activation [189]. Although the mechanism of regulation is still uncertain, Golli has been found to interact with the C-terminal domain of STIM1 following store depletion and modulated by $[Ca^{2+}]_c$ [190]. Furthermore, the protein orosomucoid-like 3 (ORMDL3) has also been reported to act as a negative modulator of I_{CRAC} and SOCE in T cells. ORMDL3 attenuates Ca^{2+} uptake by mitochondria, thus resulting in inhibition of I_{CRAC} and SOCE by SCDI [191]. More recently, filamin A has been shown as a modulator of the interaction of STIM1 with the actin cytoskeleton and Orai1. Thus, in cells lacking filamin A, as well as in cells where filamin A expression had been attenuated using siRNA, it has been reported enhancement of the STIM1-Orai1 interaction and SOCE [192]. Other SOCE modulators includes CRACR2A, a cytosolic Ca^{2+}-binding protein that modulates STIM1–Orai1 complexes [193], and STIMATE, an ER-resident protein and facilitates STIM1 puncta formation and the activation of CRAC channels [194] (Table 17.1).

SOCE is further subjected to a variety of physico-chemical conditions. For instance, Ca^{2+} influx through SOCE is subjected to the electrochemical gradient across the cell membrane, which, at physiological conditions, favours Ca^{2+} entry into the cytoplasm. However, the opening of non-selective cation channels or closure of K^+ channels might significantly alter the membrane potential and, thus, the driving force for Ca^{2+} influx. Similarly, Orai1 has been found to be sensitive to pH,

Table 17.1 Biochemical components of the STIM1-regulated Ca^{2+} channels

Channel	Components	Regulators	Refs.
CRAC	Orai1α/Orai1β, Orai2, Orai3	STIM1, STIM2.1, STIM2.2, SARAF, CRACR2A, Golli, ORMDL3, STIMATE, filamin A.	[87, 97, 103, 128, 129, 189, 191–194]
SOC	Orai1α/Orai1β, TRPC1	STIM1, STIM2.1, STIM2.2, CRACR2A, Golli, ORMDL3, SARAF, filamin A.	[87, 97, 103, 190, 191, 193]
ARC	Orai1α, Orai3	STIM1, SARAF	[32, 33, 103]

with a decrease in Orai1 function at mildly acidic pH and opposite effects observed upon mild alkalinisation [145, 195]. The expression of STIM and Orai proteins has also been found to be sensitive to hypoxia, which has been reported to upregulate the expression of STIM1, STIM2, Orai1, Orai2 and TRPC6 at the protein level [196]. Finally, reactive oxygen species (ROS) have also been reported to alter STIM1-Orai1 function. For instance, ROS have been reported to activate SOCE in human platelets [197]. Furthermore, oxidative stress in DT40 cells has been found to induce STIM1 S-glutathionylation at C^{56}, an event that has been associated to a decrease in the affinity of STIM1 for [Ca^{2+}]$_{ER}$, thus mimicking the effect of store depletion and enhancing SOCE [198]. In sum, these mechanisms, together with different covalent modifications of the key players, such as phosphorylation of STIM1 [79, 199, 200] and Orai1 [201], fine tune SOCE in order to provide functional Ca^{2+} signals.

Acknowledgements   This work was supported by MINECO (Grant BFU2016-74932-C2-1-P/2-P) and Junta de Extremadura-FEDER (IB16046 and GR18061). IJ and JJL were supported by Juan de la Cierva Program (IJCI_2015-25665) and Junta de Extremadura-FEDER (IB16046), respectively.

References

1. Berridge MJ, Bootman MD, Roderick HL (2003) Calcium signalling: dynamics, homeostasis and remodelling. Nat Rev Mol Cell Biol 4(7):517–529
2. Bobe R, Bredoux R, Wuytack F, Quarck R, Kovacs T, Papp B, Corvazier E, Magnier C, Enouf J (1994) The rat platelet 97-kDa Ca^{2+}ATPase isoform is the sarcoendoplasmic reticulum Ca^{2+}ATPase 3 protein. J Biol Chem 269(2):1417–1424
3. Wuytack F, Papp B, Verboomen H, Raeymaekers L, Dode L, Bobe R, Enouf J, Bokkala S, Authi KS, Casteels R (1994) A sarco/endoplasmic reticulum Ca^{2+}-ATPase 3-type Ca^{2+} pump is expressed in platelets, in lymphoid cells, and in mast cells. J Biol Chem 269(2):1410–1416
4. Cavallini L, Coassin M, Alexandre A (1995) Two classes of agonist-sensitive Ca^{2+} stores in platelets, as identified by their differential sensitivity to 2,5-di-(tert-butyl)-1,4-benzohydroquinone and thapsigargin. Biochem J 310(Pt 2):449–452
5. Lopez JJ, Camello-Almaraz C, Pariente JA, Salido GM, Rosado JA (2005) Ca^{2+} accumulation into acidic organelles mediated by Ca^{2+}- and vacuolar H^{+}-ATPases in human platelets. Biochem J 390(Pt 1):243–252
6. Juska A, Redondo PC, Rosado JA, Salido GM (2005) Dynamics of calcium fluxes in human platelets assessed in calcium-free medium. Biochem Biophys Res Commun 334(3):779–786
7. Brini M, Carafoli E (2011) The plasma membrane Ca(2)+ ATPase and the plasma membrane sodium calcium exchanger cooperate in the regulation of cell calcium. Cold Spring Harb Perspect Biol 3(2)
8. Cancela JM (2001) Specific Ca^{2+} signaling evoked by cholecystokinin and acetylcholine: the roles of NAADP, cADPR, and IP$_3$. Annu Rev Physiol 63:99–117
9. Choe CU, Ehrlich BE (2006) The inositol 1,4,5-trisphosphate receptor (IP3R) and its regulators: sometimes good and sometimes bad teamwork. Sci STKE 2006(363):re15
10. Lam AK, Galione A (2013) The endoplasmic reticulum and junctional membrane communication during calcium signaling. Biochim Biophys Acta 1833(11):2542–2559
11. Loh YP, Tam WW, Russell JT (1984) Measurement of delta pH and membrane potential in secretory vesicles isolated from bovine pituitary intermediate lobe. J Biol Chem 259(13):8238–8245

12. Koivusalo M, Steinberg BE, Mason D, Grinstein S (2011) In situ measurement of the electrical potential across the lysosomal membrane using FRET. Traffic 12(8):972–982
13. Catterall WA (2011) Voltage-gated calcium channels. Cold Spring Harb Perspect Biol 3(8):a003947
14. Ertel EA, Campbell KP, Harpold MM, Hofmann F, Mori Y, Perez-Reyes E, Schwartz A, Snutch TP, Tanabe T, Birnbaumer L, Tsien RW, Catterall WA (2000) Nomenclature of voltage-gated calcium channels. Neuron 25(3):533–535
15. Catterall WA, Perez-Reyes E, Snutch TP, Striessnig J (2005) International Union of Pharmacology. XLVIII. Nomenclature and structure-function relationships of voltage-gated calcium channels. Pharmacol Rev 57(4):411–425
16. Suzuki Y, Inoue T, Ra C (2010) L-type Ca^{2+} channels: a new player in the regulation of Ca^{2+} signaling, cell activation and cell survival in immune cells. Mol Immunol 47(4):640–648
17. Suzuki Y, Yoshimaru T, Inoue T, Ra C (2009) Ca v 1.2 L-type Ca^{2+} channel protects mast cells against activation-induced cell death by preventing mitochondrial integrity disruption. Mol Immunol 46(11–12):2370–2380
18. Yoshimaru T, Suzuki Y, Inoue T, Ra C (2009) L-type Ca^{2+} channels in mast cells: activation by membrane depolarization and distinct roles in regulating mediator release from store-operated Ca^{2+} channels. Mol Immunol 46(7):1267–1277
19. Matza D, Flavell RA (2009) Roles of Ca(v) channels and AHNAK1 in T cells: the beauty and the beast. Immunol Rev 231(1):257–264
20. Dionisio N, Smani T, Woodard GE, Castellano A, Salido GM, Rosado JA (2015) Homer proteins mediate the interaction between STIM1 and Cav1.2 channels. Biochim Biophys Acta 1853(5):1145–1153
21. Wang Y, Deng X, Mancarella S, Hendron E, Eguchi S, Soboloff J, Tang XD, Gill DL (2010) The calcium store sensor, STIM1, reciprocally controls Orai and CaV1.2 channels. Science 330(6000):105–109
22. Sage SO (1992) Three routes for receptor-mediated Ca^{2+} entry. Curr Biol 2(6):312–314
23. North RA (2002) Molecular physiology of P2X receptors. Physiol Rev 82(4):1013–1067
24. Kyrozis A, Goldstein PA, Heath MJ, MacDermott AB (1995) Calcium entry through a subpopulation of AMPA receptors desensitized neighbouring NMDA receptors in rat dorsal horn neurons. J Physiol 485(Pt 2):373–381
25. Salazar H, Eibl C, Chebli M, Plested A (2017) Mechanism of partial agonism in AMPA-type glutamate receptors. Nat Commun 8:14327
26. Twomey EC, Yelshanskaya MV, Grassucci RA, Frank J, Sobolevsky AI (2017) Channel opening and gating mechanism in AMPA-subtype glutamate receptors. Nature 549(7670):60–65
27. Hofmann T, Obukhov AG, Schaefer M, Harteneck C, Gudermann T, Schultz G (1999) Direct activation of human TRPC6 and TRPC3 channels by diacylglycerol. Nature 397(6716):259–263
28. Venkatachalam K, Zheng F, Gill DL (2003) Regulation of canonical transient receptor potential (TRPC) channel function by diacylglycerol and protein kinase C. J Biol Chem 278(31):29031–29040
29. Berna-Erro A, Galan C, Dionisio N, Gomez LJ, Salido GM, Rosado JA (2012) Capacitative and non-capacitative signaling complexes in human platelets. Biochim Biophys Acta 1823(8):1242–1251
30. Jardin I, Gomez LJ, Salido GM, Rosado JA (2009) Dynamic interaction of hTRPC6 with the Orai1/STIM1 complex or hTRPC3 mediates its role in capacitative or non-capacitative Ca^{2+} entry pathways. Biochem J 420:267–276
31. Togashi K, Hara Y, Tominaga T, Higashi T, Konishi Y, Mori Y, Tominaga M (2006) TRPM2 activation by cyclic ADP-ribose at body temperature is involved in insulin secretion. EMBO J 25(9):1804–1815
32. Shuttleworth TJ (2009) Arachidonic acid, ARC channels, and Orai proteins. Cell Calcium 45(6):602–610

33. Shuttleworth TJ, Thompson JL, Mignen O (2007) STIM1 and the noncapacitative ARC channels. Cell Calcium 42(2):183–191

34. Shuttleworth TJ, Thompson JL, Mignen O (2004) ARC channels: a novel pathway for receptor-activated calcium entry. Physiology (Bethesda) 19:355–361

35. Albarran L, Lopez JJ, Woodard GE, Salido GM, Rosado JA (2016) Store-operated Ca^{2+} entry-associated regulatory factor (SARAF) plays an important role in the regulation of arachidonate-regulated Ca^{2+} (ARC) channels. J Biol Chem 291(13):6982–6988

36. Rohacs T, Nilius B (2007) Regulation of transient receptor potential (TRP) channels by phosphoinositides. Pflugers Arch 455(1):157–168

37. Jardin I, Redondo PC, Salido GM, Rosado JA (2008) Phosphatidylinositol 4,5-bisphosphate enhances store-operated calcium entry through hTRPC6 channel in human platelets. Biochim Biophys Acta 1783(1):84–97

38. Putney JW (2011) The physiological function of store-operated calcium entry. Neurochem Res 36(7):1157–1165

39. Putney JW Jr (2005) Capacitative calcium entry: sensing the calcium stores. J Cell Biol 169(3):381–382

40. Putney JW Jr (1986) A model for receptor-regulated calcium entry. Cell Calcium 7(1):1–12

41. Zbidi H, Jardin I, Woodard GE, Lopez JJ, Berna-Erro A, Salido GM, Rosado JA (2011) STIM1 and STIM2 are located in the acidic Ca^{2+} stores and associates with Orai1 upon depletion of the acidic stores in human platelets. J Biol Chem 286(14):12257–12270

42. Casteels R, Droogmans G (1981) Exchange characteristics of the noradrenaline-sensitive calcium store in vascular smooth muscle cells or rabbit ear artery. J Physiol 317:263–279

43. Kwan CY, Takemura H, Obie JF, Thastrup O, Putney JW Jr (1990) Effects of MeCh, thapsigargin, and La3+ on plasmalemmal and intracellular Ca^{2+} transport in lacrimal acinar cells. Am J Physiol 258(6 Pt 1):C1006–C1015

44. Rosado JA (2006) Discovering the mechanism of capacitative calcium entry. Am J Physiol Cell Physiol 291(6):C1104–C1106

45. Abdullaev IF, Bisaillon JM, Potier M, Gonzalez JC, Motiani RK, Trebak M (2008) Stim1 and Orai1 mediate CRAC currents and store-operated calcium entry important for endothelial cell proliferation. Circ Res 103(11):1289–1299

46. Darbellay B, Arnaudeau S, Konig S, Jousset H, Bader C, Demaurex N, Bernheim L (2009) STIM1- and Orai1-dependent store-operated calcium entry regulates human myoblast differentiation. J Biol Chem 284(8):5370–5380

47. Feske S (2011) Immunodeficiency due to defects in store-operated calcium entry. Ann NY Acad Sci 1238:74–90

48. Stiber J, Hawkins A, Zhang ZS, Wang S, Burch J, Graham V, Ward CC, Seth M, Finch E, Malouf N, Williams RS, Eu JP, Rosenberg P (2008) STIM1 signalling controls store-operated calcium entry required for development and contractile function in skeletal muscle. Nat Cell Biol 10(6):688–697

49. Yoshida J, Iwabuchi K, Matsui T, Ishibashi T, Masuoka T, Nishio M (2012) Knockdown of stromal interaction molecule 1 (STIM1) suppresses store-operated calcium entry, cell proliferation and tumorigenicity in human epidermoid carcinoma A431 cells. Biochem Pharmacol 84(12):1592–1603

50. Avila-Medina J, Calderon-Sanchez E, Gonzalez-Rodriguez P, Monje-Quiroga F, Rosado JA, Castellano A, Ordonez A, Smani T (2016) Orai1 and TRPC1 proteins Co-localize with CaV1.2 channels to form a signal complex in vascular smooth muscle cells. J Biol Chem 291(40):21148–21159

51. Diez-Bello R, Jardin I, Salido GM, Rosado JA (2017) Orai1 and Orai2 mediate store-operated calcium entry that regulates HL60 cell migration and FAK phosphorylation. Biochim Biophys Acta 1864(6):1064–1070

52. Galan C, Zbidi H, Bartegi A, Salido GM, Rosado JA (2009) STIM1, Orai1 and hTRPC1 are important for thrombin- and ADP-induced aggregation in human platelets. Arch Biochem Biophys 490(2):137–144

53. Albarran L, Lopez JJ, Salido GM, Rosado JA (2016) Historical overview of store-operated Ca^{2+} entry. Adv Exp Med Biol 898:3–24
54. Roos J, DiGregorio PJ, Yeromin AV, Ohlsen K, Lioudyno M, Zhang S, Safrina O, Kozak JA, Wagner SL, Cahalan MD, Velicelebi G, Stauderman KA (2005) STIM1, an essential and conserved component of store-operated Ca^{2+} channel function. J Cell Biol 169(3):435–445
55. Zhang SL, Yu Y, Roos J, Kozak JA, Deerinck TJ, Ellisman MH, Stauderman KA, Cahalan MD (2005) STIM1 is a Ca^{2+} sensor that activates CRAC channels and migrates from the Ca^{2+} store to the plasma membrane. Nature 437(7060):902–905
56. Oritani K, Kincade PW (1996) Identification of stromal cell products that interact with pre-B cells. J Cell Biol 134(3):771–782
57. Parker NJ, Begley CG, Smith PJ, Fox RM (1996) Molecular cloning of a novel human gene (D11S4896E) at chromosomal region 11p15.5. Genomics 37(2):253–256
58. Sabbioni S, Barbanti-Brodano G, Croce CM, Negrini M (1997) GOK: a gene at 11p15 involved in rhabdomyosarcoma and rhabdoid tumor development. Cancer Res 57(20):4493–4497
59. Manji SS, Parker NJ, Williams RT, van Stekelenburg L, Pearson RB, Dziadek M, Smith PJ (2000) STIM1: a novel phosphoprotein located at the cell surface. Biochim Biophys Acta 1481(1):147–155
60. Lopez JJ, Salido GM, Pariente JA, Rosado JA (2006) Interaction of STIM1 with endogenously expressed human canonical TRP1 upon depletion of intracellular Ca^{2+} stores. J Biol Chem 281(38):28254–28264
61. Spassova MA, Soboloff J, He LP, Xu W, Dziadek MA, Gill DL (2006) STIM1 has a plasma membrane role in the activation of store-operated Ca^{2+} channels. Proc Natl Acad Sci USA 103(11):4040–4045
62. Jardin I, Lopez JJ, Redondo PC, Salido GM, Rosado JA (2009) Store-operated Ca^{2+} entry is sensitive to the extracellular Ca^{2+} concentration through plasma membrane STIM1. Biochim Biophys Acta 1793(10):1614–1622
63. Stathopulos PB, Zheng L, Li GY, Plevin MJ, Ikura M (2008) Structural and mechanistic insights into STIM1-mediated initiation of store-operated calcium entry. Cell 135(1):110–122
64. Zheng L, Stathopulos PB, Li GY, Ikura M (2008) Biophysical characterization of the EF-hand and SAM domain containing Ca^{2+} sensory region of STIM1 and STIM2. Biochem Biophys Res Commun 369(1):240–246
65. Baba Y, Hayashi K, Fujii Y, Mizushima A, Watarai H, Wakamori M, Numaga T, Mori Y, Iino M, Hikida M, Kurosaki T (2006) Coupling of STIM1 to store-operated Ca^{2+} entry through its constitutive and inducible movement in the endoplasmic reticulum. Proc Natl Acad Sci USA 103(45):16704–16709
66. Derler I, Fahrner M, Muik M, Lackner B, Schindl R, Groschner K, Romanin C (2009) A Ca^{2+} release-activated Ca^{2+} (CRAC) Modulatory Domain (CMD) within STIM1 mediates fast Ca^{2+}-dependent Inactivation of ORAI1 channels. J Biol Chem 284(37):24933–24938
67. Jardin I, Dionisio N, Frischauf I, Berna-Erro A, Woodard GE, Lopez JJ, Salido GM, Rosado JA (2013) The polybasic lysine-rich domain of plasma membrane-resident STIM1 is essential for the modulation of store-operated divalent cation entry by extracellular calcium. Cell Signal 25(5):1328–1337
68. Yuan JP, Zeng W, Dorwart MR, Choi YJ, Worley PF, Muallem S (2009) SOAR and the polybasic STIM1 domains gate and regulate Orai channels. Nat Cell Biol 11(3):337–343
69. Li Z, Lu J, Xu P, Xie X, Chen L, Xu T (2007) Mapping the interacting domains of STIM1 and Orai1 in Ca^{2+} release-activated Ca^{2+} channel activation. J Biol Chem 282(40):29448–29456
70. Muik M, Frischauf I, Derler I, Fahrner M, Bergsmann J, Eder P, Schindl R, Hesch C, Polzinger B, Fritsch R, Kahr H, Madl J, Gruber H, Groschner K, Romanin C (2008) Dynamic coupling of the putative coiled-coil domain of ORAI1 with STIM1 mediates ORAI1 channel activation. J Biol Chem 283(12):8014–8022
71. Muik M, Fahrner M, Derler I, Schindl R, Bergsmann J, Frischauf I, Groschner K, Romanin C (2009) A cytosolic homomerization and a modulatory domain within STIM1 C terminus

determine coupling to ORAI1 channels. J Biol Chem 284(13):8421–8426

72. Muik M, Fahrner M, Schindl R, Stathopulos P, Frischauf I, Derler I, Plenk P, Lackner B, Groschner K, Ikura M, Romanin C (2011) STIM1 couples to ORAI1 via an intramolecular transition into an extended conformation. EMBO J 30(9):1678–1689

73. Park CY, Hoover PJ, Mullins FM, Bachhawat P, Covington ED, Raunser S, Walz T, Garcia KC, Dolmetsch RE, Lewis RS (2009) STIM1 clusters and activates CRAC channels via direct binding of a cytosolic domain to Orai1. Cell 136(5):876–890

74. Kawasaki T, Lange I, Feske S (2009) A minimal regulatory domain in the C terminus of STIM1 binds to and activates ORAI1 CRAC channels. Biochem Biophys Res Commun 385(1):49–54

75. Yang X, Jin H, Cai X, Li S, Shen Y (2012) Structural and mechanistic insights into the activation of Stromal interaction molecule 1 (STIM1). Proc Natl Acad Sci USA 109(15):5657–5662

76. Nwokonko RM, Cai X, Loktionova NA, Wang Y, Zhou Y, Gill DL (2017) The STIM-Orai pathway: conformational coupling between STIM and Orai in the activation of store-operated Ca^{2+} entry. Adv Exp Med Biol 993:83–98

77. Jha A, Ahuja M, Maleth J, Moreno CM, Yuan JP, Kim MS, Muallem S (2013) The STIM1 CTID domain determines access of SARAF to SOAR to regulate Orai1 channel function. J Cell Biol 202(1):71–79

78. Williams RT, Manji SS, Parker NJ, Hancock MS, Van Stekelenburg L, Eid JP, Senior PV, Kazenwadel JS, Shandala T, Saint R, Smith PJ, Dziadek MA (2001) Identification and characterization of the STIM (stromal interaction molecule) gene family: coding for a novel class of transmembrane proteins. Biochem J 357(Pt 3):673–685

79. Lopez E, Jardin I, Berna-Erro A, Bermejo N, Salido GM, Sage SO, Rosado JA, Redondo PC (2012) STIM1 tyrosine-phosphorylation is required for STIM1-Orai1 association in human platelets. Cell Signal 24(6):1315–1322

80. Williams RT, Senior PV, Van Stekelenburg L, Layton JE, Smith PJ, Dziadek MA (2002) Stromal interaction molecule 1 (STIM1), a transmembrane protein with growth suppressor activity, contains an extracellular SAM domain modified by N-linked glycosylation. Biochim Biophys Acta 1596(1):131–137

81. Horinouchi T, Higashi T, Higa T, Terada K, Mai Y, Aoyagi H, Hatate C, Nepal P, Horiguchi M, Harada T, Miwa S (2012) Different binding property of STIM1 and its novel splice variant STIM1L to Orai1, TRPC3, and TRPC6 channels. Biochem Biophys Res Commun 428(2):252–258

82. Darbellay B, Arnaudeau S, Bader CR, Konig S, Bernheim L (2011) STIM1L is a new actin-binding splice variant involved in fast repetitive Ca^{2+} release. J Cell Biol 194(2):335–346

83. Luo X, Hojayev B, Jiang N, Wang ZV, Tandan S, Rakalin A, Rothermel BA, Gillette TG, Hill JA (2012) STIM1-dependent store-operated Ca(2)(+) entry is required for pathological cardiac hypertrophy. J Mol Cell Cardiol 52(1):136–147

84. Rosado JA, Diez R, Smani T, Jardin I (2015) STIM and Orai1 variants in store-operated calcium entry. Front Pharmacol 6:325

85. Soboloff J, Spassova MA, Hewavitharana T, He LP, Xu W, Johnstone LS, Dziadek MA, Gill DL (2006) STIM2 is an inhibitor of STIM1-mediated store-operated Ca^{2+} Entry. Curr Biol 16(14):1465–1470

86. Wang JY, Sun J, Huang MY, Wang YS, Hou MF, Sun Y, He H, Krishna N, Chiu SJ, Lin S, Yang S, Chang WC (2014) STIM1 overexpression promotes colorectal cancer progression, cell motility and COX-2 expression. Oncogene 34:4358–4367

87. Rana A, Yen M, Sadaghiani AM, Malmersjo S, Park CY, Dolmetsch RE, Lewis RS (2015) Alternative splicing converts STIM2 from an activator to an inhibitor of store-operated calcium channels. J Cell Biol 209(5):653–669

88. Ercan E, Chung SH, Bhardwaj R, Seedorf M (2012) Di-arginine signals and the K-rich domain retain the Ca(2)(+) sensor STIM1 in the endoplasmic reticulum. Traffic 13(7):992–1003
89. Bauer MC, O'Connell D, Cahill DJ, Linse S (2008) Calmodulin binding to the polybasic C-termini of STIM proteins involved in store-operated calcium entry. Biochemistry 47(23):6089–6091
90. Berna-Erro A, Braun A, Kraft R, Kleinschnitz C, Schuhmann MK, Stegner D, Wultsch T, Eilers J, Meuth SG, Stoll G, Nieswandt B (2009) STIM2 regulates capacitive Ca^{2+} entry in neurons and plays a key role in hypoxic neuronal cell death. Sci Signal 2(93):ra67
91. Schuhmann MK, Stegner D, Berna-Erro A, Bittner S, Braun A, Kleinschnitz C, Stoll G, Wiendl H, Meuth SG, Nieswandt B (2010) Stromal interaction molecules 1 and 2 are key regulators of autoreactive T cell activation in murine autoimmune central nervous system inflammation. J Immunol 184(3):1536–1542
92. Oh-Hora M, Yamashita M, Hogan PG, Sharma S, Lamperti E, Chung W, Prakriya M, Feske S, Rao A (2008) Dual functions for the endoplasmic reticulum calcium sensors STIM1 and STIM2 in T cell activation and tolerance. Nat Immunol 9(4):432–443
93. Darbellay B, Arnaudeau S, Ceroni D, Bader CR, Konig S, Bernheim L (2010) Human muscle economy myoblast differentiation and excitation-contraction coupling use the same molecular partners, STIM1 and STIM2. J Biol Chem 285(29):22437–22447
94. Liou J, Kim ML, Heo WD, Jones JT, Myers JW, Ferrell JE Jr, Meyer T (2005) STIM is a Ca^{2+} sensor essential for Ca^{2+}-store-depletion-triggered Ca^{2+} influx. Curr Biol 15(13):1235–1241
95. Dionisio N, Galan C, Jardin I, Salido GM, Rosado JA (2011) Lipid rafts are essential for the regulation of SOCE by plasma membrane resident STIM1 in human platelets. Biochim Biophys Acta 1813(3):431–437
96. Mignen O, Thompson JL, Shuttleworth TJ (2007) STIM1 regulates Ca^{2+} entry via arachidonate-regulated Ca^{2+}-selective (ARC) channels without store depletion or translocation to the plasma membrane. J Physiol 579(Pt 3):703–715
97. Miederer AM, Alansary D, Schwar G, Lee PH, Jung M, Helms V, Niemeyer BA (2015) A STIM2 splice variant negatively regulates store-operated calcium entry. Nat Commun 6:6899
98. Berna-Erro A, Jardin I, Salido GM, Rosado JA (2017) Role of STIM2 in cell function and physiopathology. J Physiol 595(10):3111–3128
99. Zhou Y, Nwokonko RM, Cai X, Loktionova NA, Abdulqadir R, Xin P, Niemeyer BA, Wang Y, Trebak M, Gill DL (2018) Cross-linking of Orai1 channels by STIM proteins. Proc Natl Acad Sci USA 115(15):E3398–E3407
100. Stathopulos PB, Li GY, Plevin MJ, Ames JB, Ikura M (2006) Stored Ca^{2+} depletion-induced oligomerization of stromal interaction molecule 1 (STIM1) via the EF-SAM region: An initiation mechanism for capacitive Ca^{2+} entry. J Biol Chem 281(47):35855–35862
101. Brandman O, Liou J, Park WS, Meyer T (2007) STIM2 is a feedback regulator that stabilizes basal cytosolic and endoplasmic reticulum Ca^{2+} levels. Cell 131(7):1327–1339
102. Subedi KP, Ong HL, Son GY, Liu X, Ambudkar IS (2018) STIM2 induces activated conformation of STIM1 to control Orai1 function in ER-PM junctions. Cell Rep 23(2):522–534
103. Desai PN, Zhang X, Wu S, Janoshazi A, Bolimuntha S, Putney JW, Trebak M (2015) Multiple types of calcium channels arising from alternative translation initiation of the Orai1 message. Sci Signal 8(387):ra74
104. Ambudkar IS, de Souza LB, Ong HL (2017) TRPC1, Orai1, and STIM1 in SOCE: friends in tight spaces. Cell Calcium 63:33–39
105. Chung WY, Jha A, Ahuja M, Muallem S (2017) Ca^{2+} influx at the ER/PM junctions. Cell Calcium 63:29–32
106. Jardin I, Lopez JJ, Salido GM, Rosado JA (2008) Orai1 mediates the interaction between STIM1 and hTRPC1 and regulates the mode of activation of hTRPC1-forming Ca^{2+} channels. J Biol Chem 283(37):25296–25304
107. Hoth M, Penner R (1992) Depletion of intracellular calcium stores activates a calcium current in mast cells. Nature 355(6358):353–356

108. Feske S, Gwack Y, Prakriya M, Srikanth S, Puppel SH, Tanasa B, Hogan PG, Lewis RS, Daly M, Rao A (2006) A mutation in Orai1 causes immune deficiency by abrogating CRAC channel function. Nature 441(7090):179–185

109. Vig M, Peinelt C, Beck A, Koomoa DL, Rabah D, Koblan-Huberson M, Kraft S, Turner H, Fleig A, Penner R, Kinet JP (2006) CRACM1 is a plasma membrane protein essential for store-operated Ca^{2+} entry. Science 312(5777):1220–1223

110. Zhang SL, Yeromin AV, Zhang XH, Yu Y, Safrina O, Penna A, Roos J, Stauderman KA, Cahalan MD (2006) Genome-wide RNAi screen of Ca^{2+} influx identifies genes that regulate Ca^{2+} release-activated Ca^{2+} channel activity. Proc Natl Acad Sci USA 103(24):9357–9362

111. Guo RW, Huang L (2008) New insights into the activation mechanism of store-operated calcium channels: roles of STIM and Orai. J Zhejiang Univ Sci B 9(8):591–601

112. Mercer JC, Dehaven WI, Smyth JT, Wedel B, Boyles RR, Bird GS, Putney JW Jr (2006) Large store-operated calcium selective currents due to co-expression of Orai1 or Orai2 with the intracellular calcium sensor, Stim1. J Biol Chem 281(34):24979–24990

113. Peinelt C, Vig M, Koomoa DL, Beck A, Nadler MJ, Koblan-Huberson M, Lis A, Fleig A, Penner R, Kinet JP (2006) Amplification of CRAC current by STIM1 and CRACM1 (Orai1). Nat Cell Biol 8(7):771–773

114. Prakriya M, Feske S, Gwack Y, Srikanth S, Rao A, Hogan PG (2006) Orai1 is an essential pore subunit of the CRAC channel. Nature 443(7108):230–233

115. Soboloff J, Spassova MA, Tang XD, Hewavitharana T, Xu W, Gill DL (2006) Orai1 and STIM reconstitute store-operated calcium channel function. J Biol Chem 281(30):20661–20665

116. Derler I, Plenk P, Fahrner M, Muik M, Jardin I, Schindl R, Gruber HJ, Groschner K, Romanin C (2013) The extended transmembrane Orai1 N-terminal (ETON) region combines binding interface and gate for Orai1 activation by STIM1. J Biol Chem 288(40):29025–29034

117. Palty R, Isacoff EY (2015) Cooperative binding of Stromal Interaction Molecule 1 (STIM1) to the N and C termini of calcium release-activated calcium modulator 1 (Orai1). J Biol Chem 291:334–341

118. Palty R, Stanley C, Isacoff EY (2015) Critical role for Orai1 C-terminal domain and TM4 in CRAC channel gating. Cell Res 25(8):963–980

119. Penna A, Demuro A, Yeromin AV, Zhang SL, Safrina O, Parker I, Cahalan MD (2008) The CRAC channel consists of a tetramer formed by Stim-induced dimerization of Orai dimers. Nature 456(7218):116–120

120. Maruyama Y, Ogura T, Mio K, Kato K, Kaneko T, Kiyonaka S, Mori Y, Sato C (2009) Tetrameric Orai1 is a teardrop-shaped molecule with a long, tapered cytoplasmic domain. J Biol Chem 284(20):13676–13685

121. Mignen O, Thompson JL, Shuttleworth TJ (2008) Orai1 subunit stoichiometry of the mammalian CRAC channel pore. J Physiol 586(2):419–425

122. Hou X, Pedi L, Diver MM, Long SB (2012) Crystal structure of the calcium release-activated calcium channel Orai. Science 338(6112):1308–1313

123. Thompson JL, Shuttleworth TJ (2013) How many Orai's does it take to make a CRAC channel? Sci Rep 3:1961

124. Peinelt C, Lis A, Beck A, Fleig A, Penner R (2008) 2-Aminoethoxydiphenyl borate directly facilitates and indirectly inhibits STIM1-dependent gating of CRAC channels. J Physiol 586(13):3061–3073

125. Yamashita M, Navarro-Borelly L, McNally BA, Prakriya M (2007) Orai1 mutations alter ion permeation and Ca^{2+}-dependent fast inactivation of CRAC channels: evidence for coupling of permeation and gating. J Gen Physiol 130(5):525–540

126. Endo Y, Noguchi S, Hara Y, Hayashi YK, Motomura K, Miyatake S, Murakami N, Tanaka S, Yamashita M, Kizu R, Bamba M, Goto Y, Matsumoto N, Nonaka I, Nishino I (2015) Dominant mutations in ORAI1 cause tubular aggregate myopathy with hypocalcemia via constitutive activation of store-operated Ca(2)(+) channels. Hum Mol Genet 24(3):637–648

127. Nesin V, Wiley G, Kousi M, Ong EC, Lehmann T, Nicholl DJ, Suri M, Shahrizaila N, Katsanis N, Gaffney PM, Wierenga KJ, Tsiokas L (2014) Activating mutations in STIM1 and ORAI1 cause overlapping syndromes of tubular myopathy and congenital miosis. Proc Natl Acad Sci USA 111(11):4197–4202

128. Frischauf I, Muik M, Derler I, Bergsmann J, Fahrner M, Schindl R, Groschner K, Romanin C (2009) Molecular determinants of the coupling between STIM1 and Orai channels: differential activation of Orai1-3 channels by a STIM1 coiled-coil mutant. J Biol Chem 284(32):21696–21706

129. Lis A, Peinelt C, Beck A, Parvez S, Monteilh-Zoller M, Fleig A, Penner R (2007) CRACM1, CRACM2, and CRACM3 are store-operated Ca^{2+} channels with distinct functional properties. Curr Biol 17(9):794–800

130. Ma HT, Patterson RL, van Rossum DB, Birnbaumer L, Mikoshiba K, Gill DL (2000) Requirement of the inositol trisphosphate receptor for activation of store-operated Ca^{2+} channels. Science 287(5458):1647–1651

131. Diver JM, Sage SO, Rosado JA (2001) The inositol trisphosphate receptor antagonist 2-aminoethoxydiphenylborate (2-APB) blocks Ca^{2+} entry channels in human platelets: cautions for its use in studying Ca^{2+} influx. Cell Calcium 30(5):323–329

132. Wang Y, Deng X, Zhou Y, Hendron E, Mancarella S, Ritchie MF, Tang XD, Baba Y, Kurosaki T, Mori Y, Soboloff J, Gill DL (2009) STIM protein coupling in the activation of Orai channels. Proc Natl Acad Sci USA 106(18):7391–7396

133. Prakriya M, Lewis RS (2001) Potentiation and inhibition of Ca^{2+} release-activated Ca^{2+} channels by 2-aminoethyldiphenyl borate (2-APB) occurs independently of IP(3) receptors. J Physiol 536(Pt 1):3–19

134. Wei M, Zhou Y, Sun A, Ma G, He L, Zhou L, Zhang S, Liu J, Zhang SL, Gill DL, Wang Y (2016) Molecular mechanisms underlying inhibition of STIM1-Orai1-mediated Ca^{2+} entry induced by 2-aminoethoxydiphenyl borate. Pflugers Arch 468(11–12):2061–2074

135. Lopez JJ, Jardin I, Cantonero Chamorro C, Duran ML, Tarancon Rubio MJ, Reyes Panadero M, Jimenez F, Montero R, Gonzalez MJ, Martinez M, Hernandez MJ, Brull JM, Corbacho AJ, Delgado E, Granados MP, Gomez-Gordo L, Rosado JA, Redondo PC (2018) Involvement of stanniocalcins in the deregulation of glycaemia in obese mice and type 2 diabetic patients. J Cell Mol Med 22(1):684–694

136. Fahrner M, Derler I, Jardin I, Romanin C (2013) The STIM1/Orai signaling machinery. Channels (Austin) 7(5):330–343

137. Korzeniowski MK, Manjarres IM, Varnai P, Balla T (2010) Activation of STIM1-Orai1 involves an intramolecular switching mechanism. Sci Signal 3(148):ra82

138. Stathopulos PB, Zheng L, Ikura M (2009) Stromal interaction molecule (STIM) 1 and STIM2 calcium sensing regions exhibit distinct unfolding and oligomerization kinetics. J Biol Chem 284(2):728–732

139. Ma G, Wei M, He L, Liu C, Wu B, Zhang SL, Jing J, Liang X, Senes A, Tan P, Li S, Sun A, Bi Y, Zhong L, Si H, Shen Y, Li M, Lee MS, Zhou W, Wang J, Wang Y, Zhou Y (2015) Inside-out Ca^{2+} signalling prompted by STIM1 conformational switch. Nat Commun 6:7826

140. Fahrner M, Muik M, Schindl R, Butorac C, Stathopulos P, Zheng L, Jardin I, Ikura M, Romanin C (2014) A coiled-coil clamp controls both conformation and clustering of Stromal Interaction Molecule 1 (STIM1). J Biol Chem 289(48):33231–33244

141. Stathopulos PB, Schindl R, Fahrner M, Zheng L, Gasmi-Seabrook GM, Muik M, Romanin C, Ikura M (2013) STIM1/Orai1 coiled-coil interplay in the regulation of store-operated calcium entry. Nat Commun 4:2963

142. Hirve N, Rajanikanth V, Hogan PG, Gudlur A (2018) Coiled-coil formation conveys a STIM1 signal from ER Lumen to cytoplasm. Cell Rep 22(1):72–83

143. McNally BA, Somasundaram A, Jairaman A, Yamashita M, Prakriya M (2013) The C- and N-terminal STIM1 binding sites on Orai1 are required for both trapping and gating CRAC channels. J Physiol 591(11):2833–2850

144. Gudlur A, Quintana A, Zhou Y, Hirve N, Mahapatra S, Hogan PG (2014) STIM1 triggers a gating rearrangement at the extracellular mouth of the ORAI1 channel. Nat Commun 5:5164

145. Hogan PG, Rao A (2015) Store-operated calcium entry: mechanisms and modulation. Biochem Biophys Res Commun 460(1):40–49

146. Parekh AB, Putney JW Jr (2005) Store-operated calcium channels. Physiol Rev 85(2): 757–810

147. Brechard S, Melchior C, Plancon S, Schenten V, Tschirhart EJ (2008) Store-operated Ca^{2+} channels formed by TRPC1, TRPC6 and Orai1 and non-store-operated channels formed by TRPC3 are involved in the regulation of NADPH oxidase in HL-60 granulocytes. Cell Calcium 44(5):492–506

148. Galan C, Dionisio N, Smani T, Salido GM, Rosado JA (2011) The cytoskeleton plays a modulatory role in the association between STIM1 and the Ca^{2+} channel subunits Orai1 and TRPC1. Biochem Pharmacol 82(4):400–410

149. Sabourin J, Le Gal L, Saurwein L, Haefliger JA, Raddatz E, Allagnat F (2015) Store-operated Ca^{2+} entry mediated by Orai1 and TRPC1 participates to insulin secretion in Rat beta-Cells. J Biol Chem 290(51):30530–30539

150. Sampieri A, Zepeda A, Saldaña C, Salgado A, Vaca L (2008) STIM1 converts TRPC1 from a receptor-operated to a store-operated channel: moving TRPC1 in and out of lipid rafts. Cell Calcium 44(5):479–491

151. Cosens DJ, Manning A (1969) Abnormal electroretinogram from a Drosophila mutant. Nature 224(5216):285–287

152. Hardie RC, Minke B (1992) The trp gene is essential for a light-activated Ca^{2+} channel in Drosophila photoreceptors. Neuron 8(4):643–651

153. Phillips AM, Bull A, Kelly LE (1992) Identification of a Drosophila gene encoding a calmodulin-binding protein with homology to the trp phototransduction gene. Neuron 8(4):631–642

154. Minke B (1977) Drosophila mutant with a transducer defect. Biophys Struct Mech 3(1):59–64

155. Wes PD, Chevesich J, Jeromin A, Rosenberg C, Stetten G, Montell C (1995) TRPC1, a human homolog of a Drosophila store-operated channel. Proc Natl Acad Sci USA 92(21):9652–9656

156. Zhu X, Chu PB, Peyton M, Birnbaumer L (1995) Molecular cloning of a widely expressed human homologue for the Drosophila trp gene. FEBS Lett 373(3):193–198

157. Petersen CC, Berridge MJ, Borgese MF, Bennett DL (1995) Putative capacitative calcium entry channels: expression of Drosophila trp and evidence for the existence of vertebrate homologues. Biochem J 311(Pt 1):41–44

158. Montell C, Birnbaumer L, Flockerzi V, Bindels RJ, Bruford EA, Caterina MJ, Clapham DE, Harteneck C, Heller S, Julius D, Kojima I, Mori Y, Penner R, Prawitt D, Scharenberg AM, Schultz G, Shimizu N, Zhu MX (2002) A unified nomenclature for the superfamily of TRP cation channels. Mol Cell 9(2):229–231

159. Flockerzi V, Nilius B (2014) TRPs: truly remarkable proteins. Handb Exp Pharmacol 222:1–12

160. Salido GM, Jardin I, Rosado JA (2011) The TRPC Ion channels: association with Orai1 and STIM1 proteins and participation in capacitative and non-capacitative calcium entry. Adv Exp Med Biol 704:413–433

161. Montell C (2003) The venerable inveterate invertebrate TRP channels. Cell Calcium 33(5–6):409–417

162. Montell C, Birnbaumer L, Flockerzi V (2002) The TRP channels, a remarkably functional family. Cell 108(5):595–598

163. Rosado JA, Brownlow SL, Sage SO (2002) Endogenously expressed Trp1 is involved in store-mediated Ca^{2+} entry by conformational coupling in human platelets. J Biol Chem 277(44):42157–42163

164. Gregorio-Teruel L, Valente P, Gonzalez-Ros JM, Fernandez-Ballester G, Ferrer-Montiel A (2014) Mutation of I696 and W697 in the TRP box of vanilloid receptor subtype I modulates allosteric channel activation. J Gen Physiol 143(3):361–375

165. Wedel BJ, Vazquez G, McKay RR, St JBG, Putney JW Jr (2003) A calmodulin/inositol 1,4,5-trisphosphate (IP3) receptor-binding region targets TRPC3 to the plasma membrane in a calmodulin/IP3 receptor-independent process. J Biol Chem 278(28):25758–25765

166. Dionisio N, Albarran L, Berna-Erro A, Hernandez-Cruz JM, Salido GM, Rosado JA (2011) Functional role of the calmodulin- and inositol 1,4,5-trisphosphate receptor-binding (CIRB) site of TRPC6 in human platelet activation. Cell Signal 23(11):1850–1856

167. Lee KP, Choi S, Hong JH, Ahuja M, Graham S, Ma R, So I, Shin DM, Muallem S, Yuan JP (2014) Molecular determinants mediating gating of Transient Receptor Potential Canonical (TRPC) channels by stromal interaction molecule 1 (STIM1). J Biol Chem 289(10):6372–6382

168. Hardie RC (2003) Regulation of TRP channels via lipid second messengers. Annu Rev Physiol 65:735–759

169. Huang GN, Zeng W, Kim JY, Yuan JP, Han L, Muallem S, Worley PF (2006) STIM1 carboxyl-terminus activates native SOC, I(crac) and TRPC1 channels. Nat Cell Biol 8(9):1003–1010

170. Pani B, Liu X, Bollimuntha S, Cheng KT, Niesman IR, Zheng C, Achen VR, Patel HH, Ambudkar IS, Singh BB (2013) Impairment of TRPC1-STIM1 channel assembly and AQP5 translocation compromise agonist-stimulated fluid secretion in mice lacking caveolin1. J Cell Sci 126(Pt 2):667–675

171. Lopez E, Berna-Erro A, Salido GM, Rosado JA, Redondo PC (2013) FKBP52 is involved in the regulation of SOCE channels in the human platelets and MEG 01 cells. Biochim Biophys Acta 1833(3):652–662

172. Jardin I, Lopez JJ, Salido GM, Rosado JA (2008) Functional relevance of the de novo coupling between hTRPC1 and type II IP_3 receptor in store-operated Ca^{2+} entry in human platelets. Cell Signal 20(4):737–747

173. Cheng KT, Liu X, Ong HL, Swaim W, Ambudkar IS (2011) Local Ca^{2+} entry via Orai1 regulates plasma membrane recruitment of TRPC1 and controls cytosolic Ca^{2+} signals required for specific cell functions. PLoS Biol 9(3):e1001025

174. Ong EC, Nesin V, Long CL, Bai CX, Guz JL, Ivanov IP, Abramowitz J, Birnbaumer L, Humphrey MB, Tsiokas L (2013) A TRPC1 protein-dependent pathway regulates osteoclast formation and function. J Biol Chem 288(31):22219–22232

175. Zeng W, Yuan JP, Kim MS, Choi YJ, Huang GN, Worley PF, Muallem S (2008) STIM1 gates TRPC channels, but not Orai1, by electrostatic interaction. Mol Cell 32(3):439–448

176. Zweifach A, Lewis RS (1995) Rapid inactivation of depletion-activated calcium current (ICRAC) due to local calcium feedback. J Gen Physiol 105(2):209–226

177. Zweifach A, Lewis RS (1995) Slow calcium-dependent inactivation of depletion-activated calcium current. Store-dependent and -independent mechanisms. J Biol Chem 270(24):14445–14451

178. Parekh AB (2017) Regulation of CRAC channels by Ca^{2+}-dependent inactivation. Cell Calcium 63:20–23

179. Parekh AB (1998) Slow feedback inhibition of calcium release-activated calcium current by calcium entry. J Biol Chem 273(24):14925–14932

180. Fierro L, Parekh AB (1999) Fast calcium-dependent inactivation of calcium release-activated calcium current (CRAC) in RBL-1 cells. J Membr Biol 168(1):9–17

181. Liu Y, Zheng X, Mueller GA, Sobhany M, DeRose EF, Zhang Y, London RE, Birnbaumer L (2012) Crystal structure of calmodulin binding domain of orai1 in complex with Ca^{2+} calmodulin displays a unique binding mode. J Biol Chem 287(51):43030–43041

182. Mullins FM, Yen M, Lewis RS (2016) Orai1 pore residues control CRAC channel inactivation independently of calmodulin. J Gen Physiol 147(2):137–152

183. Gilabert JA, Parekh AB (2000) Respiring mitochondria determine the pattern of activation and inactivation of the store-operated Ca^{2+} current I(CRAC). EMBO J 19(23):6401–6407

184. Palty R, Raveh A, Kaminsky I, Meller R, Reuveny E (2012) SARAF inactivates the store operated calcium entry machinery to prevent excess calcium refilling. Cell 149(2):425–438

185. Albarran L, Regodon S, Salido GM, Lopez JJ, Rosado JA (2017) Role of STIM1 in the surface expression of SARAF. Channels (Austin) 11(1):84–88

186. Albarran L, Lopez JJ, Ben Amor N, Martín-Cano FE, Berna-Erro A, Smani T, Salido GM, Rosado JA (2016) Dynamic interaction of SARAF with STIM1 and Orai1 to modulate store-operated calcium entry. Scientific Reports 6:24452

187. Albarran L, Lopez JJ, Gomez LJ, Salido GM, Rosado JA (2016) SARAF modulates TRPC1, but not TRPC6, channel function in a STIM1-independent manner. Biochem J 473(20):3581–3595

188. Maleth J, Choi S, Muallem S, Ahuja M (2014) Translocation between PI(4,5)P2-poor and PI(4,5)P2-rich microdomains during store depletion determines STIM1 conformation and Orai1 gating. Nat Commun 5:5843
189. Feng JM, Hu YK, Xie LH, Colwell CS, Shao XM, Sun XP, Chen B, Tang H, Campagnoni AT (2006) Golli protein negatively regulates store depletion-induced calcium influx in T cells. Immunity 24(6):717–727
190. Walsh CM, Doherty MK, Tepikin AV, Burgoyne RD (2010) Evidence for an interaction between Golli and STIM1 in store-operated calcium entry. Biochem J 430(3):453–460
191. Carreras-Sureda A, Cantero-Recasens G, Rubio-Moscardo F, Kiefer K, Peinelt C, Niemeyer BA, Valverde MA, Vicente R (2013) ORMDL3 modulates store-operated calcium entry and lymphocyte activation. Hum Mol Genet 22(3):519–530
192. Lopez JJ, Albarran L, Jardin I, Sanchez-Collado J, Redondo PC, Bermejo N, Bobe R, Smani T, Rosado JA (2018) Filamin A modulates store-operated Ca^{2+} entry by regulating STIM1 (Stromal Interaction Molecule 1)-Orai1 association in human platelets. Arterioscler Thromb Vasc Biol 38:386–397
193. Srikanth S, Jung HJ, Kim KD, Souda P, Whitelegge J, Gwack Y (2010) A novel EF-hand protein, CRACR2A, is a cytosolic Ca^{2+} sensor that stabilizes CRAC channels in T cells. Nat Cell Biol 12(5):436–446
194. Jing J, He L, Sun A, Quintana A, Ding Y, Ma G, Tan P, Liang X, Zheng X, Chen L, Shi X, Zhang SL, Zhong L, Huang Y, Dong MQ, Walker CL, Hogan PG, Wang Y, Zhou Y (2015) Proteomic mapping of ER-PM junctions identifies STIMATE as a regulator of Ca(2)(+) influx. Nat Cell Biol 17(10):1339–1347
195. Beck A, Fleig A, Penner R, Peinelt C (2014) Regulation of endogenous and heterologous Ca(2)(+) release-activated Ca(2)(+) currents by pH. Cell Calcium 56(3):235–243
196. He X, Song S, Ayon RJ, Balisterieri A, Black SM, Makino A, Wier WG, Zang WJ, Yuan JX (2018) Hypoxia selectively upregulates cation channels and increases cytosolic [Ca^{2+}] in pulmonary, but not coronary, arterial smooth muscle cells. Am J Physiol Cell Physiol 314(4):C504–C517
197. Rosado JA, Redondo PC, Salido GM, Gomez-Arteta E, Sage SO, Pariente JA (2004) Hydrogen peroxide generation induces pp60src activation in human platelets: evidence for the involvement of this pathway in store-mediated calcium entry. J Biol Chem 279(3):1665–1675
198. Hawkins BJ, Irrinki KM, Mallilankaraman K, Lien YC, Wang Y, Bhanumathy CD, Subbiah R, Ritchie MF, Soboloff J, Baba Y, Kurosaki T, Joseph SK, Gill DL, Madesh M (2010) S-glutathionylation activates STIM1 and alters mitochondrial homeostasis. J Cell Biol 190(3):391–405
199. Smyth JT, Petranka JG, Boyles RR, DeHaven WI, Fukushima M, Johnson KL, Williams JG, Putney JW Jr (2009) Phosphorylation of STIM1 underlies suppression of store-operated calcium entry during mitosis. Nat Cell Biol 11(12):1465–1472
200. Sundivakkam PC, Natarajan V, Malik AB, Tiruppathi C (2013) Store-operated Ca^{2+} entry (SOCE) induced by protease-activated receptor-1 mediates STIM1 protein phosphorylation to inhibit SOCE in endothelial cells through AMP-activated protein kinase and p38beta mitogen-activated protein kinase. J Biol Chem 288(23):17030–17041
201. Kawasaki T, Ueyama T, Lange I, Feske S, Saito N (2010) Protein kinase C-induced phosphorylation of Orai1 regulates the intracellular Ca^{2+} level via the store-operated Ca^{2+} channel. J Biol Chem 285(33):25720–25730

Chapter 18
Canonical Transient Potential Receptor-3 Channels in Normal and Diseased Airway Smooth Muscle Cells

Yong-Xiao Wang, Lan Wang, and Yun-Min Zheng

Abstract All seven canonical transient potential receptor (TRPC1–7) channel members are expressed in mammalian airway smooth muscle cells (ASMCs). Among this family, TRPC3 channel plays an important role in the control of the resting $[Ca^{2+}]_i$ and agonist-induced increase in $[Ca^{2+}]_i$. This channel is significantly upregulated in molecular expression and functional activity in airway diseases. The upregulated channel significantly augments the resting $[Ca^{2+}]_i$ and agonist-induced increase in $[Ca^{2+}]_i$, thereby exerting a direct and essential effect in airway hyperresponsiveness. The increased TRPC3 channel-mediated Ca^{2+} signaling also results in the transcription factor nuclear factor-κB (NF-κB) activation via protein kinase C-α (PKCα)-dependent inhibitor of NFκB-α (IκBα) and calcineurin-dependent IκBβ signaling pathways, which upregulates cyclin-D1 expression and causes cell proliferation, leading to airway remodeling. TRPC3 channel may further interact with intracellular release Ca^{2+} channels, Orai channels and Ca^{2+}-sensing stromal interaction molecules, mediating important cellular responses in ASMCs and the development of airway diseases.

Keywords Canonical transient potential receptor channel · Inositol 1 · 4 · 5-trisphosphate receptor · Ryanodine receptor · Orai channel · Stromal interaction molecule · Nuclear factor κB · Protein kinase C · Calcineurin · Airway hyperresponsiveness · Airway remodeling · Airway diseases

Authors "Yong-Xiao Wang and Lan Wang" have contributed equally for this chapter

Y.-X. Wang · Y.-M. Zheng (✉)
Department of Molecular and Cellular Physiology, Albany Medical College, Albany, NY, USA

L. Wang
Department of Molecular and Cellular Physiology, Albany Medical College, Albany, NY, USA

Department of Cardiopulmonary Circulation, Shanghai Pulmonary Hospital, Tongji University School of Medicine, Shanghai, China
e-mail: zhengy@amc.edu

© Springer Nature Switzerland AG 2020
M. S. Islam (ed.), *Calcium Signaling*, Advances in Experimental Medicine and Biology 1131, https://doi.org/10.1007/978-3-030-12457-1_18

Abbreviations

$[Ca^{2+}]i$	Intracellular Ca^{2+} concentration
ASMCs	Airway smooth muscle cells
COPD	chronic obstructive pulmonary disease
DAG	Diacylglycerol
GPCR	G protein-coupled receptor
IP_3	Inositol 1,4,5-trisphosphate
IP_3R	IP_3 receptor
IκB	Nuclear factor κB inhibitor
JNK	Jun amino-terminal kinase
mAch	Methacholine
NF-κB	Nuclear factor κB
NPo	Open probability
NSCC	Non-selective cation channel
OAG	1-oleoyl-2-acetyl-sn-glycerol
PIP_2	Phosphatidylinositol 4,5-bisphosphate
PKCα	Protein kinase C-α
PLC	Phospholipase C
RyR	Ryanodine receptor
SOCE	Store-operated Ca^{2+} entry
SR	Sarcoplasmic reticulum
STIM	Stromal interaction molecule
TNFα	Tumor necrosis factor-α
TRPC	Canonical transient potential receptor
Vm	Membrane potential

18.1   Introduction

The canonical transient potential receptor (TRP) channels are encoded by genes that most closely related resemble the *trp* gene, which was originally identified in *Drosophila* [1]. Photoreceptor cells in *Drosophila* produce a transient receptor potential in response to light. This potential is comprised of an initial rapid spike followed by a sustained phase. Both phases are mediated by TRP encoded and TRP-like channels. The first mammalian TRP gene was cloned from the human brain using an expressed sequence tag, categorized into the TRPC channel family based on its primary amino acid sequence, and thus termed TRPC1 channel [2]. The TRPC channel family is known to consist of seven members designated TRPC1–7 channels.

TRPC channels have been well investigated in a number of cell types; however, their functional roles and underlying signaling mechanisms are not well known in airway smooth muscle cells (ASMCs). A series of our recent studies using the patch clamp technique, genetically-manipulated approach and other methods have started to meticulously address whether, which and how TRPC channels play an important role in physiological and pathological cellular responses in ASMCs. In this article, we intend to provide a comprehensive overview of the major exciting findings from our research works and others, with particularly focus on the molecular expression, functional roles and underlying signaling mechanisms of TRPC3 channels in physiological cellular responses in ASMCs and airway diseases.

18.2 Multiple TRPC Channel Members Are Expressed, but TRPC3 Channel Is the Predominant Member in ASMCs

Ong HL et al. reported that TRPC1, TRPC2, TRPC3, TRPC4, TRPC5 and TRPC6 channel mRNAs are detected in primary isolated guinea-pig airway SMCs [3]. It has also been reported that TRPC1, TRPC3, TRPC4 and TRPC6, but not TRPC2 and TRPC5 channel mRNAs are present in cultured human ASMCs [4]. In cultured human ASMCs, TRPC1, TRPC3, TRPC4, TRPC5 and TRPC6, but not TRPC2 channel mRNAs have been found [5]. Prior studies detected expression of TRPC1,TRPC3 and TRPC6 channel proteins in primary isolated guinea pig airway SM cells and tissues [3, 6], TRPC1, TRPC3/6/7 and TRPC4 channel proteins in freshly isolated porcine airway SM tissues [7], TRPC6 channel protein in primary isolated guinea pig and cultured human airway SMCs [4, 8], and TRPC1, TRPC3, TRPC4, TRPC5 and TRPC6 channel proteins in cultured human ASMCs [5]. Our studies reveal that TRPC1 and TRPC3 channel mRNAs and proteins are expressed in freshly isolated mouse ASMCs [9].

Researchers have found that non-selective cation channels (NSCCs), which are permeable to both Na^+ and Ca^{2+}, with higher permeability to Na^+, are present in freshly isolated bovine and human ASMCs [10–13]. We have further characterized NSCCs in freshly isolated mouse ASMCs using the excised inside-out single channel recording [9]. The channel open probability in freshly isolated mouse ASMCs is significantly higher at positive than negative potentials, suggesting that the native NSCCs exhibit outward rectification in ASMCs. A diacylglycerol analogue, 1-oleoyl-2-acetyl-sn-glycerol (OAG), significantly increases the channel activity. The channel activity is augmented by elevating extracellular Ca^{2+} concentration and inhibited by reducing extracellular Ca^{2+} concentration. Thus, constitutively-active NSCCs in ASMCs possess diacylglycerol- and Ca^{2+}-gated properties, which are very similar to native constitutively-active NSCCs in vascular SMCs [14].

Our study also reveals that application of specific TRPC3 channel antibodies blocks the activity of native single NSCCs by ~80% in freshly isolated ASMCs [9]. TRPC3 channel gene silencing by specific siRNAs inhibits the single channel activity to a very similar extent in primary isolated cells. These patch clamp and genetic studies for the first time provide compelling evidence that native constitutively-active NSCCs are mainly encoded by TRPC3 channel in ASMCs

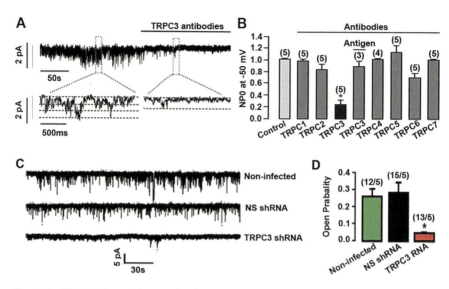

Fig. 18.1 TRPC3 channel is a predominant member of the TRPC channel family showing functional activity in ASMCs. (a) Specific TRPC3 channel antibodies inhibit the activity of single native constitutively active NSCCs in freshly isolated ASMCs (adopted from Xiao et al, *Am J Respir Cell Mol Biol*, 43: 17–25, 2010). An original recording of single NSCCs in an inside-out patch at −50 mV before and after bath application of specific TRPC3 channel antibodies (1:200 dilution). The *inserts* exhibit the channel recording at an extended time scale. (b) Summary of the effect of TRPC1 – TRPC6 antibodies on the activity of single NSCCs (unpublished data). The relative open probability (NPo) is presented as the difference in the channel activity recorded before and after application of individual TRPC antibodies (Ab). Numbers in parentheses indicate the numbers of cells from different mice examined. *$P < 0.05$ compared with control cells (without application of TRPC antibodies). (c) TRPC3 channel activity in ASMCs isolated from noninfected mice and mice receiving intravenous injection of NS and TRPC3 shRNAs (adopted from Song et al, *FASEB J*, 30: 214–29, 2016). TRPC3 channel activity in ASMCs isolated from noninfected mice (not receiving lentiviral SM22-driven shRANs) and mice receiving intravenous injection of lentiviral SM22-driven non-silencing (NS) and TRPC3 channel shRNAs. The channel activity was recorded in excised membrane patches from ASMCs using the inside-out single-channel recording. *$P < 0.05$ compared with cells from control (noninfected) mice

(Fig. 18.1), These findings also suggest that TRPC3 channel may form homomeric and/or heteromeric NSCCs with different conductance states.

18.3 TRPC3 Channel Is Important for Controlling Resting Membrane Potential and [Ca^{2+}]$_i$ in ASMCs

The resting membrane potential (Vm) in ASMCs is between −40 and − 50 mV, similar to other types of SMCs, and significantly less negative than a K$^+$ equilibrium potential of $\sim$ − 85 mV [15, 16]. Consistent with these prior reports, we have also found that the resting Vm is −44 mV in freshly isolated ASMCs [9]. More importantly, our data reveal that specific TRPC3 channel antibodies and gene silencing both result in a pronounced hyperpolarization of the resting membrane potential by approximately 14 mV. Conversely, TRPC1 channel antibodies and gene silencing have no obvious effect on the resting Vm. Thus, TRPC3 channel predominates the control of resting Vm in ASMCs, consistent with the vital role of TRPC3 channel in mediating the activity of constitutively-active NSCCs.

Comparable to the effect on the resting membrane potential, we have found that siRNA-mediated TRPC3 channel gene silencing significantly lowers the resting $[Ca^{2+}]_i$ in primary isolated ASMCs [9]. However, TRPC1 channel gene silencing does not alter the resting $[Ca^{2+}]_i$. These findings provide evidence that TRPC3, but not TRPC1 channel, play an important role in the control of the resting $[Ca^{2+}]_i$ in ASMCs. We have also shown that IP_3, an important intracellular second messenger, can activate TRPC3 channel to cause extracellular Ca^{2+} influx [17]. This novel extracellular Ca^{2+} entry route may play a significant role in mediating IP_3-mediated numerous cellular response in ASMCs.

18.4   TRPC3 Channel Is Involved in Agonist-Induced Increase in $[Ca^{2+}]_i$ in ASMCs

Using the patch clamp recording technique, we and other investigators have shown that muscarinic agonists acetylcholine and methacholine (mAch) activate NSCCs in freshly isolated canine, equine, guinea-pig and swine ASMCs [18–22]. Simultaneous measurements of membrane currents and $[Ca^{2+}]_i$ reveal that activation of NSCCs during muscarinic stimulation is always accompanied by a sustained increase in $[Ca^{2+}]_i$ due to extracellular Ca^{2+} influx [19–22]. The sustained increase in $[Ca^{2+}]_i$ induced by mACH and other agonists are largely inhibited or abolished by the general NSCC blockers Ni^{2+}, Cd^{2+}, La^{3+}, Gd^{3+} and SKF-96365 [12, 15, 20, 22–28]. Thus, functional NSCCs mediate agonist-induced Ca^{2+} influx and associated increase in $[Ca^{2+}]_i$ in ASMCs.

We have started to identify which of the TRPC channels are responsible for agonist-induced increase in $[Ca^{2+}]_i$. Our recent study reveals that TRPC3 channel gene silencing inhibits mACH-evoked increase in $[Ca^{2+}]_i$ in primary isolated ASMCs. Similarly, a previous report has shown that TRPC3 channel gene silencing blocks acetylcholine- and tumor necrosis factor-α (TNFα)-induced increase in $[Ca^{2+}]_i$ and cultured human ASMCs [5]. Moreover, OAG, a putative activator for TRPC channels [29, 30], causes a significant increase in $[Ca^{2+}]_i$ in primary isolated guinea pig ASMCs. Collectively, TRPC3 channel is vital for agonist-induced Ca^{2+} responses in airway myocytes. Relative to Ca^{2+} release from the sarcoplasmic reticulum (SR), extracellular Ca^{2+} influx through TRPC3 channel makes a smaller contribution to agonist-induced initial increase in $[Ca^{2+}]_i$, whereas TRPC3 channel-mediated Ca^{2+} signaling is persistent during agonist stimulation. This persistent Ca^{2+} signaling may be essential for maintaining cell contraction and other cellular responses, as well as refilling intracellular Ca^{2+} stores to start a new response.

Activation of muscarinic receptors or other G protein-coupled receptors can result in production of IP_3, which causes the opening of TRPC3 channel, serving as a novel mechanism for Ca^{2+} signaling to mediate cellular responses in ASMCs. In support, we have found that TRPC3 channel knockdown significantly decreases mACH-induced Ca^{2+} influx in ASMCs. The mACH-evoked contraction (cell shortening) is also inhibited by TRPC3 channel knockdown.

18.5   TRPC3 Channel Is Increased in Expression and Activity in ASMCs from Airway Diseases

TRPC3 channel in ASMCs has been shown to make an important contribution to the pathogenesis of asthma. TRPC3 channel is up-regulated in ASMCs isolated from asthma model mice [9, 31, 32] and human ASM [33], and has the ability to regulate cell proliferation of asthmatic mouse and human ASMCs in vivo and in vitro [32, 33]. In cultured human ASMCs, TRPC3 channel mRNA and protein expression is significantly increased following treatment with TNFα. Furthermore, TRPC3 channel gene silencing inhibits TNFα-induced Ca^{2+} influx, the associated increase in $[Ca^{2+}]_i$, and TNFα-mediated augmentation of acetylcholine-evoked increase in $[Ca^{2+}]_i$ [5]. In ovalbumin-sensitized/challenged asthmatic mice, TRPC1 channel protein expression level is not changed in freshly isolated asthmatic ASM tissue, but TRPC3 channel protein expression level is increased significantly [5, 9]. The asthmatic membrane depolarization is blocked by specific TRPC3 channel antibodies. We have also revealed that the asthmatic ASMCs increase in the activity of constitutively active NSCCs and depolarization in the membrane potential are both blocked by TRPC3 channel antibodies. Thus, TRPC3 channel is upregulated in its molecular expression and functional activity, which contributes to membrane depolarization and increased $[Ca^{2+}]_i$ in ASMCs, leading to airway hyperresponsiveness, remodeling and asthma [32, 34].

In a recent study, we have demonstrated that intravenous injection of lentiviral SM-specific promoter-driven TRPC3 channel shRNAs can sufficiently knockdown TRPC3 channel expression and activity in ASMC of mice in vivo and can block allergen-induced airway hyperresponsiveness and remodeling [32] (Fig. 18.2). It is interesting to point out that in-vivo administration of specific pharmacological

----

Fig. 18.2 (continued) normal and asthmatic mice following intranasal inhalation of vehicle (control) or the specific pharmacological inhibitor of TRPC3 channel Pyr3. (**c**) In-vitro airway muscle contractile responses to mAch were recorded in freshly sliced lung tissues from normal, AIAD (asthmatic), and asthmatic mice treated with NS or TRPC3 shRNAs. Scale bars represent 25 μm in length. Graph shows the quantification of airway lumen changes. (**d**) Immunohistochemistrical co-stains of α-smooth muscle actin (pink) and Ki67 (brown) in airways in normal, AIAD (asthmatic), asthmatic mice treated with lentiviral SM22-driven NS or TRPC3 shRNAs in vivo. Scale bars indicate 20 μm in length. Green arrows indicate colocalization of α-SM actin and Ki67. The insert shows an enlarged part of ASM layers, illustrating the colocalization of Ki67 and α-SM actin. Bar graph displays the quantification of ASM areas and summary of Ki67-positive cells in ASM layers in each group. Numbers in parentheses indicate the numbers of airways/mice examined. *$P < 0.05$ compared with noninfected ASMCs or normal mice. #$P < 0.05$ compared with asthmatic mice treated with NS. The figure is adopted from Song et al, *FASEB J*, 30: 214–29, 2016

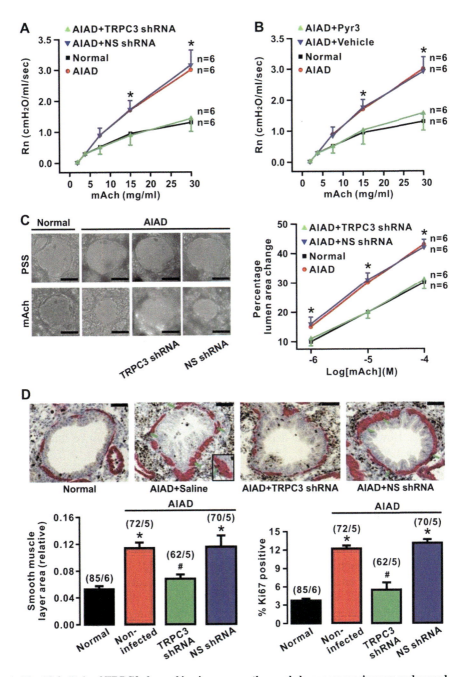

Fig. 18.2 Role of TRPC3 channel in airway smooth muscle hyperresponsiveness and remodeling in mice with allergen (ovalbumin)-induced airway disease (AIAD, asthma). (a) Inhibitory effect of SMC-specific TRPC3 channel knockdown on in-vivo airway muscle contractile responses to the muscarinic agonist methacholine (mAch) in mice with asthma (AIAD). The in-vivo airway muscle contractile responses in non-asthmatic (normal) mice, asthmatic mice, and asthmatic mice following intravenous injection of lentiviral SM22 (SMC promoter)-driven non-silencing (NS) or TRPC3 channel shRNAs were determined by assessing conducting airway (Newtonian) resistance (Rn) using an invasive FlexiVent device. (**b**) In-vivo airway muscle contractile responses (Rn) in

inhibitors of TRPC3 channel produce similar effects. These results are consistent with reports from our group and others, in which among all 7 members, only TRPC3 channel exhibits a predominant functional activity in normal ASMCs and an increase in the activity and expression in asthmatic ASMCs [5, 9, 31].

18.6   Upregulated TRPC3 Channel Can Activate NF-κB to Mediate Asthma and Other Airway Diseases Via PKCα–Dependent IκBα and Calcineurin-Reliant IκBβ Signaling Pathways in ASMCs

The recent studies have revealed that expression of protein kinase C-α (PKCα), a Ca^{2+}-sensitive PKC isoform that may activate nuclear factor-κB (NF-κB) by decreasing its inhibitor-α (IκBα) activity in cancer cells and ASMCs [32, 35] and induce cyclin D1 expression to promote proliferation in ASMCs, is increased in ASMCs from asthmatic subjects [32]. More importantly, TRPC3 channel-over expression increases PKCα expression in asthmatic and normal ASMCs, whereas TRPC3 channel knockdown blocks the increased PKC-α expression in asthmatic ASMCs with no effect in normal cells [32]. NF-κB activity in asthmatic ASMCs can also be inhibited by a PKCα antagonist. This information supports that up-regulated TRPC3 channel induces an increased Ca^{2+} influx, which stimulates PKCα, inactivates IκBα, and increase NF-κB activity, leading to cyclin expression, cell proliferation, airway hyperresponsiveness and remodeling, and ultimately asthma. However, this PKCα–dependent signaling pathway may not be functional in normal ASMCs.

We and others have found that calcineurin expression is increased in asthmatic ASMCs [32, 36–38]. The increased calcineurin expression is inhibited by TRPC3 channel knockdown, but augmented by TRPC3 channel overexpression. Similarly, specific calcineurin inhibition blocks, while its activation enhances, NF-κB activity. These findings unveil that in addition to PKCα/IκBα signaling pathway, TRPC3 channel can also stimulate calcineurin and inhibit IκBβ, leading to the activation of NF-κB and induction of cyclin D1 in asthmatic ASMCs [32]. It has been proven that calcineurin plays an important role in TRPC3 channel-mediated activation of NF-κB in neurons and myocardiocytes [39–41]. Perceptibly, PKCα–dependent IκBα and calcineurin-reliant IκBβ inhibition, the two distinct signaling pathways, are critical for TRPC3 channel-mediated increased activity of NF-κB (Fig. 18.3), which plays a vital role in the development of airway hyperresponsiveness and remodeling, eventually leading to asthma and possibly other airway diseases [32].

The upregulated TRPC3 channel has also been found to induce p-p38, p-Jun amino-terminal kinase (JNK), cleaved caspase-3, and Bcl-2 expression, as well as promote cell cycle in asthmatic mouse ASMCs. Moreover, the Ca^{2+} chelator EGTA

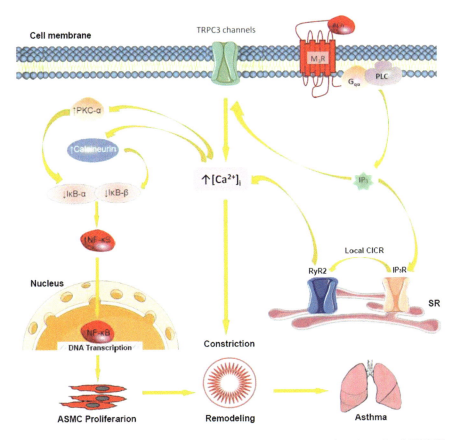

Fig. 18.3 A schematic diagram for signaling mechanisms underlying the role of TRPC3 channel in asthma. The asthmatic stimuli increase TRPC3 channel expression and activity leading to an increase in $[Ca^{2+}]_i$ in ASMCs. In addition to the direct role in mediating airway hyperresponsiveness, the increased $[Ca^{2+}]_i$ not only stimulates PKCα and then inhibits IκBα, but also stimulates calcineurin and inhibits IκBβ. The inhibited IκBα and IκBβ together result in NF-κB activation, cyclin induction, and cell proliferation, ultimately airway hyperresponsiveness, airway remodeling, and asthma. Stimulation of muscarinic receptors or other GPCRs activates PLC to produce DAG and IP$_3$. DAG causes a direct opening of TRPC3 channel. In a traditional view, IP$_3$ opens its receptors and induces Ca^{2+} release from the SR to mediate cellular responses. Notably, there is a new paradigm for the molecular functions of IP$_3$ at least in ASMCs, in which IP$_3$-induced Ca^{2+} release through IP$_3$R is significantly amplified by RyR2 due to their local interaction. In addition to Ca^{2+} release, IP may activate TRPC3 channel and cause extracellular Ca^{2+} influx to mediate cell contraction and proliferation, playing an important role in airway hyperresponsiveness, remodeling and diseases

or BAPTA significantly diminishes the effects of TRPC3 channel knockdown on the cell viability, cell cycle and the increased protein expression levels of p-p38, p-JNK, cleaved caspase-3 and Bcl-2 in asthmatic mouse ASMCs [42], which is consistent with the functional role of Ca^{2+} influx through TRPC3 channel.

18.7 TRPC3 Channel Is Regulated by Orai Channel and Ca^{2+}-Sensing Stromal Interaction Molecule

Store-operated Ca^{2+} entry (SOCE) is a vital route in controlling $[Ca^{2+}]_i$ in various types of cells including ASMCs [43, 44]. Ca^{2+} store depletion may result in activation of stromal interaction molecule 1 (STIM1) that is located on the sarco/endoplasmic reticulum [45]. The activated STIM1 is translocated from the ER to the ER-PM junctions, where it interacts with Orai1 channel that permits Ca^{2+} from the extracellular space to the cytosol [46–50]. STIM1 and Orai1 work together as mediators of SOCE in ASMCs, and SOCE has been shown to be increased in proliferative ASMCs accompanied by a modest increase in STIM1 mRNA expression and significant increase in Orai1 channel mRNA expression [47]. Gene knockdown of STIM1 or Orai1 channel by shRNAs significantly attenuates proliferation of ASMCs [47]. Moreover, STIM1 and Orai1 are involved in PDGF-mediated SOCE [51]. Overall, STIM1 and Orai1 are important contributors to SOCE in ASMCs that exhibit hyperplasia [52, 53].

It has been shown that Orai1l channel physically interacts with the N and C terminus of TRPC3 channel, and overexpressing TRPC3 channel only becomes sensitive to the store depletion in cells that also express exogenous Orai channel; thus, Orai channel interacts with TRPC3 channel to form a functional integrative unit that confers STIM1-mediated store depletion sensitivity to these two channels [54].

Interestingly, an elegant study has found that under the resting state, the N and C terminus coiled-coil domains of both TRPC3 and TRPC1 channel interact with each other to shield the STIM1 binding site, thereby preventing the role of STIM1 in regulating the activity of TRPC3 and TRPC1 channel [55]. However, cell stimulation can facilitate the formation of TRPC3/TRPC1 channel heteromultimers, enhance both interaction, and also dissociate between the TRPC1 channel N and C terminus coiled-coil domains. As such, the free TRPC1 channel C terminus coiled-coil domain interacts with the TRPC3 channel coiled-coil domains to dissociate them, making the TRPC3 channel C terminus coiled-coil domain available for interaction with STIM1, thereby allowing STIM1 to activate TRPC3 channel [55]. These novel results provide an important molecular mechanism for the regulatory interaction of STIM1 with TRPC3 channel and also potentially other TRPC channel members.

18.8 TRPC3 Channel May Interact with Ca^{2+} Release Channels

TRPC3 channel can be regulated by muscarinic receptors and other G protein-coupled receptors (GPCRs) in ASMCs [43, 56]. Stimulation of GPCRs causes activation of phospholipase C (PLC), which hydrolyzes phosphatidylinositol

4,5-bisphosphate (PIP$_2$) to produce diacylglycerol (DAG) and inositol 1,4,5-trisphosphate (IP$_3$). These two-important intracellular second messengers will possibly mediate the role of GPCRs in controlling the activity of TRPC3 channel. Indeed, we have demonstrated that the DAG analog 1-oleyl-2-acetyl-sn-glycerol (OAG) activates TRPC3 channel in ASMCs [9]. A similar effect of OAG on TRPC3 channel has also been observed in ear artery SMCs [57, 58].

IP$_3$ is generally considered to elevate [Ca^{2+}]$_i$ by activating IP$_3$ receptor/Ca^{2+} release channel (IP$_3$R) to induce intracellular Ca^{2+} release from the SR [59]. Interestingly, we have discovered that Ca^{2+} release through IP$_3$R can activate adjacent ryanodine receptor-2/Ca^{2+} reelase channel (RyR2) to induce further Ca^{2+} release from the SR, termed a local IP$_3$R/RyR2 interaction-meidated Ca^{2+}-induced Ca^{2+} release process [59–61]. Equally importantly, our more recent investigations have found that IP$_3$, can notably increase the activity of single NSCCs [17]. The effects of IP$_3$ can be fully blocked by shRNA-mediated TRPC3 channel knockdown. The stimulatory effect of IP$_3$ is also abolished by heparin, an IP$_3$R antagonist that blocks the IP$_3$-binding site, but not by xestospongin C, an IP$_3$R antagonist that has no effect on the IP$_3$-binding site. In contrast, shRNA-mediated knockdown of IP$_3$R1, IP$_3$R2, or IP$_3$R3 does not change the excitatory effect of IP$_3$ on the activity of TRPC channel. Furthermore, TRPC3 channel knockdown significantly diminishes IP$_3$-induced increase in [Ca^{2+}]$_i$. IP$_3$R1 knockdown produces a similar inhibitory effect. TRPC3 channel and IP$_3$R1 knockdown both diminish the muscarinic receptor agonist methacholine-evoked Ca^{2+} responses. IP$_3$ may also produce a stimulatory effect on the activity of TRPC3 channel in the absence of OAG in vascular SMCs [62, 63]. Taken together, IP$_3$ can not only open IP$_3$Rs to induce intracellular Ca^{2+} release, but also activate TRPC3 channel to cause extracellular Ca^{2+} influx.

It should be noted there are limited studies on the interaction of TRPC3 channel with Ca^{2+} release channels in ASMC. Further studies to determine how TRPC3 channels are regulated in normal and diseased (e.g., asthmatic) ASMCs are necessary.

18.9   TRPC3 Channel May Become a Novel and Effective Drug Target in Lung Diseases

It is well established that airway hyperresponsiveness and remodeling are the two major cellular responses in asthma [43, 45, 64–67]. We have recently demonstrated that lentivirus-based, shRNA-mediated, SMC-specific TRPC3 channel gene knockdown in vivo blocks the development of hyperresponsiveness and remodeling in mice with allergic asthma [32]. As lentivirus- and/or shRNA-mediated loss and gain of gene function embody a promising human gene therapy [68], our findings may provide a new, specific and effective therapeutic avenue in the treatment of asthma. Consistent with this view, we have further revealed that intranasal administration of the TRPC3 channel blocker Pyr3 also abolishes allergen-induced asthma in animals

[32]. It is interesting to point out that a number of TRP channels have been identified as the novel 'druggable' targets in numerous diseases; some of pharmacological blockers of these channels, although not the TRPC3 channel, have been in clinical trials [68–70]. Convincingly, the TRPC3 channel is an advantageous drug target in the therapy of asthma.

Similar to asthma, chronic obstructive pulmonary disease (COPD) is a common devastating airway disease. Indeed, this disease is the fifth leading cause of death worldwide and expected to be the third cause of mortality by 2020 [71]. Airway hyperresponsiveness develops in COPD, and thus is the rationale for using bronchodilators as the first-line therapy for COPD [72–77]. Airway remodeling has been increasingly recognized in COPD [45, 64, 67, 78–83]. Cigarette smoke is a primary factor in 80% to 90% of COPD [84–86]. Interestingly, cigarette smoking increases TRPC3 channel expression and associated Ca^{2+} influx, altering airway structure and function [87]. Presumably, lentivirus-based shRNA-mediated genetic inhibition and pharmacological inhibitors of TRPC3 channel may also become valuable therapeutics for COPD.

It has been reported that TRPC3 channel expression levels are correlated with differentiation of non-small cell lung cancer, regulate cancer cell differentiation and proliferation, and mediate the inhibitory effect of all-trans-retinoic acid [88]. In support, a recent study has found that TRPC3 methylation variation is prominent in lung cancer tissues [89]. These data suggest that in addition to asthma and COPD, TRPC3 channel may also serve as a valuable drug target in lung cancer.

18.10   Conclusion

Multiple TRPC channels are expressed in animal and human ASMCs; however, TRPC3 channel is a major functional member of the TRPC family. This channel plays an important role in controlling the resting $[Ca^{2+}]_i$ and mediating agonist-evoked increase in $[Ca^{2+}]_i$ in ASMCs. TRPC3 channel is significantly increased in expression and activity in ASMCs from subjects with asthma or other airway diseases. The increased TRPC3 channel causes excessive extracellular Ca^{2+} influx, which serves as a very important player in mediating airway hyperresponsiveness in asthma and other relevant airway disorders. In addition to this direct imperative role, the excessive extracellular Ca^{2+} influx through the upregulated TRPC3 channel can also cause the suppression of $PKC\alpha$–dependent $I\kappa B\alpha$ and calcineurin-reliant $I\kappa B\beta$ signaling pathways, activation of NF-κB, induction of cyclin expression, and cell proliferation in ASMCs, which may not only lead to airway remodeling, but can also contribute to airway hyperresponsiveness, thereby mediating the development of asthma and other airway diseases.

In-vivo administration of lentiviral SMC-specific TRPC3 channel shRNAs or specific TRPC3 channel blockers blocks airway hyperresponsiveness and remodeling, thus preventing the development of asthma. These findings promote the potential use of gene therapies or pharmacological interventions specifically target-ing at TRPC3 channel in the clinical treatment of asthma and other airway diseases.

Studies suggest that the plasmalemmal TRPC3 channel well interacts with the plasmalemmal Orai channel, cytosolic STIM, and sarcolemmal Ca^{2+} release channels in ASMCs. This coordinative network may provide a unique system in precisely controlling and regulating $[Ca^{2+}]_i$ to meet adequate physiological cellular responses, further indicating the functional significance of TRPC3 channel in ASMCs. Clearly, further investigations to determine whether, which, and how TRPC3 channel is involved in cellular responses in normal and diseased (e.g., asthmatic) ASMCs are needed, with hopes of yielding novel and important findings to enhance our understanding of the functional roles, regulatory mechanisms and signaling processes of TRPC3, as well as aid in the creation of effective therapeutic targets for the treatment of asthma and other respiratory diseases.

References

1. Montell C, Jones K, Hafen E, Rubin G (1985) Rescue of the drosophila phototransduction mutation trp by germline transformation. Science (New York, NY) 230(4729):1040–1043
2. Wes PD, Chevesich J, Jeromin A, Rosenberg C, Stetten G, Montell C (1995) TRPC1, a human homolog of a drosophila store-operated channel. Proc Natl Acad Sci U S A 92(21):9652–9656
3. Ong HL, Brereton HM, Harland ML, Barritt GJ (2003) Evidence for the expression of transient receptor potential proteins in Guinea pig airway smooth muscle cells. Respirology 8(1):23–32
4. Corteling RL, Li S, Giddings J, Westwick J, Poll C, Hall IP (2004) Expression of transient receptor potential C6 and related transient receptor potential family members in human airway smooth muscle and lung tissue. Am J Respir Cell Mol Biol 30(2):145–154
5. White TA, Xue A, Chini EN, Thompson M, Sieck GC, Wylam ME (2006) Role of transient receptor potential C3 in TNF-alpha-enhanced calcium influx in human airway myocytes. Am J Respir Cell Mol Biol 35(2):243–251
6. Ong HL, Chen J, Chataway T, Brereton H, Zhang L, Downs T et al (2002) Specific detection of the endogenous transient receptor potential (TRP)-1 protein in liver and airway smooth muscle cells using immunoprecipitation and Western-blot analysis. Biochem J 364(Pt 3):641–648
7. Ay B, Prakash YS, Pabelick CM, Sieck GC (2004) Store-operated Ca^{2+} entry in porcine airway smooth muscle. Am J Physiol Lung Cell Mol Physiol 286(5):L909–L917
8. Godin N, Rousseau E (2007) TRPC6 silencing in primary airway smooth muscle cells inhibits protein expression without affecting OAG-induced calcium entry. Mol Cell Biochem 296(1–2):193–201
9. Xiao JH, Zheng YM, Liao B, Wang YX (2010) Functional role of canonical transient receptor potential 1 and canonical transient receptor potential 3 in normal and asthmatic airway smooth muscle cells. Am J Respir Cell Mol Biol 43(1):17–25
10. Snetkov VA, Pandya H, Hirst SJ, Ward JP (1998) Potassium channels in human fetal airway smooth muscle cells. Pediatr Res 43(4 Pt 1):548–554
11. Snetkov VA, Ward JP (1999) Ion currents in smooth muscle cells from human small bronchioles: presence of an inward rectifier K^+ current and three types of large conductance K^+ channel. Exp Physiol 84(5):835–846
12. Snetkov VA, Hapgood KJ, McVicker CG, Lee TH, Ward JP (2001) Mechanisms of leukotriene D4-induced constriction in human small bronchioles. Br J Pharmacol 133(2):243–252
13. Helli PB, Janssen LJ (2008) Properties of a store-operated nonselective cation channel in airway smooth muscle. Eur Respir J 32(6):1529–1539

14. Albert AP, Large WA (2001) Comparison of spontaneous and noradrenaline-evoked non-selective cation channels in rabbit portal vein myocytes. J Physiol 530(Pt 3):457–468
15. Hirota S, Helli P, Janssen LJ (2007) Ionic mechanisms and Ca^{2+} handling in airway smooth muscle. Eur Respir J 30(1):114–133
16. Liu XS, Xu YJ (2005) Potassium channels in airway smooth muscle and airway hyperreactivity in asthma. Chin Med J 118(7):574–580
17. Song T, Hao Q, Zheng YM, Liu QH, Wang YX (2015) Inositol 1,4,5-trisphosphate activates TRPC3 channels to cause extracellular Ca^{2+} influx in airway smooth muscle cells. Am J Physiol Lung Cell Mol Physiol 309(12):L1455–L1466
18. Janssen LJ, Sims SM (1992) Acetylcholine activates non-selective cation and chloride conductances in canine and Guinea-pig tracheal myocytes. J Physiol 453:197–218
19. Wang YX, Fleischmann BK, Kotlikoff MI (1997) M2 receptor activation of nonselective cation channels in smooth muscle cells: calcium and Gi/G(o) requirements. Am J Phys 273(2 Pt 1):C500–C508
20. Fleischmann BK, Wang YX, Kotlikoff MI (1997) Muscarinic activation and calcium permeation of nonselective cation currents in airway myocytes. Am J Phys 272(1 Pt 1):C341–C349
21. Wang YX, Kotlikoff MI (2000) Signalling pathway for histamine activation of non-selective cation channels in equine tracheal myocytes. J Physiol 523(Pt 1):131–138
22. Yamashita T, Kokubun S (1999) Nonselective cationic currents activated by acetylcholine in swine tracheal smooth muscle cells. Can J Physiol Pharmacol 77(10):796–805
23. Murray RK, Kotlikoff MI (1991) Receptor-activated calcium influx in human airway smooth muscle cells. J Physiol 435:123–144
24. Parvez O, Voss AM, de Kok M, Roth-Kleiner M, Belik J (2006) Bronchial muscle peristaltic activity in the fetal rat. Pediatr Res 59(6):756–761
25. Dai JM, Kuo KH, Leo JM, Pare PD, van Breemen C, Lee CH (2007) Acetylcholine-induced asynchronous calcium waves in intact human bronchial muscle bundle. Am J Respir Cell Mol Biol 36(5):600–608
26. Dai JM, Kuo KH, Leo JM, van Breemen C, Lee CH (2006) Mechanism of ACh-induced asynchronous calcium waves and tonic contraction in porcine tracheal muscle bundle. Am J Physiol Lung Cell Mol Physiol 290(3):L459–L469
27. Gorenne I, Labat C, Gascard JP, Norel X, Nashashibi N, Brink C (1998) Leukotriene D4 contractions in human airways are blocked by SK&F 96365, an inhibitor of receptor-mediated calcium entry. J Pharmacol Exp Ther 284(2):549–552
28. Hirota S, Janssen LJ (2007) Store-refilling involves both L-type calcium channels and reverse-mode sodium-calcium exchange in airway smooth muscle. Eur Respir J 30(2):269–278
29. Nilius B, Owsianik G, Voets T, Peters JA (2007) Transient receptor potential cation channels in disease. Physiol Rev 87(1):165–217
30. Abramowitz J, Birnbaumer L (2009) Physiology and pathophysiology of canonical transient receptor potential channels. FASEB J 23(2):297–328
31. Wang L, Li J, Zhang J, He Q, Weng X, Huang Y et al (2017) Inhibition of TRPC3 downregulates airway hyperresponsiveness, remodeling of OVA-sensitized mouse. Biochem Biophys Res Commun 484(1):209–217
32. Song T, Zheng YM, Vincent PA, Cai D, Rosenberg P, Wang YX (2016) Canonical transient receptor potential 3 channels activate NF-kappaB to mediate allergic airway disease via PKC-alpha/IkappaB-alpha and calcineurin/IkappaB-beta pathways. FASEB J 30(1):214–229
33. Wylam ME, Sathish V, VanOosten SK, Freeman M, Burkholder D, Thompson MA et al (2015) Mechanisms of cigarette smoke effects on human airway smooth muscle. PLoS One 10(6):e0128778
34. Wang YX, Zheng YM (2011) Molecular expression and functional role of canonical transient receptor potential channels in airway smooth muscle cells. Adv Exp Med Biol 704:731–747
35. Vertegaal AC, Kuiperij HB, Yamaoka S, Courtois G, van der Eb AJ, Zantema A (2000) Protein kinase C-alpha is an upstream activator of the IkappaB kinase complex in the TPA signal transduction pathway to NF-kappaB in U2OS cells. Cell Signal 12(11–12):759–768

36. Hai CM (2007) Airway smooth muscle cell as therapeutic target of inflammation. Curr Med Chem 14(1):67–76
37. Walczak-Drzewiecka A, Ratajewski M, Wagner W, Dastych J (2008) HIF-1alpha is up-regulated in activated mast cells by a process that involves calcineurin and NFAT. J Immunol 181(3):1665–1672
38. Said SI, Hamidi SA, Gonzalez Bosc L (2010) Asthma and pulmonary arterial hypertension: do they share a key mechanism of pathogenesis? Eur Respir J 35(4):730–734
39. Rosenberg P, Hawkins A, Stiber J, Shelton JM, Hutcheson K, Bassel-Duby R et al (2004) TRPC3 channels confer cellular memory of recent neuromuscular activity. Proc Natl Acad Sci U S A 101(25):9387–9392
40. Nakayama H, Wilkin BJ, Bodi I, Molkentin JD (2006) Calcineurin-dependent cardiomyopathy is activated by TRPC in the adult mouse heart. FASEB J 20(10):1660–1670
41. Poteser M, Schleifer H, Lichtenegger M, Schernthaner M, Stockner T, Kappe CO et al (2011) PKC-dependent coupling of calcium permeation through transient receptor potential canonical 3 (TRPC3) to calcineurin signaling in HL-1 myocytes. Proc Natl Acad Sci U S A 108(26):10556–10561
42. Zhang X, Zhao Z, Ma L, Guo Y, Li X, Zhao L et al (2018) The effects of transient receptor potential channel (TRPC) on airway smooth muscle cell isolated from asthma model mice. J Cell Biochem 119(7):6033–6044
43. Xiao JH, Wang YX, Zheng YM (2014) Transient receptor potential and Orai channels in airway smooth muscle cells. In: Wang YX (ed) Calcium signaling in airway smooth muscle cells. Springer, Cham, pp 35–45
44. Ong HL, Ambudkar IS (2017) STIM-TRP pathways and microdomain organization: contribution of TRPC1 in store-operated Ca^{2+} entry: impact on Ca^{2+} signaling and cell function. Adv Exp Med Biol 993:159–188
45. Song T, Zheng Y-M, Wang YX (2014) Calcium signaling in airway smooth muscle Remodeling. In: YX W (ed) Calcium signaling in airway smooth muscle cells. Springer, Cham, pp 393–407
46. Dolmetsch RE, Xu K, Lewis RS (1998) Calcium oscillations increase the efficiency and specificity of gene expression. Nature 392(6679):933–936
47. Zou JJ, Gao YD, Geng S, Yang J (2011) Role of STIM1/Orai1-mediated store-operated Ca(2)(+) entry in airway smooth muscle cell proliferation. J Appl Physiol (1985) 110(5):1256–1263
48. Spinelli AM, Gonzalez-Cobos JC, Zhang X, Motiani RK, Rowan S, Zhang W et al (2012) Airway smooth muscle STIM1 and Orai1 are upregulated in asthmatic mice and mediate PDGF-activated SOCE, CRAC currents, proliferation, and migration. Pflugers Archiv Eur J Physiol 464(5):481–492
49. Chin D, Means AR (2000) Calmodulin: a prototypical calcium sensor. Trends Cell Biol 10(8):322–328
50. Feske S, Prakriya M (2013) Conformational dynamics of STIM1 activation. Nat Struct Mol Biol 20(8):918–919
51. Ogawa A, Firth AL, Smith KA, Maliakal MV, Yuan JX (2012) PDGF enhances store-operated Ca^{2+} entry by upregulating STIM1/Orai1 via activation of Akt/mTOR in human pulmonary arterial smooth muscle cells. Am J Physiol Cell Physiol 302(2):C405–C411
52. Peel SE, Liu B, Hall IP (2008) ORAI and store-operated calcium influx in human airway smooth muscle cells. Am J Respir Cell Mol Biol 38(6):744–749
53. Peel SE, Liu B, Hall IP (2006) A key role for STIM1 in store operated calcium channel activation in airway smooth muscle. Respir Res 7:119
54. Liao Y, Erxleben C, Yildirim E, Abramowitz J, Armstrong DL, Birnbaumer L (2007) Orai proteins interact with TRPC channels and confer responsiveness to store depletion. Proc Natl Acad Sci U S A 104(11):4682–4687
55. Lee KP, Choi S, Hong JH, Ahuja M, Graham S, Ma R et al (2014) Molecular determinants mediating gating of Transient Receptor Potential Canonical (TRPC) channels by stromal interaction molecule 1 (STIM1). J Biol Chem 289(10):6372–6382

56. Brightbill HD, Jeet S, Lin Z, Yan D, Zhou M, Tan M et al (2010) Antibodies specific for a segment of human membrane IgE deplete IgE-producing B cells in humanized mice. J Clin Invest 120(6):2218–2229
57. Albert AP, Piper AS, Large WA (2005) Role of phospholipase D and diacylglycerol in activating constitutive TRPC-like cation channels in rabbit ear artery myocytes. J Physiol 566(Pt 3):769–780
58. Albert AP, Pucovsky V, Prestwich SA, Large WA (2006) TRPC3 properties of a native constitutively active Ca^{2+}−permeable cation channel in rabbit ear artery myocytes. J Physiol 571(Pt 2):361–369
59. Mei L, Zheng YM, Wang YX (2014) Ryanodine and inositol trisphosphate receptors/Ca^{2+} release channels in airway smooth muscle cells. In: Wang YX (ed) Calcium signaling in airway smooth muscle cells. Springer, Cham, pp 1–20
60. Liu QH, Zheng YM, Korde AS, Yadav VR, Rathore R, Wess J et al (2009) Membrane depolarization causes a direct activation of G protein-coupled receptors leading to local Ca^{2+} release in smooth muscle. Proc Natl Acad Sci U S A 106(27):11418–11423
61. Liu QH, Savoia C, Wang YX, Zheng YM (2014) Local calcium signaling in airway smooth muscle cells. In: YX W (ed) Calcium signaling in airway smooth muscle cells. Springer, Cham, pp 107–120
62. Adebiyi A, Thomas-Gatewood CM, Leo MD, Kidd MW, Neeb ZP, Jaggar JH (2012) An elevation in physical coupling of type 1 inositol 1,4,5-trisphosphate (IP3) receptors to transient receptor potential 3 (TRPC3) channels constricts mesenteric arteries in genetic hypertension. Hypertension 60(5):1213–1219
63. Xi Q, Adebiyi A, Zhao G, Chapman KE, Waters CM, Hassid A et al (2008) IP3 constricts cerebral arteries via IP3 receptor-mediated TRPC3 channel activation and independently of sarcoplasmic reticulum Ca^{2+} release. Circ Res 102(9):1118–1126
64. Prakash YS (2016) Emerging concepts in smooth muscle contributions to airway structure and function: implications for health and disease. Am J Physiol Lung Cell Mol Physiol 311(6):L1113–L1l40
65. Panettieri RA, Pera T, Liggett SB, Benovic JL, Penn RB (2018) Pepducins as a potential treatment strategy for asthma and COPD. Curr Opin Pharmacol 40:120–125
66. Alves MF, da Fonseca DV, de Melo SAL, Scotti MT, Scotti L, Dos Santos SG et al (2018) New therapeutic targets and drugs for the treatment of asthma. Mini Rev Med Chem 18(8):684–696
67. Prakash YS, Halayko AJ, Gosens R, Panettieri RA Jr, Camoretti-Mercado B, Penn RB (2017) An official American Thoracic Society research statement: current challenges facing research and therapeutic advances in airway remodeling. Am J Respir Crit Care Med 195(2):e4–e19
68. Tiapko O, Groschner K (2018) TRPC3 as a target of novel therapeutic interventions. Cell 7(7):83
69. Nilius B, Szallasi A (2014) Transient receptor potential channels as drug targets: from the science of basic research to the art of medicine. Pharmacol Rev 66(3):676–814
70. Cui C, Merritt R, Fu L, Pan Z (2017) Targeting calcium signaling in cancer therapy. Acta Pharm Sin B 7(1):3–17
71. Vestbo J, Hurd SS, Agusti AG, Jones PW, Vogelmeier C, Anzueto A et al (2013) Global strategy for the diagnosis, management, and prevention of chronic obstructive pulmonary disease: GOLD executive summary. Am J Respir Crit Care Med 187(4):347–365
72. Vestbo J, Hansen EF (2001) Airway hyperresponsiveness and COPD mortality. Thorax 56(Suppl 2):ii11–ii14
73. Scichilone N, Battaglia S, La Sala A, Bellia V (2006) Clinical implications of airway hyperresponsiveness in COPD. Int J Chron Obstruct Pulmon Dis 1(1):49–60
74. van den Berge M, Vonk JM, Gosman M, Lapperre TS, Snoeck-Stroband JB, Sterk PJ et al (2012) Clinical and inflammatory determinants of bronchial hyperresponsiveness in COPD. Eur Respir J 40(5):1098–1105
75. Prakash YS (2013) Airway smooth muscle in airway reactivity and remodeling: what have we learned? Am J Physiol Lung Cell Mol Physiol 305(12):L912–L933

76. Fricker M, Deane A, Hansbro PM (2014) Animal models of chronic obstructive pulmonary disease. Expert Opin Drug Discovery 9(6):629–645
77. Tkacova R, Dai DLY, Vonk JM, Leung JM, Hiemstra PS, van den Berge M et al (2016) Airway hyperresponsiveness in chronic obstructive pulmonary disease: a marker of asthma-chronic obstructive pulmonary disease overlap syndrome? J Allergy Clin Immunol 138(6):1571–1579. e10
78. Hogg JC (2004) Pathophysiology of airflow limitation in chronic obstructive pulmonary disease. Lancet (London, UK) 364(9435):709–721
79. Chung KF (2005) The role of airway smooth muscle in the pathogenesis of airway wall remodeling in chronic obstructive pulmonary disease. Proc Am Thorac Soc 2(4):347–354; discussion 71–2
80. Kim V, Rogers TJ, Criner GJ (2008) New concepts in the pathobiology of chronic obstructive pulmonary disease. Proc Am Thorac Soc 5(4):478–485
81. Jones RL, Noble PB, Elliot JG, James AL (2016) Airway remodelling in COPD: it's not asthma! Respirology 21(8):1347–1356
82. Kistemaker LE, Oenema TA, Meurs H, Gosens R (2012) Regulation of airway inflammation and remodeling by muscarinic receptors: perspectives on anticholinergic therapy in asthma and COPD. Life Sci 91(21–22):1126–1133
83. Nayak AP, Deshpande DA, Penn RB (2018) New targets for resolution of airway remodeling in obstructive lung diseases. F1000 Res 7:680
84. Dewar M, Curry RW Jr (2006) Chronic obstructive pulmonary disease: diagnostic considerations. Am Fam Physician 73(4):669–676
85. Zeller M (2016) The deeming rule: keeping pace with the modern tobacco marketplace. Am J Respir Crit Care Med 194(5):538–540
86. Temitayo Orisasami I, Ojo O (2016) Evaluating the effectiveness of smoking cessation in the management of COPD. Br J Nurs (Mark Allen Publishing) 25(14):786–791
87. Vogel ER, VanOosten SK, Holman MA, Hohbein DD, Thompson MA, Vassallo R et al (2014) Cigarette smoke enhances proliferation and extracellular matrix deposition by human fetal airway smooth muscle. Am J Physiol Lung Cell Mol Physiol 307(12):L978–L986
88. Jiang HN, Zeng B, Zhang Y, Daskoulidou N, Fan H, Qu JM et al (2013) Involvement of TRPC channels in lung cancer cell differentiation and the correlation analysis in human non-small cell lung cancer. PLoS One 8(6):e67637
89. Kettunen E, Hernandez-Vargas H, Cros MP, Durand G, Le Calvez-Kelm F, Stuopelyte K et al (2017) Asbestos-associated genome-wide DNA methylation changes in lung cancer. Int J Cancer 141(10):2014–2029

Chapter 19
Pathophysiological Significance of Store-Operated Calcium Entry in Cardiovascular and Skeletal Muscle Disorders and Angiogenesis

Javier Avila-Medina, Isabel Mayoral-González, Isabel Galeano-Otero, Pedro C. Redondo, Juan A. Rosado, and Tarik Smani

Abstract Store-Operated Ca^{2+} Entry (SOCE) is an important Ca^{2+} influx pathway expressed by several excitable and non-excitable cell types. SOCE is recognized as relevant signaling pathway not only for physiological process, but also for its involvement in different pathologies. In fact, independent studies demonstrated the implication of essential protein regulating SOCE, such as STIM, Orai and TRPCs, in different pathogenesis and cell disorders, including cardiovascular disease, muscular dystrophies and angiogenesis. Compelling evidence showed that dysregulation in the function and/or expression of isoforms of STIM, Orai or TRPC play pivotal roles in cardiac hypertrophy and heart failure, vascular remodeling and hypertension, skeletal myopathies, and angiogenesis. In this chapter, we summarized the current

J. Avila-Medina · I. Galeano-Otero
Department of Medical Physiology and Biophysics, University of Seville, Sevilla, Spain

Institute of Biomedicine of Seville (IBiS), University Hospital of Virgen del Rocío/CSIC/University of Seville, Sevilla, Spain

I. Mayoral-Gonzàlez
Department of Medical Physiology and Biophysics, University of Seville, Sevilla, Spain

Institute of Biomedicine of Seville (IBiS), University Hospital of Virgen del Rocío/CSIC/University of Seville, Sevilla, Spain

Department of Surgery, University of Seville, Sevilla, Spain

P. C. Redondo · J. A. Rosado
Department of Physiology, Cell Physiology Research Group and Institute of Molecular Pathology Biomarkers, University of Extremadura, Cáceres, Spain

T. Smani (✉)
Department of Medical Physiology and Biophysics, University of Seville, Sevilla, Spain

Institute of Biomedicine of Seville (IBiS), University Hospital of Virgen del Rocío/CSIC/University of Seville, Sevilla, Spain

CIBERCV, Madrid, Spain
e-mail: tasmani@us.es

© Springer Nature Switzerland AG 2020
M. S. Islam (ed.), *Calcium Signaling*, Advances in Experimental Medicine and Biology 1131, https://doi.org/10.1007/978-3-030-12457-1_19

knowledge concerning the mechanisms underlying abnormal SOCE and its involvement in some diseases, as well as, we discussed the significance of STIM, Orai and TRPC isoforms as possible therapeutic targets for the treatment of angiogenesis, cardiovascular and skeletal muscle diseases.

Keywords Cardiac disease · Vascular disorders · Skeletal muscle · Angiogenesis

Abbreviations

CRAC	Ca^{2+}-Release Activated Ca^{2+} Channels
CREB	cAMP Response Element-Binding
DMD	Duchenne Muscular Dystrophy
EC	Endothelial Cell
EPC	Endothelial Progenitor Cell
ER/SR	Endoplasmic/Sarcoplasmic Reticulum
HF	Heart Failure
MD	Muscular Dystrophy
NFAT	Nuclear Factor of Activated T-cell
PAH	Pulmonary Arterial Hypertension
PASMC	Pulmonary Artery Smooth Muscle Cell
PDGF	Platelet-Derived Growth Factor
SERCA	Sarco/Endoplasmic Reticulum Ca^{2+} ATPase
siRNA	small interfering RNA
SOCC	Store-Operated Ca^{2+} Channel
SOCE	Store-Operated Ca^{2+} Entry
STIM1/2/1L	Stromal Interaction Molecule 1/2/1Large
TAC	Transverse Aortic Constriction
TRPC	Transient Receptor Potential-Canonical
VEGF	Vascular Endothelial Growth Factor
VSMC	Vascular Smooth Muscle Cell

19.1 General Overview of Store-Operated Ca^{2+} Entry

Store-Operated Ca^{2+} Entry (SOCE) is a ubiquitous mechanism for the entry of Ca^{2+} across ion channels located in the plasma membrane as illustrated in Fig. 19.1. SOCE is activated when the luminal Ca^{2+} concentration decreases in the intracellular Endoplasmic Reticulum (ER) Ca^{2+} stores [1]. More than a decade ago, Stromal Interacting Molecules 1 (STIM1) and Orai1 proteins were identified as the most relevant molecular components of SOCE [2, 3]. STIM1 was characterized as the ER luminal Ca^{2+} sensor that can interact and activates Orai channels upon ER

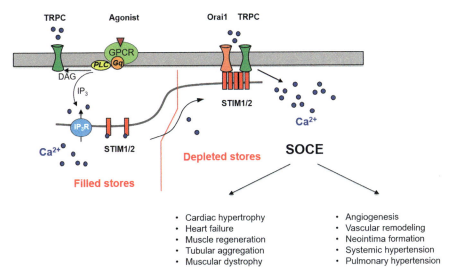

Fig. 19.1 Consensus model and standard molecular components of SOCE. Agonist binding to G-protein coupled receptor (GPCR), induces Ca^{2+} release from intracellular stores, STIM1/2 oligomerization in puncta and translocation to the plasma membrane, activation of Ca^{2+} entry through Orai1 and/or TRPCs. Agonists might also stimulate store-independent Ca^{2+} entry through others TRPCs activated by diacylglycerol (DAG). Several reports demonstrated that dysregulation in the expression and/or activation of STIM, Orai or TRPCs is associated with pathogenesis of different diseases such as cardiac hypertrophy and heart failure, muscle generation, tubular aggregation and dystrophies, angiogenesis, vascular remodeling, systemic and pulmonary hypertension

depletion [2]. Another isoform of STIM, STIM2, also senses the Ca^{2+} concentration in ER. This isoform apparently has a greater sensitivity for free intraluminal Ca^{2+} than STIM1 [4]. STIM2 can be partially active by resting ER Ca^{2+} concentration and it is further activated by small Ca^{2+} concentrations reductions in ER [4]. Along with STIM1, an alternative spliced long variant, STIM1L, has been related to SOCE especially in skeletal and cardiac muscle [5]. STIM1L is colocalized with Orai1 as well as others channels, such as Transient Receptor Potential-Canonical (TRPC) channels: TRPC1, TRPC3, TRPC4 and TRPC6 [6–8]. STIM1L is thought to form permanent clusters with Orai1 that are responsible for the rapid (<1 s) activation of SOCE in skeletal muscle cells in comparison to other cells [5].

Soon after the identification of STIM1, Orai1 was identified as the pore-forming subunit of Store-Operated Ca^{2+} Channels (SOCCs) [9, 10]. SOCE is generally mediated by the highly Ca^{2+}-selective channel, known as Ca^{2+} Release-Activated Ca^{2+} channel (CRAC) formed by homo- or hetero-multimeric Orai subunits (Orai1/2) [11, 12]. Along with Orai1, Ca^{2+} store depletion also activates other isoforms of Orai, Orai2 and Orai3 [13]. Although, Orai3 also contributes to store-independent Ca^{2+} channels, such as the Arachidonate Regulated Ca^{2+} (ARC) channels [14] and Leukotriene C4-Regulated Ca^{2+} (LRC) channels [15].

In addition to the highly selective CRAC channel, independent studies showed that SOCE can be also mediated by a non-selective SOCCs, involving Orai subunits interacting with members of TRPC channels, which are permeable to monovalent and divalent cations [16, 17]. This non-selective SOCCs has been largely described in excitable cells such as cardiac myocytes [18], Vascular Smooth Muscle Cells (VSMCs) [19] and skeletal muscle cells [20, 21]. Like Orai proteins, different members of the TRPC family can be activated by store depletion or by a variety of extracellular signals leading to changes in the Ca^{2+} concentration in spatially restricted microdomains underneath the plasma membrane, which support various Ca^{2+} dependent intracellular pathways [22].

Ca^{2+} influx via SOCE has been reported to be necessary for numerous cellular functions as VSMCs contraction and proliferation [19, 23, 24]; the contractile function in the skeletal and cardiac muscle [25, 26]; the insulin secretion [27, 28], etc. Given the physiological importance of SOCE, compelling evidences confirmed that alterations in the mechanism of activation, maintenance or inactivation of SOCE might lead to a pathophysiological phenotype. Loss- or gain-of-function gene mutations in the key components of SOCE have been reported to underlie a number of human diseases [29]. Actually, as resumed in Fig. 19.1 activation of SOCE as well as the level of expression of its molecular determinant, STIM1, Orai1 and TRPCs, have been implicated in the development of angiogenesis and several types of cancer [22], in the alteration of cardiac conduction and ventricle adverse remodeling [30], in skeletal muscle myopathy and tubular aggregation [31], and in vascular disorders [32].

In this chapter we will describe direct or indirect observations supporting the critical role of SOCE in cardiovascular and skeletal muscle disorders and angiogenesis.

19.2   SOCE in Cardiac Hypertrophy and Heart Failure

Most of the described molecular determinants of SOCE are also expressed in both excitable and non-excitable cardiac cells, such as fibroblasts, although the contribution of SOCE to normal cardiac physiology is still under debate [17, 26]. Growing set of experiments showed that stimulation of cardiomyocytes with agonists that activated G-protein coupled receptors or treatment with inhibitors of SERCA evokes persistent Ca^{2+} influx sensitive to classical blockers of SOCE. Whereas, they are insensitive to the inhibition of L-type Ca^{2+} channel [18, 33–35]. Interestingly, SOCE emerged as potential key player in Ca^{2+} dysregulation in cardiomyocytes. Therefore, most reports focused on its role in cardiac disease such as cardiac adverse remodeling, hypertrophy and consequent heart failure (HF), arrhythmia or cardiac conduction disorders [36–38]. In this chapter we will focus on the role of SOCE in cardiac hypertrophy and consequent HF. Cardiac hypertrophy can be reversible, nevertheless prolonged stress on the heart promotes deleterious remodeling of cardiomyocytes, leading to a weaker dilated heart and potentially HF [36, 39].

Cardiac hypertrophy occurs to preserve the pump function in response to a greater hemodynamic demand or to different stress such as myocardial infarction or hypertension-induced pressure overload [40, 41]. It is well known that the calcineurin-Nuclear Factor of Activated T-cells (NFAT) complex, which requires cytosolic Ca^{2+} increase, is one of the key mechanisms that switches on the fetal genes causing cardiac hypertrophy [36]. Growing evidences suggest that STIM1/Orai1-mediated SOCE provide Ca^{2+} signals necessary to initiate cardiac hypertrophy, mainly through the calcineurin/NFAT signaling [42]. Actually, the overexpression of STIM1 is associated with an increase in neonatal cardiomyocytes size and an enhanced NFAT activity, which are abolished in the presence of SOCC inhibitors [43]. Furthermore, silencing of STIM1 and Orai1 prevents phenylephrine-mediated hypertrophic neonatal cardiomyocyte growth [44]. Likewise, a more recent report found that STIM1 deletion protects the heart from pressure over-load induced cardiac hypertrophy [45]. Previously, Luo et al. [46] demonstrated that STIM1L is barely expressed in adult cardiomyocytes but it reemerges with hypertrophic agonists or in mice subjected to transverse aortic constriction during 3 weeks to trigger cardiac hypertrophy. Conversely, STIM1L levels correlate with exacerbated SOCE, and the expression of STIM1L is induced in isolated adult cardiomyocytes stimulated with phenylephrine [46].

On the other hand, independent studies demonstrated significant upregulation of TRPCs in several animal models of cardiac hypertrophy and heart failure [22, 37, 38]. In fact, experiments using pharmacological and molecular silencing or overexpression demonstrated that essential proteins of SOCE (STIM1, Orai1 and TRPC1/3/4 and 6) are involved in hypertrophy mainly through calcineurin/NFAT signaling [22, 38, 47]. Seth et al. [48] demonstrated in TRPC1 knockout mice that the calcineurin/NFAT pathway is inhibited, which reduces the hypertrophic response related to a better survival after Transverse Aortic Constriction (TAC) treatment. In this way, another study further confirmed that TRPC1/4 double knockout prevents cardiac hypertrophy and fibrotic infiltration after TAC and chronic neurohumoral stimulation [47].

Recent studies tried to demonstrate the involvement of SOCE in the transition from cardiac hypertrophy to HF. An early study demonstrated that TRPC3 protein levels are elevated in failing hearts of 20-months-old rats. This study also deter-mined, specific increase in the expression of mRNA and protein levels of TRPC5 in heart failure patients comparing to non-failing heart [49]. Whereas, other research showed that TRPC6 is also significantly upregulated in human failing heart and HF was observed in transgenic mice overexpressing TRPC6 [50]. Horton et al. [51] showed that Orai1-knock out mice submitted to TAC develop earlier loss of cardiac function and greater dilatation of the left ventricle comparing to WT mice. However, Orai1 role in HF is still unclear. Recently, Correll et al. [52] showed that STIM1 overexpression increases SOCE in heart and enhances sudden death in transgenic mice at 6 weeks of age. Authors observed that when these mice reached 12 weeks, they develop HF with prominent hypertrophy, loss of ventricular function, and pulmonary edema. Moreover, when submitted to pressure overload or neurohormonal stimulation these transgenic mice display accelerated hypertrophic [52]. In contrast, other study showed that specific STIM1 silencing in heart promotes the rapid transition from cardiac hypertrophy to HF [53]. Therefore, the presence of STIM1 is likely required for the persistence of adaptive cardiac hypertrophy.

Altogether, these studies suggest that different SOCE components play pivotal roles in the pathogenesis of heart disease, although many questions remain unanswered and require more detailed investigations.

19.3   SOCE in Vascular Disorder

Considerable interest has been focused towards the influence of ion channels plasticity and gene expression in VSMCs, which are normally quiescent and contractile. However, they can change to a proliferative and migratory state in certain conditions as arterial injury or inflammation [54]. This change is considered a characteristic step in the pathogenesis of multiple vascular diseases. Compelling evidence indicate that changes in the handling of SOCE originated by an alteration of STIM and Orai proteins is a significant contributor to the development of numerous vascular diseases [32], such as occlusive diseases, atherosclerosis or restenosis following angioplasty or those involving arterial remodeling like hypertension [15, 55–57].

Earlier studies demonstrated that vasoactive agonists and growth factors stimulate VSMCs growth, proliferation or migration through SOCE activation. In fact, angiotensin-II activates STIM1 and Orai1- dependent SOCE to stimulate aortic VSMCs growth and proliferation [58, 59]. Also, the addition of urotensin-II to aortic VSMCs activates I_{CRAC} and SOCE, stimulates Ca^{2+}/cAMP Response Element-Binding (CREB) transcription factor and promotes cells proliferation [24]. Urotensin-II mechanism requires a complex signaling pathway that involves STIM1, Orai1, and TRPC1 activation. On the other hand, Platelet-Derived Growth Factor (PDGF) activates Ca^{2+} entry and VSMCs migration involving STIM1 and Orai1 pathway, but not TRPC1/4/6 and Orai2/3 proteins [55]. Conversely, another study demonstrated that PDGF-induced proliferation VSMCs is strongly reduced in smooth muscle-specific STIM1 Knockout (sm-STIM1KO) mice, due to the diminution of NFAT activation dependent of Ca^{2+} entry through SOCE [60].

Interestingly, the use of in vivo animal model of angioplasty, performed generally by mechanical injury procedure of rat carotid arteries, helped to confirm that SOCE play a critical role in migration and proliferation of VSMCs on luminal side of injured vessels to form neointima [61]. Actually, inhibition of STIM1 and Orai1 with lentivirus attenuates neointima formation and prevents the increase of both proteins 14 days post-injury in the medial and neointimal layer. Moreover, STIM1 and Orai1 knockdown inhibits NFAT activation, which correlates with decreasing VSMCs proliferation compared to control [62]. In the same way, Guo et al. [59, 63] showed that angiotensin-II increases significantly the expression of STIM1 and Orai1 in the neointima layer. They also demonstrated that silencing of Orai1 and STIM1 reduces VSMCs proliferation and inhibits the accelerated neointimal growth induced by angiotensin-II in balloon-injured rat carotid arteries. Consistent with these data, Mancarella et al. described a significantly attenuation of neointima formation after carotid artery ligation in sm-STIM1KO mice compared to control mice [60]. Recently, a role of Homer1, a scaffolding protein that bind to several Ca^{2+}-signaling proteins [64], was characterized in VSMCs proliferation and neointima formation [65]. This study showed that Homer1 is upregulated in neointimal VSMCs after balloon-injured rat carotid arteries, interacting with TRPC1/3/4/6 channels and Orai1. Furthermore, knockdown of Homer 1 in vivo by AAV virus expressing a short hairpin RNA against Homer1, and in vitro by transfection with siRNAs, evokes a significant reduction of neointimal area and SOCE, which attenuate VSMCs proliferation and migration [65].

SOCE has also been implicated in the pathogenesis of systemic and pulmonary hypertension in different animal models. SOCE activation, associated with augmented levels of Orai1 and STIM1, is increased in aorta from spontaneously hypertensive rats compared to wild-type rats [56]. In mesenteric artery from ouabain-induced hypertensive rats compared to normotensive rats, SOCE is also increased and correlates with an enhanced expression of TRPC1, but not of TRPC4/5 [66]. Meanwhile, STIM1 upregulation was observed after hypertension development by ethanol consumption in rats [67]. Similarly, an augmented SOCE associated with an increase in TRPC channels expression was observed in Pulmonary Artery Smooth Muscle Cells (PASMCs) isolated from patients with idiopathic or hypoxic Pulmonary Arterial Hypertension (PAH) patients compared to normotensive patients [68, 69]. Likewise, others reported the participation of Orai1, STIM1 and/or STIM2 in SOCE and PASMCs proliferation in patients with idiopathic PAH [70–72]. Recently, it has been described that hypoxic conditions in PASMCs upregulated expression of STIM1/2 and Orai1/2/3 disturbing resting Ca^{2+} and SOCE, which could explain why hypoxic PAH causes pulmonary arterial vasoconstriction [73, 74].

Therefore, several players of SOCE play critical roles in VSMCs proliferation and migration making them fair therapeutic targets to mitigate vascular remodeling.

19.4   SOCE in Skeletal Muscle: Differentiation and Myopathies

The role of SOCE in the skeletal muscle contraction was neglected for a long time, however it is now commonly accepted that Ca^{2+} entry via SOCE is essential to sustain muscle contraction [25, 75, 76]. In the skeletal muscle, SOCE machinery involves a signaling complex located at the triad, the structure formed by a T-tubule with a sarcoplasmic reticulum (SR) known as the terminal cisterna, containing STIM1, STIM1L, Orai1, TRPCs and ryanodine receptors [17, 31]. These proteins control Ca^{2+} influx to refill sarcoplasmic reticulum Ca^{2+} store. This occurs not only in repetitive tetanic stimulation, but also in an immediate basis [77]. SOCE has a specific feature in the skeletal muscle related to its very fast kinetics of activation. In fact, the process leading to SOCE involving STIM1 aggregation and ER translocation to the plasma membrane and Orai1 opening takes place in less than a second [78, 79]. This feature is apparently due to the architecture of the sarcoplasmic reticulum. It is located permanently close to T-tubules at the triad allowing the proximity of STIM1 and Orai1 even before store depletion. SOCE seems also involved in the muscle contractions during tetanic stimulations in vigorous exercise and fatigue [80]. Actually, the absence of SOCE in the skeletal muscle results in mice more prone to fatigue, displaying a reduced muscle mass and force [25, 81].

SOCE also plays a role for muscle regeneration process to repair everyday life injuries or in case of myopathies partially the damage [8]. In fact, in human skeletal muscle the first event described during the differentiation process is an influx of Ca^{2+} caused by the activation of SOCE where STIM1, STIM2 and Orai1 proteins seem strongly involved [82]. Moreover, silencing of STIM1, Orai1 and Orai3 reduces the amplitude of SOCE and affects to myoblast differentiation. In contrast, the overexpression of STIM1 together with Orai1 increases SOCE and accelerates muscle differentiation [82].

An early study by Olah et al. [83] suggested that the overexpression of TRPC1 enhances SOCE in myotubes but it delays muscle differentiation leading to the formation of small myotubes. In contrast, TRPC1 overexpression enhances SOCE and accelerates the process of myogenesis in human muscle [84]. This finding was supported by recent data by Xia et al. [85] where they determined that TRPC1 downregulation by siRNA delays significantly the muscle regeneration. Similarly, a recent study demonstrated that TRPC1/4 and STIM1L are necessary for SOCE and are required for a proper skeletal muscle differentiation [6]. So far TRPC1, which is likely accepted as a channel necessary for the proper muscle differentiation both in human and murine models, can be activated by store depletion as well as stretch.

In addition to its role in muscle differentiation the alteration of the Ca^{2+} influx through SOCCs is related to different myopathies. Patients with the hereditary Severe Combined Immunodeficiency (SCID) syndrome have depressed SOCE because of loss-of-function associated to mutations of STIM1 and Orai1 genes. They manifest atrophy in skeletal-muscle fibers, hypotonia and a severe chronic pulmonary problem due to respiratory muscle weakness [29]. Interestingly, gain-of-function mutations in STIM1 or Orai1 genes results as well in muscular weakness, mainly associated with tubular aggregate myopathy syndrome [86]. Moreover, several reports suggested an important role of SOCE in Muscular Dystrophies (MD), which developed progressively accompanied with an increase of muscular weakness, atrophy and fatigue. SOCE is proposed to be a cause of Ca^{2+} overload which could initiate the characteristic cell necrosis in MD [78, 87]. Duchenne Muscular Dystrophy (DMD) is the most common form of MD in children and it is characterized by a progressive muscle degeneration and weakness [88]. DMD is caused by the absence of dystrophin, a cohesive protein located between the sarcolemma and the myofilaments in skeletal muscle cells. When the myofilaments lack of dystrophin occurs a pathological Ca^{2+} dynamic where SOCE is exacerbated [89]. Significant increase in the expression of STIM1, Orai1 and TRPC1 is associated with Ca^{2+} overload in fiber of a mouse model of DMD (*mdx*) [78, 89, 90]. Moreover, STIM1L isoform is upregulated in *mdx* fiber, meanwhile STIM2 does not participate [91, 92]. Furthermore, the inhibition of TRPC1 and TRPC4 in *mdx* recovered Ca^{2+} homeostasis suggesting a role of these TRPCs [8, 90]. On the other hand, TRPC3 is also overexpressed in MD although its function seems independent of SOCE [93, 94].

Taking everything into account, it is confirmed that STIM1/1L, Orai1 and different isoforms of TRPCs are relevant for Ca^{2+} influx and for the progression of skeletal muscle disease such as DMD. Hence, molecular determinants of SOCE could be potential therapeutic targets to attenuate the impact of different myopathies.

19.5   SOCE and Angiogenesis

Angiogenesis is understood as the growth of new blood vessels from existing vasculature [95, 96]. This process is tightly regulated and may occur in a number of situations including prolonged exercise [97], reproduction [98], physiological repair (e.g. regeneration of heart after a myocardial infarction [99–101], and different pathologies such as diabetes, endometriosis [102] and cancer [103]). Basically, angiogenesis is initiated in hypoxic environment because a variety of growth factors, including vascular endothelial growth factor (VEGF), fibroblast growth factor (FGF), and epidermal growth factor (EGF). The last one stimulates proliferation, migration and tube formation of endothelial cells (ECs), resulting in the generation of new capillary tubes [104]. Increasing evidences confirmed an expression of different SOCE proteins in ECs, which are intimately related to tumor progression and angiogenesis [105–108]. Chen et al. [106] demonstrated that STIM1 overexpression remarkably enhances tumor growth, local spread and angiogenesis; whereas, knockdown of STIM1 significantly decreases tumor growth and tumor vessels number. Independent studies showed that VEGF increases cytosolic Ca^{2+} due to SOCE in ECs [102, 105, 109]. Li et al. [102] demonstrated that specific inhibition of CRAC channels prevented VEGF-induced Ca^{2+} influx and tube formation. Using siRNAs, a dominant negative or neutralizing antibodies, they also demonstrated that Orai1 instead of TRPC1 is involved in VEGF-induced Ca^{2+} increase and angiogenesis [102]. Moreover, beside the implication of Orai1/STIM1, a recent study suggested a role for Orai3 in VEGF-induced ECs tube formation both in vitro and in vivo [110].

Although there is a consensus in the role of STIM1 and Orai1 in angiogenesis, TRPCs involvement is still under debate [111]. Indeed, knockdown of TRPC1 did not have a significant effect on the in vitro formation of human umbilical vein-derived endothelial tubes [112]. Besides, TRPC1 knockout mice developed the vasculature normally [113]. Furthermore, Antigny et al. [112] reported that silencing of TRPC3, TRPC4, or TRPC5 expression inhibited endothelial tube formation in vitro, suggesting a possible role of these TRPC channels in angiogenesis. Therefore, additional studies about the role of TRPCs in angiogenesis will be welcomed.

SOCE proteins are also expressed in Endothelial Progenitor Cells (EPCs) [114], and they are significantly enhanced in EPCs isolated from patients with renal cellular carcinoma [115]. The exacerbated SOCE correlates with an upregulation of STIM1, Orai1 and TRPC1, which controls proliferation and in vitro tubulogenesis both in normal EPCs and in their malignant counterparts [116].

Taken together, these studies suggest that STIM1, Orai1 and some TRPCs stand out as promising molecular target of anti-angiogenic therapies to prevent tumor neovascularization. Therefore, SOCE signaling pathway is worth to take in consideration as alternative strategy to hit highly vasculogenic tumors.

19.6 Conclusion and Perspectives

Over the last decades, the rapid progress in SOCE research has revealed that different proteins of the SOCE mechanism play pivotal roles not only in the regulation of basal Ca^{2+} homeostasis but also in several pathogenesis related to heart and vascular diseases, skeletal myopathies and angiogenesis. Nevertheless, many questions remain unanswered and require future investigations to decipher the physiological and pathological roles of SOCE in different systems. Given that STIM, Orai and TRPC isoforms are widely expressed in different cell types, using transgenic mouse models or tissue specific knockout of these genes will undoubtedly provide more precise information about the causative role of these proteins in the development of each disease. For this reason, many studies are now focusing on such proteins as therapeutic targets to consider for pharmacological intervention in the treatment of these pathologies.

Acknowledgements This work was supported by FEDER funds and by Spanish Ministry of Economy and Competitiveness [BFU2016-74932-C2]; Institute of Carlos III [PI15/00203; CB16/11/00431]; and by the Andalusia Government [P12-CTS-1965; PI-0313-2016].

References

1. Putney JW (1986) A model for receptor-regulated calcium entry. Cell Calcium 7:1–12
2. Roos J, DiGregorio PJ, Yeromin AV et al (2005) STIM1, an essential and conserved component of store-operated Ca^{2+} channel function. J Cell Biol 169:435–445
3. Hogan PG, Rao A (2015) Store-operated calcium entry: mechanisms and modulation. Biochem Biophys Res Commun 460:40–49
4. Brandman O, Liou J, Park WS, Meyer T (2007) STIM2 is a feedback regulator that stabilizes basal cytosolic and endoplasmic reticulum Ca^{2+} levels. Cell 131:1327–1339

5. Rosado JA, Diez R, Smani T, Jardín I (2015) STIM and Orai1 variants in store-operated calcium entry. Front Pharmacol 6:325

6. Antigny F, Sabourin J, Saüc S, Bernheim L, Koenig S, Frieden M (2017) TRPC1 and TRPC4 channels functionally interact with STIM1L to promote myogenesis and maintain fast repetitive Ca^{2+} release in human myotubes. Biochim Biophys Acta 1864:806–813

7. Horinouchi T, Higashi T, Higa T et al (2012) Different binding property of STIM1 and its novel splice variant STIM1L to Orai1, TRPC3, and TRPC6 channels. Biochem Biophys Res Commun 428:252–258

8. Saüc S, Frieden M (2017) Neurological and motor disorders: TRPC in the skeletal muscle. Adv Exp Med Biol 993:557–575

9. Feske S, Gwack Y, Prakriya M, Srikanth S, Puppel SH, Tanasa B, Hogan PG, Lewis RS, Daly M, Rao A (2006) A mutation in Orai1 causes immune deficiency by abrogating CRAC channel function. Nature 441:179–185

10. Vig M, Beck A, Billingsley JM et al (2006) CRACM1 multimers form the ion-selective pore of the CRAC channel. Curr Biol 16:2073–2079

11. Desai PN, Zhang X, Wu S, Janoshazi A, Bolimuntha S, Putney JW, Trebak M (2015) Multiple types of calcium channels arising from alternative translation initiation of the Orai1 message. Sci Signal 8:ra74

12. Vaeth M, Yang J, Yamashita M et al (2017) ORAI2 modulates store-operated calcium entry and T cell-mediated immunity. Nat Commun 8:14714

13. Lis A, Peinelt C, Beck A, Parvez S, Monteilh-Zoller M, Fleig A, Penner R (2007) CRACM1, CRACM2, and CRACM3 are store-operated Ca^{2+} channels with distinct functional properties. Curr Biol CB 17:794–800

14. Mignen O, Thompson JL, Shuttleworth TJ (2008) Both Orai1 and Orai3 are essential components of the arachidonate-regulated Ca^{2+}−selective (ARC) channels. J Physiol 586:185–195

15. Gonzalez-Cobos JC, Zhang X, Zhang W et al (2013) Store-independent Orai1/3 channels activated by intracrine leukotriene C4: role in neointimal hyperplasia. Circ Res 112:1013–1025

16. Ambudkar IS, de Souza LB, Ong HL (2017) TRPC1, Orai1, and STIM1 in SOCE: friends in tight spaces. Cell Calcium 63:33–39

17. Avila-Medina J, Mayoral-Gonzalez I, Dominguez-Rodriguez A, Gallardo-Castillo I, Ribas J, Ordoñez A, Rosado JA, Smani T (2018) The complex role of store operated calcium entry pathways and related proteins in the function of cardiac, skeletal and vascular smooth muscle cells. Front Physiol 9:257

18. Domínguez-Rodríguez A, Ruiz-Hurtado G, Sabourin J, Gómez AM, Alvarez JL, Benitah J-P (2015) Proarrhythmic effect of sustained EPAC activation on TRPC3/4 in rat ventricular cardiomyocytes. J Mol Cell Cardiol 87:74–78

19. Avila-Medina J, Calderon-Sanchez E, Gonzalez-Rodriguez P, Monje-Quiroga F, Rosado JA, Castellano A, Ordonez A, Smani T (2016) Orai1 and TRPC1 proteins co-localize with CaV1.2 channels to form a signal complex in vascular smooth muscle cells. J Biol Chem 291:21148–21159

20. Lyfenko AD, Dirksen RT (2008) Differential dependence of store-operated and excitation-coupled Ca^{2+} entry in skeletal muscle on STIM1 and Orai1. J Physiol 586:4815–4824

21. Stiber J, Hawkins A, Zhang Z-S et al (2008) STIM1 signalling controls store-operated calcium entry required for development and contractile function in skeletal muscle. Nat Cell Biol 10:688–697

22. Smani T, Shapovalov G, Skryma R, Prevarskaya N, Rosado JA (2015) Functional and physiopathological implications of TRP channels. Biochim Biophys Acta 1853:1772–1782

23. Dominguez-Rodriguez A, Diaz I, Rodriguez-Moyano M, Calderon-Sanchez E, Rosado JA, Ordonez A, Smani T (2012) Urotensin-II signaling mechanism in rat coronary artery: role of STIM1 and Orai1-dependent store operated calcium influx in vasoconstriction. Arter Thromb Vasc Biol 32:1325–1332

24. Rodriguez-Moyano M, Diaz I, Dionisio N, Zhang X, Avila-Medina J, Calderon-Sanchez E, Trebak M, Rosado JA, Ordonez A, Smani T (2013) Urotensin-II promotes vascular smooth muscle cell proliferation through store-operated calcium entry and EGFR transactivation. Cardiovasc Res 100:297–306

25. Wei-Lapierre L, Carrell EM, Boncompagni S, Protasi F, Dirksen RT (2013) Orai1-dependent calcium entry promotes skeletal muscle growth and limits fatigue. Nat Commun 4:2805

26. Bootman MD, Rietdorf K (2017) Tissue specificity: store-operated Ca^{2+} entry in cardiac myocytes. Adv Exp Med Biol 993:363–387

27. Chang H-Y, Chen S-L, Shen M-R, Kung M-L, Chuang L-M, Chen Y-W (2017) Selective serotonin reuptake inhibitor, fluoxetine, impairs E-cadherin-mediated cell adhesion and alters calcium homeostasis in pancreatic beta cells. Sci Rep 7:3515

28. Sabourin J, Le Gal L, Saurwein L, Haefliger J-A, Raddatz E, Allagnat F (2015) Store-operated Ca^{2+} entry mediated by Orai1 and TRPC1 participates to insulin secretion in rat β-cells. J Biol Chem 290:30530–30539

29. Lacruz RS, Feske S (2015) Diseases caused by mutations in ORAI1 and STIM1. Ann N Y Acad Sci 1356:45–79

30. Eder P (2017) Cardiac remodeling and disease: SOCE and TRPC signaling in cardiac pathology. Adv Exp Med Biol 993:505–521

31. Pan Z, Brotto M, Ma J (2014) Store-operated Ca^{2+} entry in muscle physiology and diseases. BMB Rep 47:69–79

32. Tanwar J, Trebak M, Motiani RK (2017) Cardiovascular and hemostatic disorders: role of STIM and Orai proteins in vascular disorders. Adv Exp Med Biol 993:425–452

33. Hunton DL, Zou L, Pang Y, Marchase RB (2004) Adult rat cardiomyocytes exhibit capacitative calcium entry. Am J Physiol Heart Circ Physiol 286:H1124–H1132

34. Kojima A, Kitagawa H, Omatsu-Kanbe M, Matsuura H, Nosaka S (2012) Presence of store-operated Ca^{2+} entry in C57BL/6J mouse ventricular myocytes and its suppression by sevoflurane. Br J Anaesth 109:352–360

35. Uehara A, Yasukochi M, Imanaga I, Nishi M, Takeshima H (2002) Store-operated Ca^{2+} entry uncoupled with ryanodine receptor and junctional membrane complex in heart muscle cells. Cell Calcium 31:89–96

36. Eder P, Molkentin JD (2011) TRPC channels as effectors of cardiac hypertrophy. Circ Res 108:265–272

37. Bartoli F, Sabourin J (2017) Cardiac remodeling and disease: current understanding of STIM1/Orai1-mediated store-operated Ca^{2+} entry in cardiac function and pathology. Adv Exp Med Biol 993:523–534

38. Yue Z, Xie J, Yu AS, Stock J, Du J, Yue L (2015) Role of TRP channels in the cardiovascular system. Am J Physiol Heart Circ Physiol 308:H157–H182

39. Ljubojevic S, Radulovic S, Leitinger G et al (2014) Early remodeling of perinuclear Ca^{2+} stores and nucleoplasmic Ca^{2+} signaling during the development of hypertrophy and heart failure. Circulation 130:244–255

40. Samak M, Fatullayev J, Sabashnikov A et al (2016) Cardiac hypertrophy: an introduction to molecular and cellular basis. Med Sci Monit Basic Res 22:75–79

41. McMullen JR, Jennings GL (2007) Differences between pathological and physiological cardiac hypertrophy: novel therapeutic strategies to treat heart failure. Clin Exp Pharmacol Physiol 34:255–262

42. Collins HE, Zhu-Mauldin X, Marchase RB, Chatham JC (2013) STIM1/Orai1-mediated SOCE: current perspectives and potential roles in cardiac function and pathology. Am J Physiol Heart Circ Physiol 305:H446–H458

43. Hulot J-S, Fauconnier J, Ramanujam D et al (2011) Critical role for stromal interaction molecule 1 in cardiac hypertrophy. Circulation 124:796–805

44. Voelkers M, Salz M, Herzog N et al (2010) Orai1 and Stim1 regulate normal and hypertrophic growth in cardiomyocytes. J Mol Cell Cardiol 48:1329–1334

45. Parks C, Alam MA, Sullivan R, Mancarella S (2016) STIM1-dependent Ca^{2+} microdomains are required for myofilament remodeling and signaling in the heart. Sci Rep 6:25372

46. Luo X, Hojayev B, Jiang N, Wang ZV, Tandan S, Rakalin A, Rothermel BA, Gillette TG, Hill JA (2012) STIM1-dependent store-operated Ca^{2+} entry is required for pathological cardiac hypertrophy. J Mol Cell Cardiol 52:136–147

47. Camacho Londoño JE, Tian Q, Hammer K et al (2015) A background Ca^{2+} entry pathway mediated by TRPC1/TRPC4 is critical for development of pathological cardiac remodelling. Eur Heart J 36:2257–2266

48. Seth M, Zhang Z-S, Mao L et al (2009) TRPC1 channels are critical for hypertrophic signaling in the heart. Circ Res 105:1023–1030

49. Bush EW, Hood DB, Papst PJ, Chapo JA, Minobe W, Bristow MR, Olson EN, McKinsey TA (2006) Canonical transient receptor potential channels promote cardiomyocyte hypertrophy through activation of calcineurin signaling. J Biol Chem 281:33487–33496

50. Kuwahara K, Wang Y, McAnally J, Richardson JA, Bassel-Duby R, Hill JA, Olson EN (2006) TRPC6 fulfills a calcineurin signaling circuit during pathologic cardiac remodeling. J Clin Invest 116:3114–3126

51. Horton JS, Buckley CL, Alvarez EM, Schorlemmer A, Stokes AJ (2014) The calcium release-activated calcium channel Orai1 represents a crucial component in hypertrophic compensation and the development of dilated cardiomyopathy. Channels Austin Tex 8:35–48

52. Correll RN, Goonasekera SA, van Berlo JH et al (2015) STIM1 elevation in the heart results in aberrant Ca^{2+} handling and cardiomyopathy. J Mol Cell Cardiol 87:38–47

53. Bénard L, Oh JG, Cacheux M et al (2016) Cardiac Stim1 silencing impairs adaptive hypertrophy and promotes heart failure through inactivation of mTORC2/Akt signaling. Circulation 133:1458–1471; discussion 1471

54. House SJ, Potier M, Bisaillon J, Singer HA, Trebak M (2008) The non-excitable smooth muscle: calcium signaling and phenotypic switching during vascular disease. Pflugers Arch 456:769–785

55. Bisaillon JM, Motiani RK, Gonzalez-Cobos JC, Potier M, Halligan KE, Alzawahra WF, Barroso M, Singer HA, Jourd'heuil D, Trebak M (2010) Essential role for STIM1/Orai1-mediated calcium influx in PDGF-induced smooth muscle migration. Am J Physiol Cell Physiol 298:C993–C1005

56. Giachini FR, Chiao CW, Carneiro FS, Lima VV, Carneiro ZN, Dorrance AM, Tostes RC, Webb RC (2009) Increased activation of stromal interaction molecule-1/Orai-1 in aorta from hypertensive rats: a novel insight into vascular dysfunction. Hypertension 53:409–416

57. Kassan M, Ait-Aissa K, Radwan E et al (2016) Essential role of smooth muscle STIM1 in hypertension and cardiovascular dysfunction. Arter Thromb Vasc Biol 36:1900–1909

58. Simo-Cheyou ER, Tan JJ, Grygorczyk R, Srivastava AK (2017) STIM-1 and ORAI-1 channel mediate angiotensin-II-induced expression of Egr-1 in vascular smooth muscle cells. J Cell Physiol 232:3496–3509

59. Guo RW, Yang LX, Li MQ, Pan XH, Liu B, Deng YL (2012) Stim1- and Orai1-mediated store-operated calcium entry is critical for angiotensin II-induced vascular smooth muscle cell proliferation. Cardiovasc Res 93:360–370

60. Mancarella S, Potireddy S, Wang Y et al (2013) Targeted STIM deletion impairs calcium homeostasis, NFAT activation, and growth of smooth muscle. FASEB J 27:893–906

61. Zhang W, Trebak M (2014) Vascular balloon injury and intraluminal administration in rat carotid artery. J Vis Exp. https://doi.org/10.3791/52045

62. Zhang W, Halligan KE, Zhang X et al (2011) Orai1-mediated I (CRAC) is essential for neointima formation after vascular injury. Circ Res 109:534–542

63. Guo RW, Wang H, Gao P, Li MQ, Zeng CY, Yu Y, Chen JF, Song MB, Shi YK, Huang L (2009) An essential role for stromal interaction molecule 1 in neointima formation following arterial injury. Cardiovasc Res 81:660–668

64. Jardin I, Albarrán L, Bermejo N, Salido GM, Rosado JA (2012) Homers regulate calcium entry and aggregation in human platelets: a role for Homers in the association between STIM1 and Orai1. Biochem J 445:29–38

65. Jia S, Rodriguez M, Williams AG, Yuan JP (2017) Homer binds to Orai1 and TRPC channels in the neointima and regulates vascular smooth muscle cell migration and proliferation. Sci Rep 7:5075

66. Pulina MV, Zulian A, Berra-Romani R, Beskina O, Mazzocco-Spezzia A, Baryshnikov SG, Papparella I, Hamlyn JM, Blaustein MP, Golovina VA (2010) Upregulation of Na^+ and Ca^{2+} transporters in arterial smooth muscle from ouabain-induced hypertensive rats. Am J Physiol Heart Circ Physiol 298:H263–H274

67. Souza Bomfim GH, Mendez-Lopez I, Arranz-Tagarro JA, Ferraz Carbonel AA, Roman-Campos D, Padin JF, Garcia AG, Jurkiewicz A, Jurkiewicz NH (2017) Functional upregulation of STIM-1/Orai-1-mediated store-operated Ca^{2+} contributing to the hypertension development elicited by chronic EtOH consumption. Curr Vasc Pharmacol 15:265–281

68. Lin MJ, Leung GP, Zhang WM, Yang XR, Yip KP, Tse CM, Sham JS (2004) Chronic hypoxia-induced upregulation of store-operated and receptor-operated Ca^{2+} channels in pulmonary arterial smooth muscle cells: a novel mechanism of hypoxic pulmonary hypertension. Circ Res 95:496–505

69. Zhang S, Patel HH, Murray F, Remillard CV, Schach C, Thistlethwaite PA, Insel PA, Yuan JX (2007) Pulmonary artery smooth muscle cells from normal subjects and IPAH patients show divergent cAMP-mediated effects on TRPC expression and capacitative Ca^{2+} entry. Am J Physiol Lung Cell Mol Physiol 292:L1202–L1210

70. Ogawa A, Firth AL, Smith KA, Maliakal MV, Yuan JX (2012) PDGF enhances store-operated Ca^{2+} entry by upregulating STIM1/Orai1 via activation of Akt/mTOR in human pulmonary arterial smooth muscle cells. Am J Physiol Cell Physiol 302:C405–C411

71. Hou X, Chen J, Luo Y, Liu F, Xu G, Gao Y (2013) Silencing of STIM1 attenuates hypoxia-induced PASMCs proliferation via inhibition of the SOC/Ca^{2+}/NFAT pathway. Respir Res 14:2

72. Fernandez RA, Wan J, Song S, Smith KA, Gu Y, Tauseef M, Tang H, Makino A, Mehta D, Yuan JX (2015) Upregulated expression of STIM2, TRPC6, and Orai2 contributes to the transition of pulmonary arterial smooth muscle cells from a contractile to proliferative phenotype. Am J Physiol Cell Physiol 308:C581–C593

73. Wang J, Xu C, Zheng Q, Yang K, Lai N, Wang T, Tang H, Lu W (2017) Orai1, 2, 3 and STIM1 promote store-operated calcium entry in pulmonary arterial smooth muscle cells. Cell Death Discov 3:17074

74. He X, Song S, Ayon RJ, Balisterieri A, Black SM, Makino A, Wier WG, Zang WJ, Yuan JX (2018) Hypoxia selectively upregulates cation channels and increases cytosolic [Ca^{2+}] in pulmonary, but not coronary, arterial smooth muscle cells. Am J Physiol Cell Physiol. https://doi.org/10.1152/ajpcell.00272.2017

75. Dirksen RT (2009) Checking your SOCCs and feet: the molecular mechanisms of Ca^{2+} entry in skeletal muscle. J Physiol 587:3139–3147

76. Stiber JA, Rosenberg PB (2011) The role of store-operated calcium influx in skeletal muscle signaling. Cell Calcium 49:341–349

77. Sztretye M, Geyer N, Vincze J et al (2017) SOCE is important for maintaining sarcoplasmic calcium content and release in skeletal muscle fibers. Biophys J 113:2496–2507

78. Edwards JN, Friedrich O, Cully TR, von Wegner F, Murphy RM, Launikonis BS (2010) Upregulation of store-operated Ca^{2+} entry in dystrophic mdx mouse muscle. Am J Physiol Cell Physiol 299:C42–C50

79. Launikonis BS, Stephenson DG, Friedrich O (2009) Rapid Ca^{2+} flux through the transverse tubular membrane, activated by individual action potentials in mammalian skeletal muscle. J Physiol 587:2299–2312

80. Allen DG, Lamb GD, Westerblad H (2008) Skeletal muscle fatigue: cellular mechanisms. Physiol Rev 88:287–332

81. Li T, Finch EA, Graham V, Zhang Z-S, Ding J-D, Burch J, Oh-hora M, Rosenberg P (2012) STIM1-Ca^{2+} signaling is required for the hypertrophic growth of skeletal muscle in mice. Mol Cell Biol 32:3009–3017

82. Darbellay B, Arnaudeau S, König S, Jousset H, Bader C, Demaurex N, Bernheim L (2009) STIM1- and Orai1-dependent store-operated calcium entry regulates human myoblast differentiation. J Biol Chem 284:5370–5380

83. Oláh T, Fodor J, Ruzsnavszky O, Vincze J, Berbey C, Allard B, Csernoch L (2011) Overexpression of transient receptor potential canonical type 1 (TRPC1) alters both store operated calcium entry and depolarization-evoked calcium signals in C2C12 cells. Cell Calcium 49:415–425

84. Antigny F, Koenig S, Bernheim L, Frieden M (2013) During post-natal human myogenesis, normal myotube size requires TRPC1- and TRPC4-mediated Ca^{2+} entry. J Cell Sci 126:2525–2533

85. Xia L, Cheung K-K, Yeung SS, Yeung EW (2016) The involvement of transient receptor potential canonical type 1 in skeletal muscle regrowth after unloading-induced atrophy. J Physiol 594:3111–3126

86. Böhm J, Chevessier F, Koch C et al (2014) Clinical, histological and genetic characterisation of patients with tubular aggregate myopathy caused by mutations in STIM1. J Med Genet 51:824–833

87. Goonasekera SA, Davis J, Kwong JQ, Accornero F, Wei-LaPierre L, Sargent MA, Dirksen RT, Molkentin JD (2014) Enhanced Ca^{2+} influx from STIM1-Orai1 induces muscle pathology in mouse models of muscular dystrophy. Hum Mol Genet 23:3706–3715

88. Brandsema JF, Darras BT (2015) Dystrophinopathies. Semin Neurol 35:369–384

89. Kiviluoto S, Decuypere J-P, De Smedt H, Missiaen L, Parys JB, Bultynck G (2011) STIM1 as a key regulator for Ca^{2+} homeostasis in skeletal-muscle development and function. Skelet Muscle 1:16

90. Vandebrouck C, Martin D, Schoor MC-V, Debaix H, Gailly P (2002) Involvement of TRPC in the abnormal calcium influx observed in dystrophic (mdx) mouse skeletal muscle fibers. J Cell Biol 158:1089–1096

91. Cully TR, Edwards JN, Friedrich O, Stephenson DG, Murphy RM, Launikonis BS (2012) Changes in plasma membrane Ca-ATPase and stromal interacting molecule 1 expression levels for Ca^{2+} signaling in dystrophic mdx mouse muscle. Am J Physiol Cell Physiol 303:C567–C576

92. Cully TR, Launikonis BS (2013) Store-operated Ca^{2+} entry is not required for store refilling in skeletal muscle. Clin Exp Pharmacol Physiol 40:338–344

93. Shirokova N, Niggli E (2013) Cardiac phenotype of duchenne muscular dystrophy: insights from cellular studies. J Mol Cell Cardiol 58:217–224

94. Millay DP, Goonasekera SA, Sargent MA, Maillet M, Aronow BJ, Molkentin JD (2009) Calcium influx is sufficient to induce muscular dystrophy through a TRPC-dependent mechanism. Proc Natl Acad Sci U S A 106:19023–19028

95. Fraisl P, Mazzone M, Schmidt T, Carmeliet P (2009) Regulation of angiogenesis by oxygen and metabolism. Dev Cell 16:167–179

96. Stapor P, Wang X, Goveia J, Moens S, Carmeliet P (2014) Angiogenesis revisited – role and therapeutic potential of targeting endothelial metabolism. J Cell Sci 127:4331–4341

97. Egginton S (2009) Invited review: activity-induced angiogenesis. Pflugers Arch 457:963–977

98. Logsdon EA, Finley SD, Popel AS, Mac Gabhann F (2014) A systems biology view of blood vessel growth and remodelling. J Cell Mol Med 18:1491–1508

99. Ingason AB, Goldstone AB, Paulsen MJ, Thakore AD, Truong VN, Edwards BB, Eskandari A, Bollig T, Steele AN, Woo YJ (2018) Angiogenesis precedes cardiomyocyte migration in regenerating mammalian hearts. J Thorac Cardiovasc Surg 155:1118–1127.e1

100. Reddy K, Khaliq A, Henning RJ (2015) Recent advances in the diagnosis and treatment of acute myocardial infarction. World J Cardiol 7:243–276

101. Melly L, Cerino G, Frobert A et al (2018) Myocardial infarction stabilization by cell-based expression of controlled vascular endothelial growth factor levels. J Cell Mol Med. https://doi.org/10.1111/jcmm.13511

102. Li J, Cubbon RM, Wilson LA et al (2011) Orai1 and CRAC channel dependence of VEGF-activated Ca^{2+} entry and endothelial tube formation. Circ Res 108:1190–1198

103. Folkman J (1971) Tumor angiogenesis: therapeutic implications. N Engl J Med 285:1182–1186

104. Kohn EC, Alessandro R, Spoonster J, Wersto RP, Liotta LA (1995) Angiogenesis: role of calcium-mediated signal transduction. Proc Natl Acad Sci U S A 92:1307–1311
105. Chen Y-F, Hsu K-F, Shen M-R (2016) The store-operated Ca^{2+} entry-mediated signaling is important for cancer spread. Biochim Biophys Acta 1863:1427–1435
106. Chen Y-F, Chiu W-T, Chen Y-T, Lin P-Y, Huang H-J, Chou C-Y, Chang H-C, Tang M-J, Shen M-R (2011) Calcium store sensor stromal-interaction molecule 1-dependent signaling plays an important role in cervical cancer growth, migration, and angiogenesis. Proc Natl Acad Sci U S A 108:15225–15230
107. Fiorio Pla A, Gkika D (2013) Emerging role of TRP channels in cell migration: from tumor vascularization to metastasis. Front Physiol 4:311
108. Martial S (2016) Involvement of ion channels and transporters in carcinoma angiogenesis and metastasis. Am J Physiol Cell Physiol 310:C710–C727
109. Dragoni S, Laforenza U, Bonetti E et al (2011) Vascular endothelial growth factor stimulates endothelial colony forming cells proliferation and tubulogenesis by inducing oscillations in intracellular Ca^{2+} concentration. Stem Cells Dayt Ohio 29:1898–1907
110. Li J, Bruns A-F, Hou B et al (2015) Orai3 surface accumulation and calcium entry evoked by vascular endothelial growth factor. Arterioscler Thromb Vasc Biol 35:1987–1994
111. Earley S, Brayden JE (2015) Transient receptor potential channels in the vasculature. Physiol Rev 95:645–690
112. Antigny F, Girardin N, Frieden M (2012) Transient receptor potential canonical channels are required for in vitro endothelial tube formation. J Biol Chem 287:5917–5927
113. Schmidt K, Dubrovska G, Nielsen G et al (2010) Amplification of EDHF-type vasodilatations in TRPC1-deficient mice. Br J Pharmacol 161:1722–1733
114. Sánchez-Hernández Y, Laforenza U, Bonetti E et al (2010) Store-operated Ca^{2+} entry is expressed in human endothelial progenitor cells. Stem Cells Dev 19:1967–1981
115. Lodola F, Laforenza U, Bonetti E et al (2012) Store-operated Ca^{2+} entry is remodelled and controls in vitro angiogenesis in endothelial progenitor cells isolated from tumoral patients. PLoS One 7:e42541
116. Moccia F, Poletto V (2015) May the remodeling of the Ca^{2+} toolkit in endothelial progenitor cells derived from cancer patients suggest alternative targets for anti-angiogenic treatment? Biochim Biophys Acta 1853:1958–1973

Chapter 20
Calcium Signaling and the Regulation of Chemosensitivity in Cancer Cells: Role of the Transient Receptor Potential Channels

Giorgio Santoni, Maria Beatrice Morelli, Oliviero Marinelli, Massimo Nabissi, Matteo Santoni, and Consuelo Amantini

Abstract Cancer cells acquire the ability to modify the calcium signaling network by altering the expression and functions of cation channels, pumps or transporters. Calcium signaling pathways are involved in proliferation, angiogenesis, invasion, immune evasion, disruption of cell death pathways, ECM remodelling, epithelial-mesenchymal transition (EMT) and drug resistance. Among cation channels, a pivotal role is played by the Transient Receptor Potential non-selective cation-permeable receptors localized in plasma membrane, endoplasmic reticulum, mitochondria and lysosomes. Several findings indicate that the dysregulation in calcium signaling induced by TRP channels is responsible for cancer growth, metastasis and chemoresistance. Drug resistance represents a major limitation in the application of current therapeutic regimens and several efforts are spent to overcome it. Here we describe the ability of Transient Receptor Potential Channels to modify, by altering the intracellular calcium influx, the cancer cell sensitivity to chemotherapeutic drugs.

G. Santoni · M. Nabissi
School of Pharmacy, Immunopathology and Molecular Medicine Laboratory, University of Camerino, Camerino, Italy

M. B. Morelli · O. Marinelli
School of Pharmacy, Immunopathology and Molecular Medicine Laboratory, University of Camerino, Camerino, Italy

School of Biosciences and Veterinary Medicine, University of Camerino, Camerino, Italy

M. Santoni
Clinic and Oncology Unit, Macerata Hospital, Macerata, Italy

C. Amantini (✉)
School of Biosciences and Veterinary Medicine, University of Camerino, Camerino, Italy
e-mail: consuelo.amantini@unicam.it

© Springer Nature Switzerland AG 2020
M. S. Islam (ed.), *Calcium Signaling*, Advances in Experimental Medicine and Biology 1131, https://doi.org/10.1007/978-3-030-12457-1_20

505

Keywords Ca^{2+} dysregulation · TRPC5 · TRPC6 · TRPM7 · TRPM8 · TRPV1 · TRPV2 · TRPV6 · Chemoresistance · Cancer

20.1 Calcium Signaling in Cancer

Intracellular calcium ions (Ca^{2+}), the most abundant and important second messenger, play a pivotal role in controlling cell proliferation, differentiation, migration and death [1–4]. Thus, it is essential to keep under tight control the Ca^{2+} signals in the form of oscillations, wave or spikes [2]. The disruption of normal Ca^{2+} signaling contributes to the development of the malignant phenotypes; in fact, cancer cells are able to modify the Ca^{2+} signaling network in order to increase proliferation, immortalization, angiogenesis, invasion, immune evasion, disruption of cell death pathways, ECM remodelling, epithelial-mesenchymal transition (EMT) and drug resistance [5, 6]. Several Ca^{2+} channels, transporters and Ca^{2+}-ATPases, as voltage-gated Ca^{2+} channel (VGCC), Transient Receptor Potential (TRP), Ca^{2+} release activated Ca^{2+} channel (CRAC), inositol 1,4,5-triphosphate receptor (IP3R) and mitochondrial Ca^{2+} uniporter (MCU) are altered in cancer. Moreover, their impairment has been found to be involved in the tumorigenesis [2] (Fig. 20.1). The aim of this chapter is to address the role of TRP channels in modulating sensitivity to chemotherapeutic drugs in different cancer types.

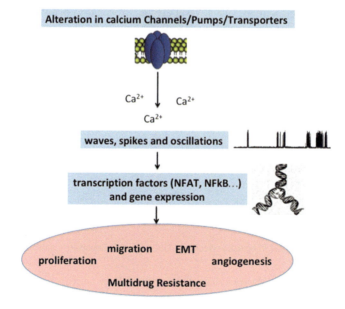

Fig. 20.1 Alterations in expression and functions of Ca^{2+} channels/pumps/transporters lead to dysregulation in calcium signaling promoting malignant phenotype and chemoresistance

20.1.1 Cation Disruption in Cancer: The Transient Receptor Potential Family

The TRP channels are non-selective cation permeable receptors localized in plasma membrane, endoplasmic reticulum, mitochondria and lysosomes [7]. They play a key role in regulating cellular Ca^{2+} concentration and membrane voltage. To date, about 30 TRPs have been identified and, on the basis of their structural homology, they are classified in: TRPC1-7, TRPV1-6, TRPM1-8, TRPP2,3,5, TRPML1-3 and TRPA1 [8]. Several findings indicate that alterations in expression and functions of TRP channels are responsible for cancer growth, metastasis and chemoresistance [9]. In particular, dysregulation of TRPC, TRPM or TRPV members has been mainly correlated with malignant growth and progression [10], so that cancer can now be considered like a "channellopathy" [11]. The central role of TRPs in cancer is to impair the Ca^{2+} homeostasis by stimulating Ca^{2+} entry or altering membrane potential. For this reason, in the recent years an increased interest in discovering agents targeting TRP channels in cancer, has been emerged and several pharmacological modulators are now used to characterize the implications of TRP channels in whole-cell membrane currents, resting membrane potential regulation or intracellular Ca^{2+} signaling [12, 13].

20.2 Drug Resistance in Cancer

Initially, cancers are susceptible to chemotherapy but over time they develop resistance by activating different strategies to limit drug efficacy eluding cell death. Thus, cancer cells become tolerant to pharmacological treatments [14]. Drug resistance can be achieved through several mechanisms involving Ca^{2+} signaling as drug inactivation, drug efflux, drug target alterations, acquisition of EMT, evasion from cell death pathways, increased DNA damage tolerance and dysregulation of critical genes (Fig. 20.2). Many chemotherapeutic agents require metabolic

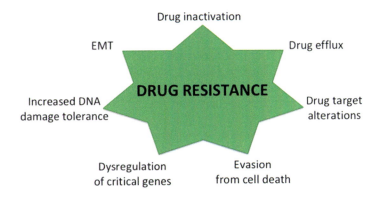

Fig. 20.2 Mechanisms promoting the acquisition of chemoresistance in cancer

conversion to be active; mutations or down-regulation of the enzymes responsible for the drug activation, as cytochrome P 450 system, glutathione-S-transferase (GST) and uridine diphospho-glucuronosyltransferase (UGT) super-families, are often present in cancer cells [15]. In addition, changes in the expression or functions of drug targets such as topoisomerase II, epidermal growth factor receptor family, Ras, Src, Raf, MEK, AKT and PTEN, lead to resistance. One of the most studied mechanisms of cancer drug resistance involves the increasing drug efflux with subsequent reduction in cellular drug concentration. ATP-binding cassette (ABC) members, multidrug resistance protein 1 (MDR1), known as P glycoprotein (P-gp), multidrug resistance-associated protein 1 (MRP1), and breast cancer resistance protein (BCRP), represent the main transporters involved in the efflux mechanism causing the non-accumulation of anti-tumoral agents in cancer cells [16].

20.2.1   TRPC5 and TRPC6 in Multidrug Resistance

TRPC5 forms homo and hetero-oligomeric complex with other TRPs and it stimulates Ca^{2+} flux in response to different stimuli as stress, growth factors, lysophospholipids, nitric oxide or thioredoxin [17]. Abnormal expression of this channel has been found to be associated with several diseases. In addition, it is well known that TRPC5 alterations, interfering with the normal Ca^{2+} homeostasis, are involved in the development of cancer progression and acquisition of chemoresistance. During therapy, the majority of cancer cell types starts to over-express P-gp, a well-known membrane efflux pump [18]. Recently [19], TRPC5 channel has been found to be overexpressed together with P-gp in adriamycin-resistant breast cancer cell line (MCF-7/ADM). As demonstrated by patch clamp, the TRPC5-dependent calcium current was higher in MCF-7/ADM cells compared to wild type indicating that the over-expressed TRPC5 is functional.

By using the TRPC5-specific blocking antibody T5E3, authors showed that the TRPC5 inhibition is associated with both marked down-regulation of P-gp expression and increase of adriamycin accumulation, demonstrating that TRPC5 is crucial for P-gp expression and chemoresistance in MCF-7/ADM cells. Moreover, the high Ca^{2+} current generated by TRPC5 activation was able to activate the nuclear factor of activated T cells cytoplasmic 3 (NFAT$_C$3) that, stimulating the transcription, promotes P-gp over-expression. In vivo studies using athymic nude mouse model of ADM-human breast tumor, showed an evident decrease of cancer growth induced by the suppression of TRPC5 channel. Thus, the TRPC5-NFATc3-P-gp signaling cascade plays an important role in promoting drug-resistance in breast cancer cells [19].

Micro RNAs (miRNAs) are single-stranded 19–25 nucleotide short RNAs that modulate gene expression at post-transcriptional stage by targeting mRNAs. It is now well accepted that they can regulate several processes closely associated with tumor progression as chemoresistance, apoptosis, cell cycle or stemness transition. In the recent years, the attention was mainly focused on miR-230a since its expression has been found to be strongly down-regulated in MCF-7/ADM cells compared to MCF-7 cells, suggesting that it is involved in the development of chemoresistance [20]. Moreover, low miR-320a expression is associated with clinical chemoresistance and poor patient outcome. As showed by Targescan and miRDB software analysis, this miRNA specifically targets TRPC5 and NFATC3 mRNAs. Therefore, its down-regulation has been found to be responsible for TRPC5 over-expression and related drug resistance in breast cancer [20].

For cancer progression, cell-cell communication in the tumor microenvironment is fundamental [21]. To this aim, cancer cells produce soluble factors and secrete membrane-encapsulated vesicles containing regulatory signals. These membrane-limited vesicles are known as Extracellular Vesicles (EVs) and they include exosomes, microvesicles and apoptotic bodies [21]. It has been recently demonstrated [22] that TRPC5, involved in growth factor-regulated local vesicular trafficking, by mediating Ca^{2+} flux, plays a role in the EVs formation and secretion in MCF-7/ADM. Since EVs membrane phospholipid bilayer is composed by the plasma membrane of the donor cells, TRPC5 channel is packaged in the developing vesicles and, in this way, transported into recipient cells where it promotes P-gp expression by increasing Ca^{2+} flux. Thus, the TRPC5-containing EVs represent a mechanism used by cancer cells to disseminate the acquisition of chemoresistance. Furthermore, immunohistochemistry analysis performed on breast cancer tissues, collected before and after the chemotherapy, showed a marked increase in the TRPC5 expression mainly in samples from not responsive patients indicating the close association between TRPC5 and chemoresistance [22]. Since TRPC5-containing EV levels correlate with acquired chemoresistance and EVs can be easily monitored in the blood of breast cancer patients [23], TRPC5-containing EVs represent a new potential diagnostic biomarker for real time measurement of chemoresistance in breast cancer.

In addition, it has been demonstrated that endothelial cells of the tumor microenvironment acquire resistance thanks to TRPC5-containing EVs released by ADM/MCF-7 [24]. As already described, the transmitted TRPC5, by activating NFATC3 in a Ca^{2+} dependent manner, stimulates the expression of P-gp.

Autophagy, an evolutionarily conserved lysosomal pathway, has been reported to show paradoxical roles in cancer: it can inhibit or promote tumorigenesis by inducing cell death or survival, respectively [25]. Since intracellular Ca^{2+} plays an important role in both basal and induced autophagy, TRP channels are now recognized as autophagy regulators [26].

Zhang and co-workers demonstrated that TRPC5 regulates the chemotherapy-induced autophagy in breast cancer cells. In fact, TRPC5, by inducing Ca^{2+} flux, initiates the autophagy via CaMKKβ/AMPKα/mTOR pathway in response to chemotherapy. Authors also showed that the TRPC5-induced autophagy functions as pro-survival mechanism promoting chemoresistance, as demonstrated by the reduction in autophagy and enhancement in ADM sensitivity in TRPC5 silenced MCF-7 cells [27].

Over-expression of TRPC5 was also found to be involved in the development of 5-Fluorouracil (5-Fu) resistance in colon rectal cancer [28]. TRPC5 is up-regulated together with the efflux pump ABC subfamily B, both at mRNA and protein levels, in resistant human HCT-8 and LoVo colon rectal cancer cells. TRPC5, by inducing intracellular Ca^{2+} flux, promotes β-catenin translocation in the nucleus, increases glycolysis and provides ATP production to avoid Ca^{2+} influx overload. Moreover it stimulates ACB and cyclin D1 expression contributing to the development of 5-Fu resistance. In fact, its suppression markedly inhibits the canonical Wnt/β-catenin signal pathway and reduces efflux pump activity reverting the chemoresistance. By contrast, the forced expression of TRPC5 results in an activated Wnt/β-catenin signal pathway and up-regulation of ABC. High expression of TRPC5 has also been found to be associated with glucose transporter 1 (GLUT1) up-regulation in colon rectal cancer cells and increased glycolysis often occurs in chemoresistance cells [29]. Taken together, these findings demonstrate in human colon rectal cancer cells an important role of TRPC5 in drug resistance via stimulating nuclear β-catenin, ABC and GLUT1 over-expression [28–30].

Cancer cells become more resistant to drugs also thanks to the EMT, a process involved in the acquisition of invasive and migratory phenotype [31]. In hepato-cellular carcinoma (HCC), it has been recently demonstrated that chemoresistance to doxorubicin occurs through up-regulation of Vimentin and down-regulation of E-cadherin and Claudin1, typical EMT markers. In fact, prolonged treatment with doxorubicin, by enhancing Ca^{2+} influx, induces EMT promoting chemoresistance. The channel involved in this process is TRPC6 that, via calcium signaling, stimulates STAT3 activation inducing the EMT [32]. The role of TRPC6 in chemoresistance was also explored in xenograft models of HCC using TRPC6-silenced and wild type Huh-7 HCC cells. Results showed that tumors, derived from the injection of TRPC6-silenced cells, grow slower than normal cells and they are more sensitive to doxorubicin [32].

20.2.2   TRPM Channels in Chemoresistance

TRPM7 is a highly Ca^{2+} and Mg^{2+} permeable member of the TRPM family activated by ATP and characterized in the C-terminal region by the presence of a kinase domain. Recent findings showed that Vacquinol-1 (Vac) promotes in glioma cells the methuosis cell death based on cell blebbing followed by rupture of the plasma membrane. This new type of cell death, caused by inefficient vacuole-lysosome fusion, is caspase-dependent and it is reverted by exogenous ATP [33]. The ATP-mediated inhibitory effect on Vac-induced cell death is due to TRPM7 activation that, by a marked Ca^{2+} influx, stimulates the phosphoinositide 3-kinase (PI3K) restoring the vacuole-lysosome fusion. Thus, the Ca^{2+} current induced by TRPM7, often found to be overexpressed in cancer cells, is responsible for the development of Vac-resistance in glioma cell lines (Fig. 20.3). It has also been demonstrated that the expression of TRPM7 is required to prevent apoptotic cell death in pancreatic adenocarcinoma [34]. The targeted silencing of TRPM7

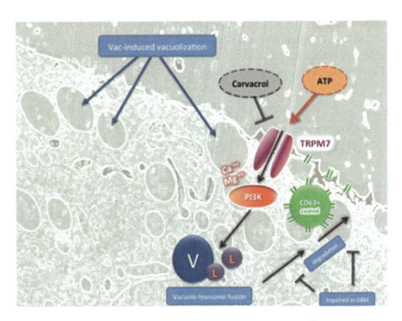

Fig. 20.3 TRPM7 activation induced by exogenous ATP stimulates Ca^{2+} influx that promotes PI3K activation increasing vesicle fusion (V) with lysosomes (L). [From reference 33]

increases the expression of senescence-associated genes inducing the replicative senescence in pancreatic cancer cells. The down-regulation of TRPM7 expression also enhances the cytotoxicity mediated by gemcitabine treatment in pancreatic cancer suggesting that its expression is strongly associated with resistance to apoptosis induction [34].

Moreover, in Lewis lung cancer cells (LLC-2), TRPM8, showing a plasma membrane and a membrane rafts localisation, is involved in the induction of proliferation, invasion and migration [35]. In addition, TRPM8, by activating Uncoupling Protein 2, contributes to the acquisition of resistance against both activated spleen CD8 T lymphocytes and doxorubicin contributing to the development of the malignant phenotype. The ability of TRPM8 to promote the acquisition of chemoresistance is also supported by data obtained in in vitro studies on prostate cancer cells [36]. By enhancing the HIF-1α protein levels, the cold-sensitive Ca^{2+} channel protein TRPM8 promotes hypoxic growth capacities and drug resistance. In particular, the TRPM8 activation induces the suppression of HIF-1α ubiquitination and enhances HIF-1 transactivation both in hypoxia- and normoxia-exposed prostate cancer cells. The potential involvement of TRPM8 channel in chemosensitivity has been also shown in osteosarcoma cells [37]. Knockdown of TRPM8 by siRNA in osteosarcoma cells leads to alterations in intracellular Ca^{2+} concentration. This Ca^{2+} imbalance induces the inhibition of several pathways as Akt-GSK-3β, ERK1/2 and FAK, promoting strong decrease in proliferation, invasion and migration.

Moreover, although TRPM8 silencing alone does not increase apoptotic cell death, it enhances the epirubicin-induced apoptosis indicating that the over expression of TRPM8 in osteosarcoma cells is associated with impaired Ca^{2+} signaling and induction of drug resistance [37].

20.2.3 Chemosensitivity and TRPV-Mediated Calcium Signaling

Calcium signaling is also required for the development of the neoplastic features in Retinoblastoma, a common intraocular pediatric cancer arising from immature cells of the retina [38]. TRPs and cannabinoid receptors are expressed in retinoblastoma cells. Their expression levels are considered useful as prognostic factors, since they correlate with tumor progression and are associated with the acquisition of etoposide-resistance. Interestingly, capsaicin, the specific agonist of TRPV1 receptor, is able to evoke Ca^{2+} influx in etoposide-resistant but not in etoposide-sensitive WERI-Rb1 retinoblastoma cells, suggesting the key role played by TRPV1-mediated calcium signaling in the acquisition of drug chemoresistance [38].

TRPV1 is also involved in the enhancement of chemosentitivity to cisplatin induced by Alpha-lipoic acid (ALA) in breast cancer cells. ALA administration, through TRPV1 activation, increases the apoptosis induced by cisplatin stimulating mitochondrial membrane depolarization, reactive oxygen species (ROS) production, lipid peroxidation, PARP1, caspase 3 and 9 expression. The ALA-dependent stimulation of TRPV1, via calcium signaling, enhances the oxidative stress making breast cancer cells more sensitive to the action of the chemotherapeutic drug [39].

Among TRPV family members, the role of TRPV2, with Ca^{2+} permeation properties, in the regulation of glioblastoma cell growth and progression, has been addressed. The aggressive behaviour of glioblastoma is mainly due to high resistance to the standard chemotherapy as Temozolomide (TMZ), Carmustine (BCNU) or Doxorubicin characterized by limited efficacy. The over-expression of TRPV2 by gene transfection in glioma cells increases the sensitivity to FAS- and BCNU-induced cytotoxicity [40, 41]. Moreover, the activation of the TRPV2 channel, induced by treatment with cannabidiol (CBD), strongly reduced the BCNU resistance in glioma cells. In fact, CBD, by generating a TRPV2-dependent Ca^{2+} influx, inhibits the Ras/Raf/MEK/ERK pathway and promotes the drug retention in glioma cells, reverting the chemoresistant phenotype and improving the apoptosis induced by TMZ, BCNU and doxorubicin [40, 41]. Mutations of the TRPV2 pore completely cancel the CBD-induced Ca^{2+} signaling demonstrating the essential role of the TRPV2 permeant cation region in chemoresistance. In addition, glioma stem-like cells represent a major problem in the treatment of glioblastoma because they maintain stem cell properties and show marked resistance to radiation and conventional drugs. Reduce their drug resistance is fundamental to increase the patient survival. At this regard, CBD, through TRPV2 activation, stimulates autophagy in glioma stem-like cells promoting cell differentiation and increasing the sensitivity to the apoptosis induced by BCNU and TMZ [42].

Furthermore, TRPV2 is expressed in multiple myeloma cells, a malignancy characterised by clonal proliferation of plasma cells and subsequent accumulation in the bone marrow [43]. Recent data demonstrated that CBD treatment induces in myeloma cells up-regulation of TRPV2 expression enhancing the sensitivity to Bortezomib, a specific proteosome inhibitor. The specific TRPV2 activation, induced by CBD, strongly reduces proliferation and improves cytotoxic effects of bortezomib by enhancing cell growth inhibition, cell cycle arrest at the G1 phase, mitochondrial and ROS-dependent necrosis mainly in TRPV2-transfected RPMI8226 and U266 multiple myeloma cells. The cell death induced by the co-administration of CBD and Bortezomib is also characterized by down-regulation of the ERK, AKT and NF-κB pathways. These findings provide a rationale for the use of TRPV2 activators (e.g., CBD) to increase the activity of proteasome inhibitors in myeloma multiple patients [43].

The TRPV6 channel, which is highly selective for Ca^{2+}, is upregulated, by a gene amplification mechanism, in breast cancer cell lines and in breast carcinoma samples compared with normal mammary gland tissue. By microarray analysis, it has been shown that the TRPV6 over-expression, associated with reduced patient overall survival, is a feature of estrogen receptor-negative breast tumors as well as HER2-positive tumors. Down-regulation of TRPV6 expression reduces the basal Ca^{2+} influx leading to decrease in cellular proliferation and DNA synthesis [44]. It has been showed, using TRPV6-transfected *Xenopus* oocytes, that tamoxifen, the most common therapy used in breast cancer treatment, inhibits the Ca^{2+} uptake regulated by this channel. In addition, tamoxifen treatment markedly reduces the expression of TRPC6 at mRNA levels in breast cancer cell lines [45]. Silencing of TRPV6 enhances the pro-apoptotic activity of tamoxifen suggesting that the increase of Ca^{2+} influx, mediated by TRPV6 over-expression in breast cancer cells, is responsible for the reduced sensitivity to tamoxifen treatment. These findings support the hypothesis that a combination therapy using tamoxifen and TRPV6 inhibitor could represent a promising strategy to improve the treatment of breast cancer [44, 46].

20.3 Conclusion

It is becoming evident that dysregulation in Ca^{2+} homeostasis plays a pivotal role in tumor progression, functioning as a driving signal in the acquisition of the aggressive phenotype. In fact, cancer cells, by changing the expression of ion channels/transporters/pumps acquire the ability to modulate Ca^{2+} intracellular concentration creating pro-survival conditions. Several evidences support the idea that Ca^{2+} signaling pathways are also involved in regulating sensitiv-

Table 20.1 Chemosensitivity and TRP roles in cancer cells

TRPs activation	Pathway mediated by Ca^{2+} dysregulation	Drug	Cancer cell line	References
TRPCS	NFATc3 activation promoting P-gp over-expression	↑Adryamycin	MCF-7	[19]
	Prosurvival autophagy via CaMKKβ/AMPKα/mTOR	↑Adryamycin	MCF-7	[27]
	ACB and cyclin D1 over-expression via β-catenin	↑5-Fluorouracil	HCT8 and LoVo	[28–30]
TRPC6	STAT3 activation inducing EMT	↑Doxorubicin	Huh-7	[32]
TRPM7	PI3K activation restoring vacuole/lysosome fusion	↑Vaquinol-1	U-87 and #12537-GB	[33]
	Prevention of non apoptotic cell death	↑Gemcitabine	BxPC3 and PANIC-1	[34]
TRPM8	Uncoupling Protein 2 activation	↑Doxorubicin	LLC-2	[35]
	Akt-GSK-3β, ERK1/2 and FAK activation	↑Epirubicin	MG-63, U2OS, SaOS2 and HOS	[37]
TRPV1	Oxidative stress	↓Cisplatin	MCF-7	[39]
TRPV2		↓Temozolomide		
	Ras/Raf/MEK/ERK activation	↓Carmustine	U-37, MZC, GSC,	[40–42]
		↓Doxorubicine		
	Inhibition of NF-kB	↓Bortezomib	RPMI3226 and U266	[43]
TRPV6	Promotes cell proliferation	↑Tamoxifen	T47D	[46]

ity to chemotherapeutic drugs. Drug resistance represents a major limitation in the application of current therapeutic regimens and several efforts are spent to overcome it. The targeting of the Ca^{2+} channels, by altering their expression and functions, has been demonstrated be effective in improving of cytotoxicity induced by the most common chemotherapeutic agents. Among Ca^{2+} channels, TRPs influence the expression and function of many drug resistance-related proteins and pathways contributing to the development of pharmacological tolerance in cancer (Table 20.1). Therefore, targeting the TRP expression and activity, can be now considered a promising and fascinating strategy to inhibit cancer growth and progression and restore/improve the sensitivity of cancer cells to chemotherapeutic drugs [10].

References

1. Giorgi C, Danese A, Missiroli S, Patergnani S, Pinton P (2018) Calcium dynamics as a machine for decoding signals. Trends Cell Biol 28(4):258–273
2. Cui C, Merritt R, Fu L, Pan Z (2017) Targeting calcium signaling in cancer therapy. Acta Pharm Sin B 7(1):3–17
3. Zhang X, Yuan D, Sun Q, Xu L, Lee E, Lewis AJ, Zuckerbraun BS, Rosengart MR (2017) Calcium/calmodulin-dependent protein kinase regulates the PINK1/Parkin and DJ-1 pathways of mitophagy during sepsis. FASEB J 31(10):4382–4395
4. Bootman MD, Chehab T, Bultynck G, Parys JB, Rietdorf K (2018) The regulation of autophagy by calcium signals: do we have a consensus? Cell Calcium 70:32–46
5. Busselberg D, Florea AM (2017) Targeting intracellular calcium signaling ($[Ca^{2+}]_i$) to overcome multiddrug resistance of cancer cells: a mini-overview. Cancers 9:48
6. Xu MM, Seas A, Kijan M, Ji KSY, Bell HN (2018) A temporal examination of calcium signaling in cancer-from tumorigenesis, to immune evasion, and metastasis. Cell Biosci 8:25
7. La Rovere RM, Roest G, Bultynck G, Parys JB (2016) Intracellular Ca^{2+} signaling and Ca^{2+} microdomains in the control of cell survival, apoptosis and autophagy. Cell Calcium 60(2):74–87
8. Nilius B, Owsianik G (2011) The transient receptor potential family of ion channels. Genome Biol 12(3):218
9. Shapovalov G, Ritaine A, Skryma R, Prevarskaya N (2016) Role of TRP ion channels in cancer and tumorigenesis. Semin Immunopathol 38(3):357–369
10. Gautier M, Dhennin-Duthille I, Ay AS, Rybarczyk P, Korichneva I, Ouadid-Ahidouch H (2014) New insights into pharmacological tools to TR(i)P cancer up. Br J Pharmacol 171:2582–2592
11. Litan A, Langhans SA (2015) Cancer as a channelopathy: ion channels and pumps in tumor development and progression. Front Cell Neurosci 9:86
12. Santoni G, Farfariello V, Amantini C (2011) TRPV channels in tumor growth and progression. Adv Exp Med Biol 704:947–967
13. Santoni G, Farfariello V (2011) TRP channels and cancer: new targets for diagnosis and chemotherapy. Endocr Metab Immune Disord Drug Targets 11:54–67
14. Housman G, Byler S, Heerboth S, Lapinska K, Longacre M, Snyder N, Sarkar S (2014) Drug resistance in cancer: an overview. Cancers (Basel) 6:1769–1792
15. Michael M, Doherty MM (2005) Tumoral drug metabolism: overview and its implications for cancer therapy. J Clin Oncol 23:205–229
16. Stavrovskaya AA (2000) Cellular mechanisms of multidrug resistance of tumor cells. Biochemistry (Mosc) 65:95–106
17. He DX, Ma X (2016) Transient receptor potential channel C5 in cancer chemoresistance. Acta Pharmacol Sin 37:19–24
18. Binkhathlan Z, Lavasanifar A (2013) P-glycoprotein inhibition as a therapeutic approach for overcoming multidrug resistance in cancer: current status and future perspectives. Curr Cancer Drug Targets 13:326–346
19. Ma X, Cai Y, He D, Zou C, Zhang P, Lo CY, Xu Z, Chan FL, Yu S, Chen Y, Zhu R, Lei J, Jin J, Yao X (2012) Transient receptor potential channel TRPC5 is essential for P-glycoprotein induction in drug-resistant cancer cells. Proc Natl Acad Sci U S A 109:16282–16287
20. He DX, Gu XT, Jiang L, Jin J, Ma X (2014) A methylation-based regulatory network for microRNA 320a in chemoresistant breast cancer. Mol Pharmacol 86:536–547
21. Xu R, Rai A, Chen M, Suwakulsiri W, Greening DW, Simpson RJ (2018) Extracellular vesicles in cancer – implications for future improvements in cancer care. Nat Rev Clin Oncol 15(10):617–638. https://doi.org/10.1038/s41571-018-0036-9
22. Ma X, Chen Z, Hua D, He D, Wang L, Zhang P, Wang J, Cai Y, Gao C, Zhang X, Zhang F, Wang T, Hong T, Jin L, Qi X, Chen S, Gu X, Yang D, Pan Q, Zhu Y, Chen Y, Chen D, Jiang L, Han X, Zhang Y, Jin J, Yao X (2014) Essential role for TrpC5-containing extracellular vesicles in breast cancer with chemotherapeutic resistance. Proc Natl Acad Sci U S A 111:6389–6394

23. Wang T, Ning K, Lu TX, Sun X, Jin L, Qi X, Jin J, Hua D (2017) Increasing circulating exosomes-carrying TRPC5 predicts chemoresistance in metastatic breast cancer patients. Cancer Sci 108:448–454

24. Dong Y, Pan Q, Jiang L, Chen Z, Zhang F, Liu Y, Xing H, Shi M, Li J, Li X, Zhu Y, Chen Y, Bruce IC, Jin J, Ma X (2014) Tumor endothelial expression of P-glycoprotein upon microvesicular transfer of TrpC5 derived from adriamycin-resistant breast cancer cells. Biochem Biophys Res Commun 446:85–90

25. Singh SS, Vats S, Chia AY, Tan TZ, Deng S, Ong MS, Arfuso F, Yap CT, Goh BC, Sethi G, Huang RY, Shen HM, Manjithaya R, Kumar AP (2018) Dual role of autophagy in hallmarks of cancer. Oncogene 37:1142–1158

26. Sukumaran P, Schaar A, Sun Y, Singh BB (2016) Functional role of TRP channels in modulating ER stress and autophagy. Cell Calcium 60:123–132

27. Zhang P, Liu X, Li H, Chen Z, Yao X, Jin J, Ma X (2017) TRPC5-induced autophagy promotes drug resistance in breast carcinoma via CaMKKβ/AMPKα/mTOR pathway. Sci Rep 7(1):3158

28. Wang T, Chen Z, Zhu Y, Pan Q, Liu Y, Qi X, Jin L, Jin J, Ma X, Hua D (2015) Inhibition of transient receptor potential channel 5 reverses 5-fluorouracil resistance in human colorectal Cancer cells. J Biol Chem 290:448–456

29. Wang T, Ning K, Lu TX, Hua D (2017) Elevated expression of TrpC5 and GLUT1 is associated with chemoresistance in colorectal cancer. Oncol Rep 37:1059–1065

30. Wang T, Ning K, Sun X, Zhang C, Jin L-f, Hua D (2018) Glycolysis is essential for chemoresistance induced by transient receptor potential channel C5 in colorectal cancer. BMC Cancer 18:207

31. Bhatia S, Monkman J, Toh AKL, Nagaraj SH, Thompson EW (2017) Targeting epithelial-mesenchymal plasticity in cancer: clinical and preclinical advances in therapy and monitoring. Biochem J 474:3269–3306

32. Wen L, Liang C, Chen E, Chen W, Liang F, Zhi X, Wei T, Xue F, Li G, Yang Q, Gong W, Feng X, Bai X, Liang T (2016) Regulation of multi-drug resistance in hepatocellular carcinoma cells is TRPC6/calcium dependent. Sci Rep 6:23269

33. Sander P, Mostafa H, Soboh A, Schneider JM, Pala A, Baron AK, Moepps B, Wirtz CR, Georgieff M, Schneider M (2017) Vacquinol-1 inducible cell death in glioblastoma multiforme is counter regulated by TRPM7 activity induced by exogenous ATP. Oncotarget 8:35124–35137

34. Yee NS, Zhou W, Lee M, Yee RK (2012) Targeted silencing of TRPM7 ion channel induces replicative senescence and produces enhanced cytotoxicity with gemcitabine in pancreatic adenocarcinoma. Cancer Lett 318:99–105

35. Du GJ, Li JH, Liu WJ, Liu YH, Zhao B, Li HR, Hou XD, Li H, Qi XX, Duan YJ (2014) The combination of TRPM8 and TRPA1 expression causes an invasive phenotype in lung cancer. Tumour Biol 35:1251–1261

36. Yu S, Xu Z, Zou C, Wu D, Wang Y, Yao X, Ng CF, Chan FL (2014) Ion channel TRPM8 promotes hypoxic growth of prostate cancer cells via an O2 -independent and RACK1-mediated mechanism of HIF-1α stabilization. J Pathol 234:514–525

37. Wang Y, Yang Z, Meng Z, Cao H, Zhu G, Liu T, Wang X (2013) Knockdown of TRPM8 suppresses cancer malignancy and enhances epirubicin-induced apoptosis in human osteosarcoma cells. Int J Biol Sci 10:90–102

38. Mergler S, Cheng Y, Skosyrski S, Garreis F, Pietrzak P, Kociok N, Dwarakanath A, Reinach PS, Kakkassery V (2012) Altered calcium regulation by thermosensitive transient receptor potential channels in etoposide-resistant WERI-Rb1 retinoblastoma cells. Exp Eye Res 94:157–173

39. Nur G, Nazıroğlu M, Deveci HA (2017) Synergic prooxidant, apoptotic and TRPV1 channel activator effects of alpha-lipoic acid and cisplatin in MCF-7 breast cancer cells. J Recept Signal Transduct Res 37:569–577

40. Nabissi M, Morelli MB, Amantini C, Farfariello V, Ricci-Vitiani L, Caprodossi S, Arcella A, Santoni M, Giangaspero F, De Maria R, Santoni G (2010) TRPV2 channel negatively controls glioma cell proliferation and resistance to Fas-induced apoptosis in ERK-dependent manner. Carcinogenesis 31(5):794–803

41. Nabissi M, Morelli MB, Santoni M, Santoni G (2013) Triggering of the TRPV2 channel by cannabidiol sensitizes glioblastoma cells to cytotoxic chemotherapeutic agents. Carcinogenesis 34:48–57

42. Nabissi M, Morelli MB, Amantini C, Liberati S, Santoni M, Ricci-Vitiani L, Pallini R, Santoni G (2015) Cannabidiol stimulates Aml-1a-dependent glial differentiation and inhibits glioma stem-like cells proliferation by inducing autophagy in a TRPV2-dependent manner. Int J Cancer 137:1855–1869

43. Morelli MB, Nabissi M, Amantini C, Farfariello V, Ricci-Vitiani L, di Martino S, Pallini R, Larocca LM, Caprodossi S, Santoni M, De Maria R, Santoni G (2012) The transient receptor potential vanilloid-2 cation channel impairs glioblastoma stem-like cell proliferation and promotes differentiation. Int J Cancer 131:E1067–E1077

44. Peters AA, Simpson PT, Bassett JJ, Lee JM, Da Silva L, Reid LE, Song S, Parat MO, Lakhani SR, Kenny PA, Roberts-Thomson SJ, Monteith GR (2012) Calcium channel TRPV6 as a potential therapeutic target in estrogen receptor-negative breast cancer. Mol Cancer Ther 11:2158–2168

45. Bolanz KA, Kovacs GG, Landowski CP, Hediger MA (2009) Tamoxifen inhibits TRPV6 activity via estrogen receptor-independent pathways in TRPV6-expressing MCF-7 breast cancer cells. Mol Cancer Res 7:2000–2010

46. Bolanz KA, Hediger MA, Landowski CP (2008) The role of TRPV6 in breast carcinogenesis. Mol Cancer Ther 7:271–279

Chapter 21
Widespread Roles of CaMK-II in Developmental Pathways

Sarah C. Rothschild and Robert M. Tombes

Abstract The multifunctional Ca^{2+}/calmodulin-dependent protein kinase type 2 (CaMK-II) was first discovered in brain tissue and shown to have a central role in long term potentiation, responding to Ca^{2+} elevations through voltage dependent channels. CaMK-II has a unique molecular mechanism that enables it to remain active in proportion to the degree (frequency and amplitude) of Ca^{2+} elevations, long after such elevations have subsided. Ca^{2+} is also a rapid activator of early development and CaMK-II is expressed and activated in early development. Using biochemical, pharmacological and genetic approaches, the functions of CaMK-II overlap remarkably well with those for Ca^{2+} elevations, post-fertilization. **Conclusion.** Activated CaMK-II plays a central role in decoding Ca^{2+} signals to activate specific events during early development; a majority of the known functions of elevated Ca^{2+} act though CaMK-II.

Keywords CaMK-II · Calcium · Calmodulin · Development · Gastrulation · Kidney · Cilia · Cardiac · Laterality · Ear · Phosphorylation

21.1 Introduction

21.1.1 Calcium Signaling During Early Development

Ca^{2+} at Fertilization Ca^{2+} elevations in living cells were first visualized in fertilized eggs from vertebrate and invertebrate species. These studies used luminescent (aequorin) or fluorescent (fura-2) indicators on species as varied as sea urchin, medaka and mice [1–4]. Multiple studies demonstrated a role for Ca^{2+} in promoting the exocytosis of cortical granules, leading to fertilization envelope elevation and the

S. C. Rothschild (✉) · R. M. Tombes
Life Sciences, Virginia Commonwealth University, Richmond, VA, USA
e-mail: chasese@vcu.edu

© Springer Nature Switzerland AG 2020
M. S. Islam (ed.), *Calcium Signaling*, Advances in Experimental Medicine and Biology 1131, https://doi.org/10.1007/978-3-030-12457-1_21

activation of early development. These studies have been summarized in excellent and comprehensive reviews [5, 6].

Ca^{2+} During Early Development Ca^{2+} signaling also influences oocyte maturation and other early developmental events [7]. The multi-functionality of Ca^{2+} in post-fertilization events may be reflected by the frequency, amplitude and location of its modulation [8] and supports its importance as a developmental controller, but also makes it challenging to identify specific roles. For instance, injection of the Ca^{2+} chelator, BAPTA, just after fertilization in mouse and zebrafish embryos, blocks all subsequent cleavages and development [9, 10], even though Ca^{2+} signals continue throughout early zebrafish development. Zebrafish Ca^{2+} elevations were observed during three stages: (1) the early rapid cleavages, (2) gastrulation and dorsal-ventral specification and (3) segmentation [10]. With enhanced spatial and temporal sensitivity, transient elevations of Ca^{2+} have been more specifically attributed to cells of the outer enveloping layer of zebrafish embryos at the blastoderm margin [11, 12], primarily on the dorsal [10] or ventral aspect [5]. In zebrafish, it was concluded [10] that Ca^{2+} signals during the discrete developmental window (6-8hpf) that corresponds to gastrulation are not necessary for specification but are important for convergent extension, through cell motility and migration [5].

Ca^{2+} at Gastrulation Coordinated inductive and morphogenetic processes generate the three germ layers and shape the embryonic body during vertebrate gastrulation. Three modes of cell migration enable these rearrangements and include (a) epiboly, (b) internalization of the presumptive mesendoderm and (c) convergent extension (CE). CE movements narrow the germ layers mediolaterally (convergence) and elongate them anteroposteriorly (extension) to define the embryonic axis. The non-canonical Wnt (ncWnt) pathway has been identified as an evolutionarily conserved signaling pathway that regulates CE cell movements during vertebrate gastrulation [13]. During development, suppression of the ncWnts, Wnt5 [14–17] and Wnt11 [18–21], leads to a shortened anterior-posterior body axis, wider dorsal structures and defects in segmentation, which are stereotypical of CE defects [22, 23]. Wnt5 and Wnt11 have been identified as essential modulators of these cellular movements and are known to cause intracellular Ca^{2+} release [24–26], whereas mutations in Wnt5 and Wnt11, reduce Ca^{2+} levels [26]. Wnt5a is known to induce prolonged Ca^{2+} elevations when injected into zebrafish embryos [24]. The zebrafish *wnt5* mutant is known as *pipetail* (*ppt*). Ppt mutant $(-/-)$ embryos exhibit altered Ca^{2+} modulation, widened somites and in some cases, split axes, indicative of defects in convergent extension [27]. Wnt11 appears responsible for the Ca^{2+} elevation that leads to directed rapid cell migration [28], whereas Wnt5/ppt acts later to influence cell intercalation. Wnt11 appears to be more important for the morphological changes associated with CE and not specification [16]. This is consistent with a cytosolic, not a nuclear role for Wnt11 in CE. CE movements in mouse embryos are also dependent on Wnt5 and Wnt11 [23].

Ca^{2+} in LR Asymmetry Laterality disorders are characterized by the misplacement of one or more organs across the left-right (LR) axis and occur as often as once in every 6000 newborn humans [29]. The positioning of internal organs in diverse vertebrate organisms is initiated by signals originating from a transient posterior structure, known as the mouse embryonic node or zebrafish Kupffer's vesicle (KV) [30]. The KV is a fluid-filled organ that forms at the posterior end of the notochord at the early somite stages of teleosts [31]. The KV, like the mouse ventral node, is lined by epithelial cells that contain motile cilia whose resultant fluid flow is necessary to establish left-right asymmetry [32, 33].

Fluid flow is believed to yield asymmetric Ca^{2+} elevations through the TRP channel, PKD2 (polycystin-2) [34, 35] and may be the earliest asymmetric event [36] responsible for developing normal laterality. Ca^{2+} elevations on the left side of the embryonic node have been detected in mice [37, 38], chick [39, 40] and zebrafish [41, 42]. Disruption of Ca^{2+} signaling causes randomization of heart and visceral organs [41].

In addition to PKD2, Ca^{2+} release has also been linked to ryanodine receptors [40, 42] and inositol phosphate dependent signals [41]. Gap junctions may enable Ca^{2+} to spread through target cells on the left side of the embryonic node [43, 44] and the H^+K^+ATPase may help maintain the driving force for Ca^{2+} elevations [40]. PKD2 targeted to endomembranes in KV cells may be more important than plasma membrane PKD2 for left-right asymmetry in zebrafish [45]. The importance of PKD2 is further supported by observations that *pkd2* morphants and mutants randomize organ placement and mis-express *spaw* (Southpaw) [34, 35]. PKD2-deficient mouse embryos also lack the normal Ca^{2+} elevation on the left side of the ventral node, fail to express *nodal* and have randomized organs [37, 46].

Ca^{2+} in Somitogenesis Zebrafish somitogenesis is accompanied by transient elevations in Ca^{2+} at the posterior end of forming somites [10, 47]. Ca^{2+} signals have also been observed in developing *Xenopus* myotome [48]. Pharmacological studies support a role for GPCR coupled PL-C activation leading to the Ca^{2+} release necessary for the formation of somites and notochord [26]. L690330, an inhibitor of Inositol monophosphatase, causes widening of somites [26] which is also observed in Wnt5 and Wnt 11 morphants [21]. These effects on somitogenesis could be due to effects on convergence and extension, as described above.

Ca^{2+} in Heart Development Ca^{2+} signals have also been implicated in the morphogenic process by which the heart tube transforms into a chambered heart. In zebrafish and mouse embryos, Ca^{2+} channel blockers, Ca^{2+} chelators, disruption of either the Na^+/Ca^{2+} exchanger, NCX, or SERCA2, the sarcoplasmic reticulum (SR) Ca^{2+} ATPase, all interfere with cardiac looping [10, 49, 50].

Ca^{2+} in Kidney Development Kidney tubules form in *Xenopus* animal caps treated with activin and retinoic acid and this is accompanied by an elevation in Ca^{2+} [51]. Activin/retinoic acid-treated animal caps do not form tubules when treated with the Ca^{2+} chelator BAPTA; the Ca^{2+} ionophore A23187 and ammonium chloride, two agents that elevate Ca^{2+} in activin-treated caps, can substitute for retinoic acid in stimulating the appearance of kidney tubules. Ca^{2+} released from IP_3 receptors during a specific time is important for the cell movements necessary to properly position zebrafish pronephric tissue [52].

Ca²⁺ in Ear Development Embryonic tissues that require motile cilia for development [32] are also sites of Ca^{2+} signaling. In addition to the KV and the kidney, the embryonic ear is also a member of this tissue category. The zebrafish inner ear contains both motile and immotile cilia; the immotile cilia, also known as kinocilia, emanate from "hair cells" and are hypothesized to contain an otolith precursor-binding factor [53]. Beating cilia appear transiently, adjacent to the kinocilia, in order to create a steady fluid flow that ensures the uniform formation of otoliths [54]. Ca^{2+} channels, such as TRP family members, have been implicated as the sensory transduction channel in hair cells [55–58]. However, well before the ear becomes a sensory organ, free Ca^{2+} is elevated in otic placode cells [10], suggesting its role in the differentiation of the ear.

21.1.2   CaMK-II Activation During Early Development

CaMK-II as a Ca²⁺ Sensor The "multifunctional" Ca^{2+}/calmodulin-dependent protein kinase, type 2 (CaMK-II) is often linked to central nervous system function, where it functions in long term potentiation (LTP) and comprises as much as 1% of the total protein in the hippocampus [59]. However, CaMK-II is evolutionarily conserved, has been found during development and in every adult mammalian tissue [60] and is expressed across all metazoan species [61]. CaMK-II is activated by Ca^{2+}-calmodulin that forms upon intracellular release of Ca^{2+} to phosphorylate protein substrates involved in functions that include transcription, secretion, cytoskeletal re-organization and ion channel regulation [59].

CaMK-II has a unique ability to oligomerize and autophosphorylate (at T^{287}) upon Ca^{2+}/CaM stimulation, distinguishing it from other CaM-dependent kinases and leading to its depiction as a "memory molecule" [62]. T^{287} autophosphorylation converts the enzyme into an "autonomous" or Ca^{2+}-independent active state where it "remembers" its activation by Ca^{2+}. Activated (P-T²⁸⁷) CaMK-II can be localized in fixed tissue as described [63] and its level has been shown to be proportional to autonomous CaMK-II enzymatic activity, including in zebrafish embryonic extracts [64]. This method can be used to provide a snapshot of Ca^{2+} signaling in zebrafish embryos as early as the 10 somite stage [64] and in locations that include the olfactory placode, apical pronephric duct and cloaca, and at the base of inner ear hair cell kinocilia [65]. These locations represent a subset of the locations where total CaMK-II is expressed such as the embryonic forebrain, olfactory placode, spinal cord, somites, ear and pronephric kidney, which are consistent with the locations of CaMK-II mRNA expression [66]. By 60hpf, CaMK-II becomes activated and located (P-T²⁸⁷) in retina, fins, anterior pituitary, neuromasts of the lateral line, with continued activation in the ear, somites, and kidney [65]. There is significant potential for using this method to define subcellular locations of Ca^{2+} signaling that act via CaMK-II in a wide variety of tissues, species and even disease states.

Two other potentially relevant and regulatory CaMK-II phosphorylation sites exist at T^{253} and T^{307}. Both of these sites are conserved across all metazoans [61]. T^{307} is an autophosphorylation site [67] that fine-tunes synaptic signaling, while T^{253} phosphorylation has been linked to G2-M progression [68]. The prevalence, overlap with P-T^{287} location and the function of these phosphorylation sites during embryogenesis has not yet been evaluated.

CaMK-II activity can be reversibly inhibited with the CaM-binding antagonist, KN-93, in mammalian cells [69, 70] and in zebrafish embryos [71]. Dominant negative CaMK-II constructs have also been used to interfere with the normal activation of CaMK-II [65].

Importance of CaMK-II at Fertilization and Resumption of Meiosis CaMK-II does not play a central role in cortical granule exocytosis or the activation of protein synthesis at fertilization, but may participate in the block to polyspermy [72]. However, in species, like humans and mice, where fertilization causes the resumption of meiosis II, CaMK-II has been strongly implicated in the release from metaphase II arrest. Such a role for CaMK-II was first proposed over 20 years ago using *Xenopus* oocyte extracts [73] and was supported by mathematical models [74]. Empirical support was also obtained in intact mouse oocytes that were pre-treated with KN-93 and then activated in vitro by ethanol [75], a treatment known to release Ca^{2+} from internal stores. Knockout or knockdown of γ CaMK-II [76, 77], but not other CaMK-IIs [78–81], interferes with mouse meiotic resumption at egg activation. Subsequent dissection of this pathway revealed a collaboration with a polo-like kinase [82, 83] in releasing spindles from metaphase II arrest. In embryos where fertilization occurs after the completion of meiosis at the pronuclear stage (sea urchins), a role for CaMK-II at fertilization has not been reported. This further supports the concept that CaMK-II is involved in meiotic resumption, not the specific Ca^{2+}-dependent activation events that accompany fertilization.

Role of CaMK-II in Cell Migration During Early Development CaMK-II is detected early in development at the time [66, 84] when important Ca^{2+} transients occur [5, 12, 85]. CaMK-II is activated by ncWnt family members [25] and can rescue mutant phenotypes of certain non-canonical Wnts (ncWnts), whose roles are to promote morphogenic cell movements [86]. CaMK-II is known to mediate cell migration [87–91] by enabling focal adhesion turnover. A reported role for CaMK-II on the ventral side of the embryo [25] is consistent with the ventral expression of zebrafish CaMK-IIs during early development [66, 92].

In zebrafish, CaMK-II is encoded by seven genes that give rise to at least two dozen splice variants in early embryos [66, 92]. CaMK-II morphants exhibit a similar phenotype to morphants and mutants of the non-canonical Wnts, Wnt11 and Wnt5. In fact, at first glance, the *camk2b1* morphant has an undulated notochord [92], exactly like the Wnt5/*ppt* morphant/mutant [21], while the *camk2g1* morphant [65] has segmentation defects similar to the Wnt11/*slb* mutant [21]. The expression patterns of *camk2b1* and *camk2g1* are consistent with tissues that exhibit CE cell movements. Previous studies have shown that *camk2b1* and *camk2g1* are expressed during gastrulation in both the mesendoderm and neuroectoderm layers where CE cell movements occur [66]. The non-variable regions encoded by *camk2b1* and *camk2g1* are 92% identical at the amino acid level, with their splice variants exhibiting different exon utilization and thus potential subcellular targeting [66]. Of the two dozen splice variants expressed from zebrafish CaMK-II genes during early development [65, 66, 92], all four that are expressed from *camk2g1* encode putative cytosolic targeting domains while two of the three that are expressed from *camk2b1* encode nuclear targeting domains. It is known that CaMK-IIs freely hetero-oligomerize [93] to form their stereotypical dodecameric structures yielding even further potential variations. Consequently, it is possible that *camk2b1* or *camk2g1* could hetero-oligomerize and target the entire complex to a location that might not be predicted based on the variants encoded by each gene.

During gastrulation, it is likely that CaMK-IIs encoded by *camk2b1* and *camk2g1* are activated by ncWnt-mediated Ca^{2+} elevations to directly enable cell migration. Oscillations of Ca^{2+} are known to occur during gastrulation [7, 94], which would activate CaMK-II to enable cell migration [91]. Interestingly, the migration of cells in culture and in embryos is compromised when CaMK-II is either hyperactivated, inhibited or eliminated [91]. A model in which ncWnts cause Ca^{2+} oscillations to transiently activate CaMK-II and enable focal adhesion turnover and thus cell migration is also consistent with findings that were described as apparently contradictory in which both Wnt5/Wnt11 gain of function and loss of function mutants inhibit convergent extension [95]. No specific focal adhesion protein has yet been identified as the target of CaMK-II phosphorylation that would enable focal adhesion turnover.

Role of CaMK-II in Cell Proliferation During Early Development CaMK-II has previously been strongly implicated in the early cell cycles of frog, sea urchin and mouse cell divisions [84, 96–100] and in cell cycle progression through cyclin D1 and cyclin dependent kinase inhibitors like p27[kip1] in cells in culture [69, 70, 84, 101]. In early sea urchin embryos, CaMK-II activity cycles with the cell cycle, like CDK1, and may actually interact with CDK1 [84]. An additional role for P-T[253] CaMK-II has also been implicated at the G2-M boundary [68].

Importance of CaMK-II in Cardiac Development CaMK-II has been extensively implicated in cardiovascular disease [102–104] and has been proposed as a therapeutic target to minimize remodeling, arrhythmias and hypertrophy [105–107]. Substrates or binding partners of CaMK-II known to influence cardiac function include phospholamban [108], L-type Ca^{2+} channels [109–111], ryanodine receptors [112] and histone deacetylases [113, 114]. Even in embryos, KN-93 reversibly slows heart rates, supporting a role for CaMK-II phosphorylation in excitation-contraction coupling [115].

During morphogenesis, the ncWnts, Wnt11 and Wnt11-R, have been implicated in cardiac specification and/or morphogenesis [116–118]. In addition, the T-box protein, Tbx5, promotes CaMK-II expression (*camk2b2*) during zebrafish cardiac development [92]. CaMK-II binding proteins or substrates, which may influence cardiac morphogenesis through cell polarity, cell migration or cell cycle control, have been identified and include Tiam1 and Flightless-I [91, 119, 120].

Importance of CaMK-II in Somitogenesis Somite broadening, notochord thickening and axis duplication were observed in Wnt5 and Wnt11 mutant embryos [21, 27] and in embryos expressing dominant negative CaMK-II [25]. While total CaMK-II is found throughout somites, activated CaMK-II is found in sarcomeres and somite boundaries [65].

Role of CaMK-II in Laterality and Kidney Development CaMK-II is a known target of polycystin2 (PKD2)-dependent Ca^{2+} signals necessary for left-right patterning but also kidney development [64, 65]. PKD2, also known as TRPP2, is a member of the TRP family of Ca^{2+} conducting channels and is mutated in patients with autosomal dominant polycystic kidney disease (ADPKD) [121]. Suppression of zebrafish PKD2 or CaMK-II causes pronephric kidney cysts and the loss of normal organ asymmetry [64, 65, 122].

A potential target of CaMK-II in this role is the histone deacetylase, HDAC4 [115]. In muscle tissue, CaMK-II directly phosphorylates HDAC4, retaining it in the cytosol to upregulate MEF2C target genes [114, 123, 124]. HDAC4 is the only class II HDAC that bears a specific CaMK-II docking site [125]. However, HDAC4 and HDAC5 can form hetero-dimers, rendering HDAC5 responsive to CaMK-II and enabling export and retention of HDAC5 in the cytosol [123].

HDAC4 and HDAC5 are known to influence MEF2C-dependent gene transcription [114, 123, 126]. MEF2C is a Ca^{2+}-dependent mediator of differentiation and development that is known to couple signaling to transcription [127]. When HDACs are exported from the nucleus, p300/CBP can then transcriptionally activate MEF2C-dependent genes through histone acetylation [127]. MEF2C target genes are not just involved in myogenesis and include *MTSS1 (MIM)*, which is necessary for ciliogenesis and actin cytoskeletal organization [128, 129]. MEF2C, and MIM are also important signaling molecules in kidney disease; loss of MEF2C or MIM leads to polycystic kidneys [130]. In addition, HDAC5 suppression or treatment with the pan-specific HDAC inhibitor, TSA, partially reverses cystogenesis in PKD2 mutants [130].

It stands to reason that CaMK-II provides a previously unknown linkage between Ca^{2+} signals and HDAC family members in ciliated embryonic tissues such as the kidney. CaMK-II may serve to refine the action of HDACs in cells where Ca^{2+} is elevated. Activated CaMK-II sequesters HDAC4 in the cytosol, presumably enabling transcription of target genes that are necessary for kidney morphogenesis and cilia stability.

Role for CaMK-II in Ear Development While the location of activated CaMK-II (P-T^{287} CaMK-II) provided insight into many of the potential roles of CaMK-II during early development, the intensity and location of P-T^{287} CaMK-II in the inner ear [65] suggested that CaMK-II could be essential for translating external stimuli into an intracellular response.

The organization of sensory epithelial cells must occur at the appropriate time and place during development to ensure normal otolith biomineralization. Alterations in this process lead to malformed otoliths and therefore impaired hearing. Signaling molecules necessary for inner ear sensory epithelial cell patterning include Delta-Notch family members, as demonstrated by the *mindbomb* (*mib*) mutant, which leads to supernumerary hair cells [131]. The *mib* locus encodes an E3 ubiquitin ligase, which ubiquitylates and then internalizes Delta ligands [132, 133]. CaMK-II (*camk2g1*) is responsible for the patterning of inner ear sensory cells through the Delta-Notch pathway. Like *mib* mutants, suppression of *camk2g1* also leads to supernumerary hair cells, which may contribute directly or indirectly to ectopic or malformed otoliths.

The enrichment of CaMK-II in the hair cells, but not in the surrounding support cells is consistent with CaMK-II influencing the Delta ligand and not the Notch receptor. Upon Delta binding to Notch, the Delta-Notch extracellular domain undergoes transendocytosis causing the Notch receptor to be proteolytically cleaved and the Notch intracellular domain (NICD) to enter the nucleus and activate gene expression [134]. Delta ligands are ubiquitylated, internalized, and degraded, but in the *mib* mutant, DeltaD is not endocytosed and degraded, causing an upregulation of DeltaD mRNA expression and increased DeltaD protein localization to the membrane with retention in endocytic vesicles [133, 135]. Suppression of *camk2g1* also causes an increase of DeltaD in particles that appeared to represent intracellular vesicles, but do not accumulate at the cell surface. These results suggest that the DeltaD protein is being synthesized in *camk2g1* morphants, but is not being transported to the membrane, therefore accumulating in secretory vesicles in the cytosol.

CaMK-II is known to phosphorylate proteins important in trafficking, docking and fusion of secretory vesicles. Substrates include synapsin I [136], synaptotagmin [137] and synaptobrevin, a vesicle associated membrane protein [138]. Although these proteins are essential in the secretion of neurotransmitters, they also function in non-neuronal tissues [139]. In the zebrafish KV, CaMK-II may be necessary for the secretion of Southpaw (Spaw) to the left lateral plate mesoderm, enabling the expression of left sided genes and therefore left-right organ asymmetry [64]. Likewise, a role for CaMK-II in promoting recycling and trafficking of the Delta ligand would explain its role in enabling Delta signaling to Notch expressing cells. In the absence or reduction of Delta ligands at the plasma membrane, expression of key genes would be inhibited, causing alterations in sensory epithelial cell patterning. Delta-Notch signaling has been linked to the differentiation of ciliated embryonic cells in other organisms as well [140]. While CaMK-II has previously been shown to activate Notch signaling in Notch expressing cells [141, 142], this role for CaMK-II in Delta ligand processing is distinct and acts through an undetermined protein target.

Protein Targets of CaMK-II During Development For a multifunctional protein kinase like CaMK-II, it stands to reason that there are many different potential substrates of phosphorylation during early development. However, histone deacetylase,

HDAC4, is a known substrate of CaMK-II and may be responsible for many of these developmental processes through epigenetics [115]. Nonetheless, there are other potential substrates that have also been identified. For Wnt signaling, Flightless-I may be a protein that is not itself phosphorylated, but binds to autophosphorylated CaMK-II and thus impacts downstream pathways, such as cell migration and gene expression [120]. While phospholamban has not been explicitly evaluated in the embryonic heart, its role as a target of cyclical phosphorylation by CaMK-II during excitation-contraction coupling [108] makes it an ideal candidate to explain the effect of reversible CaMK-II inhibition on heart rates. As described above, there are multiple CaMK-II substrates, including synapsin, that could mediate the effect of CaMK-II on secretion and thereby explain both its role in Delta-Notch signaling in the ear and in Southpaw secretion in the embryonic node. A characterization of these substrates in early development would be justified.

Other Protein Targets of Calcium During Development Among alternative Ca^{2+} targets including PK-C, calcineurin and CaM kinases, only CaMK-II has the molecular capability of decoding Ca^{2+} of different amplitude, duration and frequency [74, 143, 144]. In addition, it is represented by a family of genes whose alternatively spliced products are differentially targeted to membranes and subcellular compartments [61]. While PKC-delta has been implicated in convergent extension in Xenopus [145] and PK-C lambda in cell polarity [146], neither PK-C or calcineurin have been implicated as targets of the asymmetric Ca^{2+} signal [5].

21.2 Conclusion

Ca^{2+} signals have long been known to play an important role during early development. CaMK-II is a multifunctional, ubiquitously expressed protein kinase that built its reputation as regulating long-term potentiation (LTP) in the central nervous system. The features of CaMK-II that make it structurally and enzymatically attractive in the CNS are also advantageous for frequency decoding and sustained activation in a variety of cellular and tissue settings during early development.

So far, evidence is strong for a role for *camk2b1* in gastrulation and the nervous system [66], while *camk2g1* is pleiotropic and has been linked to functions found primarily in ciliated cells, such as the development and function of the ear, kidney and KV, all of which rely on cilia [64, 65, 71]. Interestingly, investigators have been stymied in their quest to prepare a knockout *camk2g1* CRISPR mutant to this clearly important early developmental gene due to a poorly understood, but gene-specific repair mechanism, which appears to act on a limited number of critical genes, including *camk2g1* [147]. While this suggests an important role for *camk2g1* in early zebrafish development, it has made it difficult to prepare *camk2g1* mutants. Activated CaMK-II plays a central role in decoding Ca^{2+} signals to activate specific events during early development. A majority of the known Ca^{2+} elevations act though CaMK-II and not other known Ca^{2+} targets (Table 21.1). Not

Table 21.1 Roles for calcium and calcium targets during development

Event	Calcium	PK-C	Calcineurin	CaMK-II
Meiotic resumption	Essential	NO	NO	YES via PLK
Gastrulation	Elevates ventrally	YES, PK-C delta	YES	Wnt/Ca pathway
Somitogenesis	Preceding somites	NO	NO	YES, compression
Heart Development	Elevates at looping	YES, "Heart and Soul" zebrafish mutant	NO	YES, by Tbx5
Kidney Development	Ventrally high	NO	NO	YES, with PKD2
LR Asymmetry	Transient on left side	NO	NO	YES, on left
Ears	TRP channel dependent	NO	NO	YES, Delta-Notch

only do Ca^{2+} signals and CaMK-II functionally overlap during meiotic resumption, gastrulation, somitogenesis, heart morphogenesis and function, kidney development and laterality and ear development, but the localization of Ca^{2+} signals, whether they be transient or sustained coincides with the location and sometimes intensity of activated CaMK-II. Linkages of CaMK-II with upstream (Tbx5, PKD2) and downstream (HDAC4) partners that are associated with human disease further supports these relationships.

References

1. Steinhardt R, Zucker R, Schatten G (1977) Intracellular calcium release at fertilization in the sea urchin egg. Dev Biol 58:185–196
2. Gilkey JC, Jaffe LF, Ridgway EB, Reynolds GT (1978) A free calcium wave traverses the activating egg of the medaka, Oryzias Latipes. J Cell Biol 76:448–466
3. Ridgway EB, Gilkey JC, Jaffe LF (1977) Free calcium increases explosively in activating medaka eggs. Proc Natl Acad Sci U S A 74:623–627
4. Cuthbertson KSR, Whittingham DG, Cobbold PH (1981) Free Ca^{2+} increases in exponential phases during mouse oocyte activation. Nature 294:754–757
5. Whitaker M (2006) Calcium at fertilization and in early development. Physiol Rev 86(1): 25–88
6. Stricker SA (1999) Comparative biology of calcium signaling during fertilization and egg activation in animals. Dev Biol 211(2):157–176
7. Webb SE, Miller AL (2003) Calcium signalling during embryonic development. Nat Rev Mol Cell Biol 4(7):539–551
8. Berridge MJ, Lipp P, Bootman MD (2000) The versatility and universality of calcium signalling. Nat Rev Mol Cell Biol 1(1):11–21
9. Tombes RM, Simerly C, Borisy GG, Schatten G (1992) Meiosis, egg activation and nuclear envelope breakdown are differentially reliant on Ca^{2+}, whereas germinal vesicle breakdown is Ca^{2+}-independent in the mouse oocyte. J Cell Biol 117:799–812
10. Creton R, Speksnijder JE, Jaffe LF (1998) Patterns of free calcium in zebrafish embryos. J Cell Sci 111(Pt 12):1613–1622
11. Reinhard E, Yokoe H, Niebling KR, Allbritton NL, Kuhn MA, Meyer T (1995) Localized calcium signals in early zebrafish development. Dev Biol 170(1):50–61

12. Gilland E, Miller AL, Karplus E, Baker R, Webb SE (1999) Imaging of multicellular large-scale rhythmic calcium waves during zebrafish gastrulation. Proc Natl Acad Sci U S A 96(1):157–161

13. Kühl M, Sheldahl LC, Park M, Miller JR, Moon RT (2000) The Wnt/Ca^{2+} pathway: a new vertebrate Wnt signaling pathway takes shape. Trends Genetics 16(7):279–283

14. Lin S, Baye LM, Westfall TA, Slusarski DC (2010) Wnt5b-Ryk pathway provides directional signals to regulate gastrulation movement. J Cell Biol 190(2):263–278

15. Zhu S, Liu L, Korzh V, Gong Z, Low BC (2006) RhoA acts downstream of Wnt5 and Wnt11 to regulate convergence and extension movements by involving effectors Rho kinase and Diaphanous: use of zebrafish as an in vivo model for GTPase signaling. Cell Signal 18(3):359–372

16. Kilian B, Mansukoski H, Barbosa FC, Ulrich F, Tada M, Heisenberg CP (2003) The role of Ppt/Wnt5 in regulating cell shape and movement during zebrafish gastrulation. Mech Dev 120(4):467–476

17. Moon RT, Campbell RM, Christian JL, McGrew LL, Shih J, Fraser S (1993) Xwnt-5A: a maternal Wnt that affects morphogenetic movements after overexpression in embryos of Xenopus laevis. Development 119(1):97–111

18. Matsui T, Raya A, Kawakami Y, Callol-Massot C, Capdevila J, Rodriguez-Esteban C et al (2005) Noncanonical Wnt signaling regulates midline convergence of organ primordia during zebrafish development. Genes Dev 19(1):164–175

19. Marlow F, Topczewski J, Sepich D, Solnica-Krezel L (2002) Zebrafish Rho kinase 2 acts downstream of Wnt11 to mediate cell polarity and effective convergence and extension movements. Curr Biol 12(11):876–884

20. Heisenberg CP, Tada M, Rauch GJ, Saude L, Concha ML, Geisler R et al (2000) Silberblick/Wnt11 mediates convergent extension movements during zebrafish gastrulation. Nature 405(6782):76–81

21. Lele Z, Bakkers J, Hammerschmidt M (2001) Morpholino phenocopies of the swirl, snailhouse, somitabun, minifin, silberblick, and pipetail mutations. Genesis 30(3):190–194

22. Wallingford JB, Fraser SE, Harland RM (2002) Convergent extension: the molecular control of polarized cell movement during embryonic development. Dev Cell 2(6):695–706

23. Andre P, Song H, Kim W, Kispert A, Yang Y (2015) Wnt5a and Wnt11 regulate mammalian anterior-posterior axis elongation. Development 142(8):1516–1527

24. Slusarski DC, Corces VG, Moon RT (1997) Interaction of Wnt and a Frizzled homologue triggers G-protein-linked phosphatidylinositol signalling. Nature 390(6658):410–413

25. Kühl M, Sheldahl L, Malbon CC, Moon RT (2000) Ca^{2+}/calmodulin-dependent protein kinase II is stimulated by Wnt and Frizzled homologs and promotes ventral Cell fates in xenopus. J Biol Chem 275:12701–12711

26. Westfall TA, Hjertos B, Slusarski DC (2003) Requirement for intracellular calcium modulation in zebrafish dorsal-ventral patterning. Dev Biol 259(2):380–391

27. Westfall TA, Brimeyer R, Twedt J, Gladon J, Olberding A, Furutani-Seiki M et al (2003) Wnt-5/pipetail functions in vertebrate axis formation as a negative regulator of Wnt/beta-catenin activity. J Cell Biol 162(5):889–898

28. Ulrich F, Concha ML, Heid PJ, Voss E, Witzel S, Roehl H et al (2003) Slb/Wnt11 controls hypoblast cell migration and morphogenesis at the onset of zebrafish gastrulation. Development 130(22):5375–5384

29. Peeters H, Devriendt K (2006) Human laterality disorders. Eur J Med Genet 49(5):349–362

30. Hirokawa N, Tanaka Y, Okada Y, Takeda S (2006) Nodal flow and the generation of left-right asymmetry. Cell 125(1):33–45

31. Essner JJ, Amack JD, Nyholm MK, Harris EB, Yost HJ (2005) Kupffer's vesicle is a ciliated organ of asymmetry in the zebrafish embryo that initiates left-right development of the brain, heart and gut. Development 132(6):1247–1260

32. Kramer-Zucker AG, Olale F, Haycraft CJ, Yoder BK, Schier AF, Drummond IA (2005) Cilia-driven fluid flow in the zebrafish pronephros, brain and Kupffer's vesicle is required for normal organogenesis. Development 132(8):1907–1921

33. Lee JD, Anderson KV (2008) Morphogenesis of the node and notochord: the cellular basis for the establishment and maintenance of left-right asymmetry in the mouse. Dev Dyn 237(12):3464–3476
34. Bisgrove BW, Snarr BS, Emrazian A, Yost HJ (2005) Polaris and Polycystin-2 in dorsal forerunner cells and Kupffer's vesicle are required for specification of the zebrafish left-right axis. Dev Biol 287(2):274–288
35. Schottenfeld J, Sullivan-Brown J, Burdine RD (2007) Zebrafish curly up encodes a Pkd2 ortholog that restricts left-side-specific expression of southpaw. Development 134(8): 1605–1615
36. Tabin CJ, Vogan KJ (2003) A two-cilia model for vertebrate left-right axis specification. Genes Dev 17(1):1–6
37. McGrath J, Somlo S, Makova S, Tian X, Brueckner M (2003) Two populations of node monocilia initiate left-right asymmetry in the mouse. Cell 114(1):61–73
38. Tanaka Y, Okada Y, Hirokawa N (2005) FGF-induced vesicular release of Sonic hedgehog and retinoic acid in leftward nodal flow is critical for left-right determination. Nature 435(7039):172–177
39. Raya A, Kawakami Y, Rodriguez-Esteban C, Ibanes M, Rasskin-Gutman D, Rodriguez-Leon J et al (2004) Notch activity acts as a sensor for extracellular calcium during vertebrate left-right determination. Nature 427(6970):121–128
40. Garic-Stankovic A, Hernandez M, Flentke GR, Zile MH, Smith SM (2008) A ryanodine receptor-dependent Ca(i)(2+) asymmetry at Hensen's node mediates avian lateral identity. Development 135(19):3271–3280
41. Sarmah B, Latimer AJ, Appel B, Wente SR (2005) Inositol polyphosphates regulate zebrafish left-right asymmetry. Dev Cell 9(1):133–145
42. Jurynec MJ, Xia R, Mackrill JJ, Gunther D, Crawford T, Flanigan KM et al (2008) Selenoprotein N is required for ryanodine receptor calcium release channel activity in human and zebrafish muscle. Proc Natl Acad Sci U S A 105(34):12485–12490
43. Levin M, Mercola M (1999) Gap junction-mediated transfer of left-right patterning signals in the early chick blastoderm is upstream of Shh asymmetry in the node. Development 126(21):4703–4714
44. Hatler JM, Essner JJ, Johnson RG (2009) A gap junction connexin is required in the vertebrate left-right organizer. Dev Biol 336(2):183–191
45. Fu X, Wang Y, Schetle N, Gao H, Putz M, von Gersdorff G et al (2008) The subcellular localization of TRPP2 modulates its function. J Am Soc Nephrol 19(7):1342–1351
46. Pennekamp P, Karcher C, Fischer A, Schweickert A, Skryabin B, Horst J et al (2002) The ion channel polycystin-2 is required for left-right axis determination in mice. Curr Biol 12(11):938–943
47. Webb SE, Miller AL (2000) Calcium signalling during zebrafish embryonic development. BioEssays 22(2):113–123
48. Ferrari MB, Spitzer NC (1999) Calcium signaling in the developing Xenopus myotome. Dev Biol 213(2):269–282
49. Porter GA Jr, Makuck RF, Rivkees SA (2003) Intracellular calcium plays an essential role in cardiac development. Dev Dyn 227(2):280–290
50. Ebert AM, Hume GL, Warren KS, Cook NP, Burns CG, Mohideen MA et al (2005) Calcium extrusion is critical for cardiac morphogenesis and rhythm in embryonic zebrafish hearts. Proc Natl Acad Sci U S A 102(49):17705–17710
51. Leclerc C, Webb SE, Miller AL, Moreau M (2008) An increase in intracellular Ca^{2+} is involved in pronephric tubule differentiation in the amphibian Xenopus laevis. Dev Biol 321(2):357–367
52. Lam PY, Webb SE, Leclerc C, Moreau M, Miller AL (2009) Inhibition of stored Ca^{2+} release disrupts convergence-related cell movements in the lateral intermediate mesoderm resulting in abnormal positioning and morphology of the pronephric anlagen in intact zebrafish embryos. Develop Growth Differ 51(4):429–442

53. Stooke-Vaughan GA, Huang P, Hammond KL, Schier AF, Whitfield TT (2012) The role of hair cells, cilia and ciliary motility in otolith formation in the zebrafish otic vesicle. Development 139(10):1777–1787

54. Colantonio JR, Vermot J, Wu D, Langenbacher AD, Fraser S, Chen JN et al (2009) The dynein regulatory complex is required for ciliary motility and otolith biogenesis in the inner ear. Nature 457(7226):205–209

55. Amato V, Vina E, Calavia MG, Guerrera MC, Laura R, Navarro M et al (2012) TRPV4 in the sensory organs of adult zebrafish. Microsc Res Tech 75(1):89–96

56. Corey DP (2006) What is the hair cell transduction channel? J Physiol 576(Pt 1):23–28

57. Shin JB, Adams D, Paukert M, Siba M, Sidi S, Levin M et al (2005) Xenopus TRPN1 (NOMPC) localizes to microtubule-based cilia in epithelial cells, including inner-ear hair cells. Proc Natl Acad Sci U S A 102(35):12572–12577

58. Sidi S, Friedrich RW, Nicolson T (2003) NompC TRP channel required for vertebrate sensory hair cell mechanotransduction. Science 301(5629):96–99

59. Hudmon A, Schulman H (2002) Neuronal Ca^{2+}/calmodulin-dependent protein kinase II: the role of structure and autoregulation in cellular function. Annu Rev Biochem 71:473–510

60. Tobimatsu T, Fujisawa H (1989) Tissue-specific expression of four types of rat calmodulin-dependent protein kinase II mRNAs. J Biol Chem 264:17907–17912

61. Tombes RM, Faison MO, Turbeville C (2003) Organization and evolution of multifunctional Ca^{2+}/CaM-dependent protein kinase (CaMK-II) genes. Gene 322:17–31

62. Swulius MT, Waxham MN (2008) $Ca^{(2+)}$/calmodulin-dependent protein kinases. Cell Mol Life Sci 65(17):2637–2657

63. Rothschild SC, Francescatto L, Tombes RM (2016) Immunostaining phospho-epitopes in ciliated organs of whole mount zebrafish embryos. J Visual Exp JoVE 108:53747

64. Francescatto L, Rothschild SC, Myers AL, Tombes RM (2010) The activation of membrane targeted CaMK-II in the zebrafish Kupffer's vesicle is required for left-right asymmetry. Development 137(16):2753–2762

65. Rothschild SC, Francescatto L, Drummond IA, Tombes RM (2011) CaMK-II is a PKD2 target that promotes pronephric kidney development and stabilizes cilia. Development 138(16):3387–3397

66. Rothschild SC, Lister JA, Tombes RM (2007) Differential expression of CaMK-II genes during early zebrafish embryogenesis. Dev Dyn 236(1):295–305

67. Lou L, Schulman H (1989) Distinct autophosphorylation sites sequantially produce autonomy and inhibition of the multifunctional Ca^{2+}/calmodulin-dependent protein kinase. J Neurosci 9(6):2020–2032

68. Hoffman A, Carpenter H, Kahl R, Watt LF, Dickson PW, Rostas JA et al (2014) Dephosphorylation of CaMKII at T253 controls the metaphase-anaphase transition. Cell Signal 26(4):748–756

69. Tombes RM, Grant S, Westin EH, Krystal G (1995) G1 cell cycle arrest and apoptosis are induced in NIH 3T3 cells by KN-93, an inhibitor of CaMK-II (the multifunctional Ca^{2+}/CaM kinase). Cell Growth Differ 6(9):1063–1070

70. Rasmussen G, Rasmussen C (1995) Calmodulin-dependent protein kinase II is required for G1/S progression in HeLa cells. Biochem Cell Biol 73:201–207

71. Rothschild SC, Lahvic J, Francescatto L, McLeod JJ, Burgess SM, Tombes RM (2013) CaMK-II activation is essential for zebrafish inner ear development and acts through Delta-Notch signaling. Dev Biol 381(1):179–188

72. Gardner AJ, Knott JG, Jones KT, Evans JP (2007) CaMKII can participate in but is not sufficient for the establishment of the membrane block to polyspermy in mouse eggs. J Cell Physiol 212(2):275–280

73. Morin N, Abrieu A, Lorca T, Martin F, Dorée M (1994) The proteolysis-dependent metaphase to anaphase transition: calcium/calmodulin-dependent protein kinase II mediates onset of anaphase in extracts prepared from unfertilized Xenopus eggs. EMBO J 13(18):4343–4352

74. Dupont G (1998) Link between fertilization-induced Ca^{2+} oscillations and relief from metaphase II arrest in mammalian eggs: a model based on calmodulin-dependent kinase II activation. Biophys Chem 72(1–2):153–167

75. Tatone C, Iorio R, Francione A, Gioia L, Colonna R (1999) Biochemical and biological effects of KN-93, an inhibitor of calmodulin-dependent protein kinase II, on the initial events of mouse egg activation induced by ethanol. J Reprod Fertil 115(1):151–157

76. Backs J, Stein P, Backs T, Duncan FE, Grueter CE, McAnally J et al (2010) The gamma isoform of CaM kinase II controls mouse egg activation by regulating cell cycle resumption. Proc Natl Acad Sci U S A 107(1):81–86

77. Chang HY, Minahan K, Merriman JA, Jones KT (2009) Calmodulin-dependent protein kinase gamma 3 (CamKIIgamma3) mediates the cell cycle resumption of metaphase II eggs in mouse. Development 136(24):4077–4081

78. Silva A, Stevens C, Tonegawa S, Wang Y (1992) Deficient hippocampal long term-potentiation in a-calcium-calmodulin kinase II mutant mice. Science 257:201–206

79. Silva AJ, Paylor R, Wehner JM, Tonegawa S (1992) Impaired spatial learning in alpha-calcium-calmodulin kinase II mutant mice. Science 257(5067):206–211

80. van Woerden GM, Hoebeek FE, Gao Z, Nagaraja RY, Hoogenraad CC, Kushner SA et al (2009) betaCaMKII controls the direction of plasticity at parallel fiber-Purkinje cell synapses. Nat Neurosci 12:823–825

81. Backs J, Backs T, Neef S, Kreusser MM, Lehmann LH, Patrick DM et al (2009) The delta isoform of CaM kinase II is required for pathological cardiac hypertrophy and remodeling after pressure overload. Proc Natl Acad Sci U S A 106(7):2342–2347

82. Liu J, Maller JL (2005) Calcium elevation at fertilization coordinates phosphorylation of XErp1/Emi2 by Plx1 and CaMK II to release metaphase arrest by cytostatic factor. Curr Biol 15(16):1458–1468

83. Hansen DV, Tung JJ, Jackson PK (2006) CaMKII and polo-like kinase 1 sequentially phosphorylate the cytostatic factor Emi2/XErp1 to trigger its destruction and meiotic exit. Proc Natl Acad Sci U S A 103(3):608–613

84. Tombes RM, Peppers LS (1995) Sea urchin fertilization stimulates CaM kinase-II (multifunctional (type II) Ca^{2+}/CaM Kinase) activity and association with $p34^{cdc2}$. Dev Growth Differ 37(5):589–596

85. Creton R (2004) The calcium pump of the endoplasmic reticulum plays a role in midline signaling during early zebrafish development. Brain Res Dev Brain Res 151(1–2):33–41

86. Kohn AD, Moon RT (2005) Wnt and calcium signaling: beta-catenin-independent pathways. Cell Calcium 38(3–4):439–446

87. Pfleiderer PJ, Lu KK, Crow MT, Keller RS, Singer HA (2004) Modulation of vascular smooth muscle cell migration by calcium/calmodulin-dependent protein kinase II-delta 2. Am J Physiol Cell Physiol 286(6):C1238–C1245

88. Bilato C, Curto KA, Monticone RE, Pauly RR, White AJ, Crow MT (1997) The inhibition of vascular smooth muscle cell migration by peptide and antibody antagonists of the alphavbeta3 integrin complex is reversed by activated calcium/calmodulin- dependent protein kinase II. J Clin Invest 100(3):693–704

89. Bouvard D, Block MR (1998) Calcium/calmodulin-dependent protein kinase II controls integrin alpha5beta1-mediated cell adhesion through the integrin cytoplasmic domain associated protein-1alpha. Biochem Biophys Res Commun 252(1):46–50

90. Lundberg MS, Curto KA, Bilato C, Monticone RE, Crow MT (1998) Regulation of vascular smooth muscle migration by mitogen-activated protein kinase and calcium/calmodulin-dependent protein kinase II signaling pathways. J Mol Cell Cardiol 30(11):2377–2389

91. Easley CA, Brown CM, Horwitz AF, Tombes RM (2008) CaMK-II promotes focal adhesion turnover and cell motility by inducing tyrosine dephosphorylation of FAK and paxillin. Cell Motil Cytoskeleton 65(8):662–674

92. Rothschild SC, Easley CA, Francescatto L, Lister JA, Garrity DM, Tombes RM (2009) Tbx5-mediated expression of Ca^{2+}/calmodulin-dependent protein kinase II is necessary for zebrafish cardiac and pectoral fin morphogenesis. Dev Biol 330(1):175–184

93. Lantsman K, Tombes RM (2005) CaMK-II oligomerization potential determined using CFP/YFP FRET. Biochim Biophys Acta 1746(1):45–54

94. Webb SE, Miller AL (2006) Ca^{2+} signaling and early embryonic patterning during the Blastula and Gastrula Periods of Zebrafish and Xenopus development. Biochim Biophys Acta 1763:1192–1208

95. Kuhl M, Geis K, Sheldahl LC, Pukrop T, Moon RT, Wedlich D (2001) Antagonistic regulation of convergent extension movements in Xenopus by Wnt/beta-catenin and Wnt/Ca^{2+} signaling. Mech Dev 106(1–2):61–76

96. Baitinger C, Alderton J, Poenie M, Schulman H, Steinhardt RA (1990) Multifunctional Ca^{2+}/calmodulin-dependent protein kinase is necessary for nuclear envelope breakdown. J Cell Biol 111:1763–1773

97. Knott JG, Gardner AJ, Madgwick S, Jones KT, Williams CJ, Schultz RM (2006) Calmodulin-dependent protein kinase II triggers mouse egg activation and embryo development in the absence of Ca^{2+} oscillations. Dev Biol 296(2):388–395

98. Markoulaki S, Matson S, Ducibella T (2004) Fertilization stimulates long-lasting oscillations of CaMKII activity in mouse eggs. Dev Biol 272(1):15–25

99. Markoulaki S, Matson S, Abbott AL, Ducibella T (2003) Oscillatory CaMKII activity in mouse egg activation. Dev Biol 258(2):464–474

100. Johnson J, Bierle BM, Gallicano GI, Capco DG (1998) Calcium/calmodulin-dependent protein kinase II and calmodulin: regulators of the meiotic spindle in mouse eggs. Dev Biol 204(2):464–477

101. Morris TA, DeLorenzo RJ, Tombes RM (1998) CaMK-II inhibition reduces cyclin D1 levels and enhances the association of p27^{kip1} with cdk2 to cause G1 arrest in NIH 3T3 cells. Exp Cell Res 240:218–227

102. Zhu W, Woo AY, Yang D, Cheng H, Crow MT, Xiao RP (2007) Activation of CaMKIIdeltaC is a common intermediate of diverse death stimuli-induced heart muscle cell apoptosis. J Biol Chem 282(14):10833–10839

103. Hagemann D, Bohlender J, Hoch B, Krause EG, Karczewski P (2001) Expression of Ca^{2+}/calmodulin-dependent protein kinase II delta-subunit isoforms in rats with hypertensive cardiac hypertrophy. Mol Cell Biochem 220(1–2):69–76

104. Zhang T, Maier LS, Dalton ND, Miyamoto S, Ross J Jr, Bers DM et al (2003) The deltaC isoform of CaMKII is activated in cardiac hypertrophy and induces dilated cardiomyopathy and heart failure. Circ Res 92(8):912–919

105. Zhang R, Khoo MS, Wu Y, Yang Y, Grueter CE, Ni G et al (2005) Calmodulin kinase II inhibition protects against structural heart disease. Nat Med 11(4):409–417

106. Grueter CE, Colbran RJ, Anderson ME (2007) CaMKII, an emerging molecular driver for calcium homeostasis, arrhythmias, and cardiac dysfunction. J Mol Med (Berlin, Germany) 85(1):5–14

107. Yang Y, Zhu WZ, Joiner ML, Zhang R, Oddis CV, Hou Y et al (2006) Calmodulin kinase II inhibition protects against myocardial cell apoptosis in vivo. Am J Physiol 291(6):H3065–H3075

108. Baltas LG, Karczewski P, Krause EG (1995) The cardiac sarcoplasmic reticulum phospholamban kinase is a distinct delta-CaM kinase isozyme. FEBS Lett 373(1):71–75

109. Grueter CE, Abiria SA, Dzhura I, Wu Y, Ham AJ, Mohler PJ et al (2006) L-type Ca^{2+} channel facilitation mediated by phosphorylation of the beta subunit by CaMKII. Mol Cell 23(5):641–650

110. Hudmon A, Schulman H, Kim J, Maltez JM, Tsien RW, Pitt GS (2005) CaMKII tethers to L-type Ca^{2+} channels, establishing a local and dedicated integrator of Ca^{2+} signals for facilitation. J Cell Biol 171(3):537–547

111. Lee TS, Karl R, Moosmang S, Lenhardt P, Klugbauer N, Hofmann F et al (2006) Calmodulin kinase II is involved in voltage-dependent facilitation of the L-type Cav1.2 calcium channel: identification of the phosphorylation sites. J Biol Chem 281(35):25560–25567

112. Zalk R, Lehnart SE, Marks AR (2007) Modulation of the ryanodine receptor and intracellular calcium. Annu Rev Biochem 76:367–385

113. Backs J, Song K, Bezprozvannaya S, Chang S, Olson EN (2006) CaM kinase II selectively signals to histone deacetylase 4 during cardiomyocyte hypertrophy. J Clin Invest 116(7):1853–1864

114. Little GH, Bai Y, Williams T, Poizat C (2007) Nuclear calcium/calmodulin-dependent protein kinase IIdelta preferentially transmits signals to histone deacetylase 4 in cardiac cells. J Biol Chem 282(10):7219–7231

115. Rothschild SC, Lee HJ, Ingram SR, Mohammadi DK, Walsh GS, Tombes RM (2018) Calcium signals act through histone deacetylase to mediate pronephric kidney morphogenesis. Dev Dyn 247(6):807–817

116. Garriock RJ, D'Agostino SL, Pilcher KC, Krieg PA (2005) Wnt11-R, a protein closely related to mammalian Wnt11, is required for heart morphogenesis in Xenopus. Dev Biol 279(1):179–192

117. Eisenberg CA, Eisenberg LM (1999) WNT11 promotes cardiac tissue formation of early mesoderm. Dev Dyn 216(1):45–58

118. Pandur P, Lasche M, Eisenberg LM, Kuhl M (2002) Wnt-11 activation of a non-canonical Wnt signaling pathway is required for cardiogenesis. Nature 418(6898):636–641

119. Fleming IN, Elliott CM, Buchanan FG, Downes CP, Exton JH (1999) Ca^{2+}/calmodulin-dependent protein kinase II regulates Tiam1 by reversible protein phosphorylation. J Biol Chem 274(18):12753–12758

120. Seward ME, Easley CA, McLeod JJ, Myers AL, Tombes RM (2008) Flightless-I, a gelsolin family member and transcriptional regulator, preferentially binds directly to activated cytosolic CaMK-II. FEBS Lett 582(17):2489–2495

121. Tsiokas L (2009) Function and regulation of TRPP2 at the plasma membrane. Am J Physiol Renal Physiol 297(1):F1–F9

122. Obara T, Mangos S, Liu Y, Zhao J, Wiessner S, Kramer-Zucker AG et al (2006) Polycystin-2 immunolocalization and function in zebrafish. J Am Soc Nephrol 17(10):2706–2718

123. Backs J, Backs T, Bezprozvannaya S, McKinsey TA, Olson EN (2008) Histone deacetylase 4 confers CaM kinase II responsiveness to histone deacetylase 5 by oligomerization. Mol Cell Biol 28(10):3437–3445

124. Zhang T, Kohlhaas M, Backs J, Mishra S, Phillips W, Dybkova N et al (2007) CaMKIIdelta isoforms differentially affect calcium handling but similarly regulate HDAC/MEF2 transcriptional responses. J Biol Chem 282(48):35078–35087

125. Di Giorgio E, Brancolini C (2016) Regulation of class IIa HDAC activities: it is not only matter of subcellular localization. Epigenomics 8(2):251–269

126. Di Giorgio E, Clocchiatti A, Piccinin S, Sgorbissa A, Viviani G, Peruzzo P et al (2013) MEF2 is a converging hub for histone deacetylase 4 and phosphatidylinositol 3-kinase/Akt-induced transformation. Mol Cell Biol 33(22):4473–4491

127. McKinsey TA, Zhang CL, Olson EN (2002) MEF2: a calcium-dependent regulator of cell division, differentiation and death. Trends Biochem Sci 27(1):40–47

128. Bershteyn M, Atwood SX, Woo WM, Li M, Oro AE (2010) MIM and cortactin antagonism regulates ciliogenesis and hedgehog signaling. Dev Cell 19(2):270–283

129. Saarikangas J, Mattila PK, Varjosalo M, Bovellan M, Hakanen J, Calzada-Wack J et al (2011) Missing-in-metastasis MIM/MTSS1 promotes actin assembly at intercellular junctions and is required for integrity of kidney epithelia. J Cell Sci 124(Pt 8):1245–1255

130. Xia S, Li X, Johnson T, Seidel C, Wallace DP, Li R (2010) Polycystin-dependent fluid flow sensing targets histone deacetylase 5 to prevent the development of renal cysts. Development 137(7):1075–1084

131. Haddon C, Jiang YJ, Smithers L, Lewis J (1998) Delta-Notch signalling and the patterning of sensory cell differentiation in the zebrafish ear: evidence from the mind bomb mutant. Development 125(23):4637–4644

132. Haddon C, Mowbray C, Whitfield T, Jones D, Gschmeissner S, Lewis J (1999) Hair cells without supporting cells: further studies in the ear of the zebrafish mind bomb mutant. J Neurocytol 28(10–11):837–850

133. Itoh M, Kim CH, Palardy G, Oda T, Jiang YJ, Maust D et al (2003) Mind bomb is a ubiquitin ligase that is essential for efficient activation of Notch signaling by Delta. Dev Cell 4(1):67–82

134. Kandachar V, Roegiers F (2012) Endocytosis and control of Notch signaling. Curr Opin Cell Biol 24(4):534–540
135. Matsuda M, Chitnis AB (2009) Interaction with Notch determines endocytosis of specific Delta ligands in zebrafish neural tissue. Development 136(2):197–206
136. Matsumoto K, Fukunaga K, Miyazaki J, Shichiri M, Miyamoto E (1995) Ca^{2+}/calmodulin-dependent protein kinase II and synapsin I-like protein in mouse insulinoma MIN6 cells. Endocrinology 136(9):3784–3793
137. Nielander HB, Onofri F, Valtorta F, Schiavo G, Montecucco C, Greengard P et al (1995) Phosphorylation of VAMP/synaptobrevin in synaptic vesicles by endogenous protein kinases. J Neurochem 65(4):1712–1720
138. Popoli M (1993) Synaptotagmin is endogenously phosphorylated by Ca^{2+}/calmodulin protein kinase II in synaptic vesicles. FEBS Lett 317(1–2):85–88
139. Bustos R, Kolen ER, Braiterman L, Baines AJ, Gorelick FS, Hubbard AL (2001) Synapsin I is expressed in epithelial cells: localization to a unique trans-Golgi compartment. J Cell Sci 114(Pt 20):3695–3704
140. Marcet B, Chevalier B, Luxardi G, Coraux C, Zaragosi LE, Cibois M et al (2011) Control of vertebrate multiciliogenesis by miR-449 through direct repression of the Delta/Notch pathway. Nat Cell Biol 13(6):693–699
141. Ann EJ, Kim HY, Seo MS, Mo JS, Kim MY, Yoon JH et al (2012) Wnt5a controls Notch1 signaling through CaMKII-mediated degradation of the SMRT corepressor protein. J Biol Chem 287(44):36814–36829
142. Mamaeva OA, Kim J, Feng G, McDonald JM (2009) Calcium/calmodulin-dependent kinase II regulates notch-1 signaling in prostate cancer cells. J Cell Biochem 106(1):25–32
143. Bayer KU, De Koninck P, Schulman H (2002) Alternative splicing modulates the frequency-dependent response of CaMKII to Ca^{2+} oscillations. EMBO J 21(14):3590–3597
144. De Koninck P, Schulman H (1998) Sensitivity of CaM kinase II to the frequency of Ca^{2+} oscillations. Science 279(5348):227–230
145. Kinoshita N, Iioka H, Miyakoshi A, Ueno N (2003) PKC delta is essential for Dishevelled function in a noncanonical Wnt pathway that regulates Xenopus convergent extension movements. Genes Dev 17(13):1663–1676
146. Horne-Badovinac S, Lin D, Waldron S, Schwarz M, Mbamalu G, Pawson T et al (2001) Positional cloning of heart and soul reveals multiple roles for PKC lambda in zebrafish organogenesis. Curr Biol 11(19):1492–1502
147. Gagnon JA, Valen E, Thyme SB, Huang P, Ahkmetova L, Pauli A et al (2014) Efficient mutagenesis by Cas9 protein-mediated oligonucleotide insertion and large-scale assessment of single-guide RNAs. PLoS One 9(5):e98186

Chapter 22
Calcium Signaling and Gene Expression

Basant K. Puri

Abstract Calcium signaling plays an important role in gene expression. At the transcriptional level, this may underpin mammalian neuronal synaptic plasticity. Calcium influx into the postsynaptic neuron via: N-methyl-D-aspartate (NMDA) receptors activates small GTPase Rac1 and other Rac guanine nucleotide exchange factors, and stimulates calmodulin-dependent kinase kinase (CaMKK) and CaMKI; α-amino-3-hydroxy-5-methyl-4-isoxazolepropionic acid receptors that are not impermeable to calcium ions, that is, those lacking the glutamate receptor-2 subunits, leads to activation of Ras guanine nucleotide-releasing factor proteins, which is coupled with activation of the mitogen-activated protein kinases/extracellular signal-regulated kinases signaling cascade; L-type voltage-gated calcium channels activates signaling pathways involving CaMKII, downstream responsive element antagonist modulator and distinct microdomains. Key members of these signaling cascades then translocate into the nucleus, where they alter the expression of genes involved in neuronal synaptic plasticity. At the post-transcriptional level, intracellular calcium level changes can change alternative splicing patterns; in the mammalian brain, alterations in calcium signaling via NMDA receptors is associated with exon silencing of the CI cassette of the NMDA R1 receptor (*GRIN1*) transcript by UAGG motifs in response to neuronal excitation. Regulation also occurs at the translational level; transglutaminase-2 (TG2) mediates calcium ion-regulated crosslinking of Y-box binding protein-1 (YB-1) translation-regulatory protein in TGFβ1-activated myofibroblasts; YB-1 binds smooth muscle α-actin mRNA and regulates its translational activity. Calcium signaling is also important in epigenetic regulation, for example in respect of changes in cytosine bases. Targeting calcium signaling may provide therapeutically useful options, for example to induce epigenetic reactivation of tumor suppressor genes in cancer patients.

B. K. Puri (✉)
CAR, Cambridge, UK
e-mail: bpuri@cantab.net

© Springer Nature Switzerland AG 2020
M. S. Islam (ed.), *Calcium Signaling*, Advances in Experimental Medicine
and Biology 1131, https://doi.org/10.1007/978-3-030-12457-1_22

Keywords Alternative splicing patterns · AMPA receptors · Calcium signaling ·
Epigenetic regulation · Gene expression · Gene reactivation · L-type
voltage-gated calcium channels · NMDA receptors · Transcription · Translation

22.1 Introduction

Gene expression is the process by which the information in a DNA sequence in
a gene is used to biosynthesize an RNA or polypeptide [1]. In turn, this involves
transcription, that is, the synthesis of an RNA copy from a DNA template, and, in
the case of polypeptides, translation, that is, protein synthesis on a messenger RNA
(mRNA) template [1]. It has recently become increasingly apparent that calcium
signaling is relevant to the regulation of eukaryotic transcription, alternative splicing
patterns, and translation. In this chapter, the roles of calcium ion signaling in these
processes and in the regulation of epigenetic mechanisms will be discussed.

Calcium ion binding, and associated phosphorylation, are associated with
changes in protein electrical charge, conformation and interactions; phosphate
moieties can be removed by protein kinases from adenosine-5′-triphosphate (ATP)
and attached covalently to the three common amino acid residues which have free
hydroxyl groups, namely the polar amino acids serine, threonine and tyrosine [2,
3]. Thus, calcium ions and phosphate ions can effect signal transduction [2, 3].
Aside from its role in gene expression, calcium ion signaling, both intercellular and
intracellular, has numerous other important functions, ranging from mitochondrial
functioning and innate immunity to apoptosis and cell death pathways [2, 4, 5].
Other chapters of this work deal with many of these. An excellent review from the
year 2000 which considers the versatility and universality of calcium signaling is
that of Berridge, Lipp and Bootman [6], while Putney and Tomita review the role
of phospholipase C signaling and calcium influx [7]; in this chapter, the focus is on
the role of calcium ion signaling in respect of gene expression.

22.2 Pre-translation

22.2.1 Eukaryotic Transcription

Eukaryotic transcription occurs on a chromatin template (unlike the case for
prokaryotes, in which a DNA template is used for transcription); the following three
classes of RNA polymerase are involved: RNA polymerase I, which transcribes
18S/28S ribosomal RNA (rRNA); RNA polymerase II, which transcribes mRNA
and certain small RNAs; and RNA polymerase III, which transcribes transfer RNA
(tRNA), 5S rRNA and certain small RNAs [1].

22.2.2 Calcium-Related Transcriptional Regulation

Calcium-dependent gene expression regulation at the transcriptional level is thought to underlie animal neuronal synaptic plasticity and thereby mediate learning and adaptation to the environment [8]. In mammalian neurons, such regulation involves a complex cascade of signaling molecules, beginning with influx of calcium ions into the postsynaptic neuron via N-methyl-D-aspartate (NMDA) receptors (for glutamate), α-amino-3-hydroxy-5-methyl-4-isoxazolepropionic acid (AMPA) receptors (also for glutamate), or L-type voltage-gated calcium channels (VGCCs) [9–11]. Each of these three possibilities will be briefly considered in turn.

Calcium ion influx through NMDA receptors activates small GTPase Rac1 (also known as Ras-related C3 botulinum toxin substrate 1), which acts as a pleiotropic activator of actin, and also activates other Rac guanine nucleotide exchange factors (GEFs) such as kalirin-7 and betaPIX (βPIX) [12, 13]. It also stimulates calmodulin-dependent kinase kinase (CaMKK) and CaMKI, which in turn phosphorylates βPIX [13]. Kalirin-7 interacts with AMPA receptors, controlling their synaptic expression [12]. While most AMPA receptors are calcium impermeable, those lacking the glutamate receptor-2 (GluR2) subunits do allow calcium ion flow. Calcium ion influx through such calcium-permeable AMPA receptors leads to activation of Ras guanine nucleotide-releasing factor (RasGRF) proteins, which in turn is coupled with activation of the mitogen-activated protein kinases/extracellular signal-regulated kinases (MAPK/ERK; also known as the Ras-ERK or Ras-Raf-MEK-ERK) signaling cascade [14]. Finally, calcium ion influx through L-type VGCCs appears to activate signaling pathways involving CaMKII, downstream responsive element antagonist modulator (DREAM), distinct microdomains (MD-I and MD-II), and possibly the distal C-terminal (dCT) fragment of the L-type receptor and beta subunits [15]. These consequences of calcium ion influx through NMDA, AMPA receptors and VGCCs are summarized in Table 22.1.

In turn, key members of the above signaling cascades, such as CAMKII, nuclear factor kappa-light-chain-enhancer of activated B cells (NF-κB), MAPK/ERK, GTP-Rac, DREAM, MD-I, MD-II, and possibly dCT and β4c, cross from the cytoplasm into the nucleus [8, 15]. Here, they alter the expression of, amongst others, the non-

Table 22.1 Primary activated molecules following calcium ion influx through NMDA and AMPA receptors and VGCCs

Type of calcium ion receptor or channel	NMDA receptors	AMPA receptors	VGCCs
Primary activated molecules	Small GTPase Rac1	RasGRF	CaMKII
	Kalirin-7		DREAM
	βPIX		MD-I
	CaMKK		MD-II
	CaMKI		dCT
			β subunits

coding RNA (ncRNA) miR-132 (which is a microRNA), *CREM* (which encodes the protein cyclic adenosine monophosphate (cAMP) responsive element modulator), *BDNF* (which encodes brain-derived neurotrophic factor), the proto-oncogene *c-Fos*, *PDYN* (which encodes a preproprotein which, following proteolysis, gives rise to several opioid peptides), *WNT2* (wingless-type MMTV integration site family, member 2; encoding signaling proteins relating to the Wnt signal transduction pathways), *BCL2* (encoding B-cell lymphoma 2 or Bcl-2), *SOD2* or *MnSOD* (encoding superoxide dismutase 2, mitochondrial), *XIAP* (X-linked inhibitor of apoptosis family of proteins), *NR4A1* or *Nur77* (nuclear receptor subfamily 4 group A member 1 or nerve growth factor IB), *ARC* (which encodes activity-regulated cytoskeleton-associated protein), HOMER1 (Homer scaffold protein 1 or Homer1a), *SLC8A1* or *NCX1* (solute carrier family 8 member A1 or sodium/calcium exchanger), and *SLC8A3* or *NCX3*. These are involved in synaptic development, dendritic growth, and neuronal plasticity; changes in their expression, as well as mutations in some of these loci, may be associated with neurocognitive disorders [8, 15]. Furthermore, EphB receptor tyrosine kinases, localized at excitatory synapses, cluster with NMDA receptors and modulate the function of the latter during early synaptogenesis [16].

A similar picture exists in respect of the mammalian heart, from which efflux of calcium ions normally takes place via plasma membrane calcium ATPases (PMCAs). Sustained increase in intracellular calcium ion concentration in cardiac cells activates the calcineurin moiety of PMCA4, which in turn dephosphorylates nuclear factor of activated T-cells (NFAT), which then translocates to the nucleus where it activates genes involved in cardiac hypertrophy [17].

The above examples have been drawn from animal cells. Calcium-related transcriptional regulation has also been shown to be important in plants. This has been studied in the unicellular green alga *Chlamydomonas reinhardtii*, which has a relatively short life-cycle and a fully sequenced genome [18–20]. In chloroplasts of this alga, calcium ion signaling and the calcium ion-binding protein CAS, acting in response to cues such as biotic and abiotic stress and carbon dioxide concentrating mechanisms, ultimately act upon a number of nuclear targets, including: *APX* (encoding ascorbate peroxidase); flg22 (flagellin 22); *HSF*s (heat shock transcription factors); *HSP*s (heat shock proteins); and *LHCRS3* (light-harvesting complex stress-related protein 3) [21]. These result in changes in basal defense responses and carbon dioxide concentration mechanisms [21].

22.2.3   Changes in Alternative Splicing Patterns

At the post-transcriptional, but pre-translational, level, intracellular calcium ion level changes can also alter gene expression by causing changes in alternative splicing patterns, whereby the same pre-mRNA generates mRNAs (post-splicing) which have different exon combinations [1].

In the mammalian brain, it has been shown that alterations in calcium ion signaling via NMDA receptors is associated with exon silencing of the CI cassette (exon 19) of the NMDA R1 receptor (*GRIN1*) transcript by UAGG motifs in response to neuronal excitation [22]. CI mediates targeting of NMDA R1 to the plasma membrane, has an endoplasmic reticulum retention signal, and contains a binding site for calcium/calmodulin [23, 24]. This may offer a powerful strategy for neuronal adaptation to hyperstimulation and may explain the diverse properties of NMDA receptors in different groups of neurons [22, 24].

Mechanical stimulation of hair cells of the basilar papilla of the avian inner ear, which is homologous to the organ of Corti, or spiral organ, of mammals, is associated with changes in intracellular calcium ion concentration via changes in the kinetic properties of calcium-ion-activated potassium ion channels; in turn, changes in calcium concentration have been found to be associated with alternative mRNA splicing patterns which tune individual hair cells to specific auditory frequencies [25–27].

In a similar vein, it is also noteworthy that GH3 pituitary cell depolarization has been shown to repress *KCNMA1* or *STREX* (potassium calcium-activated channel subfamily M alpha 1, previously stress-axis regulated exon) exon splicing in BK (big potassium, also known as Maxi-K, Kcal.1 or slo1) potassium ion channel transcripts via CaMKs [28].

Mammalian VGCCs are able to be activated over a relatively wide range of electrical potential differences, whereas the activation voltage dependence of calcium channel isoforms found in different tissues are tuned to their specific corresponding physiological functions. For example, the type known as 1.1 is the VGCC least responsive to depolarization and it has been found to achieve this electrical property through alternative splicing [29]. It acts both as a calcium ion channel in embryonic muscle and as a sensor of electrical potential difference in mature skeletal muscle for excitation-contraction coupling, and its relative lack of responsiveness to depolarization serves these functions well [29, 30]. On the other hand, the type of VGCC known as 1.2, which is the main type found in the brain and the cardiovascular system, is more responsive to depolarization; interestingly, the adjustment of its optimum activation voltage-dependency has recently been shown not to result from alternative splicing, showing that more than one mechanism is involved in fine tuning VGCCs [30].

22.3 Translation

Calcium regulation of gene expression at the translational level has been demonstrated in human cultured cells. The peptide transforming growth factor beta (TGFβ) controls cell proliferation in many tissues, including connective tissue [31]. In particular, repair of mammalian tissue injury can be initiated by TGFβ1 receptor signaling [32–35]. Indeed, poor regulation of this process may lead to dysfunctional cardiopulmonary fibrosis and chronic myofibroblast differentiation [36–38]. In

2013, it was shown, by Willis and colleagues, that the protein cross-linking enzyme transglutaminase-2 (TG2) mediates calcium ion-regulated crosslinking of Y-box binding protein-1 (YB-1) translation-regulatory protein in TGFβ1-activated myofibroblasts; YB-1 binds smooth muscle α-actin (*SMαA*) mRNA and regulates its translational activity [39].

22.4 Epigenetics

22.4.1 Epigenetic Mechanisms

Tollefsbol has defined epigenetic processes as 'changes of a biochemical nature to the DNA or its associated proteins or RNA that do not change the DNA sequence itself but do impact the level of gene expression' [40]. These biochemical changes are reversible and include DNA methylation, modifications in chromatin, nucleosome positioning, and ncRNA profile alterations [41]. The study of epigenetics is a rapidly developing field of research, which is of relevance to the study of diseases and, at a fundamental level, to a deeper understanding of intracellular communication [40–42]. It has recently become increasingly clear that calcium ion signaling plays an important role in epigenetic regulation.

22.4.2 Calcium-Related Epigenetic Regulation

A few recent examples are given to illustrate the important role of calcium signaling in epigenetic regulation.

Regarding DNA methylation, it has been shown that changes in the calcium content of murine diets can induce methylation changes in DNA cytosine bases. For example, a calcium-deficient diet in pregnant and nursing rats is associated with hypomethylation of the pup hepatic *HSD11B2* promoter region; this gene encodes the NAD^+-dependent enzyme corticosteroid 11-β-dehydrogenase isozyme 2 (also known as 11-β-hydroxysteroid dehydrogenase 2), and such pups have higher serum corticosterone levels than matched control pups from mothers fed a normal diet [43].

Raynal and colleagues tested a number of drugs which re-activate silenced gene expression in human cancer cells [44]. They found 11 newly identified pharmacological agents, such as cardiac glycosides, which induce methylated and silenced CpG island promoters which drive *GFP*, the gene for green fluorescent protein, and endogenous tumor suppressor genes in cancer cell lines. Surprisingly, rather than causing local DNA methylation changes or global histone changes, all 11 agents were found to alter calcium ion signaling and trigger CaMK activity; in turn, this released methyl CpG binding protein 2 (MeCP2), a methyl-binding protein, from silenced promoters, thus causing gene activation [44–46]. Given that epigenetic changes are, in principle, reversible, this suggests that a potential therapeutic approach to the treatment of cancer might involve targeting calcium signaling in order to induce epigenetic reactivation of tumor suppressor genes [44].

It has been pointed out that the calcium ion influx through postsynaptic NMDA receptors and VGCCs mentioned above, which can lead to changes in *BDNF* expression, for example, also cause epigenetic changes such DNA hypomethylation (unmethylated cytosines) and histone acetylation; indeed, histone modification and changes in DNA methylation appear to be important features of the mediation of the risk of the development of major depressive disorder [47].

It should also be noted that epigenetic changes can also regulate calcium ion homeostasis. For example, epigenetic modification of the promoter region of *SERCA2a*, which encodes sarcoplasmic reticulum Ca^{2+}-ATPase and which is rich in CpG islands, changes the expression of this gene and is associated with alterations in calcium ion homeostasis; indeed, it has been suggested that demethylation in this promoter region, induced by the hydrazinophthalazine antihypertensive pharmacological agent hydralazine, may lead to modulated cardiomyocytic calcium homeostasis and consequent improved cardiac functioning [48].

22.5 Discussion

The examples given above have shown that calcium signaling has an important role in gene expression. This may involve regulation at the level of gene transcription; it may involve the regulation of alternative splicing; it may occur at the level of gene translation; and it may entail epigenetic mechanisms. Furthermore, these regulatory processes are bidirectional, in that changes in gene expression can themselves affect calcium ion homeostasis and calcium ion signaling. These findings offer important potential therapeutic avenues for the treatment of numerous diseases.

References

1. Krebs JE, Goldstein ES, Kilpatrick ST (2018) Lewin's genes XII. Jones & Bartlett Learning, Burlington
2. Clapham DE (2007) Calcium signaling. Cell 131(6):1047–1058
3. Westheimer FH (1987) Why nature chose phosphates. Science 235(4793):1173–1178
4. Morris G, Puri BK, Walder K, Berk M, Stubbs B, Maes M et al (2018) The endoplasmic reticulum stress response in neuroprogressive diseases: emerging pathophysiological role and translational implications. Mol Neurobiol 55(12):8765–8787
5. Puri BK, Morris G (2018) Potential therapeutic interventions based on the role of the endoplasmic reticulum stress response in progressive neurodegenerative diseases. Neural Regen Res 13(11):1887–1889
6. Berridge MJ, Lipp P, Bootman MD (2000) The versatility and universality of calcium signalling. Nat Rev Mol Cell Biol 1:11

7. Putney JW, Tomita T (2012) Phospholipase C signaling and calcium influx. Adv Biol Regul 52(1):152–164
8. Greer PL, Greenberg ME (2008) From synapse to nucleus: calcium-dependent gene transcription in the control of synapse development and function. Neuron 59(6):846–860
9. Berridge MJ (1998) Neuronal calcium signaling. Neuron 21(1):13–26
10. Berridge MJ, Bootman MD, Lipp P (1998) Calcium–a life and death signal. Nature 395(6703):645–648
11. Jonas P, Burnashev N (1995) Molecular mechanisms controlling calcium entry through AMPA-type glutamate receptor channels. Neuron 15(5):987–990
12. Xie Z, Srivastava DP, Photowala H, Kai L, Cahill ME, Woolfrey KM et al (2007) Kalirin-7 controls activity-dependent structural and functional plasticity of dendritic spines. Neuron 56(4):640–656
13. Saneyoshi T, Wayman G, Fortin D, Davare M, Hoshi N, Nozaki N et al (2008) Activity-dependent synaptogenesis: regulation by a CaM-kinase kinase/CaM-kinase I/betaPIX signaling complex. Neuron 57(1):94–107
14. Tian X, Feig LA (2006) Age-dependent participation of Ras-GRF proteins in coupling calcium-permeable AMPA glutamate receptors to Ras/Erk signaling in cortical neurons. J Biol Chem 281(11):7578–7582
15. Naranjo JR, Mellström B (2012) Ca^{2+}-dependent transcriptional control of Ca^{2+} homeostasis. J Biol Chem 287(38):31674–31680
16. Takasu MA, Dalva MB, Zigmond RE, Greenberg ME (2002) Modulation of NMDA receptor-dependent calcium influx and gene expression through EphB receptors. Science 295(5554):491–495
17. Cartwright EJ, Oceandy D, Austin C, Neyses L (2011) Ca^{2+} signalling in cardiovascular disease: the role of the plasma membrane calcium pumps. Sci China Life Sci 54(8):691–698
18. Jain M, Shrager J, Harris EH, Halbrook R, Grossman AR, Hauser C et al (2007) EST assembly supported by a draft genome sequence: an analysis of the Chlamydomonas reinhardtii transcriptome. Nucleic Acids Res 35(6):2074–2083
19. Merchant SS, Prochnik SE, Vallon O, Harris EH, Karpowicz SJ, Witman GB et al (2007) The Chlamydomonas genome reveals the evolution of key animal and plant functions. Science 318(5848):245–250
20. Misumi O, Yoshida Y, Nishida K, Fujiwara T, Sakajiri T, Hirooka S et al (2008) Genome analysis and its significance in four unicellular algae, Cyanidioschyzon [corrected] merolae, Ostreococcus tauri, Chlamydomonas reinhardtii, and Thalassiosira pseudonana. J Plant Res 121(1):3–17
21. Rea G, Antonacci A, Lambreva MD, Mattoo AK (2018) Features of cues and processes during chloroplast-mediated retrograde signaling in the alga Chlamydomonas. Plant Sci 272:193–206
22. An P, Grabowski PJ (2007) Exon silencing by UAGG motifs in response to neuronal excitation. PLoS Biol 5(2):e36
23. Black DL, Grabowski PJ (2003) Alternative pre-mRNA splicing and neuronal function. Prog Mol Subcell Biol 31:187–216
24. Zukin RS, Bennett MV (1995) Alternatively spliced isoforms of the NMDARI receptor subunit. Trends Neurosci 18(7):306–313
25. Black DL (1998) Splicing in the inner ear: a familiar tune, but what are the instruments? Neuron 20(2):165–168
26. Navaratnam DS, Bell TJ, Tu TD, Cohen EL, Oberholtzer JC (1997) Differential distribution of Ca^{2+}−activated K^+ channel splice variants among hair cells along the tonotopic axis of the chick cochlea. Neuron 19(5):1077–1085
27. Rosenblatt KP, Sun ZP, Heller S, Hudspeth AJ (1997) Distribution of Ca^{2+}−activated K^+ channel isoforms along the tonotopic gradient of the chicken's cochlea. Neuron 19(5):1061–1075
28. Xie J, Black DL (2001) A CaMK IV responsive RNA element mediates depolarization-induced alternative splicing of ion channels. Nature 410(6831):936–939

29. Tuluc P, Yarov-Yarovoy V, Benedetti B, Flucher BE (2016) Molecular interactions in the voltage sensor controlling gating properties of CaV calcium channels. Structure 24(2):261–271
30. Coste de Bagneaux P, Campiglio M, Benedetti B, Tuluc P, Flucher BE (2018) Role of putative voltage-sensor countercharge D4 in regulating gating properties of CaV1.2 and CaV1.3 calcium channels. Channels (Austin) 12(1):249–261
31. Canney PA, Dean S (1990) Transforming growth factor beta: a promotor of late connective tissue injury following radiotherapy? Br J Radiol 63(752):620–623
32. Sun X, Liu W, Cheng G, Qu X, Bi H, Cao Z et al (2017) The influence of connective tissue growth factor on rabbit ligament injury repair. Bone Joint Res 6(7):399–404
33. Toomey D, Condron C, Wu QD, Kay E, Harmey J, Broe P et al (2001) TGF-beta1 is elevated in breast cancer tissue and regulates nitric oxide production from a number of cellular sources during hypoxia re-oxygenation injury. Br J Biomed Sci 58(3):177–183
34. Wang S, Denichilo M, Brubaker C, Hirschberg R (2001) Connective tissue growth factor in tubulointerstitial injury of diabetic nephropathy. Kidney Int 60(1):96–105
35. Desmouliere A, Geinoz A, Gabbiani F, Gabbiani G (1993) Transforming growth factor-beta 1 induces alpha-smooth muscle actin expression in granulation tissue myofibroblasts and in quiescent and growing cultured fibroblasts. J Cell Biol 122(1):103–111
36. Gabbiani G (2003) The myofibroblast in wound healing and fibrocontractive diseases. J Pathol 200(4):500–503
37. Grotendorst GR, Rahmanie H, Duncan MR (2004) Combinatorial signaling pathways determine fibroblast proliferation and myofibroblast differentiation. FASEB J 18(3):469–479
38. Chen G, Grotendorst G, Eichholtz T, Khalil N (2003) GM-CSF increases airway smooth muscle cell connective tissue expression by inducing TGF-beta receptors. Am J Phys Lung Cell Mol Phys 284(3):L548–L556
39. Willis WL, Hariharan S, David JJ, Strauch AR (2013) Transglutaminase-2 mediates calcium-regulated crosslinking of the Y-Box 1 (YB-1) translation-regulatory protein in TGFβ1-activated myofibroblasts. J Cell Biochem 114(12):2753–2769
40. Tollefsbol TO (2016) An overview of medical epigenetics. In: Tollefsbol TO (ed) Medical epigenetics. Elsevier, Amsterdam, pp 3–7
41. Kanwal R, Gupta K, Gupta S (2015) Cancer epigenetics: an introduction. Methods Mol Biol 1238:3–25
42. Huang B, Jiang C, Zhang R (2014) Epigenetics: the language of the cell? Epigenomics 6(1):73–88
43. Takaya J, Iharada A, Okihana H, Kaneko K (2013) A calcium-deficient diet in pregnant, nursing rats induces hypomethylation of specific cytosines in the 11beta-hydroxysteroid dehydrogenase-1 promoter in pup liver. Nutr Res 33(11):961–970
44. Raynal NJ, Lee JT, Wang Y, Beaudry A, Madireddi P, Garriga J et al (2016) Targeting calcium signaling induces epigenetic reactivation of tumor suppressor genes in cancer. Cancer Res 76(6):1494–1505
45. Martinowich K, Hattori D, Wu H, Fouse S, He F, Hu Y et al (2003) DNA methylation-related chromatin remodeling in activity-dependent BDNF gene regulation. Science 302(5646):890–893
46. Chen WG, Chang Q, Lin Y, Meissner A, West AE, Griffith EC et al (2003) Derepression of BDNF transcription involves calcium-dependent phosphorylation of MeCP2. Science 302(5646):885–889
47. Nagy C, Vaillancourt K, Turecki G (2018) A role for activity-dependent epigenetics in the development and treatment of major depressive disorder. Genes Brain Behav 17(3):e12446
48. Kao YH, Cheng CC, Chen YC, Chung CC, Lee TI, Chen SA et al (2011) Hydralazine-induced promoter demethylation enhances sarcoplasmic reticulum Ca^{2+}-ATPase and calcium homeostasis in cardiac myocytes. Lab Investig 91(9):1291–1297